A NATURAL HISTORY OF THE MAMMALIA

This is a volume in the Arno Press collection

BIOLOGISTS AND THEIR WORLD

See last pages of this volume
for a complete list of titles

A

NATURAL HISTORY

OF THE

MAMMALIA

G[eorge] R[obert] Waterhouse

Volume II

ARNO PRESS
A New York Times Company
New York • 1978

Editorial Supervision: MARIE STARECK

Reprint Edition 1978 by Arno Press Inc.

BIOLOGISTS AND THEIR WORLD
ISBN for complete set: 0-405-10641-6
See last pages of this volume for titles.

Manufactured in the United States of America

Publisher's Note: Pages 481-484 are a foldout in the original volume and have been reprinted as four pages of text in this edition.

Library of Congress Cataloging in Publication Data

Waterhouse, George Robert, 1810-1888.
A natural history of the Mammalia.

(Biologists and their world)
Reprint of the 1846-1848 ed. published by H. Baillière, London.
Includes index.
1. Marsupialia. 2. Rodentia. I. Title.
II. Series.
QL737.M3W37 1978 599.2 77-83843
ISBN 0-405-10739-0

A

NATURAL HISTORY

OF THE

MAMMALIA.

BY

G. R. WATERHOUSE,

CORRESPONDING MEMBER OF THE PHILOMATIC SOCIETY OF PARIS,

ASSISTANT IN THE DEPARTMENT OF MINERALOGY AND GEOLOGY OF THE BRITISH MUSEUM.

VOL. II.

CONTAINING THE ORDER

RODENTIA, OR GNAWING MAMMALIA;

WITH

22 ILLUSTRATIONS ENGRAVED ON STEEL, AND ENGRAVINGS ON WOOD.

LONDON:

HIPPOLYTE BAILLIÈRE, PUBLISHER, 219, REGENT STREET;

FOREIGN BOOKSELLER TO THE ROYAL COLLEGE OF SURGEONS,

AND TO THE ROYAL MEDICO-CHIRURGICAL SOCIETY.

PARIS: J.-B. BAILLIÈRE, LIBRAIRE DE L'ACADÉMIE ROYALE DE MÉDECINE.

LEIPSIC: T. O. WEIGEL.

1848.

LONDON:
WILSON AND OGILVY, 57, SKINNER STREET, SNOWHILL.

ORDER—GLIRES, OR RODENTIA.

Mammalia, having the jaws provided in front with long, curved, cylindrical (or nearly cylindrical) teeth, the exposed ends of which are bevelled off on the inner surface, so that they terminate in a sharp, cutting edge; these teeth, two in number in each jaw, (sometimes four in the upper jaw), separated by a wide toothless space from the molars, which have the crowns flat, or studded with blunt and but little elevated tubercles: the condyle of the lower jaw longitudinal or rounded, and inclosed by the glenoid cavity in such a manner as to permit very little lateral motion to the jaw, which, however, moves freely in the longitudinal direction: surface of the brain smooth, or very nearly so; anterior pair of the optic lobes (the nates) the largest; cerebellum exposed.

The Rodentia, so called from their gnawing propensities, form one of the most clearly defined groups of the Mammalia; a group which has representations in all parts of the world, and the species of which are very numerous. They feed upon vegetable substances, and are of small size, few exceeding the Common Hare in bulk. The most striking characters of the Rodents are those furnished by the teeth; the long vacant space which separates the incisors in front, here adapted for gnawing, from the masticating teeth behind. The gnawing teeth in most Rodents are two in number in each jaw, but in some few, forming the family *Leporidæ*, there are four of these teeth in the upper jaw; the superadded pair, however, are small, and hidden behind the principal pair. They are long, curved, and more or less cylindrical, but often their depth is greater than their width, and the sides are somewhat flattened: such is the case in most of the species of the

Squirrel and Rat families. Sometimes the width of the incisor is very great, and exceeds the depth; the Rodents, which burrow, and live almost entirely under ground, present this form of incisors; these powerful teeth being no doubt used to gnaw through the roots, which would otherwise obstruct their subterranean course. Other modifications in the form of the incisor teeth will be hereafter pointed out in the descriptions of the various species; I shall merely here observe, that those of the upper jaw are always shorter than those of the lower, and usually describe about three parts of a circle. The longer incisors of the lower jaw form a smaller segment of a larger circle.

Owing to the unequal density of different parts of the incisor teeth, the wear by their mutual attrition is unequal; the softer hinder part wears most rapidly, and the hard enamel in front least so; hence a sloping, chisel-like edge is always preserved. In proportion as the exposed end of the incisor is thus worn, additional dentary matter being supplied to the tooth, it continues to grow from the base, and thus preserves an uniform size.

As not only the sharp edge of the Rodent's incisor, but likewise the proper length of the tooth, is maintained by the constant attrition of an opposing incisor, should one of these teeth receive an injury, and be thrown out of its proper direction, the tooth which it should be opposed to will continue to grow, and retaining the curve given to it by the form of the socket in which it is implanted, will sometimes describe a complete circle: the power of mastication is destroyed, and the death of the animal necessarily ensues[1].

[1] Prof. Owen, in the 104th plate of his Odontography, has figured the upper jaw of a Rabbit in which, from some such cause, one of the upper incisors has assumed an unnatural length; its recurved point penetrates the upper jaw. The same author also calls attention to a lower jaw of a Beaver, preserved in the Museum of the College of Surgeons, in which, from un-

Eight is the smallest number of molar teeth found in any animal of the present group, and that occurs only in one species — the Hydromys of Australia; $\frac{3-3}{3-3}$ is the normal, though not constant number, in a very extensive section (the *Muridæ*), and $\frac{4-4}{4-4}$ is the constant dental formula of another large section (*Hystricidæ*): this number is exceeded in most of the species of the Squirrel family (*Sciuridæ*), where there is an extra small tooth in front of each row in the upper jaw, making $\frac{5-5}{4-4}$; and the highest number of molars is found in the Hare section, where the molars are $\frac{6-6}{5-5}$. The teeth under consideration are rooted, or rootless; in the former case, as the roots are formed by the decrease in the quantity of dental pulp —sudden at first, immediately after the formation of the body of the tooth, and gradual afterwards, when it is supplied only from certain points determining the number of fangs—they discontinue growing after a certain period; when, in fact, the animal has attained maturity; but in the latter case the molar teeth, like the incisors, continue to grow at the root in proportion as they wear at the opposite extremity.

The molars are composed of three substances of different degrees of density, of which the chief are called dentine and enamel: the dentine forms the main portion of the tooth, whilst the enamel, a much harder substance, sometimes merely forms an outer layer; but in most Rodents it likewise forms loops or folds, which penetrate deeply into the body of the tooth. External to the enamel is the third substance—cement, which enters very little into the structure of the rooted molar, but in the rootless, forms an important part, filling the interstices of the folds of enamel, and sometimes

checked growth, the incisor has described a complete circle, and in its course has passed through the masseter muscle, and entered the back part of the mouth.—Odontography, p. 411.

serving to tie together the different parts of a tooth where it is completely subdivided by the enamel, as in the molars of the Capybara.

The enamel being harder than the other two substances of which the tooth is composed, wears less rapidly, and presents sharp ridges on the crown of the tooth, which for the most part have a nearly transverse direction, whilst, as has been pointed out by Cuvier, in connection with this structure, the motions of the lower jaw are longitudinal. The patterns formed by the folds of enamel are very variable, and have been seized upon by mammalogists to assist in defining the minor groups of the order. Very frequently there are two principal folds indenting the outer side of the upper molars, and one on the inner; and the lower molars present the opposite condition in this respect. The Rodentia agree with most other placental Mammalia in never having more than three true molars on either side of each jaw; for where the series is composed of more than three, it is found that the additional teeth are pre-molars, having replaced deciduous, or milk teeth. In some species the deciduous teeth are shed before birth, as in the Guinea-pig; in the Hares they are shed when the animal is about eighteen days old, whilst in the Beaver I have found the milk tooth, which is afterwards replaced by the foremost of the four permanent molars, in the skull of an animal which was about half grown.

Deciduous incisors have not hitherto been detected in animals of the group under consideration, excepting in the Hares and Rabbits[1], in which, a few days after birth, six incisors are found in the upper jaw: they are small and pointed, and packed closely together, so as to form two longitudinal series: of these, four are deciduous, the anterior pair only

[1] I have sought in vain for deciduous incisors in young *Marsupialia*; if they exist, they must be shed at a very early period in these animals.

being permanent. In the lower jaw the two permanent incisors only are found.

The intestinal canal, in most Rodents, is very long, and the division of large and small intestines is marked by the presence of a cœcum, which is often very voluminous, and more or less divided into cells by internal septa: it is largest in those species which have rootless molars, and is of moderate size, and more simple in form, in the Rats, which subsist less exclusively upon vegetable substances, and have rooted, tubercular molars. In one small group, the *Myoxina*, or Dormouse section, the cœcum is entirely wanting, and here, moreover, the intestinal canal is very short. In most Mammalia the small intestines greatly exceed the large in length: the difference, however, is less marked in the herbivorous section; and in the division which now occupies our attention we find the small intestines occasionally even surpassed in length by the large, and, as a general rule, not only is the difference of length of the two divisions of the canal less than other Mammals, but the large intestines do not retain the character from which they derived their name, that of having a greater diameter than the small intestines: if we except a short dilated portion which joins the cœcum, they scarcely exceed the small in diameter.

The stomach varies much in its general contour in the Rodentia, but its transverse almost invariably exceeds its vertical diameter; the œsophagus usually enters near the middle of the upper, or concave surface; the left, or cardiac portion, is sometimes nearly globular, but most frequently ovate, or nearly conical, and is strongly recurved. The pyloric portion is generally much inflated, and separated from the cardiac half, by a more or less strongly marked constriction; its parietes are thicker, and more muscular than those of the cardiac, which are clothed with an epidermis. The two halves of the stomach are in general nearly equal in calibre. In the definition of the order I have already alluded to a very characteristic feature in the skull of the Rodent—the longitudinal direction of the glenoid cavity. The frequent occurrence of a distinct interparietal bone, even in the adult animals of this order, the great development of the inter-maxillary bones, and the large size which the ant-orbital opening often attains, are other points which must not remain unnoticed in connection with the skull. The eye cavities are lateral, often very large, and are never perfectly separated from the temporal fossæ: these latter are in many species of the group very small, especially in those which have a large ant-orbital opening[1].

The usual number of dorsal vertebræ is thirteen, but in many species there are twelve; some few have as many as fourteen, and in the Prehensile-tailed Porcupine there are sixteen. Of lumbar vertebræ, six is the most common number, but seven not unfrequently are found, and five occur in some few species. Many Rodents have three sacral vertebræ; five, and two occur, but are very rare, whilst four is the most frequent number. Nearly all Rodents are claviculated; in the Hares, however, the clavicles are very small, and the Cavies form an exception to the general rule, in being destitute of clavicles. The bones of the forearm are applied closely together, but rarely anchylosed. I have seen them partially anchylosed in the skeleton of the Paca (*Cœlogenys Paca*); and, in a skeleton of a small species of Sciuropterus,

[1] Of the peculiarities of the skull and skeleton presented by the different subdivisions of the present order, we shall have to speak hereafter (as well indeed as of other parts), in conjunction with the classification of the group, which it will be most convenient to take into consideration when the plates are before the reader, and can be referred to in illustration of the points advanced. The geographical distribution of the order will at the same time be treated of.

in the College of Surgeons, these two bones are perfectly joined at the lower extremity, as are the tibia and fibula in the same animal. The fore foot is almost always provided with four well-developed toes, and a rudimentary inner toe or thumb. Excepting in the Hare tribe and the Cavies, the fore-foot is used, like a hand, to convey food to the mouth. All the ordinary bones of the wrist are distinct, and the os magnum is generally divided.

The tibia and fibula are distinct bones in very many Rodents, but in the whole of the *Muridæ* and *Leporidæ* the latter bone is joined to the former, so that the two bones appear to form but one on the lower half of the shank. In the Cavies, Agoutis, and Gerboas (*Dipus*, as now restricted[1]), the hind foot is provided with but three toes; some few Rodents have four, but usually five toes are present. The *os naviculare* is often divided, and a supernumerary bone is very frequently met with on the inner side of the tarsus joining the os naviculare.

The structure of the terminal part of the muzzle offers some well-marked characters in the different sections of Rodents, but unfortunately the stuffed specimens in museums present a condition very unfavourable for the examination of this part: having, however, drawings and notes of the part in question of many species, taken from the animals immediately after death, or from specimens preserved in spirits, I will here point out some of the leading modifications.

A distinctly cleft upper lip, combined with a small naked muffle, divided by a vertical groove, and separating the nostrils, is characteristic of nearly the whole of the great Murin and Sciurine sections[2]. In these animals the nostril open

[1] The three metatarsals, which bear the toes, are united in this genus into one bone, having three articular surfaces at its distal extremity.

[2] I speak of these sections as they have been defined by myself in the Magazine of Natural History, and in the Annals of Natural History.

ings are lateral, rounded in front, suddenly contracted behind the rounded part by a descending fleshy lobe, and as they run backwards they curve upwards: they are, in fact, shaped like a comma. In several species of both these groups I have observed a ridge or fold crossing the naked part of the muffle from the upper angles of the nostrils, and where this fold occurs it indicates that the animals have the power of drawing the skin of the hairy upper portion of the muzzle down over the naked portion: such I know to be the case in many Squirrels and Rats, as well as in the Gerbilli and Gerboas. When the skin of the muzzle is folded over, the nostrils are closed, and the naked and sensitive portion of the muffle protected. The principal exceptions which I have noticed occur in the *Saccomyina*[1], where the upper lip is not cleft, presenting only a small notch immediately above the incisors. The same structure is observable likewise in the genus *Hydromys*; and in some of the *Arvicolina* the upper lip is also but partially cleft. The next modification in the muzzle is presented by the Hystricine division of Rodents. The common character of this group is to have an obtuse muzzle, the nostrils linear, and forming a slight S-like curve, and the muffle clothed (often rather sparingly) with fine velvet-like hairs. The upper lip is rarely cleft, and there is very seldom any groove between the nostrils. Sometimes the nostrils are nearly round, but run into a point above; the muzzle recurved, enormously swollen, and covered through-

[1] A group of Rodents found in North, and Central America, and in some of the West India Islands, all the species of which possess cheek pouches, opening externally: they have $\frac{4-4}{4-4}$ molar teeth. In some the teeth are rootless, and the tail is short; they constitute the genus *Geomys;* in others the tail is long, and the molars are rooted, as in the genera *Heteromys*, *Saccomys*, and *Perognathus*. *Dipodomys* no doubt also belongs to this section, which I provisionally form for genera which there appear to be good reasons thus to unite.

out with velvet-like hairs; no groove separates the nostrils, and the upper lip is merely notched at the margin. The Prehensile-tailed Porcupine (*Hystrix prehensilis,* Linn.) presents this structure of muzzle[1]. With the muffle similarly clothed with hairs, and destitute of vertical groove, I find the nostril openings assuming a nearly linear form, and a cleft upper lip, in the Fasciculated Porcupine of Africa (*Atherura Africana*, Gray). In the Hares the muffle is hidden by the folding over of the skin of the muzzle, and the upper lip is cleft. The Cavies approach the Hares most nearly in the structure of the muzzle, but they retain the common (I do not say constant) character of the Hystricine group, in having the upper lip entire.

In many Rodents the hairy upper lip laps over the anterior palatal portion of the mouth.

LEPORIDÆ, or HARE FAMILY.

Rodents, with incisors, $\frac{4}{2}$, and rootless molars, $\frac{6-6}{5-5}$ or $\frac{5-5}{5-5}$; skull with the two optic foramina united.

The Hare section is less numerous in species than other families of Rodents, and offers many exceptions to the general, or normal, characters of the order. The large size of the openings in the skull, combined with a very imperfect condition of the palate; the perforations in the nasal process of the superior maxillary bone, large orbits meeting in the mesial line of the cranium, the small temporal fossæ,

[1] When the animal is alive this great fleshy muzzle is constantly in motion. I may observe, there is a great tendency *generally* in the Hystricine group to this incrassated form of muzzle.

and the increased number of incisor and molar teeth, are among the more striking characters presented by the skull. The extra pair of incisors in the upper jaw are small, and placed behind the principal pair, and these latter are grooved in front. The incisors, both of the upper and under jaw, are shorter, *i. e.* less deeply implanted in the jaw than in other Rodents, and they are always white. The molar teeth are always rootless.

The stomach is simple, or partially divided internally; the cœcum, which is very long, is divided into numerous cells by tendinous bands, the partitions being indicated externally by a corresponding number of constrictions: similar constrictions are observable on the first part of the colon, but the colon soon contracts to a diameter which is nearly equal to that of the small intestines. At the point of junction of these latter with the cœcum, is a small glandular cul-de-sac.

The spine of the blade-bone, or scapula, terminates in a moderately long acromion process, which, near its extremity, sends down a long branch at right angles with the spine. The fore feet are provided with five well-developed toes, and the hind feet with four. At the lower half of the shank the two bones (tibia and fibula) are always anchylosed.

The inner surface of the cheeks is clothed with small hairs; at least, there is a moderately broad band of skin which is thus clothed, running backwards from the angle of the mouth. The tail is either short and bushy, and carried erect, or exists only in a rudimentary condition.

Examples of this family are found both in the Old and New World, but they are chiefly confined to the northern hemisphere; some, like the *Lepus glacialis*, extend into the polar regions. About 35° S. Lat. is the most southern range of the family in either hemisphere. The great continent of South America has yielded but a single species, and that does not occur south of the Rio de la Plata, whilst, on the other hand, it is in North America (extent of territory being considered) that the species are proportionately most numerous. In the Old World, the group, in like manner, almost disappears in the tropical portions; it reappears, however, south of the tropics, there being several species found at the Cape of Good Hope. But one species has been discovered in the Indian Islands[1], and in Australia there are no Hares; indeed, all the Rodentia found in that continent belong to the family *Muridæ*.

Fossil remains of the present family occur only in the newer tertiary deposits, and are referrible to the same genera as the existing species.

The family *Leporidæ* is composed of two genera only; one of them constitutes Cuvier's

Genus, *Lagomys*.

Leporidæ, with no visible tail, short and rounded ears, short hind legs, and molars, $\frac{5-5}{5-5}$.

The tailless Hares, or Pikas, as the animals of this division have been called, are of small size, the largest not exceeding the Common Guinea-pig in bulk. To Pallas we are indebted for the first account of certain members of this genus, discovered by him in his travels in Central Asia, one of which has since been found to extend into Europe, occurring in the southern districts of the Volga. A new species of the group was discovered by Dr. Richardson in the Rocky Moun-

[1] Of these islands it is in Java only that any Hare has been found, and that is specifically identical with an animal inhabiting continental India, and which occurs likewise in the Mauritius.

tains of North America, and more recently several species of Pikas have been found in the Himalayas.

Although in certain spots the Pikas are found in considerable numbers, they do not appear to be, strictly speaking, gregarious. They occur only in alpine or sub-alpine districts, where they form burrows in the ground, or sometimes take shelter amongst the loose stones. Occasionally, when the weather is cloudy, they will quit these retreats in quest of food during the day but the night is their ordinary time of feeding. Their food consists of various kinds of herbage, and, as in the high and cold regions which they inhabit, the herbage is covered with snow during the winter months, they must then starve, were it not that their instincts lead them to lay up a stock for this season. Large quantities of dried grasses, and other vegetable productions, are collected by the Pikas for their winter's consumption; these they pile up during the autumn, like small hay-stacks, which gradually disappear as the spring approaches, unless, as not unfrequently happens, these stores are robbed by the sable-hunters to feed their horses.

The Pikas, when feeding, frequently utter a chirping or whistling noise: in the *Lagomys pusillus* the sound is said to resemble the call of a Quail, and to be remarkably loud for so small an animal.

Beyond the chief distinguishing characters of the genus *Lagomys*, already pointed out, there are others which are worthy of notice, more especially in the structure of the skull. Upon comparing the skull of a Pika with that of a Hare[1], the most striking differences are the depressed form of the former; its being more dilated behind, the contracted interorbital space, and absence of supra-orbital process, the upward direction of the orbits, the great prolongation of the

[1] See Plate 2.

malar bone beyond the zygomatic process of the temporal, extending backwards, as it does, nearly to the opening of the ear-chamber. In lieu of the numerous perforations found in the nasal process of the superior maxillary bone of the Hares, there is but one chief opening in the same bone of the *Lagomys*[1]. Viewing the palatal portion of the skull of the *Lagomys*, we find that, owing to the depressed form of the cranium, the bodies of the sphenoids and the vomer are brought nearer to the plane of the palate than in the true Hares. The posterior sphenoid wants the mesial perforation, and the openings on either side of this bone are smaller; the vomer, moreover, joins the anterior sphenoid, whilst in the Hare an oblong opening separates the bodies of these two bones; the incisive openings are sometimes separated from the middle or chief palatine openings, but in some of the Pikas the openings are confluent, as in the true Hares. The occiput is broader than high, and the zygomatic arch is remarkably short. In the lower jaw (Plate 2, fig. 1. *b*.), the chief differences consist in the smaller antero-posterior diameter of the angular portion, and in the condyloid portion being sloped less backwards: the long thin plate representing the coronoid process in the true Hares, is in the Pikas replaced by a small tubercle, and there is a second, still smaller tubercle, placed below the one just mentioned, and, in fact, but little removed from the posterior molar teeth. The situation of the mental foramen is remarkable, since, instead of being placed in the fore part of the jaw, we here find it near the middle of the outer surface of the horizontal ramus.

The principal incisor teeth of the upper jaw are broad, but have a very small antero-posterior diameter; each of these teeth has a deep vertical groove on the outer side, and ter-

[1] A small bony branch sometimes crosses this opening—perhaps it is constant, but has been destroyed in some of the crania before me.

minates in two points, there being a strong notch at the extremity. The lower incisors are simple, and smaller than the upper. The upper molars can scarcely be said to differ in their structure from those of the true Hares; they present the same three transverse ridges of enamel, but on the hinder part of the last molar a small extra loop is visible—it is placed nearer to the inner, than to the outer angle of the tooth: this molar corresponds to the penultimate molar in the Hare's skull. The molar teeth in the lower jaw differ from the corresponding teeth in the Hares, in having the groove on the outer, as well as that on the inner surface, more deep; the body of the tooth is therefore more contracted in the middle, and the salient angles of the two halves of the tooth are more prominent; the foremost molar, as in the Hares, has two grooves on the outer surface; the last molar has but one salient external and internal angle, and in bulk does not equal half of one of the preceding molars. Small naked pads at the ends of the toes have been noticed amongst the characters which distinguish the Pikas from the Hares; these pads are generally distinct, but in two species (*L. pusillus* and *L. Ogotona*) they are exceedingly small, and can scarcely be traced; excepting these pads, the soles of the feet are throughout densely clothed with fur.

LAGOMYS ALPINUS.

Alpine Lagomys.

Lepus alpinus	**PALLAS, Novæ Species Quadrupedum e Glirium Ordine, p. 52, Pl. 2. 1778.—Reise durch verschiedene Provinzen des Russischen Reiches, tom. ii. p. 701, Tab. A. 1773.**
" "	**SCHREBER, Säugthiere, iv. p. 911, Tab. 238.**
Lagomys alpinus.	**CUVIER, Règne Animal (ed. 1829), i. p. 219.**

General hue of the upper parts grey-brown, of the under parts grey, tinted with yellow; feet pale, suffused with brownish yellow; ears with dirty yellowish white hairs internally, dusky towards the margin, and with the margin white.

The Alpine Lagomys, or Pika, inhabits Siberia, from the River Irtish eastwards, and is said to extend into Kamtschatka. Its fur is long and soft, and of a deep grey hue next the skin; on the upper parts of the body the hairs have a whitish ring beyond the middle, and from this part they become gradually darker to the point, first passing into brownish yellow, and then into black; on the sides of the body the brownish yellow hue is more conspicuous than on the back, for here the black is less extended on the exposed ends of the hairs. The hairs on the under parts of the body are of a yellow hue at the point, but the deep grey colour of the lower part of each hair is not entirely hidden when the fur is in its ordinary condition, and hence the general tint of these parts is greyish yellow. Over the hinder part of the haunches, and on the sides of the chest, the fur assumes a richer hue than elsewhere, being much suffused with rusty yellow. The dense fur on the soles of the feet is brown. The hairs of the moustaches are exceedingly long, and of a black hue. The ears are large and rounded, and clothed with dusky black hairs externally, excepting at the base, where they are greyish; on the inner side they are of a dirty yellowish white colour, but tipped with blackish; remote from the margin they are almost entirely white, but towards the margin the dark hue prevails; the margin itself is dirty white, but the pale edge is very narrow.

	British Museum.		From Pallas.	
	Inches.	Lines.	Inches.	Lines.
Total length	8	9	9	7
Length of ear	0	10	1	0
Width of ditto	0	11½	1	0½
Length of fore foot and nails ...	0	9½		10
" of hind foot and nails ...	1	5	1	5

The dimensions in the second column, taken from Pallas's account of *L. alpinus*, are, I do not doubt, from a fresh specimen, and more accurate than those contained in the first column, which are from a stuffed specimen in the British Museum—the original of my description. In the British Museum collection is a second specimen, which is entirely black; the skull, which has been removed from this latter animal, agrees very closely with Pallas' figure of the skull of *L. alpinus.* The skull (Pl. 2, fig. 1) is larger than that of other species of the genus; the interorbital portion of the cranium is broader than in the skull of *L. Ogotona* (which approaches it most nearly in size); the orbital fossæ are much smaller; the principal palatal opening, which is very broad, is not so long, and the incisive opening is separated by a long interval from the principal opening by the uniting of the superior maxillary bones in the mesial line; the lateral opening in the superior maxillary bone is smaller. The condyloid portion of the lower jaw, though not higher, has a considerably greater transverse diameter than the corresponding part of the lower jaw of *L. Ogotona;* the angle of the jaw, moreover, is less recurved.

LAGOMYS OGOTONA.

Ogotona Lagomys.

Lepus Ogotona.	PALLAS, Nov. Spe. Glir. p. 64, Pl. 3.
" "	SCHREBER, Säugthiere, iv. p. 915, Pl. 239.
Lagomys Ogotona.	CUVIER, Règne Animal, p. 219.

Ears and feet very densely clothed with fur: upper parts of the body of a pale ashy grey, with a slight admixture of yellow; feet, ears, and under parts of the body, white, tinted with cream-colour, or pale yellow.

Inhabits the subalpine districts beyond Lake Baikal, as well as of the Daurien, and the Mongolean Desert, and extending into China.

The fur of this animal is long, very soft, and of a deep grey colour next the skin: on the upper parts of the body the hairs are yellowish white, or cream-coloured, near the point, and grey, or some of them, black, at the point; the dark grey hue of the inner parts of the hairs is completely hidden by the pale terminal part, which is chiefly of a cream colour, but the extreme points of the hairs being dusky, the mixture produces an ashy grey hue; on the sides of the body the hairs are less distinctly tipped with dusky, and here a yellowish cream colour predominates: two very large patches behind the ears are of the same tint; the hinder parts of the haunches are slightly suffused with rufous; the under parts of the body are cream colour, and so are the feet, but the toes are white, and the under surface of the hind foot is brown. The ears are densely clothed with long hairs, and these are for the most part of an uniform white; those covering the fore part of the outer surface, however, are somewhat dusky, and the extreme points of many of the hairs on the inner surface of the ear are likewise dusky, more especially those near the margin.

The above description is taken from a specimen in the British Museum collection: the skull resembles that represented by Pallas of his *L. Ogotona,* not only in size, but in having the interorbital portion elevated; this part is much more contracted than in other species of Lagomys I am acquainted with. The superior maxillary bones do not quite meet in the mesial line, so as to separate the incisive opening from the principal palatal opening.

	Brit. Museum. Inches.	Lines.
Total length	8	0
Length of ear		$8\frac{3}{4}$
Width of ditto	1	0
Length of fore foot		11
" of hind foot	1	$5\frac{1}{2}$

LAGOMYS PUSILLUS.

Pigmy Lagomys, or Sulgan.

(*Plate* 1, Fig. 2.)

Lepus pusillus.	PALLAS, Nov. Comm. Petrop. xiii. p. 531, Pl. 14, f. 1; Nov. Spec. Glir. p. 37, Pl. 1.
" "	SCHREB., Säugthiere, iv. p. 906, Pl. 237.
Lagomys pusillus.	DESM. Mammalogie, p. 353.
" "	CUVIER, Règne Animal, p. 219.

General tint brown, the fur rather strongly pencilled with black and brownish yellow; feet and under parts yellow-white; ears small and rounded, with a distinct white margin, and internally with a submarginal black band, within which the hairs are yellow.

Inhabits the southern districts of the Volga, and the southern portions of the Ouralian Mountains, and occurs likewise in South Siberia, extending eastward to the River Obi.

Lagomys pusillus may be distinguished from the two preceding species (*L. Ogotona* and *L. alpinus*), by its smaller size, and, from these and other species described, in having the ears proportionately smaller, with a broader white margin, and with a distinct submarginal black band on the inner surface: the dark band, however, is confined to the hinder and lower parts of the ear: within the dark band the hairs assume a rich yellow hue: externally the ears are black in front, and grey on the hinder part, where the hairs are very long. The general brown tint of the upper parts of the body is produced by the mixture of black and brownish yellow, the hairs being annulated towards the point with the latter colour, and black at the point: next the skin they assume a blackish grey hue; on the under parts of the body the deep grey hue is but partially hidden by the overlapping pale yellowish points of the hairs. The feet are nearly white above, but slightly tinted with yellow; the dense fur on their under surface is dusky brown. Naked pads beneath the toe-nails can scarcely be said to exist, or at least they are smaller than in other species, if we except the *L. Ogotona*.

The foregoing description is from a specimen in the British Museum, the dimensions of which agree very closely with those given by Pallas, some of which I have added in the second of the subjoined columns.

	British Museum.		From Pallas.	
	Inches.	Lines.	Inches.	Lines.
Total length	7	0	6	9
Ear		8½		9
Fore foot and nails		8		8
Hind foot and nails	1	1½	1	1

The skull of *L. pusillus* agrees with that of *L. alpinus* in being comparatively broad between the orbits—much broader and more depressed at this part than in *L. Ogotona*. In the form of the nasal bones it differs from both species, inasmuch as these bones are as broad behind as in front (or very nearly so); the skull is considerably smaller than in either of the species just mentioned, and the lateral openings in the superior maxillary bones are larger. The incisive and principal palatine openings are confluent, and the angle of the lower jaw is very little recurved.

LAGOMYS RUFESCENS.

Rufous Lagomys.

Lagomys rufescens. GRAY, Annals and Magazine of Nat. Hist. No. 65 (vol. x.), December, 1842, p. 266.

General hue pale sandy red, darker on the upper surface of the head, as well as on the shoulders and fore part of the back;

two very large patches behind the ears, the feet, and the under parts of the body, are of a pale buff yellow: ears moderately large, subovate, and well clothed with hairs; rusty yellow on the inner surface, as well as on the fore part of the outer surface, but pale buff yellow on the under part: soles of the feet obscurely suffused with rufous.

Inhabits the Rocky Hills of Cabul.

The hairs of the fur of this animal are of a dark slate-grey colour, tipped with pale yellow on the under parts of the body, and on the upper parts they are broadly ringed with pale rufous near the point, and black, or dusky, at the point. The large pale patches behind the ears are formed of hairs which differ from those of other parts of the body in having the grey occupying but a small part of their extent, being chiefly confined to the root. The hairs of the moustaches are very long, and for the most part brown; some few are brown-white. The toes, both of fore and hind feet, are terminated with distinct naked pads of a blackish colour.

	Inches.	Lines.	Inches.	Lines.
Total length	8	0	7	0
Length of ear		8½		8
Width of ditto		9		9
Length of fore foot and nails		10½		9½
" of hind foot and nails	1	5	1	3¼

The skull of *Lagomys rufescens* is considerably broader than that of *L. nepalensis;* it has the nasal bones narrower, and distinctly contracted behind, whilst in the Nepal animal, the nasal bones are nearly as broad behind as in front; the interorbital space is narrower, and there is a marked difference in the palatine openings, as I have pointed out in the account of *L. nepalensis*. The only skull of *L. rufescens* which I have had an opportunity of examining[1] is imperfect, wanting

[1] This skull was removed from the smaller specimen referred to in my description, the dimensions of which are given in the second column.

the hinder parts; the existing parts are larger than in *L. nepalensis*, and smaller than in *L. Ogotona*, and, if the skull were perfect, its entire length, no doubt, would be very nearly $1\frac{3}{4}$ inches. Although narrower between the orbits than in *L. nepalensis*, this part is not so contracted as in *L. Ogotona*: the lateral openings in the nasal portion of the superior maxillary bones are distinctly smaller than in the skull of either of the animals just mentioned.

The above description is taken from one of two specimens in the British Museum; the second specimen is smaller, and differs somewhat in its colouring, having the back of a sandy hue, very obscurely tinted with rufous, and finely pencilled with brown, and the whole of the back of the neck yellowish white. Other specimens are contained in the museum of the India House, having been sent by Dr. Griffith, who states that the animal is "not uncommom in rocky grounds about Cabool, and perhaps generally between an elevation of 6,000 or 8,000 feet[1]."

Mr. Blyth, speaking of a species of Lagomys found in the western part of the Himalaya district (the *L. Hodgsonii*), states that it is probably the same as the small species noticed by Capt. Thomas Hutton in Afghanistan; this latter is, no doubt, the animal just described, and, upon comparing it with Mr. Blyth's description of his *Hodgsonii*, it would appear to differ in its general colouring being paler, and in having either the whole, or the greater part, of the back of the neck of a yellowish white tint; the superior breadth of the skull, moreover, would distinguish the *L. rufescens* from the *L. Hodgsonii*: on the other hand, the skull of Mr. Blyth's animal, judging from the figures and dimensions given, agrees

[1] See Dr. Griffith's Report in the tenth volume of the Journal of the Asiatic Society of Bengal.

most closely with that of the *L. nepalensis*, and, as in the description of its colouring I can perceive no important point of distinction, I am strongly inclined to believe the two are specifically identical. The principal characters of Mr. Blyth's species I shall proceed to extract from that gentleman's account.

Lagomys Hodgsonii.

Lagomys Hodgsonii. BLYTH, Journal of the Asiatic Society of Bengal (1841), vol. x. p. 816, and Plate at p. 854.

General colouring nearly resembling that of the Common Water-Rat, having a distinct rufous tint; the longer hairs of the fur, which are three-quarters of an inch in length, have the basal half, or rather more, of a slaty black colour; above this they are first pale brown, then rufous brown, and, finally, at the point, they are dusky; the under parts are of a dirty white hue; the feet brown, with a faint tinge of rufous on the upper surface; the ears are of a rounded oval form, nearly naked externally, but with a narrowish border of brown hairs on the fore part; internally there is likewise a border of brown hairs, but this is comparatively broad; the margin is obscurely fringed with white: upper surface of the head, and the region of the shoulders, of a richer rufous tint than other parts.

This species is said to be common in Lahoul, Ladakh, and Kooloo.

	Inches.	Lines.
Total length	6	0
Hind foot and nails	1	3
Length of ear		10½
Width of ditto		9
Length of skull	1	7½
Width of ditto		10½
From front of palate to inferior margin of occipital opening		10½ (nearly)
Height of cranium, with the lower jaw *in situ*	0	13½

LAGOMYS NEPALENSIS.

The Nepal Lagomys.

Lagomys nepalensis. **HODGSON, Journal of the Asiatic Society of Bengal (1841), vol. x. p. 854, and Plate at p. 816.—Annals and Mag. of Nat. Hist. for Sept. 1842, vol. x. p. 76.**

Fur very soft, that on the head and fore parts of the body of a rich chestnut red, on the back brown, and on abdomen white, more or less suffused with rufous or pale chestnut; feet whitish, tinted with rufous above; ears rather large, brownish internally, and with an indistinct pale margin; a whitish patch behind each ear.

Inhabits the northern hilly districts of Nepal, and Mr. Hodgson is informed that the species is likewise found in Thibet.

The skull of *L. nepalensis,* like that of *L. pusillus,* has a comparatively broad interorbital space, and the nasal bones are of very nearly equal width throughout; but these bones are longer than in *L. pusillus.* The most marked difference existing between the skull of the Nepal Lagomys and that of the *L. pusillus* (or, indeed, of that of any other species which I have examined), consists in the perfect confluence of the incisive and principal palatine openings, a character which approximates the Nepal Pika more nearly to the true Hares. In *L. alpinus* the superior maxillary bones unite so as to separate the two openings in question by a longish space, and in other species of *L. Lagomys* the same bones are approximated more or less closely: in *L. Ogotona* they nearly meet, and in *L. pusillus* and *L. rufescens,* though they are more distinctly separated, and the two openings are more confluent, yet there is a decided contraction between them, serving to mark their natural boundaries, whilst in *L. nepalensis* no contraction at the part in question exists; there is but one large opening, commencing close behind the incisive teeth, where it is very narrow, and becoming gradually broader to its termination, which is in a line with the space between the second and third molars. The groove in the upper incisor teeth is rather broader than usual.

Beyond the distinctions furnished by the structure of the skull, the *L. nepalensis* differs from other species in having the head and fore parts of the body of a deep and rich chestnut red. In *L. rufescens* the top of the head, and the shoulders, are reddish, but paler than in the Nepal animal, which has but a small pale patch behind each ear. The fur on the back has a general brownish hue, but it is rather distinctly pencilled with black, and pale yellowish brown, the hairs (which are as usual of a deep slate-grey for more than half their length), being annulated with the pale colour towards the point, and black at the point. On the under parts of the body the dark grey of the inner part of the fur is but partially hidden by the pale points of the hairs; they are either tipped with yellowish white, or pale chestnut red; on the chest the latter colour is distinct. The ears are moderately well clothed with short hairs[1], the general hue of which is brown on the inner surface; and on the outer they are nearly black on the fore part, and pale on the hinder: these organs are large and of a rounded form; they are rather broader than in *L. rufescens*. The toes are terminated with distinct naked pads, and there is a small naked tubercle on the under side of the wrist. The dense fur on the under surface of the feet is of a dusky brown hue.

[1] Mr. Hodgson states that the ears are "nearly nude, except on the anterior and incurved edge of the helix, where very short hairs are pretty closely set." This descriprion agrees with one only of several specimens in Mr. Hodgson's collection which I have examined, and in that it appears to me the hairs have been rubbed off of the ears.

	British Museum. Inches.	Lines.	Leyden Museum. Inches.	Lines.
Total length	8	0	7	6
Length of the ear		8¾		8½
Width of ditto		11½		9½
Length of fore foot and nails, about		10		10
" of hind foot and nails ...	1	4	1	3½

LAGOMYS ROYLII.

Royle's Lagomys.

Lagomys Roylii. OGILBY, Memoir on the Mammalogy of the Himalayas, in Royle's Illustrations of the Botany, &c. of the Himalayan Mountains, p. lxix, Pl. 4.[1] 4to. London, 1839.

" " IS. GEOFFROY, in the Voyage dans l'Inde, par Jacquemont, Mammifères, p. 62.

General hue of the fur greyish brown ; on the under parts of the body dirty white, tinted with yellow : ears small and rounded, with a very narrow pale margin.

The specimen upon which Mr. Ogilby founded the *L. Roylii* was discovered by Mr. Royle in the Choor Mountain, at an elevation of 11,500 feet, and, according to M. Jacquemont, the same species is likewise found in the valley of the Yurpo, and in Cashmere, in the high valley where the Sind and Gombur divide.

The *Lagomys Roylii* most nearly resembles the *L. pusillus*, having the same small and rounded ears, but these organs have but an indistinct narrow margin, and they want the decided black mark on the inner surface. The comparatively small size of its ears, and the total absence of any rufous colouring about the head and shoulders, at once dis-

[1] The animal is here named *Lagomys alpinus*, a mistake which no doubt gave rise to the statement made by Dr. Wagner (Schreb. Säug. Suppl. iv. p. 120), that the *L. alpinus* is found on the Choor Mountain.

tinguish it from the other species of Northern India. The fur of the present animal, although soft, is rather less so than in the *L. nepalensis ;* as in other species, it is of a deep slate-grey hue next the skin. On the upper parts of the body the hairs have a whitish ring towards the point, are black at the point, and between the white and black parts they assume a brownish yellow colour; the mixture of these colours produces a general hue, which may be described as greyish brown. On the under parts of the body the hairs are tipped with very pale yellow. The ears have the usual dusky patch on the fore part of the external surface, and internally they are tolerably well clothed with brownish hairs, which are dark towards the margin, and become paler more remote from that part; the margin of the ear is very narrowly edged with dirty white. The toes of the fore foot terminate with distinct naked pads, and there is a small naked tubercle on the under surface of the wrist. The hind feet are wanting in the specimen before me, which is likewise the original of Mr. Ogilby's description. The length of the skin, which is partially stuffed, and in bad condition, is 8 inches; of the ear, $\frac{7}{12}$ths of in inch, and of the fore foot, $9\frac{2}{3}$ lines.

The skull, according to the figure in Dr. Royle's work, is about equal in size to that of *Lagomys nepalensis ;* its upper surface, it would appear, is more arched, and the palatine opening smaller; the incisive opening is confluent with the principal opening, and the figure does not represent the sudden contraction of the fore part of the great opening formed by the junction of the two foramina mentioned, in which respect the skull of *L. Roylii* presents a marked difference when compared with that of *L. pusillus*.

On the whole, the materials at my disposal, as it appears to me, fully justify the conclusion that there are three distinct species of *Lagomys* in the northern parts of India, and that these species are neither of them identical with either of the three Russian species described by Pallas in his "Glires."

LAGOMYS PRINCEPS.

The Rocky Mountain Lagomys.

Lepus (Lagomys) princeps. RICHARDSON, Zoological Journal, No. 12, p. 520, March, 1828.—Fauna Boreali Americana, p. 227, Pl. 19. 1829.

General hue of the fur on the upper parts of the body greyish, but these parts are strongly pencilled with black and yellowish white; on the sides of the body yellowish brown, and on the under parts dirty white, considerably suffused with pale brown-yellow; feet white, tinted with yellow on the upper surface; ears rather large, with a distinct white margin.

Dr. Richardson has been enabled to trace this little animal, known as the "Little Chief Hare," along the Rocky Mountains, from the 52d to the 60th parallel of latitude, and we learn from Dr. Bachman that it has a still more extensive range on this great North American chain, it having been found by Mr. Nuttall as low as 42° N. Lat. According to the observations of Mr. Drummond, quoted by Dr. (now Sir John) Richardson, the *L. princeps* frequents heaps of loose stones, through the interstices of which it makes its way with great facility. It is often seen after sunset, mounted on a stone, and calling to its mates by a peculiar shrill whistle. On the approach of man it utters its feeble cry, like the squeak of a Rabbit when hurt, and instantly disappears, to reappear in a minute or two at the distance of twenty or thirty yards, if the object of its apprehension remains stationary. Mr. Drummond never found any burrows, and is of opinion that they do not make any, but that they construct their nests amongst the stones. They do not make their appearance during the winter months, and hence we may infer that, like other members of the group, they lay up a store of provisions for these months; such stores, however, have not been observed.

The Rocky Mountain Lagomys is about equal in size to the *L. pusillus*, but is readily distinguished from that animal by its colouring, and the larger size of its ears: its fur is long and soft, and of a darkish grey next the skin; on the back each hair has a broad subterminal yellow-white ring, and a black point; on the sides of the body the visible portions of the hairs are of a yellow-brown hue, and on the under parts they are tipped with pale yellow: a tolerably large whitish patch is observable behind each ear: the ears are moderately well clothed on the inner side with short hairs, which are partly white, and partly black, the black prevailing near the margin, which is pure white; on the outer side the hairs covering the ears are long and fine, black on the fore part, and grey-white on the hinder part. The feet are pale, and the naked pads at the ends of the toes are distinct.

	Inches.	Lines.
Total length	7	0
Length of ear		8
Width of ditto		10
Length of fore foot and nails		8½
" of hind foot and nails	1	2½

The above description is from a specimen in the British Museum, it having been presented to the National Collection by Sir John Richardson. A second specimen, presented to the Museum by the Hudson's Bay Company, differs from the one described in having the hairs on the inner surface of the ears long, and much suffused with black.

LAGOMYS HYPERBOREUS.

The Polar Lagomys.

Lepus hyperboreus. PALLAS, Zoographia Rosso-Asiatica, i. p. 152, published 1831.

Lagomys " WAGNER, in SCHREB., Säugth. Suppl. vol. iv. p. 121. 1844.

Fur soft and dense, on the upper parts of the body grey-brown, on the crown of the head tinted with reddish; chest, sides of the head, neck, and body, suffused with rufous; chin and throat white; ears rounded; white at the margin; feet whitish; toes with a naked, black fleshy pad at the extremity: the fur of a lead-grey at the root.

The Polar Lagomys is described by Pallas from specimens received by him from the country of the Tschouktschis, in the north-east portion of Siberia. It is said to be about equal in size to the Norway Lemming, and hence is considerably smaller than either of the other known species. Its head is longer than in *L. alpinus*, and in form nearly resembles that of the Common Rat (*Mus decumanus*).

	Inches.	Lines.
Total length	5	3
Length of head	1	5
" of ear	0	7
" of fore foot	0	7
" of hind foot	0	10½

The subjoined dimensions are taken from the skulls of the various species of Lagomys contained in the British Museum collection.

	Lagomys alpinus.		Lagomys Ogotona.		Lagomys pusillus.		Lagomys spelæus.		Lagomys rufescens.		Lagomys nepalensis.	
	Ins.	Lines.	Ins.	Lines.	Ins.	Lines.	Ins.	Lines.	Ins.	Lines.	Ins.	Lines.
Total length of skull	2	1	1	10	1	7					1	$8\frac{2}{3}$
Width	1	0		$11\frac{1}{2}$		$9\frac{2}{3}$		$9\frac{2}{3}$		11		$10\frac{1}{2}$
From apex of intermaxillaries to posterior edge of glenoid cavity ...	1	$2\frac{2}{3}$	1	$2\frac{1}{2}$		$11\frac{1}{3}$		$11\frac{1}{3}$	1	$1\frac{1}{4}$	1	1
Length of nasal bones		$8\frac{1}{3}$		8		$6\frac{1}{2}$				$6\frac{2}{3}$		7
Width of ditto behind		$2\frac{1}{4}$		$2\frac{5}{6}$		$2\frac{1}{2}$		2		$1\frac{5}{6}$		$2\frac{1}{3}$
" " in front		3		$2\frac{2}{3}$		$2\frac{5}{8}$		$2\frac{1}{3}$		$2\frac{2}{3}$		$2\frac{3}{4}$
" of interorbital space		$2\frac{5}{6}$		$1\frac{1}{2}$		$2\frac{3}{8}$		2		2		$2\frac{1}{3}$
Longitudinal diameter of the joined orbital and temporal cavities		$5\frac{1}{2}$		$6\frac{1}{4}$		5		$4\frac{3}{4}$		6		$5\frac{1}{2}$
Vertical diameter of ditto		$4\frac{1}{2}$		5		4		$3\frac{3}{4}$		$4\frac{5}{6}$		$4\frac{1}{2}$
Length of palatine opening		$3\frac{2}{3}$		$4\frac{1}{3}$		$3\frac{1}{6}$		3		6		6
Width of ditto behind		$2\frac{2}{3}$		$2\frac{2}{3}$		$2\frac{5}{8}$		$2\frac{1}{3}$		$2\frac{1}{2}$		2
From back of principal upper incisor to posterior margin of palatine opening		8		$7\frac{3}{4}$		$5\frac{1}{2}$		$5\frac{1}{2}$		7		$6\frac{1}{2}$
Antero-posterior diameter of the narrow bridge of the palate		$1\frac{1}{3}$		$1\frac{1}{4}$		$0\frac{7}{8}$		$0\frac{7}{8}$				1
Width of palate, taken opposite the last molar		$4\frac{1}{4}$		4		$3\frac{1}{3}$		$3\frac{1}{3}$		$3\frac{1}{2}$		4
Length of the five upper molars taken together		$4\frac{1}{2}$		$4\frac{1}{2}$		$3\frac{5}{8}$		$3\frac{3}{4}$		4		4
" of lower jaw	1	4	1	3	1	$0\frac{1}{4}$			1	$1\frac{1}{3}$	1	$1\frac{1}{3}$
Greatest height behind		9		9		$7\frac{2}{3}$				8		$8\frac{1}{3}$
Transverse diameter of the angular portion		$3\frac{3}{4}$		$3\frac{3}{4}$		$2\frac{7}{8}$						$2\frac{3}{4}$

Fossil species of Lagomys.

Lagomys Corsicanus, BOURDET.

Founded upon a nearly entire skull discovered in the osseous breccias of Corsica, and described and figured in Cuvier's Ossemens Fossiles, 4to. ed. vol. iv. p. 199, Pl. 14, figs. 4, 5, and 6.

This skull very nearly resembles that of the *Lagomys alpinus*, having the same broad space between the orbits, but presents some slight differences in the proportions of certain parts, and is a trifle larger.

Lagomys Sardus, WAGNER.

Also founded upon some remains figured and described by Cuvier (see Oss. Foss. iv. p. 203, Pl. 15, figs. 16, 17, and 18). These remains were found in the breccias of the neighbouring island—Sardinia. The principal fragment is a portion of an upper jaw containing the alveoli of the five molar teeth. According to Cuvier, the animal must have been larger than the recent species, *Lagomys Ogotona*, and smaller than the *L. alpinus*, and consequently likewise inferior in size to the fossil species of Corsica. *Lagomys* ——— *?* M. Pictet refers a humerus found in the cavern at Mialet (Cévennes) to the genus Lagomys.—See his Traité Elémentaire de Paléontologie.

Lagomys Œningensis.

The examination of a crushed skeleton from the Pliocene, lacustrine formation, at Œningen, contained in the British Museum, leaves no doubt on my mind that the animal was a species of *Lagomys*. The five upper molar teeth are distinct, and the angular portion of the lower jaw preserves the characteristic form of that part in the species of Lagomys. It is very probably specifically identical with the animal whose skeleton is figured by Cuvier, and which was received from the same quarter.—See Oss. Foss., vol. v. pt. 1, Pl. 3, fig. 18, p. 62.

Lagomys spelæus, OWEN.

A British species founded by Prof. Owen upon the anterior portion of a skull preserved in the collection of the British Museum, and which is from Kent's Hole. The dimensions of this specimen are given with those of the skulls of the recent species of Lagomys, upon referring to which, it will be seen that the fossil corresponds very closely with the *Lagomys pusillus*; there are, however, some slight differences, which, if they do not warrant us in considering the two animals as distinct, are at least sufficient to caution us against assuming their specific identity. The fossil alluded to, is figured by Prof. Owen in his History of British Fossil Mammalia, p. 213.

Genus, *Lepus*.

Lepus. LINNÆUS, Systema Naturæ (12th ed.), i. p. 77. 1766.

Leporidæ, with molars, $\frac{6-6}{5-5}$; large, and elongated ears (as long, and often longer, than the head); a short and very bushy tail; the hinder legs powerful, and much longer than the fore legs.

The Common Hare and Rabbit, together with numerous other species for the most part confined to the northern hemisphere, are comprised under the generic title *Lepus*. The species being numerous, and often very difficult to distinguish, naturalists have sought for characters to subdivide the group, and more especially to distinguish the Rabbits from the Hares; but although the young of the former are blind and naked when born (at least such is the case in the Common Rabbit), and the latter are clothed with hair, and have the eyes open, and in their habits there is this difference, that the Rabbits burrow, whilst the Hares make a "form," or kind of nest on the surface nf the ground on which they squat,—no constant structural distinction has, that I am aware, been discovered[1]. By far the greater number of the species of Lepus agree with the Hare in the habits noticed, and that animal may therefore be regarded as the type of the genus, and we shall use the term Hare in a generic sense.

Besides the difference in the number of the molar teeth, the possession of a tail, the large size of the hind legs, and the large and elongated ears, which distinguish the Hares from the species of *Lagomys* (the only other genus of Rodents in which the upper jaw is provided with more than two incisors), there are many points in the structure of the animals under consideration which require to be noticed.

Destitute of means of defence, the Hares are timid, have remarkable powers of flight, and, to warn them of danger, their senses of hearing, seeing, and smelling, are unusually developed. The eyes are very large and prominent, and,

[1] Upon comparing the bones of the feet of the Rabbit with those of the Hare, I have observed differences, some of which will probably be found to have a constant relation to the habits, but these differences would scarcely warrant a subgeneric separation of the animals which present them. The terminal phalanges of the toes in the Rabbits, for instance, differ from those in the Hare in being less deep, and less compressed, and, like the same bones in several other Mammals having burrowing habits, they are cleft—on the upper surface at least. The hind foot of the Hare possesses three bones more than is found in the Rabbit, viz. a tolerably large supernumerary bone with a convex outer surface, which is attached to the back of the os naviculare; a distinct internal cuneiform, and, attached to this, a rudimentary metatarsal of the inner toe: neither of these do I find in the Rabbit's foot. In the form of some of the bones there are marked differences. The great expanded process on the hinder, or under, part of the os naviculare, in the Rabbit's foot, is not found in that of the Hare, whilst, on the other hand, we find a large protuberance on the under surface of the external, or third cuneiform, partly taking its place. The inner metatarsal, or that which corresponds to the second in man, articulates with the second cuneiform bone by a simple and nearly flat surface in the tarsus of the Hare, whilst in that of the Rabbit the inner metatarsal sends up a process on the inner side, which, enclosing the second cuneiform, runs up to articulate with the naviculare; this process, indeed, occupies the place of the first cuneiform. In the form of the proximal extremity of the outer metatarsal, the two animals, moreover, differ considerably.

being placed laterally, enable the animal to see in all directions, or nearly so, at the same time; the nasal portion of the skull[1] is very large in proportion to the part devoted to the protection of the brain. The upper surface of the skull is strongly arched in the longitudinal direction, its highest point being rather behind the centre of the orbit; from this point it descends rather suddenly to the occiput, but forms nearly a straight line in front; the cranial portion is also convex in the transverse direction, but the upper surface of the facial part is flattened. The broad nasal bones send down from the inner surface some thin bony laminæ, of which the principal one is longitudinal, and are somewhat loosely attached to the skull. The frontal bones are remarkable for the large supra-orbital processes which are thrown out from them, a character which is almost peculiar to the Hares; the supra-orbital processes, after leaving the frontal bone, are greatly produced, both in the anterior and posterior directions, the extremities being free, and they form the whole upper boundary of the orbits. A slight depression in the comparatively small temporal bone marks the boundaries of the diminutive temporal muscle. The nasal process of the superior maxillary bone is remarkable for being pierced by numerous perforations; the lachrymal bone is situated almost entirely within the orbital cavity, but throws out a small process which projects beyond the boundary of the orbit: the infra-orbital opening is small, and serves for the transmission of the nerve only; the zygomatic arches are compressed, moderately deep, straight, and parallel to each other, or very nearly so; no suture marks the anterior boundary of the malar bone, which, as in the species of *Lagomys*, projects in the form of a pointed process beyond the hinder boundary of the glenoid cavity, but is less prolonged than in those animals. The

[1] Plate 2, fig. 2.

occipital surface of the skull is small, and slopes inwards from the upper edge, which thus overhangs the lower part; the supra-occipital bone presents a singular conformation, the upper surface being divided into three nearly equal parts—a square raised platform in the middle, and a wing-like process on either side; and there are numerous minute perforations in these parts. The auditory bullæ are of moderate size, and the petro-tympanic bone is but loosely attached to the cranium, but is held in its place by a little bony band thrown out from the temporal bone. The opening for the spinal chord is very large. Viewing the under surface of the cranium[1], the most striking point is the very imperfect condition of the palate; the large incisive openings, and the deep posterior emargination, contract the palate, so that it forms only a narrow bridge, crossing opposite the anterior molars. Owing to the contracted form of the palate, the sphenoid bones are exposed; the bodies of these bones are perforated and hollow within[2]. The lower jaw (Plate 2, fig. 2 *b.*) is remarkable for its flatness, and the great height and width of the ascending ramus: the coronoid process is represented by a thin plate, of but little width, which runs along the anterior edge of the ascending ramus: a perforation passes through this ascending branch immediately behind the last molar. The angular portion of the jaw is on the same plane with the horizontal and ascending rami, is very large, and approaches to a semi-circular form, but presents two obtuse angles, as will be seen in the figure.

The upper and lower incisors are about equal in width, and

[1] Plate 2, fig. 2 *a*.

[2] Not only are the bodies of two of the cranial vertebræ perforated, but the corresponding parts in nearly the whole of the vertebræ in the Hares are perforated; usually the openings, two in number, are confined to the inner surface of the body, but many of the lumbar vertebræ likewise present two perforations on the under surface.

their diameter is less in the antero-posterior, than in the transverse direction ; those of the lower jaw have the anterior surface smooth and flat, and those of the upper jaw have a strong groove in front, which is placed rather nearer to the inner than to the outer side of the tooth : the extra pair of incisors are small, and placed behind the principal pair[1]. The molar teeth are small in proportion to the skull ; those of the upper jaw, with the exception of the first and last, have a vertical groove, both on the outer and inner surface ; and the foremost broader tooth has three grooves ; the crowns of these teeth are broader than long, and present three transverse ridges ; the foremost is somewhat smaller than either of the four following teeth (which are of equal size), wants the mesial ridge, and has a small fold of enamel indenting its anterior surface ; the hindermost molar is very small, and in the form of a compressed cylinder. The crowns of the molars of the lower jaw have their longitudinal and transverse diameters nearly equal ; like those of the upper jaw, they present three transverse ridges of enamel, and the shaft of each tooth is grooved on the outer and inner side, but the grooves are much stronger, and so indent the tooth that its crown presents two salient angles on each side : the hindermost molar is considerably smaller than the rest. The motion of the jaws must be chiefly lateral in the Hares, for the crowns of their molar teeth are never worn flat, as in other Rodents which have rootless molars.

With regard to the vertebræ of the trunk, the points most worthy of notice are, that the foremost, or atlas vertebra, has but one perforation on either side, besides the usual one in the transverse process, and this passes from the outer to the inner surface, and has its outlet immediately in front of the

[1] See Plate 2, fig. 2 *a*, and 2 *c* ; the latter figure represents the upper incisors viewed from the side.

transverse process. The posterior dorsal and the lumbar vertebræ have the anterior articular processes much produced in the vertical direction, in some of these vertebræ rising as high as the spinous process; the three lumbar vertebræ are remarkable for having a spine on the under surface of the body. The scapular bone is much elongated; the humerus is perforated immediately above the lower articular surface, which presents two strongly-raised and acute ridges, and in this respect the humerus of the Hare differs from that of most other Rodents. The femur, which is long and slender, has a third trochanter in the form of a ridge, which is produced in the middle into an obtuse angle; it is situated unusually near the upper extremity of the femur. The fibula joins the tibia rather above the middle of that bone.

LEPUS TIMIDUS.

The Common Hare.

Lepus timidus.	LINN., Syst. i. p. 77.
" "	SCHREB., Säugth. iv. p. 865. Pl. 233 A.
" "	PALLAS, Zocgraphia Rosso-Asiatica, i. p. 148.
" "	DESM., Mammalogie, Part 2. p. 347.
" "	WAGNER, SCHREB., Säugth. Suppl. iv. p. 75.
" "	BELL, British Quadrupeds, p. 333.
" *Europæus.*	PALLAS, Glires, p. 30.
Lièvre.	BUFFON, Hist. Nat. vi. p. 247, Pl. 38 : DAUBENT. p. 264, Pl. 39—49—Anatomy.

Hase, of the Germans.

Ears longer than the head ; fur mottled with black and brownish ochre ; back of the neck and the limbs externally suffused with rufous; abdomen, inner side of limbs, and tail, white ; the upper surface of the latter black ; ears black at the point.

Excepting in the more northern parts, the Hare is found throughout Europe, extending southward to the Mediterranean, being found in Italy and Greece. According to Pallas, it occurs in Russia below the 55th parallel of latitude, and there, is the only species found. It is a native of Poland, Lithuania, and the Crimea; is found about the Caucasus, the lower part of the Volga, and the southern extremity of the Ural chain of Mountains, and extends into Persia. In Norway and Sweden, as well as in Siberia, it is wanting.

The Hare feeds by night, and during the day lies in its seat, or form; a mere depression in the ground made by the animal, usually near some bush. When crouched on its seat it is with difficulty distinguished, and it will often allow a person to approach without quitting its position. Timid, and swift, the Hare has been an object of the chace from the time of the Romans. It swims well, and readily takes to the water, when, by crossing a stream, or even a river, a more plentiful supply of food can be obtained.

The female Hare has ten mammæ, and brings forth from two to five young at a birth.

A full-grown Hare weighs eight or nine pounds, but notices of Hares of twelve pounds in weight are on record, and in the Magazine of Natural History mention is made of a Hare attaining even thirteen pounds in weight.

Black or white varieties occasionally occur, but the Common Hare does not change its colour in the winter months, as is the case with many of the species inhabiting the more northern latitudes. Its fur, on the back, is white at the root, and each hair has a broad black space in the middle, which is followed by an equally broad ring of a rufous yellow hue, the extreme points being black; the yellow ends of the hairs lap over the lower black parts, but do not entirely hide them, and hence the mottled black and yellowish appearance of the upper parts of the body. On the sides of the body the fur is of a pale grey next the skin, and chiefly rufous yellow externally, and on the under parts they are white, slightly tinted with reddish, and at the root they are grey-white; the limbs are whitish internally, but on the outer surface they are rufous, or of a yellowish rust colour, as are also the chest and back of the neck, but the latter part is of a less bright hue than the limbs, and is whitish near the roots of the ears. The ears are white externally, excepting on the fore part, where they are grizzled with black and yellow, and at the tip, which is black; the inner surface of the ear is for the most part clothed with yellow hairs; hairs of this colour form a conspicuous ridge on the anterior margin; the hinder margin is fringed with whitish hairs, and within the latter margin, but near to it, is a long and broad mark formed of hairs, which are partly black and partly yellow; the black prevails near the margin: the black at the tip of the ear extends downwards about half an inch, but on the posterior margin it runs down about one-third of the entire length of the organ. The chin, throat, and orbit of the eye, are white, and a broad whitish mark runs along the side of the face from the muzzle to the ear. The outer surface of the thigh is always greyish: the tail is white below, and at the point, and black above; the dense fur on the under surface of the feet is of a dirty brownish colour. but it is very pale on the hinder half of the tarsus.

The collection of the British Museum contains an English specimen of the Common Hare which is worthy of notice, its prevailing hue being sooty brown.

The subjoined admeasurements are taken from a tolerably fine male Hare:—

	Inches.	Lines.
Length from tip of nose to root of tail (measuring along the curve of the back) ...	23	0
" of ear	4	10

	Inches.	Lines.
Length of tarsus to the end of the nails ...	5	8
" of tail, including the hair, about ...	4	0
" of skull[1]	3	10½
Width of ditto	1	9½
" between the orbits	1	2
Length of nasal bones	1	8
Width of ditto behind		10½
" " in front		8
Length of incisive opening	1	0
" of palate		3¾
Width of upper incisors		3
From back part of principal incisors to molars	1	2
Length of the six upper molar teeth, taken together		7¾
" of lower jaw	2	9
Height of ditto	1	8
Transverse diameter of the angular portion ...	1	0½

The scapula measures 3″ 5½‴ in length, and 1″ 9‴ in width behind; humerus, 4″ 0‴; ulna, 5″ 2‴; femur, 5 2½; tibia, 5 11½.

As many of the species of Hares approach, in their general appearance, very nearly to the Common Hare, we shall use that animal as an object of comparison, and hence it has been described more in detail than otherwise would have been necessary. All the Hares have a fringe of long hairs on the anterior margin of the ears, and this, for brevity sake, we shall call *the fringe;* the outer surface has the anterior portion densely clothed with short, adpressed hairs, and these being of a darker colour than elsewhere, form a dark band on that part—this will be called the *external band;* and, for a similarly clothed part on the inner surface of the ear, which runs along the hinder margin, the terms *internal band* will be used. The remaining inner portions of the ear are naked, or very nearly so, and need not be noticed. The bands of dark coloured hairs are very conspicuous in the variable Hare, when in its winter dress, for they keep their dark hue when almost every other part is white.

[1] This skull is represented on Plate 2, fig. 2; the nasal bones are here foreshortened.

LEPUS MEDITERRANEUS.

The Sardinian Hare.

Lepus Mediterraneus. WAGNER, in Wiegmann's Archiv für Naturgeschichte, 7th year, Pt. 1, 1841, p. 136; SCHREB. Säugth. Suppl. vol. iv. p. 77, Tab. 233 c. 1844.

Much smaller than the *L. Timidus ;* ears longer than the head, nearly naked in the middle, and black at the apex; the back of the neck and limbs of a reddish ochre colour; tail black above, white beneath; a white mark behind the eyes.

Dr. Wagner regards as distinct from the common species, the Hare which inhabits Sardinia, and which there is some reason to believe occurs likewise in Gibraltar. It approaches the *L. timidus* in many respects, but is fully one-third smaller; its ears are proportionably rather longer, and they are more sparingly clothed with hairs; the hinder part of the outer surface of the ear is of a less pure white; the occiput is of a more uniform rusty red hue, and there is a greater admixture of black in the colouring of the upper parts of the body. The under parts of the body are white, but tinted with yellow.

	Inches.	Lines.
Length	15	5
Head	3	0
Ear	4	3

In the British Museum are specimens of a small Hare, brought from Tunis, and from Kirkany, a small island situated very near the coast, by Mr. Fraser, which agree so closely with Dr. Wagner's description, as to leave little doubt of their specific identity with the Sardinian species. Its skull is not longer than that of the Common Rabbit, and beyond the difference of size, offers many points of distinction when compared with that of the Common Hare; the most striking are, the contracted form of the nasal bones in front, their being truncated at a right angle behind, where, however, they are separated in the mesial line by a small and nearly semi-oval projection of the frontals, the small size of the supra-orbital processes, and the small depth of the zygomatic arch. The nasal bones are shorter than those of the Rabbit's skull (which the Tunis Hare much more nearly resembles), and, mereover, differ in being truncated behind; the cerebral portion of the skull is longer; the supra-orbital process shorter, the zygomatic arch less deep; the incisive opening, although of the same length, is much broader behind; the palate shorter, and the incisor teeth are rather narrower. The most strking difference, however, and one in which the present species differs from all others of the genus hitherto examined, consists in there being but five molar teeth on either side of the upper jaw; the small sixth molar being absent; in this respect the Tunis Hare resembles the species of *Lagomys*, but in no other character that I can perceive[1]. Its dimensions are—

	Inches.	Lines.
Length	3	$0\frac{1}{2}$
Width	1	$4\frac{3}{4}$
Length of nasal bones	1	$1\frac{3}{4}$
Width behind		$7\frac{2}{3}$
" in front		$4\frac{3}{4}$
" between orbits		$10\frac{1}{4}$
Depth of zygomatic arch		$2\frac{1}{3}$
Length of incisive opening		$9\frac{1}{2}$
Width of ditto behind		$4\frac{1}{3}$
Antero-posterior diameter of palate		$2\frac{2}{3}$
Width of upper incisors		2
Length of the five upper molars, taken together		$5\frac{3}{4}$

[1] With such a singular exception, one is anxious to examine other specimens of this species of Hare, to ascertain whether the difference in the number of molar teeth is not peculiar to the individual, for I have seen but one skull. It presents another peculiarity, and that is in having two small oblong openings in the frontal bones, placed near the mesial line, and close to the parietal suture.

The animal differs from the Common Hare in its general colouring, being much paler: the limbs and back of the neck are of a delicate yellowish rufous, and the sub-terminal rings on the hairs of the back are almost cream colour, whilst these hairs are very distinctly terminated with black; the feet are mottled with whitish, and the rufous colour on the tibia is confined to the hinder part, the fore part being pure white; a large rufous patch is observable immediately in front of the thigh: the ears are margined with brown at the extremity internally, and with brown black externally; the dark colour not extending downwards on the middle part of the ear, as in the Common Hare, where it forms a large patch of three-quarters of an inch in depth. The tail is sooty black along the mesial part of the upper surface.

The pale hue of this species approximates it to the *Lepus Ægyptius*, but in that animal the ears are much longer[1].

	Inches.	Lines.
Length, about	16	0
" of ear	4	3
" of tarsus	4	1

LEPUS HYBRIDUS.

The Russak Hare.

Lepus hybridus.	(Pallas) Desmarest, Mammalogie, Pt. 2, p. 349. 1822.
" *medius.*	Nilsson, Skandin. Faun. i. p. 224.
" *aquilonius.*	Blasius, in den Verh. der Vers. d. Naturf. z. Braunschw., 1842, p. 88.
" "	Wagner, Schreb., Säugth. Suppl., iv. p. 78. 1844.
" *altaicus.*	(Eversm.) Gray, in List of the Mammalia in the British Museum, p. 126. 1843.

[1] The skull has not been removed from the specimen from which this description is taken, and I find the incisor teeth rather broader than in the skull described.

Lepus variabilis, hybridus. (Pallas) Schreb., Säugth., iv. Pl. 235 c. 1792.

Russak. Pallas, Nov. Spec. Quad. e Glirium Ordine, p. 5, 1778; Zoographia Rosso-Asiatica, i. p. 147. Edit. 1831.

Ears rather exceeding the head in length, with a black patch at the apex, which is much extended on the outer surface of the ear: fur long and silky; on the back mottled with black and pale yellow; on the sides of the body and the haunches nearly white; on the under parts, and inner side of the limbs, pure white: the fore legs in front, and the tarsi, suffused with very pale rusty yellow: tail black above, white below.

This animal is found in Russia, generally, according to Blasius, between the 55th and 63d parallels of latitude; a specimen sent to the British Museum as the *Lepus Altaicus* of Eversman, being from the Altai Mountains, shows that it ranges a little further south, and, if I am right in identifying this animal with the *L. medius* of Nilsson, it occurs likewise in Zealand.

From *L. variabilis* and *L. Tolai* the present species is readily distinguished by its superior bulk, and the comparatively great extent of the black on the outer surface of the ears; from *L. variabilis* it differs, moreover, in having longer ears, and the upper surface of the tail black.

As there appears to be considerable difficulty in determining the European species of Hares, I shall proceed to point out the characters of the animal under consideration more in detail; first observing, that the specimen upon which my notes were made, is in the Museum at Leyden[1].

As regards the colouring, I have to observe, in addition to the points already noticed, that the fur on the upper parts of

[1] It is labelled, "*Lepus russatus*, F. Cuvier; Russac en Russie, mentionné par Pallas, donné par H. F. Boie, tué le 3 Mars, 1824, environs de Petersbourg."

the body has a very mottled appearance, perhaps owing to its undergoing a slight change in the winter and summer months: the hairs are pure white at the root, have a sooty black ring about the middle, a broader, delicate yellow ring beyond, and the extreme point black. On the sides of the body, which are almost white, there is a faint tinge of yellow, and the hairs on this part are brown in the middle: the cheeks are white, and there is a yellowish patch on the sides of the muzzle. The ears have the external and internal bands composed of party-coloured, black and yellow hairs; the black patch at the tip, externally, is extended in the vertical direction for about one inch, or one-fifth of the entire length of the organ; but on the inner side the black runs downwards from the point scarcely a quarter of an inch. The throat is white, and the chest is nearly white, but here the hairs are of a dirty rust-colour in the middle.

	Leyden Museum.		British Museum.	
	Inches.	Lines.	Inches.	Lines.
Length from tip of nose to root of tail	26	0	26	0
From tip of nose to ear, about ...	4	4	4	10
Length of tail, including the hair, about	6	0	5	9
" of ear	5	0	4	8
Width of ditto	2	6	2	8
Length of tarsus	6	8	6	6

A skull, also in the Leyden Museum, labelled as that of the Russak, agrees more nearly with the skull of *L. timidus* than with that of *L. variabilis:* although considerably larger than the skull of *L. timidus*, the teeth are very nearly of the same size; the incisive openings are relatively larger, and the palate has a smaller antero-posterior diameter. Its dimensions are as follows:—

	Inches.	Lines.
Total length of skull	4	1[1]
Width	2	0

[1] Perhaps a trifle more; the hinder part of the skull is mutilated.

	Inches.	Lines.
Width between the orbits	1	$2\frac{2}{3}$
Length of nasal bones	1	11
Width of ditto behind		$11\frac{1}{2}$
" " in front		$9\frac{3}{4}$
Antero-posterior diameter of palate		3
Width of upper incisors		$3\frac{1}{3}$
From principal upper incisors to molar teeth ...	1	$4\frac{1}{2}$
Length of six upper molars taken together ...		$7\frac{1}{3}$

The specimen already alluded to as being in the British Museum collection, does not differ in its colouring from the one described : its dimensions are given in the second column.

LEPUS TOLAI.

The Baikal Hare.

Lepus Tolai. PALLAS, Nov. Spe. Glir., p. 20 ; Zoographia Rosso-Asiatica, i. p. 149.
" " SCHREB. Säugth. iv. p. 878, Pl. 234.
" " LICHT. in Eversm. Reise, p. 118.
" " DESM., Mamm., p. 349.
" " WAGNER, Schreb. Säug. Suppl., iv. p. 84.
The Baikal Hare. PENNANT, Synopsis of Quadrupeds, p. 253. 1771.

Fur long, dense, and soft ; upper parts of the body mottled with a delicate yellow (almost cream-colour), and dusky, or brownish black ; sides of the body grey-white ; ears about equal to the head in length, margined with black at the apex ; back of the neck of a very pale and delicate rufous tint.

Inhabits the regions beyond Lake Baikal, and is found throughout the Mongolian Desert, where likewise *Lepus variabilis* occurs.

Unlike the *L. variabilis,* this species does not become white in the winter months : the general colouring of its fur is very nearly the same as in *L. hybridus,* but from that species, as well as from the Varying Hare, it is at once distinguished by the black on the ears being confined to the margin ; the delicate rufous tint of the occiput and back of the neck offers another point of distinction. Its size is about equal to that of the *L. variabilis*, and consequently inferior to the Russak Hare, with which, in some other respects, it is more likely to be confounded.

The following description is drawn up from a specimen in the British Museum, which agrees so closely with the accounts of the *L. Tolai* given by Pallas and Lichtenstein, as to leave no doubt that it is correctly named in our collection.

The fur on the back is of a very delicate grey, or grey-white, next the skin, and each hair is of an equally delicate rufous tint in the middle, has a black ring beyond the middle, is then of a very pale rufous yellow (nearly cream-colour), and terminated with dusky black. The chest and sides of the body are of a pale ashy grey ; the hairs on these parts are dirty white at the point, of a pale rufous brown in the middle, and of a more distinct grey hue next the skin than are those of the back. On the under parts of the body the hairs are uniformly white. The legs are slightly tinted with yellowish externally, and the feet are white, or very nearly so, but slightly mottled with pale rufous. An indistinct whitish mark is observable on the sides of the head. The ears are well clothed, and the external and internal coloured bands consist of hairs which are freckled with cream-colour and black; on the inner surface of the ear the black and cream-colour are about equal in proportion, whilst the cream-colour prevails on the outer surface. The apical portion of the ear is fringed with black; and on the outer surface a dusky hue is observable on the parts immediately adjoining the margin ; there is, however, no distinct black patch such as is observed in the *Lepus variabilis* and other nearly allied species. A narrowish yellow band follows the course of the apical black fringe on the inner surface of the ear. The upper surface of the nose is of a delicate yellow tint. The general hue of the animal is very pale.

	Brit. Mus.		From Pallas. Male.		From Pallas. Female.	
	Ins.	Lines.	Ins.	Lines.	Ins.	Lines.
Length from tip of nose to root of tail ...	22	0	21	5	28	4
" of tail, with the hair, about ...	4	0	4	1	3	6
" from nose to ear, about	4	0				
" of ear	4	0	4	10	4	8
Width of ditto	2	5	2	6	2	5
Length of tarsus	5	6	5	3	5	0

The skull of *Lepus Tolai* is much smaller than that of the Common Hare ; has the nasal bones relatively smaller, more contracted in front, and depressed behind ; the supraorbital processes are also relatively smaller, the zygomatic arch less deep, and the incisive opening narrower. Upon comparing it with the cranium of *Lepus variabilis*, it presents many points of distinction, the most striking of which are the smaller size of the supraorbital processes, and the compara tively small depth of the zygomatic arch ; the incisive opening is less dilated in the middle. The skull, moreover, is more depressed, and consequently less strongly arched above than that of *L. variabilis*[1] ; the upper incisor teeth are less deeply grooved, and the molars are smaller.

	Inches.	Lines.
Total length of skull[2]	3	6
Width	1	$8\frac{1}{2}$

[1] The height of the skull in a vertical line above the third molar tooth, is 1 inch in *L. Tolai ;* and, in a skull of *L. variabilis*, the length of which does not surpass that of the *L. Tolai*, the height in the same part is 1 inch $1\frac{1}{2}$ lines ; the zygomatic arch in the same skull is $4\frac{1}{4}$ lines in depth, whilst in *L. Tolai* it is only 3 lines.

[2] A *very* small portion of the hinder part of the skull is wanting, and has been allowed for in the above dimensions.

	Inches.	Lines.
Width between orbits		$11\frac{1}{2}$
Length of nasal bones	1	$4\frac{1}{4}$
Width of ditto behind		$8\frac{1}{4}$
" " in front		$5\frac{3}{4}$
Length of incisive opening		$10\frac{1}{3}$
Width of ditto behind		$4\frac{2}{3}$
Length of six upper molar teeth, taken together		$7\frac{1}{4}$

This skull is in the collection of the British Museum.

LEPUS VARIABILIS.

Varying Hare.

Lepus variabilis. PALLAS, Glires, pp. 1 and 30.
" " SCHREB., Säugth. iv. p. 885, Pl. 235 A, B.
" " DESM., Mammalogie, p. 349.
" " FLEMMING, British Animals, &c. p. 22. 8vo. Edinburgh, 1828.
" " (Varying Hare). BELL, British Quadrupeds, p. 343. 1837.
" " and *borealis.* NILSSON, Skandin. Fauna, Plates 16 and 19, Pt. 6, p. 1, and Pt. 7, p. 2. 1832.
" *albus.* (BRISSON) JENYNS, British Vertebrate Animals, p. 35. 1835.
Alpine Hare. PENNANT, British Zool. (4th edit.), i. p. 102, Pl. 10, No. 21, 8vo. 1776-7.
Varying Hare. SHAW, General Zoology, ii. p. 201. 1802.
Lepus Hibernicus. BELL, Brit. Quad. p. 341.
" *timidus,* var. *β.* JENYNS, Brit. Vert. An. p. 35.
Irish Hare. YARRELL, Proceedings of the Zoological Society for July, 1833, p. 88.

Ears about equal to the head in length, with a smallish, but distinct black patch at the tip. *Summer fur*—general hue rusty brown, being finely pencilled with black and rufous yellow; under parts impure white: tail white; greyish above. *Winter fur*—white; ears with a black tip, and the external and internal bands freckled with black and yellow; upper surface of the nose generally yellow.

Inhabits Ireland and Scotland, and occasionally extends into Cumberland: on the continent, it is found throughout Scandinavia, extending northwards to the Arctic Ocean: thence it is traced through Siberia, north of the 55th parallel of latitude, into Kamtschatka. It appears to be for the most part wanting in the central portions of Europe, but reappears again on the Alps of Switzerland, and extends eastwards as far as Salzburg. It is said, moreover, to occur in Bavaria.

The Varying Hare differs from the Common Hare in being rather smaller, in having the ears proportionately smaller, and with the black less extended at the point, especially on the outer surface: the tail, moreover, is shorter, and wants the black on the upper surface; the pale and dark annulations of the hairs of the fur are narrow, and the animal does not present that mottled appearance which is observed in the colouring of the Common Hare; the white cheek mark is likewise wanting. In the winter months the two animals are easily distinguished, since the present species becomes white, whilst the *Lepus timidus* undergoes no perceptible change. The change to white colour in the Varying Hare is distinctly traced to the effects of the low temperature to which that animal is subjected during the winter months, in the northern regions which it inhabits, and on the elevated parts of the mountains to which it appears to confine itself when found in more temperate parts of Europe or Asia. The colouring matter leaves the hairs when thus subjected to the effects of cold, for the summer fur is not cast off, and replaced by new white hairs [1].

In Scotland the Varying Hare begins to change its colouring towards the end of autumn, and by the middle of November it has usually become wholly white, with the exception of the upper surface of the nose, the tips of the

[1] I am strongly inclined to believe that the extreme cold, in such cases as the present, by checking the plasmatic circulation, not only bleaches, but ultimately destroys the vitality of the hairs, and that this is the reason why they are after a time cast off, to be replaced by new and coloured hairs.

ears, and the bands of the ears. These parts I have generally found, in Scotch specimens, retain the summer hue more or less perfectly, but in Swedish specimens, every part becomes white, with the exception of the point of the ears, which invariably retains the black hue.

A specimen now before me, which was purchased in the London market in the month of December, and no doubt is from Scotland, is perfectly white, with these exceptions—that there are a few scattered black hairs on the back, and similar black hairs on the chest, which being here tolerably numerous, give to the part a greyish tint. The tip of the ears is black, and the external and internal bands are composed of hairs which are partly black and partly yellow; the upper surface of the muzzle is yellow, and the crown of the head has very many party-coloured black and yellow hairs. I have seen many specimens like this, from Scotland: its dimensions are given in the first column, and those of its skull and some other parts of the skeleton, are likewise appended to this account.

A specimen of the Varying Hare, in its summer fur, which I observed in the Leyden Museum, had the general hue of the fur brown, obscurely tinted with rufous; the outer surface of the legs was nearly of the same tint as the body; the tibia, however, was somewhat tinted with soot colour, and the tarsi were whitish; the cheeks rufous yellow; the chest of the same colour as the sides of the body, but a trifle paler; the bands of the ears dark, being pencilled with black and yellow; the margin white, excepting at the point, where there was a conspicuous black patch, having a diameter of about half an inch in the vertical direction, on the outer surface of the organ, and a trifle less extended on the inner surface. The throat and under parts of the body white; the tail also white, but somewhat suffused with soot colour on the upper surface.

A specimen in the British Museum, sent by M. Sundevall, and which it appears to me is not quite in its summer fur, differs from the above in being of a somewhat less rich hue; and the tail is merely greyish above. The tips of the ears have a similar black patch, which is neither more nor less extended; indeed, a similar patch I have invariably found on the ears of all specimens which have come under my notice.

The Irish Hare agrees in all essential particulars with the above specimens, but is of a richer colour: its fur is shorter than that of the Common Hare, and, as it will not felt, like the hare of that animal, it is valueless. The general tint of the fur is rufous brown, a tint produced by the admixture of black and reddish yellow, the exposed ends of the hairs being of these colours. The hairs on the back are of a very pale grey colour at the root, rather narrowly annulated with rufous-yellow towards the point, and black at the point. The head is of a brighter colour than other parts, and the sides of the body are less distinctly pencilled with black than the back; the hind feet are somewhat mottled with white; the haunches greyish; the chin, throat, and abdomen, are white, but not very pure; the tail is also white, excepting on the upper surface, which is greyish. The ears have a black patch at the tip, which is rather more extended on the outer surface than on the inner, having a vertical diameter of about half an inch on the former.

Mr. Thompson, of Belfast, who has carefully compared the Irish Hare with specimens of the *Lepus variabilis* from the continent, is of opinion that these animals are specifically identical[1]—an opinion in which I agree.

The Irish Hare only occasionally becomes white in the winter.

[1] Prof. Blasius, who has likewise compared these animals, has arrived at the same conclusion.

	From Scotland[1].		From Scotland.		From Scotland.		From Lapland.		From Sweden.		Irish Hare.	
	Ins.	Lines.	Ins.	Lines	Ins.	Lines.	Ins	Lines.	Ins.	Lines.	Ins.	Lines.
Length from tip of nose to root of tail ...	20	0	22	6	21	9	25	6	21	0	23	0
" from tip of nose to ear	3	9					4	3	4	2	3	11
" of ditto to eye	2	1										
" of ear	3	8	3	6	3	6	3	5	4	3	3	6
Width of ditto	1	11							2	2	2	0
Length of fore leg and foot, from the elbow joint	7	0										
" of the fore foot and nails	3	0										
" of hind leg and foot, from the knee joint	10	6										
" of hind foot and nails	6	0	6	4	6	3	6	6	6	6	6	0
" of tail, including the fur, about ...	4	3					4	3	4	6	4	0
" of ditto without the skin	3	0										

Two skulls, from specimens of the *Lepus variabilis* bought in the London market, agree very closely with each other, and differ from a skull of the Common Hare, in being smaller and proportionately rather broader; in having the nasal bones

[1] These dimensions are taken from a fresh specimen, the weight of which was 5½ lbs. The admeasurements in the other columns are from stuffed specimens.

smaller and more contracted in front, the incisive opening also more contracted in front, and the palate shorter, owing to the antero-posterior diameter of the palatine portion of the palatine bone being considerably less. The incisor teeth, moreover, are distinctly narrower.

	Inches.	Lines.	Inches.	Lines.
Total length of skull	3	5½	3	7
Width	1	9½	1	9½
" between orbits	1	1	1	2⅓
Length of nasal bones	1	3¾	1	4¼
Width of ditto behind		8¾		8¾
" " in front		6⅓		6½
Distance between upper incisor teeth and the molars	1	0¾	1	1½
Antero-posterior diameter of the palate ... .:		3		2¾
Length of the six upper molar teeth, taken together		7½		7½
Width of the incisors		2½		2½
Length of scapula	2	11		
Greatest width of ditto	1	5½		
Length of humerus...	3	11		
" of ulna	4	7½		
" of radius	4	0		
" of bones of fore foot taken together	2	9		
" of femur	4	8		
" of tibia	5	6		
" of bones of hind foot, taken together	5	6		

[1] The dimensions in this column are from the skull and other parts of the skeleton of the animal, the admeasurements of which are given in the first column in the preceding series of dimensions. Upon preparing this skeleton I was surprised to find that the structure of the tarsus did not agree with my description of that of the Common Hare at p. 34 of the preceding part of this work, and was hence led to re-examine the specimens from which my description was taken, and ultimately to discover that the foot alluded to, and supposed to be that of the Hare, is, in fact, the foot of an *Agouti*. The note at the bottom of p. 34 should be either expunged, or altered.

LEPUS CANESCENS.

The Ashy-grey Hare.

Lepus canescens. NILSSON, ARCH. Skand. Beitr. i. p. 172.
" *borealis.* var. NILSSON, Illuminerade Figurer till Skandinaviens Fauna, 6th Part, p. 1, Pl. 22. 1838.

Fur long and soft, ashy-grey on the upper parts of the body, white on the under parts; ears about equal to the head in length, with the apex black, and a black fringe extending about half way along the hinder margin; tail white, slightly tinted with grey on the upper surface.

Inhabits Scandinavia.

This animal was formerly regarded by Nilsson as a variety of the *L. variabilis,* but is now recognized by that naturalist, as well as by Sundevall, as a distinct species. It appears to be chiefly confined to the southern parts of Scandinavia, is found throughout Gothland, and extends northwards to Jaemtland.

From the examination of a specimen sent by Sundevall to the British Museum, I am induced to agree with the two naturalists above mentioned, with regard to the distinctness of this species. This specimen is in its winter fur, which, unlike that of the *Lepus variabilis,* is of a pale ashy-grey hue. The fur on the back is composed of hairs which are of an ashy white colour at the root, very pale rufous brown beyond, followed by a broad grey-white ring, and a dusky point: the longest hairs are for the most part dusky. The sides of the body, and the limbs, are chiefly of an ashy white tint; the feet are white, but in parts suffused with rufous; the fore feet most distinctly so. The whole of the under parts, including the chest (an uncommon circumstance), are white, and so is the tail, but this latter is slightly tinted with grey on the upper surface. The crown of the head is faintly suffused with rufous, and the nose is rufous-yellow above. The ears are well clothed; the external band is formed of party-coloured, pale rufous, and black hairs, and so is the internal band, but here the black is more conspicuous: the black at the point of the ear extends down about half an inch on the inner side, and about three quarters of an inch on the outer; and, on the margin of the ear, is a fringe of black hairs, which, on the fore part of the ear, extends about an inch from the point, whilst on the hinder margin the black is continued to below the middle of the ear.

	Inches.	Lines.
Length from tip of nose to root of tail ...	21	0
" from nose to ear	4	3
" of ear	4	0
Width of ditto	2	1
Length of hind foot and nails	6	0
" of tail, including the fur	5	0

LEPUS TIBETANUS.

Tibet Hare.

Lepus Tibetanus. Waterhouse, Proceedings of the Zoological Society for January, 1841, p. 7.

? " *Oiostolus.* Hodgson, Journ. of the Asiatic Soc. of Bengal, vol. ix. p. 1186, Feb. 1841.

Ears longer than the head, margined with yellow-white internally; externally with the apex edged with black, and with a narrow edging of black, extending about half way down the hinder margin: general hue palish ashy grey; the back mottled with dusky and yellowish white; back of neck pale rufous brown; legs rather long, and chiefly white; tail white, with the upper surface grey-black.

Inhabits Little Tibet.

The above description is taken from a species of Hare

presented to the Zoological Society by Mr. Vigne, relating to which that gentleman furnishes us with the following particulars [1]:—

"The Hares of Little Tibet, which is a barren country, are found almost exclusively within, or within reach of, sandy valleys, through which a stream flows, creating on its flat banks just verdure enough for them to feed upon. They lie in 'forms' under rocks and stones; sometimes, when pursued, they will take to any natural hollows beneath them, and are occasionally, and, in fact, frequently, sheltered by the thick bushes of Tartarian furze, with which the valleys of Little Tibet abound, more or less. They do not burrow, but scratch away a deep 'form' under a stone or rock."

In its pale grey hue, having a slight yellow tint only on the back, this species agrees most nearly with the *L. Tolai* and the *L. hybridus;* it is, however, a much smaller animal than the last mentioned species, and differs from both in having proportionately larger ears, and in wanting the black patch on the inner surface of the apical portion of those organs. The same characters distinguish the Tibet Hare from the *L. canescens* (also a pale grey species), and *L. variabilis.* Other differences present themselves when the skull of the present species is compared with that of either of the animals just mentioned. In the ears being pale at the apical margin internally, the Tibet Hare agrees with the *L. ruficaudatus* of India, but besides the differences in the colouring of these two species, the Tibet animal is readily distinguished from the Indian, by its longer legs and tarsi.

The fur is moderate as to length and texture, of a very pale grey next the skin, on the upper parts and sides of the body, and pale brownish ash colour externally on the latter parts; on the back, each hair of the ordinary fur is dusky

[1] See the Zoological Society's Proceedings as above quoted.

beyond the middle, has a cream-coloured subterminal ring, and a black point: the hidden parts of the longer hairs are almost entirely white. The back of the neck is of a very pale rufous brown, inclining to white immediately below the ears, and the chest is very nearly of the same tint. The fur on the under parts of the body is white, obscurely suffused with cream-yellow; the orbit of the eye is white, and the legs and feet are very nearly white; a pale ashy grey tint is observable on the outer surface of the hind legs, as well as on the fore part of the anterior pair, and the fore foot is somewhat suffused with yellow; the dense fur on the under surface of the feet is of a dirty brown colour. The ears, which are considerably longer than the head, are for the most part clothed with white hairs internally; the band on the hinder part is of an ashy grey colour, and somewhat indistinct; at the apical margin the hairs are yellow-white, but the black fringe of the outer surface projects slightly beyond the pale fringe of the inner one; the ear, externally, is dusky immediately below the black edging at the tip, and on the fore margin a narrow black line is seen to descend about an inch from the point, whilst on the hinder margin (as in *L. canescens*) a similar black line runs full half way down the ear. The tail is of moderate length; white, excepting on the upper surface, which is greyish black. The groove in the upper incisor very nearly divides the tooth into two equal parts, and is very deep.

	Inches.	Lines.
Length from tip of nose to root of tail ...	18	0
" from nose to ear	3	8
" of ear	4	7
Width of ditto	2	4½
Length of tarsus, including the nails ...	5	0
" of tail, including the fur	4	6

A skull of a Tibet Hare brought home by Mr. Vigne, (which, it is necessary to state, does not belong to the same individual as that from which my description is taken), much more closely approximates in size to the skull of the Rabbit than to that of the Common Hare; in the comparative breadth of the palate, the larger size of the nasal bones, and of the incisive opening, and the greater space between the sphenoid alæ, as well as in the form of the anterior molar tooth, however, it will bear a closer comparison with the Hare's skull than with that of the Rabbit. From the skull of the Common Hare, or from that of the *L. variabilis*, it differs in having the interorbital space considerably more contracted, the raised platform, formed by the supraoccipital bone, narrower and more elongated, the basi-occipital bone distinctly more contracted in the middle, and the auditory bullæ larger[1]. The skull is narrower, and the nasal bones are more elongated than in *L. variabilis*, but in having a very small antero-posterior diameter to the palatine bone, *L. Tibetanus* agrees with the last mentioned animal. The incisive opening is very nearly the same in both.

	Inches.	Lines.
Length of skull	3	4
Width of ditto	1	7½
" between orbits		11¼
Length of nasal bones	1	5
Width of ditto behind		8
" " in front		6¾
From superior incisor to foremost molar ...	1	0¼
Length of the six molars taken together ...		6½
" of palate		3
" of lower jaw	2	5
Height of ditto	1	6

Lepus Oiostolus. Hodgs.

"Fur, consisting almost wholly of wool, considerably curved, and interspersed rarely with very soft hairs; slaty grey-blue for

[1] The transverse diameter of the auditory bullæ is nearly 5¼ lines, whilst in the much larger skull of the *L. timidus* it is but 5 lines; in *L. variabilis* it is also 5 lines.

the most part and internally; but externally fawn-tinted above, and whitish below, and on the limbs; some hairs on the back tipped with black beneath a sub-rufous ring. Tail white, with a grey-blue stripe towards the back."—*Hodgs.*

Mr. Hodgson states that he had only been able to procure some "wretched remains" of this animal at the time that he drew up the above description. From these he was led to infer that the animal was about the same size as the *Lepus macrotus*—a species which is rather less than the Common Hare. The habitat given is, the snowy region of the Himalayas, and perhaps also Tibet.

In a subsequently published part of the Journal[1], already quoted, we are informed by Mr. Hodgson that the *L. Oiostolus* is common in Tibet, near the Hemâchal, but not so abundant in the central and eastern provinces of Utsâng and Khân, as the much larger species, *Lepus pallipes.*

A Hare in the British Museum collection, named *Lepus Oiostolus,* and presented by Mr. Hodgson, is evidently an extremely young animal. Whether it be the young of the *Lepus Tibetanus* I will not pretend to determine—its ears are similarly coloured.

Lepus pallipes.

Lepus pallipes. HODGSON, Journal of the Asiatic Society of Bengal, vol. xi. p. 288, with Plate. 1842.

Fur long, dense, and soft; of a pale ochre colour, but, on the back of the animal, pencilled with black; haunches greyish; under parts white; chest of a delicate yellow-rufous tint; the front of the fore legs, and the fore feet, nearly of the same hue: tarsus almost white, but somewhat suffused with rufous in front: tail white, excepting along the middle portion of the upper surface, where it is grey.

Inhabits the central and eastern portions of Tibet.

[1] Vol. vi. p. 288. 1842.

In Mr. Hodgson's collection, presented by that gentleman to the British Museum, is an imperfect skin of the *Lepus pallipes;* one of those, I believe, upon which that gentleman's account of the species is drawn up. From this skin the above description is taken; the head being very imperfect, and destitute of ears, prevents my describing those parts, but Mr. Hodgson's description states that the ears are equal to the head in length, whilst the latter is said to measure 4¾ inches. The whole length of the animal, according to Mr. Hodgson, is from 22 to 23 inches, and the tail, without the hair, 4 inches, or, with the hair, 6 inches.

The general colouring of this animal agreeing very nearly with that of *Lepus Tibetanus,* combined with its habitat, would lead us to suspect the two animals were specifically identical, but the tarsus of *L. pallipes* being absolutely shorter, whilst the bulk of the animal is greater than in *L. Tibetanus,* causes me to hesitate in linking them together. There must, moreover, be a difference in the proportions which the ears bear to the head, in these animals.

LEPUS CUNICULUS.

The Rabbit, or Cony.

Lepus Cuniculus.	LINN., Syst. ed. 12, i. p. 72.
" "	DESM., Mamm. p. 348.
" "	JENYNS, British Vertebrate Animals, p. 35.
" "	BELL, British Quadrupeds, p. 348.
" *Magellanicus.*	LESSON et GARN. in Bullet. des Sci. Nat. viii. p. 96.
Rabbet.	PENNANT, British Zoology, i. p. 104, No. 22, t. x.
"	SHAW, General Zoology, ii. Pt. 1, p. 204, Pl. 164.
Le Lapin.	BUFFON, Hist. Nat. vi. p. 303, Tab. 50.
Kannichen.	KEYS et BLASIUS, Die Wirbelthiere Europas, i. 30.

Ears about equal to the head in length; tarsus short: general colour grey-brown; under parts white; back of neck rufous; tail white below, blackish above, but pencilled with dirty white.

Inhabits the southern and western parts of Europe, including the greater part of Germany; and occurs likewise in North Africa.

The Rabbit, or Cony[1], as it was more anciently called, is readily distinguished from the Hare, not only by its smaller size and greyish hue, but likewise by its comparatively short feet and ears, and the absence of the distinct black patch at the extremity of the latter. The short feet, indeed, distinguish this animal from very many of its congeners. Its burrowing habits have already been alluded to. Dry, sandy situations are the favourite haunts of Rabbits, but they by no means confine themselves to such spots, apparently having a constitution fitted to bear great variety both of soil and climate. The flesh being of good flavour, and the animal easily domesticated, it has been transported to various parts of the globe[2]. It is said to be originally from Spain.

The Rabbit is a very prolific animal, having several litters in a year, and bringing forth from five to eight young at a litter. Unlike the young of the Hare, these are at first blind and naked, but to compensate for their helpless condition they are amply protected by the warm and secure nest made by the parent, consisting as it does of fur from her own body, and being deposited at the bottom of a separate burrow formed for the purpose.

Like many other animals, when domesticated the Rabbit

[1] The word Rabbit is from the Dutch *Robbeken*, whilst that of Cony is from the Latin *Cuniculus*, and has its analogue, as Mr. Bell has pointed out, in many of the European languages. The Italian *Coniglio*, Spanish *Conejo*, Portuguese *Coelho*, German *Caninchen*, Danish and Swedish *Kaning*, Belgic *Konin*, and Welsh *Cwningen*, are clearly all from the same root.

[2] The Rabbit has been introduced into the Falkland Islands, and, according to Mr. Darwin (Journal of the Voyage of the Beagle, vol. iii. p. 248), these animals abound over large tracts of the eastern island. Black, grey, and piebald varieties are there found mixed together, and it is upon a black variety found in these islands that MM. Lesson and Garnot established the supposed species, *Lepus Magellanicus*. The animal does not occur, as these naturalists imagined, on the main land, near the Strait of Magellhaen: a small cavy (*Cavia Kingii*) there found being spoken of by Magellan under the name "Conejos," gave rise to the error.

is subject to much variety of colouring, and attains a large size. Black, white, and sandy-yellow, are the prevailing colours (as is also the case in the domestic Guinea-Pig), and wild Rabbits of the latter colour are occasionally met with. A silvery-grey domestic variety—the tint being produced by the hairs being party-coloured white and black—is known in England by the name *Silver Rabbit,* and is the "var. B. *Lepus cuniculus argenteus*" of Desmarest. Another variety, remarkable for the length and softness of its fur, known to us under the names *Russian Rabbit,* and *Angora Rabbit,* is the "var. C. *L. cuniculus angorensis*" of the same author.

Angora Rabbit.

Besides these, our Lop-eared Rabbits should be noticed, as being among the more striking of the domestic varieties. These animals are sometimes of enormous size, as compared with the wild Rabbit, and are remarkable for having the ears more or less bent downwards from the base[1].

The general hue of the wild Rabbit is grey-brown; the hairs on the back are greyish white at the root; above this they are grey, then pale brown, shaded through darker brown into dusky, followed by a ring of sandy white, and a black point. The under parts of the body are white, but the fur on these parts is grey next the skin: the chin aud throat are also white, as well as the region of the angle of the mouth. The back of the neck is rufous; the ears brown-grey externally, towards, and at the fore part; of a paler hue on the hinder part, and darker towards the point, which is edged with black; the hairs on the inner surface of the ears are for the most part greyish white, but assume an ashy-brown hue towards the hinder margin; the flesh of the ear is of a bluish white colour. The admeasurements of an average sized individual are as follows:

	Inches.	Lines.
From tip of nose to root of tail	16	6
" " to ear	3	4
Length of ear	3	1
Width of ditto	1	9
Length of fore leg and foot, from the elbow	4	9
" of the fore foot and nails	2	1
" of the hind leg and foot, from the knee	6	10
" of the foot	3	11
" of the tail, including the hair ...	3	0

Among the more striking differences which present themselves upon comparing the skull of the Rabbit with that of the Hare, may be noticed the comparatively contracted form of the muzzle and interorbital space in the former; and the

[1] These are known to the fanciers by the names, Horn-lop, Half-lop, Oar-lop, and Perfect-lop.—See Bell's British Quadrupeds, pp. 351 and 2. In the latter page is a vignette, in which these varieties are represented.

incisive openings are consequently narrow; the post-palatal channel is also considerably narrower in proportion in the Rabbit than in that of the Hare.

Two skulls of the Lop-eared variety of the domestic Rabbit, in the British Museum collection, are about four times larger than those of the wild race, but they retain the characteristics of the species; the *general* proportions being the same. The bones of these skulls want the compact texture observable in those of the wild animals, being coarse and spongy—many of them very rough and full of perforations—and there is a general want of symmetry. The bones which enclose the brain are very thick, so that, notwithstanding their superior magnitude, the brain cavity in these skulls is apparently but little larger than in the wild Rabbit. The teeth are not large in proportion to the skull, the molars being, in fact, but little larger than those of the wild animal, and they are absolutely smaller than in the skull of the Wild Hare, whilst the skulls of the Domestic Rabbit under consideration are much larger than the Hare's skull. These skulls measure respectively 4 inches 5 lines, and 4 inches 7 lines, in length; and both of them, 2 inches in width.

The subjoined dimensions are from skulls of the Wild Rabbit; those in the two last columns are from specimens obtained in a small island near the coast of Tunis: the remainder are from English specimens.

These skulls, it will be perceived, are smaller than those of English specimens—one of them is very small; but their proportions appear to be exactly the same, judging by the eye. Mr. Fraser, who brought these skulls, together with skins of the animals, to England, likewise found the Rabbit on the main land, at Tunis.

	Ins.	Lines.	Ins.	Lines.	Ins.	Lines.	Ins.	Lines.	Ins.	Lines.
Total length of skull	3	1	3	1¼	3	0½	2	11	2	8⅔
Width	1	6¾	1	6⅔	1	6¼	1	5½	1	5
" between orbits		8		9¼		8		8⅔		7¾
Length of nasal bones	1	5⅓	1	6	1	5	1	3	1	1¾
Width of ditto behind		6⅔		7¼		7		6⅔		5¾
" " in front		5⅓		5¾		4¾		4⅔		4¼
From upper incisors to molar teeth ...		11½		11¾		11		10		9
Length of six upper molars, taken together		6⅔		6¾		6¼		6½		6¼
" of incisive opening		9½		9¾		9¼		9		8¼
Width of ditto behind		3⅔		3⅓		3¼		3		2¼
" of post-palatal channel		2⅓		2⅓		2		1⅚		1⅚
Length of lower jaw	2	2	2	2½	2	2	2	0⅓		
Height of ditto	1	4	1	4	1	3¾	1	3		
Transverse diameter of angular portion		10½		10⅔		10¾		10		

LEPUS BRACHYURUS.

The Japan Hare.

Lepus brachyurus. TEMMINCK, in Siebold's Fauna Japonica, Pl. 11.

Ears small; shorter than the head: tail very short; brown, with the upper surface dusky: general hue of the animal rufous brown; throat white; abdomen rufous white.

Inhabits Japan.

A very distinct species of Hare, but approaching, in some respects, the Black-necked Hare of India; its ears, however, are much smaller, and it differs in its general colouring, as well as in the want of the black back to the neck. The head is more uniformly coloured than in most Hares, there being neither any white cheek mark, nor any white round the eye. The fur on the back of the neck is of the same general hue as that of the back, the hairs being pencilled in the same manner.

The fur of the Japan Hare is short, its texture is moderate (less harsh than in *L. nigricollis*), and its colouring, at a little distance, appears more uniform than most other Hares. This arises from the fur being very finely pencilled with brownish and pale rufous, instead of the exposed ends of the hairs having a long space of pale colour, and an equally extended space of black, as is more commonly the case, and which gives the mottled appearance to the fur of those species which have the hairs thus coloured. The fur is of an ashy-grey tint next the skin, excepting on the abdomen, where the hairs are uniformly coloured. The head is short, and the ears are unusually small; the latter are of the same general tint as the body, but at the apex they are narrowly margined with brown-black internally; externally they are also brown-black at the point, but here the black is more extended, and a narrow black line runs down the outer margin so as to occupy two-thirds of the length of the ear. Four specimens which have come under my notice (three in the Leyden Museum, and one in the British Museum), have a white spot on the crown of the head. The hind feet are of a richer rufous brown than other parts.

	Inches.		Lines.
Length from tip of nose to root of tail ...	20	—	0
" from ditto to ear, about	3	—	7
" of ear	2	—	10
Width of ditto	1	—	9
Length of tail, including the fur, about ...	2	—	0
" of the hind foot and nails	4	—	10

The skull of the Short-tailed, or Japan Hare, is remarkable for the small size of the supra-orbital process; the notch in front of this process is very small; the palate is longer than usual, and the antero-posterior diameter of the palatine bone, and that of the palatine portion of the superior maxillary, are more nearly equal. All the bones of the skull display more than common strength, arising from the comparatively large size of the teeth. The molars and incisors are fully as large as those of the Common Hare, although the skull is very much smaller. The lower jaw is short and strong, and the coronoid process is less elevated than in most Hares.

	Leyden Museum. Inches.	Lines.	British Museum. Inches.	Lines.
Length of the skull to the hinder margin of the parietal bones	3	5	3	1?[1]
Width of ditto	1	8⅓	1	7
Length of nasal bones	1	6		?
Width of ditto behind		9		8¼
" " in front		7		5¾
" between orbits	1	0½		9½
From upper incisors to the molar teeth	1	0⅔	1	0
" incisors to hinder margin of palate	1	5	1	4
Antero-posterior diameter of palate ...		4		3¾
Length of the six upper true molars taken together		7½		7½
Width of upper incisors		3		3
Length of lower jaw			2	3½
Height of ditto			1	4
Transverse diameter of angular portion				11½

This completes the list of the Europeo-Asiatic species of Hares: we shall now proceed to the

[1] The whole length of of the perfect skull would no doubt be *very* nearly 3″ 4‴.

Indian Hares.

LEPUS NIGRICOLLIS.

Black-necked Hare.

Lepus nigricollis. F. CUVIER, Dictionnaire des Sciences Naturelles, tom. xxvi. p. 307, 1823.—(*Lièvre à cou noir*), Mammifères, No. 69. 1837.

" " Is. GEOFFROY, Dict. Classique d'Hist. Nat. ix. p. 386. 1826.

" " OGILBY, in Royle's Illustrations of the Botany, &c. of the Himalayan Mountain, vol. i. p. lxviii. 1839.

" " WAGNER, SCHREB., Säugth. Suppl., iv. p. 88, Pl. 233 F. 1844.

" *melanauchen.* TEMMINCK, Discours Préliminaire, &c. in SIEBOLD'S Fauna Japonica, p. xiii. 1835.

" *Kurgosa* (The Sassa or Khargosh). GRAY, in List of the Mammalia in the Collection of the British Museum, p. 128. 1843.

Ears rather shorter than the head: fur rather short, and harsh; on the upper parts of the body mottled with black and rufous yellow; the back of the neck sooty black; chin and throat white; ears black at the apex externally, and yellow-white at the point internally: abdomen white, suffused with rusty yellow; feet rufous yellow: tail white beneath, and pencilled with black and pale rufous above.

Inhabits Java and the Mauritius, and likewise continental India—Bengal, the Deccan, and Carnatic.

Upon comparing my descriptions taken from F. Cuvier's *Lepus nigricollis*, in the Paris Museum, and from the *Lepus melanauchen* of M. Temminck, in the Leyden Museum, I find them to agree so closely as to leave no doubt on my mind as to the specific identity of these animals; the *L. Kurgosa*[1] of the British Museum collection is also clearly the same species.

The Black-necked Hare is well named by M. F. Cuvier,

[1] Mr. Gray quotes this name (it should be *Hurgosa*) from Buchanan's Mysore, i. p. 169. I find no description of the present animal in this work. The name *Lepus Hurgosa* merely occurs in a list of game animals.

for I am acquainted with no other species in which the back of the neck is black. The harsh and short fur is mottled, or strongly pencilled, with black and sandy yellow on the whole of the upper parts, if we except the occiput and back of the neck, which are sooty black; the sides of the body are of a brownish hue, or sometimes inclining to grey; the under parts white, but suffused more or less with rufous yellow; the chest, and foot of the fore leg, are rufous, as are likewise the hind feet, but these latter are very pale; the hind legs externally are greyish, and internally whitish: the orbit of the eye, the chin, and the throat, are white; the ears are brownish on the inner surface, but fringed with white, and margined with yellow-white at the apex; on the outer surface they are grizzled with black and yellow on the fore part, whitish on the hinder, and edged with black at the point: the tail is white beneath, dusky above, and generally more or less pencilled with rufous or yellow on this part.

	Leyden Museum. From Java.		Paris Museum.	
	Inches.	Lines.	Inches.	Lines.
Length from nose to root of tail ...	19	9	18	0
" " to ear	3	6	3	3
" of ear	3	4	3	6
Width of ditto	1	10		
Length of tail, with the fur, about	3	9		
" of tarsus	4	1		

With regard to the skull of this species, I may observe that the supra-orbital processes are smaller than in the generality of the Hares (in this respect resembling the Japan Hare), and the interorbital space is narrower. The nasal bones are rather long. The upper incisor teeth have each two longitudinal grooves, placed very closely together, and not very distinct; the inner half of the tooth projects so as to form a somewhat prominent ridge. The following admeasurements are taken from two skulls of the *L. nigricollis* in the Leyden Museum, and are from Java specimens.

	Inches.	Lines.	Inches.	Lines.
Total length of skull	3	5	3	5
Width	1	$6\frac{1}{4}$	1	$6\frac{1}{2}$
" between orbits		9		
Length of nasal bones	1	$5\frac{1}{2}$	1	5
Width of ditto behind		8		9
" " in front		6		6
From upper incisors to molar teeth ...		$11\frac{1}{2}$		$11\frac{2}{3}$
Length of six upper molars taken together		$6\frac{1}{2}$		$6\frac{1}{2}$
Antero-posterior diameter of palate ...		3		3
Length of lower jaw	2	$4\frac{3}{4}$	2	$5\frac{1}{2}$
Height of ditto	1	5	1	6
Transverse diameter of angular portion		$11\frac{1}{2}$	1	0

LEPUS RUFICAUDATUS.

Rufous-tailed Hare.

Lepus ruficaudatus. Is. Geoffroy, Dictionnaire Classique d'Histoire Naturelle, tom. ix. p. 381 (1826); in Bélanger's Voy. aux Indes Orientales, Zoologie, p. 156 (1834); Guérin's Magaz. de Zoologie, 1832.

" " Ogilby, in Royle's Illustrations of the Botany, &c. of the Himalayan Mountains, vol. i. p. lxix. (1839).

" *macrotus.* Hodgson, Journal of the Asiatic Society of Bengal, vol. ix. p. 1183 (1841).

Ears longer than the head: fur short, and somewhat harsh to the touch; on the upper parts of the body strongly pencilled with black and brownish rufous; on the under parts, white, suffused with bright rust colour; chest, back of neck, and limbs, for the most part of a brownish rust colour: ears margined with yellowish internally at the apex, and with black externally, the black running down the posterior margin: tail rufous above; white below.

Inhabits the northern parts of India, the plains of the Ganges, extends into Nepal, and westward to Delhi.

Lepus ruficaudatus is the Common Hare of the northern parts of India, and is apparently replaced in the south by the Black-necked Hare: it greatly resembles the common European species, but is rather smaller, and has the hind feet proportionately shorter; the colouring of the ears and tail, moreover, furnish distinguishing characters, the former being destitute of the black patch on the inner surface of the apical portion, whilst the tail differs in having its upper surface of a brownish rufous tint. Its fur is shorter than in the *L. timidus*, and somewhat harsh to the touch; on the upper parts of the body it is strongly pencilled with black and brownish rufous; the sides of the body assume a more uniform tint, being but little pencilled with black, and on the haunches a greyish tint is perceptible. The hairs on the back are of a very pale grey tint at the root; beyond the grey, each hair assumes a pale rufous tint, which is followed by a broadish black ring, then a less extended space of brownish rufous, and lastly black, which is the colour of the point. On the abdomen the fur is white, but the exposed ends of the hairs are generally more or less tinted with a rusty yellow colour. The chest, back of the neck, and the limbs (for the most part) are of a brownish rust colour; the fore legs, however, of a brighter hue than the hinder. The chin, throat, and orbit of the eye, are white, and the white is considerably extended immediately in front of the eye. The ears are long, rather sparingly clothed, have the posterior edge white, the fringe in front composed of white and yellow hairs, and the apex rather broadly margined with yellowish: this applies to the inner surface of these organs; externally they are margined at the extremity with black, and a black line runs about halfway down the hinder margin, commencing at the apex of the ear. The tail is white below, and its general hue above is brownish rufous; but there is an admixture of black hairs on this part.

I have seen specimens from the mountains which are of a paler hue than those of the plains.[1] The dense fur on the under surface of the feet is generally of a dirty brown hue; a specimen in the British Museum, however, has the under parts of the feet of a deep red colour.

To the admeasurements contained in the two first of the subjoined columns are added some dimensions given by Mr. Hodgson, and which are, no doubt, taken from fresh specimens :—

	Dark var.		Pale var.		From the Plains.				Hills.	
					MALE.		MALE.		FEMALE.	
	Ins.	Lns.	Ins.	Lns.	Ins.	Lns.	Ins.	Lns.	Ins.	Lns.
Length from tip of nose to root of tail	22	0	19	6	18	0	18	0	19	0
" from nose to ear	3	9	3	10						
" of ear	4	4	4	4	4	9	4	9	5	0
Width of ditto	2	7	2	6						
Length of tail with the fur	4	6	4	6	7	10½	5	0	5	0
" without the fur ...	—		—		3	6	3	7½	3	7½
" of hind foot and nails ..	4	8	4	7						
From elbow to wrist	—		—		3	8	3	7½	3	8
" wrist to end of foot, including the nails	—		—		2	0	2	0	2	1
Length of tibia	—		—		4	10½	4	11⅓	5	0⅔

[1] A specimen of the Rufous-tailed Hare in the museum of the Zoological Society, and which is labelled as coming from the Himalaya, is considerably

According to Mr. Hodgson this species of Hare weighs from five to seven pounds.

The skull of *Lepus ruficaüdatus* is rather smaller than that of the Common Hare, has the nasal bones narrower, especially in front; the supra-orbital processes less developed, the zygomatic arch less deep, and the post-palatal channel narrower. There is a distinct crest, in adult skulls, at the union of the parietal bones, and the crest is continued forwards for some little distance on to the frontals. In the structure of the superior incisor teeth there exists a tangible distinctive character between the present species and the Common Hare. In this latter animal the antero-posterior diameter of the incisor is equal, or *very* nearly equal, whether taken near the outer or the inner side of the tooth, whilst in *L. ruficaudatus* there is a very distinct decrease in the diameter of the tooth in passing from the inner to the outer edge; and, moreover, the superior incisor in the Indian Hare has the external groove less deep, and placed nearer to the inner edge of the tooth, which is sharp and prominent. Sometimes there are two parallel grooves, placed closely together. The lower jaw differs, in having relatively less space between the incisor and molar teeth, and the horizontal ramus thicker, and presenting a distinctly convex outline below the molars, instead of being straight at that part, as in *L. timidus;* the angular portion of the jaw descends less, but has a proportionately greater transverse diameter.

paler in its general tint than a second individual in the same collection, presented by Mr. Hodgson, and which is probably from the Sub-Himalaya. In the British Museum collection the darker and paler (greyish-brown) varieties will also be found, being presented by the zealous and public-spirited naturalist just mentioned, to whom we are indebted for most of our information relating to the animals of Nepaul.

	Ins.	Lines.	Ins.	Lines.
Length of skull	3	8	3	8
Width	1	$8\frac{1}{2}$	1	8
" between orbits	1	1	1	0
Length of nasal bones	1	$7\frac{1}{2}$	1	$6\frac{2}{3}$
Width of ditto behind		10		10
" " in front		7		7
From upper incisor to molar teeth	1	$0\frac{3}{4}$	1	$0\frac{3}{4}$
Length of six molar teeth, taken together		$7\frac{1}{3}$		$7\frac{1}{3}$
Antero-posterior diameter of palate		$3\frac{4}{5}$		4
Width of post-palatal channel ...		$3\frac{2}{3}$		4
Length of lower jaw	2	$6\frac{1}{2}$	2	$6\frac{1}{2}$
Height of ditto	1	$4\frac{3}{4}$	1	$5\frac{3}{4}$
Transverse diameter of the angular portion	1	$2\frac{1}{2}$	1	$2\frac{1}{3}$

LEPUS HISPIDUS.

Assam Hare.

Lepus hispidus. PEARSON, in Bengal Sporting Magazine for August, 1843, p. 131.

Carpolagus hispidus. BLYTH, Journal of the Asiatic Society of Bengal, No. 160 (76 of New Series), p. 247, with Plates. 1845.

Ears much shorter than the head; fur harsh, on the upper parts of the body pencilled with black and yellow-brown; on the lower parts paler, and inclining to white; toes yellowish white: tail reddish, excepting at the root beneath.

According to Mr. Blyth, who has given a detailed description of this animal in the work above quoted, its length, measured in a straight line, is about fifteen inches and a half; the ears two inches; the tail, with the hair, scarcely an inch and a half; the tarsus, to the end of the nails, three and three quarter inches; the length of the skull also three inches and three quarters. The head is large, and the eyes small; the whiskers slender, and by no means conspicuous; the ears are very short as compared with those of most other species, as is likewise the tail; the limbs small, and less unequal in length than usual; the claws straight, sharply pointed, and remarkably strong. The fur is remarkable for an animal of the Hare section, on account of its harshness, but this harsh texture is due to the longer hairs of the fur only, the under fur being soft: this latter is of an ashy hue, whilst the outer fur consists of hairs, some of which are annulated with black and yellowish brown, whilst others (which are longer) are entirely black and glossy. On the under parts of the body the fur is paler, being of a dingy white tint. The fur of the tail is soft, and of a rufous tint throughout, with the exception of the basal part beneath.

Beyond these points Mr. Blyth draws attention to some peculiarities observable in the skull, which that gentleman remarks "is much more solid and strong than in any *Lepus*, with every modification that should contribute to increased strength, but upon the same subtypical model of confirmation: dentition also similar, but the grinders broader and more powerful, and the incisors proportionably much larger. The palatal foramina are reduced, so that the bony palate is as long as broad; the ant-orbital foramina are nearly closed by obliquely transverse bony spiculæ, corresponding to the open bony network observable in *Lepus*. The nasal bones are broad, with an evenly arched transverse section, and are less elongated backwards than in the true Hares; the maxillaries and intermaxillaries corresponding to their greater width and solidity. The zygoma fully twice as long as in *Lepus;* the supra-orbital processes continued forwards uninterruptedly, the anterior emargination, seen in the Hares, being quite filled up with bone, while the posterior is also much less deep."

These aberrant modifications found in the skull of the Assam Hare, combined with certain other peculiarities of the animal already pointed out, have induced Mr. Blyth to remove it from the genus *Lepus,* and to found upon it the new genus *Carpolagus.* Strongly marked, however, as these distinctions are, if the Assam Hare be compared with the Common Hare, they are less so when that animal is compared with the Indian Hare (*Lepus ruficaudatus*), and *much* less so, when it is compared with the *Lepus brachyurus* of Japan. This last mentioned animal has the short ears and tail of the *Lepus hispidus,* and the same large molar, and incisor teeth, combined with a powerfully formed skull, but in this skull the notch which sets free the fore part of the supra-orbital process, is not absent, as in *Lepus hispidus* : it agrees in having the patch unusually long, but differs from the skull of *L. hispidus* (as it would appear from Mr. Blyth's figures) in having the zygomatic arches straight and parallel as in other Hares ; the Assam species having the zygoma somewhat arched outwards. The peculiarities which I have pointed out as distinguishing the lower jaw of the *Lepus ruficaudatus* from that of the *L. timidus,* are also found in the lower jaw of *L. hispidus,* but here the angular portion has a still greater transverse diameter.

The specimen upon which Mr. Pearson drew up his description, and founded the present species, was killed on the right bank of the river Teestah, close under the Saul Forest, and about six miles north of Jalpee Goree. In this place the animal is said to be rare, but towards the hills it appears to be found in greater abundance. There are reasons to believe the animal burrows.

Lepus Sinensis.

Lepus Sinensis. GRAY, Illustrations of Indian Zoology.

Of a species of Hare which inhabits China, a well executed figure will be found in the work above quoted. It represents an

animal about fourteen or fifteen inches in length, with ears of the same length as the head, and pale on the inner surface, to the point; the body of a rich rufous colour, but much suffused with black on the back, and the tail with the lower surface white, and the upper, partly black and partly rufous. Beyond these points, gleaned from the figure mentioned, I am unable to furnish any further particulars, excepting that the fur of the animal, as Mr. Gray informs us, was very harsh.

Hares of Africa, Arabia, and Syria.

Of the several species of Hares described as natives of the northern parts of Africa, as well as of Syria and Arabia, two only have come under my notice, namely, the Egyptian Hare, and the Isabelline Hare; the former has a palish but mottled colouring, and the latter is almost of an uniform delicate fawn yellow; both are much smaller than the Common Hare, and are remarkable for having very large ears. The remaining species are described by Ehrenberg, and do not appear to be generally known, excepting by the descriptions. One other species must be included in the North African Fauna, the *Lepus Mediterraneus,* which is already described amongst the European species, since it occurs in Sardinia as well as in North Africa.

LEPUS SYRIACUS.

Syrian Hare.

Lepus Syriacus. HEMPR. et EHR. Symbolæ Phys. Mamm. Pt. ii. Pl. xiv. fig. 2.

Head and tail short; limbs long; ears about one-third longer than the head: fur somewhat harsh, its general hue yellowish, but on the upper parts of the body variegated with black:

throat and abdomen white; tail white beneath, black above; a large black patch at the apex of the ear externally.

Inhabits Syria.

The Syrian Hare, we learn from Ehrenberg, is about equal in size to the Egyptian species, but differs in having a coarse fur, as well as in its general colouring and proportions. A considerable portion of the hairs of the back are ashy grey at the root; above this they are yellowish white; then follows a broad dusky brown space, and the exposed portions are yellow, but with a fine black point: the longer and coarser scattered hairs are entirely black, except at the root, which is grey, and there are others which are black, with a subterminal narrow yellow ring. On the sides of the body the hairs are ashy grey at the root, black in the middle, and yellow at the apex: the back of the neck is of a rusty hue, but whitish near the root of the ear: the abdomen and throat are white, and a white mark passes through the eye. The posterior surface of the ear, externally, is pure white, and at the apex is a large black patch, about one inch and a quarter in depth; the fringes of the ears are whitish. The tail is white, excepting along the mesial line above, which part is black. The feet are densely clothed beneath with bright rufous hairs. Ehrenberg states that this species of Hare was met with on Mount Lebanon, not far far from Beirout, and again between Beirout and Tripoli[1]. The following are its dimensions according to the author just mentioned.

[1] It is known to the inhabitants of Mount Lebanon by the Arabic name *Erneb*: a second species of the genus is said to occur in Syria, which, from the burrowing habits attributed to it, may be the Common Rabbit; it is called by the natives *Tabsum*, a name which, Ehrenberg remarks, corresponds very nearly with the Hebrew term *Taphza*. The name *Erneb* or *Aerneb* (corresponding to the Hebrew Arnebet of Moses), it must be observed, is applied indiscriminately to the following North African or Arabian species:—*L. Ægyptius*, *L. isabellinus*, *L. Arabicus*, and *L. Sinaiticus*.

	Inches.	Lines.
From tip of nose to root of tail	20	5
From ditto to the hinder part of the root of the ear	4	$1\frac{1}{3}$
Length of tail, including the fur	3	$9\frac{1}{2}$
" of ditto without the fur	2	$3\frac{1}{2}$
" of ear	5	5
" of fore foot and nails	2	$5\frac{3}{4}$
" of tarsus	5	6

LEPUS SINAITICUS.

Sinai Hare.

Lepus Sinaiticus. **HEMPR. et EHR., Symb. Phys. Mamm. ii. Pl. xiv. fig. 1. Oct. 1832.**

Ears one-fifth larger than the head; tarsi nearly equal to the head in length: upper parts of the body yellow brown, variegated with black; ears with a distinct black tip; tail black above, white below.

Inhabits Arabia Petræa, in the valleys near Mount Sinai.

The fur in this species is grey at the root; each hair is yellow in the middle, brownish above, and black at the point; the longer scattered hairs, however, are white at the extremity. The general hue of the upper parts is brown, variegated with black, as is the case in the *Lepus Ægyptius*. A white mark passes through the eye, and there is a white spot on the neck immediately behind each ear. The ears are fringed with white, and have a black patch at the apex. The tail is black in the middle of the upper surface, but white in the remaining parts; the soles of the feet tawny. The colouring of a second individual is described by Ehrenberg as reddish ash, variegated with blackish: the orbit of the eye and the throat white; the throat, lower part of the breast, and front of the legs, also white; the cheeks, upper part of the chest, and the under parts of the feet, rufous : ears brownish ash-colour, margined with white and black at the tip.

	Inches.	Lines.
Length from tip of nose to root of tail ...	20	5
" from ditto to the hinder part of the root of the ear	4	7
" of tail, with the fur	3	7
" of ditto, without the fur	3	0
" of ear	5	4¾
" of fore foot and nails	2	1½

The author has never met with this species of Hare in the collections which he has visited.

LEPUS ARABICUS.

Arabian Hare.

Lepus Arabicus. Hempr. et Ehr. Symbolæ Physicæ, Mammalia, Part 2.

Head and ears elongated, the latter one-fifth longer than the head; tibiæ long; tarsi about equal to the head in length; teeth stout: fur ashy grey, or tawny, variegated with black and rufous; on the under parts of the body white; back of the neck tawny yellow; ears immaculate: tail white, with upper surface brown black.

Inhabits Arabia.

The Hares which inhabit the opposite coasts of the Red Sea, Ehrenberg observes, have a great affinity with each other, but present certain points of distinction, and hence have been regarded as species, but more complete observations are required to determine the value, or importance, of the differences noticed. The present species was met with at Confodah (a little south of Mecca), and likewise in Arabia Felix, near Toheia.

The fur on the upper parts of the body, in the Arabian

Hare, is ashy white at the root, and each hair is brown in the middle, has a subterminal, broad yellow-white ring, and a blackish, ashy, or rufous point; the longer interspersed hairs are entirely black, excepting at the root, which is greyish. The general hue of the upper parts is ashy grey, but with white pencillings; the back of the neck is pale yellow. A white mark passes through the eye, and the throat and abdomen are white. The ears have no black patch at the extremity; their anterior surface is ashy grey, whilst the hinder external part is paler; the apex is margined with black; the inner surface yellowish. The feet are brownish yellow, but assume a paler hue at the extremities of the toes; the fur on their under parts is brown. The tail is brown-black above, and white beneath.

The fur was short in the specimens examined by Ehrenberg, arising, it is supposed, from their being killed when changing their winter, for the summer fur.

The following dimensions (as well, indeed, as the other parts of this description), are taken from thc "Symbolæ Physicæ," of MM. Hemprich and Ehrenberg, the author having had no opportunity of examining the animal.

	Inches.	Lines.
Length from tip of nose to root of tail ...	22	$0\frac{3}{4}$
" from ditto to the hinder part of the root of the ear	4	6
" of tail, including the fur	3	$3\frac{1}{2}$
" of ditto, without the fur	2	$2\frac{1}{3}$
" of ear	5	5
" of fore foot and nails	1	$11\frac{1}{4}$
" of hind foot	4	5

LEPUS ÆGYPTIUS.

Egyptian Hare.

Lepus Ægyptius. GEOFFROY, Description de l'Egypte, Hist. Nat. tom. ii. p. 739, Atl. Pl. 6, fig. 2 (1812).

Lepus Ægyptius. Desm. Mammalogie, Part ii. p. 350.
" " Ehrenberg, Symbolæ Physicæ, Pl. 15, fig. 1.
" " Wagner, Schreb. Säugth. Suppl. iv. p. 90.

Ears much longer than the head; fur moderately long, and soft, its general hue ashy yellow; on the upper parts of the animal mottled with black; chin, throat, and abdomen, white; ears rather broadly margined with black externally, at the apex; tail white, with the upper surface black.

Inhabits Egypt.

The Egyptian Hare is considerably smaller than the common species; has proportionately larger ears, and its general hue is paler. It appears to be found throughout Egypt, and we are informed by Messrs. Hemprich and Ehrenberg, that it confines itself chiefly to the deserts, being rare in the fields. It is a delicately made animal, and having long legs, and especially a long hind foot, we should imagine, was very fleet. Its soft and moderately long fur is of a grey-white hue next the skin, on the upper parts of the body, and here the hairs have a sub-terminal broad cream-coloured ring, below which is a broad dusky grey space, and above the pale ring they are blackish. The chin, throat, and abdomen, are white; the chest, of a very pale dirty yellowish hue. The orbit of the eye is whitish: the ears, which are large, are clothed with pale hairs internally, and have a yellow-white fringe on the anterior margin; externally they are of a yellowish ash colour in front, white behind, and rather broadly margined with black at the apex. The back of the neck is pale rufous, but whitish immediately below the ear: the legs are very pale, inclining to white, but somewhat suffused with a dirty yellow tint.

	Leyden Museum.		British Museum.	
	Inches.	Lns.	Inches.	Lns.
Length from tip of nose to root of tail	19	9	18	0
" from nose to ear	3	2 ?	3	7
" of tail	3	6	4	4
" of ear	5	0	5	6

	Leyden Museum. Inches.	Lns.	British Museum. Inches.	Lns.
Width of ditto	2	11	2	11
" of fore leg and foot ...			6	0
" of fore foot			2	0
" of tarsus	4	6	4	3

	Inches.	Lines.
Length of skull		?
From the anterior extremity to the posterior root of the zygoma	2	6
Width	1	8¼
Length of nasal bones	1	3½
Width of ditto behind		9¼
" " in front		6¾
" of interorbital space	1	0¾
Between incisors to molar teeth		11
Length of palate		3¼
" of six true molars, taken together		6½
Width of second upper true molar ...		2½
" of upper incisors		2⅚
Length of lower jaw	2	6⅓
Height of ditto	1	6¼
Transverse diameter of angular portion ...		11

Of this species I have examined the specimens in the Paris and Leyden Museums, as well as those in the British Museum. Those in the Leyden Museum were procured in Egypt by Dr. Rüppell. The feet in the Egyptian Hare are more densely clothed with fur than in the *L. isabellinus.*

Lepus Habessinicus.

Lepus Habessinicus. HEMPR. et EHR. Symb. Phys. Mamm. Part ii. Plate 15, fig. 2.

Ears about equal to the head in length; tarsi rather longer than the ears, muzzle in adult animals rather arched: upper parts of the body brownish ash colour, finely pencilled with black, under parts white; ears externally with a black patch at the extremity.

Inhabits Abyssinia.

It would appear from Ehrenberg's description, from which this account is taken, that the Abyssinian Hare approaches the Egyptian species in size, but differs in having the ears comparatively shorter, and the muzzle more arched. It is stated to be of a brownish ash colour on the upper parts of the body, and pencilled with black on these parts; yellowish on the back of the neck; white on the under parts of the body as well as on the throat; an indistinct white mark passes through the eye; the ears are of a brownish ash colour, with a large blackish patch at the apex externally, about three quarters of an inch in depth; the fringe of the ear, as well as the posterior margin, white, and there is a white patch at the base of this organ externally. The feet and chest are yellowish.

	Inches.	Lines.
Length from tip of nose to root of tail	20	5
" from ditto to back of ear ...	4	0¼
" of tail, with the fur	3	10
" of ditto, without the fur ...	2	9
" of ear	4	2½
" of hind foot	4	3⅔

LEPUS ISABELLINUS.

Nubian Hare.

Lepus isabellinus. CRETZSCHMAR, Atlas zu der Reise im Nordlichen Africa von Rüppell, p. 52, Pl. 20 (1826).

" " WAGNER, Schreb. Säugh. Suppl. iv. p. 95, Pl. 233, D.

" *Æthiopicus.* HEMP. et EHRENB. Symbolæ Physicæ, Mamm. ii. Pl. 13 (1832).

Form slender; limbs moderately long; ears about one-third longer than the head: fur moderate, its general hue fawn yellow; on the back slightly pencilled with black, under parts of the animal white; ears margined with black at the apex externally: tail white below; partly brown-black, and partly yellow above.

Inhabits Nubia.

An elegant little Hare, discovered by Dr. Rüppell, the enterprising Abyssinian traveller, who states that it frequents districts which are overgrown with shrubs, and is common in the Steppes south-west of Ambukol (on the Nile 18° N. lat.), where it is hunted with greyhounds: it extends southwards into Sennaar, but in Abyssinia is replaced by another Hare, which, by Cretzschmar, is regarded as specifically identical with the Egyptian Hare, but which, according to Ehrenberg, forms a third species, and is described by that author under the name *Habessinicus*.

The Nubian Hare is readily distinguished from its congeners by its almost uniform delicate fawn yellow colouring, combined with a small stature, very large ears, and slender and moderately long legs. The fur on the back of this animal is of an ashy-white colour next the skin; each hair is brownish in the middle, and has the exposed end of a pale yellow tint; the longer and coarser hairs are black at the point. On the back of the neck, chest, and sides of the body, there is no black pencilling; the former parts are slightly tinted with rufous, but on the back of the neck it is chiefly in the mesial line that this hue is visible; for immediately behind the ears, and for some distance backwards from that point, the fur is whitish. The throat and abdomen, as well as the orbit of the eye and inner surface of the limbs, are white; the tail is white beneath, sometimes faintly tinged with yellow; and above, it is brown-black along the mesial line, but yellow at the sides: the dense fur on the soles of the feet is of a rufous colour. The ears have the external and internal bands of the same yellow hue as the body; the fringe on their anterior margin is yellow-white, and the hinder margin is edged with pure white; on the inner surface the ears are margined with yellow at the apex, and on the outer, they are rather broadly margined with black.

The above description is taken from a specimen in the

British Museum, which was sent from Sennaar. This specimen agrees most closely (as I perceive, upon comparing my notes) with specimens in the Paris and Leyden Museums. Two specimens in the last-mentioned Museum were received from Dr. Rüppell.

	Brit. Mus.		Paris Mus.		Leyden Mus.		From Ehrenberg.			
	Ins.	Lines.	Ins.	Lines.	Ins.	Lines.	Ins.	Lines.	Ins.	Lines.
Length from tip of nose to root of tail .	17	6	18	0	19	0	19	7	16	2
" from nose to ear, about ...	3	2	3	4	3	1	4	1½	3	10[1]
" of ear	4	6	4	8	5	0	4	7	4	2½
Width of ditto	2	9			2	10				
Length of fore leg and foot, about ...	5	3								
" of the fore foot and nails ...	1	9							1	11
" of the hind foot and nails ...	4	0			4	1	3	10	4	1½
" of the tail, including the fur, about	3	6			4	0				
" of ditto, not including the fur .							2	7	2	3½

Upon comparing the skulls of *Lepus Ægyptius* and *L. isabellinus*, in the Leyden Museum, I found much general

[1] Here measured to the *back* of the ear, my own measurements were taken to the fore part of that organ.

resemblance between them ; they have the nasal bones rather short and very flat behind. The skull of *L. Ægyptius* was rather larger than that of *L. isabellinus*, and had a greater antero-posterior diameter to the palatine bone ; the incisor teeth, moreover, were proportionately broader. The upper incisors in both animals had each a double groove, the grooves being placed closely together. The following are the dimensions of the cranium of the Isabelline, or Nubian Hare.

	Inches.	Lines.
Length of skull to the hinder margin of the parietal bones	2	3½
Width of ditto	1	6
Length of nasal bones	1	2¾
Width of ditto behind		8½
" " in front		6
" of interorbital space		11
From back part of upper incisor to molar teeth		10¾
" ditto to hinder margin of palate... ...	1	2
Antero-posterior diameter of palate		2⅔
Length of six upper molar teeth, taken together		6¼
Width of upper incisors		2½
Length of lower jaw	2	3¼
Height of ditto	1	6
Transverse diameter of angular portion ...		10½

Hares of South Africa.

Much confusion has arisen with respect to the Cape Hares; many nominal species have been formed, and one of the species has been very generally confounded with the Egyptian Hare. Having fortunately had an opportunity of examining most of the specimens upon which the various species have been founded, I have been enabled to satisfy myself that there exist in South Africa, three well-marked species of Hares, perfectly distinct from those of other parts, and easily distinguished from each other. Two of these are very nearly equal in size to the Common Hare (*L. timidus*), and have very large ears, but one attains a larger size than the other, and has ears proportionately larger; has a rufous back to the neck, a white throat, and its limbs are of a brown-grey hue, as well as the sides of the body; this is the *L. saxatilis* of F. Cuvier. The second is distinguished from the first by the absence of the rufous tint on the back of the neck, by the limbs and sides of the body being of a more or less bright rust colour, and the throat yellow: this is the *Lepus Capensis* of F. Cuvier. The third is a small animal, about equal in size to the Common Rabbit (*L. Cuniculus,*) and is further distinguished from the large species, by its having short ears, and short feet and tail; by its deep red-brown colouring, and by the tail being for the most part of the same red-brown colour as the body, often more or less dusky, however, at the point; the tail in the larger species being white, with the upper surface black. This smaller species is the *L. crassicaudatus* of Is. Geoffroy.

According to the Leyden Museum, the first species is known to the Dutch colonists, by the name '*Berg-haas*' (or Mountain Hare); the second, is the '*Vlakte-haas*' (or Plain Hare) of the colonists; and the third is the '*Roode-haas* (or Red Hare) of the Dutch settlers. Besides these, there is a very small Hare in the Paris Museum, upon which M. Is. Geoffroy established his *L. arenarius;* this I feel almost certain is a very young animal, and, moreover, I suspect it will prove to be the young of *L. Capensis*.

LEPUS SAXATILIS.

Rock Hare.

(*Plate* 1, Fig. 1.)

Lepus saxatilis. F. Cuvier, Dictionnaire des Sciences Naturelles, Tom. xxvi. p. 309. 1823.

Lepus saxatilis. Is. Geoff. Dict. Classique d'Hist. Nat. ix. p. 382. 1826.
" *rufinucha.* A. Smith, Zoological Journal, Vol. iv. p. 440. 1829.
" *longicaudatus.* Gray, Magazine of Nat. Hist. Vol. i. (new series), p. 586. 1837.
" *fumigatus.* Wagner, in Schreb., Säugth. Suppl. vol. iv. p. 98. 1844.
Berg-haas of the Dutch colonists.

Ears much longer than the head; tail long: fur moderate; general hue grey-brown, on the upper parts of the body mottled with brownish black, and brown-white; under parts of body, white; back of neck, rufous; ears with a large blackish patch externally at the apex; internally fringed with black at the apex; tail black above, and white below, as well as at the point.

Inhabits the Cape of Good Hope — "rocky and mountainous situations."—Dr. A. Smith.

The Rock Hare of the Cape is about equal in size to the Common Hare, and is remarkable for the enormous size of its ears, and the comparatively great length of its tail. From the other Cape Hares it may be distinguished not only by its superior bulk, and the relatively greater size of its ears and tail, but likewise by the absence of any rufous tint on the limbs and sides of the body.

Dr. A. Wagner, in his able Supplement to Schreber's great work upon Mammalia, has recently described the present species under the name *Lepus fumigatus*, not without expressing a suspicion, however, that it might prove to be identical with F. Cuvier's *L. saxatilis*, the description of which he did not regard as sufficiently detailed to enable him to satisfy himself upon this point. My notes made upon M. Cuvier's animal in the Paris Museum, describe it as fully as large as the Common Hare, with the ears immensely large, and the general hue of the fur grey-brown—on the back of the animal mottled with black and brown-white; the back of the neck of a rusty yellow tint; the abdomen and inner surface of the limbs, as well as the throat and orbit of the eye, white; the ears externally with the fore part clothed with party-coloured black and dirty white hairs, the hinder part whitish, and the apical portion black; internally for the most part clothed with dirty yellow-white hairs, but becoming blackish near the hinder margin, which is edged with white; the fringe on the anterior margin is composed of dirty yellowish hairs; fore legs pale brown; the tarsi mottled with pale brown and white; the dense fur on the soles of the feet dusky brown. Length $22\frac{1}{2}$ inches; ear, about $5\frac{1}{3}$ inches; nose to ear, about $4\frac{1}{12}$ inches,

The specimens in the British Museum and the Leyden Museum agree in all essential particulars with this description. In all, the back of the neck is of a palish rufous tint, and I may add that this hue extends on to the basal portion of the ears behind; they have the outer surface of the limbs of the same brown hue as the sides of the body; the tarsi brownish, more or less mottled with white; the upper lip, lower part of cheeks, orbit of eye, throat, abdomen, and the inner surface of the limbs, white; the ears brownish or dusky near the posterior margin internally; the remaining inner portions whitish, or, towards the point, yellowish; the fringe on the hinder margin is white, and that on the anterior margin is of a dirty yellowish colour; the apical third of the ear is fringed with black, and there is a large dusky patch on the apical portion of the ear externally. The fur on the back is of an ashy grey hue next the skin, and each hair is shaded through brown into brown-black, in the middle; has a broad subterminal brown-white ring, and a dusky point. On the chin there is sometimes a brown spot, and the throat is sometimes a greyish white. The tail is long, white below and at the point, and black above.

	British Museum.		Leyden Museum.	
	Inches.	Lines.	Inches.	Lines.
Length from tip of nose to root of tail	23	0	24	0
" from nose to ear, about ..	4	6	4	6
" of ear	6	10	7	0
Width of ditto	3	6	3	6
Length of tarsus	5	5	5	6
" of tail, including the fur, about	6	0	5	6

The skull of the *Lep. saxatilis* is considerably larger than that of *L. Capensis,* and a large lobe of the frontal bones is wedged in between the nasal bones behind. The following admeasurements are taken from a skull in the Leyden Museum.

	Inches.	Lines.
Total length of skull	4	2
Width of ditto	1	10
" between orbits	1	2
Length of nasal bones	1	9
Width of ditto behind	1	0⅓
" in front		8¼
Between upper incisors and molar teeth ...	1	3¼
From ditto to hinder part of palate ...	1	7¼
Antero-posterior diameter of palate ..		3½
Length of the six upper true molars, taken together		7⅔
Width of upper incisors		3¼
Length of lower jaw	2	8½
Height of ditto	1	6
Transverse diameter of angular portion ...	1	0

LEPUS CAPENSIS.

Cape Hare.

Lepus Capensis. LINN. Syst. Nat. i. p. 76.

" " F. CUVIER, Dictionnaire des Sciences Naturelles, xxvi. p. 308. 1823.

" " IS. GEOFF., Dictionn. Classique d'Hist. Nat. ix. p. 382. 1826.

" *ochropus.* A. WAGNER, in Schreb. Säugth. Suppl. ii. p. 96. 1844.

? " *Capensis.* THUNBERG, Mammalia Capensia Recensita et Illustrata, in Mem. de l'Acad. Imper. des Sciences, de St. Petersbourgh, vol. iii. p. 309. 1811.

? " *arenarius.* IS. GEOFF., in Dict. Class. d'Hist. Nat. ix. p. 383.

Ears distinctly longer than the head; tail rather long: general hue grey brown, distinctly mottled with black on the upper parts of the body; back of neck brown-grey; throat fawn yellow; sides of body, as well as the limbs, of a more or less bright rust colour; abdomen white, more or less suffused with rust colour.

Inhabits the Cape of Good Hope.

The Cape Hare bears considerable resemblance to the Common Hare (*L. timidus*), but may be distinguished by its longer ears and tail, the brighter colour of its limbs, and the yellow tint of the throat; it wants, moreover, the rufous tint on the back of the neck which is observable in the Common Hare. Although large, the ears are proportionately rather smaller in the Cape Hare than in the Egyptian: the weight of this latter animal, when full grown, is probably not more than half that of the Cape species; the two species, moreover, differ in their colouring.

The fur on the upper parts of the body, in *L. Capensis*, is mottled with black and brownish white (sometimes having a very faint rufous tint): the hairs on these parts are nearly white at the root, dusky in the middle, have a black point, and a broadish sub-terminal brown-white ring (or the ring is white, with a faint vinous red tint): the flanks and the limbs are of a bright rust colour; the abdomen is suffused with the same colour, but here the hairs are white below the points, and along the mesial line of the abdomen, as well as on the inner surface of the hind legs, the hairs are usually entirely white; the back of the neck is of the same grey-brown hue as the upper parts of the body, or nearly so, but wants the black pencilling; the chest is of a brownish rufous tint, by no means dark, and the throat is always of a yellowish hue, usually of a dirty fawn-yellow. The region of the eye and the sides of the muzzle are tawny, but bordering the eye is a white ring. The ears are finely freckled with black and tawny on the fore part of the outer surface, and they have a large dusky patch at the apex, nearly an inch in vertical diameter; the point is fringed with black; the fringe of the ear is of a dirty yellow, or rufous white colour; the hinder margin is white. The feet are clothed with shortish hairs, and hence the nails of the toes are not hidden;[1] the denser fur on the soles of the feet is of a dirty brown colour. The tail is rather long; white beneath and at the point, and black above.

	British Museum.		Leyden Museum.	
	Ins.	Lines.	Ins.	Lines.
Length from tip of nose to root of tail	20	0	21	0
" from nose to ear	3	8	4	3
" of ear	5	4	5	6
Width of ditto	2	10		
Length of hind foot	4	11	5	6
" of tail, including the fur, about	5	0	6	0

The skull of *L. Capensis* is about equal in size to that of *L. timidus;* it has the nasal bones broad behind, and obliquely truncated at this part, where a lobe of the frontal bones separates them as usual, but this lobe is much smaller than in the *L. saxatilis.* The subjoined admeasurements, taken from a skull in the Leyden Museum, will serve to convey an idea of the proportions of the parts:—

	Inches.	Lines.
Total length of skull	3	5
Width	1	6¾
" between orbits		?
Length of nasal bones	1	6
Width of ditto behind		9
" " in front		5¼
Distance between upper front incisors and molar teeth	1	0½
" " ditto and posterior margin of palate	1	4½
Antero-posterior diameter of palate		3

[1] In *L. saxatilis* the feet are more densely clothed.

	Ins.	Lines.
Length of the six upper molar teeth, taken together		$6\frac{1}{2}$
Width of upper incisors		$2\frac{2}{3}$
Length of lower jaw	2	$5\frac{1}{2}$
Greatest height of ditto	1	$4\frac{1}{2}$
Transverse diameter of the angular portion ...		10

The limbs and sides of the body, it must be observed, are not always of an equally bright rust colour in this species; and sometimes they are of a yellow hue.

In the British Museum collection is a specimen in which these parts are of a remarkably bright yellow colour. The animal described by Dr. Wagner, under the specific name *ochropus*, would appear to be an intermediate variety between this last mentioned animal, and the specimens having the limbs of a rusty red hue. In all the individuals of the *L. Capensis* which have come under my notice (and they are numerous) the throat was yellow. From the local name 'Vlakte-haas' we are led to suppose this species differs from the *L. saxatilis*, in having the habit of frequenting the plains.

Lepus arenarius. Is. Geoffroy.

I am of opinion that the above name has been applied to specimens of a species of Hare in an immature condition, because their comparatively small size is accompanied by peculiarities in the fur, and also in the teeth, which are usually (if not always) characteristic of young Hares. In the young animal there will generally be found interspersed with the ordinary fur, numerous very long and slender hairs, of which those on the sides of the body, from having their exposed ends whitish, are very conspicuous: and, in a small Cape Hare in the British Museum collection, which closely resembles the specimens upon which M. I. Geoffroy drew up his description, I find the incisor teeth wedge-shaped, another characteristic of the young animal; in the adult, the teeth being always of equal width throughout. The general colouring of the animal under consideration is such as, from analogy, I imagine the young of *L. Capensis* would present; the back of the neck is of a greyish hue, as in that animal, and there is a slight trace of the rufous tint on the limbs, and sides of the body. The proportionately smaller size of the ears in the *L. arenarius*, (if it be a young animal, as I suppose) affords no valid objection to its specific identity with the *L. Capensis*. According to my description taken from the *L. arenarius* in the Paris Museum, it has the—

Fur long and soft, and of a general greyish brown hue, the limbs and sides of the body of a pale brownish rufous; the back of the neck greyish; the belly white, and the tail also white, excepting on the upper surface, which is black; the sides of the muzzle and chin are yellow-white; the ears have a black patch at the apex externally; the long interspersed hairs on the body are dusky on the back, and whitish on the sides of the body. Length, measuring along the curve of the back, about 13 inches; the ear rather more than 2¼ inches.

The specimens are labelled as being from Port Natal.

LEPUS CRASSICAUDATUS.

Thick-tailed Hare.

Lepus crassicaudatus. Is. Geoff., in Guérin's Magasin de Zool. 1832, Pl. 19—Etudes Zoolog., No, 1, Pl. 9.

" " Wagner, in Schreber's Säugth. Suppl. iv. p. 99. 1844.

" *rupestris.* A. Smith, in South African Quarterly Journal, No. 2, Pt. 3, p. 86, March 1834.

" *melanurus.* Rüppell, in Museum Senckenbergianum, vol. iii. p. 137. 1842.

"*Roode-haas*" of the Dutch colonists.

Ears broad, rounded at the extremity, and rather short; legs and feet short: fur dense, and of a dark hue, mottled with black and brown-white, on the upper parts, rufous white on the under, and dark brownish rufous on the limbs; tail short and thick, of a rufous brown colour at the base, and dusky at the point.

Inhabits the Cape Colony, and extends to Port Natal.

The general proportions of this animal approximate it closely to the Common Rabbit. It is subject to some slight variations in its colouring, which is more or less intense, though always, comparatively speaking, dark; and the tail in some specimens is entirely of a rusty brown hue, though usually it is dusky or blackish at the point, and I feel satisfied it is upon a specimen in which the black is more extended than usual, that Dr. Rüppell has founded his *Lepus melanurus*, an animal which clearly retains all the essential characters of M. Geoffroy's previously described *L. crassicaudatus*.

The fur of the Thick-tailed Hare is dense and soft, mottled with black and brownish white on the back, but of a pale grey hue next the skin; the limbs are of a deep brownish rust colour, and so is the back of the neck; the abdomen rufous white, the hairs here being pale rufous in the middle, and whitish at the root as well as the point; the chin grey-white; the small hairs covering the inner surface of the ear (which is broad in proportion to its length, and rounded at the extremity) are for the most part of a pale greyish brown hue; the hairs forming the fringe on the anterior margin, which are shorter than usual, are whitish with dusky points; the hinder margin is dirty white; the apical portion of the ear externally is dusky, and the external band is formed of party-coloured black and grey-white hairs; the tail is brown-black, and rufous at the root.

	Leyden Museum.		British Museum.	
	Inches.	Lines.	Inches.	Lines.
Length from tip of nose to root of tail	17	9	14	0
From nose to ear, about	2	10	2	8
Length of ear	3	4	3	2
Width of ditto	2	5	1	11
Length of tarsus	3	2	2	10
" of tail, including the fur, about	3	0	2	6

My notes, made upon the *L. crassicaudatus* in the Paris Museum, describe the animal as having the tail entirely of a rusty brown hue, the legs of an uniform bright rusty red; the top of the muzzle rusty brown, and in other respects agreeing in colouring with the animal from which my description is taken. A specimen in the British Museum has the tail rufous brown at the base, and of a dark brown hue beyond.

The skull of the *Lepus crassicaudatus* is of the same size as that of the Common Rabbit; the nasal bones are narrower than in other Cape Hares, and are remarkable for being separated behind by a very large lobe of the frontals; this lobe, wedged in between the nasal bones, terminates at the commencement of the posterior third part of the bones just mentioned.

	Inches.	Lines.
Length of skull from the apex of the intermaxillaries to the posterior root of the zygomatic arch	2	4
Width of ditto	1	5¼
Length of nasal bones	1	5
Width of ditto behind		7½
" " in front		5
Length between the superior incisor, and the molar teeth	1	0½
Antero-posterior diameter of palate		3¼
Length of six upper true molars taken together		5¾
Width of upper incisors		2¼
Length of lower jaw	2	3½
Height of ditto	1	2⅓
Transverse diameter of angular portion ...		10¼

North American Hares.

We shall commence our account by noticing those *American Hares which occur in the more northern latitudes, and which become white during the winter months.*

LEPUS GLACIALIS.

The Polar Hare.

Lepus glacialis.	LEACH, in Ross's Voyage, 8vo. ed. Lond. 1819, vol. 2, App. iv. p. 170 (*L. arcticus*, l. c. p. 151).
" "	SABINE, Suppl. to Parry's First Voyage, p. clxxxviii. (1821).
" "	RICHARDSON, Appendix to Parry's Second Voyage, p. 321 (1825); Fauna Boreali Americana, p. 221 (1829).
" "	HARLAN, Fauna Americana, p. 194 (1825).
" "	GODMAN's Natural Hist. vol. ii. p. 162 (1826).
" "	BACHM. Journal of the Acad. of Nat. Sci. of Philadelphia, vol. 7, Part 1, p. 285, Pl. 21 (1837).
" "	AUDUBON and BACHMAN, Quadrupeds of America, Vol. 1, p. 242: No. 7, Pl. 32 (Summer fur) of the Illustrations.
" *timidus.*	FABRICIUS, Fauna Grœlandica, p. 29 (1780).
Varying Hare.	PENNANT, Arctic Zoology, i. p. 94 (1784).

Ears about equal to the head in length; hind foot large; tail short, and very bushy; incisor teeth narrow. *Summer fur*—General hue sooty-grey (not dark); the ears black, with a white fringe to the greater part of the hinder margin. *Winter fur*—Pure white (the hairs white to the root); tip of ear with a small black patch.

Inhabits the northernmost parts of North America, having been found as far north as the N. Georgian Islands, in lat. 75°; occurs on both sides of Baffin's Bay, and, on the east side of America, extends through Labrador into Newfoundland[1]. In the interior of the continent its southern limit is about 62° N. lat.[2]

There is scarcely a spot in the Arctic regions, the most desolate and sterile that can be discovered, Capt. Sir James

[1] Dr. Bachman is inclined to believe, from statements which have been made to him, that the Polar Hare extends into Nova Scotia, and the northern parts of Maine.

[2] The Polar Hare "is not found in wooded districts; hence it does not come further south on the line of the Mackenzie and Slave Lake, than lat. 64°."—*Rich.*, Faun. Bor. Am.

Ross observes, in which this animal is not to be found, and that, too, throughout the winter; nor does it seek to shelter itself from the inclemency of the weather, by burrowing in the snow, but is found generally sitting solitary under the lee of a large stone, where the snow-drift, as it passes along, seems in some measure to afford a protection from the bitterness of the blast that impels it, by collecting around, and half burying the animal beneath it; or, according to Dr. (now Sir John) Richardson, it seeks the sides of hills, where the wind prevents the snow from lodging deeply, and where, even in winter, it can procure the berries of the Alpine Arbutus, the bark of some dwarf willows, or the evergreen leaves of the Labrador tea-plant (*Ledum*[1]). "The Polar Hare," Sir John Richardson proceeds to observe, "is not a very shy animal, and on the approach of a hunter it merely runs to a little distance, repeating this manœuvre as often as its pursurer comes nearly within gun-shot, until it is thoroughly scared by his perseverance, when it makes off."

A specimen captured by Sir John Ross's party became sufficiently tame to feed out of the hand, and was allowed to run loose about the ship's cabin. Although constantly in a temperature never much below the freezing point, its fur, it is stated, assumed the white colour as early as those which were running wild. Its winter coat was cast early in May, and was not replaced by the darker hairs of the ordinary summer fur, but retained the pure white colour[2].

The Polar Hare appears to be a more bulky animal than our common European species, but this arises from the greater length and density of its fur; its size, in fact, being

1 "On the barren coast of Winter Island, the Hares went out on the ice to the ships, to feed upon the tea-leaves thrown overboard by the sailors."—Quoted by Richardson from Lyon's Private Journal.

2 Appendix to Ross's Second Voyage—Nat. Hist., p. xvi.

nearly the same. In its winter fur it can scarcely be distinguished from the white specimens of *L. variabilis,* having, like that animal, but a small black patch at the extremity of the ears; whilst, however, the Polar Hare is somewhat larger than the Varying Hare, its hind foot is absolutely shorter[1]. In its summer fur it is readily distinguished from the *L. variabilis* (and, indeed, from all other Hares here described), by its almost uniformly black ears.

In the British Museum are specimens, from Labrador, exhibiting the animal both in its summer and winter dress. A specimen in its summer fur most closely resembles the individual described and figured by Dr. Bachman, and which that gentleman allowed me to examine. Its general hue is grey, suffused with sooty black, but so faintly, that the prevailing tint is by no means dark: the back of the animal is rather finely pencilled with greyish white, and the hairs on this part are whitish at the root, then of a dirty yellowish hue followed by a brownish tint, a greyish white, subterminal ring, and a dusky point; on the under parts of the body the hairs are uniformly white; the orbit of the eye, and the tail, are also white. The feet are very densely clothed with fur; that on their under surface is of a dirty brownish yellow hue. The ears are densely clothed, and almost entirely black, slightly inclining to grey on the hinder part of the outer surface and on the middle part of the inner surface; and

[1] The average length of the hind foot, in four specimens, I found to be 5″ 11¼‴, the average length of the animals being 25″ 1‴; whilst in the *L. variabilis*, taking the average of six specimens, the foot was 6″ 3⅙‴, and the length of the body 22″ 3½‴. When, with this point of distinction, are taken into consideration other differences observable in the structure of the skull and teeth, and in the colouring of the animals in their summer fur, it must be admitted that there are ample grounds for regarding the Polar Hare of the Western Hemisphere as specifically distinct from what may be termed the Polar Hare (*L. variabilis*), of the Eastern Hemisphere; two animals which, by many naturalists, were formerly confounded.

somewhat brownish at the point; the hinder margin is fringed with white hairs, but the white hairs do not extend either to the point or to the base of the ear, being replaced by a black fringe at those parts.

With regard to the winter fur, it is only necessary to state that it is of a pure white colour on all parts, excepting on the soles of the feet and point of the ears; it is long, soft, and very dense: at the apex of the ear the black patch is small, being not more than half an inch in diameter, and composed almost entirely of the long hairs which spring from the margin of the ear.

Dr. Bachman's specimen was killed in the middle of August, at which time the change from the summer to the winter fur had begun to take place, for on the hinder part of the back was a large white patch, and some few white spots were visible on the sides of the body.

A nearly mature fœtal specimen of the Polar Hare, examined by Sir J. Richardson, was of a blackish brown colour on the upper parts of the body and outer surface of the extremities. Some specimens examined by Sir James Ross, and which were apparently less advanced in development, were of a dark grey colour.[1] A specimen, three months old (measuring 17 inches in length, and with the ears 3½ inches in length), is described by Sir J. Richardson as having the general tint of the upper parts hoary, the hairs of the fur being black at the root and white at the point; the abdomen, feet, and tail, entirely white; the ears similar in colour to the back, but with the proportion of black hairs rather greater, their margins white, and their points with a brownish black spot.

[1] Four, six, and eight young have been found *in utero*.

	Brit. Museum. Summer fur. Labrador.		Brit. Museum. Winter fur. Labrador.		Brit. Museum. Winter fur. Baffin's Bay.		Summer fur. Newfoundland.		From Bear Lake. Dr. Richardson.	
	Ins.	Lns	Ins.	Lns	Ins.	Lns	Ins.	Lns	Ins.	Lns
Length, tip of nose to root of tail	25	0	26	0	26	0	24	0	22	6
" " ditto to ear	4	2	4	3	4	3	4	1	4	6[1]
" of ear	4	0	4	3	4	0	4	0	4	6
Width of ditto	2	1								
Length of hind foot	6	0	6	0	5	9	6	0	6	6
" of tail, including the fur, about	2	6	3	0			3	0	3	6

The skull of the Polar Hare is about equal in size to that of the Common Hare, but its proportions differ considerably; the most striking difference being the greater width of the interorbital portion of the cranium in the former animal: the upper surface of the frontal bones have the anterior half rather strongly depressed, and the hinder half very convex; the cranial portion of the skull is broader, and its outline descends more suddenly from before backwards; the supra-orbital processes are very large; the zygoma is very deep, and shorter than in the Common Hare: the sudden contraction of the fore half of the incisive opening, and the structure of the upper incisor teeth, furnish well-marked distinguishing characters, when the skull of the Polar Hare is compared with that of the *L. timidus*, or of the *L. variabilis*. In both these last-named species the principal upper incisor has the transverse diameter greater than the longitudinal, whilst in the Polar Hare, these teeth (which are remarkable for their narrowness) have the antero-posterior diameter equal

[1] This last dimension was taken to the back of the head; the skull of the same specimen measured (according to Dr. Richardson) 3 inches and 10 lines in length.

to the transverse one: the anterior surface is convex; the groove is less deep than usual, and two or three small longitudinal striæ may be observed in this groove. The lower incisors, like those of the upper jaw, are very narrow, being rather narrower than the corresponding teeth in a skull of *L. variabilis* before me, although that skull is considerably smaller than the skull of the Polar Hare; their front surface is somewhat concave.

	Melville Island. Leyden Museum.		Greenland. Leyden. Museum.		Greenland. British Museum.		Greenland. British Museum.	
	Ins.	Lines.	Ins.	Lines.	Ins.	Lines.	Ins.	Lines.
Total length of skull	4	3½		?	3	10½	3	6¼
Width of ditto	2	1	2	0¾	1	11	1	10½
Length of nasal bones	1	9	1	7½	1	8	1	4⅓
Width of ditto behind		11¼		11		10½		9½
" " in front		9½		9		8¾		7¾
" between orbits	1	7	1	7½	1	5⅓	1	4
From upper incisor to molar teeth	1	5	1	4¼	1	3⅓	1	1½
From front of upper incisor to hinder margin of palate ...	1	9⅓	1	8½	1	7½	1	5⅔
Antero-posterior diameter of palate		3⅓		3¾		3½		3¼
Length of the six upper true molars, taken together		9		9		8¾		8
Width of the two front upper incisors		2¾		2¾		2⅔		2½
Length of lower jaw	3	2½	3	2	2	11	2	7
Height of ditto	1	10½	1	8½	1	10	1	8½
Transverse diameter of angular portion	1	2¼	1	2	1	1½	1	0½

The skull, the dimensions of which are given in the third column, presents all the same general characters which I observed in the other three skulls—having the same strongly-convex swelling of the frontal bones, the broad interorbital space, deep and short zygoma, contracted incisive opening, and narrow incisors, convex in front; but it is remarkable for the shortness of the facial portion, a part of the skull which is more subject to variation in the Hares than other parts.

LEPUS AMERICANUS.

American Hare.

Lepus Americanus. ERXLEBEN, Systema Regni Animalis, p. 330 (1777).
" *Hudsonius.* PALLAS, Nov. Spe. Glir. p. 30 (1778).
" *Americanus.* RICHARDS., Appendix to Parry's Second Voy., p. 324, (1825).—Fauna Borealis Americana, p. 217 (1829).
" " BACHM., Journ. of the Acad. of Nat. Sci. of Philad. vii. p. 403 (1837); vol. viii. p. 76.
" " AUDUB. and BACHM., Quadr. of N. America, i. p. 93. Illustr. No. 3, Plates 11 and 12.
" *Virginianus* (The Northern Hare). BACHM., l. c. vii. p. 301.
" " HARLAN., Fauna Americana, p. 196 (1825).

Ears nearly equal to the head in length; well clothed with longish hairs; tarsus long, and densely clothed with fur; tail very short. Colour, *in winter,* white, but with the hairs on the upper parts of the body grey at the root, of a palish yellow-brown in the middle, and with the exposed parts, only, white. *Summer fur* with the general hue reddish brown; pencilled with black on the upper parts of the body; throat, abdomen, and under surface of the tail, white.

Inhabits the northern parts of North America, but does not extend into such high latitudes as the *Lepus glacialis*[1]; according to Dr. Richardson, it ranges northward as high as lat. 64° 30′; and Dr. Bachman gives its southern limit at about lat. 51°; it is the Common Hare about Hudson's Bay, is found in Canada, Newfoundland, all the New England States, and in the northern portions of New York, Pennsylvania, and Ohio. It is apparently confined to the eastern portions of the continent.

Dr. Bachman informs us that the present species is the

[1] The mean range of the *Lep. glacialis* is about lat. 61°, and that of *L. Americanus* about eight or nine degrees more south.

most fleet and active of the American Hares, and that excellent observer confirms the statement made by Lewis and Clarke, of its possessing the power of leaping the enormous distance of twenty-one feet at a bound. It is almost hopeless to hunt this animal, for when chased, it winds and doubles through the entangled paths of hemlock and spruce, until it fairly wears out its pursuers; and unless the hunter should appear to stop its course with the gun, it is certain to escape. In the northern portions of the United States it avoids the open grounds, confining itself to the forests, and more particularly preferring dry and elevated grounds, covered with Pines and Firs.

Never having had an opportunity of examining the Northern Hare in its full summer fur, the following is extracted from Dr. Bachman's description :—" In summer, the whole of the upper surface of the animal is reddish brown, formed by hairs which are at their roots, and for two-thirds of their length, of a bluish ash colour, then reddish yellow, succeeded by a narrow line of dark brown, the part next the tips, or points, reddish brown, but nearly all the hairs tipped with black—this colour predominating towards the rump. Whiskers mostly black; a few of the hairs white, the longest reaching beyond the head: ears brown, with a narrow black border on the outer margin, and a slight fringe of white hairs on the inner. In some specimens there is a fawn, and in others a light coloured edge, around the eye, and a few white hairs on the forehead. The iris of the eye is light silvery yellow. The point of the nose, chin, and throat, white; neck yellowish brown. Inner surface of legs, and under surface of the body, white. The tail brown above, and white below. The summer fur is assumed in April, and remains without much change till about the beginning of November, in the latitude of Quebec, and till the middle of the same month in the State of New York, and western parts of Pennsylvania; after which season the animal gains its winter pelage. During the winter, in high northern latitudes, it becomes nearly pure white, with the exception of the black edge on the outer borders of the ears. In the latitude of Albany, New York, it has always a tinge of reddish brown."

In the collection of the Zoological Society, and in the British Museum, are several specimens of the Northern Hare having the winter fur, which, though white externally, differs from the winter fur of the Polar Hare in always being of a pale grey next the skin on the upper parts of the body, and each hair on these parts is moreover of a yellow-brown hue in the middle. The hairs on the fore part of the ears are likewise party-coloured, being yellowish or dusky at the roots; towards the point, externally, they are blackish at the root, and have white points; at the edge sometimes wanting the white point, they form a black fringe, but very commonly even here the black is more or less hidden by the points being white.

The National Collection also contains a specimen of the *L. Americanus*, which is interesting as having been procured in the month of November, a time in which the fur is undergoing its change from summer to winter colouring. This animal is white, with the exception of the upper surface of the head, the cheek, back, and chest; here the party-coloured, rufous-yellow, and black hairs of the summer dress, remain. On the back, the coloured hairs are considerably longer than on the sides of the body, where they are white, and it is evident that a considerable portion of the summer hairs have been shed from the back, since those remaining are somewhat scanty, and, upon separating them, we find an abundance of white hairs, which are evidently new, and had not attained their full length at the time that the animal was killed. Were the long and somewhat scanty coloured hairs to be plucked from this animal, it would present the same condition as other specimens which have attained their full winter pelage; with these exceptions, that its back would not be of so pure a white, for the yellowish colour, which at all times is seen in the middle portion of each hair, would be but partially hidden, the white points of the hairs being less elongated; and the fur, generally, would be shorter. On the whole, I feel convinced from the examination of this specimen, that the white of the winter pelage is not always (as good observers state it to be in our *Lepus variabilis*), due to a change of colour only, but, in some cases (as in the case before us), is accompanied by a shedding of the summer fur—or rather by a partial shedding of the fur; for, judging from the conditions presented by different specimens of the Northern Hare, now before me, I am strongly inclined to believe that the longer and coarser hairs of the fur only, are shed and replaced by equally long, but white-pointed hairs, and that the dense under fur undergoes a change of colour, and becomes longer.

	British Museum.		From fresh specimens. Dr. Bachman.			
	Ins.	Lines.	Ins.	Lines	Ins.	Lines.
Length from tip of nose to root of tail ...	19	6	19	3	16	0
" from nose to ear	3	7				
" of ear	3	4	3	6	3	6
Width of ditto, about	2	0				
Length of tail (vertebræ)			1	3	1	6
" of ditto, with the fur	2	9	2	3	2	6
" of hind foot to end of nails ...	5	6	5	6	5	3

The average weight of full grown specimens, according to Dr. Bachman, is about $5\frac{1}{2}$ lbs.

Hares confined, for the most part, to the middle and southern portions of North America:—

LEPUS AQUATICUS.

Swamp Hare.

Lepus aquaticus. BACHM. Journal of the Acad. of Natural Sciences of Philadelphia, vol. vii. p. 319, Plate 22, No. II.—(Read March, 1837).

" " AUDUBON and BACHMAN, Quadrupeds of North America, vol. i. p. 287 (1847); Illustr. No. VIII. Plate 37.

" *Douglasii*, var. 1. GRAY, Magaz. of Nat. Hist. for Nov. 1837; vol. i. p. 586.

Ears but little shorter than the head; tail and tarsus moderate as to length: fur coarse, glossy, and somewhat closely applied to the body; on the upper parts of the animal of a brownish yellow colour, much mottled with black (chiefly black in the winter); behind the ear, rufous; chin and abdomen, white; ears edged with black externally at the apex; tail yellow-brown above, white beneath; legs and feet suffused with rusty red.

This species is common in the western part of the State of Alabama; still more abundant in the State of Mississippi, and in the lower part of Louisiana, and extends into Texas.

The Swamp Hare greatly resembles the two species next to be described (the *L. palustris* and *L. sylvaticus*), but may be distinguished by its much larger size,[1] by its ears, tarsi, and tail being proportionately longer, and its fur coarser and more closely applied to the body.

Although the present species is occasionally seen on high grounds in the dense forest, it prefers (as we are informed by Dr. Bachman) low and marshy situations, or the neighbourhood of streams and ponds of water. It swims with great

[1] "The Swamp Hare is a third larger than the Marsh Hare; the largest specimen of the latter, in more than fifty that we measured, was only 14 inches long, whilst the largest Swamp Hare was 22 inches; and we are informed that it is often much larger."—AUD and BACHM.

facility, and will sometimes dive, when danger threatens, to the distance of eight or ten feet. When chased by dogs the Swamp Hare runs with great swiftness, and is able to escape from them without difficulty; but it almost invariably directs its flight to the water, where it commonly swims to some spot where hollows are made in the banks by the water, or where some overhanging roots of trees afford a shelter.[1] It subsists on the roots of various kinds of aquatic plants, especially on a species of iris growing in the water.

The young of this Hare are frequently found in nests formed of leaves and grasses, placed on hillocks in the swamps, or in the hollow of some fallen tree. The female is said to produce from four to six young, and to have at least two broods in a season.

The following dimensions (from the work of Messrs. Audubon and Bachman) were taken by Dr. Lee, from a specimen of the Marsh Hare, in the flesh :—

	Inches.	Lines.
Length from the point of the nose to the insertion of the tail	20	0
" of head	4	3
" of ears posteriorly	3	$10\frac{1}{2}$
Height to shoulder	11	0
Length of hind foot	4	6
" of the tail vertebræ	2	$1\frac{1}{2}$
" tail, including the fur	3	0

The weight of a female specimen, in bad condition, was found to be six pounds.

Lepus Douglasii. Var. 1, Gray.

Ears rather shorter than the head; tail and tarsus moderate as to length; the nails stout and large: fur rather harsh, its general hue of a rich brownish yellow; on the upper parts

[1] Messrs. Audubon and Bachman refer to a specimen of the Swamp Hare (presented by them to the Academy of Natural Sciences of Philadelphia), which, upon being pursued, swam across the Alabama River.

of the body strongly pencilled with black; on the sides of the body with but little black pencilling, and on the lower parts of the flanks, as well as on the shoulders, the colour is uniform, and the chest likewise is of an uniform yellowish colour, or very nearly so; throat and abdomen rufous yellow, inclining to white: ears moderately well clothed with yellow hairs internally, which become richer in tint, and almost rufous at the point of those organs; the band indistinctly grizzled with dusky; externally the ear-band is more distinctly grizzled with dusky and yellowish, and the remaining portion, like the back of the neck, is of a bright rusty yellow, excepting at the apex, where there is a smallish, but distinct, black patch; and, moreover, there is a black line running down the hinder margin, extending from the tip of the ear about half way towards the base; the extreme edge of the ear (in front, behind, and at the apex) is rufous yellow: the limbs are of a somewhat darker hue than the sides of the body, being suffused with rust colour: the tail is pencilled with dusky, and rufous-yellow above, and with white and dirty yellow below. The hairs of the back are of a very pale grey at the root, then pale brownish yellow, followed by black, a sub-terminal broad, rich brown yellow ring, and a black point; on the abdomen the hairs are white, with a very indistinct trace of grey next the skin.

From Texas (?): in the British Museum.

Dr. Bachman, who examined this animal when in England, recognised in it a specimen of his *Lep. aquaticus;* nevertheless (although I do not doubt this view is correct), I have thought it desirable to introduce the above description of its character, since by so doing I am enabled to notice several points in addition to those given in the description gleaned from Dr. Bachman's account of *L. aquaticus*, some of which will probably furnish constant distinguishing characters of this species when compared with the two very nearly allied animals, the *L. sylvaticus* and *L. palustris*. For instance, the uniform rufous-yellow hue of the hinder half of the ear externally, combined with the distinct black spot at the apex, and the black line next the hinder margin, as well as the hairs of the abdomen being white at the root, are characters in which the animal before me differs from either the Wood Hare or the Marsh Hare, both these last mentioned species having the fur of the abdomen of a distinct grey hue below the exposed ends of the hairs—which, by the by, is a rare circumstance in the Hares: in the latter (*L. palustris*) there is no black edging to the ear, and in the *L. sylvaticus* the black edging at the point of the ear is narrow, and I always find a black line running along the anterior margin, instead of along the posterior one, as in *L. aquaticus*. On the other hand, the greater harshness of the fur, and the tail being pure white beneath—characters which are noticed as distinguishing the *L. aquaticus* from *L. palustris*—do not in the specimens in the British Museum present points of distinction, the tail being coloured beneath, and the fur being somewhat less harsh in the *L. aquaticus* than in the *L. palustris*. This probably arises from our specimen being immature. Differences of proportion in the three above-mentioned animals will be perceived upon comparing the following dimensions:—

	L. Douglasii. Var. 1. GRAY. *L. aquaticus.* BACHM.		*L. palustris.* Brit. Museum.		*L. sylvaticus.* Brit. Museum.	
	Ins.	Lns.	Ins.	Lns.	Ins.	Lns.
Length from tip of nose to root of tail ...	17	0	17	0	16	6
" from tip of nose to ear	3	7	3	4	3	3
" of ear	3	3	2	6	2	10
Width of ditto	2	3	1	10	1	9
Length of hind foot and nails	4	0	3	5	3	7
" of tail, including the fur	3	0	1	10	?	

LEPUS SYLVATICUS.

The Wood Hare.

Lepus sylvaticus. BACHM. Journal of the Acad. of Nat. Sci. of Philadelphia, vol. vii. p. 403; (*L. Americanus*), l. c. p. 326.

" " AUDUB. et BACHM. Quadrup. of North Amer. p. 173; Illustr. No. 5, Pl. 22.

" *Americanus.* DESMAREST, Mammalogie, Part II. p. 351.[1]

" *nanus.* DE KAY, Nat. Hist. of New York.

" " SCHREB. (?), Säugth. vol. iv. p. 881, Pl. 234, B.

Grey Rabbit of the Americans.

Ears rather shorter than the head; limbs and tail short; feet clothed with shortish hairs, which do not hide the nails: fur somewhat harsh to the touch; on the upper parts of the body mottled with black and pale brownish rufous; on the top of the head suffused with reddish chesnut; on the back of the neck of a bright rust colour; sides of body greyish; throat and abdomen, white; legs suffused with rusty red, the tarsi whitish above: ears, with the apical portion, clothed with pale hairs internally, margined with brown-black at the apex externally, and with a blackish line along the anterior margin; tail brown above, white beneath.

Found almost throughout the United States. According to Dr. Bachman's observations, the *L. sylvaticus* does not extend in the north direction beyond the southern counties of New Hampshire: it abounds in the sandy districts, covered with pine-trees, west of the city of Albany, and from thence it is traced southwards to the northern parts of Florida, occurring in the intermediate States of Pennsylvania, New Jersey,

[1] From my notes I find that the present species is the *Lepus Americanus* of the Paris and Leyden Museums; in the former the name *Lepus nanus* of Schreber is attached to the animal as a synonym. Schreber's description, under the name just mentioned, certainly, as it appears to me, for the most part applies to the *Lepus sylvaticus*, but is rendered inaccurate by his having introduced into his account statements made by other authors which refer to the Northern Hare.

Maryland, Virginia, and North and South Carolina. It is a native likewise of the southern States west of Florida, and very abundant on the upper Missouri River to nearly 1000 miles above Saint Louis.[1]

The three species of Hare, first clearly described and distinguished by Dr. Bachman under the specific names *aquaticus, sylvaticus,* and *palustris,* so closely resemble each other, that it is somewhat difficult to convey, in words, the points of distinction. They all agree in having the feet clothed with comparatively short and stiff hairs; their ears and limbs are shorter than in many of the species of *Lepus,* and their fur is by no means soft, as compared with that of the Common Hare or Rabbit. The *L. sylvaticus* is intermediate in size between the *L. aquaticus* and the *L. palustris,* and with regard to the proportions of the ears and hind feet, it also holds a middle station. The reddish chestnut hue of the upper surface of the head, a bright rusty-red colour on the back of the neck, and extending to the shoulders, the sides of the body (and especially the haunches), being greyish, the tarsi inclining to white on the upper surface, the ears edged with black, or brown-black, externally, both on the fore and apical margins, are points in the colouring which, combined, will serve to distinguish the Wood Hare. To these points it is only necessary to add, that the orbits of the eyes are whitish; the ears are clothed with dirty white hairs externally, or with yellow hairs (generally of a distinct yellow cast at the apex of the ear), and present but an indistinct trace of an inner band of party-coloured hairs. The hairs of the back are of a pale grey hue at the root, from which part, proceeding towards the point, they are gradually shaded into black, have a broad, subterminal, pale brownish, rufous ring: on the abdomen, about two-thirds of each hair

[1] See Audubon and Bachman's Quad. of North America, i. p. 179.

is white, and the remaining, basal, portion is grey. The tail is white beneath, and brown above, but many of the hairs on this part have a yellow point. The fore legs are of a bright rufous colour; the feet somewhat paler; those of the hind foot almost white on the upper surface, and dirty brown on the under; the back of the hind legs is of a rusty red hue.

	Zoological Society.		Zoological Society.		Leyden Mus.		From Dr. Bachman.	
	Ins.	Lines.	Ins.	Lines.	Ins.	Lines.	Ins.	Lines.
Length from tip of nose to root of tail	18	0	16	0	16	9	15	0
" from ditto to ear ...	3	2	3	0	3	3	3	5
" of ear	3	0	2	11	2	8	3	0
Width of ditto	1	7	1	7	1	6		
Length of hind foot and nails ...	3	11	3	8	3	7	3	7
" of tail, including the fur, about	2	6	2	6	3	0	2	2
" of vertebræ of ditto ...							1	2

Weight of adult male, according to Dr. Bachman, 2 lbs 7 oz.

The above description is drawn up from specimens presented to the Zoological Society by Dr. Bachman; according to whom, the *Lepus sylvaticus* frequents woods, and even the densest thickets; is fond of spots which are overgrown with young pines, thickly crowded together, or thickets of the high bush-blackberry (*Rubus villosus*). If alarmed, it runs with great swiftness, but usually for a short distance only, to take refuge in some hollow tree, or amongst the entangled roots. When hunted by dogs, it sometimes makes its escape by winding and doubling in thickets which are too dense to be penetrated by a larger animal. In the northern and middle states, the burrows of the Maryland Marmot afford a safe retreat in times of danger, unless, as sometimes happens, a skunk has already taken possession of the burrow. Like others of its genus, the Wood Hare has the habit of stamping with its hind feet, when alarmed at night, and also when engaged in combat with another of its own species. It has many enemies, falling a prey to the Weasels, Wild Cats, Foxes, and Snakes, but is, nevertheless, an abundant animal in most parts of the United States, being very prolific. Unlike the Northern Hare, the present animal never becomes white in the winter: sometimes the extreme points of the hairs become bleached in the winter months, but this, it appears, is by no means constantly the case.

LEPUS PALUSTRIS.

The Marsh Hare.

Lepus palustris. BACHM., Journal of the Acad. of Nat. Sc. of Philadelphia, vol. 7, p. 194 (May 1836), and p. 336.
" " AUDUBON and BACHM., Quadrupeds of North America, vol. 1, p. 151. Illustr. No. 4, Pl. 18.
" *Douglasii,* var. 2. GRAY, Magazine of Nat. Hist. for Nov. 1837, vol. 1 (New Series), p. 586.

Limbs short; ears shorter than the head: fur harsh, on the upper parts of the body mottled with black and rich brownish yellow, on the throat and under parts white; fore legs and tarsi deep brown-rufous; orbit of eye yellowish; back of neck of a dirty rufous tint; ears with yellow hairs on the inner surface, and without any distinct black edging externally; tail very short, rufous-brown above, dirty yellowish beneath.

Inhabits the southern and western parts of the United States, and extends into Texas; is found in the southern parts of North Carolina, more abundantly in South Carolina; abundant in Georgia, Alabama, and Louisiana, and ranges to the southern point of Florida.

The habits of the Marsh Hare, as detailed by Dr. Bachman, are very similar to those of the *Lepus aquaticus*, nearly confining itself to marshy situations, and taking freely to the water. It is an excellent swimmer, but, as would be supposed from the shortness of its limbs, its powers of running and leaping are far inferior to most other Hares.

The nest of the Marsh Hare has usually an arched roof, and is not unfrequently formed of a large species of rush (*Juncus effusus*), which the animal cuts into pieces of about a foot in length. These nests are often partially surrounded by water. The number of young found in the nests varies from five to seven, and the animal is said to have several litters in a season, the first of which is fully a month later than that of the Wood Hare (*L. sylvaticus*). *Lepus palustris* may be distinguished from the *L. sylvaticus* by its ears, feet, and tail, being proportionately shorter, and in the specimens which have come under my notice (among which are some presented by Dr. Bachman to the Zoological Society), I have observed certain differences in the colouring, which are perhaps constant. For instance, they want the black edging to the anterior and apical margins of the ear, which is present in all the specimens of *L. sylvaticus* I have seen; the back of the neck is of a dirty rufous tint, and this colour does n t extend backwards on to the shoulders, whilst in the Wood Hare, the back of the neck, and upper part of the shoulders, are of a bright rust colour. Another difference is somewhat striking, when the two animals are laid side by side; and that is, that the flanks and haunches want the grey hue which distinguishes *L. sylvaticus*, being of nearly the same colours as the back. The fur in the present species, moreover, is coarser, and this is more strikingly the case with respect to the fur of the abdomen. The fur on the upper parts of the body is of a deep grey colour next the skin, and each hair has a black point, and a broad ring of rich yellow-brown below the point; on the throat and abdomen the hairs are also of a dark grey at the root, but their exposed ends are white: the chest is of a somewhat paler hue than the sides of the body, and is sometimes whitish in the mesial line; the fore legs and feet, as well as the tarsi, and the hinder part of the tibiæ, are of a deep brown-rufous tint; the orbit of the eye is a buff yellow colour[1]. The ears are clothed for the most part with yellowish hairs internally; and towards, and at the point, the hairs assume a decided yellow colour; externally they are rather sparingly clothed behind with dirty yellowish hairs, or, towards the hinder margin, with white hairs, and at the root of the ear they assume a rufous tint; the band on the fore part of the ear is composed of party-coloured dusky and yellow hairs. The tail is of a rufous-brown hue above (where, however, the hairs are dusky grey at the root), and of a dirty yellow colour beneath—sometimes white along the middle line. Of this species I have examined five specimens, three of which are in the Museum of the Zoological Society,—two being presented to that Society by Dr. Bach-

	Museum Zoological Society. Specimen from Mr. Douglas.		Museum Zoological Society. Specimen from Dr. Bachman.		Dimension from a fresh specimen. Dr. Bachman.	
	Ins.	Lines.	Ins.	Lines.	Ins.	Lines.
Length from tip of nose to root of tail ...	16	0	16	0	13	0
" from ditto to ear	3	5	3	4		
" of ear	2	7	2	6	2	6
Width of ditto	1	9	1	10		
Length of hind foot and nails	3	6	3	5	3	0
" of tail, including the fur			1	6	1	6
" of tail, without the fur					1	0
From shoulder to extremity of fore foot					7	0

[1] In three specimens before me, such is the case; whilst in three specimens of *L. sylvaticus* the orbit is white.

man, and the third formed part of a collection obtained from Mr. David Douglas, and supposed to be from California[1], and two in the British Museum collection; these latter were described by Mr. Gray under the name *L. Douglasii.*

Hares of the Western and South-western portions of North America.

Of these, the first three species described are of very small size, and have short ears; the others are of moderate, or large size, and most of them have remarkably long ears.

LEPUS NUTTALLII.

Nuttall's Hare.

Lepus Nuttallii. Bachm., Journ. of the Acad. of Nat. Sci. of Philad. vii. p. 345, Pl. 22, No. 1.
" " Audubon and Bachman's Quadrupeds of N. America, Illust. No. 19, Pl. 94.

Ears broad, but very short—much shorter than the head: tail also very short: fur soft; on the upper parts of the body pencilled with buff-yellow, and black; under parts pale yellowish grey; legs and feet of a rufous-yellow colour, and a large patch on the back of the neck, which extends on to the shoulders, is of the same colour: tail white beneath.

Inhabits the west side of the Rocky Mountains in the neighbourhood of the Columbia and Shoshonee Rivers.

From all the known Hares the present species may be distinguished by its diminutive size, its bulk being nearly equal to that of the Common Norway Rat. From the *Lepus Bachmani*, which approaches it most nearly in size, Nut

[1] As there may be some doubt on this point, I have not introduced California as amongst the parts of North America inhabited by this species.

tall's Hare differs in having much shorter ears, and in having a very large rufous patch covering the back of the neck, and extending on to the shoulders, and likewise in having the limbs rufous. The ears are described by Dr. Bachman as broad and rounded; the feet thickly clothed with soft hair, which completely hides the nails, and the tail concealed (in the dried skin) by the long fur of the hinder part of the body. The fur soft; on the upper parts of the body, presenting a mixture of buff and dark brown, and on the under parts, of a light buff-grey colour. The fur on the back is of a lead-grey colour for three-fourths of its length from the roots, then light ash, mixed with buff; and the long interspersed hairs are tipped with black. The ears are tolerably well clothed, both externally and internally, with ash-coloured hairs, have a black line on the fore part, and are edged with white. A large buff-coloured patch extends from behind the ears to the back, and the limbs are of the same colour, but with a mixture of rufous, which is extended upwards to the shoulders, and on to the thighs. The soles of the feet are yellowish brown, and the tail is white on the under surface.

	Inches.	Lines.
Length from tip of nose to root of tail ...	6	9
" of head	2	1½
" of ear	1	6
" of hind foot	2	0
" of tail (vertebræ)		9
" of ditto, including the fur	1	3

This little Hare was discovered by Mr. Nuttall, whose researches in the botany of the Rocky Mountains are well known, and, according to that gentleman's note which accompanied a specimen presented to Dr. Bachman, it was met with west of the Rocky Mountains, where it inhabits thickets by the banks of several of the small streams which flow into the Shoshonee and Columbia Rivers.

LEPUS BACHMANI.

Bachman's Hare.

Lepus Bachmani. WATERH., Proceedings of the Zoological Society for August, 1838, p. 103.

Ears rather longer than the head; tarsi short, and well clothed with fur; fur moderately soft, its general hue brown (on the back pencilled with black and pale brown); sides of body grey-brown; abdomen white, slightly suffused with rufous; occiput brown: ears with a narrow indistinct dusky edging at the apex externally.

Inhabits the south-west portions of North America—perhaps California.

The specimen upon which the *Lepus Bachmani* was founded is *possibly* not full grown, but the condition of the incisor teeth forbid our supposing it very young, since they are well developed, and of equal width throughout, and, in fact, the only ground for suspecting the animal may not be adult, is its small size. It has the short feet which characterize the *L. palustris*, and as the last mentioned animal was received by the Zoological Society in the same collection as the *Lepus Bachmani*, I have more than once compared the two animals very carefully together. Notwithstanding that the Hare, which I have named after my friend Dr. Bachman, is not above one-third of the size of the *L. palustris*, its ears are absolutely longer; and another important difference is observable in the feet, which, instead of being clothed with short and adpressed hairs, which do not conceal the nails, are, in Bachman's Hare, clothed with long and woolly hairs—those from the toes projecting about a quarter of an inch beyond the points of the nails. The fur in the present species is much softer than in *L. palustris*, and differs in its colouring; the incisor teeth are strongly grooved, and the division of the tooth which is on the outer side of the groove is less broad than in the Marsh Hare. The general brown hue of the back is produced by a mixture of black and pale yellow brown, the visible portions of the hairs being of this colour, but each hair has its lower half of a deep grey colour, and the hairs are almost black immediately below the subterminal pale ring: the hairs on the abdomen are of a distinct grey colour at the root, and white beyond, but slightly suffused with rufous at the point; and, towards the sides of the body, the rufous tint is much more conspicuous. The legs and feet (if we except the upper surface of the tarsus, which is very pale) are very nearly of the same brown tint as the sides of the body, these parts wanting the red tint which is seen in the *L. palustris*, *L. aquaticus*, and *L. sylvaticus*, which, like the present species, have short feet. The ears are clothed with grey-white hairs internally, but near the hinder margin they are grizzled with black, and at the apical margin there is a very narrow band of brownish yellow hairs; on the front edge of the ear is a small white fringe; externally, the band of the ear is grizzled with black and yellow; there is a narrow dusky edging at the apex, and the hinder half is grey-white, excepting at the root of the ear, where a rufous tint is visible. The back of the neck is brown, but towards the shoulders slightly tinted with rufous. The tail is white below, and dusky above, with pencillings of dirty yellow.

	Inches.	Lines.
Length from tip of nose to root of tail ...	11	0
" from ditto to ear	2	5½
" of ear	2	8
Width of ditto	1	8
" of the hind foot	2	10
" of tail, including the fur, about ...	1	3

LEPUS ARTEMISIÆ.

Wormwood Hare.

Lepus artemisiæ.	BACHM., Journ. of the Acad. of Nat. Sci. of Philad. vol. viii. p. 94.
" "	AUDUB. and BACHM., Quadr. of N. America, Illustr. No. 18, Pl. 88.
" *artemisiacus.*	WAGNER, Schreb. Säugth. Suppl. iv. p. 115.

Ears about equal to the head in length; tarsi tolerably long, and well clothed; tail short: fur soft; on the upper parts of the animal pencilled with black and brownish white; on the under parts, white; back of the neck, and limbs, for the most part, of a pale rusty-fawn colour: ears margined with black at the apex externally: tail above, coloured like the back; beneath, white.

Inhabits the Rocky Mountains.

The Wormwood Hare is a very small species (much smaller than the Common Rabbit), being about equal in size to the *Lep. Bachmani*, from which it differs in having a longer tarsus, as well as in its colouring.

The following description was taken from a specimen brought to England by Dr. Bachman, which that gentleman kindly allowed me to examine.

The fur is soft, and the general hue is somewhat pale; the hairs on the back are of a pale grey at the root, shaded into brown in the middle, annulated with brown-white towards the point, and black at the point; on the abdomen the hairs are also pale grey at the root, but they are white externally, as are also the hairs on the throat: the inner surface of the limbs are white; on the sides of the neck, and on the chest, the hairs are grey, tipped with brown-white, but with a faint yellow hue; the whole of the back of the neck—commencing immediately behind the ears, and extending to the back—is of a pale rusty-fawn colour; and the legs, externally, are of the same hue; the feet are of a dirty yellow-brown hue beneath, and the tarsus is very pale along the middle part of the upper surface. The tail is coloured like the back of the animal on its upper surface, but assumes a grey-black tint towards the tip; beneath, it is white. The bands of the ears are finely freckled with dusky and dirty white; the hinder surface, externally, is dirty white, as well as the innner surface; the margins are whitish, but on the outer surface are edged with black at the apex. The upper incisors are deeply grooved.

	Inches.	Lines.
Length from tip of nose to root of tail ...	12	0
" from ditto to ear	2	7
" of ear	2	8
" of tarsus	3	2

LEPUS CAMPESTRIS.

Prairie Hare.

Lepus campestris. BACHM., Jour. of the Acad. of Nat. Sci. of Philad. vol. vii. p. 349 (1837).
" *Townsendii.* BACHM., l. c. vol. viii. Pt. 1, p. 92, Pl. 2 (1839).
" " AUDUB. and BACHM., Quadr. of N. America, vol. i. p. 25 (1847), No. 1, Pl. 3 of the Illustrations.
" *Virginianus.* RICHARDS., Faun. Bor. Amer. Pt. 1, p. 224 (1829).
Prairie Hare of the Fur Traders.

Ears longer than the head; tarsi long, and densely clothed with fur: fur very soft, the prevailing hue pale ashy grey, obscurely tinted with yellow; throat and under parts of the body, and the tail, white, the latter with a few dark hairs on the upper surface; back of neck dirty white; ears with a large black patch at the apex externally.

Found on both sides of the Rocky Mountains in the region of the Columbia River, and ranges eastwards, on the Missouri, beyond the Yellowstone River[1]—range of *L. Townsendii* of Bachman. The *L. Virginianus* of Richardson, according to that author, inhabits the plains through which the north and south branches of the Saskatchewan flow, and which extend westward as far as the Winnipegoos Lake, and the southern extremity of Lake Winnipeg; it is not known to range further north than the 55th parallel.

With regard to the names associated at the head of this account, it is necessary to state that Dr. Bachman gave the name *Lepus campestris* to the animal described by Dr. Richardson under the name *L. Virginianus*, because this latter name had been previously applied by Dr. Harlan to a distinct species of Hare. The *Lepus Townsendii* was founded by Dr. Bachman upon specimens of a Hare which he was induced to believe would prove distinct from the *L. campestris*, from his having received information that it did not become white in the winter months, like that animal; more recently, however, Dr. Bachman has been assured by residents in Missouri, that the *L. Townsendii* is subject to the same change of colour. As, on the other hand, the animals described under the names *campestris* and *Town-*

[1] The accounts before us, from which we have to ascertain the geographical range of this animal, are extremely vague; one of them gives "more than 1200 miles east of the Rocky Mountains," as a locality for the species! From the context, however, there can be no doubt that the traveller from whose notes the quotation is made, was on the Missouri (though it is not so stated in plain terms), and it seems probable that the 1200 miles mentioned may include the windings of that river. Mr. Nuttall, as quoted by Dr. Bachman, states that our Hare, though seen frequently on the plains of the Platte, is still more common to the west of the Rocky Mountains.

The habitats of the *L. Virginianus* of Richardson are kept distinct, since it is just possible that the animal may prove a different species to the *L. Towsendii.*

sendii agree, according to the descriptions, in the colouring of the summer fur, and in size, and are both inhabitants of the western prairies, there are no grounds for regarding them as distinct.

A specimen of the Prairie Hare, which Dr. Bachman allowed me to examine, and which is probably one of the specimens referred to in the "Quadrupeds of North America," was very nearly equal in size to the Common European Hare, and likewise agreed very closely with that animal in the proportions of its limbs, but its ears were distinctly longer, and its colouring differed from that of most Hares in being, on the upper parts of the body, of a palish ashy grey hue, in which a faint yellow tinge was perceptible; the hairs on the back were grey-white at the root, brown-grey beyond, almost black below the cream-coloured sub-terminal ring, and black at the point; the fur on the abdomen white to the skin; throat and inner side of the limbs also white; the tail white, having, however, a greyish hue at the base, on the upper surface, owing to there being an admixture of black hairs at that point: the tarsi white above, and of a pale brownish colour beneath; the sides of the muzzle suffused with cream-colour; the orbit of the eye white; back of the neck nearly white, especially near the root of the ear: the coloured bands of the ears composed of small hairs, partly blackish and partly cream-coloured; the hairs forming the fringe on the anterior margin, very long, and of a cream yellow, the hinder margin edged with white; the hinder half of the outer surface of the ear pure white, excepting at the point, where the hairs were black, and formed a moderate sized patch. The fur was dense, and very soft, and the feet were densely clothed, though rather less so than in the Polar Hare.

			Adult Male, From Upper Missouri. Dr. Bachman.		Adult Female. Dr. Bachman.		Young. Dr. Bachman.	
	Ins.	Lines.	Ins.	Lines.	Ins.	Lines.	Ins.	Lines.
Length from tip of nose to root of tail	21	0	21	6	21	0	12	0
" of head			4	7			2	9
" of ear	5	2	5	6	5	3	2	7½
" of hind foot ...	5	6	5	7½			3	7½
" of tail, including the fur ...	5	0	4	9	4	6	2	1½
" of ditto, without the fur ..			3	1½	3	0	1	3
From shoulder to the extremity of the fore foot					13	6	7	1½
Weight			6½ lbs.					

The above description agrees so closely with that of *Lepus Townsendii* of Dr. Bachman, that I have only thought it necessary to add the dimensions of the specimens examined by that naturalist, whose correspondents inform him that the species exclusively inhabits the plains, and more especially those which are overgrown with the aromatic Wormwood (*Artemisia*). It is very abundant in the neighbourhood of Fort Walla-Walla, but is said to disappear, together with the wormwood bushes, in proceeding westwards towards the sea.

Lepus Virginianus. Richardson.

This animal, which is said to be common on the plains through which the north and south branches of the Saskatchewan flow, is described as being entirely white, with the exception of the ears, which have a brownish black tip, the black about as much extended as in the Polar Hare, and has the fore part of the outer surface, and the hinder part of the inner surface, of a fawn colour. Although white externally, the hairs of the fur are found to be coloured below the exposed parts: on the back they are white for one-third of their length from the root, and pale brownish yellow in the middle; on the cheeks the hairs are grey at the roots, and on the neck they are grey at the root, then buff colour: on the abdomen the hairs are white to the root. Dr. Richardson's measurements, which are from a mutilated skin, give 22 inches as the entire length of the animal, and 4 inches for the length of the ears.

According to Messrs. Lewis and Clark's account, quoted by Dr. Richardson, the animal weighs from seven to eleven pounds; its head, neck, back, shoulders, and the outer parts of the legs and thighs, are of a lead colour; the tail white: the fur long, dense, and soft. The animal assumes the above colouring from the middle of April till the middle of November; during the rest of the year it is of a pure white, with the exception of the black, and reddish brown of the ears, which never changes.

LEPUS CALIFORNICUS.

Californian Hare.

Lepus Californicus. Gray, Magazine of Natural History for November, 1837, vol. 1 (New Series), p. 586.

" *Richardsonii.* Bachm., Journ. of the Acad. of Nat. Sci. of Philad., vol. viii. p. 88.

" " Wagner, Schreb. Säugth. Suppl. iv. p. 111 (1844).

" *Bennettii.* Gray, Zoology of the Voy. of H.M.S. Sulphur—Mammalia, p. 35. Pl. 14 (1843).

Ears very large, much longer than the head; limbs long and slender; tail long: fur soft, moderate as to length; on the upper parts of the body pencilled with black and pale rufous-brown; on the chest, sides of the body, and outer surface of limbs, pale brown, more or less tinted with rufous; abdomen white, suffused with bright rust colour, or with fawn-yellow tail dirty brown-yellow below; black above: ears with a large black patch at the apex, externally: back of neck grey-white, brownish in the mesial line.

Inhabits California—"in the open hilly country which surrounds the harbour of St. Diego" (lat. 32° 04′).

I have before me, notes of ten nominal species of Hares, supposed, or known to be, inhabitants of the south-western portions of North America. Of these, one, which was described from a specimen in the Museum of the Zoological Society, from which the label had become detached and lost, was merely surmised to be a native of this part of America, as it was likewise of the southernmost part of South America, and was described under the name *Lepus longicaudatus*: this proves to be a South African animal, being, without doubt, the *L. saxatilis*. Of the remaining, so-called species, there are two of which I have only very short descriptions—they are the *L. cunicularius* and *L. Mexicanus* of the Berlin Museum[1]. Of the others I have detailed descriptions; and, in most cases, I have examined the specimens from which the original descriptions were drawn up. They appear to me to be clearly referrible to three species only; and the earliest names imposed upon them respectively, were *Californicus, Texianus,* and *callotis*. They are all nearly

[1] For notes upon these two species I am indebted to my friend Dr. Bachman. The note relating to the *L. Mexicanus* is brief, but leaves scarcely a doubt on my mind that that animal is specifically identical with the *Lepus callotis*. Judging from the description of the

Lepus cunicularius: LICHTENSTEIN,

that animal must be nearly allied to the short-eared Hares of the United States. Its fur is said to be short, and coarse; the general hue of the upper parts of the body, brown-black; the hairs on these parts being of a pale ashy grey at the root, light brown in the middle, annulated with yellowish white towards the point, and with a long black point: ears, internally clothed with equal in size to each other, and to the Common European Hare, and have long ears and long limbs. The *Lep. callotis* is readily distinguished by the terminal portion of the ear, externally, being white, or yellowish; in *L. Californicus* and *L. Texianus* the same part being marked with a large black patch: the points of distinction between the two last-named species are noticed in the account of *Lep. Texianus*. The Californian Hare is subject to considerable variation in its colouring, and more especially with regard to the colour of the flanks and limbs; in some specimens the rufous tint is tolerably bright and distinct, whilst in others there is scarcely any trace of the reddish hue on these parts. The upper parts of the body are strongly pencilled with black, and pale rufous-brown; the chest, the sides of the body, and the limbs, are rufous, slightly tinted with brown; the throat yellow-white; chin brownish, and abdomen white, suffused with bright, but pale, rust colour; the fore legs are considerably pencilled with black in front, especially towards the feet; the fore feet are brown-white; there is less of the red tint on the outer surface of the hind legs than on the sides of the body, and on the tarsus it is replaced by pale brown; the soles of the feet are dusky brown. The ears, which are very large, have the fore part of the outer surface, and the hinder part of the inner, clothed

short cream-coloured hairs; externally, pale grey-brown, but assuming a blackish hue towards the tip: back of the neck yellow; orbit of the eye, throat, and abdomen, white.

	Inches.	Lines.
Length from tip of nose to root of tail ...	15	6
" of tail	2	9
" of ear	3	5
" of tarsus	4	0

Inhabits Mexico.

I may here mention, that in the Leyden Museum I examined an animal labelled as coming from Mexico, which appeared to me to differ in no respect from the common Wild Rabbit of Europe.

with short, party-coloured, dusky and brown-white hairs; the fringe on the anterior margin is composed of yellow-white hairs, having dusky points; the hinder edge of the ear is cream-coloured, or, in parts, white; the tip is distinctly edged with black, and the black runs down from the point, on the hinder margin, for about half the length of the ear; on the fore margin, the black is much less extended: on the outer surface of the ear, at the apex, is a large dusky black patch, which is about an inch and a half in vertical diameter; the hinder half of the ear, below the black part, is of a dirty white colour. The back of the neck is brown in the middle line, and greyish white behind the ears. The tail is moderately long, of a dirty brownish yellow colour below, and black above to the point. The hairs on the back are grey-white at the root, have a broad black space in the middle, a sub-terminal ring of rufous brown (but very pale), and a black point: on the abdomen the hairs are white at the root.

This description is from the specimen upon which Mr. Gray founded the *Lepus Californicus;* its admeasurements are given in the first of the four columns of dimensions. Two other specimens, which, together with the one just noticed, are in the collection of the Zoological Society, and are the originals, I believe, of Dr. Bachman's descriptions, under the name *Lepus Richardsonii*[1]. They differ slightly from each other, and from the specimen described by Mr. Gray, but, after a careful comparison, I cannot doubt that the three animals are specifically identical. The tail in all is of a dirty yellowish colour beneath; and the back of the neck is grey-

[1] I regret that up to the present moment I have been unable to obtain the Eighth Volume of the Journal of the Academy of Natural Sciences of Philadelphia, in which Dr. Bachman describes the *Lepus Richardsonii;* the account, however, as extracted by Dr. Wagner in his Supplement to Schreber's Säugthiere, leaves little doubt on my mind that I am correct in regarding the specimens referred to, as those upon which that account was drawn up.

white, but more or less brown in the middle, and they have a large black patch on the outer surface of the ear. One of the specimens of *L. Richardsonii* differs from the animal from which my detailed description is taken, in having less of the rufous tinge on the chest and sides of the body; these parts are of a palish brown colour, with a faint rufous tint. The region of the shoulders is brown, and destitute of black pencilling. The second specimen differs in its general tint, being less brown—inclining to grey: its belly is yellow-white, and there is a very slight rusty tint about the base of the legs, and lower parts of the sides of the body. The legs are brown.

	Lepus Californicus.		*Lepus Richardsonii.*				*Lepus Bennettii.*	
	Ins.	Lines.	Ins.	Lines	Ins.	Lines.	Ins.	Lines.
Length from tip of nose to root of tail	22	6	21	0	21	0	19	6
" from nose to ear	4	6	4	3	3	10	4	0
" of ear	5	6	5	6	5	5	5	2
Width of ditto	3	0	3	0	3	0	2	10
Length of tail, including the fur	4	8	5	0			4	8
" of tarsus	4	10	4	10	4	10	4	9

The *Lepus Bennettii* of Mr. Gray, in its colouring, resembles the second specimen referred to in the foregoing account; that is to say, it is less rufous on the sides of the body than the specimen described under the name *L. Californicus.* Its dimensions are given in the fourth column. The following are the admeasurements of its skull:—

	Lepus Bennettii. British Museum.	
	Inches.	Lines.
Length of skull from tip of intermaxillary bones to lower part of occiput[1]	3	5

[1] A very small portion of the occiput only is remaining in this skull, which, if perfect, would no doubt be about 3″ 7‴, or 3″ 8‴, in total length.

	Lepus Bennettii. British Museum. Inches.	Lines.
Width of skull	1	$7\frac{2}{3}$
Length between orbits		$11\frac{1}{2}$
Length of nasal bones	1	$4\frac{1}{2}$
Width of ditto behind		$8\frac{1}{3}$
" " in front		$5\frac{2}{3}$
" of upper incisors		$2\frac{2}{3}$
From upper incisors to molar teeth	1	$0\frac{1}{2}$
Length of six molar teeth, taken together ...		7
" of palate		3
" of lower jaw	2	$6\frac{1}{2}$
Height of ditto	1	6
Transverse diameter of angular portion ...		$11\frac{1}{3}$

LEPUS TEXIANUS?

Texian Hare?

Lepus Texianus. Aud. and Bachm., Quadrupeds of North America?

Legs long; ears very large—more than one-third longer than the head; tail long: fur long and soft; the general hue of the animal pale, inclining to ashy grey, but strongly mottled with black and brownish white, on the upper parts of the body; the throat and abdomen white; haunches, and outer surface of the legs, grey; tarsus nearly white; back of neck pale ashy grey; tail black above, and greyish beneath; a large black patch on the rump; ears with a large black patch at the apex externally.

Inhabits Texas?

A well-marked species of Hare, in the collection of the Zoological Society, of which the history is not known, is recognized by Mr. J. W. Audubon as a species with which he is well acquainted, and which that gentleman informs me will shortly be published in the great work on the North American Quadrupeds, which I have already frequently quoted. According to Mr. Audubon, to whom we are indebted for the splendid plates which illustrate the work alluded to, the animal inhabits Texas, and has been named *Lepus Texianus*. In its colouring, the specimen alluded to, agrees more closely with the *L. callotis* than with the Californian Hare, but it resembles the latter in having a large black patch at the tip of the ear, and in having the back of the neck pale: its limbs are apparently proportionately shorter, and less slender than those of the animals just mentioned, and it differs from them, in a marked degree, in the relative size of its ears. The black patch, on the hinder part of the back, and which is hidden by the recurved tail, it seems, is found likewise in the *Lep. callotis*. The hairs on the upper parts of the body are white (*very* slightly tinted with grey) at the root, brownish towards the middle, black above the middle, have a broad sub-terminal pale ring, which is almost white, but shaded into reddish near the black with which the hair is pointed. The throat and abdomen are white, but a slight yellow tint is observable on the latter part. The feet are dusky grey beneath. The fore legs are of a pale ashy grey colour in front, and obscurely pencilled with dusky. The fore part of the ears, externally, and the hinder part, internally, are clothed with party-coloured, dusky, and yellow-white hairs; the fringe on the anterior margin is composed of long dirty yellow hairs; the hinder margin is white; with the exception of the part already noticed, the inner surface of the ear is clothed with pale hairs, yellowish near the point, and white below that part; the tip of the ear is fringed with brown-black hairs, and a similar dark edging runs down the hinder margin for about two inches from the point; the hinder half of the ear, externally, is white, if we except the apical portion, where there is a large dusky black patch of about an inch or more in vertical diameter. The tail is black above, and at the point, and dirty greyish white below: here the hairs are grey at the root. The orbit of the eye is whitish.

	Inches.	Lines.
Length from tip of nose to root of tail ...	20	0
" from nose to ear, about	4	0
" of ear	6	6
Width of ditto	3	0
Length of tarsus	5	0
" of fore leg and foot, about	5	0
" of tail, including the fur	4	0

LEPUS CALLOTIS.

Mexican Hare.

Lepus callotis.	WAGLER, Natürliches System der Amphibien, p. 23, (1830): Isis, 1831, p. 511.
" "	WAGNER, Schreb. Säugth. Suppl., iv. p. 106, Tab. 233 E (1844).
" *nigricaudatus.*	BENNETT, Proceedings of the Zool. Soc. for March, 1833, p. 41.
" *flavigularis.*	WAGNER, in Mus. Monac.; Schreb. Säugth. l. c.
" *Mexicanus.*	LECHTENSTEIN, in the Berlin Museum. RICHARDSON, Report on North American Zoology.—Sixth Report of the British Association, 1836, p. 150.

Ears very large, about one-fifth longer than the head; limbs long, and very slender: fur soft, on the upper parts of the body of a delicate sienna yellow colour, mottled with black; lower half of the sides of the body, limbs, and rump, white: tail black above, and at the point; white beneath; back of neck, and base of ear externally, black, the remaining outer surface of the ear (with the exception of the band in front) pale—no black patch at the apex.

Inhabits Mexico, and the adjoining part of California.

The Mexican Hare is a well-marked species, very nearly equal in size to the Common Hare of Europe, but has longer ears than that animal, and its limbs are much more slender. In its colouring the most striking points are, the dusky black back to the neck, the absence of any black patch at the tip of the ear externally, and the mottled, black, and yellowish colouring of the upper parts of the animal, terminating abruptly about half way down the sides of the body, and also on the hinder part of the back at some distance from the root of tail, and being replaced by a white fur on the lower parts of the animal.

I have seen but one specimen of this species—the specimen which Mr. Bennett described in the Proceedings of the Zoological Society under the name *Lepus nigricaudatus*, and which was received from that part of California which adjoins Mexico. Dr. Wagner has evidently examined several specimens, and states that upon two occasions he has received the animal from Mexico. All the individuals examined, present the leading characters above pointed out, but I must observe, that on the sides of the body there are a few black hairs interspersed with the white fur, and these become more numerous towards the root of the tail, above which there is a dusky patch—sometimes, however, very indistinct; and the back of the neck assumes a brownish hue in the middle line, and sometimes many of the hairs on this part are pointed with whitish; sooty black, however, is the prevailing hue of the back of the neck, and this colour is extended on to the root of the ear.

Lepus flavigularis. WAGNER.

This specimen, to which Dr. Wagner gave the name *flavigularis*, differs from others in having the upper parts of the body of a richer tint, the yellow parts of the hairs being of an ochreous hue; the whole fore part of the neck being of an uniform ochre colour, and the hinder half of the outer surface of the ears being yellow to the point;—a large pure white patch is usually found at the point of the ear, externally[1].

[1] The same part is *black* in the Texian Hare, which has stouter limbs and longer ears than the Mexican Hare, and wants the black on the back of the neck.

Lepus nigricaudatus. BENNETT.

In applying the name Black-tailed Hare to the present animal, Mr. Bennett merely wished to call attention to the fact, that its tail had more black than usual in its colouring; in very many Hares the black is confined to the upper surface of the part in question, whilst in the *L. nigricaudatus*, the tip and the sides of the tail are also sooty black, the white of the under surface being almost confined to the root of the tail. As the *L. nigricaudatus* presents all the general features which have been noticed as appertaining to the *L. callotis*, I shall merely add, that the fur on the upper parts of its body is composed of long and soft hairs, which are pale grey at the root; above the grey, a long space on each hair, is of a pale rufous colour, this is followed by a dusky or blackish space, a broad, pale rufous-yellow ring, and a black point: the sides of the face are chiefly yellow; the chest yellow; the back part of the tibia dusky grey, and the dense fur on the soles of the feet is of a dirty blood-red.

	Lepus callotis. From Wagner.		*Lepus flavigularis.* From Wagner.		*Lepus nigricaudatus.*		*L. Mexicanus.* Licht. From Bachman.	
	Ins.	Lines.	Ins.	Lines.	Ins.	Lines.	Ins.	Lines.
Length from tip of nose to root of tail	18	6	21	9	19	9	18	0
" from ditto to ear	4	0	4	0	3	11		
" of ear	5	2	5	2	4	10	5	7
Width of ditto					2	8		
Length of fore foot and leg to the elbow joint, about ...					7	0		
" of the tarsus	4	8	4	9	4	7	4	9
" of tail, including the fur	2	9	3	0	3	0	2	2

The brief note relating to the *L. Mexicanus* of the Berlin Museum, furnished me by Dr. Bachman, describes that animal as having the back of the neck black; the white of the under parts of the body extending high up on the flanks, and, indeed, in all other respects agreeing with the characters of *L. callotis.*

South American Hares.

Hitherto but one species of Hare has been found in South America.

LEPUS BRAZILIENSIS.

Brazilian Hare.

Lepus Braziliensis.	LINN., Syst. Nat., 12th Ed., i. p. 78.
" "	SHREB., Säugth. iv. p. 902.
" "	DESMAREST, Mammalogie, Pt. 2, p. 352.
" "	WAGNER, Shreb. Säugth. Suppl. iv. p. 116.
" *Tapeti.*	PALLAS, Nov. Spec. Glir. p. 30.
" *Tapiti.*	BUFFON, Histoire Naturelle, tom. xv. p. 162.
"	AZARA, Essais sur l'Histoire des Quadrupèds de la Province de Paraguay, 8vo. Paris, 1801, tom. ii. p. 57.

Ears small; tail very short; tarsus short: fur of moderate length, and rather coarse; upper parts of the body distinctly mottled with black and ochreous-yellow; region of nostrils, orbits of eye, throat, and abdomen, white: legs rust coloured; tail brown above, rufous-yellow beneath; ears with yellow hairs internally, and chiefly brown externally; back of neck bright rufous.

This species is scattered throughout Brazil, and is found likewise in Peru, Bolivia, and Paraguay.

The Brazilian Hare, in affinity, approaches most nearly to the *Lepus palustris*, and certain other North American species in which the ears and feet are short, and the latter are less densely clothed with fur than usual; it bears, likewise, a considerable general resemblance to the *Lepus nigricollis* of India, but is smaller than either of these animals, and is readily distinguished by its extremely short tail, which is almost hidden by the fur of the haunches. The bright rust colour of the back of its neck, moreover, will prevent its being confounded with the Black-necked Hare. Its fur is by no means long, and is much less soft than in the Common Rabbit; the hairs on the back are grey at the root, and brownish above; and each hair is shaded into black near the sub-terminal pale ring, which is partly ochreous white, but assumes a rich ochre colour towards the point, which is black; on the upper parts of the body, where the black terminal part of each hair is long, the fur is strongly mottled, or variegated with black and ochreous yellow, but on the sides of the body there is much less black in the colouring. The white in the region of the eye is considerably extended near the angles, and is very conspicuous, as is also the white mark which surrounds the nostril opening. The throat and abdomen are white, and the hairs on these parts are uniform in colour to the root, or, an extremely faint grey tint is sometimes seen on the hairs of the abdomen next the skin. The limbs are of a tolerably bright rust colour, but the fore parts of the hind leg, and the upper surface of the tibia, are almost white; and the hinder surface of the tibia is sometimes brown. The ears are clothed with rusty yellow hairs internally, and there are no party-coloured hairs near the hinder margin as is usually found in the species of the present group: the outer surface is clothed for the most part with brownish hairs; those on the fore part are, however, pencilled with black. The tail is rufous yellow beneath, and the hairs on the upper surface are coloured like those of the upper parts of the body.

The above description is taken from specimens brought

from Bolivia by Mr. Bridges, and which are now in the British Museum. An extremely young specimen from the same collection, measuring about four inches in length, is almost entirely black on the upper parts of the body: the ears are entirely of a bright rust colour internally, and black externally at the point, and along the anterior margin; the cheeks are chiefly black, and the upper surface of the head is chestnut-brown; throat and abdomen white; orbit yellow-white; legs rufous.

	Inches.	Lines.	Inches.	Lines.
Length from tip of nose to root of tail	13	0	15	0
" from nose to ear, about ...	2	6	2	6
" of ear	2	0	2	0
Width of ditto	1	4	1	4
Length of hind foot	2	10½	3	0
" of tail, including the fur, about	1	3		

The skull removed from one of these specimens, and a second skull in the British Museum, sent from Para, agree most closely with each other, as will be seen upon comparing the subjoined dimensions. Like the skull of the Common Rabbit, they have the inter-orbital space narrow, as compared with the same part in the more typical Hares, owing to the supra-orbital process being but little developed in the lateral direction; a small notch marks their anterior boundary, and the produced free hinder portion is less elongated than in the Rabbit's skull. The peculiarities of the skull of the Brazilian Hare will perhaps be rendered more evident, by our continuing a comparison between it and the skull of the Common Rabbit. Viewing the upper surface, the most striking distinctions are as follows:—the facial part of the cranium are much shorter; the nasal bones are not only shorter from this circumstance, but likewise from the fact that they terminate posteriorly at a point more advanced, or more removed from the angle of the orbit: in the Rabbit the nasal bones terminate in a line, which, if carried out, would be rather behind the point of the orbital process of the superior maxillary bone; whilst in the Brazilian animal, in which the skull is considerably smaller, the nasal bones terminate more than $\frac{1}{12}$ of an inch in front of the corresponding part of the superior maxillary bone. The upper surface of the skull is less arched, and the frontal and parietal bones are remarkable for having small punctures more or less thickly scattered over their surface, which are visible, in a much less degree, in the Common Rabbit. The nasal bones, on either side, at the termination of the anterior third part of their length, have a smallish concave area, in which several perforations are visible. Besides these points, I shall merely notice, that the palate is relatively longer than in the Rabbit, and the post-palatal channel is less contracted. The lower jaw, besides being proportionately shorter, has the ascending ramus less directed backwards, and its antero-posterior diameter is considerably smaller, arising chiefly from the ridge, corresponding to the coronoid process, being much less developed. Now, in all these characters, and in several other minute points—in the form of the auditory bullæ, and the palatine opening, &c.—a fossil skull from the caverns of Minas Geraes most perfectly agrees. This fossil skull, indeed, has *all* the peculiarities of that of the living species of the country, but differs in being about one-third larger, and the same difference is likewise observable in parts of two other fossil skulls, from the same caverns, and in several lower jaws, which are more or less imperfect. I will arrange the dimensions of the fossil skull next to those of the recent.

	Recent. Bolivia.		Recent. Para.		Fossil.	
	Ins.	Lines.	Ins.	Lines.	Ins.	Lines.
Total length of skull	2	6½	2	7	2	11
Width	1	2¾	1	3¼	1	5 [1]
" between orbits		7⅓		7¼		
Length of nasal bones	1	0		11⅔		
Width of ditto behind		6		6		6¾
" " in front		4		4¼		4½
From front of intermaxillary bone to hinder margin of nasal bones		11⅔		11	1	0
Length of palatine opening		8		7½		9
Width of ditto behind		3		3¼		3⅔
Depth of zygomatic arch		2		2		
Antero-posterior diameter of palate ...		3		3		3¼
Length of six upper molar teeth taken together		6		6		7¼
Width of upper incisors		2⅙		2⅕		
Length of lower jaw	1	11	1	11		
Height	1	1½	1	2½	1	4
Antero-posterior diameter of ascending ramus, taken in the middle		4¾		4 ¾		5⅔

Fossil species of Lepus.

Already have been noticed some remains, found in the caverns of Brazil, appertaining to a Hare which is exceedingly close to the recent species—*Lepus Braziliensis.* The bones referrible to the Hare genus found in the European deposits, and chiefly in the caverns, in like manner agree for the most part very closely with the recent species of the country.

I have before me several lower jaws, and some other parts of the skeletons of Hares from Kirkdale cavern, and from

[1] The width of two other fossil skulls, according to portions preserved in the British Museum, is respectively—1·4½ and 1·3⅔ ; length of the six molars, taken together, 6½ lines.

Kent's Hole, Torquay; amongst others, in the British Museum collection, is the lower jaw figured by Dr. Buckland in the "Reliquiæ Diluvianæ" (Plate 13, fig. 8); it agrees with the lower jaw of a recent Hare (*Lep. timidus*), except in being a trifle larger, and proportionately rather deeper. This jaw is from Kirkdale: the same differences are observable in two lower jaws found in the cave at Torquay, called Kent's Hole: all three of the jaws measure 8 lines in depth below the second molar tooth, whilst a recent Hare's jaw is 7 lines in depth at the same part. One of the fossil jaws from Kent's Hole[1] measures, from the anterior extremity to the back part of the last molar, 1 inch $9\frac{1}{2}$ lines, and the second, 1 inch 9 lines; the recent jaw being 1 inch $8\frac{1}{4}$ lines, measured from the same parts. The molar teeth, in all the fossils, agree in size and form with those of the recent species, and the same remark applies to an imperfect tibia from Kent's Hole, and, according to Cuvier[2], to the heel-bone figured by Dr. Buckland, in the work above quoted—Pl. 10, fig. 14. A metatarsal bone on the same plate (figs. 15 and 16) Cuvier found, upon comparison, to be nearly equal in length to that of the Common Hare, but proportionately stouter.

Three lower jaws from Kent's Hole are not distinguishable from those of the Common Rabbit—*Lep. Cuniculus.*

Remains of the Hare genus have been found in the caverns of Liège, Montpelier, Lunel-Viel, and Aude[3]; and in the tertiary strata of the Puy de Dôme. In the osseous breccia of Gibraltar, Cette, and Pisa, remains of *Lepus* have likewise been found: some of the fossils resemble the bones of the Common Rabbit, and some appertain to a species of

[1] It is represented in Owen's British Fossil Mammalia, p. 210, fig. 80.

[2] See "Ossemens Fossiles," 4to. ed. 1823, p. 55.

[3] See Marcel de Serres, Essai sur les Cavernes à Ossements, &c. 8vo. 1838; Recherches sur les Ossemens Humatiles des Cavernes de Lunel-Viel, 4to. 1829, and Comptes-Rendus, 1842, p. 388.

Lepus about one-third smaller than that animal. The lower jaw, from Gibraltar, represented in Plate 13, fig. 4, of Cuvier's Ossemens Fossiles, (Vol. IV.) differs much from that of the Rabbit or Hare, not only in being smaller, but in having the ascending ramus nearly at right angles with the horizontal one; and the angular portion of the jaw was apparently narrower in the transverse direction, as in the species of *Lagomys*. The situation of the mental foramen (which is not indicated in the figure) would assist in determining to which of the two genera of *Leporidæ* the jaw should be referred.

FAMILY HYSTRICIDÆ.

Rodents with $\frac{4-4}{4-4}$ molar teeth; the muffle, or terminal portion of the muzzle, clothed with small hairs; the skull with a large ant-orbital opening, through which a portion of the masseter muscle passes, as well as the infra-orbital nerve; the lower jaw with the angular portion joined to the outer (not the under) surface of the bony covering of the inferior incisor; tibia and fibula distinct.

The family *Hystricidæ* is essentially a South American group: it may be divided into six minor sections, or sub-families, viz.—*Hystricina, Dasyproctina, Echimyina, Octodontina, Chinchillina,* and *Caviina;* of these, the most highly organized division, which includes the Porcupines (*Hystrix* of Linnæus), has a wide geographical range, having representatives in the four quarters of the globe: with the exception of two species[1], the whole of the species forming the remaining five sub-families are exclusively found in the New World, being chiefly confined to South America,—two

[1] *Aulacodus Swinderianus*, and *Petromys typicus*; the former found in Western Africa, and the latter in South Africa.

or three species in the West Indian Islands, and about the same number in central America, form the exceptions. The members of the sub-families *Hystricina, Dasyproctina,* and *Echimyina,* increase in number towards the tropical portions of South America; whilst, in the southern parts of that continent, the less highly organized species only are found, these being members of the remaining three sub-families. On the western side of the Southern Andes but one of the sub-sections alluded to has representatives—the *Octodontina.*

The relations which the sub-families of the *Hystricidæ* bear to each other may be expressed by arranging them as follows:—

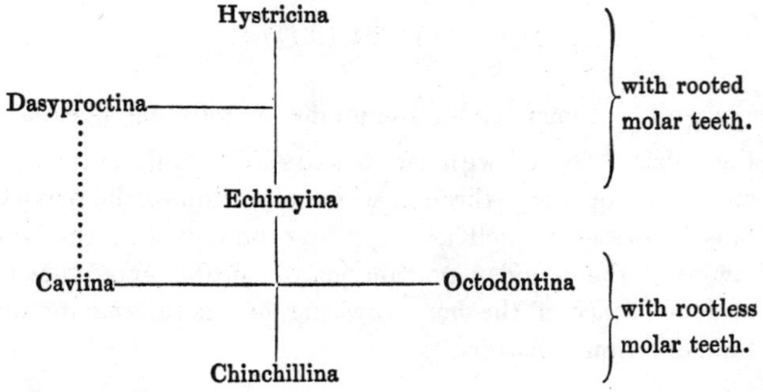

By this mode of placing the groups, we wish to express that the Agoutis (*Dasyproctina*), in their characters, partake of those of the three groups to which they are approximated; that certain members of the section *Echimyina* lead off to the *Hystricina,* whilst others approach more nearly to the Chinchillas; and, lastly, that the Octodons are nearly related, on the one hand, to the *Echimyina,* and on the other, to the Chinchillas: they bear the same kind of relationship to the *Echimys* group as do the *Arvicolæ* to the true Rats. That such is the case we shall endeavour to shew when treating of the groups separately.

With regard to my definition of the family *Hystricidæ,* it

is necessary to observe that, in the Cavies, where the lower incisor is short, the angular portion of the jaw cannot strictly be said to spring from the outer surface of the bony covering of that tooth, though the direction of the incisor is such that, were it prolonged backwards, the alveolus of the tooth, and the angular portion of the jaw, would hold the same relative positions as in the other *Hystricidæ*[1]. Usually the incisors in this family are large and long; and, on viewing the under surface of the lower jaw, a longitudinal groove separates the dentary and angular portions. The latter is almost always of a triangular form, and usually terminates behind in an acute angle. Commonly the masseteric ridge is well developed, and gives a somewhat expanded, flattened surface to the lower edge of the jaw, which is broadest near the middle, being dilated at that part so as to form an obtuse angle on the inner side of the ramus. This angular projection, which is not found in other groups of Rodents, is very distinct in the *Echimyina* and *Octodontina*[2], but either disappears, or becomes indistinct, at the extremes of the family, as, for example, in most of the *Hystricina* and *Caviina*. Besides these points, the small size of the coronoid process is very characteristic of the Hystricine family, having relation to

[1] The transitional forms which link the structure of jaw which is characteristic of the Cavy, with that which is usual in the *Hystricidæ*, are found in the lower jaw of the animals belonging to the genera *Dolichotis*, *Lagotis*, and *Lagostomus*. In *Dolichotis* the incisors are somewhat more elongated than in the typical Cavy; in *Lagotis*, the incisors, which are still comparatively short, yet extend as far back as the line of the springing of the angular portion, or descending ramus of the jaw—here we begin to trace the angle on the inner side of the ramus; and, in the closely allied genus, *Lagostomus*, the angle is distinct, the incisors longer, and the angular portion of the jaw is attached to the outer side of its alveolus. In the *Sciuridæ* and *Muridæ* the corresponding part of the jaw invariably springs from the *under* surface of the dental portion.

[2] Excepting the genus *Habrocoma*, where the angle in question is very obtuse; it is unusually far back in *Octodon* (see fig. 2 *b*, of Plate 8), and is very prominent in *Ctenomys* (fig. 6 *b*, Plate 8).

the small development of the temporal muscle: the condyloid portion of the jaw is, moreover, but little elevated, though rather deep from front to back, and the condyle is generally narrow. As exhibiting all these characters of the lower jaw in a well-balanced degree, we might select the jaw of *Capromys Fournieri;* and the skull of the same animal furnishes an excellent model for comparison in describing the various forms of crania found in the present family. The facial part of this skull is large, the interorbital space broad, the frontal bones being much developed; the depression, marking the course of the temporal muscle, small; the zygomatic arch deep; the malar bone is traversed, on the outer surface, by a well-marked ridge, running obliquely downwards and forwards, from the posterior extremity of the bone, which is, moreover, remarkable for having the lower edge produced into an acute angle, opposite which, on the upper edge of the zygoma, is a second angular projection, marking the hinder boundary of the orbit; the ant-orbital opening very large, and enclosed externally by a process of the superior maxillary bone; the zygomatic process of the superior maxillary bone thrown out horizontally from the plane of the palate; the palate deeply emarginated behind, and much contracted between the foremost molar teeth; the series of molars of opposite sides of the jaw converging in front.

This description of the leading characters of the skull of *Capromys Fournieri* will apply, in almost every particular, equally well with a large portion of the *Hystricidæ,* viz. to the *Echimyina* and *Octodontina.* With regard to the other sub-families, the most important deviation from this type to be noticed is the absence of the angular projection on the under surface of the malar bone, and the want of the oblique ridge on the outer surface of that bone. In some species the molar teeth do not converge and contract the palate in front; but this is a distinction of minor importance, since, where these exceptions occur, we find the species which present them very closely connected in affinity with others in which the palate is contracted. We may safely state, that there is a *general tendency*, in the present group, to a contracted form of palate, arising from a convergence of the molar teeth in front, and that, in other groups of Rodents, it is extremely rare to meet with this character: it is most strongly marked in the species which we place at the end of the Hystricine section—the Cavies and Chinchillas, and disappears in some of the species of the higher divisions of the family in which the molar teeth are rooted.

As the remarkable character of having a portion of the masseter muscle passing through the anterior root of the zygomatic arch is peculiar to the Rodents, and constant in the present family, the two heads on Plate 6 A have been engraved to show this, and the other masticatory muscles—*a*, figs. 1, 2, and 3, is the masseter muscle: in fig. 3, the superficial layers of this muscle have been removed to show an inferior layer (*a**), which arises from the sides of the muzzle, and after passing through the ant-orbital opening, suddenly bends downwards, and becomes tendinous, and is inserted near the middle of the outer surface of the lower jaw: *b*, is the temporal muscle; *c, c*, fig. 4, are the pterygoid muscles; and *d*, the digastric, which, in the Rodents, preserves its normal character of being tendinous in the middle; *e*, is the infra-orbital nerve—a nerve of sensation which supplies the upper lip. On the same plate will be found representations of the terminal portion of the muzzle of some of the genera of the *Hystricidæ*, together with a view of the same part in one of the *Muridæ*, added for comparison[1].

It will be seen that the Capromys (figs. 5 and 5*a*), which has been associated with the Rats, presents a very different struc-

[1] These figures are from drawings taken from the animals immediately after death.

ture of muzzle to that of fig. 9, which represents the common form of that part in the *Muridæ;* and that, notwithstanding *Capromys* and *Lagostomus* are genera of distinct and well-marked sections of the *Hystricidæ,* yet in the structure of the terminal part of the muzzle, or muffle, there exists a far greater similarity in these genera than exists between either of them, and the Murine Rodent; not only in the expanded form of the muffle, but in the fact that that part is clothed throughout with velvet-like hairs, whilst in the Rats it is naked.

Beyond these points, I will merely here call attention to the form of the ear, which always, or nearly always, has the hinder edge emarginated in the present group of animals.

SUB-FAMILY CAVIINA.

Hystricidæ with rootless molar teeth, divided by folds of enamel so as to form lobes having acute angles; the series of molars on opposite sides of the upper jaw converging, and nearly meeting, in front; incisor teeth comparatively short, those of the lower jaw not being extended backwards as far as the springing of the angular portion, or descending ramus; four toes to the fore feet, and three to the hind; tail wanting, or rudimentary; upper lip entire: a strongly-developed horizontal ridge on the outer surface of the lower jaw: the angular portion of this jaw produced much beyond the condyloid portion, descending below the level of the dental portion, and with a curved lower margin: clavicles wanting.

The Cavies have almost invariably been associated with the Agoutis (*Dasyprocta* and *Cœlogenys*), and frequently the animals of both divisions have been linked together, under the sectional name *Subungulata,* by the more recent writers, who thus adopt Illiger's name for the section, and his views regarding it[1]. I am satisfied that the nature of the affinity which exists between these two groups is by no means so near as we are generally led to suppose.

Both groups belong to the same great family; and, besides the characters which are peculiar to this family, the species of the genus *Dasyprocta* may be said to approach the Cavies, from the circumstances of their having the same number of toes to the feet, and being almost (or entirely) destitute of a tail; but this I maintain to be weak evidence in favour of the supposition that there exists a very near affinity between the genera associated in this work under the name *Caviina*, and those arranged under the head *Dasyproctina;* such an affinity, in fact, as is indicated by the arrangement of these two groups in a section apart. A certain degree of importance must be given to the characters just referred to, when we find them *combined* with numerous other peculiarities exhibited in the structure of the teeth, skull, and skeleton, as in the *Caviina*, but evidence of a distinct minor type, in the structure of these parts, will be educed when we treat of the *Dasyproctina*.

A remarkable structure of the fauces, or entrance to the throat, has been pointed out by Mr. Morgan in the Capybara

[1] The characters of the section *Subungulata*, given by Illiger, will apply to the whole of the great Hystricine section of Rodents (*Hystricidæ*), as defined in this work, if we except one phrase only, viz. "*Corpus pilis solis tectum;*" and, with regard to this phrase, I have only to observe, that in one of the genera of the *Hystricidæ* (*Nelomys*), there are species, some of which are clothed with spines, and others in which the clothing consists of soft hairs only. I object to the name *Subungulata*, because it is only applicable to some of the species of the group, as Illiger was well aware, since in his definition he says—"*Ungues falculares, paucis postici subungulæformes;*" because it is more applicable to some other groups of Mammalia; and, lastly, because other, more convenient and less objectionable, names for families, than those framed upon the same principle as the name under consideration, have been proposed and generally adopted—as, for instance, names taken from the leading genus in the family: such names have many advantages; amongst others, they at once give an approximate idea of the contents of the group.

and Guinea Pig, and is probably constant in the present subfamily: it consists in the fleshy lining of the mouth, at this part, being continued so as to form a funnel-shaped cavity, which passes backwards into the pharynx, into which it opens by a very small aperture, incapable of admitting the passage of any but well masticated food. The walls of this funnel-shaped cavity, moreover, are supplied by muscles by which its circumference may be dilated or contracted, and when produced, this conical prolongation of the *velum palati* passes over the epiglottis, so that the entrance of any particle of the food into the larynx is prevented[1]. The stomach is simple, but the cœcum is large and complicated in the Cavy group. The molar teeth are generally divided by a deep fold of enamel into two principal lobes; those of the upper jaw have the entering fold of enamel on the inner side, whilst those of the lower jaw have the corresponding fold on the outer side. The palatine portion of the skull in front of the molars is much contracted, and the incisive opening is small; between the molars, the palate assumes a triangular form, being contracted in front by the convergence of the molars to that point; the post-palatal emargination is usually deep, and leaves exposed the anterior sphenoid bone; the auditory bullæ are moderate; the occipital condyles very narrow, and the occipital opening is comparatively large. In the structure of the lower jaw is found a well-marked character, which distinguishes the species of the present section from all other Rodents, and that is a peculiar ridge (shewn in the figures on Plates 4 and 6) extending along the outer side of the lower jaw, commencing opposite the first molar tooth, and at first running horizontally backwards, but recurved towards the hinder extremity, to run up to the condyloid portion of the

[1] For a more detailed account of the structure of the fauces in the Capybara, the 16th Volume of the Linnean Transactions must be consulted—see p. 465.

jaw: it is much protruded in all the Cavies, and most so in the more typical species, where the upper surface of the protruded part is strongly concave in the transverse direction. The condyloid portion of the jaw is remarkable for being but little elevated above the crowns of the molar teeth, and the extremely small coronoid process reminds us of the condition of the same part in the Tailless Hares (*Lagomys*): indeed, the Cavies approach the Hares in many particulars, most of which are more strongly marked in the Patagonian Cavy.

Genus *Dolichotis.*

Dolichotis. DESMAREST, Mammalogie, Part 2, p. 360. 1822.
Mara. LESSON, Centurie Zoologique, p. 113. 1830.

Caviina with the limbs long; the ears fully half as long as the head, pointed, broad at the base, and deeply emarginated behind; tail very short, and recurved; metatarsus with the anterior half clothed with hairs on the under surface; the hinder half, as well as the heel, naked; molar teeth small, compared with the bulk of the skull; the three foremost of the upper jaw, and the three hindermost of the lower, divided by folds of enamel, each into two equal lobes; the last molar of the upper jaw three-lobed, as well as the foremost of the lower jaw (Plate 4, fig. 1 B).

The comparatively large size of the ears, and the long legs of the Patagonian Cavy, (the only species contained in this genus) give to it a very different aspect to that of the ordinary Cavies, and these characters, combined with the presence of a distinct, though short, tail, form the chief generic distinctions.

The structure of the skull (Plate 4, fig. 1) of the Patagonian Cavy more nearly resembles that of the *Cavia rupestris* than other species of the Cavy group[1]; and these two

[1] The skull of the Patagonian Cavy more nearly resembles that of the *Cavia rupestris* than that of the *Cavia Cobaya*, in its more elongated form, and more especially in the greater length of the facial portion; in the greater depth between the orbits; the shorter and more rounded form of the hollow formed by the junction of the orbital and temporal cavities; in the smaller size of the

animals agree in having the angular portion of the lower jaw deeper and less developed than in the more typical Cavies. The projecting ridge on the outer surface of the lower jaw is less sharp in the Patagonian Cavy than in the species of the genus *Cavia*, and, being very little recurved, the upper surface of the ridge is but slightly concave. In the structure of the teeth, the *Dolichotis Patachonica* agrees very closely with the *C. rupestris*, but in the former animal the principal fold of enamel of the molars is met by a distinctly indenting fold from the opposite side of the tooth; and the supplementary lobe of the last upper molar is broader than in *C. rupestris*, being nearly of an oval form: the notch which separates it from the second lobe is much more deep and distinct.

The Cavies, generally, evince an approximation to the Hares in the comparative shortness of their incisor teeth, the imperfect condition of the palate, which, from being short, and deeply emarginated behind, leaves exposed the sphenoid bones, the narrow form of the bodies of the sphenoids, and the small brain cavity. The facial part of the skull, however, is not so large in proportion as in the Hares, and the skull is more depressed, and wants the great supra-orbital processes, and the large incisive openings, which form some of the many characteristics of the Hare's skull. With these evidences of an approximation between the Cavies in general, and the Hares

portion of the palate which lies between the molar teeth, and in the malar bone being produced into a distinct angle above, forming a post-orbital process. The malar bone, however, rises vertically to form this angle, in the Patagonian Cavy, immediately above its point of union with the zygomatic process of the temporal bone, whilst in the *C. rupestris* the angle is rather more advanced in position: no such angle exists in the malar bone of *C. Cobaya*. The ridge forming the upper boundary of the ant-orbital opening is less prolonged, in the animal under consideration, than in the typical Cavies, terminating, as it does, in the same vertical line as the fore part of the root of the zygoma. The lachrymal bone is here larger than usual, descends so as nearly to join the malar bone, and is interposed between the upper and lower processes of the superior maxillary bone, which, in the true Cavies, generally meet and form a perfect ring enclosing the ant-orbital opening.

(which, nevertheless, are separated by many important differences), it is interesting to find, in the Patagonian Cavy, a link uniting the two groups still more closely together—here the cerebral portion of the skull is short, deep and arched, and the facial portion immensely developed, the nasal bones very large, and, unlike the other Cavies, very convex above; the orbits large, approximated in the mesial line of the skull, and protected above by a strongly-developed supra-orbital ridge, presenting a distinct notch above the eye, and with the hinder extremity separated from the body of the frontal bone by a fissure; the palate so little extended, that not only are the sphenoids exposed, but even the vomer comes into view. These points, which approximate the Patagonian Cavy to the Hares, however, it must be observed, are none of them so strongly marked, in the animal under consideration, as in the *Leporidæ*.

In the scapula of the Patagonian Cavy, we find not only the descending branch of the acromion process, as in the genera *Cavia* and *Lepus*, but an equally long ascending branch also exists: the spine is separated from the body of the scapula, in front, by a very deep emargination, as in the Hares and typical Cavies, and, indeed, the deep emargination at this part is found in most of the *Hystricidæ*, and forms a striking point of distinction when the scapula of the true Rat is compared with that of the Rat-like animals of South America, forming the genus *Echimys;* or when the scapula of the Beaver of the northern hemisphere is compared with that of the Coypu (*Myopotamus*) of South America, which by many naturalists is erroneously associated with the Beaver.

DOLICHOTIS PATACHONICA.

The Patagonian Cavy.

(*Plate* 3, fig. 1).

Cavia Patachonica.	SHAW, General Zoology, vol. ii. Part i. p. 226, Pl. 165 (1801).
" "	WATERH., Zool. of H.M.S. Beagle, Mammalia, p. 89 (1839).
Dasyprocta Patachonica.	DESMAREST, Journal de Physique, 1819, tom. 88, p. 205 ; Mammal. Part 2, p. 358 (1822).
Chloromys Patagonicus.	LESSON, Manuel de Mammal. p. 301 (1827).
Mara Magellanica.	LESSON, Centurie Zoologique, p. 113, Pl. 42 (1830).
Dolichotis Patagonica.	WAGNER, Schreb. Säugth Suppl. iv. p. 66. (1844).
Patagonian Cavy.	PENNANT, Hist. of Quadr., ed. 1781, vol. ii. p. 363, Plate No. 234.
Lièvre Pampa.	AZARA, Essais sur l'Hist. Nat. des Quadr. de la Province du Paraguay, French transl. vol. ii. p. 51 (1801).

Fur dense and crisp ; on the upper parts of the head and body grey, on the sides of a yellow-rust colour ; chin, throat, and abdomen, white ; rump black, but with a broadish white transverse band crossing immediately above the tail : limbs for the most part rusty yellow, but greyish in front.

Inhabits Patagonia, ranging from about 48° 30′ in the South, to 37° 30′ on the East coast, and extending into La Plata as far north as Mendoza.

" The Patagonian Cavy," Mr. Darwin observes, " is found only where the country has a desert character. It is a common feature in the landscape of Patagonia, to see in the distance two or three of these Cavies hopping one after the other in a straight line over the gravelly plains, thinly clothed by a few thorny bushes, and a withered herbage. Near the coast of the Atlantic, the northern limit of the species is formed by the Sierra Tapalguen, in latitude 37° 30′, where the plains rather suddenly become greener, and more humid. The limit certainly depends upon this change, since near Mendoza (33° 30′), four degrees further northward, where the country is very sterile the animal again occurs. Its southern limit is between Ports Desire and St. Julian, about 48° 30′." From the same source[1] we learn the following particulars relating to the habits of this Hare-like Cavy. It is a burrowing animal, but when found in the same districts with the Viscacha it will avail itself of the excavations of this latter animal for a retreat. The Patagonian Cavies wander at times to great distances from their homes, and usually two or three are seen together on these occasions. The animal, in its mode of running, more nearly resembles the Rabbit than the Hare, and, though its limbs are long, it does not run very fast: it seldom squats after the manner of a Hare, is very shy and watchful, and feeds by day; and, in connexion with this circumstance, I may mention that the eyes, like those of the Kangaroos, are defended from the direct rays of the sun, by well-developed eye-lashes, which is not the case with the ordinary Cavies. It generally produces two young at a birth, and brings forth in its burrow: its flesh is white when cooked, but rather dry and tasteless.

The long legs, and tolerably long erect ears, combined with the general form of the head, and the short recurved tail, have caused this animal to be very generally mistaken for a Hare, by casual observers, and we have pointed out certain other characters in which the resemblance is further borne out; it nevertheless is a Cavy in all the more essential details of structure. In size, it greatly surpasses the Common Hare, full-grown individuals weighing from twenty to thirty-six pounds, and, indeed, the Patagonian Cavy must

[1] Zoology of H.M.S. Beagle.

rank amongst the largest of the Rodent tribe, though far surpassed by its aquatic congener, the Capybara. The head is large, and terminates in a blunt muzzle, which is entirely clothed with hairs, even in the mesial line, if we except an exceedingly narrow space, of about half an inch in length, between the muffle and the upper lip, which is naked. The upper lip merely presents a slight angular notch above the incisor teeth. The ears are very broad at the base, and pointed at the opposite extremity, and, being rather deeply emarginated behind, the upper half is thus somewhat suddenly contracted in width. Unlike other Cavies, the hind feet are in a great measure clothed with hair beneath; the under surface of the toes, however, is naked, and there is a long callous pad on the heel. The fur of the animal is moderately long, somewhat crisp to the touch, and very dense; on the upper parts of the head and body, it is of a grey hue, being finely freckled with black and dirty yellow-white; on the sides of the head and body, as well as on the chest, the fur is of a bright yellowish rust colour, varying in intensity in different parts, and is most rich on the lower part of the cheeks, where the hairs are very long. The prevailing hue of the legs is rusty yellow—paler than on the sides of the body—and inclining to ashy grey on the fore part of the fore legs. The tarsi are of a pale brownish colour above, but rufous at the sides, and a rich rufous hue is observable on the hinder part of the posterior limbs, above the heel: the outer surface of the thigh is greyish; the terminal portion of the muzzle, the chin, throat, and abdomen, are white; on the last mentioned part not very pure, being slightly suffused with rusty yellow. The rump is black, but on the haunches is a transverse, broad, white patch; and a narrow white band, crossing immediately above the tail, connects the two patches of the opposite sides of the body. The hairs of the moustaches are black, and very long. The ears are tolerably well clothed with hairs, and these are for the most part of a rusty yellow colour, but near, and at the hinder margin—on the inner surface of the ear—spring numerous long and very slender black hairs, and on the fore part of the ear externally, as well as at the point, are similar hairs. The short and thick tail is rather sparingly clothed with rusty yellow hairs at the base, and is quite naked at the point.

The above description is taken from a specimen in the British Museum collection; other specimens in the same collection show that the colouring of the animal varies somewhat in intensity in different individuals; the rusty yellow hue on the sides of the body being sometimes paler, and less bright, than in the specimen described: and the hairs on the inner surface of the ears being occasionally very pale.

	British Museum.		Zoological Society.	
	Inches.	Lines.	Inches.	Lines.
Length from tip of nose to root of tail	30	0	34	0
" from ditto to ear	4	10	5	0
Height of the animal at the shoulders, about	13	0		
Length of the ear	3	4	3	2
Width of ditto at the base	2	11		
Length of fore leg and foot, about ...	9	0		
" of fore foot (without including the nails)	2	0		
" of the longest nail		6		
" of hind foot, exclusive of the nails	7	2	7	3
" of the nail of the middle toe		10		
" of the tail, about	2	0	1	4
" of skull	5	4	4	9½
Width of ditto	2	7	2	4¾
" between orbits	1	9	1	6½
Length of nasal bones	1	10	1	5?
Width of ditto behind		10½		8⅔
" " in front		9½		8⅔
" of upper incisor teeth ...		3¾		3⅓
Between incisor and molar teeth—upper jaw	1	9	1	6⅔

	British Museum. Inches.	Lines.	Zoological Society. Inches.	Lines.
Length of four upper molars, taken together	1	1		$11\frac{1}{4}$
Average width of upper molars ...		$2\frac{2}{3}$		$2\frac{1}{3}$
Width between hinder molars ..		$11\frac{1}{4}$		
" between front molars		1		
Length of lower jaw	3	9	3	5
Height of ditto, behind	1	7	1	5

The dimensions in the second column are taken from a skull in the Leyden Museum ; those in the first, are from the skull of the animal which is the original of my description.

Genus, *Cavia*.

Cavia. KLEIN, Quadrupedum Dispositio, p. 49 (1751).
Anœma. F. CUVIER, Ann. du Mus. xix. p. 292 (1812).
Les Cobayes. CUVIER, Règne Animal (ed. 1817) ; ed. 1829, p. 220.

Caviina with short limbs, and the ears likewise short ; feet naked beneath ; molar teeth nearly of equal size ; each molar with two principal lobes.

The short-footed Cavies, having the molar teeth nearly equal in size, present two well-marked modifications in the structure of these teeth, the one exhibited in the Common Guinea-Pig, and the other in the *Cavia rupestris*. Perceiving these differences, F. Cuvier, in his *Dents des Mammifères*[1], instituted a new genus, under the name *Kerodon,* for the reception of the latter animal, which was found to differ from the *Cavia Cobaya,* and its allies, in the structure of its toe-nails, as well as in the teeth. Subsequently, new species of Cavies were discovered, having the same type of dentition as the *C. rupestris,* but in which the claws retained the form found in the Common Guinea-Pig, and it then became a question, to which of the characters—that furnished by the teeth, or that observed in the structure of the claws or nails—we should give the greater weight. I confess I cannot regard

[1] P. 151, Pl. 48 (1825).

either of the differences as of sufficient importance for the establishment of a genus, but am inclined to consider the structure of the teeth as of higher value than that of the nails, and think it desirable to adopt M. F. Cuvier's division, as a section of the genus *Cavia*, only.

Section, or Sub-genus, *Cerodon*[1].

Caviæ with the two lobes of the molar teeth equal, or very nearly so; the hinder lobe of the upper molars without any distinctly indenting fold of enamel.

Here the upper molar teeth have each a deep fold of enamel on the inner side, which penetrates the body of the tooth so deeply, that it touches the enamel on the opposite, or outer side; in this latter part the enamel merely forms a waved line (viewing the crown of the tooth), being indistinctly bent inwards in the middle of each lobe (as in *C. rupestris*), or forming a slight notch in, or near the middle of, each tooth, in some other species, as in *C. Australis*. The hindermost upper molar differs from the rest in having a small posterior lobe, or process thrown backwards from the second principal lobe. The molars of the lower jaw resemble those of the upper, but are reversed in position, have a more or less deeply indenting fold of enamel (marked *a*, in figs. 1, 2, 4, &c.) in the hinder lobe, running parallel with, and close to, the principal fold (in this respect differing from the molars of *Cavia* proper—Pl. 6, fig. 9); and it is in the foremost molar that a small extra lobe is found, projecting from the front of the tooth. The molars are relatively smaller in *Cerodon* than in *Cavia* proper. Figs. 2 *b*, of Pl. 4, and figs. 3, 5, and 7, represent the second upper molar tooth of various species of the *Cerodon* section; and the two foremost

[1] From Κέρας, a horn, bow, &c.; and ὀδοὺς, a tooth.

molars, of the lower jaw of the Cerodons, are represented on Pl. 6, figs. 1, 2, 4, 6, and 8.

CAVIA (*Cerodon*) RUPESTRIS.

Rock Cavy.

Cavia rupestris.	PR. MAXIMILIAN, Isis, 1820, p. 43 ; Beiträge zur Naturgeschichte Brasiliens, ii. p. 466 (1826) ; Abbildungen zur Naturgesch. Bras. 1822—31.
Kerodon Moco	F. CUVIER, Dents des Mammifères, p. 151 (1825); Dictionnaire des Sci. Nat.
" *sciureus.*	Is. GEOFFROY, Dict. Class. d'Hist. Nat., ix. p. 120, (1826).
Cerodon rupestris.	WAGNER, Schreb. Säugth. Suppl. iv. p. 69 (1844).
Rock Cavy.	PENNANT, History of Quadr., ii. p. 362 (1781).

Nails of the toes short, obtusely pointed, and scarcely projecting beyond the large fleshy pads with which the toes are terminated ; fur soft, its general hue grey, tinted with rufous on the hinder parts of the back ; throat white ; chest whitish ; abdomen white, suffused with pale ochreous yellow ; sides of face tinted with rufous ; hinder part of the posterior limbs suffused with chestnut-red : fore legs and tarsi whitish, tinted with rufous.

Inhabits Brazil, in rocky situations.

The Rock Cavy, according to Prince Maximilian, inhabits the interior of Brazil, and is confined to rocky districts, where it seeks its retreat in holes amongst the fragments of the rocks. It is frequently met with at Belmonte, Rio Pardo, and Rio de St. Francisco (all nearly under the 16th parallel of south latitude), being found near the rivers, but always in the higher parts of their course. Its flesh is said to be well flavoured, on which account it is sought by the Indians, to whom it is known by the name *Hokê*, and by the Portugese it is called *Mokó*.

The *Cavia rupestris* is superior in size to most other Cavies, stands higher on its legs, and is remarkable, in its group, for the comparative softness of its fur, and the structure of the nails of the toes. The feet are entirely naked beneath, and the toes are terminated by rather large fleshy pads, on which lie the short and blunt nails, which scarcely project over the fleshy portion of the toe: a large trilobed fleshy cushion is situated at the root of the toes of the fore feet, which, like those of the hind feet, have four or five transverse incisions on the under surface, and there is a large, naked, and rounded carpal pad. A nearly similar tri-lobed pad is found at the roots of the toes of the hind feet, behind which is a narrow transverse groove, and the remaining portion of the sole of the foot is protected by a smooth, flat, and hardened skin. The nail of the inner toe of the hind foot differs from that of the other toes in being larger, and in projecting considerably beyond the fleshy portion of the toe; it is thin, and hollow beneath, and, viewed from above, it presents a straight outline on the side next the other toes, and is rounded on the opposite side: this nail, I feel no doubt, is used by the animal to clean its dense and soft fur. The general hue of the upper parts is grey, the fur on these parts being distinctly pencilled with black and white; they are of a grey-brown colour at the root, black at the point, and annulated with greyish white near the point; on the crown of the head the black prevails, but on the back the black and pale colour are about equal in proportions; the hinder part of the back assumes a brownish tint, owing to the pale sub-terminal ring of each hair being of a rufous white hue at this part; and the rump, and hinder part of the hind legs, are of a reddish chestnut colour. The throat is white, the chest whitish, and the abdomen white, suffused with pale ochreous yellow; the hairs, however, are of a grey-brown colour at the root on this last mentioned part; between the hind legs is a white patch.

A rufous yellow tint prevails on the sides of the head, which is most bright in the region of the eye, and between that organ and the ear; the ears are clothed for the most part with pale yellowish hairs; the fore legs are whitish, suffused with pale rufous; the inner surface of the wrist is clothed with very short silvery white hairs, which are closely applied to the skin: the tarsi are whitish above, silvery white on the inner side, and suffused with rufous, or pale rust colour. The hairs of the moustaches are unusually long, and of a black colour.

The above description is taken from a specimen in the Museum of the Zoological Society, which was procured in Bahia. I have before me notes made upon specimens of the same animal preserved in the Leyden Museum, from which it appears that the abdomen is sometimes white, and that the chest is tinted with rufous in some individuals, and yellowish in others.

	Zoological Society Museum.		Leyden Museum.	
	Inches.	Lines.	Inches.	Lines.
Total length, measuring along the curve of the back	14	0	13	6
Length from tip of nose to ear, about	2	7	2	6
" of ear		7		6
Width of ditto	1	0½		
Length of fore foot	1	2½	1	2
" of hind foot	2	4	2	5

The dimensions of the skull of the specimen from which my description is taken will be found, with those of other species of *Cavia*, at the end of the descriptions of the animals of this genus. This skull (Plate 4, fig. 2), is readily distinguished from that of other Cavies by its more elongated form, the greater breadth of the interorbital space, and the comparatively short zygomatic arch: this is distinctly less deep beneath the anteror angle of the eye than usual; and, indeed, its vertical diameter, at this part, is scarcely greater than at a point somewhat behind the middle of the malar bone: here, however, it is suddenly contracted in width by a deep emargination in the upper edge, leaving a distinct post-orbital angle to the malar bone. In other Cavies the zygomatic arch is most deep beneath the anterior angle of the eye, and decreases in vertical diameter more or less gradually from that point to the hinder root of the zygoma; generally, however, an angle (either obtuse or acute) is observable on the upper edge, but this forms part of the zygomatic process of the superior maxillary bone, and in none is there any angular post-orbital projection to the malar bone, as in the present species. The molar teeth are relatively smaller than in any other species of Cavy, and so is the portion of the bony palate which lies between the molar teeth: the palate in front of the molars is remarkable for its compressed form, and the incisive opening is exceedingly narrow.

Besides the species just described—which cannot be confounded with any other member of the present genus, if attention be paid to the structure of the claws, and the texture and colouring of the fur—I am acquainted with six species of *Cavia*, all of which may be readily distinguished by characters furnished by the skull and teeth; in their colouring and proportions points of distinction likewise present themselves. To these, *perhaps*, may be added a seventh species of the group having long and pointed nails to the toes—the *Cavia Cutleri;* but notwithstanding the striking difference which the colouring of the fur of this animal presents, I strongly suspect that it is a black variety of the *Cavia Aperea*.

It is to be regretted that the authors who have described the various species of Cavies have made no mention of the structure of the skull and molar teeth—probably they had no opportunity of examining these parts—but have confined their descriptions chiefly to the colouring of the fur, which being subject to some slight variation, I have been unable, in some cases, to convince myself of the specific identity of the animals already described with those I am about to notice, since these latter do not quite agree, in this respect, with the published descriptions.

Of the six above-mentioned species, the leading points of distinction may be thus expressed :—

A. With molar teeth very nearly agreeing with those of *Cavia rupestris*—see Pl. 4, fig. 2 *b* ; and Pl. 6, figs. 1 to 8, = section CERODON.

1. General hue of the fur rich rufous-brown ; incisor teeth yellow. *Cavia* A. 1 = *C. flavidens*, Brandt ? (Inhabits Brazil).

2. General hue of the fur grey.

a, with yellow incisors,

* with white spot behind each ear, and a whitish mark behind the eye ... ... *C. Spixii*, Wagler. (Brazil).

** without any pale spot behind the ear, or eye ... ... *C. Boliviensis.* (Bolivia).

b, With white incisor teeth: head very short. *C. Australis*, Is. Geoff. (Patagonia).

B. With molar teeth resembling those of the Domestic Guinea-Pig—Pl. 4, fig. 4, and Pl. 6, fig. 9 : incisor teeth white.

1. Prevailing hue of the fur brown-black ; the hairs narrowly ringed with rufous-brown ... *Cavia* B. 1 = *C. fulgida*, Wagler ? (Brazil).

2. Prevailing hue brownish — distinctly freckled with black and yellow ... ... *Cavia* B. 2 = *C. Aperea*, Auct. ? (La Plata).

To which may be added—

3. Fur black (brown-black on the sides and under parts of the body) ; no pale annulations to the hairs .. *C. Cutleri*, Bennett. (Peru ?).

CAVIA (*Cerodon*), No. I.

Rufous-brown Cavy

? *Cavia flavidens*. BRANDT, Mém. de l'Acad. de Petersb. (1835), iii. Pt. 2, p. 439, Pl. 17 ; Mammalium Exoticorum, p. 103, Pl. 17.

? *Cavia nigricans.* WAGNER, Schreb. Säugth. Suppl. iv. p. 64.
? *Kerodon obscurus.* (LICHT.) RÜPPELL, Verz. der in dem Museum der Senckenb., p. 33.

General hue rich rufous brown; throat and abdomen pale rusty yellow; incisors yellow.

Inhabits Bahia, Brazil.

The Rufous-brown Cavy is larger than the Domestic Guinea-Pig, and has the head relatively narrower and longer: in the texture of the fur it very nearly resembles that animal. The rich brown hue of the upper parts of the body is due to the hairs being rather broadly annulated towards the point, with rich rufous-yellow, and at the point of a brown-black colour: the hidden parts of the hairs are of a pale brownish grey; the coarser hairs are almost white at the root. On the sides of the body the hairs are pointed with brown, instead of brown-black, and hence these parts are rather less dark than the back. The throat and abdomen are of a pale rusty yellow—inclining to a golden yellow; but the hairs on these parts are greyish at the root: the legs and feet are of the same colour as the sides of the body. Behind each ear is a pale rufous patch; the ears are sparingly clothed with yellowish hairs, and on the cheeks is an indistinct dusky mark, running towards the ear from under the eye.

	Inches.	Lines.
Total length, measuring along the curve of the back	12	0
From nose to ear, about	2	4
Height of ear		6
Greatest width of ditto		11
Length of fore foot, including the nails ...	1	1½
" of hind foot, including the nails ...	2	2
" of nail of middle toe		3½

The skull of this animal (Plate 6, fig. 10), is relatively narrower than that of the *Cavia Cobaya*, and has the facial portion more elongated; here, if the skull be divided into two equal parts, the centre will be found in a line with the small supra-orbital notch, but the length of the cerebral portion of the skull in *C. Cobaya* (measuring from the same point) considerably exceeds the facial portion. The ant-orbital opening is more extended, in the direction of the zygoma, and its outer angle is consequently more remote from the intermaxillary suture — $10\frac{3}{4}$ lines in the Rufous-brown Cavy, and $8\frac{1}{2}$ lines in the Guinea-Pig. The lachrymal bone is interposed between the upper and lower processes of the superior maxillary, which, unlike those in the *C. Cobaya,* do not meet to form the outer boundary of the ant-orbtial opening. Upon viewing the under surface of the skull, the most striking points of distinction between the two animals consist in the narrowness of the molar teeth of the Rufous Cavy, and the greater size of the palatine bone; in *C Cobaya* the post-palatal emargination contracts the palatine bone, so that no horizontal portion remains opposite the hinder part of the posterior molar teeth, whilst in the Rufous Cavy, the lateral branches of the same bone, which here continue the plane of the palate, are fully $\frac{1}{12}$th of an inch in width. The teeth of this animal, when compared with those of *Cavia rupestris,* present scarcely any points of distinction worthy of notice: the incisors are rather broader, the hinder lobe of the last upper molar is rather more elongated, and the indenting fold on the inner side of the molars of the lower jaw is deeper. See Plate 6, fig. 2, which represents the first and second molar teeth of the lower jaw, magnified.

A Cavy in the Leyden Museum, there regarded as the *Cavia Aperea,* I feel little doubt is identical with the animal just described, though its dimensions are somewhat less. Its general hue, according to my notes, is rich brown, the fur on the upper parts of the body being pencilled with black and rusty yellow. The hairs on the back are brownish grey at the root, black at the point, and with a subterminal rusty-yellow ring; the throat and abdomen are pale rusty yellow; the feet dark brown; the ears clothed with bright rust coloured hairs, both internally and externally, but somewhat dusky on the fore part of the outer surface. Incisor teeth yellow.

	Inches.	Lines.
Total length	10	6
From nose to ear	2	0
Height of ear, about		4½
Length of tarsus, including the nails	2	2

A skull in the same collection, which is likewise labelled as that of the *Cavia Aperea*, most closely agrees (judging from my description) with the skull of my *Cavia*, A. 1. Elsewhere, it appears that the animal under consideration is regarded as the *Cavia Aperea;* erroneously, as I think, since the British Museum has received a skeleton, *evidently* of my Rufous-brown Cavy, and which was labelled "*Cavia Aperea.*"

Cavia flavidens. Brandt.

Incisors yellow; upper parts of the body of mixed brownish yellow and palish brown colours, the prevailing hue inclining to yellow; upper surface of the head, back of the neck, and an oblong patch, extending from beneath the eye to the neck, dusky; throat and abdomen yellowish white.

Inhabits Brazil.

The *C. flavidens* is said to exceed the *C. Aperea* in size, and to have a shorter head and shorter ears than that animal; the fur is likewise described as being somewhat shorter: the hairs on the back, pale brownish grey at the root, pale brownish yellow in the middle, and brown-black at the point; the hairs on the abdomen, pale brownish grey at the root, and white, slightly suffused with yellow, externally: a small white spot on the throat, formed by a whirl of short white hairs[1].

	Inches.	Lines
Total length	10	0
From nose to ear	2	0
From ditto to eye	1	1

Cavia nigricans. WAGNER.

Fur somewhat glossy; general hue dark brown; abdomen and mesial line of the throat of a dirty yellow colour: the hairs on the upper parts of the animal, blackish brown at the point, paler at the root, and with a narrow rusty red, or rusty yellow, subterminal ring—or, some few of them, with the exposed ends entirely dark: nails long, curved, and pointed, and projecting considerably beyond the fleshy portion of the toes. Length, 12 inches; hind foot and nails, 2″ 2‴.

Inhabits Brazil.

The general colouring of this animal is evidently darker than that of my *Cavia* (*Cerodon*) No. 1, and I should have been led, on this account, to associate it with my Cavia B. 1, but from its size, combined with the facts that the same animal

[1] In the carefully executed figure which accompanies M. Brandt's paper, the *C. flavidens* is represented as of a rufous-brown colour, tolerably dark on the upper parts of the head and body, and paler, and inclining to yellow, on the flanks; and, in fact, this figure closely resembles the animal I have described under the name of the Rufous-brown Cavy. The shortness of the head of *Cavia flavidens* would no doubt distinguish M. Brandt's animal from my Cavia A. 1, but in skins or stuffed specimens the measurement of this part, we know, is not always to be depended upon. I have described the head of the Rufous-brown Cavy as longer than that of the *C. Aperea*, not because it is longer in the stuffed specimen, but because the *skull* is longer. The small whirl of white hairs on the throat, somewhat dwelt upon as one of the characteristics of the *C. flavidens*, I have sought for in my Cavia A. 1, but, unfortunately, the skin is in part destitute of hairs on the mesial line of the throat in the only specimen I have had an opportunity of examining; I may observe, however, that a similar whirl occurs in some other species of Cavies.

is placed in the genus *Cerodon* by Lichtenstein and Rüppell (being by the former naturalist named *Kerodon obscurus*), and that Dr. Wagner states that, according to the structure of its skull, it should be arranged with the Cerodons, I am led to believe it will prove to be a variety of the Cavia described by me under the name of Rufous-brown Cavy.

CAVIA (*Cerodon*) SPIXII.

Spix's Cavy.

Cavia Spixii. Wagler, in Isis, 1831, p. 512.
" " Brandt, Mammalium Exoticorum Novorum, &c. p. 106.
" " Wagner, Schreb. Säugth. Suppl. iv. p. 62, Pl. 173 A, fig. 2.

Incisor teeth yellow: general colour of the animal grey, with a somewhat indistinct brownish tint on the back; the space between the eye and the ear whitish, a white patch behind each ear, and the throat and abdomen likewise white.

Inhabits Brazil, Rio de Janeiro, Bahia, and the neighbourhood of the Amazon.

The yellow colour of the incisor teeth, and the general grey hue of the fur, combined with a distinct white patch behind, and a second in front of each ear, are the leading characteristics of this species.

A Cavy in the Zoological Society's Museum agrees so perfectly with the published descriptions of *Cavia Spixii*, that I shall proceed to describe it under this head. It is somewhat larger than the Common Guinea-Pig, and its fur is shorter and rather softer. The hairs of the fur, on the upper parts of the body, are of a very pale grey hue at the root, dusky at the point, and have a moderately broad subterminal ring of pale brownish yellow; the admixture of the darker and lighter colours producing a general brown-grey tint—by no means dark; on the sides of the body, the hairs being less dark at the point, and the subterminal ring almost white (scarcely tinted with yellow), the general tint may be described as ashy grey: the throat, abdomen, and inner surface of the limbs are white, and the white of the under parts extends somewhat on to the sides of the body. Immediately above the eye is a narrow white mark; a less pure white band runs from the eye to the ear, and there is a somewhat conspicuous white patch behind each ear. The feet are nearly white; the fore feet rather shorter than usual, as well as the nails of the toes of these feet.

	Inches.	Lines.
Total length	11	6
From nose to ear	2	3
Length of ear		6
Width of ditto	1	1
Length of hind foot, and nails	2	1

The skull of *Cavia Spixii* (Pl. 6, fig. 11) is smaller than that of *C. Cobaya;* viewing this skull from the side, the outline of its upper surface does not become straight, or slightly concave, between the orbits, as in many Cavies, but forms a gentle, convex curve from the tip of the nasal bones to the hinder part of the parietals, where it descends somewhat suddenly towards the occiput: the nasal bones are distinctly longer than in *C. Cobaya;* the auditory bullæ larger, and these, encroaching upon the body of the basi-occipital bone, contract it in front, so that it is very narrow at that part as compared with the Guinea-Pig's skull; the auditory bullæ, moreover, encroach rather more on the occipital surface of the skull: the lachrymal bone is interposed between the superior and inferior zygomatic processes of the superior maxillary bone, as in *C. rupestris,* and some other species of the genus. The palatine bone is larger than in *C. Cobaya,* broader opposite the last molar tooth, and horizontal at this part as in *C. Boliviensis*: the incisor teeth bend more suddenly downwards than in *C. Cobaya;* the thin plate of bone which extends between the condyloid process and the coronoid of the lower jaw is scarcely emarginated above, forming at this part almost a straight line. The molar teeth are much smaller than in *C. Cobaya*. Fig. 3, of Plate 6, represents the second molar of the upper jaw; and fig. 4 represents the two foremost molars of the lower jaw.

CAVIA BOLIVIENSIS.

Bolivian Cavy.

? *Galea musteloïdes*. **Meyen, Act. Acad. Nat. Cur. xvi. Part 2, p. 598. 1833.**

Incisor teeth of an orange yellow colour; general hue of the fur grey, with a very faint yellow tinge; throat and abdomen white; feet whitish.

Inhabits Bolivia, occurring only at great elevations.

This little Cavy so greatly resembles the *Cavia australis* of Patagonia, that for some time I mistook it for that animal; having, however, recently had an opportunity of examining the skulls removed from several specimens of the Bolivian Cavy, I found well-marked distinguishing characters in that part, and hence was led to make a careful comparison of the external parts of the two animals. In the colouring there is very little difference; the general hue of the *C. Boliviensis* is rather darker, and more brown on the upper parts of the body; in the proportions of the head and feet there exist more important points of distinction, the feet being proportionately shorter, and the head longer in the Bolivian animal than in the Patagonian species; I may likewise add, that the Bolivian Cavy has the ears rather larger, and its incisor teeth are somewhat broader, and invariably of a yellow or orange colour in front[1]; those of the *C. australis* being white.

The *Cavia Boliviensis* is rather smaller than the *Cavia Aperea;* the texture of its fur is nearly the same as in that animal, but the hairs are rather shorter, and the general tint of the animal is decidedly paler; this arises from the black at the points of the hairs on the upper parts of the body being less pure, and the pale rings below the points being of a less decided yellow tint. The hairs on the back are of a very pale grey colour at the root, dusky externally, and have a very pale brownish yellow ring towards the point; the general hue produced by the admixture of colours is grey, slightly tinted with brown, and by no means dark. The sides of the body are of a palish ashy grey tint; here the subterminal pale ring of the hairs is almost white, having but a very faint yellow tinge, and the points of the hairs are dusky grey: the throat and abdomen are white—not quite pure. The feet are whitish; on the outer surface of the tarsus the hairs are slightly pencilled with greyish and yellow-white. The eye-lids are whitish, and the ears are clothed with white, and in parts with yellowish hairs. Sometimes the sides of the body have a distinct yellow tint, and the upper parts are more brown; and in one specimen before me the general hue is very pale; on the sides of the body dirty yellow-white, with scarcely a trace of the dark pencilling: the back is chiefly ashy brown. The specimens described are in the British Museum collection.

[1] I have examined about twenty specimens of the Bolivian animal.

			Darker variety.		Paler variety.	
	Ins.	Lines.	Ins.	Lines.	Ins.	Lines.
Length	9	6	9	6	9	6
From nose to ear, about	1	9	1	8	1	$9\frac{1}{2}$
Height of ear	0	$6\frac{1}{4}$		$6\frac{1}{2}$		(
Width of ditto, about		$9\frac{3}{4}$		$9\frac{1}{2}$		9
Fore foot and nails		$9\frac{1}{4}$		$9\frac{3}{4}$		10
Hind foot and nails	1	8	1	$7\frac{1}{2}$	1	8

The skull of this animal (Pl. 6, fig. 3) is considerably smaller than that of *Cavia Cobaya*, has the incisive openings narrower, and the palatine bone with the posterior emargination much narrower; the upper and lower zygomatic processes of the superior maxillary bone do not meet in front of the lachrymal bone: these points link the skull of *Cavia Boliviensis* more closely with that of my *Cavia* (*Cerodon*) No. 1, or with that of *Cavia Spixii*, and it agrees much more closely with those two animals in the structure of its molar teeth than with the *Cavia Cobaya*. As its comparatively small size approximates it most nearly to the *C. Spixii*, I will point out the principal distinctions which present themselves in the skulls of these two grey coloured Cavies, both of which have yellow incisors[1]. The upper surface of the skull of the Bolivian Cavy is less arched than that of the *C. Spixii*; viewing it from the side, the outline is nearly straight, from the tip of the nasal bones backwards to a point a little in front of the fronto-parietal suture, but descends slightly between the orbits, and towards the apex of the nasal bones;

[1] I should observe that the white of the abdomen is less pure in *Cavia Boliviensis* than in *C. Spixii*, and does not extend so much on to the sides of the body: although the fur is rather paler behind the ear, in the present species, than on the other parts of the head, we do not observe the distinct white spot at that part which characterizes the *C. Spixii*, which has shorter and weaker nails to the toes, than the Bolivian animal.

the inter-orbital space is rather narrower; the nasal bones are narrower: the hinder part of the zygomatic arch is thrown more boldly outwards; and a distinct post-orbital process, terminating in an acute angle, projects from the zygomatic process of the superior maxillary bone. I find the grinding surface of the molar teeth, in all the skulls of the *C. Boliviensis*, very little raised above the plane of the palate, as in *Cavia Aperea*, but in the *C. Spixii*, the molars, as in *C. rupestris*, have their crowns raised much above the level of the palate: the molar teeth in *C. Boliviensis* are relatively larger than in *C. Spixii;* those of the upper jaw have two small, but distinct, indenting folds of enamel near the middle of the outer surface, whilst in *C. Spixii* there is but one distinct indentation, though it has a very slight tendency to divide; here the hinder lobe of the last molar is equal in size to the anterior one, whilst in the Bolivian animal the hinder lobe of this tooth is the larger. Lastly, I may notice a conspicuous difference in the palatal portion of the skull of these two animals, which, from the similarity of their colouring, might perhaps be confounded. In *C. Boliviensis*, the portion of the palate which lies between the molar teeth is nearly flat, whilst in the *C. Spixii* it is strongly concave, especially in the fore part; here the palatine bone occupies nearly the whole of the space between the molar teeth; the palatine portion of the superior maxillary bone forms a small raised platform between the anterior pair of molars, and then runs back from that part, in the form of a very narrow ridge, on either side, to form the socket for the molars. The palatine bone in *C. Spixii,* on the other hand, has a very small diameter in the longitudinal direction, and if measured in the mesial line; but its lateral branches are more expanded than in *C. Spixii,* and the post-palatal emargination is consequently narrower than in the skull of the animal last mentioned.

Mr. Bridges, who brought many specimens of the present species from Bolivia, procured them on the high table land between Cochabamba and La Paz, but that gentleman informs me it is found throughout the higher portions of the country, usually at elevations of from ten to twelve thousand feet.

Galea musteloïdes. MEYEN.

Meyen informs us that the lofty plains on the Pass of Tacna, to the lake of Titicaca, as well as those of Tajori (on the Andes, 18° S. Lat.), are inhabited by immense numbers of a small animal, which burrows, and so undermines the ground, that every step of the horse is attended with danger. The animal is so shy that he was unable to procure a specimen; and, indeed, it was only seen from a distance. It is described as having a long and slender body, of a brownish red colour on the upper parts, and pale on the under. Its length, without including the tail, appeared to be about eight inches. To this animal Meyen gives the name *Galea musteloïdes.*

The characters of the genus *Galea* are derived from a skull found by Meyen at the entrance of one of the burrows of the little animals just referred to, and, judging from Meyen's figures and description[1], I feel no doubt it belongs to the Rodent I have named *Cavia Boliviensis;* but, if this be the case, it cannot be the skull of the rufous-brown, burrowing animal, as Meyen conjectured; an animal which, it appears, had a tail, and which I feel little doubt is a species of *Ctenomys;* and, moreover, I think it is probably identical in species with a *Ctenomys* found by Mr. Bridges in the same district as the *Cavia Boliviensis,* a district, it will be observed, which is not far distant from Tacna and Tijori, the habitats of the *Galea musteloïdes.*

[1] Nova Acta Nat. Cur., xvi. Pt. 1, p. 597, Pl. 42, figs. 4—7, and fig. 12.

CAVIA (*Cerodon*) AUSTRALIS.

The Southern Cavy.

(*Plate* 3, fig. 2).

Cavia australis. Is. Geoff., in Guerin's Magazin de Zoologie, 3me Année, Classe 1, Pl 12. 1833.

" " D'Orbigny, Voyage dans l'Amerique, 79me livr. Pl. 18, figs. 1—4.

Kerodon Kingii. Bennett, Proceedings of the Zoological Society for December, 1835, Pt. 3, p. 190.

" " Waterh., Zoology of the Voy. of H.M.S. Beagle—Mammalia, p. 88.

" " Gray, List of the Mammalia in the Brit. Mus. 1843, p. 126.

Incisor teeth white; head short; tarsi long; fur rather soft, its general hue grey—by no means dark; the eye edged with white; a whitish spot behind each ear; the chin and edge of the lower jaw likewise white.

Inhabits Patagonia, from about the 39th parallel of latitude down to the Strait of Magalhaen.

This little grey Cavy, according to Mr. Darwin, is common along the coast of Patagonia, from the Rio Negro to the Strait of Magalhaen. It is very tame, commonly feeds by day, and is said to bring forth two young at a birth. At the Rio Negro it frequents, in great numbers, the bottoms of old hedges, and at Port Desire it takes up its abode amongst the ruins of the old Spanish buildings. M. D'Orbigny observed the Southern Cavy as far north as 39° S. Lat., which it appears is its extreme range in that direction; and this author states that it lives in families, and digs deep burrows in the ground, preferring sandy declivities which are more or less overgrown with bushes: its burrows are said to have several outlets.

The food of the Southern Cavy consists of seeds and green herbage, and it has been observed to ascend trees to feed upon their fruits.

The *Cavia australis* is smaller than the Common Guinea-Pig, and distinguished from others of its group by its very short head: the ears are rather smaller than usual, and the tarsi somewhat longer, in proportion to the bulk of the animal, than in most other Cavies ; they are likewise narrower than usual in the group, and the hairs which cover them encroach somewhat on the under surface, so as to contract the naked sole in the middle. Its fur is shorter, and less harsh, than in the *Cavia Aperea*, and, indeed, is somewhat soft. The general hue of the upper and under parts of the body is almost uniformly grey ; the abdomen being very little paler than other parts ; here the hairs are of a darker grey at the root than externally ; and, on the back, the hairs are grey-brown at the root, rather broadly annulated with very pale-brownish yellow towards the point, and dusky grey at the point : on the sides of the body the pale annulations of the hairs are brownish white. The chin is white, and so is the edge of the lower jaw ; the eyes are edged with white, both above and below, and there is a small dirty white patch behind each ear. The nails of the toes are tolerably long, and pointed, more especially those of the hind feet.

	Inches.	Lines.
Total length	9	6
From nose to ear, about	1	8
Height of ear		4
Width of ditto		7
Length of fore foot and nails		10
" of hind foot and nails	1	11

The skull of the *Cavia australis* (Plate 6, fig. 13) is remarkable for its short and rounded form, and for the nasal bones being distinctly broader behind than in front. The upper surface of the skull is strongly arched; the orbits very large, the ant-orbital opening less oblique in its direction than in other Cavies: here the upper boundary of the bony ring which encloses this opening is more remote from the tip of the muzzle than the lower boundary, whilst in other Cavies the upper portion of the ring projects over the lower. The auditory bullæ are larger than usual, and they encroach upon, and contract, the occipital bones more than in any other species of Cavy[1]. The incisive opening is proportionately larger, here assumes a triangular form, and is about 1¾ lines in width behind. The palate is strongly indented between the molar teeth, as in *C. Spixii*, but in *C. australis* the palatine bone forms a very small portion of the surface of the palate, being composed of two narrow branches widely separated behind, and where they meet in front they are but little produced in the antero-posterior direction. The anterior portion of the bony ring which encloses the ant-orbital opening (as in *C. rupestris* and *C. Aperea*) is entirely formed of the superior maxillary bone, and is not broken through by the lachrymal, as in *Cavia Boliviensis*, *C. Spixii*, and *C.* A 1. Immediately below the lachrymal bone, the ring in question is remarkable for its slenderness. The posterior upper molar tooth has the terminal lobe relatively wider than in other Cerodons. The incisor teeth are very narrow.

[1] The *Cavia Spixii* approaches the *C. australis* most nearly in this respect: in the *Cavia Aperea*, and the *C.* A 1, the bulla in question presents less of its surface when the skull is viewed from behind; and in the *Cavia rupestris* the tympanic bullæ form no part of the occipital surface of the cranium.

	Cavia rupestris. Zoological Society.		*Cavia (Cerodon),* No. 1. British Museum.		*Cavia (Cerodon),* No. 1. Zoological Society.		*Cavia Spixii.* Zoological Society.		*Cavia Boliviensis.*		*Cavia Boliviensis.*		*Cavia australis.*		*Cavia australis.*	
	Ins.	Lines.	Ins.	Lines.	Ins.	Lines.	Ins.	Lines.	Ins.	Lines.	Ins.	Lines.	Ins.	Lines.	Ins.	Lines.
Total length of skull	2	11	2	9½	2	7½	2	3	2	1	2	1	1	10	1	9⅔
Width	1	5	1	5½	1	4¼	1	3	1	2¾	1	3	1	2	1	1½
" between orbits		9½		7¼		7¼		6		5½		5⅓		5		5¼
Length of nasal bones		11¾	1	1¼	1	0		10⅓		8¼		9		7		7¼
Width of ditto behind		4¼		3¾		3⅔		3⅓		2¾		3½		3⅙		3⅙
" " in front		5		4¾		4¼		4⅙		3¼		3½		2⅚		2¼
From incisors to molar teeth		11		9⅔		8¼		7		6⅔		6¾		5½		5
Length of four upper molars, taken together ...		6¾		7		6⅚		5¾		5½		6		5¼		5
Width of lateral branch of palatine bone, opposite the last molar tooth		1		1⅙		1⅙		0⅚		1 4/6		1¼		0⅓		0⅓
" of post-palatal emargination		3		3		2⅔		2½		1 4/6		1⅔		2¼		2¼
From outer angle of ant-orbital opening to inter-maxillary		10¾		11		10¾		9		8¼		8		6½		6¼
Length of zygomatic arch	1	1½	1	2½	1	1¼	1	0¼	1	0		11		10⅓		10
From lower edge of lachrymal bone to lower edge of zygoma		4⅔		5½		4⅚		4½		4¼		4¼		3½		4¼
Length of lower jaw			2	0	1	11	1	6¾	1	7¼	1	8½	1	5¼	1	4
Height of ditto from condyle		8¾		9		8¼		7		6¾		6¾		6		5⅓
Length of angular portion		9½		11½		10		8½		8½		9		7¾		7¼

Section 2. *Cavia proper.*

Caviæ with the hinder lobe of each of the molar teeth larger than the front one: the hinder lobe of the upper molars with a tolerably deep, indenting fold of enamel on the outer side, and the corresponding half of the lower molar with a deep fold of enamel on the inner side.

The difference in the structure of the molar teeth, as I have before remarked, I regard as of very minor importance—it is only one of degree; and yet, if attended to, will be found useful in determining the species. Here the molars are broader than in the Cerodons—a distinction, however, which is rather less marked, if the species of the present section be compared with the *C. Boliviensis;* but I would more particularly direct attention to the modification of the structure observable in the lower molars of the species of *Cavia proper.*

Upon referring to fig. 9 of Pl. 6, it will be seen that the fold of enamel (*a*) which penetrates the hinder lobe of these teeth, runs in from the middle of the lobe, and, sloping obliquely forwards, meets the deep fold of the outer surface of the tooth, and encloses the small area, marked *. No such area is visible in the tooth of a *Cerodon.* In having the molar teeth more approximated to the hinder root of the zygoma, and the crowns of these teeth descending but little below the level of the glenoid cavity, the *Cavia Aperea* (which may be regarded as the type of the present section) carries to an extreme a characteristic feature of the Cavy group. In the *Cavia rupestris* the molars are considerably more remote from the hinder root of the zygoma, but here, as in nearly all the other members of the sub-family *Caviina*, the dental portion of the superior maxillary bone is carried so far back that it touches the temporal bone [1].

[1] The only exception I am acquainted with is the *Cavia Patachonica*; in the skull of which animal, an extremely narrow strip of the palatine bone

CAVIA APEREA.

Restless Cavy, or Guinea-Pig.

Wild.

? *Aperea Braziliensibus.*	MARCGRAVE, Historiæ Rerum Naturalium Brasiliæ, p. 244 (1648).
? *Cavia Aperea.*	LINN., Syst. Nat. ed. Gmel. i. p. 122.
? " "	KUHL, Beiträge zur Zoologie, p. 65.
" "	F. CUVIER et GEOFFR., Hist. Nat. des Mammif. fasc. 48.
? " "	PR. MAXIMILIAN, Abbild. zur Naturgesch. Bras.; Beiträge zur Nat. Bras. ii. p. 462.
" "	RENGGER, Naturgesch. der Säugethiere von Paraguay, p. 274.
" "	WAGNER, Shreb. Säugth. Suppl. iv. p. 58, Pl. 173 A, fig. 3.

Domesticated.

Cavia Cobaya.	MARCGR. Bras. p. 224.
" "	SCHREB., Säugth. iv. p. 617, Pl. 173.
" "	DESMAREST, Mammalogie, p. 356.
" "	(Guinea-Pig). JENYNS, Manual of British Vertebrate Animals, p. 36.
" "	(Restless Cavy). BELL, British Quadrupeds, p. 353.
" "	(Meerschweinchen). WAGNER, Schreb. Säugth. Suppl. iv. p. 65.
Mus Porcellus.	LINN., Syst. Nat. 12th ed., i. p. 79.
Variegated Cavy.	SHAW, General Zoology, ii. p. 17, Pl. 126.
Cochon d'Inde.	BUFFON, Hist. Nat. viii. p. 1, Pl. 1.

Wild Guinea-Pig. Incisors white : fur long, and somewhat coarse ; on the upper parts and sides of the body distinctly pencilled with black and dirty yellow ; chest grey-brown ; throat and abdomen pale dirty yellow, or, sometimes, brown-grey.

Inhabits the banks of the Rio Plata, and extends northwards into Paraguay, Bolivia, and Brazil.

separates the superior maxillary bone behind, from the temporal bone. In other Rodents we find the two bones in question more or less widely separated by the sphenoid, and often by a portion of the palatine bone as well as the sphenoid.

I have examined specimens of this Cavy from Maldonado, La Plata, and from Bolivia; the former brought home by Mr. Darwin, and the latter collected by Mr. Bridges. According to Mr. Darwin, the animal is known by the name "Aperea," and is exceedingly common in the neighbourhood of several towns which stand on the banks of the Rio Plata; sometimes frequenting the sand hillocks, or the hedge-rows formed of the Agave and Opuntia; but marshy places, covered with aquatic plants, appear to be preferred. Where the soil is dry, it makes a burrow, but where otherwise, it lies concealed amidst the herbage. It generally comes out to feed in the evening, and if the day be gloomy, it will likewise make its appearance in the morning. In Bolivia, Mr. Bridges informs me, this animal is confined to the low lands, and in this respect differs from the *Cavia Boliviensis,* which is always found at a considerable elevation. It is not uncommon in fields, in the vicinity of Chuquisaca and Cochabamba, and takes shelter amongst the loose stones of the walls by which the fields are enclosed.

Dr. Rengger informs us that the Aperea is found throughout Paraguay, and extends southwards as far as lat. 35°; he has himself not only met with the animal (with which he is well acquainted) in those parts, but likewise in Brazil, both in Bahia and Pernambuco. In Paraguay it generally frequents moist situations, and near the borders of the forests, but it never occurs either in the forests, or in the open fields. It lives in little societies (according to the same author) of from six to fifteen individuals, in the impenetrable groves of Bromelias; and its presence is detected by the numerous little beaten paths which it forms amongst these plants. It feeds early in the morning, and after sunset in the evening, but never strays far from its home. The Aperea breeds but once in the year, and then has but one or two young.

A specimen from Maldonado, presented to the Zoological

Society by Mr. Darwin, has the fur long, and somewhat coarse; on the upper parts, and sides of the head and body, it is distinctly pencilled with black and dirty yellow, the black prevailing on the back and upper surface of the head, but, on the sides of the body, the black and yellow colours are in about equal proportions. The general tint is darkish brown, slightly inclining to grey, being totally destitute of any red hue. The throat and abdomen are of a dirty yellow colour, but very pale; here the hairs are of a very pale grey colour at the root: the hairs on the upper parts of the body are of a dusky grey colour at the root, and black, or brownish black, beyond, with the exception of a moderately broad subterminal yellow ring: some of the longest hairs are destitute of the pale ring. The hairs on the ears are for the most part of a dirty yellow colour: the feet are pencilled with yellow and dusky, and on the inner side of the wrist (as is usual in this group) there is a space which is rather sparingly clothed with minute silvery white hairs. On the back of the neck the fur is very long.

	Maldonado.		Bolivia.		La Plata?		Domestic Guinea-Pig.		Domestic Guinea-Pig.	
	Ins.	Lines.	Ins.	Lines	Ins.	Lines.	Ins.	Lines.	Ins.	Lines.
Length	9	0	9	9	10	0	9	6	10	9
From nose to ear ...	2	3	2	1	2	0	2	1	2	2
Height of ear, about ...		5½		6		6		8		9
Width of ditto, about .		10	1	0		11	1	1½	1	1½
Length of fore foot and nails	1	0		11		10		10	1	0
" of hind foot and nails	1	11	1	8¾	1	9	1	8½	1	11

An old male, killed at Maldonado, weighed, according to Mr. Darwin, one pound three ounces.

A specimen in the British Museum collection agrees with the above description, excepting that a brown mark runs from the chin along the middle line of the throat, and joins the broad grey-brown band which crosses the chest: it is from Bolivia. A second specimen, in the same collection, differs in having the throat and abdomen of a pale brownish grey colour, very indistinctly suffused with dirty yellow.

It has been supposed that the Aperea is the wild race from which the Cobaya, or Domestic Guinea-Pig, sprang, but observing several differences in the structure of the skull and teeth of the two animals, Dr. Rengger arrives at an opposite conclusion, and considers that his view is further supported by the facts, that the Aperea, by choice, frequents damp situations, and can bear the cold with impunity, whilst the Cobaya, or Domestic Guinea-Pig, shuns both cold and moisture; that the wild animal produces but once in the year, and then has but one or two young, whilst the Cobaya has many young at a birth, and breeds two or three times in the year: that he could not succeed in pairing tamed individuals of the Aperea with the Guinea-Pig; and, lastly, that he had observed no differences in the colouring of the Aperea, even after having been kept in confinement for several generations[1].

Amongst other points of distinction, Dr. Rengger notices that in the *C. Aperea* the facial portion of the skull is more pointed in front than in the *C. Cobaya*, and that the nasal bones, in the former animal, differ from those of the latter in being pointed behind. In two out of three skulls of the *C. Aperea*, now before me, the terminal portion of the muzzle is distinctly narrower than the same part in four skulls of the *C. Cobaya*, but in a third skull of *C. Aperea* this difference

[1] Dr. Rengger observes that he saw fourteen tame Apereas, whilst on a journey to Villa Rica (in Paraguay), at the house of a countryman, who had reared five or six generations from a single couple, which he had captured when they were very young, seven years previously. They agreed with the wild animal in their colouring and mode of life, and, as in the wild condition, the females brought forth but once in the year, and then had but one or two young.

does not exist; here, however, as in the other skulls of *C. Aperea*, the nasal portion of the muzzle appears to be more elongated, and more attenuated than in *C. Cobaya*; this arises from the slope of the ridge, above the ant-orbital opening, being more gradual than in the skulls of the *C. Cobaya*. The hinder part of the nasal bones terminates in a point in the crania of *C. Aperea*, and are more or less truncated in those of Cobaya. Another difference which may be noticed is, that there is a greater extent of the petro-tympanic bone visible above the *meatus auditorius* in the Aperea than in the Cobaya[1]. Now, in the *C. Boliviensis* I found well-marked points of distinction in the skull and teeth, separating that animal from other Cavies; but though the skulls of different specimens of this species agreed amongst each other in possessing these peculiarities, they likewise differed, and the parts which were found to vary were precisely those in which dissimilarity of structure has been observed upon comparing the skulls of the *C. Aperea* and *C. Cobaya*; with this exception, that I have perceived no difference in the petro-tympanic bone analogous to that pointed out in the crania last referred to; I have, in fact, seen but two skulls of the Bolivian Cavy in which these bones were not wanting—the skulls being imperfect. In the form of the nasal bones they present the same kind of difference as that which distinguishes the *C. Aperea* from *C. Cobaya*, being sometimes truncated, or very nearly so, and sometimes prolonged in the mesial portion, behind, even to a greater degree than in the Aperea; and the ridge in front of the supra-orbital opening

[1] Dr. Rengger alludes to a difference in the curvature of the incisors, in the relative length and breadth of the molars, and in the colour of the cement of these teeth, as well as to a difference in the direction of the plane of the occiput (if I correctly interpret his expressions); moreover, he did not find the proportionate height of the lower jaw the same in the animals under consideration. In none of these points do I find any distinction in the skulls before me.

differs fully as much, and in the same manner, as has been pointed out in the skulls of the Domestic and Wild Guinea-Pig. On the other hand, if the skulls of *C. Aperea* and *C. Cobaya* be studied with reference to those parts which are little subject to variation in individuals of the same species of the genus *Cavia*, their points of resemblance become very striking. They agree most closely in the form of the zygomatic arch, in having no distinct angular process on its upper edge, the palate slightly concave, the palatine bone small, with the lateral or pterygoid branches externally narrow, especially opposite the last molar tooth; these branches do not extend the plane of the palate in the horizontal direction, as in many other species of Cavy, but slope into the post-palatal channel. The ant-orbital opening is entirely enclosed in front by a portion of the superior maxillary bone. Then, in connection with these points of agreement in the structure of the skull, we have to add that the animals themselves agree in size and proportion, and that a third, the *C. Cutleri,* has a skull presenting all the same general resemblances, and is intermediate between the skulls of *C. Cobaya* and *C. Aperea,* in certain points in which differences have been observed; having the broad and truncated nasal bones of the former, and the less abruptly projecting upper anterior root of the zygoma, of the latter; and, lastly, I shall be able to show that the structure of the molar teeth, which, as I have already noticed, presents peculiarities linking these three animals together, may be essentially the same as in these animals, and yet combined with some well-marked differences in the structure of the skull, and in other parts. Taking all these points into consideration, I have the conviction forced upon me, that the *C. Aperea, C. Cobaya,* and *C. Cutleri,* do not present peculiarities of structure which can justify our regarding them as distinct species.

Agouti. Guinea-Pig.

Cavia Cobaya of authors.

Colours—black, white, and rich fulvous yellow; or of two or one of these colours only.

The Common Guinea-pig[1], or Restless Cavy, has been engraved on the same block with an Agouti, and the cut will therefore serve to show the difference in the general form which exists in these animals. The former is known, through the paintings of Aldrorandus, to have attained the varied colouring, such as we now find it, within half a century of the discovery of South America, from whence, whatever doubts there may be regarding the original race of the species, it undoubtedly was brought, and hence it is supposed that the animal must have been previously domesticated by the

[1] Perhaps so called by mistake for *Guiana-Pig*, as it is by no means improbable the animal may have been originally brought from Guiana.

natives of America. I think it more probable, however, that some pretty variety [1]—some unusual condition presented by the little *C. Aperea*, so common in many parts of South America, —had attracted the attention of the earliest European settlers in the New World, and given rise to its capture and domestication, more especially as the harmless disposition and pretty colouring of the Common Guinea-Pig appear to be the only claims of interest which are attached to it, unless it really possesses the one useful property attributed to it, that of banishing the rats from its vicinity.

Cavia Cutleri. BENNETT.

General hue black, the sides and under parts of the body inclining to brown, especially the latter.

[1] Besides black or white varieties, which occasionally occur in wild animals, there is a yellowish or yellow-brown variety, which is not very rare. Amongst the Rodents, I have seen such a variety in the Beaver, the Muskwash, Mouse, in more than one species of Rat and Squirrel, in the Rabbit, and in the *Octodon Degus*. Since there are many grounds for believing in the existence of a very close affinity in the conditions which produce all such variations as those noticed in the Guinea-Pig, I can conceive it highly probable that the offspring of an animal presenting an abnormal condition in its colouring, would, if paired with one in the usual state, be subject to throw out other varieties, and, in fact, that from such a stock, variations like those presented by the Guinea-Pig, might in a short period be obtained, whilst, on the other hand, many generations might be reared from a pair in which there was no difference in colouring, without producing varieties. Whether my conjecture be well founded or not, I think that Dr. Rengger's opinion of the non-specific identity of the *C. Aperea* and *C. Cobaya*, derives very little support from that part of the argument which relates to the absence of variety in colouring in the former animal, for it is difficult to conceive that that species should be exempt from the effects of peculiar conditions which, in so many other species of animals, produce such varieties—species not confined to the class Mammalia; for it is impossible not to be struck with the great similarity which exists between the variations of colour in the domestic animal under consideration, and those of the Gold-fish, which may be summed up in the same terms, merely substituting "silver" for "white," and "golden red" for "fulvous-yellow." If certain conditions superinduce a similar change in animals so widely separated—if they operate so far in the animal scale—may they not extend to the highest?

The habitat of this animal is not known, but in Capt. King's note relating to the specimen upon which the species was founded, it is stated that the animal was known by the name of the *Peruvian Cavy.* In size and proportions it closely resembles the *C. Aperea*, but the fur differs from that of all other wild Cavies in being black, or nearly so—the hairs having no trace of a sub-terminal pale ring; on the back they are brown-grey at the root, and black externally; and those on the sides of the body are grey-brown, with a black point, and on the abdomen they are of a very deep brown.

	Inches.	Lines.
Length, about	10	0
" of ear		7
" of fore foot and nails	1	0
" of hind foot and nails	1	10

The proportions of this animal being essentially the same as those of the *C. Aperea*, and the skull presenting all the characteristics of that species, I cannot regard the colouring as furnishing good grounds for a specific distinction: in fact, I firmly believe it to be a black variety of the *C. Aperea.*

CAVIA (B. 1).

? *Cavia fulgida.* Wagler, in Isis, for 1831, p. 512.
? " " Wagner, Schreb. Säugth. Suppl. iv. p. 59.
? " *rufescens.* Lund. Det K. Danske, Vidensk. Selks., viii. 282.

Incisor teeth white: fur very long, and glossy; prevailing hue brown-black, obscurely and finely freckled with rich yellow-brown; abdomen brown-yellow.

Inhabits Minas Geraes, Brazil.

This is a small species—smaller than the *C. Cobaya*, and differs from other members of the genus, which have come under my notice, in being of a darker hue; the visible portions of the hairs on the upper parts of the body being black, with the exception of a very narrow rich yellow-brown ring, at some little distance below the point; the mixture producing a brown-black hue: at the root, the hairs, on the parts just mentioned, are grey, as are also those on the abdomen, but these latter are of a brown-yellow colour at the point. There is no pale mark near the eye, nor behind the ear. The fur is glossy and long, especially that on the hinder part of the back.

	Inches.	Lines.
Length	9	0
" from nose to ear, about	1	11
" of ear		5
Width of ditto		9
Length of fore foot and nails		11
" of hind foot and nails	1	10

The skull of the specimen from which my description is taken, and which is in the Museum of the Zoological Society, is imperfect, having lost the portions behind the zygomatic arch, but the remaining parts shew it to have been equal in size to the crania of the *C. Cobaya* or *C. Aperea*, and relatively rather broader. Viewing the upper surface, the most striking difference which presents itself, when compared with the skulls of the animals just mentioned, is in the form of the zygoma; this is less arched, and the skull is more distinctly widest if measured at the hinder root of the zygomata. The nasal bones are broad, and rounded behind. The upper edge of the zygoma is produced into a distinct and acute post-orbital process, which springs from the maxillary portion, immediately above its union with the malar bone. The palatine bone is larger, and the pterygoid processes extend the plane of the palate, and are fully double the width, measuring the palatal surfaces; and the post-palatal channel on which they encroach is proportionately narrow. The molar teeth are distinctly smaller in proportion to the skull; the hinder lobe of the last molar is smaller, and the fold of enamel, entering from the outer side of the tooth, which separates the lobe from the body, is deeper. A small longitudinal ridge of bone projects from the upper surface of the zygomatic process of the maxillary, and forms the outer boundary of a groove for the infra-orbital nerve: a similar ridge is found in the skulls of all the other species of *Cavia* I have examined, with the exception of *C. rupestris*, *C.* A. No. 1, and *C. Spixii*: it is wanting in the genera *Dolichotis* and *Hydrochœrus*.

Cavia fulgida. WAGLER.

According to Dr. Wagler's description, the *C. fulgida* has the same long glossy fur and white incisor teeth as my *Cavia* B. 1, but it would appear to be of a richer colour, having the pale annulations near the points of the hairs more distinct. I feel little doubt, however, that the two animals are the same. The upper parts and sides of the animal are said to be pencilled with rusty-red, and black; on the flanks, gradually passing into the rusty-ochreous tint which covers the whole of the abdomen and inner surface of the limbs. The hairs on the back are grey at the root: some of them entirely black beyond, but the greater portion, with pale rusty points. The ears are margined with black, and on the inner surface clothed with rusty red hairs. The hairs of the moustaches are dark brown, or some of them, pale yellow. Length, nine inches and a quarter.

Dr. Wagner's description is drawn up from the same specimen as that described in the Isis by Wagler: it was procured by Spix whilst on his journey on the Amazon River. The former of these authors says that Lund's *Cavia rufescens* is unquestionably of the same species, in which statement I agree, and my *Cavia* B. 1, being from the same district as Lund's *C. rufescens*, adds further grounds for supposing it to be a mere variety of the *C. fulgida*.

	C. Aperea. Maldonado, La Plata.		*C. Aperea.* Maldonado, La Plata.		*C. Aperea.* Maldonado, La Plata?		Domestic Guinea-Pig. Female.		Domestic Guinea-Pig.		Domestic Guinea-Pig.		*Cavia leucogaster.* Leyden Museum.		*Cavia Cutleri.*		*Cavia*, B. 1.	
	Ins.	Lines	Ins.	Lines.	Ins.	Lines.	Ins.	Lines.	Ins.	Lines.	Ins.	Lines.	Ins.	Lines.	Ins.	Lines.	Ins.	Lines.
Total length of skull	2	6	2	4¼	2	3¼	2	3⅔	2	5½	2	7½	2	6	2	7½		
Width	1	4	1	3⅓	1	2	1	3⅛	1	4¼	1	5	1	5¼	1	6	1	3¾
" between orbits		6¾		5⅔		5½		6		6		6½		6½		6¼		6
Length of nasal bones		10¼		9		9		9		9½				10¼		10¼		9
Width of ditto behind		4		3¾		3⅙		3¼		3⅔		3⅔		4¼		3¾		4
" " in front		4⅓		3⅔		3 4/6		4¼		4⅓		4½		4½		4⅓		4¼
From upper incisors to molar teeth ..		8¼		7⅔		7½		7¼		8¼		8⅓		8½		8¾		8
Length of four upper molars, taken together		7		7		6⅚		6¾		7		7		7⅔		7⅔		5¾
Width of lateral branch of palatine bone, opposite the last molar tooth ...		0⅓		0⅓		0⅓		0⅓		0⅓		0⅓				0½		0⅚
" of post-palatal emargination. ...		2⅙		2⅓				2½		3		3				2½		1¾
From outer angle of ant-orbital opening to the intermaxillary suture		8½		8¼		8		7		7⅔		8½				9		7¾
Length of zygomatic arch	1	1½	1	0¾	1	0¾	1	1⅔	1	1¼	1	2¼			1	3	1	0½
From lower edge of lachrymal bone to lower edge of zygoma		4¾		4⅔		4⅔		4¼		4¾		5				5¼		4½
Length of lower jaw	2	1	1	11½			1	10	1	10½			2	0½				
" of ditto, measuring from the hinder side of the condyle	1	7¼	1	6¾	1	6¼	1	6	1	7					1	8½	1	6
Height of ditto, from apex of condyle ...		7¼		7		7½		7¼		7¼				8		8½		6⅔
Length of angular portion		11⅓		11				10¾		9¼								

I must now notice one or two species of Cavy with which I am only acquainted from descriptions.

CAVIA LEUCOPYGA.

The White-rumped Cavy.

Cavia leucopyga. BRANDT, Mem. de l'Acad. de Petersb. (1835), iii. Pt. 2, p. 436, Pl. 16; Mammalium Exoticorum, p. 100.
" " WAGNER, Schreb. Säugth. Suppl. iv. p. 63.

Incisor teeth white: fur on the upper parts of the animal of mixed brown-black, and brownish yellow colours; throat, chest, and abdomen, white.

Inhabits Brazil.

The *C. leucopyga* is said to be equal in size to the *C. rupestris,* and therefore superior to the *C. Aperea;* its head is more elongated, and somewhat more convex, and the ears rather smaller, than in the latter animal. The hairs of the fur harsh (rather more so than in *C. Aperea*); those on the upper parts of the body of an ashy grey colour at the root, brownish black in the middle, black at the point, and with a subterminal ring of palish yellow, or inclining to white, with a faint brownish yellow tint: some of the hairs, however, want this pale ring. On the throat and under parts of the body the hairs are likewise greyish at the root, but their exposed portions are white, or yellowish white. The claws are shorter and of a less dark colour than in *C. Aperea.*

	Inches.	Lines.
Total length	11	0
From nose to ear	2	6
From ditto to the angle of the eye	1	3

To the above, which is extracted from Brandt's very detailed description, I may add, that in the figure which accompanies the account, the animal is represented as of a grey colour on the upper parts of the body, and that the figure greatly resembles the species I have described under the name of *Spixii*, from which it, however, may be distinguished by its incisor teeth being white, and by the absence of any white spot either behind the ear or the eye. The structure of the skull and molar teeth is not alluded to.

Cavia Cutleri. Tschudi.

Cavia Cutleri. **Tschudi, Fauna Peruana, Part 5, p. 195 (1845).**

Upper parts of the body blackish brown, pencilled with clay-yellow; each hair from the root to the middle blackish brown, with a broad whitish yellow ring, and a blackish point: muzzle, and the region immediately in front of the eye, somewhat brighter than other parts: outer surface of the ear thinly clothed with short blackish hairs, the inner surface with longer yellowish hairs: sides of the body of a brighter hue than the upper parts: throat, breast, and abdomen, as well as the inner surface of the limbs, of a bright clay-yellow, which colour is very rich (particularly in young individuals): feet somewhat darker, the soles blackish grey, and the nails black. Total length of the animal from 12 to 13 inches.

Inhabits Peru, on the western side of the Andes, in the neighbourhood of the town of Ica, South Lat. 14° 15′.

Dr. Tschudi, who was acquainted with but one wild species of Cavy in Peru, concludes that species to be identical with the animal described by Mr. Bennett (see p. 191 of this work), which is supposed (certainly not upon very satisfactory grounds) to be a native of Peru: the much larger size of Dr. Tschudi's animal, however, renders it highly probable that it is a distinct species. The difference in the colouring I think less important, for, as I have before stated, I strongly suspect the *C. Cutleri* of Mr. Bennett to be an accidental variety—in fact, a case of melanism. Dr. Tschudi states, with regard to the present species, that it is found in extraordinary numbers near the hedges and walls in the vicinity of the town of Ica, and that it makes its appearance early in the morning; and, in the evening, after sunset.

Dr. Lund, in his papers upon the Fossil Mammalian Remains found in the Caverns of Brazil, published in the Transactions of the Royal Danish Society[1], enumerates three recent species as occurring in the district of Minas Geraes, viz., *C. Aperea*, *C. rufescens*, and *C. saxatilis*; and four fossil species—*C. Apereoides*, *C. gracilis*, *Cavia* allied to *C. saxatilis*, and *C. bilobidens*. The characters of some of these are briefly noticed, and illustrated by figures representing the structure of the teeth.

Cavia saxatilis. Lund.

Judging from the figure of the last upper molar tooth (Pl. 25, fig. 18, in the 8th volume of the work quoted), I strongly suspect this animal will prove to be the same as my *Cavia* A. 1.

Cavia rufescens. Lund.

Said to be smaller than the *C. Aperea*; of a reddish colour above, with a yellow-grey abdomen, and white incisor teeth.

The last upper molar tooth, figured by Dr. Lund (Pl. 25, fig. 15), corresponds most closely with the same tooth in my

[1] Det Kongelige Danske Videnskabernes Selskabs Naturvidenskabelige og Mathematiske Afhandlinger.

Cavia B. 1: the indentation on the outer surface of that tooth, and which separates the small third lobe from the body of the tooth, as in my *Cavia* B. 1, is deeper and narrower than in *C. Aperea.* The colouring of Dr. Lund's animal would appear to be paler and brighter than in the specimen described in this work, and to which I have just referred.

Of the fossil species, the

Cavia bilobidens, LUND,

is evidently nearly allied to the *C. Boliviensis.* The molar teeth represented in Dr. Lund's Plate 21, fig. 6, and Pl. 25, fig. 17, agree very nearly with the corresponding teeth in the Bolivian animal: as in that species, the second lobe of the last upper molar has a distinctly greater transverse diameter than the first lobe; the third lobe of this tooth, however, is decidedly longer and narrower in the recent species than in the fossil.

Cavia Apereoides. LUND.

Whilst, according to my views, the preceding species would belong to the Cerodon section, the hinder molar tooth of *C. Apereoides* (Pl. 25, fig. 16, of the Danish Transactions) presents the essential structure of the typical Cavies, and, like the species of that section, it has the interspace between the lobes of the teeth filled with cement, there being little or no cement between the lobes of the molars in the Cerodons. Compared with the corresponding tooth in *C. Aperea*, this fossil tooth differs in having the front lobe relatively larger in the antero-posterior direction; the hindermost lobe longer, separated by a distinct notch on the inner side of the tooth, and with the apex curved somewhat inwards.

Genus, *Hydrochærus*[1].

Hydrochærus. BRISSON, Regnum Animale, p. 117 (1756).

Caviæ with the upper incisor teeth having a broad and shallow groove in front; the molars unequal in size; the three foremost of the upper jaw, each with two principal lobes, and the last, which is much larger than the rest, with many narrow transverse lobes (see Pl. 4, fig. 3 *b*): the two foremost molars of the lower jaw, each with three principal lobes; the third and fourth, each with four lobes, of which two, forming the middle part of the tooth, are simple, narrow, and transverse; the hindermost molar exceeding either of the three preceding teeth in size (see Pl. 4, fig. 3 *c*): feet short and broad, and semi-palmated; the toes terminating in broad and depressed nails: ears small.

HYDROCHŒRUS CAPYBARA.

The Capybara.

(*Plate* 5, fig. 2.)

Hydrochærus Capybara.	ERXLEBEN, Systema Regni Animalis, p. 193.
" "	DESM., Mamm. Part 2, p. 355.
" "	CUVIER, Règne Animal, ed. 1829, vol. i. p. 220.
" "	PR. VON NIEUW., Beitrage, ii. p. 475.
" "	RENGGER, Naturgesch. der Säugth. von Paraguay, p. 268.
" "	DARWIN, Voy. of H.M.S. Beagle — Mammalia, p. 91.
Cavia Capybara.	SCHREB., Säugth. iv. p. 620, Pl. 174.
Sus hydrochærus.	LINN., Syst. Nat., 12th ed. i. p. 103.
Cabiai	BUFFON, Hist. Nat., xii. p. 348, Pl. 49—50.

Head large, the muzzle obtuse; eyes small; ears likewise small, emarginated behind; body very stout, and clothed, rather scantily, with long and coarse hairs, of a rufous-brown colour on the upper parts of the animal, and dirty brownish yellow on the under parts.

[1] From ὕδωρ, water; and χοῖρος, a hog.

Inhabits the eastern portions of South America, from Guayana in the North, down to the Rio de la Plata in the South, and extends westwards into the lowlands of Peru and Bolivia.

As, in the class Mammalia, the largest known species are aquatic, so, in the minor divisions of the class, we find, *as a general rule*, the largest species have aquatic habits. Few species of the order Rodentia attain a size approaching to that of the Beaver or Coypu, whilst in the Capybara we have presented to us by far the largest species of the group—a gigantic Water Cavy; its stout body, scantily clothed with long and coarse hairs, gives to it a certain resemblance to the Pigs, with which, indeed, it has been arranged. Its head is large, remarkably flat on the upper surface, the muzzle broad and obtuse, the upper lip slightly notched above the incisor teeth, the eye small, placed considerably nearer to the ear than to the end of the nose, and protected above by distinct eye-lashes; the ears relatively small, rounded above, and emarginated behind; the limbs short and stout; the feet short, broad, and naked beneath; the toes joined by a web, which, however, does not extend to their extremities, the toe-nails short, broad, depressed, and by no means acutely pointed. A small horny protuberance represents the tail, and, according to Dr. Rengger, there is a small naked spot immediately beneath the nose in the male sex, which does not exist in the female[1]. The hairs covering the body are coarse and long (five or six inches in length on the hinder part fo the back in a large specimen in the British Museum), and are not sufficiently abundant completely to hide the skin. The colouring varies very little; in the large specimen just referred to, the hairs on the upper parts, and sides of the body, are of a rufous-brown colour, by no means dark; on the back many of the hairs are dusky at the point, and on the hinder parts they assume a rusty hue; the hairs on the under

[1] A specimen now living in the Menagerie of the Zoological Society has a large oil-secreting gland on the upper surface of the muzzle.

parts are of a pale dirty brownish yellow hue The hairs of the moustaches are dusky.

A second, and much smaller specimen, in the same collection, differs only in having the upper parts of the body more distinctly pencilled with dusky.

	Inches.	Lines.	Inches.	Lines.
Total length	51	0	44	0
From nose to ear	9	0	10	0
Length of ear	2	0	2	6
Width of ear	2	0	1	6
Length of fore foot and nails ...	5	9		
The nail of the middle toe		6		
Its width		7		
Length of hind foot and nails	9	0		
" of nail of middle toe ...	1	3		
Width of ditto	1	1		
Medium height	20	0	18	6

The dimensions in the first column are from a remarkably large specimen in the British Museum; those in the second column are furnished by Dr. Rengger, and are said to be from a fine male specimen. Residing in Paraguay for a considerable time, where the Capybara is common, Dr. Rengger no doubt took his dimensions from a recent specimen[1].

The skull of the Capybara (Plate 4, fig. 3) is remarkable for the flatness of its upper surface; the outline of this part, viewed from the side, presents very nearly a straight line from the tip of the nasal bones to the middle of the parietal bones, from which latter point it descends slightly to the occiput. The nasal bones are very broad, flat behind, and convex, in the transverse direction, on the fore part; their diameter is nearly the same at each extremity. The zygomatic arch is very deep, and differs from the corresponding part in the skulls of other Cavies, in presenting but little difference in the vertical diameter, when measured in front, beneath the orbit,

[1] The difference in the relative proportions of width and length in the ear perhaps arises from Dr. Rengger's having taken the width of the ear in a straight line, whereas my dimension gives the width as measured on the outer surface, and taking the curve.

or behind. A small, obtusely pointed, post-orbital process, rises from the malar bone: the lachrymal bone is large, and remarkable for the large and pointed process which projects outwards from its posterior upper angle. But, viewing the side of the skull, the most striking peculiarity is the enormous size of the mastoid process which is thrown downwards from the occipital bone. The auditory bullæ are proportionately smaller than in other Cavies. With regard to the palatine portion of the cranium, as that is represented on the plate, I will only call attention to the increased size of the palatine bone, as compared with that of other *Caviina*: here, this bone is much less deeply emarginated behind than usual. The lower jaw is represented (fig. 3 *a*), as viewed from the outer side: the horizontal protuberance on this part, although very prominent, is but slightly concave on the upper surface towards the fore part, and again towards the hinder part. The incisor teeth, which are white, are rather broader than deep, and have a broad and shallow channel on the anterior surface; those of the under jaw as well as those of the upper. The molar teeth, the grinding surfaces of which are represented in figs. 3 *b*, and 3 *c*, are remarkable for the large proportion of cement which enters into their composition, and the Capybara present a rare exception, amongst Rodents, in having the body of these teeth, in parts, completely traversed by the plates of enamel: in other Rodents the enamel almost always enters into the body of the tooth in the shape of folds or loops, but in the Capybara each tooth consists of two or more parts, composed of dentine enclosed in enamel, and these separate parts are united by the cement. The three foremost molars of the upper jaw (fig. 3 *b*) are each composed of two lobes separated by the cement, and each lobe has a deeply indenting fold of enamel on the outer side. The last upper molar has nine or ten simple, narrow plates, placed transversely, and joined to each other, and to an anterior, and posterior lobe, by layers of cement of corresponding width; the anterior lobe of the tooth resembles one of the lobes of the preceding molars, terminating in an acute angle on the inner side, and broad, and divided by a deeply indenting fold of enamel externally: the hinder lobe is composed of two transverse plates, which are joined on the outer surface of the tooth. In the tooth figured there are nine of the intermediate simple plates, which so strongly remind us of the plates of which the Elephant's tooth is composed, and which are united in a similar manner by cement; but in the skull of a young animal there exist ten of the simple plates in the hinder molar. In this same skull I found certain points in which the molars of the lower jaw do not agree with those represented in my plate (fig. 3 *c*); for instance, in the young skull the hindermost plates of the last molar are united on the inner side of the tooth, and the two foremost are joined on the outer side, and in the front molar the enamel is continuous, entering and returning on either side of the tooth without interruption. The teeth figured are those of an aged animal, and the differences just noticed arise from the circumstance that some portions of the outer surface of the teeth have been broken off.

The dimensions of the skull of the large animal from which my description is taken, are as follows:—

	Inches.	Lines.
Total length of skull	10	0
Width	5	5
" between orbits	2	9
Length of nasal bones	3	4
Width of ditto behind	2	0½
Ditto in front	1	8½
Length of the mastoid processes	2	2
Width of space between foremost molars ...		2
Ditto between hindermost molars	1	9
" of incisors	1	0
Length of four upper molars, taken together	3	1
" of the last upper molar	1	7

	Inches.	Lines.
Greatest width of the last upper molar ...		8
Total length of lower jaw	?	
Length of lower jaw to hinder part of condyloid portion	7	1
Antero-posterior diameter of condyloid process	1	4
Length of the four molar teeth, taken together	3	5
" of the last lower molar		11
Greatest width of ditto		$7\frac{1}{2}$

A specimen of the Capybara shot by Mr. Darwin at Monte Video, weighed ninety-eight pounds: its length, from the end of the snout to the stump-like tail, was three feet two inches, and its girth, three feet eight. "These great Rodents," says Mr. Darwin, "are generally called 'Carpinchos:' they occasionally frequent the islands in the mouth of the Plata, where the water is quite salt, but are more abundant on the borders of fresh-water lakes and rivers. Near Maldonado three or four generally live together. In the day-time they either lie among the aquatic plants, or openly feed on the turf plain. When viewed at a distance, from their manner of walking, as well as from the colour, they resemble Pigs; but when seated on their haunches, and attentively watching any object with one eye, they reassume the appearance of their congeners, the Cavies. These animals, at Maldonado, were very tame; by cautiously walking, I approached within three yards of four old ones. This tameness may probably be accounted for by the Jaguar having been banished for some years, and by the Guacho not thinking it worth his while to hunt them. As I approached nearer and nearer, they frequently made their peculiar noise, which is a low abrupt grunt, not having much actual sound, but rather arising from the sudden expulsion of the air: the only noise I know at all like it is the first hoarse bark of a large dog. Having watched the four, from almost within arm's length (and they me) for several minutes, they rushed into the water at full gallop, with the greatest impetuosity, and emitted at the same time their bark. After diving a short distance, they came again to the surface, but only just showed the upper parts of their heads. When the female is swimming in the water, and has young ones, they are said to sit on her back. These animals are easily killed in numbers, but their skins are of trifling value, and the meat is very indifferent. I never heard of the Carpincho being found south of the Plata, but as I see in a map that there is a Laguna del Carpincho high up the Rio Salado, I suppose such must have occurred. On the islands of Rio Parana they are exceedingly abundant, and afford the ordinary prey of the Jaguar[1]." In some parts of Paraguay they are said to be likewise very numerous, as many as twenty and more individuals being sometimes seen together. The female has from one to four young. Mr. Bridges met with the Capybara on the banks of the Mamorè, as well as in the vicinity of some other rivers of Bolivia; and, that it occurs in the rivers in the lowlands of Peru, we learn from Dr. Tschudi. It has long been known as a native of Brazil. Dr. Lund found amongst the fossil remains from the caverns of the last mentioned country, some of which are apparently referrible to the Capybara, and others which he regards as indicating the former existence of a second species, to which he gives the name *Hydrochœrus sulcidens*.

SUB-FAMILY, CHINCHILLINA.

Chinchillidæ. BENNETT, Transactions of the Zoological Society, vol. i. p. 58; Proceedings of the Zoological Society for May, 1833, Pt. 1, p. 57.

[1] Voyages of the Adventure and Beagle, vol. iii. being the Journal of Charles Darwin, M.A. Naturalist to the Beagle, p. 57.

Chinchillidæ. WATERH., Proc. Zool. Soc. for April, 1839, Pt. 7, p. 61; and for Nov. 1839, p. 172.

Chinchillina. WAGNER, Wiegm. Archiv. fur Naturgesch. 1841, p. 120; Schreb. Säugth. Suppl. iii. p. 299 (1843).

Hystricidæ with rootless molar teeth, having parallel, or nearly parallel, transverse plates of enamel; the series of molars on either side of each jaw converging in front; tail long, or of moderate length, recurved, and bushy. Clavicles perfect; slender: fibula very slender.

Inhabit the mountains of Peru and Chili, and one species occurs in the plains of La Plata.

The sub-family *Chinchillina* forms a well-marked group among the Rodentia, and has furnished the subject of several monographs, each adding materials towards its elucidation, but at the same time towards the confusion of the nomenclature of the species. This latter was to a certain extent unavoidable, since several papers upon the Chinchillas were published nearly at the same time, and at different widely separated places. But it is to Mr. Bennett that we are indebted not only for the first strict definition of the group, but likewise for the most important details relating to the species.

The Chinchillas present certain characters which link them to the Cavies, Octodons, and Hares; and we find likewise considerable approximation of one of the *Echimyina* (the Capromys) to the present group. In the shortness of the brain cavity, the general imperfection of the palatine portion of the cranium, the narrowness of the sphenoids, the large size of the openings in their lateral wings, as well as the large size of the occipital opening, the narrowness of the occipital condyles, and, in the fact, that a very considerable portion of the supra-occipital bone enters into the composition of the roof of the skull, we find characters which justify the approximation of the Chinchillas to the Hares. The palate, in the Chinchillas, is still more imperfect than in the Cavies, and the incisive openings are larger; and the parallel arrangement of the enamel plates of the molar teeth, combined with the greater height of the condyloid portion of the lower jaw, are points in which the Chinchillas evince a nearer affinity to the Hares than do the Cavies[1]. As, however, the Chinchillas and Hares are essentially of two very distinct types of the Rodent structure, and the characters just alluded to are for the most part characters indicative of a low grade of organization, it would seem that the amount of resemblance which exists between these two groups, or between the Cavies and the Hares, rather arises from these lowest members of the *Hystricidæ* being nearly parallel in rank, in the animal scale, with the *Leporidæ*, and does not indicate an affinity of a very near degree.

As regards the resemblances which the Chinchillas bear to the other groups mentioned, I will only here state that, in addition to those characters which are common to the *Hystricidæ*, the members of the present section show their nearer affinity to the Cavies, Octodons, and *Capromys*, than to others of the family, in their having the molar teeth rootless. The short and triangular form of the portion of the palate which lies between the molar teeth, the comparatively great space which separates the hinder molars, the curvature of these teeth, and in the form of the zygomata, are characters which link the Chinchillas to the Cavies more nearly than to the other sections just mentioned. Here, as in the Cavies, the zygoma has no descending process on its under surface, and its anterior root, in leaving the sides of the muzzle, is almost immediately directed backwards as well as outwards, and even has a slight tendency to descend, as in the Cavies[2]. On the

[1] In these remarks I allude more especially to the Chinchillas forming the sections *Lagidium* and *Chinchilla*; the Viscacha furnishes some exceptions, which will be pointed out.

[2] The zygoma, however, descends much more in the Cavies than in the

other hand, in the comparatively small size of the lachrymal bone, in having the ala of the sphenoid wedged in between the maxillary and the temporal bones, as well as in the form of the lower jaw, the Chinchillas depart from the Cavies, and resemble the Octodons.

The sub-family Chinchillina contains species fitted by their structure to live on the mountains, forming the genera *Chinchilla* and *Lagidium;* and one species, which is an inhabitant of the plains: it forms the genus *Lagostomus.*

Genus, *Lagostomus.*

Lagostomus. Brookes, in Linnean Transactions, vol. xvi. Part 1, p. 102. 1829.
" Bennett, Transactions of the Zoological Society, vol. i. p. 59.
" Wagner, Schreb. Säugth. Suppl. iii. p. 309.

Chinchillina with the molar teeth composed of two narrow lamellæ; the hindermost upper molar with three lamellæ; fore feet with four toes, armed with shortish, arched, and pointed nails; hind feet with three toes, provided with long, compressed, and sharply-pointed nails: upper lip with a vertical groove; muffle broad and expanded.

The genus *Lagostomus* is easily distinguished from the other *Chinchillina* by the reduced number of toes to its hind feet, and the comparatively long nails with which they are provided; the nails of the fore feet, moreover, although they are by no means large, are still considerably stronger and more produced than in the mountain Chinchillas. Another striking difference is observable in the structure of the molar teeth, these (with the exception of the hinder-

Chinchillas; indeed so much that its lower edge, in the middle of the arch, lies considerably below the level of the crowns of the molar teeth: in the Octodons the molar teeth descend far below the level of the zygomatic arch, and in the Chinchillas they descend but little below.

most molar of the upper jaw) having but two plates composed of dentine and surrounded by enamel. The three foremost molars of the upper jaw approach to a square form, but have the antero-posterior diameter about one-third less than the transverse one, and each tooth is divided into two very nearly equal parts by a transverse plate of enamel—at least such appears to be the case; but upon removing one of these teeth from its socket, and examining its base, it will be seen that this mesial plate is composed of two layers of enamel which enter from the inner side of the tooth, and, running parallel with each other, are not united till they reach the opposite side: here the small loop formed by their junction is met by a slight indentation of the enamel of the outer surface of the tooth. The hindermost upper molar differs from the rest in having three transverse laminæ, and in being rather longer than broad. The molars of the lower jaw resemble the foremost of those of the upper, excepting that their transverse diameter is relatively greater. All the molars are deeply implanted in the jaw, and distinctly curved; those of the upper jaw have the convex surface directed forwards and slightly inwards, and those of the lower jaw are reversed as regards the relative positions of the convex and concave parts. The incisor teeth are moderately broad, and have the depth and width nearly equal: their enamelled surface presents several indistinct, small, longitudinal furrows. The upper incisors are less strongly arched in front than usual, and the lower pair are but little curved as compared with the incisors of most other Rodents; they terminate behind, nearly beneath the hinder margin of the last molar.

The genus *Lagostomus* contains but one species—the Viscacha—and the remaining points of distinction are noticed in the description of that animal.

LAGOSTOMUS TRICHODACTYLUS.

The Viscacha.

Plate 5, fig. 1.

Lagostomus trichodactylus.	BROOKES, Linn. Trans. xvi. p. 95, Pl. 9.
" "	LESSON, Illustr. de Zool. Pl. 8.
" "	BENNETT, Trans. of the Zool. Soc. i. p. 60.
" "	WAGNER, Schreb. Säugth. Suppl. iii. p. 310.
Callomys Viscacia.	IS. GEOFF. et D'ORBIGNY, in Annales des Sci. Nat. for Nov. 1830, xxi. p. 291.
Lagostomus Viscacha.	MEYEN, Nov. Act. Bonn. xvi. p. 584.
Dipus maximus.	(DE BLAINV.) DESM. Nouv. Dict. d'Hist. Nat. xiii. p. 117; Mammalogie, Pt. 2, p. 315.
La Vizcache.	AZARA, Essais sur l'Hist. Nat. des Quadr. de Paraguay, ii. p. 41.
Marmot Diana.	GRIFF., Trans. of Cuvier's Animal Kingdom, iii. p. 170, with Plate.

Body stout; limbs powerful; tarsi long; ear nearly half as long as the head, broad at the base, narrow at the opposite extremity, being distinctly emarginated behind: fur soft, and moderately long; general hue of the upper parts of the animal grey, somewhat mottled with dusky, and distinctly pencilled with black; the whole of the under parts white, or yellow-white: a broad dusky, or black band, extends, on either side, from the muzzle to the back part of the cheek; an equally broad white band crosses the muzzle, and terminates on each side beneath the eye, and a third narrow band passes across the forehead, which is of a dusky hue: tail about half the length of the body, for the most part of a dusky brown or blackish colour, and clothed with long hairs on the upper surface.

Inhabits La Plata.

The Viscacha is well described by Azara in his work upon the Quadrupeds of Paraguay, but was so little known to the European naturalists that it was not recognized when a living specimen was brought to this country thirteen years after the appearance of that work. MM. De Blainville and F. Cuvier both made notes upon this specimen, which was exhibited in the Menagerie at Exeter Change, and the former of these authors gave to the animal the name *Dipus maximus,* supposing it to be a gigantic member of the Gerboa group. The same specimen is figured under the name *Marmot Diana,* in Griffith's Animal Kingdom. After its death it became the property of Mr. Brookes, by whom many important points in its structure (more especially relating to the teeth and skeleton) were for the first time made known to us. This able anatomist soon perceived that there were many striking points of distinction between the Viscacha and Gerboa, or, indeed, between that animal and any other then known Rodent; and, in his Memoir in the Linnean Transactions, founded upon it the new genus *Lagostomus,* and substituted the specific name *trichodactylus* for that of *maximus* previously given; the latter name not being applicable to the only known species of a genus.

The stout-bodied and strong-limbed Viscacha would, upon a cursory inspection, scarcely be regarded as a member of the same small group as the delicately made and active little Chinchilla, and yet in all essential points of structure these animals are most clearly framed on the same model, and the differences which they present remind us of the modifications of structure observable in the Squirrels and Marmots; the former light and active, and the latter of comparatively large size, strong and powerful, and fitted for burrowing habits: they are, in fact, burrowing Squirrels, whilst the Viscacha (which has been mistaken for a Marmot) is a burrowing Chinchilla.

The head of the Viscacha is large, and flattened above; the eyes large, and placed nearer to the ears than to the end of the muzzle; the ears are about half as long as the head, or rather less, and clothed (rather sparingly) with remarkably fine, and moderately long hairs; those springing from the anterior margin are very long, and form a fringe; they are some of them, of a pale grey-brown colour, whilst others are white, as are likewise the hairs on the inner surface, and on the hinder part of the outer surface of the ear; near the anterior margin externally, they are brownish. The terminal portion of the muzzle is very blunt, and the muffle is clothed throughout with velvet-like hairs of a brown colour; it is nearly twice as broad as high, and almost flat; presents no distinct mesial groove, but has two principal folds which diverge from the centre of the lower edge of the muffle, and external to these diverging grooves, are the long and narrow nostril openings, which take the same direction, and are capable of being closed at the will of the animal by a lappet of skin which lies above them. There is a vertical fold in the upper lip, extending from the muffle to the incisor teeth, but the upper lip cannot be said to be cleft in the same manner that we find it in the Hares—it is merely notched above the incisors; no part, that I can perceive, is destitute of hairs. The palms of the fore feet are naked, and of a flesh colour: a large semicircular pad at the base of the toes is separated by a transverse groove from a still larger pad behind, which is nearly of a triangular form. The fore toes are short and stout, and differ but little in length; the nails with which they are terminated are about a quarter of an inch in length, concave beneath, and arched and somewhat keeled above; like those of the hind feet, they are nearly white. The tibia is long, and so is the tarsus; the under surface of the latter presents a long, callous, flat pad behind, but the anterior half of the metatarsus is clothed with hair beneath; the toes are naked beneath, long, and powerful; the middle toe presents a remarkable swelling at the extremity, on the upper surface, and this is thickly beset with extremely stiff, curved, white bristles, which are for the most part about a quarter of an inch in length: similar bristles are observable at the base of the nail, and springing from the upper surface, of the inner toe. They are, no doubt, used by the animal to cleanse its fur. The nails are long, but little curved, solid, and keeled on the upper edge; the nail on the middle toe, which is the longest, is usually about three-quarters of an inch in length. The fur of the animal is soft, and moderately long, and varies somewhat in its colouring in the different specimens of the Viscacha which are found in our collections. The colour of the upper parts of the body is usually greyish; sometimes suffused with brown, and sometimes with a yellowish, or even a faint rufous cast, but always somewhat mottled with dusky, and more or less distinctly pencilled with black; the black pencilling is due to numerous long and comparatively coarse hairs mingled with those of the ordinary fur, the exposed ends of which are black. One of two specimens brought by Mr. Darwin from Buenos Ayres, has the whole of the under parts of the body of a delicate yellow, but in five other specimens which I have examined, these parts were pure white, as well as the inner surface of the limbs. The feet are usually white, but the tarsi are usually suffused with brown externally, becoming darker towards the lower and under part. A broad white band crosses the muzzle in the middle, and passes downwards and backwards on either side, to terminate beneath the eye; and immediately behind this band a dusky mark (more or less distinct) is observed to cross the forehead; it bends slightly forwards in the middle; above each eye is a distinct white spot, and numerous longish, but not very coarse, black hairs spring from this part. The hairs of the moustaches are numerous, and remarkably long and stout, some of them as much as six inches in length. The sides and under surface of the tail are clothed with coarse and shortish hairs, sometimes entirely of a dusky brown colour, but usually this tint is found on the under surface, and the sides are whitish: the upper surface of the tail is densely clothed (excepting near the root) with very long and somewhat coarse hairs: those at the extremity of the tail are the longest, measure from two to three inches in length, and form a large brush, or tuft.

Azara mentions, that a female Viscacha which he examined, differed from a male which he describes, in wanting certain coarse bristles on the cheeks; and I may mention that a specimen which died in the menagerie of the Zoological Society, and which was found to be a female upon dissection,[1] wanted the coarse bristles in question, and there are two specimens in the British Museum collection where they are likewise wanting. In these specimens there are a few coarse long black hairs on the cheeks, but these are flexible, and unlike the bristles which I observe in three other specimens, which it may therefore be presumed are males. Here broad black band is observed passing along the side of the muzzle, and terminating on the hinder part of the cheek, and this band is formed of long, and extremely stiff, black bristles, which rattle like the quills of the Porcupine when the hand is passed over them: they are directed downwards and backwards. The same specimens have scattered long black bristly hairs on the sides of the neck and across the shoulders; and they differ, moreover, from the specimens in which the cheek bristles are wanting, in having the chin and throat covered with spiny hairs, the same part in the other

[1] This animal was dissected by Prof. Owen, who communicated a paper on its anatomy to the Zoological Society, which will be found in the Proceedings of that Society for Decemher, 1839. The same specimen was examined by myself immediately after its death, and it is from my notes then made that many of the details of the above description have been extracted. In it I found but four mammæ, and the two on either side of the body were remarkable for their forward position, the first being an inch and a half behind the base of the fore leg; the second was about two inches behind this; they were very distinct, long, and slender. The dimensions of this specimen are given in the first column of admeasurement, which are seen in the next page.

individuals referred to, being clothed with exceedingly soft fur.

	Female. Zoolog. Soc.		Male? from Buenos Ayres.		Male? Zoolog. Soc.		Male? British Mus.		Female? British Mus.	
	Ins.	Lines.	Ins.	Lines.	Ins.	Lines.	Ins.	Lines.	Ins.	Lines.
From nose to root of tail	19	0[1]	26	0	23	0	24	0	23	0
" " to ear ...	4	3	5	0	4	9	4	3	4	3
Length of ear	1	10	1	9	1	9	1	9	1	9
" of tail, without the hair ...			8	0	7	0	8	0	6	0
" of ditto, including the hair					8	6	10	6	8	6
" of fore foot, without the nails			1	9	1	9	1	8	1	5
" of nail of longest toe				5		4½		3½		
" of hind foot, without the nails	4	5	4	8	4	8	4	10	4	5
" of nail of middle toe				9		9½		9		
Height at shoulder, about			9	0			8	0		

The skull of the Viscacha differs from that of the Mountain Chinchillas in the great strength of its parts, the less contracted form of the bodies of the sphenoids, as well as of that of the occipital bone, the smaller size of the incisive foramen, and much less inflated auditory chamber: here no part of the petro-tympanic bone is visible on the upper surface of the skull, which terminates behind in a well-developed occipital ridge, up to which the parietal bones are extended; the zygomatic arch is stouter, but less deep, especially beneath the fore part of the orbit. The malar bone here, as in

[1] Measured in a straight line; in the other dimensions, taken from stuffed specimens, the length of the animals is given as measured over to curve of the back.

the species of *Lagidium*, runs up in front to join the lachrymal bone. The nasal bones are broader than in the Mountain Chinchillas, and in the Viscacha's skull there is a bony plate projecting from the anterior root of the zygoma, and which forms the outer boundary of a tolerably deep channel for the infra-orbital nerve: this plate does not exist either in *Chinchilla* or *Lagidium*. The coronoid process of the lower jaw is more forward in its position and less elevated, and the ridge on the outer surface of the ramus, which runs obliquely downwards and forwards from the condyle, is very strongly developed; it is the same ridge being carried more forward, which gives such a peculiar form to the lower jaw of the Cavy.

	Female.		Sex unknown.		Male.	
	Ins.	Lines.	Ins.	Lines.	Ins.	Lines.
Total length of skull	4	0	4	2	4	7
" width near the hinder root of the zygoma	2	7	2	6	2	9
Width of the occiput	1	9	1	8½		
" between orbits	1	2	1	3		
Length of nasal bones	1	7	1	9	1	10
Greatest width of ditto, which is rather in front of the middle		8		8½		9
Width of ditto behind		5½		6½		6½
From incisors to molar teeth	1	3	1	3¼	1	5¾
Length of four upper molars taken together		11½		11¾	1	0
Width of upper incisor teeth		4½		4½		4½
" of palate between foremost molars ...		1¾		1½		2⅓
Ditto between the hindermost		8		8		9
Length of lower jaw			3	2½	3	7
Height of ditto, measuring from the condyle			1	3½	1	7

The following interesting account of the habits of the Viscacha, from Mr. Darwin's Journal, leaves little to be desired.

" The Viscacha is well known to form a prominent feature

in the zoology of the Pampas. It is found as far south as the Rio Negro, in lat. 41°, but not beyond. It cannot, like the Agouti (*Dolichotis Patachonica*), subsist on the gravelly and desert plains of Patagonia, but prefers a clayey or sandy soil, which produces a different and more abundant vegetation. Near Mendoza, at the foot of the Cordillera, it occurs in close neighbourhood with the allied Alpine species. It is a very curious circumstance in its geographical distribution, that it has never been seen, fortunately for the inhabitants, in Banda Oriental, to the eastward of the river Uraguay; yet in that province there are plains which appear admirably adapted to its habits. That river has formed an insuperable obstacle to its migration, although the broader barrier of the Parana has been passed, and the Viscacha is common in Entre Rios (the province between the two rivers), directly on the opposite shore of the Uraguay. Near Buenos Ayes these animals are exceedingly common. Their most favourite resort appears to be those parts of the plain which during one half of the year are covered with great thistles, to the exclusion of other plants. The Gauchos affirm that it lives on roots, which, from the great strength of its gnawing teeth, and the kind of localities frequented by it, seems probable. As in the case of the Rabbit, a few holes are commonly placed together. In the evening the Viscachas come out in numbers, and there quietly sit on their haunches. They are at such times very tame, and a man on horseback passing by, seems only to present an object for their grave contemplation. They do not wander far from their burrows. They run very awkwardly, and when hurrying out of danger, from their elevated tails and short front legs, much resemble great rats. Their flesh, when cooked, is very white and good, but it is seldom used.

"The Viscacha has one very singular habit; namely, dragging every hard object to the mouth of its burrow. Around each group of holes many bones of cattle, stones, thistlestalks, hard clumps of earth, dry dung, &c. are collected into a heap, which frequently amounts to as much as a wheelbarrow would contain. I was credibly informed, that a gentleman, when riding on a dark night, dropped his watch; he returned in the morning, and by searching in the neighbourhood of every Viscacha hole on the line of road, as he expected, soon found it. This habit of picking up whatever may be lying on the ground anywhere near its habitation, must cost much trouble[1]. For what purpose it is done I am quite unable to form even the most remote conjecture: it cannot be for defence, because the rubbish is chiefly placed above the mouth of the burrow, which enters the ground at a very small inclination."

The chief distinguishing characters of the alpine Chinchillas may be summed up as follows:—

Chinchillina with the molar teeth each composed of three narrow transverse lamellæ; ears large; tail long and bushy; tarsi entirely naked beneath; the nails of the toes short; auditory bullæ large.

This little group is confined to the Andes of Chili, Bolivia, and Peru, and the species which it contains are remarkable for the extreme softness of their fur. In connexion with the large size of the external organs of hearing, we find the bony chamber of the ear of unusually large size in these animals: the petrous, tympanic, and mastoid portions of the temporal bone here unite to form a large and irregular bulb, which not only increases the occipital surface of the cranium, but likewise, to a more or less degree, enters into the composition of the upper part. With regard to the molar teeth, I may here

[1] In the Satin-birds (*Ptilonorhyncus*) of Australia, Mr. Gould noticed a similar propensity to collect together various bright-coloured or glittering substances; these it arranges around a kind of bower, formed by sticks and other materials; for what purpose, is not known.

observe, that the foremost of the three narrow plates of which each tooth is composed (and which is formed of dentine and surrounded by enamel), is rather less extended in the transverse direction than the other two, in the grinders of the lower jaw; whilst, in the molars of the upper jaw, the hindermost plate is the shortest. Some slight differences are observable in the form of the last upper, and foremost lower molars, which are hereafter alluded to.

The fore limbs are small and weak; the hind limbs toes; and relatively strong. The fore feet have four or five long, the hind feet have four toes, of which the outermost is much shorter and smaller than the rest, and terminates at the commencement of the anterior third of the foot: of the three large toes, the central one is the longest, and the inner one the shortest. All the toes are terminated by a very large and somewhat compressed fleshy pad, above which the small nail is placed: the nails are rather strongly arched, pointed, and hollow beneath, and scarcely project beyond the fleshy portion of the toe: they are more or less hidden by longish hairs which spring from the end of the toe, and the nail of the inner toe of the hind foot is completely hidden by a tuft of very stiff and curved bristles. The naked soles of the feet are black: a large, hardened, flat pad, covers the hinder half of the tarsus.

To the three species of Alpine Chinchillina described by Mr. Bennett in his monograph of the group, in the Transactions of the Zoological Society, a fourth has been added—the *Eriomys Chinchilla* of Lichtenstein, which there are grounds to believe is a good species. They agree so closely together in most of the details of their structure, that it would have been well to have associated them under one generic title, and to have formed sub-generic divisions only, upon the distinctions furnished by the number of the toes of the fore feet. We then could have distinguished, by the rank of the divisions, the greater amount of difference which characterizes the genus *Lagostomus*, from the minor distinctions which separate *Lagidium* from *Chinchilla*—two sections which have been formed of the Alpine *Chinchillina*, and which are regarded as genera.

Genus, *Lagidium*.

Lagidium. MEYEN, Nova Acta Acad. Nat. Cur. Bonn. vol. xvi. Part 2, p. 576, March, 1833.
" WAGNER, Schreb. Säugth. Supp. iii. p. 305.
Lagotis. BENNETT, Proceedings of the Zool. Soc for May, 1833, Part 1, p. 58.—Trans. Zool. Soc. vol. i. p. 59.

Alpine Chinchillas with four toes to the fore feet ; ears elongated.

LAGIDIUM CUVIERI.

Cuvier's Chinchilla.

Plate 7, fig. 1.[1]

Lagotis Cuvieri.	BENNETT, Proc. Zool. Soc. for May, 1833 ; Trans. Zool. Soc. vol. i. p. 46.
Lagidium Cuvieri.	WAGNER, Schreb. Säugth. Suppl. iii. p. 306 (1843).
? *Lagidium Peruanum.*	MEYEN, Nova Acata. Acad. Nat. Cur. Bonn. xvi. Part 2, p. 578.
" "	TSCHUDI, Fauna Peruana, Pt. 4, p. 164 (1845).
Callomys aureus.	IS. GEOFF. et D'ORBIGNY, Ann. des. Sci. Nat. xxi. p. 291 (1830).

Ears rather shorter than the head ; fur very soft, its general hue ashy grey, suffused with yellowish, and somewhat pencilled with black ; under parts of a very pale yellow, but somewhat richer towards the sides of the body : feet grey ; tail black beneath, and with very long hairs above, some of which are black : a narrowish black line runs along the back, but does not extend to the root of the tail.

Inhabits the Andes of Chili, Bolivia, and Peru.

[1] *Lagidium Peruanum* on the Plate.

The animal described by Mr. Bennett under the name *Lagotis Cuvieri*, is smaller than most other specimens I have examined, and which appear to me to be referrible to the same species. Its general tint may be described as ashy grey, but with a mixture of dirty yellow, and somewhat pencilled and mottled with dusky: under parts of the animal pale yellow; towards the sides of the body, and root of the tail, of a richer yellow hue; sides of muzzle, palish yellow; cheeks suffused with yellow. The ears are of an elongated oval form, scantily clothed with grey-white hairs, but on the outer surface, towards the tip, as well as near the anterior margin, somewhat dusky. The hairs of the moustaches stout, and immensely long, the longest measuring as much as seven inches; they are entirely black. The feet are densely clothed with hairs, the general hue of which is grey, but the hairs are brownish at the root, and most of them very pale at the point; the strong bristles on the inner toe of the hind foot are black: the nail beneath these bristles differs from the nails of the other toes in being broad and truncated. The tail, at the root, is clothed with fur like that on the body; this, however, extends but a short distance—about an inch and a half; the sides, and under surface of the remaining portions, are well covered with stiff hairs of about half an inch in length; those on the under part of the tail are black, and those on the side for the most part dirty white; the long bushy hairs are confined to the upper surface of the tail, and gradually increase in length as they approach the point of the organ, where their average length is about three inches; along the mesial line of the tail, and at the point, they are almost entirely black, but towards the sides, the greater portion of them are dirty white. The hairs on the back, taken singly, are pale grey at the root, dirty yellow-white at the point, and grey-brown below this pale part; many of them are slightly tipped with dusky, and there are longer interspersed hairs which have the visible extremity black. Along the mesial line of the back, all the hairs are black at the point, and hence there is a black dorsal line ; this is about half an inch in width, and is gradually obliterated towards the hinder part of the back. The hairs on the under part of the body are ashy grey at the root, and pale yellow at the point.

	Ins.	Lines.
Length from tip of nose to root of tail ...	16	6
" from ditto to ear, about	3	2
" of ear	2	8
Greatest width of ditto	1	10
Length of tail, not including the hair, about	11	0

The dimensions of the skull will be found with those of other crania about to be noticed. The total length of the skeleton, according to Mr. Bennett, is twenty-four inches and six-tenths; of which the caudal vertebræ occupy one foot and $\frac{4}{10}$ths of an inch; the length of the fore foot to the end of the middle finger is $1\frac{4}{10}$ths of an inch; of the hind foot, $3\frac{6}{10}$ths inches.

The locality from which this specimen was obtained is not known; it was brought alive to this country, and lived for a considerable time in the menagerie of the Zoological Society.

Amongst numerous specimens of *Lagidium* brought to this country by Mr. Bridges were four, agreeing very nearly with Mr. Bennett's animal, having the same yellow-grey hue: one of these (now in the collection of the British Museum) was procured on the east side of the Andes, between Villavicencia and Uspallata, at an elevation of between four and five thousand feet[1]; the remaining three were shot in Bolivia, in the

[1] According to the various accounts of the habits of these animals, which I have read, it is very unusual to meet with them at so small an elevation. This specimen may be described as having the general colour palish ashy grey, slightly suffused with yellow on the sides of the body, and somewhat brownish on the back: the dorsal stripe is dusky brown: the sides of the muzzle, lower part of cheeks, the fore legs, and the whole of the under parts, of a distinct

vicinity of Potosi. Other specimens in Mr. Bridges' collection were procured, as that gentleman informs us, to the west of La Paz, on the east side of the Andes, at an elevation of perhaps fifteen thousand feet. They differ considerably in their colouring from the animal described by Mr. Bennett, and are, moreover, of large size. The following description is from one of these specimens, now in the British Museum.

Incisor teeth white: hairs of the fur about an inch in length on the back; rather longer on the hinder parts; its general hue on the upper parts of the animal, bluish grey; the hairs pale grey at the root, gradually shaded into brownish grey externally, and grey-white at the point; the longer interspersed hairs, black externally: chin, throat, chest, and abdomen, as well as the under side of the tail at the root, and the inner surface of the limbs, of a bright yellow colour. Ears rather sparingly clothed with grey-white hairs; or, near the fore part, and at the tip, externally, with brown-black hairs. Fore feet nearly white, but with the longish hairs springing from the ends of the toes, dusky; hind feet ashy grey, rather dark on the outer side, and pale on the inner; the longer hairs at the end of the toes, as well as the stiff bristles which spring from the extremity of the inner toe, blackish. Tail as long as the body, and very bushy; the long hairs, however, confined to the upper surface; at the end of the tail, where they are longest, measuring four inches and upwards in length; they are some of them dirty white and

yellow; the fore feet almost white; the hind feet greyish, but with the inner surface, as well as the heel externally, dirty pale yellow: ears, externally, of a pale brown-ash colour, but with a narrowish white band along the anterior margin, and a similar one at the hinder margin, the tip somewhat rufous: a large rufous patch on the crown of the head: tail brown-black beneath; the long hairs on the upper surface are some of them brown-black, with whitish points, but very many of them are dirty white throughout.

some black, on the mesial line of the tail; chiefly dirty white near the sides, and black at the end of the tail; the shorter hairs on the under surface, black. Hairs of moustaches, black. A black mark runs along the back, commencing in a line with the shoulders, and terminating over the loins.

	Brit. Museum.		Zool. Society.		Yellow-grey variety. Brit. Museum.	
	Ins.	Lines.	Ins.	Lines.	Ins.	Lines.
Length from tip of nose to root of tail ...	19	0	20	6	16	6
" from ditto to ear	3	8	3	7	3	4
" of ear	3	0	2	11	2	8
Greatest width of ditto	1	11	2	0	1	8
Length of fore foot	1	8	1	10	1	7
" of hind foot	4	2	4	4	3	7
" of tail, without the hair	12	6	12	0	12	0
" of ditto, including the hair, about	16	6	16	0	16	0

I have added the dimensions of a second specimen, from Mr. Bridges' collection, contained in the Zoological Society's museum; it agrees with the specimen just described in its colouring. The dimensions of the skull of the animal last referred to are given below, in the second column of admeasurements; those in the first being from the animal described by Mr. Bennett, and those in the third column are from a skull in the British Museum, which was removed from one of Mr. Bridges' specimens, which agrees with the above description.

	Yellow-grey variety. Mr. Bennett's specimen.		Blue-grey variety. British Museum.		Blue-grey variety. Zoolog. Society.	
	Ins.	Lines.	Ins.	Lines.	Ins.	Lines.
Total length of skull	3	1	3	8	3	5
Width at the occiput	1	2⅔	1	4¼	1	3⅔
" measured from outer surface of zygomatic arches	1	8	1	9½	1	9¾
" between orbits		10		9½		9½
Length of nasal bones	1	0½	1	6½	1	4
Width of ditto in front		4⅓		5¼		5⅓
" " behind		3¾		4¼		4
Space between upper incisors and molar teeth		10¾	1	0½	1	0
Length of the four upper molars taken together		8½		10		9⅓
Width of incisors		2⅔		3		2¾
Length of lower jaw	2	1½	2	8	2	4⅓
Height, at the condyle		10⅓	1	0	1	0¼
Length of angular portion	1	0	1	3	1	1½

A notice of the habits of this animal will be found in the account of the *Lagidium pallipes*.

Lagidium Peruanum. Meyen.

This animal is usually supposed to be identical with Mr. Bennett's *Lagotis Cuvieri*, but I have never seen a specimen of the *L. Cuvieri* in which the under parts of the body are white, as described by Dr. Meyen, and in which the dorsal black stripe is wanting, as would appear to be the case in the *L. Peruanum*, since it is not mentioned in the account of that animal. The stripe is sometimes wanting in the *L. pallipes*, and this leads me to suppose the *L. Peruanum* may be of the same species as that animal, which is more generally diffused in Peru than the *L. Cuvieri*: moreover, the length of the *L. Peruanum*, according to Dr. Meyen, is about sixteen inches, which is inferior to that usually found in *L. Cuvieri*, and more nearly accords with *L. pallipes*. Unfortunately the length of the ears is not mentioned in Dr. Meyen's description.

Callomys aureus. Is. Geoff. et D'Orb.

Founded upon a skin in which there were neither ears, feet, nor tail. The fur is said to be yellow, with a greenish cast on the back of the animal; slightly undulated with black, and with a black dorsal stripe: the under parts golden yellow. Length, about twenty inches. It is most probably a variety of the *Lagidium Cuvieri*.

LAGIDIUM PALLIPES.

The Pale-footed Chinchilla.

Lagotis pallipes. Bennett, Proceedings of the Zoological Society for May 1835, Part 3, p. 67; Trans. Zool. Soc. Vol. 1, p. 331, Pl. 42.
Lagidium pallipes. Wagner, Schreb. Säugth. Suppl. iii. p. 308 (1843).
" " Tschudi, Fauna Peruana, Part 4, p. 165 (1845).

Ears about one-third shorter than the head; fur on the upper parts of the animal of an ashy grey colour, much suffused with dusky brown; the whole of the under parts of a fawn-yellow, as well as the feet and inner side of the limbs; ears dusky brownish in front, externally; tail, with the long hairs springing from the mesial line of the upper surface, of a dusky brown colour at the root, and pale rufous brown at the point.

Inhabits the Bolivian and Peruvian Andes, extending northwards to the Republic of Ecuador.

The animal upon which Mr. Bennett founded the *Lagotis pallipes*, differs from the *L. Cuvieri* in being rather smaller, and in having the ears relatively shorter. The general hue of its fur, moreover, is darker; arising from the longer dark-pointed hairs, mixed with those of the ordinary fur, being much more numerous, and likewise from the circumstance that the hairs of the ordinary fur (which are shorter than those of *L. Cuvieri*) being many of them dusky at the point. On the back is a trace of a dark dorsal stripe; but it is so indistinct, that it would scarcely be perceived, were we not led to look for it on account of the presence of such a mark in the *L. Cuvieri*. The fur of the back is grey next the skin, and each hair is of a dirty yellow-white colour at or near the point; and, immediately below the pale ring, the hairs are somewhat darker than at the root. The hairs of the throat, chest, and abdomen, are of an ashy grey colour at the root, and palish fawn-yellow externally; the yellow assumes a richer hue on the belly, and on the fore and hinder part of the fore legs. The feet are yellowish white; the tarsi slightly mottled with brownish on the outer surface. The sides of the head and neck are pale ashy grey; the lower part of the cheeks, and the sides of the muzzle, yellow. The ears are rather sparingly clothed with grey-white hairs internally, and likewise on the hinder part externally, but, on the fore part, they are dusky brown. The hairs of the moustaches are, for the most part, black; but some few are brown at the root and whitish at the point. The tail is clothed with somewhat coarse and short hairs at the sides and on the under surface; those at the sides are of a dirty yellow-white colour, but dusky brown at the root; and those on the under part are blackish brown. The long hairs on the upper surface are, many of them, of a dirty yellow-white colour; whilst others (chiefly confined to the mesial line) are blackish brown at the root, and rufous brown beyond; the longest of these hairs are about two and a half inches in length; these spring from the end of the tail, and are entirely black. The bristly hairs which cover the extremity of the inner toe of the hind foot are silvery white. The feet are less densely clothed with fur than in *L. Cuvieri.*

	Inches.	Lines.
Length from tip of nose to root of tail ...	14	0
" from ditto to ear	3	0
" of ear	2	2
Width of ditto	1	5
Length of fore foot	1	2½
" of hind foot	3	0
" of the tail, not including the hair, about	9	6

Mr. Bennett's specimen of the *Lagotis pallipes* is supposed to have come from the Chilian Andes. An animal now in the British Museum, and which was procured by Mr. Bridges in the Bolivian Andes, in the neighbourhood of La Paz, at an elevation of about twelve thousand feet, I feel little doubt is specifically identical with Mr. Bennett's species just referred to, since it has the same comparatively short ears; but it differs in several respects from the specimen I have just described. The general colour of the upper and under parts of the body is the same, or very nearly so; but the fur is somewhat longer, its average length being full three-quarters of an inch; whilst that in Mr. Bennett's specimen is about half an inch. On the back is a dusky stripe; but this is less strongly defined than in the *L. Cuvieri.* The hind feet are dusky grey externally, and whitish on the inner side; and the bristles on the inner toe, instead of being silvery white, as in Mr. Bennett's specimen, are dusky. The ears are, for the most part, dusky on the outer side, but there is a narrowish white band along the fore margin, and a second at the hinder margin. The tail is covered at its base, for a distance from the root of about two inches, with soft fur like that on the body, as is also the case in *L. Cuvieri;* but in that animal this part of the tail is yellow beneath like the body, whilst here the soft fur on the tail is entirely grey:

this soft hair is followed by coarser hairs, and these at first are entirely black, the black being extended in the longitudinal direction about two inches; the hairs then begin to lengthen considerably on the upper surface of the tail, and increase in length to the point, where they are about three inches long; they are brown (some of them inclining to white, at the point), and black-brown at the root: the shorter hairs on the under surface of the tail are brown-black.

	British Museum.		*Lagotis pallipes.* From Dr. Tschudi.		*Lagotis Cuvieri (Peruana).* From Dr. Tschudi.	
	Ins.	Lines.	Ins.	Lines.	Ins.	Lines.
Length, from tip of nose to root of tail ...	18	0	14	0[1]	15	0
" from ditto to ear	3	4				
" of ear	2	2	2	3	3	0
Width of ditto	1	3	1	3	1	4
Length of fore foot	1	3				
" of hind foot	3	9	3	4		
" of tail, not including the hair, about	11	6	10	0	11	0
" of ditto, including the hair ...	15	6	13	0	14	0

Dr. Tschudi, whose observations were made in the native country of the Chinchillas, or Viscachas, as the species of Lagotis are called (as well as the animal which inhabits the plains of La Plata), is quite satisfied that the present species is distinct from the *L. Cuvieri,* and notices the shorter ears and moustaches, the darker colour of the fur and its being tinted with yellow brown, and having numerous black waves, combined with the rufous feet, and distinctly distichous tail,

[1] I have not reduced Dr. Tschudi's dimensions to English measure, for I do not know what measure he uses; he may have stated in his work, but that I am unable to refer to.

as its principal distinguishing characters. Dr. Tschudi, moreover, states that the geographical range of the two species is different, and affords another proof of their specific distinction. Whilst the *L. Cuvieri* occurs only on the South Peruvian and Bolivian high lands, the *L. pallipes*, according to the author just mentioned, extends over the Puna and Cordillera regions of Middle and Northern Peru, as far as the Republic of Ecuador. The Puna and Cordillera regions[1], are alone inhabited by the Viscachas (*Lagotis*), which are usually met with at a height of from 12 to 16,000 feet above the sea; lower than 11,000 feet they rarely descend, and when seen as low as 9 or 10,000 feet, it is only in pairs or single individuals. The highest point at which the Viscachas were noticed by our informant, was at 15,800 feet of elevation, on the eastern slope of the Huijracocha[2]. In the province of Jauja (nearly in the latitude of Lima), Dr. Tschudi met with several Viscachas, which were entirely black: he was unable to procure a specimen, but is of opinion they were mere varieties of the *L. pallipes*, which was met with in abundance not many thousand paces from the spot in which the black specimens were seen.

The Viscachas commonly live in communities on the naked rocks, where (to use Dr. Tschudi's own words) a dry and scanty vegetation is scarcely to be seen. They quit their abodes, which are amongst the loose pieces of rock, shortly after sunrise and rather before sunset, and then display the most extraordinary activity, leaping from rock to rock; in a moment, as it appears, they may be seen to ascend an almost precipitous rock of twenty or thirty feet in height, and if fired at, they disappear as if by magic, having retreated to the

[1] The Puna region is the high table-land of about 12,000 feet of elevation, and the Cordillera region is on the ridges of the Andes themselves.

[2] I cannot find this place in the maps—perhaps it should be Huarmicocha, which is on the Andes, S. lat. about 13° 25′.

holes and crevices. The sunshine they scrupulously avoid, and it is therefore only in shaded spots they are seen, unless the day be cloudy. Their food consists of grasses, dry roots and mosses, and to procure this, they often have to wander far from their homes. Whilst eating they sit upon their haunches with the tail extended, and the food is held in the fore paws. They have many enemies, but appear to be very prolific, and Dr. Tschudi says he has met with the young at all periods of the year. The birds of prey (especially the condor) and foxes, as well as the shepherds' dogs, are their principal enemies. The Indians, by means of snares, capture large numbers of the Viscachas, both for their skins and flesh, the latter being well-flavoured; and the former is deprived of its fine and soft fur, which is mixed with wool, and woven into warm scarfs, &c., so much needed in the icy cold regions inhabited by these animals. The furs are moreover shipped to Europe, but are less prized than that of the true Chinchilla.

Genus, *Chinchilla*.[1]

Chinchilla. Bennett, Gardens and Menagerie of the Zoological Society, Part I. p. 1 (1829): Transactions of the Zoological Society, Vol. i. p. 59.
" Gray, Spicilegia Zoologica, Part II. p. 11, August 1830.
" Rousseau, Annales des Sciences Naturelles, xxvi. p. 337 (1832).
Eriomys. Lichtenst, Darstellung neuer oder wenig bekannter Säugth. Tab. 28, 1829?
" Van der Hoeven, in Bijdragen tot de Naturkundige Wetenschappen, vi. p. 115, 1831.
" Wagner, Schreb. Säugth. Suppl. iii. p. 301 (1843).

Alpine Chinchillas, with large and rounded ears; five toes to the fore foot, and four to the hind.

The teeth of the species of *Chinchilla* can scarcely be said to differ from those of *Lagidium*, the only distinction

[1] The names *Chinchilla* and *Eriomys* were applied to the present genus, it appears,

being, that whilst the alternate plates of enamel and dentine in *Lagidium* are rather distinctly curved, they are straight in *Chinchilla*, excepting it be that the second plate of the upper molar is somewhat curved backwards on the inner side of the tooth; and again, the hindermost lobe of the last molar in *Chinchilla* is less narrow than in *Lagidium*, being of a somewhat irregular oval form; it has a slight indentation on the outer side. The foremost lobe of the front lower molar is not completely separated from the body of the tooth by a transverse plate of enamel, as in *Lagidium;* the tooth, indeed, might be described as having but two lobes; one, narrow and transverse, forming the hinder part of the tooth; and the second, likewise transverse, but with the antero-posterior diameter broader than in the first, and with an indenting fold of enamel on the inner side, which, though deep, does not reach the opposite side of the tooth.[1] This same tooth is relatively smaller than in *Lagidium*. The small and delicate fore foot of the Chinchilla has five perfect toes, each provided with a strongly arched and pointed nail, which is concave beneath, and in the living animal certainly cannot project beyond the large fleshy pad with which the toes are terminated: the

nearly at the same time; but considering that the former name was accompanied by a very complete account of the characters of the animal to which it was applied, and the latter by one which left the dentition unnoticed, and was incorrect in an important particular—the number of the toes to the feet, I feel I am justified in giving the preference to the name *Chinchilla*. The date of the Part of Lichtenstein's work containing the account of the Chinchilla is given as 1829 by several authors, but the only copy which I can refer to bears the date 1831 on the cover: the cover for the first part of the work in question is used for the remaining parts, but altered in ink, as regards the date and number of the part. The first of the part of "The Gardens and Menagerie" is quoted by Van der Hoeven as bearing the date October, 1829. The volume is dated, as usual, from the time of its completion (1830).

[1] Possibly a transverse section of the foremost lower molar of *Lagotis* would, when placed under the microscope, show that the difference consists only in the indenting fold on the inner side of the tooth of that animal being deeper than in *Chinchilla*.

inner toe, or thumb, is very small, scarcely extending beyond the base of the metatarsal bone of the second toe.[1] The three large toes of the hind foot are each provided with a small tuft of bristles at the extremity, but those on the inner toe are the longest and most stiff.

[1] In the skeleton of a Chinchilla now before me, I find the wrist bones to present the ordinary number, but the *os scaphoides* and *os lunare* are joined to form one bone, whilst the normal number is made up by the presence of an extra small bone wedged in between this and the *os magnum*. The ulna and radius are in close contact for their whole length, but not anchylosed, as Mr. Brookes found the corresponding bones in the Lagostomus.

CHINCHILLA LANIGERA.

The Smaller Chinchilla.

Chinchilla lanigera.	BENNETT, Gardens and Menagerie of the Zoological Society, Part I. p. 1; Transactions of the Zoological Society, Vol. i. p. 59.
" "	GRAY, Spicilegia Zoologica, Part II. p. 11.
Cricetus laniger.	DESM. Mamm. Part II. p. 313.
Callomys laniger.	Is. GEOFF. et D'Orb. Ann. des Sci. Nat. p. 291.
Mus laniger.	MOLINA, Saggio sulla Storia Naturale del Chila, p. 267, (1782).
Chinchille.	ACOSTA, Hist. Nat. des Indes Occid. p. 199.
Chinchilla.	F. CUV. et GEOFF. Mammifères, livr. 64.

Ears broad, and equal in length to about two-thirds the length of the head; incisor teeth of a bright orange colour in front; fur about an inch in length, and extremely soft: it general hue grey, but strongly mottled with dusky, or black; under parts, of an impure yellow-white; feet dirty white: tail bushy, for the most part black beneath (where the hairs are short);

black likewise above, along the mesial line, as well as at the point; and dirty white at the sides.

Inhabits the Andes of Chili, Bolivia, [and Peru; extending northwards, on the east side of the mountains, as far as the 9th parallel of latitude, and having a vertical range of from about eight to twelve thousand feet[1]].

Although the furs of the Chinchillas have long formed an important article of commerce, the animal itself was almost unknown to naturalists until the year 1829, in the early part of which, Mr. Yarrell first pointed out the chief characteristics of its dentition[2], which, however, being examined under unfavourable circumstances, left to be desired a more detailed description. In the same year the animal was figured and described by Lichtenstein, and by Mr. Bennett, in his "Gardens and Menagerie," where the structure of the teeth, as well as the external peculiarities of the animal, are most fully described. The animal was again figured, and described in detail, by Mr. Gray, in the following year[3]; and Mr. Yarrell's notes on the internal anatomy of the Chinchilla, which appeared in the Proceedings of the Zoological Society for February 1831, has rendered our knowledge of the structure of the animal almost complete. The skull has been figured and described by Van der Hoeven, and subsequently by M. Rousseau, in the works quoted at the head of this account, and to Mr. Bennett we are indebted for a figure of the skeleton, stomach, and cœcum[4].

Dr. Tschudi informs us that Chinchillas are plentiful on

[1] It is possible that the remarks on the habitat and range, enclosed within brackets, refer to the next species.

[2] See the Zoological Journal for January, 1829, Vol. iv. p. 317.

[3] In his Spicilegia.

[4] See Plates 6 and 7 of Vol. 1 of the Transactions of the Zoological Society. The sub-family *Chinchillina*, scarcely known to Cuvier at the time of the publication of his last edition of the Règne Animal, is now more perfectly worked out than any other group of Rodents. Those who require further details than it is desirable to introduce into a general work, will find a full account of the internal and external peculiarities of two of the genera of the group in the

the east declivity of the Cordillera throughout Peru up to the 9th parallel of latitude, but he is not aware that they occur on the inner Cordilleras. Further south, Mr. Bridges met with the species here described both on the Bolivian and Chilian sides of the Cordillera; and I learn from this gentleman that the animal is very plentiful at Cobija. In its habits it greatly resembles the species of the preceding genus. The native hunters, according to Mr. Bridges, train a species of weasel (the Quique of Molina—*Galictis vittata* of Bell) to enter into the crevices of the rocks, where the Chinchillas lie hidden during the day, and capture them as they endeavour to make their escape.

The *Chinchilla lanigera* is about equal in size to our common squirrel, but appears to have a stouter form, owing to the greater length and density of its fur. Its muzzle is somewhat pointed; the eye large: the ears very large, with the length and breadth nearly equal; they are rounded at the extremity, and their broadest part is towards the base; the hairs with which they are clothed are moderately long and extremely slender—they are not sufficiently numerous to hide the skin of the ear. The fur is dense and extremely soft, about three-quarters of an inch in length on the back, and an inch or more on the hinder parts and sides of the body; the upper parts of the animal are of a very pale grey colour, or grey white, and more or less mottled with dusky, or black: the under parts, as well as the feet, are white, with a faint greyish, or yellowish tinge. The hairs of the moustaches, which are very long, are some of them white, and some black; the flesh of the ears is sometimes brown, and sometimes nearly black in preserved specimens, and the hairs which cover the anterior portion of their outer surface, and those near the margins, on the inner surface, are grey-brown: on

Transactions just quoted; and for the third genus, we must refer to Mr. Brookes' paper, chiefly on the skeleton, of the only known species, in the Linnean Transactions; and to Professor Owen's notes on its internal anatomy, contained in the Proceedings of the Zoological Society, Part VII. p. 175.

the other parts of the ears the hairs are silvery white. The tail, excepting at the base, where it is covered with fur like that on the body, is clothed with coarse hairs; those on the upper surface average about an inch and a half in length, and those on the sides and under surface are most of them about half an inch in length; the long hairs on the upper part of the tail are dirty white at the root and point, and brown-black in the middle, the dark colour being much extended; the sides of the tail are dirty white, and the under surface brown. All the fur on the body of the animal is of a deep slate-grey colour next the skin.

	No. 1.[1]		No. 2.		No. 3.		No. 4.	
	Ins.	Lines.	Ins.	Lines.	Ins.	Lines	Ins.	Lines.
Length from tip of nose to root of tail	10	6	9	2	9	0	10	3
" from ditto to ear	2	5	2	3	2	2	2	3
" of ear	1	10	1	9	1	7½	1	9
Width of ditto	1	10	1	7½	1	8	1	10
Length of fore foot		10		9½		9½		10
" of hind foot	2	4½	2	2	2	1½	2	2½
" of tail, without the hair..	5	0	5	9	5	0	5	9
" of do., including the hair			7	6	7	0	7	6

The animal, of which the admeasurements are given in column No. 4, differs from other specimens in our collections

[1] The dimensions in column No. 1 are taken from a specimen in the Zoological Society's museum, and which lived in the Society's menagerie for some time; it was presented to the Society by Lady Knighton, and is a specimen referred to by Mr. Gray, Mr. Bennett, and Mr. Yarrell, in their accounts of the Chinchilla. It is said to be from Coquimbo. The admeasurements of column No. 2 are from a specimen in the collection of the British Museum, to which collection it was presented by Mr. Collie, and is alluded to by Mr. Gray in his "Spicilegia Zoologica." I mention this because Dr. Wagner imagines that the species described by Mr. Gray is distinct from that described by Mr. Bennett. All the specimens in the British Museum and in that of the Zoological Society I am satisfied belong to the smaller of the two species described by Dr. Wagner. The dimensions in the third column are from a specimen in the British Museum, procured either in Chili, or Bolivia, by Mr. Bridges.

in having the upper parts of the body suffused with brownish; the points of the hairs here are dusky, but with an admixture of red. The sides of the body are of a very pale ashy grey hue, and the under parts are white, slightly tinted with cream colour. It is a most elegant little animal, and forms part of the series in the British Museum collection. The skull, removed from this specimen, has the muzzle more elongated than in three other Chinchilla skulls which I have before me. Its dimensions are given in the first column of the following series:—

	British Museum.		Zoological Society[1].		Zoological Society.		British Museum.		Leyden[2].	
	Ins.	Lines.	Ins.	Lines.	Ins.	Lines.	Ins.	Lines.	Ins.	Lines.
Total length of skull ...	2	5½	2	3½	2	3½	2	2½	2	2½
Width behind	1	2¼	1	2	1	1⅔	1	1½		
" from the outer surface of the zygomatic arches .	1	3	1	3¼			1	2	1	1½
Length of nasal bones .		10¾		9⅔		9⅓		9		
Width of ditto behind .		3⅓		3¼		3		3		
" " in front .		3½		3⅔		3½		3		
" between orbits .		5¼		5⅔		5½		4¾		5⅓
Vertical diameter of the auditory bullæ .		11½		11¼		11¼		11		
Antero-posterior diameter of ditto ...		11		10¼		10		9¾		
From incisor to molar teeth		7		6⅔		6¾		6½		6⅓
Width of incisors ...		2		2⅙		2		2		1⅚
Length of the four molars, taken together		6		6		6		5⅔		5¾
Length of lower jaw ...	1	8¾	1	8½			1	7		
Height of ditto, behind		8⅓		8⅔				7½		
Length of angular portion		10½		10¾				9		

[1] The dimensions in this column are from the skull belonging to the skeleton described by Mr. Yarrell and Mr. Bennett. The skull of which the admeasurements are given in the fourth column differs from others, in having the incisor teeth white; in all other specimens which I have seen, these teeth are of a rich orange colour in front.

[2] From the skull figured and described by M. Van der Hoeven.

CHINCHILLA BREVICAUDATA.

Short-tailed Chinchilla.

Eriomys Chinchilla. LICHT. Darstell. Part 6, Pl. 28.
" " WAGNER, Schreb. Säugth. Suppl. iii. p. 302.
" " Of the Leyden Museum.

Ears oblong, rounded at the extremities; fur on the back about an inch and a quarter in length; of a deep blue-grey at the root; and, on the upper parts, each hair has a broad white ring near the point, and is dark grey at the point: the general tint silvery grey, suffused with black, especially on the back; under parts of the body, inner surface of the limbs, and the feet, white; tail with two dark bands on the upper surface.

Inhabits Peru.

The above is extracted from Dr. Wagner's description of the *Eriomys Chinchilla* of the Berlin Museum. The admeasurements of this animal, as given by the same author, are as follows:—

	From Wagner.		Leyden Museum.		Leyden Museum.	
	Ins.	Lines.	Ins.	Lines.	Ins.	Lines.
Length of the body	14	0	14	0	15	0
" of tail, without the hair	3	0				
" of ditto, including the hair ...	5	6	5	6	5	0
Length of the ear	1	6	1	8	1	7
Width of ditto			1	7	1	6
Length of the hind foot	2	6	2	7	2	6½

On my visit to the museum at Leyden during the last autumn, I made a note, to the effect that two Chinchillas in that museum appeared to me to be decidedly larger, and to have the ears and tail relatively shorter than the Chinchillas in our English collections; that such is actually the case, will be seen upon comparing the admeasurements which I have arranged with those given by Dr. Wagner, with the admeasurements of the *Chinchilla lanigera,* and moreover it will be evident that the two specimens in question agree most closely in size and proportions with Lichtenstein's animal. Their colouring, moreover, is essentially the same; the general hue of the upper parts of these specimens being (according to my note) of a very pale bluish grey, but mottled with slaty-black; the under parts, as well as the feet, white. Tail becoming gradually more bushy to the end; the hairs greyish white, and grey-black, in about equal proportions. Fur on the back grey, and each hair with a broad subterminal white ring, and a grey-black point.

I feel little doubt that further investigation will prove this to be a distinct species from the *Chinchilla lanigera.* Wiegmann and Dr. Wagner have arrived at the same conclusion.

Sub-Family, OCTODONTINA.

Octodontidæ. Waterh., in the Zoology of the Voyage of H.M.S. Beagle, Mammalia, p. 83 (1839); Proc. Zool. Soc. for Nov. 1839, Pt. 7, p. 172.

Psammoryctina (part). Wagner, Schreb. Säugth. Suppl. iii. p. 312; Wiegm. Archiv. vii. 1841.

Hystricina with rootless molar teeth, having but a single indenting fold of enamel on either side, or, rarely, with an extra fold on the inner side of the molars of the lower jaw: zygomatic arch with an angular process on the lower edge: hind feet with five toes; fore feet likewise with five toes, or, sometimes, with four.

The species of this section inhabit the middle and southern parts of South America, occurring on both sides of the Andes:

they are of small size, the largest known species scarcely surpassing the Common Squirrel in bulk: some live almost entirely under ground, and have the fore feet strong, and armed with powerful claws, fitted to their burrowing habits; others are found at the roots of hedges, occasionally climbing amongst the branches of the shrubs of which the hedges are formed, or even (though rarely, I suspect) ascending trees.

The *Octodontina* bear the same kind of relationship to the *Echimyina*, as do the *Arvicolina*, of the Northern Hemisphere, to the true Rats; they agree with the *Echimyina*, in fact, in all essential points, but differ in having the molar teeth rootless, and almost always of a more simple structure. Accompanying the rootless molar teeth, we find that the incisors are almost always broader, and relatively less deep, from front to back; and connected with these differences, the structure of the skull denotes increased powers in the muscles of mastication, the zygomatic arch being stronger, and the temporal fossæ more extended: generally, in proportion as the incisor teeth increase in size, the cranium assumes a broader and shorter form, is more contracted between the orbits, and has the cerebral cavity proportionately smaller[1]; and it is in those species which burrow, and live almost entirely under ground, that these last mentioned peculiarities are carried to the extreme.

From the other two sub-families of *Hystricidæ* which are characterized by rootless molar teeth—the *Caviina* and *Chinchillina*—the Octodons may be distinguished by the hind feet being always provided with five well-developed toes: the

[1] The same kind of differences are observable in the skulls of the *Arvicolæ*, when compared with those of the true Rats; but in each group the general type remains evident amidst all the various modifications: hence, as the skull of an *Echimys* may be distinguished from that of the true Rat, amongst other characters, by the greater depth of the zygoma, and the superior breadth of the frontal bones between the orbits, so are the Octodons' skulls distinguished from those of *Arvicolæ* by similar differences.

presence of a distinct tail will prevent the Octodons from being confounded with the Cavies; and this organ being clothed with short hairs in the Octodons gives to these animals a very different aspect to that of the bushy-tailed Chinchillas. The skull and dentition likewise furnish good distinguishing characters. The palate—as will be seen upon comparing the figures on Plate 8 with those of the skulls of Cavies and Chinchillas—is less contracted between the foremost molar teeth: the malar bone has a distinct angular projection on its under side, as in the *Echimyina*[1] (*a* fig. 3, Pl. 12), and there is the same strong muscular ridge running obliquely forwards and downwards from the hinder root of the zygoma on the outer surface of the malar bone. The molar teeth are usually indented on each side by a single fold of enamel, and the crown of one of these teeth approaches more or less nearly to the form of the figure 8: the most marked exception is found in the lower molars of the genus *Habrocoma*—Pl. 8, fig. 1.

Five genera, having well-marked distinguishing characters, are included in the present sub-family: their chief external peculiarities are as follows:—

I. Fore feet with four toes; ears very large ... ... *Habrocoma.*
II. Fore feet with five toes.
  A, Ears large or of moderate size.
    a, tail as long as the body; slightly bushy at the extremity ... ... *Octodon.*
    b, tail shorter than the body; clothed throughout with small adpressed hairs ... ... *Schizodon.*
  B, Ears very small, almost entirely hidden by the fur of the head.
    a, incisor teeth moderately broad; nails to the toes of the fore feet shorter than the toes ... *Spalacopus*
    b, incisor teeth very broad; nails to the toes of the fore feet longer than the toes ... ... *Ctenomys.*

[1] This angular process is very large in the species of Ctenomys, but unfortunately was broken off in the skull—fig. 5, Pl. 8.

The skull of the species of *Octodon* and *Schizodon* differs from that of other species of the present sub-family, in having a separate groove for the infra-orbital nerve ; a small vertical plate, rising from the anterior root of the zygoma above, forms the outer boundary of this groove.

Genus, *Habrocoma*[1].

Abrocoma. WATERH., Proceedings of the Zoological Society for February, 1837, p. 30 ; Zool. of H.M.S. Beagle, Mammalia, p. 83.

Octodontina with the incisor teeth very narrow ; the upper molar teeth with a deeply indenting fold of enamel on either side dividing each tooth into two equal, transverse lobes ; the hindermost molar with an extra small lobe behind ; lower molars, each with a single fold on the outer side, and two folds on the inner side ; all the folds of enamel deep, and running obliquely forwards ; the lobes of the teeth narrow, terminating externally in very acute angles ; ears very large ; feet naked beneath, and with the soles covered with small fleshy tubercles ; the nails small ; fore feet without any inner toe : tail moderate, clothed throughout with small adpressed hairs.

Inhabit Chile.

In the large size of the ears, and in having long and extremely soft fur, the species of the genus *Habrocoma* evince an approach to the Chinchillas. The skull likewise presents some points of resemblance in the two groups ; as, for instance, in the general weakness of the bones, the large size of the auditory bullæ[2], the narrowness of the basi-occipital bone, and of the bodies of the sphenoids ; in the compressed, and somewhat elongated form of the nasal portion of the cranium, and the large size of the orbits : the incisor teeth, likewise, are very narrow in both groups. The form of the

[1] From Ἁβρος, soft ; and Κομη, hair.

[2] In Habrocoma, as in *Chinchilla*, the petro-tympanic bulb is visible on the upper surface of the cranium.

zygomatic arch, and that of the lower jaw, as well as the structure of the molar teeth, on the other hand, present strong points of distinction, when these parts in *Habrocoma* and *Chinchilla* are compared. The zygomatic arch in the Octodons, generally, is thrown out more boldly at its anterior root, than in the Chinchillas and Cavies, and, in this respect, *Habrocoma* is the furthest removed from the last mentioned sections: here, the skull, measured at the fore part of the zygomatic arches, is as broad as at the hinder root of those arches; the malar bone sends down, from its under surface, a large and distinct angular process, and there is a post-orbital process on the upper edge. The palatal portion of the cranium in *Habrocoma* differs from that in the other genera of *Octodontina*, in being perfectly smooth, though strongly concave in the transverse direction, and in having a single, tolerably large opening opposite the second molar tooth; the incisive foramina are relatively larger, and more elongated. The rami of the lower jaw present a distinct longitudinal groove on the under surface, separating the dental from the angular portion, as in the other sections of the *Octodontina*, and in this respect differing from the Chinchillas,—but this groove is deeper than usual. The fore part of the jaw is elongated and very narrow, and the condyloid, and angular process, are much elongated; a deep, and very nearly semi-circular emargination, separates the two portions behind; the angular portion of the jaw, moreover, presents a very narrow surface when viewed from beneath, and the angular projection on the inner edge[1] is very slightly marked, as is the case in the Chinchillas.

The upper molar teeth may be described as being composed each of two transverse lobes, which are joined in the middle of the tooth by a narrow ridge, formed by the uniting of the external with the internal indenting fold of enamel; the

[1] Corresponding to the part marked *a* in fig. 4, Pl. 12.

inner fold runs inwards, and slightly backwards: the last molar has an extra lobe, projecting backwards, and slightly outwards from the back of the tooth. The lower molars much resemble the grinding teeth in the Arvicolæ, and they will likewise bear a close comparison with the lower molars in Capromys; the loops of enamel are narrow and deep, those which enter from the outer surface of the teeth meeting those which enter from the opposite side; each tooth has a single fold, or loop, on the outer side, and two folds on the inner side; in fact, resembling the teeth, fig. *b*, Pl. 12, excepting that the folds of enamel are rather more oblique, and the lobes of the teeth terminate in very acute angles: and, moreover, the interspaces are not filled with cement, as in the teeth of Capromys. The feet are rather small; the toes short, the nails weak, and hollow beneath; the nail on the second toe of the hind foot is broader than the other nails, and obliquely truncated at the extremity: the soles of the feet are naked, but covered with minute fleshy tubercles, as are also the whole under parts of the toes: the fore feet are remarkable for not exhibiting, externally, even a rudiment of a thumb: rigid hairs cover the nails of the three middle toes of the hind feet.

The skeleton of *Habrocoma Bennettii* is remarkable for possessing a greater number of ribs, and consequently of dorsal vertebræ, than has been observed in any other Rodent[1], there being no less than seventeen pairs of ribs, the hindermost of which are nearly an inch in length: the lumbar vertebræ are five in number, and there are four anchylosed sacral vertebræ, two of which only join the ilium; of caudal vertebræ there are about twenty-three[2].

[1] In the Capromys and Prehensile-tailed Porcupine we find the nearest approach to this number of ribs, there being sixteen pairs in these animals.

[2] This is three more than exists in the skeleton before me, which has evidently lost about that number of vertebræ. I may here mention, that I have examined the skeletons of four out of the five genera contained in the section

The intestines were remarkably long in a specimen of *Habrocoma Cuvieri*, which I dissected; the small intestines measuring full five feet in length, and the large intestines, the diameter of which was distinctly greater than that of the small, measuring three feet: the cœcum was very voluminous, and measured eight inches in length.

HABROCOMA BENNETTII.

Bennett's Habrocoma.

(*Plate* 7, fig. 2).

Abrocoma Bennettii. WATERH., Proceedings of the Zool. Soc. for February, 1837, p. 31; Zool. of H.M.S. Beagle—Mammalia, p. 85, Pl. 28.

Habrocoma helvina. WAGNER, in Wiegmann's Archiv. für Naturgesch. 1842, p. 7.

Ears but little shorter than the head; tail shorter than the body: fur long, and extremely soft; its general hue ashy grey, slightly tinted with yellow; abdomen pale dirty yellow; feet white; tail, pale grey above, grey-white beneath.

Inhabits Chile.

Octodontina, and that I do not find any increase above the ordinary number of thirteen dorsal vertebræ in the genera *Octodon*, *Spalacopus*, and *Ctenomys*. In all, there are well-developed clavicles; the scapula has the spine very deeply emarginated in front: in fact, the spine of the scapula in these animals rises only in the form of a compressed process, of but little antero-posterior diameter, on the hinder half of the scapula, and thence sends forwards a long acromion process, which projects over the articular cavity, and is more or less expanded, and bifurcated, at the extremity: the coracoid process is well developed. The humerus in all has a strong deltoid process; the femur in *Habrocoma*, *Octodon*, and *Ctenomys*, has a slight trace of a third trochanter, represented by a ridge projecting on the outer side of the shaft about midway between the extremities (or rather above that point), and forming an obtuse angle: it is the least prominent in *Ctenomys*, and in *Spalacopus* can scarcely be traced: the tibia and fibula are distinct bones in *all* the genera of the *Octodontina*, for this part of the skeleton I have examined in *Schizodon* as well as the other genera mentioned. The spine of the second dorsal vertebra is stronger and more elevated than the spines of the other vertebræ, and is laterally expanded at the extremity, or even slightly bifurcated, as is the case in the spine in Habrocoma: this spine is very large and broad in *Ctenomys*. In Habrocoma all that exists of the thumb is a short metatarsal bone.

This animal was discovered by Mr. Darwin amongst some thickets in a valley on the banks of the Cordillera, near Aconcagua, and that gentleman informs us that he saw two specimens of the same animal, on the elevated plain, near the town of Santa Rosa, in front of the same part of the Andes: they were crawling up the trunk of an Acacia tree with so much facility as to induce the belief that they habitually ascended trees. More recently, numerous specimens of the *Habrocoma Bennettii* have been sent to England by Mr. Bridges from Chile, from which I perceive that the animal sometimes attains a larger size than the individual described by me in the "Voyage of the Beagle," and, as might be expected, the different specimens exhibit some slight variations in their colouring.

The *Habrocoma Bennettii* is somewhat larger than the Common Rat (*Mus decumanus*), has much larger ears, and these are nearly round, if we except the hinder edge, which is straight in the middle portion: the tail, in length, is about equal to half that of the head and body, and rather thick at the base; it is densely clothed throughout with short hairs, averaging about one-eighth of an inch in length. The fur of the animal is long, and remarkably soft; the long hairs, which are very abundant, are extremely slender. The general hue of the fur is ashy grey, with a slight admixture of yellow, and sometimes having a faint brownish hue on the back; the hairs on all parts of the body are of a deep slate-grey below the exposed ends; those on the back have the exposed end chiefly of a pale, impure ochreous yellow tint, but with the point dusky: on the sides of the body the hairs differ from those of the back in wanting the dusky point, and in having the yellowish tint very pale; on the abdomen the points of the hairs are white, more or less suffused with pale dirty yellow or cream-colour, especially in the mesial line. The feet are white; the tail pale ashy grey above; the hairs brown, dark at the root, and dirty white beneath. The hairs of the moustaches are very numerous, slender, and of moderate length; some of them dusky, and others silvery white. The ears are sparingly clothed; the incisor teeth yellow, or pale orange colour in front.

	No. 1[1].		No. 2.		No. 3.		No. 4.	
	Ins.	Lines.	Ins.	Lines.	Ins.	Lines.	Ins.	Lines.
Length from tip of nose to root of tail	9	6	9	9	10	0	8	9
" from nose to ear	1	9	1	11	1	10	1	8
" of ear				10		11		11
Width of ditto			1	0½	1	2	1	3
Length of fore foot and nails ...		8		9		9½		9½
" of hind foot and nails ...	1	2½	1	4	1	5	1	4⅓
" of tail	5	0	5	0	5	9	5	0
Diameter of ditto at the base ...						4		4⅓
Length of skull	2	0½						
Width of ditto	1	0						
Length of nasal bones		9¼						
Width of ditto behind		2¼						
" in front		2⅝						
" of interorbital space ...		4¼						
From incisor teeth to molars ...		7½						
Width of incisors		1½						
Length of four upper molars, taken together ...		5¼						
Space between the foremost molars		1						
" between the last pair ...		2¼						
Antero-posterior diameter of auditory bullæ		7¼						
Transverse diameter of ditto ...		5½						
Length of lower jaw	1	6¼						
" of angular portion of ditto		9¼						
Height of lower jaw, measuring from the condyle ...		7⅔						

[1] The dimensions in column No. 1 are from a stuffed specimen; those in column No. 2 are from the specimen described in the Zoology of the Beagle, and those in column No. 3 are from a specimen in the British Museum, which is more suffused with brown than usual, and has the cheeks and sides of the body distinctly tinted with pale rufous: in column No. 4 are the admeasurements of a male specimen preserved in spirits.

HABROCOMA CUVIERI.

Cuvier's Habrocoma.

Abrocoma Cuvieri. **Waterh., Proceedings of the Zoological Society for February, 1837, p. 32; Zool. of H.M.S. Beagle—Mammalia, p. 86, Pl. 29.**

Ears slightly emarginated behind; tail for the most part black: general hue of the animal ashy grey; abdomen whitish, or tinted with cream colour.

Inhabits Chile—near Valparaiso.

I gave the name *Cuvieri* to two specimens of a small Habrocoma forming part of Mr. Darwin's collection, relating to which that gentleman's note states—"This species is abundant on the dry hills, partly covered with bushes, near Valparaiso."

These two specimens are very much smaller than the specimens of *Habrocoma* named *Bennettii*—indeed their bulk is scarcely more than one-third of that of the individuals last mentioned; they both have the ears slightly emarginated behind, and the tail entirely of a dusky black colour, with the exception of the under part of the basal portion, which is pale. With regard to their small size, it is necessary to observe that the skull, which has been removed from one of them, shews that the animal was not adult, since the hindermost molar in each jaw is scarcely protruded from its socket: the three developed molars are very much smaller than the corresponding teeth in the *H. Bennettii;* they occupy, in fact, the same space, in the longitudinal direction, as is occupied by the two foremost molars in the larger animals[1]; the skull is relatively shorter, and the hinder portion is more arched.

	Inches.	Lines.
Length from nose to root of tail	6	2
" from ditto to ear	1	4½
" of ear		7
Width of ditto		7
Length of fore foot and nails		7
" of hind foot and nails	1	0½
" of tail	3	0
" of skull	1	4⅔
Width of ditto		9
Length of nasal bones		6¼
Width of ditto behind		2
" in front		2
From incisor teeth to molars		4½
Length of three foremost molars, taken together		2¾
" of the four molars, about		3¾
Antero-posterior diameter of auditory bullæ		6
Length of lower jaw		11½
" of angular portion		5½
Height of lower jaw, measured from the condyle		4¾

Genus, *Octodon.*

Octodon. BENNETT, Proceedings of the Zool. Soc. for March, 1832, p. 46; Trans. Zool. Soc. vol. ii. p. 81.

Dendrobius. MEYEN, Nova Acta Acad. Nat. Cur. xvi. p. 600. 1833.

Octodontina with moderately large ears; tail nearly as long as the body, with long hairs at, and near the point; five toes to the fore or hind feet; the claws small: incisor teeth moderate as to width, or somewhat narrow; convex in front:

[1] I have before me the skull of a young Coypu (*Myopotamus Coypus*) in which the molar teeth are in precisely the same condition as in the *Habr. Cuvieri*, and, upon comparing the three foremost molars of this skull with the same teeth in an adult animal, I find they measure a trifle more in the longitudinal direction in the young than in the old animal, where each tooth has been slightly worn by the friction of the adjoining tooth or teeth.

molar teeth each with a single indenting fold of enamel on either side ; the hinder half of the upper molars much narrower than the anterior portion : a small vertical plate (to protect the infra-orbital nerve) within the ant-orbital opening of the skull.

The species of this genus inhabit Chile and the Bolivian Andes, and extend into Peru ; they are readily distinguished from the members of the other genera of the *Octodontina* by their tail being tufted at the extremity ; on the basal portion of the tail the hairs are very short, and somewhat scanty, only partially hiding the scaly skin from which they spring ; about the middle of the tail they begin to increase in length, and those which spring from the point are the longest. The molar teeth of the lower jaw are indented on either side by a tolerably deep fold of enamel ; the folds of the opposite sides, however, do not meet in the middle of the tooth : in form, the crown of one of these teeth nearly resembles the figure 8, but each lobe of the tooth is broader than long, and is placed obliquely with respect to the jaw—see Plate 8, figs. 2 and 3. The upper molars differ from the lower (as will be seen in the figures just referred to) in having the hinder lobe much narrower, in the transverse direction, than the front lobe.

OCTODON DEGUS.

Cuming's Octodon.

(*Plate* 11, fig. 2).

Sciurus Degus.	MOLINA, Saggio sulla Storia Naturale del Chile, 4to. ed. p. 251 (1810); 8vo. ed. 1782, pp. 303 and 342.
" "	GMEL. Linn. Syst. i. p. 152.
Octodon Cumingii.	BENNETT, Proc. of the Zool. Soc. for March, 1832, Pt. 2, p. 47 ; Trans. Zool. Soc. vol. ii. p. 81, Pl. 16.
" "	WAGNER, Schreb. Säugth. Suppl. iii. p. 317.

Octodon pallidus.	WAGNER, in Wiegm. Archiv. 1845, Pt. 3, p. 33.
Cumingii.	TSCHUDI, Fauna Peruana, Säugth., Pt. 14, p. 171, Pl. 12.
Dendrobius Degus.	MEYEN, Acta Acad. Nat. Cur. xvi. p. 601, Pl. 44, (1833).
Chilian Squirrel.	SHAW, Gen. Zool. ii. p. 148.

Fur long, and moderately soft; upper parts of the body pencilled with black and pale brownish yellow; sides of the body chiefly of the latter colour; abdomen dirty yellow; feet white, faintly tinted with yellow; tail dusky above, dirty white beneath; the long hairs on the apical third, black, or dusky.

Inhabits Chile.

According to Mr. Darwin, this species of Octodon may be seen by hundreds in the hedgerows and thickets in the central parts of Chile, where they make burrows close together, leading one into another. They feed by day in a fearless manner, and are very destructive to fields of young corn; when disturbed, they all run together towards their burrows, in the same manner that Rabbits in England do when feeding outside a covert. When running they carry their tails elevated; and often they may be seen seated on their haunches, like the Squirrel. According to Molina, they lay up a store of food for the winter, and do not become dormant. According to more than one author, the Octodon is seen occasionally to climb amongst the branches of the bushes. Mr. Bridges informs us[1] that Cuming's Octodon has a very extended range, that gentleman having met with it as far north as lat. 28°, and southwards to the 35th parallel. In the province of Coquimbo, where the country is sterile, the little animal takes up its abode amongst the loose stones on the sides of the mountains, and is frequently met with in the holes of the Chinchillas. Its food consists of the herbage

[1] See Proceedings of the Zoological Society for 1843, p. 130.

which grows at the roots of the hedges; in the winter months, when pressed by hunger, it feeds upon the tender bark of the *Mimosa Cavenia*, and also on that of the *Cestrum Palqui*. Mr. Bridges is inclined to believe that it breeds but twice in the year, and states that it has from four to six young at a birth.

The *Octodon Cumingii* is rather smaller than the Common Water-Rat, has moderately large ears, which are distinctly emarginated behind, and the tail rather shorter than the body. Its long and soft fur is of a very deep grey colour next the skin, indeed almost black, but, on the upper parts of the body, the hairs are rather broadly annulated towards the point with pale brownish yellow, and at the point they are dusky: numerous long hairs are interspersed with those of the ordinary fur; the visible portion of those which spring from the back are black, whilst those on the haunches are white at the extremity. On the sides of the body these hairs differ from those of the back in being almost destitute of any dusky hue at the point, and here, therefore, the prevailing tint is brownish yellow: the hairs of the abdomen are of a pale yellow colour at the point. The feet are almost white, and so is the under surface of the tail, as well as the sides, the upper parts being dusky, with a very slight pencilling of yellow; the apical third of the tail is clothed with coarsish black, or dusky hairs, averaging about half an inch in length. The skin of the ears is apparently black; they are rather sparingly clothed for the most part with small pale hairs; those on the fore part of the outer surface, however, are of a dusky hue; the orbit of the eye is pale yellow, and there is a pale spot behind each ear. The hairs of the moustaches are very numerous, and somewhat slender; some of them are black, others are white, and some few are black at the base, and white beyond. The nails of the toes are black; the naked soles of the feet are apparently also black; they are covered with minute tubercles;

besides which there are the ordinary fleshy pads at the base of the toes and on the summit of each of these is a tubercle of larger size than elsewhere: on the under surface of the toes are several transverse incisions. The inner toe of the fore foot is very small, and has a short and blunt nail; the small nails of the other toes are compressed, curved, sharply pointed, and hollow beneath: long and stiff white hairs project over the nails of the hind feet. The female Octodon has eight mammæ; the three anterior pairs of which are remarkable for their position, being almost on the sides of the body[1].

	From stuffed specimen.		From stuffed specimen.		From specimen preserved in spirit. Female.		*O. pallidus.* Pale variety.		From Tschudi.	
	Ins.	Lines.	Ins.	Lines.	Ins.	Lines.	Ins.	Lines.	Ins.	Lines.
Length from tip of nose to root of tail	8	0	7	6	7	6	7	6	7	3
" of tail, not including the hair	4	2	4	4	4	8	3	10	4	3[2]
" from nose to ear	1	8¼	1	7	1	6	1	6	1	6
" of fore foot and nails		8¾		8½		9⅔		8¾		9
" of hind foot and nails ..	1	4¼	1	6	1	6	1	5	1	3
" of ear		8½		8¾		8		7		9
Width of ditto ...				11	1	0		9½		

[1] I have found mammæ similarly placed in the Viscacha and Coypu. The foremost nipple on either side of the body, in the Octodon, is placed on a line with the base of the fore leg, and is about half an inch behind the limb; the third nipple is placed close in front of the femur, and the second is found midway between these two; the remaining two nipples are placed between the hind legs.

[2] Including the tuft.

Octodon Cumingii, var. *Peruana*. TSCHUDI.

Dr. Tschudi regards as a variety of Cuming's Octodon, an animal which he found in Peru, but in one spot only, namely—"in the Quebrada of San Mateo, in the neighbourhood of the village of San Juan de Matucanas, at an elevation of about 9000 feet above the sea[1]." Its general hue is stated to be of a yellowish grey-brown on the upper parts of the body, but with pencillings of black; the hairs, however, are of a dark slate-grey colour at the root; the under parts yellowish white; the head and neck suffused with rufous, and with less admixture of black than on the back; the region of the ear yellowish white, which colour is expanded into a tolerably large patch near the posterior angle of the ear; the feet dirty white; nails blackish brown; tail blackish brown at the base above; the tuft at the apex black, passing into red-brown; the hairs of the moustaches are some of them entirely black, others silvery white, and some of them black at the root and white at the point: the ears blackish grey, and clothed with short hairs, of which those on the anterior margin are white. Long and coarsish hairs are interspersed with those of the ordinary fur; those on the back are black, whilst those on the sides of the body, and near the root of the tail, are white. The dimensions of the animal are given with those of the *Octodon Degus*, in the fourth column. Dr. Tschudi, who has compared his Octodon with the specimen in the Berlin Museum, which is the original of Meyen's description under the name *Dendrobius Degus*[2], states that it is of a brighter hue, both

[1] I find in the maps that Matucana and San Mateo are situated on the west slope of the Cordillera, very nearly in the latitude of Lima.

[2] Dr. Tschudi states that the length of this specimen is (measured from the tip of the nose to the root of the tail) 7″ 1‴, and not 5½″, as stated by Meyen; and that its hind foot is 1″ 3‴.

on the upper and under parts of the body, than the Chilian animal, and that its abdomen is of a yellow-white colour, and not dingy grey, as in the *Octodon Cumingii* of Mr. Bennett. To this statement I have to remark, that I have seen at least sixty specimens of the *Oct. Cumingii* from Chile, and find that they vary as to the brightness of their colouring, and that the abdomen is almost invariably tinted with yellow. It is to be desired that the skulls of the Peruvian Octodon be compared with that of the Chilian, and likewise with the species which inhabit the high lands of Bolivia.

Octodon pallidus. Wagner.

Three specimens of an Octodon from Chile, contained in the British Museum collection, are undoubtedly identical with the *Octodon pallidus* of Wagner, and, I have little doubt, are simple varieties of the *Oct. Degus*. Mr. Darwin, when in Chile, noticed that piebald, or albino varieties of this little animal, were by no means uncommon, and Mr. Bridges, who has likewise studied these animals in their native country, informed me that the specimens, about to be noticed, were varieties of Cuming's Octodon.

The three specimens in question (and I have seen several others like them) agree with the *Octodon Degus* in the size and proportions of the skull, as well as other parts, but differ in being almost of an uniform pale sienna yellow colour: the abdomen is of a paler hue than the upper parts of the body; and the tail and feet are white, or white slightly tinted with yellow. I shall have to notice similar varieties of several other species of Rodents, as well as of Mammals of other orders: they present a condition which is intermediate between the true albino and the normal state.

OCTODON BRIDGESII.

Bridges' Octodon.

Octodon Bridgesii. Waterh., Proceedings of the Zoological Society for October, 1844, p. 153.

Fur very long, and moderately soft; upper parts of the animal strongly pencilled with brown-yellow, and black,—the black prevailing on the back: abdomen white, suffused with pale brown-yellow, or cream coloured; tail indistinctly tufted; dusky, excepting at the base beneath, where it is dirty white: feet greyish white.

Inhabits Chile—the Province of Colchagua.

This species—which has been named after Mr. Bridges, who has not only enriched our collections with the animals and plants of Chile, but who has likewise furnished us with excellent observations on the habits and ranges of the former[1] —may be distinguished from the *Octodon Degus* by its large size; its tail being relatively longer and less distinctly tufted, and the ears rather smaller. Its general hue, moreover, is darker—somewhat inclining to brown—there being a greater admixture of black on the upper parts of the body, and the hairs being annulated with brownish yellow[2]: there are likewise differences in the skull and dentition, which are hereafter pointed out.

The fur in Mr. Bridges' Octodon is dense, very long, and but moderately soft; the interspersed coarse hairs are very abundant, and their exposed ends are black on the upper

[1] The contents of various letters, which Mr. Bridges has been so kind as to address to me on these subjects, having been communicated to the Zoological Society, will be found in the "Proceedings"—so often quoted in this work.

[2] In colouring, this animal is exceedingly like the Common Rat.

parts of the animal, and white on the sides of the body, as well as over the haunches, where they are very conspicuous. The hairs of the ordinary fur are of a slate-grey colour at the root, and, indeed, for the greater portion of their length; less dark than in *Oct. Degus*, and rather broadly annulated with brownish yellow towards the point, which is black; on the sides of the body the yellow hue prevails, whilst the black is more prominent on the back. The hairs on the under parts of the body are sometimes of a cream-colour at the point, and sometimes a brownish yellow hue is tolerably distinct: the throat, and inner side of the legs at the base is white; the feet greyish white. The tail is dusky black above; below, and at the sides, it is white, or dirty white: the hairs on this organ are rather scanty throughout, and hence do not entirely hide the skin; from the tip of the tail they project about half an inch, or less, and from that point towards the base they gradually decrease in length. The ears are clothed with small hairs, which are pale, excepting on the fore part externally, where they are dusky brown. The hairs of the moustaches resemble those of *Oct. Degus*: the nails of the toes differ in being brown, or horn-colour, and are shorter than in *O. Degus*.

	Zoological Society.		British Museum.		British Museum.	
	Ins.	Lines.	Ins.	Lines	Ins.	Lines.
Length from tip of nose to root of tail ..	8	4	8	3	8	6
" of tail, not including the hair ...	5	2	5	4	5	0
" from nose to ear			1	7	1	6
" of ear		7¼		8		7½
Width of ditto				11		10
Length of fore foot and nails		10		10½		10
" of hind foot and nails	1	6	1	7	1	7

The skull of *Octodon Bridgesii* (Pl. 8, fig. 3) differs from that of *Oct. Degus* (Pl. 8, fig. 2[1]) in being larger, and less arched; somewhat narrower between the orbits; in having the zygoma relatively deeper, and the nasal bones longer and narrower: the palate is rather more contracted between the anterior molar teeth, and more deeply emarginated behind: the basi-occipital bone is broader. The lower jaw is less deeply emarginated behind, and has the coronoid process broader in the antero-posterior direction. The upper molar teeth have the transverse diameter relatively greater, and the indenting fold of enamel on the inner side distinctly deeper; in *O. Degus* the fold on the inner side of the last molar indents the tooth so slightly as merely to produce a waved outline, whilst this same tooth in *O. Bridgesii* has the inner fold of enamel fully as deep as that on the inner side of the foremost molars of *O. Degus*. The molars of the lower jaw, like those of the upper, are proportionately broader.

[1] The lower jaw on this plate, marked 2 *b*, should have been 3 *b*, since it belongs to the *Oct. Bridgesii*, and not to *Oct. Degus*, as is made to appear by the lettering of the plate.

	O. Bridgesii.		O. Bridgesii.		O. Bridgesii.		O. Degus.		O. Degus.		O. Degus.		O. Degus. Var. pallidus.	
	Ins.	Lines.	Ins.	Lines.	Ins.	Lines.	Ins.	Lines.	Ins.	Lines.	Ins.	Lines.	Ins.	Lines.
Total length of skull	1	$9\frac{1}{2}$	1	$9\frac{3}{4}$	1	$10\frac{1}{4}$	1	6	1	7	1	$6\frac{1}{4}$		[1]
Width	1	$0\frac{1}{2}$		$11\frac{1}{3}$		$11\frac{2}{3}$		$10\frac{1}{2}$		$10\frac{3}{4}$		$10\frac{2}{3}$		$10\frac{2}{3}$
" between orbits				$4\frac{1}{4}$		$4\frac{1}{4}$		$4\frac{1}{2}$		$4\frac{2}{3}$		$4\frac{3}{4}$		$4\frac{2}{3}$
Length of nasal bones		$8\frac{1}{2}$		$8\frac{1}{2}$		$8\frac{1}{2}$		7		7		7		$7\frac{1}{2}$
" of zygomatic arch		11		$10\frac{1}{2}$		$10\frac{1}{2}$		$8\frac{1}{2}$		9		$8\frac{3}{4}$		9
From incisors to molar teeth		$5\frac{1}{4}$		5		$4\frac{4}{5}$		$4\frac{1}{4}$		$4\frac{1}{2}$		$4\frac{1}{3}$		$4\frac{1}{2}$
Width of palate between the last molars		$2\frac{2}{3}$		$2\frac{1}{2}$		$2\frac{3}{4}$		2		$2\frac{1}{5}$		$2\frac{1}{5}$		$2\frac{1}{4}$
Length of four upper molars, taken together		$5\frac{1}{4}$		5		$5\frac{1}{5}$		$4\frac{1}{3}$		$4\frac{1}{2}$		$4\frac{1}{3}$		$4\frac{1}{2}$
Width of incisors		$1\frac{3}{4}$		$1\frac{3}{4}$		$1\frac{3}{4}$		$1\frac{3}{5}$		$1\frac{1}{3}$		$1\frac{1}{2}$		$1\frac{1}{3}$
Length of the lower jaw, measured from the hinder part of condyle	1	$2\frac{1}{4}$	1	2	1	2		$11\frac{3}{4}$	1	$0\frac{1}{2}$			1	$0\frac{1}{2}$
Height of ditto		7				$6\frac{1}{2}$		$5\frac{2}{3}$		$5\frac{3}{4}$				6

[1] The hinder portion of this skull is wanting, but measuring to the extremity of the posteriorly produced portion of the temporal bone, the length is 1″ 5‴; the skull, of which dimensions are given in the sixth column, gives the same length when measured in the same manner. Having now examined many specimens of ***Oct. Bridgesii***, and the skulls of four specimens, and finding them always to differ from ***O. Degus*** in the points mentioned, I am perfectly satisfied that they form two distinct species.

OCTODON GLIROIDES.

Dormouse Octodon.

Octodon Gliroides. D'ORBIGNY et GERVAIS, Revue Zoologique, Année 1844, vol. vii. p. 123 ; D'Orbigny's Voyage dans l'Amérique.

Fur soft, ashy grey on the upper parts of the body, and white on the under ; tail, brown-black beneath to the extremity, which is somewhat tufted ; feet, white above.

The animal upon which Messrs. D'Orbigny and Gervais found the *Octodon gliroides*, is said to inhabit the summit of the Bolivian Andes at La Paz, where it feeds upon the Cactus plants. Its proportions are said to be nearly the same as in *O. Degus*, but, besides the dissimilarity of colouring in the two animals, the *Oct. gliroides* differs from *O. Degus* in having the molar teeth rather less elongated ; the fourth molar has the folds of enamel less oblique ; the upper molars are of a more triangular form, and the lower more regularly 8-shaped, with the exception of the hindermost, which is of the form of a comma, and has the indenting fold on the outer side, and not on the inner, as in *O. Cumingii*. In the last mentioned species, the last molar of either jaw differs less from the preceding molars, both in size and form, than do the same teeth in *O. Degus*.

The above is extracted from the description given by Messrs. D'Orbigny and Gervais : I am unacquainted with the animal.

Genus, *Schizodon*[1].

Schizodon. WATERH., Proceedings of the Zoological Society for November, 1841, p. 91.

[1] From σχίζω, to cut, divide ; and ὀδοὺς, a tooth.

Octodontina with moderate sized ears; short tail, clothed throughout with short hairs; fore feet strong, the claws about equal to the toes in length: incisor teeth stout; molars each with a single deep fold of enamel on either side, dividing the tooth into two oval-shaped lobes, the long diameters of which are placed transversely; the lobes equal in size, or very nearly so, excepting in the hindermost molar of each jaw, in which the posterior lobe is considerably smaller than the rest: skull strong, rather short and broad, with a separate small canal for the infra-orbital nerve. Toes 5—5.

Inhabits the Southern Andes—eastern side.

The genus *Schizodon* in many respects is intermediate between the genera *Octodon*, *Ctenomys*, and *Spalacopus*; in its dentition it differs from either, but, on the whole, approaches most nearly to the last-mentioned genus. To the naked eye, the series of molars appear to be made up of a succession of cylinders, compressed from before backwards, and these scarcely presenting any difference in size in the three foremost molars in each jaw; but under a lens, each molar is seen to have a deep external, and a correspondingly deep internal, fold of enamel; the folds of opposite sides meeting in the mesial line of the tooth. In the upper jaw the hindermost molar is somewhat smaller than either of the three preceding teeth, and has the posterior lobe smaller than the anterior one; in the lower jaw the corresponding tooth is of the same form, but is of smaller size, and it is so placed that the inner side of the tooth inclines inwards and forwards. Beyond these points, I have only to remark, that each lobe of the molars is slightly angular, both on the outer and inner sides of the tooth. Compared with the molars of *Octodon* and *Ctenomys*, the corresponding teeth in *Schizodon* are at once distinguished by the general equality of the lobes; and from the molars of *Spalacopus* they differ in having the folds of enamel of opposite sides meeting in the middle of the tooth. The incisor teeth are very large; broader than in the *Octodons*, but with the anterior surface convex, as in the incisors of those animals, and their direction is the same. The skull in *Schizodon* (Pl. 8, fig. 4) agrees with that of the *Octodons* in having a separate, small canal for the infra-orbital nerve; in all other respects it is intermediate between the genera *Octodon* and *Ctenomys*. The lower jaw (Pl. 8, fig. 4 *a* [1]), in having the angular portion thrown very boldly outwards, and in other respects, agrees more nearly with the lower jaw of *Ctenomys* than with that of *Octodon*.

SCHIZODON FUSCUS.

The Brown Schizodon.

(*Plate* 11, fig. 2).

Schizodon fuscus. WATERH., Proceedings of the Zool. Soc. for November, 1841, p. 91.

Fur tolerably long, moderately soft, and somewhat glossy; its general hue deep brown—pencilled black and yellow-brown: abdomen of a pale dirty yellow hue; feet dusky brown.

Mr. Bridges, who discovered this animal, informs me that he met with it in the Valle de las Cuevas, on the eastern side of the Andes, about six leagues from the Volcano of Peteroa (S. lat. about 75°), at an elevation of from five to seven thousand feet. Its favourite abode is stated to be in grassy situations near the mountain streams. In certain parts the ground is completely undermined by the burrows of these little animals, and whilst riding over these, the horses of Mr. Bridges party frequently plunged into the ground almost up

[1] The molar teeth of the lower jaw are much curved, and in the jaw figured on the Plate, the root of the penultimate molar has protruded slightly through the ramus, as is indicated by the two round dark spots in the figure.

to the hock. The valleys inhabited by the Schizodon are so elevated that they are covered with snow at least four months during the year, and Mr. Bridges is inclined to believe that, like the Spalacopus, the Schizodon lays up a store of food for the winter season. It is a nocturnal animal, and, it would appear, lives almost entirely under ground.

The general colouring of the *Schizodon fuscus* nearly resembles that of the Common Rat (*Mus decumanus*), but it is somewhat darker, and the fur rather softer, and more glossy: it is of a very dark slate grey next the skin, both on the upper and under parts of the body; on the latter parts, the hairs are of a pale dirty yellow colour (sometimes brown-yellow) at the point, and on the upper parts of the animal they are black at the point, and annulated with yellow-brown towards the point; on the sides of the body the hairs are merely tipped with dusky, and the yellow-brown hue prevails; the long, interspersed, coarser hairs on the back of the animal, have the exposed ends black, but those on the sides of the body have dirty white points, or, some of them, especially those near the tail and over the haunches, are broadly annulated with yellowish white near the point, and black at the point. The ears, which perhaps should be described as rather small, are emarginated behind, and somewhat sparingly clothed with small hairs, which are for the most part of a dirty white colour, but are dusky brown on the fore part of the outer surface. The hairs of the moustaches are, the greater portion of them, dusky at the root, and white beyond: the feet are of a dusky brown hue, but pencilled with grey-white; the long bristly hairs which cover the nails of the three longest toes of the hind feet are white. The tail is tolerably well clothed with very short and stiff hairs; dusky brown, or brown, on the upper surface, and brown beneath—pale near the root.

	Zoological Society. Inches.	Lines.	Brit. Museum. Inches.	Lines.
Length from tip of nose to root of tail	9	0	7	6
" of tail	1	8	2	6
" from nose to ear			1	7
" of ear		$5\frac{1}{2}$		$6\frac{1}{2}$
Width of ditto				$9\frac{1}{2}$
Length of fore foot and nails ...				9
" of the nail of the middle toe				$3\frac{1}{4}$
" of the hind foot and nails ...	1	$5\frac{3}{4}$	1	3
" of skull	1	10	1	8
Width of ditto	1	0	1	$0\frac{1}{2}$
Length of nasal bones		8		$7\frac{1}{2}$
Width of ditto behind		2		2
" " in front		3		3
Length of zygomatic arch		10		$9\frac{2}{3}$
Greatest depth of ditto		$2\frac{1}{2}$		$2\frac{1}{2}$
Width between orbits		$4\frac{1}{2}$		$4\frac{3}{4}$
" of palate, between the anterior molars		1		$1\frac{1}{5}$
" of ditto between the posterior molars		$2\frac{2}{3}$		$2\frac{1}{2}$
From incisors to molar teeth ...		$5\frac{3}{4}$		$5\frac{1}{4}$
Width of incisors		$2\frac{1}{2}$		$2\frac{1}{2}$
Length of the four molars, taken together		$5\frac{1}{2}$		$4\frac{3}{4}$
" of lower jaw	1	5		
Height of ditto		$7\frac{1}{3}$		

Genus, *Spalacopus*.

Spalocopus. WAGLER, in Isis. 1832, p. 1219.
Poephagomys. F. CUVIER, Ann. des Sci. Nat. i. (2d Series), p. 321 (1834).
Psammoryctes. POEPPIG, in Wiegm. Archiv für Nat. 1835, i. p. 252.
Psammomys. POEPPIG, Reise in Chile, Peru, &c. i. p. 166 (1835).
Bathyergus. POEPPIG, in Froriep's Notizen aus dem Gebiete der Natur- und-Heilkunde, xxiii. No. 18, p. 279 (1829).

Octodontina with rudimentary ears; tail short, and clothed with short hairs; the nails of the toes of the fore feet rather shorter than the toes; incisor teeth moderately broad, those of the upper jaw distinctly directed forwards as well as downwards; molar teeth with the crowns shaped like the figure 8, the anterior and posterior lobes being equal, excepting in the last tooth—the indenting fold of enamel of one side of each tooth not meeting that of the opposite side; hindermost molar in each jaw smaller than the rest, and with the posterior lobe smaller than the anterior one.

Inhabit Chile. Live almost entirely under ground[1].

The very small ear, and the folds of enamel on the opposite sides of the molars being separated from each other in the mesial line of the tooth, distinguishes the genus *Spalacopus* from *Schizodon:* the skull, moreover, in the present genus, is much shorter than in the last-mentioned one; the

[1] Dr. Burmeister doubts the propriety of arranging the present genus in the section *Octodontina*, and has removed it into the group *Cunicularia*, and Dr. Wagner has expressed his approval of this view (see Wiegmann's Archiv, 1845, p. 171); the latter author likewise places the genus *Ctenomys* in the *Cunicularia*. This arrangement involves principles of classification totally at variance with my own, and can only be upheld by the assumption that Rodents, whose structure is modified so as to be fitted for burrowing habits, are *necessarily* allied to each other, since over and above those characters which are most clearly adapted to their mode of life, the *Cunicularia* differ much in the structure of the skull, and other parts of the skeleton, from the *Octodontina*, whilst in all the peculiarities of skull and skeleton, *Spalacopus* and *Ctenomys* essentially agree with Octodon: they are, moreover, linked with Octodon by a recently discovered genus (*Schizodon*), which in most of its characters presents an intermediate condition; and, in connexion with these resemblances between the different minor sections of the *Octodontina*, it is interesting to observe that they are all found together in a quarter of the globe in which the *Cunicularia* do not occur. I have further to remark, that the peculiarities which distinguish *Ctenomys* from *Octodon* are less strongly marked in the smaller species of the former genus than in the larger; that such is the case, so far as the skull and lower jaw are concerned, will be seen upon comparing the skull of *Ctenomys* (fig. 5) with that of the larger species (fig. 6), and with the skull of Octodon (fig. 3); and, lastly, I know that the distinctive characters in question almost disappear in the young *Spalacopus* or *Ctenomys*. Dr. Burmeister's observations upon the classification of the Rodentia, are contained in the "Allgemeine Literatur-zeitung" for August, 1843, p. 552, and I am happy to find that so accurate an observer has arrived at the same *general* views upon this subject as myself. Although Dr. Burmeister's views were published long after my own, they appear to have been in no way influenced by the observations contained in the numerous papers I had written on the subject.

ant-orbital opening is relatively smaller, and there is no separate canal in this opening for the infra-orbital nerve. The upper incisor teeth, upon quitting their sockets, instead of being bent suddenly downwards at a right angle with the plane of the skull, or nearly so, as in other *Octadontina*[1], here, are distinctly directed forwards as well as downwards; they, in fact, in their curve, form a smaller segment of a larger circle than usual; their roots, with their bony covering, diverge much posteriorly, and in their passage backwards they encroach upon the ant-orbital opening, and terminate close to the crown of the third molar tooth, on either side. The lower incisor teeth are very long, running back nearly to the condyle of the jaw. Stiff hairs cover the nails of the toes of the hind feet, less distinctly on the outer and inner toes than on the others. The thumb of the fore foot has a short truncated nail.

SPALACOPUS POEPPIGII.

Poeppig's Spalacopus.

(*Plate* 9, fig. 1).

Spalacopus Poeppigii.	WAGLER, in Isis. 1832, p. 1219.
Poephagomys ater.	F. CUV. Ann. des Sci. Nat. tom. i. (2d Series), p. 321, Pl. 13 (1834).
" "	WATERH. Zool. of H. M. S. Beagle, Mammalia, p. 82. 1839.
" "	EDOUX et GERVAIS, Voy. de la Favorite, Zool. i. p. 17, Pl. 7 (1839).
Psammoryctes noctivagus.	POEPPIG, in Wiegm. Archiv, (1835), i. p. 252.
" "	WAGNER, Schreb. Säugth. Suppl. iii. p. 318 (1843).
Psammomys.	POEPPIG, Reise in Chile, Peru, &c. i. p. 166 (1835).
? *Mus cyanus.*	MOLINA, Saggio sulla Storia Naturale del Chile, 4° ed. p. 248; 8° ed. p. 308.

Fur soft, and very glossy; black, with an admixture of purple brown; the brown hue most distinct on the under parts of

[1] With one exception only, found in the *Ctenomys leucodon*.

the animal: incisor teeth pale yellow, or yellow-white in front.

Inhabits Chile.

The *Mus cyanus* of Molina has been noticed in all works upon Mammalia for upwards of sixty years past, but I believe has never been identified with any animal received from Chile by the various Museums of Europe; I feel little doubt, however, that Mr. Bridges is correct in the identification suggested in the following observations, communicated to me in a letter:—

"The *Poephagomys ater*, or *Spalacopus Poeppigii*, is undoubtedly the animal alluded to by Molina under the name *Mus cyanus*[1]; his long description of its habits agrees in most respects with the habits of this little animal; but I have never yet heard it called by the natives 'Guanque:' it is generally known in Chile by the name of Cururo and *Cuyeita;* Guanque is the vernacular name of a species of *Dioscorea* on which the 'Cururo subsists. Molina is perfectly correct in saying that it stores up a considerable quantity of provisions, which consist of the *Dioscorea, Conanthera, Ornithoagalum, Brodiæa,* and other bulbs and tubers which abound in the country. The poorer class of inhabitants being aware of its habits, sound the caves or burrows, and rob them of their store, which they eat. The jaws of the Cururo are capable of extraordinary expansion, and by this provision of Nature it is enabled to carry bulbs and tubers of a large size to its granary.

[1] Molina describes the colouring of the *Mus cyanus* in the following terms—"*corpore cæruleo, subtus albido.*" It is by no means improbable that the blue gloss which a fur of the same texture and colour as that of the Spalacopus presents, when viewed in a strong light, may have given rise to the use of one of these terms; but, upon the supposition that the Spalacopus is identical with *Mus cyanus*, we must further take for granted that Molina had a white-bellied variety of the animal before him, or that his description in this, as in many other cases, is incorrect.

"The work of this little animal would surprise a person unacquainted with its habits: I have frequently seen a considerable surface of ground completely undermined by its burrows. It generally selects the slopes of hills and mountains, where bulbs are found, especially in the interior parts of the country: its caves are carried in a horizontal course, at the depth of eight or ten inches; or rather, about the depth in which they meet their food.

"The Cururo may be considered nocturnal, since it seldom, if ever, makes its appearance during the day; those which I procured were obtained by waiting for them in the evening, and shooting them when their head scarcely emerged from their caves."

With regard to the specific characters of this animal I need add very little to the few lines with which this account commences. The fur is slightly grey next the skin, most distinctly so on the abdomen: the general hue of the animal is sometimes black brown, sometimes nearly a perfect black, but usually black, with an admixture of purple-brown. The ears are slightly emarginated behind, and tolerably well clothed with black hairs. The fore foot has a fringe of silvery grey hairs on the inner side; the nails of the second and third toes of the hind foot are rather broader than the other nails; the nail of the second toe is the broadest, and is obliquely truncated at the extremity: stiff hairs project over these nails, as in other species of the section *Octodontina*. The hairs of the moustaches are numerous, but by no means strong. The following admeasurements are from stuffed specimens in the British Museum:—

	Ins.	Lines.	Ins.	Lines.	Ins.	Lines.
Length from tip of nose to root of tail ...	6	2	6	0	6	6
" of tail	1	10	1	9	1	10
" from nose to ear	1	3	1	3½	1	3½
" of ear		3		2¾		2¾
" fore foot and nails[1]		10		9		9½
" of hind foot and nails	1	1¾	1	1½	1	1½
Length of skull	1	4½				
Width of ditto		10				

Genus, *Ctenomys*[2].

Ctenomys. De Blainville, Bulletin de la Société Philomatique, for April, 1826, p. 62 ; Ann. des Sci. Nat. x. p. 97.
" Lesson, Manuel de Mammalogie, p. 252. 1827.
" Wagner, Schreb. Säugth. Suppl. iii. 375.

Octodontina with rudimentary ears ; small eyes ; short tail ; fore feet large and powerful, and armed with nails, which exceed the toes in length ; incisor teeth very broad, the upper pair indistinctly convex in front, the lower pair flat at the same part : molars with two unequal lobes ; the hindermost molar of each jaw much smaller than the rest, nearly cylindrical : skull very strong, short, and broad ; the occipital portion much dilated : no separate canal for the infra-orbital nerve.

Geographical range ; from Brazil westwards into Bolivia, and southwards to the Straits of Magalhaen. Live underground.

The enormous size of the incisor teeth, and the unequal lobes of the molars, distinguish this genus from *Spalacopus ;*

[1] The nail of the middle toe of the fore foot is from 2¼ to 2½ lines in length in these specimens.

[2] From κτεὶς, a comb ; and μῦς, a mouse—having allusion to the comb-like apparatus (formed of stiff bristles), with which the toes of the hind feet are furnished.

the claws of the fore feet, moreover, are relatively larger; they are but little arched, and keeled beneath, excepting at the extremity, the keel or ridge being formed by the sides of the nail being bent inwards, and meeting in the middle line: the thumb nail is tolerably large, and hollow beneath. The nails of the toes of the hind feet are smaller, and nearly straight, and entirely hollow beneath: two rows of exceedingly stiff bristles spring from the ends of the toes, and partly cover the nails; and a fringe of long and tolerably stiff hairs projects from the outer side of the foot. The muscular ridge which crosses the malar bone is here remarkably strong, as are likewise the upper and lower angular processes of this bone, which have before been noticed[1].

CTENOMYS BRASILIENSIS.

Brazilian Ctenomys.

Ctenomys Braziliensis.	DE BLAINVILLE, in Bulletin de la Société Philomatique for April, 1826, p. 62.
" "	EDOUX et GERVAIS, Voyage de la Favorite, v. Pl. 8, fig. 2 (skull).
" "	DARWIN, Voyage of the Beagle—Mammalia, p. 79.
" "	D'ORBIGNY, Voy. dans l'Amerique, Pl. 17.
" *torquatus.*	LICHT., Darstellung, 6tes Heft, Pl. 31, f. 1.

Fur soft, fine, and rather short; of a deep slate-grey next the skin, and bright rusty brown externally; on the under parts of the animal, however, the hairs are rusty white at the point: the hairs on the feet short and harsh, as well as those

[1] How different is the form of the malar bone in the *Ctenomys* and the Bathyergus, or Cape Mole-Rat! this bone being small and slender in the latter animal, as is almost invariably the case in the Muridæ; and, where exceptions occur, we do not find this part then taking the form which I have pointed out as characteristic of a great mass of the South American Rodents. Another well-marked modification of the malar bone is seen in the Beaver and Squirrels, where it runs up in front to be wedged in between the lachrymal bone and the zygomatic process of the superior maxillary.

of the tail, which latter are of a blackish brown colour: hairs of the moustaches tolerably long.

Inhabits Brazil, Paraguay, La Plata, and Bolivia.

The above specific characters are extracted from M. De Blainville's original description, which is from a specimen procured in Minas Geraes. I cannot doubt that the animal described by Lichtenstein is specifically identical with the *Ctenomys Braziliensis* of De Blainville; and specimens of a species of Ctenomys brought from Bolivia by Mr. Bridges, together with others procured at Maldonado, La Plata, by Mr. Darwin, appear to me to be clearly of the same species.

Ctenomys torquatus, LICHTENSTEIN,

Is described as of a bright yellow-brown colour on the upper parts of the body, passing into dark brown along the mesial line; sides of the head, as well as the whole under parts, dirty white: a white band extends from behind the ear to the throat.

This animal, which, according to Lichtenstein, inhabits the southern provinces of Brazil and the banks of the Uraguay, is supposed to be a distinct species from the *Ct. Braziliensis* of De Blainville, on account of the latter being of smaller size, its having a shorter tail, wanting the comb-like bristles on the fore feet, its uniform colouring, and the want of the band on the neck[1].

[1] I cannot admit that these differences furnish good grounds for separating the two animals. In the first place, there is an evident mistake with regard to the size of the *Ctenomys Braziliensis*: it is true, De Blainville (who does not give dimensions) states that the figures on his plate are of the natural size, but if this be true of the figures of the skull, it cannot be so with the figure of the animal, the head of which, as represented, being one-third smaller than the skull; and it will be seen that on the plate in question, there is a mark intended, apparently, to denote that the figure of the animal is one-third of the natural size. The skull figured agrees most closely in size and proportions

Ctenomys, from Madonado, La Plata.

General hue bright rufous brown; abdomen of a pale reddish yellow colour; and a band of the same colour passes down to the throat from behind the ear: tail brown above, and white beneath: ears with a narrow fringe of rufous and dusky coloured hairs on the hinder margin: hairs of the moustaches for the most part dirty white, with dirty points: fur of a deep slate-grey at the root.

Ctenomys, from Bolivia,

Differs from the above in having the general hue of the fur paler; on the sides and under parts of the body, of an extremely pale rufous yellow tint; and on the upper parts, pale rufous brown: upper surface of the muzzle dusky; a pale vertical mark on the neck; tail pale dirty yellow, more or less suffused with brown above; the apical half with the hairs longer in the mesial line, both above and below, than elsewhere—sometimes more than a quarter of an inch in length; these hairs are harsh, and form a slight crest on each surface of the tail, and cause the orangan to appear compressed; they are usually white, or nearly so: feet nearly white; nails horn-coloured: fur, both on the upper and under parts of the body, of a very dark slate-grey next the skin.

Mr. Bridges brought to England numerous specimens, from the neighbourhood of the town of Potosi, agreeing with the above description; in the same district, however, were found others in which the colouring was somewhat richer and more nearly agreeing with the animal brought from Maldonado by Mr. Darwin; others, again, with the general hue

with skulls before me, which were removed from specimens of Ctenomys received from Bolivia, and which agree in all essentials with Lichtenstein's animal. With regard to the pale mark on the neck, I may observe, that in some of the specimens last alluded to, the mark is distinct, whilst in others it is so faint that it might be overlooked. In all the species of Ctenomys which I have seen, this mark can be traced more or less distinctly. The bristles of the feet are not unfrequently almost entirely wanting, being very apt, it would appear, to be worn off. The difference in the length of the tail alluded to must be very slight—if there really exists any such difference.

very pale, had the upper surface of the head of a black hue, and the hinder part of the back slightly suffused with dusky; and one was of a brown colour, darker than the rest, and very obscurely tinted with rufous. Having examined skulls removed from specimens exhibiting each of these variations in colouring, I am enabled to state that this part presents no differences which would lead us to suppose there was more than one species amongst the Potosi specimens. The skull (Pl. 8, fig. 6) is from one of these specimens; of three others I have given the admeasurements.

Mr. Bridges informs me this animal is called by the natives "Toco," and "Tofo."

	From Maldonado.		From Bolivia.		From Bolivia.		From Bolivia.		From Bolivia.		*Ct. torquatus*, Lichtenst.	
	Ins.	Lines.	Ins.	Lines.	Ins.	Lines.	Ins.	Lines	Ins.	Lines.	Ins.	Lines.
Length from tip of nose to root of tail ...	8	2	9	0	9	6	8	3	8	8	9	0
" of tail	2	5	3	0	3	6	2	6	2	6	2	9
" of fore foot, without the nails ...		8		8		8½		8½		8½		
" of ditto, including the nails ...		11½	1	0	1	0¾	1	1½	1	1¼		
" hind foot, not including the nails			1	4½	1	5½	1	4	1	3½	1	3
" of ditto, including the nails ...	1	5	1	7½	1	9	1	6	1	7		

The following account of the habits of the Ctenomys is abstracted from Mr. Darwin's notes published in the "Voyage of the Beagle." The species found in Maldonado is there known by the name Tucotuco, the name being in imitation of the sound the animal emits. It is exceedingly abundant, Mr. Darwin states, in the neighbourhood of Maldonado, but is difficult to be procured, and rarely seen. The Tucotuco lives almost entirely under ground, and prefers a sandy soil with a gentle inclination, but it sometimes frequents damp places, even on the borders of lakes. The burrows are said not to be deep, but of great length. They are seldom open; the earth being thrown up at the mouth into hillocks, not quite so large as those made by the mole. Considerable tracts of country are completely undermined by these animals. They appear to a certain degree to be gregarious, for the man who procured my specimens had caught six together, and he said this was a common occurrence. They are nocturnal in their habits, and their principal food is afforded by the roots of plants, which is the object of their extensive and superficial burrows. Azara states that they lay up magazines of food within their burrows.

The Tucotuco is universally known by a very peculiar noise which it makes when beneath the ground. A person, when he first hears it, is much surprised, for it is not easy to tell whence it comes. The noise consists in a short, but not rough, nasal grunt, which is repeated about four times in quick succession; the first grunt is not so loud, but a little longer and more distinct than the three following: the musical time of the whole is constant, as often as it is uttered.

Of specimens kept in confinement by Mr. Darwin, several were quite tame the first day, neither attempting to bite nor to run away: their movements were slow and clumsy, which apparently arose from the outward action of the hind legs;

and they are quite incapable of jumping even the smallest vertical height[1]. Mr. Darwin was informed that the Tucotuco is often found blind. This gentleman met with indications of the presence of the Ctenomys, or of an animal having similar habits, in several parts south of the Plata; in fact, at Bahia Blanca, Rio Negro, and Cape Negro. At Bahia Blanca (lat. 39°), an animal burrows in the same manner as the *Ct. Braziliensis*, and its noise is of the same general character, but instead of being repeated twice at short intervals, it is single, and is uttered at equal intervals, or in an accelerating order. At Rio Negro (lat. 41°), where burrows were again met with, the noise was likewise heard; here the sound was only repeated twice, instead of three or four times, as with the La Plata animal. The sound, moreover, was louder and more sonorous, and closely resembled that made in cutting down a small tree with an axe.

CTENOMYS BOLIVIENSIS.

Bolivian Ctenomys.

Incisor teeth very broad: fur soft, very glossy, and rather short; its general hue bright rufous brown; upper surface of head and muzzle blackish brown, and the same colour is continued in a broad, but not well defined band, along the back of the neck and on to the fore part of the back; under parts of a bright rusty yellow colour, with the exception of the space between the hind legs, and a large patch covering the fore part of the abdomen, where the hairs are entirely white: tail dark brown above, pale brown beneath: nails of fore feet immensely long.

Inhabits the plains of Santa Cruz de la Sierra.

[1] A specimen brought home in spirits by Mr. Darwin was found by Mr. Reid, who dissected it, to be destitute of a ligamentum teres to the thigh-bone, and hence, it is supposed, arises the awkward movements of the hinder extremities; this ligature, however, is certainly not always absent, since in a skeleton from a Bolivian specimen now before me, the ligamentum teres remains attached to the head of the thigh-bone.

A few specimens of a large species of Ctenomys were procured by Mr. Bridges in the above-mentioned locality, which not only differ in their size and colouring from other specimens of Ctenomys found in the high table lands of Bolivia, but present some strongly marked peculiarities in the structure of the skull, which lead me to believe it will prove a distinct species.

The fur of the animal, unlike that of other species which have come under my notice, is but slightly tinted with grey next the skin on the under parts of the body: on the upper parts the hairs are of the usual dark slate grey at the root. The sides of the body are of a rufous brown; the back is of a darker rufous brown, becoming gradually deeper towards the fore part, where the dark colour is contracted into a band which runs along the neck, and joins the black-brown colour of the upper surface of the head: from behind the ear descends a broad mark of somewhat paler hue than on the sides of the body. The hairs of the moustaches are for the most part dirty white. The feet are sparingly clothed with pale brown hairs; or, at the sides, with whitish hairs. The tail is, as usual, clothed for the most part with very short hairs, but those which spring from the point are comparatively long, being fully half an inch in length; they are of a dirty white colour, as are likewise nearly equally long hairs which spring from the mesial line of the tail both above and below, but on the apical half of the organ only.

The incisor teeth are proportionately broader than in *Ct. Braziliensis;* and the claws of the fore feet longer. The skull is remarkable for the great dilatation of the middle part of the muzzle; has the nasal bones broader at the extremity than those of *Ct. Braziliensis;* the auditory bullæ narrower, and the portion of these bulbs which enters into the plane of the occiput is considerably less extended in the vertical direction. The ridge which traverses the outer surface of the malar bone is here immensely developed. The molar teeth are broader in proportion to their length; whilst the length of the foremost molar in *Ct. Boliviensis* does not exceed that of the same tooth in *Ct. Braziliensis:* I find the transverse diameter in the former $1\frac{1}{5}$ lines, and the latter $1\frac{1}{2}$ lines. I have examined but one cranium of this species, but though I have seen several skulls (evidently adult) of the *Ct. Braziliensis,* I have found no trace in them of the extraordinarily expanded form of the muzzle which characterizes the skull of *Ct. Boliviensis.*

	Inches.	Lines.
Length from tip of nose to root of tail ...	12	0
" of tail	3	0
" of fore foot, without the nails ...		11
" of ditto, including the nails	1	$5\frac{1}{2}$
" of hind foot, without the nails ..	1	$7\frac{1}{2}$
" of ditto, including the nails	1	11

A Ctenomys in the collection of the British Museum agrees in its general colouring with the large animal just described, excepting in wanting the large white patch on the chest, but is of smaller size; not being larger than a full-grown *Ct. Braziliensis;* its incisor teeth, however, are considerably broader, being $3\frac{1}{2}$ lines in width, whilst in the largest specimens of the last-mentioned species they do not measure more than three lines. These circumstances lead me to believe it is an immature individual of the *Ct. Boliviensis.* This specimen was brought from Bolivia by Mr. Bridges, but I regret I neglected to ascertain from what part[1].

[1] As a circumstance which somewhat militates against the conclusions I have arrived at with regard to the two species of Ctenomys described in the preceding pages, I must mention that the larger species, which I have named *Boliviensis,* is from a part of Bolivia, the Fauna of which is completely Brazilian, and where one might expect to find Geoffroy's species, whilst the Bolivian animal, which I have supposed to be specifically identical with the *Ct. Braziliensis,* is from the high lands—procured, indeed, as Mr. Bridges informs me, "at an elevation of 12,000 feet, in sandy slopes and valleys, at no great distance from water"—a part of Bolivia, where nearly all the mammals are distinct from those of Brazil.

CTENOMYS LEUCODON.

White-toothed Ctenomys.

Incisor teeth white, those of the upper jaw distinctly directed forwards as well as downwards: nails of the toes relatively narrow: fur brown, slightly inclining to grey on the upper parts, and faintly tinted with rufous on the sides of the body: abdomen of a pale dirty rufous tint; muzzle dusky above.

Inhabits Bolivia, in the department of La Paz.

Mr. Bridges procured a large number of specimens of this species from San Andres de Machaca, a little south of the lake of Titicaca: they all agreed in having the incisor teeth white, or yellow white, and in this respect differed from all other specimens of Ctenomys which I had examined, they having the incisors of a deep orange colour in front: a more important point of distinction, however, is found in the direction of these teeth. Whilst in other species of Ctenomys the upper incisors descend somewhat suddenly upon leaving the jaw, in *Ct. leucodon* these teeth are very distinctly directed forwards as well as downwards: in both cases the incisors form a segment of a circle, but in the white-toothed animal, they form a smaller segment of a larger circle than in the other species; they terminate posteriorly, on the outer side of the second molar tooth, whilst in *Ct. Braziliensis* and *Ct. Boliviensis* the incisors terminate on the outer side of the first molar. Another difference, which I find constant in the specimens, is, that the nails of the toes of the fore feet are smaller, and relatively more slender; in *Ct. leucodon* the ridge on the under side of the nails is continued but a little distance from the base, whilst in *Ct. Braziliensis* it terminates a little short of the point of the nail. There was scarcely any variation in the colouring in the specimens of the White-toothed Ctenomys. The fur, as in *Ct. Braziliensis*, is of a deep slate grey at the root, both on the upper and under parts of the body: its general hue is brown, slightly inclining to grey, on the upper part of the body, and a trifle paler, and with a faint rufous tint on the sides. The tail is greyish brown above, dirty white below, and generally has a slight crest of longer whitish hairs both on the upper and under surface of the apical half. The feet are of a very pale rufous hue, but the fringe of long hairs on the outer side of the hind feet is white.

	Ins.	Lines.	Ins.	Lines.	Ins.	Lines.
Length, from tip of nose to root of tail ...	8	0	8	6	8	0
" of tail	2	4	2	6	2	4
" of fore foot, not including the nails		7½		8		8
" of ditto, including the nails ...		11	1	0		11
" of hind foot, without the nails ...	1	1	1	1	1	1½
" of ditto, including the nails ...	1	4½	1	4	1	4½

In two skulls of this species, which I have had an opportunity of examining, I found a slight sagittal crest formed by the junction of the temporal ridges, which in other species were invariably separated

CTENOMYS MAGELLANICUS.

Magellanic Ctenomys.

(*Plate* 9, fig. 2.)

Ctenomys Magellanicus. BENNETT, Proceedings of the Zool. Soc. for Dec. 1836, p. 190; Trans. Zool. Soc. vol. ii. p. 84, Plate 17.

General tint of the fur ashy grey, faintly suffused with yellow, and on the back brownish; abdomen pale ochreous yellow; tail very pale brown: the fur is moderately long, very soft, and of a deep slate grey colour at the root.

Inhabits Port Gregory, Strait of Magalhaen.

But one specimen, that I am aware, of the species of Ctenomys which inhabits the southernmost parts of Patagonia, has been brought to Europe—the specimen described by Mr. Bennett, and which was presented by Capt. P. T. King to the Zoological Society. The small size of the skull of this animal, and its proportionate narrowness, quite satisfies me that it is a distinct species[1], as these differences cannot in this case be attributed to immaturity, all the teeth being fully developed. The nails of the toes are smaller than usual in the genus.

	Inches.	Lines.
Length from tip of nose to root of tail ...	8	0
" of tail	2	6
" of fore foot and nails		$10\frac{1}{3}$
" of the longest nail		$2\frac{2}{3}$
" of hind foot and nails	1	$3\frac{1}{2}$

[1] The skull and lower jaw is represented, of the natural size, on Plate 8, fig. 5.

	Ctenomys Braziliensis. British Museum.		*Ctenomys Braziliensis.* From Geoffroy's Plate.		*Ctenomys Braziliensis.* From Bolivia.								*Ctenomys leucodon.*			
	Ins.	Lines.	Ins.	Lines.	Ins.	Lines.	Ins.	Lines.	Ins.	Lines.	Ins.	Lines.	Ins.	Lines.	Ins.	Lines.
Total length of skull	2	2½	1	11¼	2	1	2	1	1	11	1	10	2	0	1	10½
Width of occiput	1	6½	1	1¼	1	3	1	3¼	1	2	1	0⅔	1	1⅔	1	1¾
" measured from the outer side of the zygomatic arches	1	7⅔	1	2¼	1	3⅚	1	4¼	1	2⅔	1	1¼	1	4½	1	0¾
" " between orbits		5¾				5½		5⅔		5		4		5		5¼
Length of nasal bones		10				9		8				8¾		8⅓		7¾
Width of ditto behind		1¾				2¼		2⅓		2		2⅓		2		2
" " in front		5				3⅔		3¾		3½		3⅚		4		3¾
Length of zygomatic arch	1	0⅔		11		11⅔		11½		11		10½		11		10¾
Width of muzzle, measured in the middle of the inter-maxillary bones		9		5½		5⅔		5½		5⅕		5⅓		6		5
From incisor teeth to molars		8		6¾		7		6¾		6¼		6⅕		7⅓		7⅓
Width of incisors		3¾		2⅔		3		3		2⅚		2¾		3		3
Length of four upper molars, taken together		6		5		5⅔		5½		5⅕		5		5½		5¼
" of lower jaw, to apex of angular portion ...	1	11	1	7	1	8	1	7	1	6	1	5	1	7½	1	6¼
Height, measuring from condyle to lower edge of angular portion		10¼				9⅓		9		8¼		6				8¼
Greatest expanse of the two rami	2	0½	1	5¼	1	8	1	7½	1	5¾	1	4½	1	8½	1	5½

With the exception of the admeasurements in the second and third columns, the above dimensions are from crania contained in the British Museum collection. The skull of the *Ct. Braziliensis*, according to the figure in D'Orbigny's work, measures 1″ 10‴ in length; and (?) 1″ 5‴ in width. The average length of the six skulls of the *Ct. Braziliensis* would be about 1″ 11‴; and the width about 1″ 3‴.

Fossil Octodontina.

Amongst the remains of Mammalia found by Mr. Darwin at Bahia Blanca, are certain fragments which Prof. Owen refers to the genus *Ctenomys:* they consist of a portion of the skull with the first and second molars in situ, and containing the sockets of the last two molars; and a fragment of the lower jaw, with the incisor, and one of the molar teeth. The lower incisor is distinctly narrower than the corresponding tooth in the *Ct. Braziliensis;* whilst Prof. Owen states that the upper molars are rather longer than in that animal, and have the external longitudinal groove deeper. These remains are from Monte Hermoso, and, according to Mr. Darwin, are most probably contemporaneous with the Megatherium, Mylodon, and other extinct Mammals found at Punta Alta (also in Bahia Blanca); they are figured in the Zoology of the Voyage of H.M.S. Beagle — Fossil Mammalia, Pl. 32, fig. 6—11, and described at p. 109.

M. D'Orbigny also figures and describes a fragment of a lower jaw of a species of *Ctenomys* (see Voy dans l'Amer. Merid. Pl. 9, figs. 7 and 8). The fragment, being the fore portion of both rami of the jaw, contains the incisor teeth and first and second molars. M. Laurillard justly remarks with regard to this fossil, that its condition does not permit of any safe conclusions relating to the specific or non-specific identity of the animal with the recent species of the genus.

Sub-Family, ECHIMYINA.

Echymidæ. Bonaparte, Catalogo Metodico dei Mammiferi Europei, 4°. Milano (1845), p. 5.

Psammoryctina (part). Wagner, Schreb. Säugth. Supp. iii. p. 312.

Hystricidæ with complicated molar teeth, and these (with one or two exceptions only) rooted; malar bone with a distinct angular process on the lower edge: both fore and hind feet with five toes.

Geographical distribution.—One species of this section is found at the Cape of Good Hope[1], and a second in Western Africa[2]: with these exceptions the group is confined to South America and the West Indian Islands. In South America the species of *Echimyina* are only known to occur in the northern and central portions; Paraguay appears to be their southern limit. On the west side of the Cordilleras none have been found.

Genus, *Capromys*.

Isodon. Say, in Journ. of the Acad. of Nat. Sciences of Philadelphia, Vol. ii. p. 333 (Nov. 1822).

Capromys. Desmarest, in Mém. de la Soc. d'Hist. Nat. i. p. 43 (Dec. 1822).

Echimyina with rootless molar teeth; each upper molar with a single deep fold of enamel on the inner side, and two deep folds on the outer; incisor teeth narrow, and convex in front: muffle broad, with a mesial longitudinal groove; upper lip slightly cleft: ears moderate; tail of moderate length, and somewhat sparingly clothed with hairs, which do not hide the scaly skin: feet naked beneath, and covered with small tubercles; nails of the toes large, and much curved: pupil of the eye vertical.

[1] *Petromys typicus.* [2] *Aulacodus Swinderianus.*

But two species of this genus are known ; they inhabit the Island of Cuba.

By arranging the present genus at the end of the *Echimyina*, adjoining the Octodon section, its principal characters will be indicated. Thus, the species of Capromys agree with the Octodons in having rootless molars, but these teeth, in their more complicated structure, resemble the molars of the Spiny Rats (*Echimys*) ; and the more elongated form of the skull is another point which induces me to arrange the genus Capromys in the Echimys' division, with which it is more distinctly connected by other genera, than with the Octodons[1].

CAPROMYS PILORIDES.

The Short-tailed Capromys.

Isodon pilorides. SAY, Journ. of the Acad. of Nat. Sci. of Philadelphia, vol. ii. p. 333, with Plate (Nov. 1822).

Capromys Fourneiri. DESMAREST, Mém. de la Soc. d'Hist. Nat. i. p. 43.

" " RAMON DE LA SAGRA, Hist. Phys. de l'Ile de Cuba, Mammifères, p. 11 ; Atlas, Plates 3, 4, 6, and 7.

" " WAGNER, Schreb. Säugth. Suppl. iii. p. 322.

Tail shorter than the body : fur long, very harsh to the touch, strongly pencilled with black and rich yellow ; or, on the hinder part of the body, with yellowish rust colour ; ears, muzzle, throat, and chest, white, or nearly white ; abdomen rusty yellow ; toes whitish, or white : tail well clothed at the

[1] It is true that the molar teeth of the lower jaw in Habrocoma are as complicated as in Capromys ; those of the upper jaw, however, have the same simple form as in other *Octodontina ;* moreover, the cranium of Habrocoma, in the extreme narrowness of the sphenoids, the large openings for the nerves, the large occipital opening, and, indeed, in its general structure, is more removed from Capromys than are the true Octodons.

root with rust-coloured hairs, sparingly clothed with brown hairs beyond.

The *Capromys pilorides* inhabits the forests of Cuba, where, according to M. Ramon de la Sagra, it is seen to climb the trees with great activity, both for safety when danger threatens, and to seek its food, which not only consists of fruits and the leaves and bark of trees, but likewise the flesh of animals, and especially of a species of Lizard, of the genus Anolis, which it hunts with great perseverance. It is readily tamed.

The specimens of Capromys in our English collections appear to me to be all referrible to the present species. In the museum of the Zoological Society are two adult specimens and a skeleton, which I will proceed to describe:—

One of these specimens, which lived in the Zoological Society's museum for some time, has the head almost entirely pure white, the crown and back part only being coloured like the back; a few dark hairs, however, are seen scattered with the others, but chiefly on the cheeks. The chin, throat, and chest, are pure white; the abdomen rusty yellow; on the sides and upper parts of the body, the hairs are black, excepting quite at the root, where they are greyish brown; and near the point, where they have a moderately broad ring of rich yellow: on the hinder parts of the body the yellow becomes deeper, and approaches to a rust colour, and at the root of the tail the hairs are entirely of a bright rust colour; about two inches of the tail is densely clothed with similar hairs; on the remaining portion, the hairs are of a deep brown colour; short, and being by no means abundant, they do not hide the scaly skin. The ears are tolerably well clothed with longish white hairs. The outer surface of the limbs is chiefly black, being but slightly pencilled with yellow; the fore legs are rusty-white internally, and the hind legs have the inner side clothed with bright rusty yellow hairs.

like those on the abdomen. The hind feet are almost entirely black, but from the end of the toes spring long white hairs: the fore feet are white. The dimensions of this specimen are given in the first column, which likewise includes the admeasurements of its skull.

The second specimen referred to, differs from the first in having only the fore part of the muzzle white; the fore legs slightly suffused with chestnut brown on the outer surface, and the hind legs and haunches almost entirely of a rich chestnut red, as well as the root of the tail; the throat and chest dirty white; the abdomen of a pale rust colour, but with many interspersed white hairs; the feet dusky brown; the toes for the most part clothed with dirty white hairs; the long hairs projecting over the nails of the hind feet are, however, most of them dusky; the hairs on the inner surface of the ears are some of them rufous, and others dirty white.

The skull figured in plate 12, was removed from a specimen in the British Museum, which I conceive is an immature individual of the present species: its fur is less harsh (though by no means soft) than that of the two adult animals already described; and on the upper parts and sides of the body the animal is pencilled with black and dirty yellow; all the hairs are, however, of an ashy grey tint at the root. On the head and cheeks the hairs are for the most part black at the extremity, and annulated with dirty yellowish white; the tip of the muzzle is grey white, as are also the chin and throat, whilst the abdomen is dirty yellow, and rather pale. The feet are brown, obscurely pencilled with a pale colour; the hairs springing from near the roots of the nails are white; the nails dark horn-coloured. The tail is clothed with hairs which average about half an inch in length, and which, though tolerably abundant, do not quite hide the scaly skin, excepting at the root of the organ; here they are of a rusty red hue, and beyond, the hairs are almost entirely of a pale rusty yellow hue, obscurely tinted with brownish on the upper part of the apical portion of the tail. The ears are clothed with greyish white hairs internally, and the hairs of the moustaches are white at the root, and black beyond. The length of the animal, from the tip of the nose to the root of the tail, is 12″ 0″; of its tail, 7″ 3‴; of the ear, 9‴; of the fore foot and nails, 1″ 5½‴; and of the hind foot and nails, 2″ 10½‴.

A specimen of *Capromys pilorides*[1], represented on Plate 3 of M. Ramon de la Sagra's work, agrees with the animal which furnishes the subject of my first description, with the exception that the fore part of the head is not white, as in the Zoological Society's specimen—a peculiarity which may probably be attributed partly to age, and partly to the circumstance of the animal having been long in confinement; and on Plate 4 of the same work is represented an almost uniformly sandy yellow variety of the animal.

	Zoological Society.		Zoological Society.	
	Inches.	Lines.	Inches.	Lines.
Length from tip of nose to root of tail	20	9	22	6
" of tail			11	6
" from nose to ear	3	3	3	5
" of ear	1	0	1	0
Width of ditto	1	3	1	4
Length of fore foot and nails ...	2	5	2	6
" of ditto, not including the nail	2	0½	2	3
" of hind foot and nails ...	4	0	4	0
" of ditto, without the nails ...	3	5	3	6
Height of animal, about	8	0	8	0
Length of skull	3	8	3	10
Width of ditto	1	11	1	10½
" of ditto between orbits ...	1	0½	1	0½
" of ditto at occiput	1	3½	1	2¾

[1] Named *Capromys Furnieri* on the Plate. I may here mention that the earliest name given to this animal (*Isodon pilorides*) has not been adopted by mammalogists: the reason given for rejecting the generic term is, that it had been previously used for a different genus of Mammalia. I am aware, however, of no objection to Say's specific name, and therefore retain it.

	Zoological Society.		Zoological Society.	
	Inches.	Lines.	Inches.	Lines.
Length of nasal bones	1	2	1	2
Width of ditto behind		$4\frac{3}{4}$		$3\frac{3}{4}$
" " in front		$7\frac{2}{3}$		6
Length of zygomatic arch	1	$8\frac{2}{3}$	1	$8\frac{1}{2}$
Greatest depth of ditto		$7\frac{1}{4}$		$6\frac{1}{3}$
Width of upper incisors		$2\frac{5}{6}$		3
From incisor to molar teeth	1	0	1	0
Length of four upper molars taken together		11		$9\frac{1}{4}$
Width of foremost molar		$2\frac{1}{2}$		2
Length of lower jaw	2	9	2	$4\frac{1}{2}$
Height of ditto behind	1	$3\frac{2}{3}$	1	$1\frac{3}{4}$
Length of angular portion	1	$7\frac{1}{2}$	1	3

The skeleton of the *C. pilorides* (Pl. 14), which is described by Prof. Owen[1], and well figured by M. Ramon de la Sagra, agrees in all essential points with the skeletons of the *Octodontina*, but has sixteen pairs of ribs, and consequently sixteen dorsal vertebræ (*d. v.* on the Plate). Of lumbar vertebræ (*l. v.*), there are seven; of sacral (*s. v.*), four; and of caudal vertebræ (*c d. v.*): I find twenty, in a skeleton in the Zoological Society's museum; one or two of the terminal vertebræ are, however, wanting in the skeleton. The last two or three of the cervical vertebræ (*c. v*) have slightly elevated spines; the foremost dorsal vertebra is furnished with a short spinous process, and the third, and several following dorsal vertebræ, differ from the corresponding bones in the Octodons, in having the spinous process as much produced (though not so stout) as that of the second dorsal. The femur (*f*) wants the prominent ridge on the outer surface, which is found in *Octodon;* in other parts I see no differences worthy of note. The admeasurements of the skull belonging to the skeleton referred to, are given in the second

[1] Proceedings of the Zoological Society, Part 2 (1832), p. 100: other parts of the anatomy of the Capromys will be found described in detail by Prof. Owen in the same Part of the Proceedings—see pp. 68 to 76.

column of dimensions in the preceding page; the principal admeasurements of other parts of the skeleton are as follows:—

	Inches.	Lines.
Total length (measuring from the atlas vertebra)	22	6
Length of scapula (Pl. 14 *s.*)	2	1
Greatest width of ditto	1	9
Length of humerus (*h.*)	2	7
" of ulna (*u.*)	3	1
" of radius (*r.*)	2	6
" of fore foot	2	0
" of femur (*f.*)	3	3
" of tibia (*t.*)	3	3½
" of hind foot	3	7
" of tail, about	10	0

CAPROMYS PREHENSILIS.

Prehensile-tailed Capromys.

Capromys prehensilis. POEPPIG, Journ. of Acad. of Nat. Sci. of Philadelphia, iv. Part 1, p. 11. 1824.
" " GUERIN, Magaz. de Zool. 1834, cl. 1.
" " RAMON DE LA SAGRA, Hist. Phys. de l'Ile de Cuba, Mammifères, p. 12; Atl. Pl. 5 (and Pl. 8, figs. 2 and 3—skull). 1840.
" *Poeyi.* GUERIN, Mag. de Zool. 1834, cl. 1, Pl. 15.

Tail nearly equal to the head and body in length; fur moderate as to texture; on the upper parts, of a mixed grey and rufous tint; neck yellowish; forehead, cheeks, and throat yellow-white; chest and abdomen white; tail rufous at the base; toes clothed with whitish hairs.

Habitat, Cuba.

The *Capromys prehensilis* is known in Cuba by the name *Hutia Carabali,* and is said to confine itself to parts of the forests which are remote from the habitations of man, and to be more shy and less tameable than the *Hutia Congo,* or *C. pilorides.* Like the last mentioned species, it is an expert climber, and, according to M. Ramon de la Sagra, it confines itself to the uppermost branches of the trees. Its tail is prehensile at the extremity; and the animal, assisted by this organ, is enabled to cling with security to the small twigs of the trees, or to the parasitic plants with which they are overgrown, and amongst which it usually conceals itself.

From *C. pilorides*, the present species is chiefly distinguished by the greater length of its tail, the less harsh nature of its fur, and its smaller size. The hairs of the fur are soft next the skin, but rigid at the point; on the upper parts of the body they are black at the root, grey in the middle, and rufous at the point. The tail is naked beneath at the extremity; the nails of the toes white.

	Inches.	Lines.
Total length, from tip of nose to end of tail	25	0
Length of tail	12	3
" of fore foot	1	7
" of hind foot and nails	3	8
" of ear		7

In the preceding description is incorporated the chief characteristics of the *C. prehensilis*, as pointed out by Dr. Poeppig. The skull of this animal, according to the well executed figures in M. De la Sagra's work, differs from that of *C. pilorides* in being considerably smaller, in having the cranial portion relatively narrower, and the zygomata more arched: the foremost molar of the lower jaw, it would appear, is more elongated, and wants the small third notch on the fore part of the inner side. Its principal admeasurements (taken from the figures alluded to) are as follows:

	Inches.	Lines.
Total length of cranium	3	0
Width from outer surface of zygomatic arches	1	6½
" of occipital portion		10½
Length from upper incisor to molar teeth ...		11
" of four upper molars, taken together		7

Capromys Poeyi. GUERIN.

M. Guerin describes the *C. Poeyi* as follows :—

Fur soft and flexible ; of a chestnut colour, pencilled with rufous and yellow ; the hairs of a deep chestnut tint next the skin ; a great portion of them rufous at the point, whilst some are yellowish at that part : the head brown, with the exception of the forehead and cheeks, which are of a palish rufous yellow colour ; throat and abdomen white ; hairs of the moustaches white at the root, and of a chestnut brown hue beyond ; feet of a palish chestnut colour ; tail clothed *throughout* with long rufous hairs. Total length (including the tail) 26″ 4‴ ; of tail, 13″ 2‴.

MM Desmarest and Gervais, after having compared the *C. Poeyi* with two specimens of the *C. prehensilis,* arrived at the conclusion that the differences which were perceptible did not warrant the specific separation of these animals. The points of distinction observed were these—

C. prehensilis.	*C. Poeyi.*
Fur coarse, of a deep chestnut colour ; dirty white on the abdomen, and slightly suffused with yellow : head of a pale golden brown, with some rust-coloured hairs.	Fur soft and flexible, of a chestnut colour, pencilled with rusty yellow ; abdomen whitish.
Hairs of the moustaches whitish at the root, and rusty at the point.	Hairs of the moustaches whitish at the root, and of a deep chestnut colour at the point.
Forehead yellowish white.	Forehead and cheeks rusty yellow.
Toes covered with whitish hairs; nails deep brown.	Toes covered with ferruginous hairs ; nails whitish.
Tail rather more than half the length of the body : thickly beset with hairs ; the extremity naked beneath.	Tail nearly as long as the body, entirely covered with long, brushy hairs.

Genus, *Plagiodontia.*

Plagiodontia. F. CUVIER, in Annales des Sciences Naturelles, vi. (Second Series), p. 347 (1836).

Echimyina with rootless molars, those of the upper jaw with a single deep fold of enamel on the inner side running obliquely forwards and inwards from near the posterior angle of the tooth, and a second deep fold entering from the outer side, near the anterior angle, and running in an opposite direction; molars of the lower jaw each with two deep folds on the inner side, and with one less deep fold on the outer side: toes 5—5: ears small: tail short, destitute of hair, and with a scaly skin.

But one species of this genus is known, an animal inhabiting the Island of St. Domingo, and which is evidently very nearly allied to the Capromys of the neighbouring island. Its short and naked tail, combined with certain differences observable in the structure of the molar teeth, furnish the chief distinguishing characters. Compared with the molars of the Capromys, those of the Plagiodontia are shorter in proportion to their width—in fact, the crowns of these teeth are very nearly equal in length and breadth, and approach to a rounded form: the foremost molar of the lower jaw, however, furnishes an exception, being decidedly longer than broad: and, with regard to the indenting folds of enamel, there is a well-marked difference in the upper molars of the Capromys and Plagiodontia; these teeth, in the latter animal, having but one external or internal fold, whilst in the former they have two folds on the outer side; and, moreover the indenting folds of the molars in Plagiodontia are remarkable for their oblique direction.

PLAGIODONTIA ÆDIUM.

(*Plate* 13, fig. 2).

Plagidontia ædium. F. Cuvier, Ann. des Sci. Nat. vi. (New Series), p. 347, Pl. 17.

Capromys (*Plagiodontia*) *ædium.* Wagner, Schreb. Säugth. Suppl. iii. p. 325.

The *Plagiodontia ædium* is described by M. F. Cuvier as a large Rat-like animal, somewhat less than the Common Rabbit, standing low on the legs, with small ears, and a short naked tail, having a scaly skin. Its fur is composed of two kinds of hair; the one fine and silky, which forms the chief clothing of the animal, and the other, longer and coarser: the finer hairs are grey at the root, and of a fawn colour externally, the latter colour occupying one-fourth of the entire length of each hair; the coarser hairs (which are tolerably abundant) are black, excepting on the under parts of the body, where they are whitish; here the softer hairs are of a pale yellowish brown hue.

	Inches.	Lines.
Length from tip of nose to root of tail ...	13	2
" of tail	5	5

This animal feeds upon roots and fruits; and being esteemed very good eating, it is so much sought after, that it has become very rare.

Genus, *Myopotamus*[1].

Myopotamus. (Commerson) Geoffroy, Annales des Muséum, vi. p. 81, (1805).

Mastonotus. Wesmael, Bullet. Roy. des Sci. de Bruxelles, Année 1841, 2me Partie, p. 61.

[1] From μῦς, a rat; and ποταμὸς, a river.

Echimyina with semi-rooted molar teeth, of which the hindermost in each jaw are the largest; each upper molar with two external, and two internal deep folds of enamel; the lower molars with three internal folds, and one indentation entering from the outer side: incisor teeth very large: palate much contracted between the front molar teeth: hind feet webbed: tail of moderate length, sparingly clothed with short hairs.

MYOPOTAMUS COYPUS.

The Coypu.

(*Plate* 15, fig. 1).

Mus Coypus.	MOLINA, Saggio sulla Storia Naturale de Chili, 8vo. ed. (1782), p. 287; 4to. ed. (1810), p. 239.
" *casteroides.*	BARROW, in Linnean Transact. xi. p. 168 (1812).
Myopotamus Coypus.	(COMMERSON) GEOFF., Ann. du Mus. vi. p. 81, (1805).
Hydromys "	GEOFF., l. c.
" "	DESM., Mamm. p. 296 (1822).
Potamys Coypou.	DESM., Dict. des Sci. Nat. xliv. p. 491.
Myopotamus Bonariensis.	RENGGER, Naturgesch. der Säugth. von Paraguay, p. 237 (1830).
Mastonotus Popelairi.	WESMAEL, Bullet. de l'Acad. Roy. des Sci. de Bruxelles, Année 1841, 2me Partie, p. 61.
Guillinomys Chilensis.	LESSON, Nouv. Tab. du Règne Anim. p. 126, (1842).
Quouiya.	AZARA, Essais sur les Quadr. de Paraguay, ii. p. 1.
Coypu Rat.	SHAW, Gen. Zool. ii. Pt. 1, p. 48.

Ears of moderate size; tail nearly equal to the body in length: fur long: the under fur very dense and soft; upper parts of the animal pencilled with dusky and brownish yellow, in about equal proportions; sides and under parts with the prevailing tint brown-yellow; tip of muzzle and chin white; a yellow patch immediately beneath the ear-opening: feet dusky brown.

Immature specimens—General hue rich brown; on the sides of the body, inclining to yellow.

The Coypu is very nearly equal in size to the Beaver, and bears a considerable superficial resemblance to that animal; its tail, however, is cylindrical, has a scaly skin, and is scantily clothed with short stiff hairs, like that of the Rat. It inhabits the rivers and streams of a great portion of South America, occurring on both sides of the Andes. On the eastern side it extends from Peru[1] southwards, to the Rio Chupat, in 43° 20′[2]. In the eastern portions of Brazil, the Coypu was not met with by the Prince of Nieuwied. On the west side of the Andes, this animal ranges, according to Mr. Darwin, from the valleys of Central Chile, (Lat. 33°) to 48° south, or perhaps even somewhat further, but does not extend to Tierra del Fuego. In the Chonos Archipelago, Mr. Darwin states, these animals, instead of inhabiting fresh water, live exclusively in the bays and channels which extend between the innumerable small islets of that group. They make their burrows within the forest, at a short distance from the rocky beaches. The inhabitants of Chiloe, who sometimes visit this Archipelago for the purpose of fishing, state that the Coypus here do not live solely on vegetable matter, as is the case with those inhabiting rivers, but that they sometimes eat shell-fish. The Coypu is said to be a bold animal, and to fight fiercely with the dogs employed in chasing it. Its flesh is white and well flavoured. At Buenos Ayres an extensive trade in the skins of the Coypus, there called Nutrias, or Otters[3], is carried on.

I learn from Mr. Bridges, that in the southern parts of Chile the Coypu abounds more in the lakes than in rivers, and more especially frequents those waters in which the *Typha latifolia* and *Scirpus* are plentiful, and serve to give them shelter. The Coypu is most abundant, according to Mr. Bridges' observations, on the borders of the River Maypo,

[1] Tschudi, Fauna Peruana. [2] Darwin, Voy. of the Beagle.
[3] Voy. of the Beagle, pp. 78-9.

near Santiago, and in the lakes of Aculeo and Quintero. In the breeding season, which is in the months of September and October, the Coypus emit a mournful cry, which much resembles that of a child when in distress. The female Coypu evinces much attachment for her young, and swims with them on her back, until they are sufficiently large to follow the parent. This habit helps to explain the singular position of the nipples noticed in the female Coypu. Of these, four were found, by M. Lereboullet, on each side of the body, and situated rather above the mesial line of the flanks, the foremost being placed behind the shoulder, and the hindermost in front of the thigh[1].

A series of specimens of the Coypu in the British Museum collection exhibits the differences of age in this animal from an early period upwards. A fine large specimen of two feet in length, without including the tail, has the upper parts of the body pencilled with black and brown-yellow, in about equal proportions; on the sides of the body the yellow hue slightly prevails; and the under parts can scarcely be said to differ from the flanks in the colouring. The long, dense, and soft under fur is of a brown-black hue on the upper parts of the animal, dusky brown on the sides of the body, and of a less deep brown on the under parts. Of the long hairs, which are tolerably abundant, and by no means harsh, the longest are entirely black on the upper parts of the body; and others, which are more numerous,

[1] See M. Lereboullet's Paper on the Anatomy of the Coypu, in the Mémoires de la Société du Muséum d'Histoire Naturelle de Strasbourg, iii. Pt. 3 (1843). Mr. Christie had previously called the attention of naturalists to the position of the mammæ in the Coypu, which he describes as being "situated extremely high up on the sides"—see Proc. of the Zool. Soc. for 1835, p. 182. This position of the nipples, rather on the sides than the under part of the body, perhaps will be found a common circumstance in the Hystricine division of Rodents, since I have found them so situated in the genera *Lagostomus*, *Octodon*, *Habrocoma*, and *Nelomys*: here, however, they are perhaps placed somewhat less high on the flanks than in the Coypu.

are black at the point, and annulated with brownish yellow towards the point. The feet are dark brown: the naked soles are black. The fore feet present no points of structure to arrest our attention, there being the usual number of toes, and a short thumb, with a curved and truncated nail of moderate size. The hind feet are remarkably large, and four of the long toes are united by an ample web: the outer toe is free, but has a narrow band of loose skin along its inner surface. The nails of the toes are long, hollow beneath, and by no means deep, nor strongly compressed: those of the hind feet, which are the longest, are but little arched. The tip of the muzzle is clothed with small white hairs throughout (very distinctly so in the young animal), with the exception of a narrow space adjoining the nostrils. The chin is white, and so are the portions of the lip which overlap the palate anterior to the molar teeth. The upper lip is not cleft, being merely notched immediately above the large incisor teeth, which latter are of a deep orange colour in front. The tail is nearly as long as the body—not including the head—and sparingly clothed with short stiff hairs, which spring from between the scales with which the skin is covered, as in the Common Rat. The admeasurements of this specimen are given in column No. 1.

A second specimen of rather smaller size differs only in having the flanks considerably suffused with yellow[1]. Other specimens, still smaller, are almost entirely of a rich brown colour, somewhat inclining to yellow on the sides and under parts of the body: and a very young specimen, of about eight inches in length, differs from these last, in having the tail well covered with bright reddish brown hairs. All the specimens have dark brown feet, the tip of the muzzle and chin white, and a large yellow patch immediately beneath the ear opening.

[1] Its dimensions are in column No. 2.

	No. 1.		No. 2.		No. 3.	
	Ins.	Lines.	Ins.	Lines.	Ins.	Lines.
Length from tip of nose to root of tail ...	24	6	18	6	17	0
" of tail	17	6	12	9	10	0
" from tip of nose to ear	4	8	3	4	3	3
" of ear		8½		7½		7½
Width of ditto	1	4	1	1½	1	1½
Length of fore foot and nails	2	9	2	3	2	2
" of the nail of the middle toe of ditto		6		5		5
" of hind foot and nails	5	6	4	7	4	5
" of the nail of the middle toe ...		6		5		5½

In the structure of its teeth the Coypu will bear a tolerably close comparison with the Beaver; the molars of the Coypu, however, are rooted, the foremost in either jaw distinctly so; the hindermost are more deeply implanted in their sockets, and have short roots[1]. With regard to the folds of enamel, they are disposed as follows in the teeth of the adult animal: each upper molar has a loop or fold entering the crown in the middle, and on the inner side, and extending half way across the tooth, and on the outer half of the crown, are seen three transverse areas enclosed by enamel, each extending from the outer side of the tooth to the middle, or very nearly so; the hindermost molar, however, differs from the rest in having but one of these isolated areas; that corresponding to the second area in the other teeth here assumes the form of a loop, entering from the outer side of the tooth, and the third area in like manner is replaced by a fold of

[1] The molars of the Coypu do not become rooted at so early a period as in the Rats and Echimys group, nor does the body of one of these teeth begin to divide into roots so near the crown as in those animals: hence (with the exception of the foremost molar of each series) we cannot perceive that they possess roots until they are removed from the jaw: in the Rats the roots are visible when the tooth is in the socket. Molars presenting the conditions of those of the Coypu I distinguish by the term semi-rooted.

enamel, but this enters from the inner side of the tooth. The molars of the lower jaw have one moderately deep fold entering from the middle of the outer side, and three deeper folds on the inner side, except it be in the two foremost molars, where the inner folds have assumed the form of isolated areas, corresponding to those on the outer half of the upper molars. In the half-grown animal, where but three molars of each series are developed, the foremost molar has one internal and three external folds of enamel; the second has two internal and two external folds, and the third molar has three loops of enamel, two external and one internal, and a transverse lobe behind, which is entirely surrounded by enamel.

The skeleton of the Coypu is essentially the same as in the genera *Echimys, Capromys,* and *Octodon,* and therefore differs considerably from that of the Beaver, with which animal the Coypu has been associated. The incisor teeth of the animal before us being very large, the zygomatic arch is deep and strong, and the temporal ridges meet to form a slight crest: the palate is moderately broad behind, but is exceedingly contracted between the foremost molar teeth—these teeth, indeed, almost meet. The form of the occiput is worthy of notice, since, besides the ordinary processes which correspond to the mastoid, and which are very large, there are two other processes; they are smaller than the first, above which they are placed, but are still very prominent: a small portion of the petro-tympanic bone separates the two processes on either side of the occiput. The lower jaw is chiefly remarkable for the small size of the coronoid process (a mere tubercle, which projects but little above the crowns of the molar teeth), and for the angular portion being thrown boldly outwards, and thus giving a great width to the space between the rami behind. The lower edge of the angular portion is much expanded, and produced into an angle on the inner side. The vertebræ are—dorsal, 13 ; lumbar, 6 ; sacral, 4 (anchylosed); caudal, 26 [1].

	Inches.	Lines.
Total length of skeleton	32	8
of which the skull is	4	8
Width of skull	2	9
Length of nasal bones	1	9
Width of ditto in front		9⅓
" " behind		6
" between orbits		11
From incisor to molar teeth	1	4
Width of incisors		6
Length of four upper molars, taken together	1	2
" of lower jaw	4	0
Width behind	3	2
Length of scapula	2	6
" of humerus	2	8
" of ulna	3	9
" of radius	3	0
" of fore foot	2	4½
" of femur	3	5
" of tibia	4	1
" of hind foot	5	0
" of caudal vertebræ, taken together ..	16	7

[1] Mr. Martin, who gives a detailed account of the anatomy of the Coypu in the Proceedings of the Zoological Society for November, 1835, expresses his regret that he was unable to examine the bones of the extremities, since Cuvier notices a peculiarity in those of the *carpus*. I may mention that in two carefully prepared fore feet of the Coypu now before me, I can perceive nothing abnormal—they present the ordinary Rodent structure in this part; the true carpal bones are eight in number; three in the upper range, viz. the pisiform (Pl. 14, fig. 2 *ps*) and cuneiform (*c*), and a large transverse bone formed by the junction of the scaphoid and lunare (*sc. l*); and, besides the ordinary four bones in the lower range, there is a small triagular bone (*) wedged in between the os magnum (*m*), trapezoides (*ts*), and lunare. Beyond these are two other bones, which are less constant, a large flat bone (†) on the inner side of the wrist attached partly to the scaphoid and partly to the trapezium (*tm*); and a somewhat smaller flat bone on the outer side of the wrist, which joins the pisiform.—The bone *un*, in the figure referred to, is the unciform; and the bones *mc* 1 to *mc* 5, are the five metacarpals; *mc* 1 being the metacarpal bone of the thumb, or inner toe. The tarsus of the Coypu is represented on the same Plate, fig. 3: *cm*, is the calcaneum; *a*, the astragalus; *n*, the naviculare; *cb*, the cuboid; *c* 1, the internal cuneiform; *c* 2, the middle cuneiform; and *c* 3,

Genus, *Cercomys*[1].

Cercomys. F. Cuvier et Geoff., Mammifères, 60me livr. Pl. 60 (Sept. 1829); Nouvelles Annales du Muséum, i. p. 449 (1832).
" Wagner, Schreb. Säugth. Suppl. iii. p. 349.

Echimyina with the molar teeth of equal size, or very nearly so; the crowns of a rounded form, those of the upper jaw with a deep fold of enamel entering from the inner side, and with three transverse areas surrounded by enamel, and occupying the outer half of the tooth; molars of the lower jaw like those of the upper, but with the areas and enamel fold, reversed in position: feet normal; tail long and scaly; ears moderately large: fur moderate as to texture.

Habitat, Brazil.

The skull of the *Cercomys cunicularius*, figured by M. F. Cuvier[2], greatly resembles that of the species of Echimys, but the molar teeth of the animal agree very closely in their structure with those of the Coypu, differing, however, in being of equal size: beyond this, the genus *Cercomys* may be distinguished from the *Myopotamus* by the hind feet being destitute of web, showing that the habits of the animal are dissimilar; and from the species of *Echimys* generally, it differs in having the body clothed with fur of the ordinary kind—destitute of bristles, or spines. In stating that the feet are of normal structure in the Cercomys, I wish it to be understood that they are provided with five toes, and that the thumb, or inner toe of the fore foot, is small, and has a small nail.

the external cuneiform bones: *mt* 1 to *mt* 5, are the five metatarsal bones, that of the inner toe being marked *mt* 1.

[1] So named from its having a tail like that of the Rat.

[2] Nouv. Ann. du Mus. i. Pl. 19, figs. 1 and 2.

CERCOMYS CUNICULARIUS.

The Cercomys.

Cercomys cunicularius (Cercomys du Brézil). F. CUVIER, Mamm. iii. Pl. 60; Nouv. Ann. du Mus. i. 449, Pl. 18, fig. 1, molar teeth of upper jaw; Pl. 19, figs. 1 and 2, cranium.
" " WAGNER, Schreb. Säugth. iii. p. 350.

Ears moderately large; tail about equal to the head and body in length; tarsus long: fur moderate as to texture, its general tint deep brown, the sides of the body of a paler hue, and the under parts impure white.

The animal upon which M. F. Cuvier founds his genus *Cercomys* was brought by M. Auguste de Saint-Hilaire from the province of Minas Geraes in Brazil, and is the only known species of the genus. In size and general appearance, it bears considerable resemblance to the Common Rat (*Mus decumanus*), but, according to M. Cuvier, its head is more arched, and it has larger ears: its nostrils are very small, and surrounded only by a narrow naked space; the hind feet are long. Although the general hue of the fur is deep brown, the hairs taken separately have their visible ends either black or fawn colour; the hairs of the ordinary fur being grey at the base, and of a fawn colour at the point, whilst the longer and coarser interspersed hairs are black at the extremity. The dimensions of the animal are not given by M. F. Cuvier, but it is said to be represented of the natural size on his Plate; according to which it measures, in a straight line, from the tip of the nose to the root of the tail, about 5¾ inches; its tail, 6½ inches; ears, 8 lines; and tarsus (including the nails), 1 inch and 10 lines. The skull (according to the figure) is in—

	Inches.	Lines.
Total length	1	11⅓
Width		10¼
" between orbits		5⅓

Genus, *Petromys*.

Petromys. SMITH, South African Quarterly Journal, No. 5, p. 2 (1831); and id. Second Series, vol. i. p. 146.

Echimyina with the molar teeth very nearly equal in size, and rooted; their crowns of a quadrate form, placed obliquely in the jaw, so that the inner front angle is the most forward; each tooth with a single indenting fold of enamel on either side, the folds of opposite sides meeting in the mesial line of the tooth: incisor teeth small and compressed, rather deeper than wide, and flat in front: body clothed with fur of the ordinary kind; ears rather small—hairy: feet small; toes 5—5, the thumb exceedingly small, but provided with a small nail: tail about equal to the body in length, well covered with moderately long hairs, which gradually increase in length towards the apex of the organ, where they are longest.

Habitat, South Africa.

The little animal on which the genus *Petromys* is founded is remarkable amongst the Old World Rodents for presenting characters which are almost peculiar to the species of South America. In having a brushy tail, and a separate small groove within the ant-orbital opening for the infra-orbital nerve, as well as in having but one indenting fold of enamel on either side of each molar tooth, it agrees with the Octodons, but it differs from those animals, and resembles the *Echimyina*, in having the molar teeth rooted. The extremity of the muzzle is clothed with small hairs, with the exception of a very narrow space around the nostrils, which is naked: the upper lip is merely notched immediately above the incisor teeth. The feet are rather small, and naked (and apparently smooth) beneath; the toes are provided with small, curved, and compressed claws. The stomach is oblong, curved, and of nearly equal diameter at each extremity; the cœcum is voluminous[1].

The skull of the Petromys is remarkably depressed; the cerebral portion short and broad; the facial portion narrow, and slightly elongated; the upper surface almost flat; the ant-orbital opening large; the zygomatic arch short and deep, the malar bone produced, above, into an obtuse post-orbital process; and, below a distinct ridge which crosses its outer surface almost horizontally—descending slightly, however, towards the fore part of the zygoma—is a narrow plate of bone, which, commencing near the anterior root of the zygoma, runs back, and terminates abruptly near the middle of the malar bone. The incisive foramen is large, extending from a point opposite the front edge of the foremost molar, to within about $\frac{1}{12}$th of an inch from the incisors, and with a width of $1\frac{1}{3}$ lines. The palate between the molar teeth is narrow, being about $1\frac{1}{4}$ lines in width at the widest part, which is behind, and scarcely one line in width between the front molars. The post-palatal emargination is V-shaped, the apex of the angle being rather behind the last molar teeth. The auditory bullæ are tolerably large, and very convex. The occipital bone sends out two branches on either side, as in Echimys: the one, directed outwards and downwards, terminates a little behind the ear-opening; the second is widely separated from the first by a portion of the petro-tympanic bone, and descends vertically, forming the ordinary mastoid process of the ex-occipital bone. The coronoid process of the lower jaw is small, and the condyloid process is but little elevated. The angular projection on the inner side of the lower edge of the angular portion is but little marked.

[1] These parts are beautifully figured, together with other details, in Dr. Smith's "Illustrations," &c. No. ix. Pl. 21.

PETROMYS TYPICUS.

The Petromys.

(*Plate* 17, fig. 1)

Petromys typicus. SMITH, South African Quarterly Journal, No. 5, p. 2; Illustrations of the Zoology of South Africa, No. ix. Mammalia, Pls. 20 and 21.

Fur moderately long and soft; prevailing colour rufous-brown; head (with the exception of the muzzle, which is rufous) and fore parts of the body greyish brown; throat inclining to white; abdomen palish rufous-yellow: tail of the same colour as the body at the root; the remaining portion black.

The Petromys, we learn from Dr. Smith, is found on the rocky hills which occur towards the mouth of the Orange River. It seems to feed principally upon vegetable substances, and was seen to eat the flowers of certain syngenesious plants, particularly those of a species of *Senecio.* When feeding it showed little fear of man, and, unless closely approached, continued actively devouring flower after flower, although it could not have been insensible to its being observed. When approached, however, too closely, it retreated amongst the loose stones, or in a crevice in the rock.

The hairs of the fur of this animal are almost entirely of one kind, and on the back and flanks, the outer surface of the hind legs, and basal portion of the tail, the fur is almost of an uniform bright brownish rufous colour; at the root, however, the hairs are of pale brown-grey, and the extreme points of many of the hairs are dusky. The upper surface of the head, and the cheeks, as well as the neck and shoulders, are clothed with ashy, or brownish grey hairs, which are finely annulated with rufous near the point: the tip of the muzzle and orbit of the eye are bright rufous: the moustaches, which are very long, are entirely black. The ears are rather small, emarginated behind, and moderately well clothed with small hairs, which are for the most part of a rufous tint, but from the edges of the ears spring numerous, longish, and extremely fine dusky hairs. The chin is rufous; the throat rufous white, with a slight mixture of grey, the hairs being of the latter colour at the roots. The hairs on the abdomen are of a palish rufous yellow colour externally, and ashy grey at the root: the feet are of the same bright brown-rufous, or brown-rust, as the back; their naked soles are black, as are also the nails of the toes. The tail is well clothed with somewhat stiff brown-black hairs, having an average length of about half an inch; near the root they are distinctly shorter, and at the opposite extremity they are most of them about three-quarters of an inch in length; about one inch of the tail, at the root, is clothed with fur like that on the back, both in texture and colour.

	Inches.	Lines.	Inches.	Lines.
Length from tip of nose to root of tail	7	6	7	5
" of tail (not including the hair)	5	8	4	10
" of ear		4¾		4
Width of ditto		7¾		6½
Length of fore foot		8⅔		8
" of hind foot	1	3⅔	1	3½
" length of skull	1	8¼	1	7⅔
Width of ditto	1	0	1	0
" between orbits		5½		5
Length of nasal bones		7		6⅚
Width of ditto behind		1½		1¼
" " in front		2⅚		2⅔
Length of zygomatic arch		8		8
Greatest depth of ditto		2⅙		
From incisors to molar teeth ...		4½		4½
Length of four upper molars, taken together		4⅓		4½
Width of incisors		1⅓		1⅓
Length of lower jaw	1	1	1	0¾
Height of ditto (measuring from the condyle)		6		5

Genus, *Dactylomys*.

Dactylomys. Is. Geoffr., Magasin de Zoologie, par M. Guérin-Méneville, Année 1840, Mammifères, p. 27.
" Wagner, Schreb. Säugth. Suppl. iii. p. 347 (1843).

Echimyina with but four toes to the fore feet, and these armed with short, convex nails; the intermediate pair of toes long: the series of molar teeth of opposite sides of the upper jaw converging and nearly meeting in front; all the molars relatively large; those of the upper jaw each composed of two separate lobes, which are contracted and pointed on the inner side of the tooth; and broad, and with a deep fold of enamel, on the outer side: the foremost molar of the lower jaw with one external and two internal folds of enamel, and the remaining molars each with two lobes, of which, one—forming the hinder part of the tooth—is narrow and transverse; the other is broad on the inner side of the tooth, where it has a deeply indenting enamel fold.

The genus *Dactylomys* is founded by M. Isidore Geoffroy St.-Hilaire upon an animal which had long previously been known, having been described by Desmarest and others, under the name *Echimys dactylinus*. From Echimys it is readily distinguished by the total absence of a thumb to the fore foot, the inequality of the toes of the same foot, the middle pair being considerably longer than the outer toes, and by these toes being provided with short, convex, and but little pointed nails. Beyond these points, it may be observed that the fur of the animal is destitute of the spines which are generally characteristic of the Echimys group. The ears are small; the hind feet short and broad; the tail long, exhibiting rings of scales like the Rat's tail, and (excepting at the base) having only short and stiff hairs springing from between them. Recently a second species has been added to this genus by Dr. Wagner.

DACTYLOMYS TYPUS.

The Typical Dactylomys.

Echimys dactylinus.	(GEOFF.) DESMAREST, in Nouv. Dict. d'Hist. Nat. tom. x. p. 57 (1817); Mammalogie, p. 291 (1822).
" "	F. CUVIER, in Nouv. Annales du Muséum, i. p. 450, Pl. 18, fig. 3 (teeth); and Pl. 19, figs. 5 and 6 (skull).
Loncheres dactylina.	FISCHER, Synopsis Mammalium, p. 306.
Dactylomys typus.	IS. GEOFF. in Magas. de Zool. Année 1840, Mammif. pp. 29 and 47, Pl. 20 (animal), and Pl. 28, figs. 1—3, molar teeth.
" "	WAGNER, Schreb. Säugth. Suppl. iii. p. 348.

Hairs of the fur long, and somewhat harsh; upper parts of the body freckled with black and yellow; under parts, as well as the head and feet, dirty yellow-white; the hairs on the upper surface of the head very long, especially towards the hinder part: tail longer than the head and body taken together.

The only specimen of this species which is known to us exists in the Paris collection, and was originally brought from the Museum at Lisbon. No doubt it is a native of South America, and is most probably from Brazil. The specimen in question I examined a few years since, and the following notes of its chief characters were then inserted in my memorandum-book. Ears small; tail longer than the head and body taken together; about two inches of its length—being the basal portion—clothed with fur like that on the body, the remaining portion with rings of small scales, as in the Rat's tail, and with small, but stiff hairs, springing from the interstices: fore feet with the middle toes long, far exceeding the outer toes in length; the thumb wanting; hind feet broad. Fur long and somewhat harsh; on the upper parts of the animal the hairs are black towards the point, yellowish white at the point, and with the hidden parts of a deep rust colour: the feet and under parts of the animal are of a dirty yellow-white hue. The head is clothed for the most part with long dirty yellow-white hairs; those on the hinder part of the head, which are the longest, are about an inch in length. The occiput is tinted with rust colour. The hairs of the moustaches are dirty white at the root, and brown at the opposite extremity. The tail is brown, but becomes whitish at the apex. The hairs on the middle line of the back are about an inch and a quarter in length.

	Inches.	Lines.
Length from tip of nose to root of tail ...	14	0
" of tail	16	6
" of hind foot	2	3

DACTYLOMYS AMBLYONYX.

Hairy-tailed Dactylomys.

Dactylomys amblyonyx. (NATTERER) WAGNER, in Wiegm. Archiv für Naturgesch. 1845, Pt. 2, p. 146.

This species is briefly described by Dr Wagner as being of a yellow hue sprinkled with black, with the under parts of a beautiful ochre colour; the nails of the toes dilated, and the tail clothed with hairs throughout.

	Inches.	Lines.
Length of the head and body	9	6
" of tail	12	0

Inhabits Ypanema, in Brazil.

Genus, *Loncheres.*

Loncheres. ILLIGER, Prodromus Systematis, Mamm. &c. p. 90 (1811).
" FISCHER, Synopsis Mammalium, p. 306 (1829).
" WAGNER, Schreb. Säugth. Suppl. iii. p. 329 (1843).

Nelomys. JOURDAN[1], Annales des Sci. Nat. Dec. 1837, viii. p. 370.
" Is. GEOFF., Magas. de Zool. Année 1840, p. 36.
" PICTET, Mém. de la Soc. de Phys. et d'Hist. Nat. de Genève, viii. p. 201 (1842).
Phyllomys. LUND, in K. Danske Vidensk. Selsk. Naturv. Afhandl. 1841, viii. p. 243.
Isothrix. WAGNER, in Wiegm. Archiv für Naturgesch. 1845, Part 2, p. 145.
Echimys (part). GEOFFROY St. HILAIRE.

Echimyina with short and broad feet ; ears somewhat small; molar teeth rather large, and longer than broad ; those of the upper jaw each with two principal transverse folds of enamel, often completely dividing the tooth ; and those of the lower jaw with one external, and two internal folds of enamel. Palate narrow — most so between the foremost molars.

Habitat, Brazil and Guayana.

Like the species of *Echimys* (the genus next to be defined), the members of the present section usually have numerous elongated lance-shaped spines mixed with the hairs of the ordinary fur—often, however, chiefly confined to the upper parts of the body ; but they may be distinguished by their shorter and broader feet, and by the toes being much less unequal in length. Thus, in *Loncheres Blainvillii*, the nail of the inner toe terminates in a line with the middle portion of the adjoining toe, whilst in *Echimys Cayennensis*, the

[1] To M. Jourdan is due the credit of having clearly pointed out the principal characters of the present genus ; and to M. Is. Geoffroy we are indebted for a more detailed enumeration of the points of distinction between this and other genera, as well as for an excellent Monograph of the genus *Echimys*, as originally instituted by his father, and which included several species of the present section. As, however, Illiger had, as early as the year 1811, founded a genus (*Loncheres*) upon an animal named by him *Loncheres paleacea*, and which is in every particular a true member of M. Jourdan's genus (instituted in 1837), the law of priority obliges me to use the older name. It is true that Illiger's definition is imperfect, but if such are considered sufficient grounds for discarding his generic term, very many of the genera of Mammalia may have their names changed upon a similar plea.

smaller inner toe terminates opposite the base of the toe next to it; this latter (the second toe) has a very broad, hollow, and obliquely truncated nail, whilst in the *Echimys* the corresponding nail can scarcely be said to differ from the nails of the other toes, which are compressed, curved, and pointed. The muzzle is more blunt in the species of *Loncheres* than in those of Echimys, the ears are smaller, and the tail is usually thicker. The molar teeth are relatively larger, and of a more complicated structure. The upper molars in Loncheres have one internal, and three external folds. The inner fold enters the tooth about midway between the extremities; and of the three external folds, the central one enters the tooth opposite the inner fold, and, in traversing the crown of the molar, is indistinctly curved, so as to pass slightly beyond, and behind the inner fold, which it touches. Whilst these central folds or loops of enamel by their meeting divide the crown into two very nearly equal parts, the remaining two external folds traverse the tooth so as again to divide each half. Such is the structure of the first and second upper molars in a specimen of *Loncheres cristata* now before me. In the third molar, which is rather less worn, this difference is observable,—that the mesial folds do not yet show the looped ends, and the body of the tooth is completely divided by a transverse groove. The fourth molar, which presents a still earlier condition, is divided in the middle like the third, but the hinder half of the tooth is perfectly divided by a transverse groove. In *Loncheres Blainvillii,* I find that the hindermost of the external folds, owing to the molars being much worn, assumes the form of a small isolated transverse area, or (in the second and third molars) of two such areas. The molar teeth of the lower jaw, both in *Loncheres cristata* and *L. Blainvillii,* have each one external and two internal loops of enamel; they enter deeply into the body of the tooth, especially the internal loops, and, in passing inwards, incline obliquely forwards. The foremost molar differs from the others in having an extra triangular lobe in front, in the middle of which is a small circular or oval area enclosed by enamel.

The first and second upper molars of *Loncheres semivillosa* are represented by M. Isidore Geoffroy as having each two external and two internal loops of enamel[1].

Dr. Wagner separates from the genus *Loncheres* certain species in which the fur is of the ordinary kind—having no admixture of spines; these are associated under the sectional name *Isothrix*: we shall first describe those species which may be called

Loncheres proper.

Upper parts of the body with numerous, bristly, flattened spines[2], mixed with the hairs of the ordinary fur.

LONCHERES CRISTATA.

White-crested Loncheres.

Echimys cristatus. (Geoff.) Desm., Nouv. Dict. d'Hist. Nat. tom. x. p. 55 (1817); Mammif. Pt. 2, p. 291.

Loncheres paleacea. (Illiger) Licht., Abh. der K. Akad. der Wissensch. in Berlin, 1818-19, p. 191.

" " Kuhl, Beitr. zur Zool. p. 72 (1820).

Nelomys cristatus et *paleaceus.* Is. Geoff., Magas. de Zool. 1840, p. 49, Pl. 21.

[1] See Guérin's Magasin de Zoologie, année 1840—Mammifères, Pl. 28, fig. 7.

[2] These, which I have termed spines, are of course merely modified hairs: they are formed nearly like a lance-head, but are much elongated. The broadest part of these spinous hairs is near the root, and below this part they become suddenly contracted into a very slender stalk which is inserted in the skin: from a short distance above the root they gradually taper to terminate in a sharp point; they are nearly flat, but have the under surface slightly convex, and the upper surface concave, in a corresponding degree; on this surface, moreover, the margins are thickened so as to form a narrow, elevated ridge on either side of the hair.

Loncheres paleacea et *cristata.*	WAGNER, Schreb. Säugth. Suppl. iii. p. 332. (1843).
Myoxus chrysurus (Gilt-tailed Dormouse).	SHAW, Gen. Zool. ii. Pt. 1, p. 169 (1801).
Hystrix chrysuros.	SCHREB. Säugth. Supp. Pl. 170 B.
Lérot à queue dorée.	(ALLAMAND) BUFFON, Hist. Nat. vii. Suppl. p. 283, Pl. 72 (1789).
Echimys à queue dorée.	CUV. Règne Animal, ed. 1829, i. p. 199.

Upper parts and sides of the body covered with broad spines; general colour brown; dark on the back, reddish on the sides of the body, and pale on the abdomen; upper surface of the head white; the hairs on this part stiff, and very long, especially towards the occiput; region of eye and occiput black; tail black, with the apical half (or more, or less) white.

Inhabits Guayana and Para.

In giving Para as a habitat for this species as well as Guayana, I must observe, that some authors regard a certain specimen of *Loncheres,* named by Illiger *paleacea,* as distinct from the *L. cristata,* and if this view be correct, Guayana only must be given as the habitat of the latter animal. That there are no good grounds, however, for regarding the *L. paleacea* and *L. cristata* as distinct species, I shall hereafter show.

In my note-books I have descriptions of three specimens of the *L. cristata;* one from a specimen in the Paris Museum, which I have good reason to believe is the individual named by M. Geoffroy; it is from Cayenne. The second description is from a specimen in the Leyden Museum, received from Surinam; and the third is from a specimen in the British Museum, captured in British Guayana. Upon comparing these descriptions, I find the three animals agree very closely, excepting that the white is more or less extended on the tail. Their general hue is rufous-brown; dark on the back, brighter on the sides of the body, and palish on the abdomen: the upper surface of the head and muzzle, as well as the sides of the latter at the tip, and the chin, are white; the white hairs on the head are for the most part flattened and bristly, and towards the occiput they are very long, and form a conspicuous, depressed tuft, projecting over the occiput, which is black, as is also a considerable space around each eye. The cheeks are dusky; some of the stiff hairs on these parts being dusky, whilst others are brown. The tail is clothed with coarse hairs, for the most part averaging rather more than a quarter of an inch in length, but which become longer at the apex of the organ, where they are full an inch in length; they are not sufficiently numerous entirely to hide the scaly skin. The root of the tail is of the same colour as the body; the terminal portion white, and the intermediate part black. In the Paris specimen about one-half of the tail is white; in the Leyden individual, about two-thirds of the organ is covered with white hairs; whilst in that in the British Museum, one-third of the tail is white. The back and flanks are almost entirely covered with flattened spiny hairs, averaging an inch or more in length, and about one line in width at the broadest part; they are brown-white at the root, dusky brown towards the point, and many of them rufous at the point. Mixed with these spines are dirty rust-coloured hairs, which are very scanty on the back of the animal, but become more numerous on the flanks.

	Paris Mus.		Leyd. Mus.		Brit. Mus.	
	Ins.	Lines.	Ins.	Lines.	Ins.	Lines.
Length from tip of nose to root of tail ...	13	0	11	9	12	0
" of tail, about	13	4	13	0	8	8
" of ear, about				4		4½
" of tarsus and nails	2	1	2	0?	1	11½
" fore foot and nails					0	11
" of the four upper molar teeth, taken together						6⅔
" of the lower jaw, measuring from the back of the condyle ...					1	5¼

Hystrix chrysurus. Schreb.

The *Hystrix chrysurus* of Schreber, or *Myoxus chrysurus* of Shaw, are founded upon the animal described by Allamand in Buffon's great work, under the name "*Lérot à queue dorée.*" With regard to this animal, it is only necessary to state that the original description was drawn up from a very young animal preserved in spirit, and that the yellow colour of the mark on the upper part of the head, as well as the yellow end to the tail, are almost without doubt due (as M. Is. Geoffroy suggests) to the hairs on these parts being stained by the spirit; they having originally been white. Beyond the yellow colouring of the parts in question, and the small size, the animal does not appear to differ from the *E. cristata.*

Loncheres paleacea. (Illiger) Lichtenstein.

The animal thus named is only known to us by Lichtenstein's description and figure; from an examination of which, M. Is. Geoffroy concludes that it forms a species distinct from the *L. cristata,*—an opinion in which Dr. Wagner coincides. The points of distinction, according to M. I. Geoffroy, are, that in the *L. paleacea* the back of the head and cheeks are of the same general hue as the other parts of the body, and not black, as in *L. cristata;* that the hairs of the moustaches are whitish, whilst in *L. cristata* they are black; and, lastly, that the tail is of the same colour at the base as the body, and that the remaining portion, being three-fifths of the entire length, is white; whilst in *L. cristata* two-fifths of the tail only are white![1]

[1] With regard to the colour of the hairs of the moustaches, I have to remark, that in the British Museum animal already referred to, some of the hairs of the moustaches only are black, those nearest the tip of the muzzle being dirty white. This same specimen has but one-third of the tail white, whilst in the Leyden specimen two-thirds of the tail are white, and in the Paris individual one-half of the tail is clothed with white hairs; thus showing that the white is more or less extended on this organ. Lastly, as Lichtenstein, in his very short description, does not absolutely state that the sides of the head and the occiput are of the same colour as the body of the animal, it is scarcely safe to conclude that such

LONCHERES BLAINVILLII.

The Black-tailed Loncheres.

Nelomys Blainvillii.	JOURDAN, Annales des Sciences Naturelles, (Second Series) viii. p. 371 (1837).
" "	Is. GEOFFROY, Magas. de Zool. Année 1840—Mammifères, pp. 19, 41, and 49, Pl. 22; and for the dentition, Pl. 28, figs. 10—12.
Loncheres Blainvillei.	WAGNER, Schreb. Säugth. Suppl. iii. p. 334.

General hue of the animal bright sienna-yellow; chin, throat, and, in fact, the whole of the under parts, pure white; fore feet also white; hind feet yellow: tail nearly as long as the head and body together, and with black hairs, excepting at the root, where the hairs are of the same yellow hue as those of the body.

Inhabits Brazil, in the vicinity of Bahia, and is likewise found in a small island (I. of Deos) off the coast of Bahia.

A well-preserved specimen in the museum of the Zoological Society, and which is from Bahia, presents the following characters. Limbs rather short and stout; feet short; tail nearly as long as the head and body taken together; ears small, slightly emarginated posteriorly. On the upper parts of the body, flattened spinous hairs predominate; they average about an inch in length, and $\frac{1}{24}$th of an inch in width at the widest part: the sides of the body are clothed with spines and hairs in about equal proportions, and the under parts are covered with stiff hairs; these latter are pure white to the root. The spines on the sides of the body are grey-white at the root, dusky in the middle, and of a rich yellow colour

is the case from the inspection of the somewhat indifferent figure which accompanies his account; and even if these parts are brown in Lichtenstein's animal, the difference between it and the *L. cristata* is but slight, since in the latter animal they vary from black to brown-black.

(inclining to rufous) at the point; many of the spines on the back are similarly coloured, but some are dusky at the point, and hence a slight pencilling of this dark hue is seen mixed with the rich sienna-yellow, which is the prevailing colour. The ears are provided with a tuft of extremely fine hairs immediately behind the opening, some of which hairs are nearly an inch in length. The fore feet are white, but tinted with yellow on the wrist; the hind feet are yellow, and the toes are white at the sides; the nails of the toes are white. About one inch of the tail, being the basal portion, is clothed with spiny hairs, coloured like those of the adjoining part of the body; the remaining portion is well covered with hairs, averaging about half an inch in length, and which are almost entirely black; the exceptions being, that towards the base they gradually assume a brownish tint, and on a considerable portion of the under surface they are dusky brown. The hairs of the moustaches are black: a white spot is observable in front of the ear.

	Zoological Society. Inches.	Lines.	From Wagner. Inches.	Lines.
Length from tip of nose to root of tail	11	6	8	0
" of tail, about...	8	6	7	9
" from nose to ear	1	11		
" of ear, about		$4\frac{1}{2}$		
Width of ditto		$8\frac{1}{2}$		
Length of fore foot and nails ...		10		
" of hind foot and nails	1	6	1	5
" of skull	2	$1\frac{1}{2}$		
Width of ditto	1	$0\frac{1}{4}$		
" between orbits		$6\frac{1}{4}$		
Length of nasal bones		$7\frac{3}{4}$		
Width of ditto behind		$1\frac{3}{4}$		
" " in front		$2\frac{5}{6}$		
From incisor to molar teeth ...		$5\frac{1}{2}$		
Length of four upper molars, taken together		$5\frac{5}{6}$		
Space between foremost molars ...		$0\frac{3}{4}$		
" between hindermost ditto ...		2		
Width of incisors		$1\frac{2}{3}$		

LONCHERES ARMATA.

Strong-spined Loncheres.

Mus hispidus.	LICHT., Darstellung der Säugth. &c. Tab. 35, fig. 2.
Nelomys armatus.	IS. GEOFF., Annales des Sciences Naturelles, 1838, tom. x. p. 125; Magas. de Zool. 1840, Mammif. pp. 12, 42, and 51.
Loncheres armata.	WAGNER, Schreb. Säugth. Suppl. iii. p. 335.

Upper parts of the body armed with broad and strong spines; tail about equal to the body, only, in length, with spines on the upper surface at the root, and scattered, small, whitish hairs beyond: general hue brown, but pencilled with yellow; under parts dirty pale yellow; muzzle rufous.

Inhabits Cayenne.

This species was first described by Dr. Lichtenstein, who, having the impression that it was identical with Geoffroy's *Echimys setosus*, gave to it the same specific name. That it is a perfectly distinct species has been proved by M. Isidore Geoffroy. It is said to be armed on the upper parts of the body with spines of about three-quarters of an inch in length, and three-quarters of a line in width, and mixed with these are ordinary hairs, which, however, are not sufficiently numerous to hide the spines; these latter are not confined to the back, for they extend on to the hinder portion of the head, as well as on the upper surface of the tail at the root; about one inch of the tail is thus armed; the remaining portion has small scattered white hairs. The moustaches are black; the nails of the toes white; the feet whitish.

	Inches.	Lines.
Length from tip of nose to root of tail ...	8	0
" of tail	6	0
" of ear		4
Width of ditto		4
" of hind foot and nails	1	5

Loncheres didelphoides.

Echimys didelphoides. (GEOFF.) DESMAREST, Nouv. Dict. d'Hist. Nat. x. p. 58; Mammalogie, p. 292.
Loncheres " FISCHER, Synopsis Mammalium, p. 307.
Nelomys " IS. GEOFF., Magas. de Zool. année 1840, pp. 41 and 50, Pl. 24.

Ear small; tail rather shorter than the head and body taken together; scaly, and with very short, scattered, brown hairs, excepting at the root, where it is well clothed: long, narrow spines cover the upper parts, and sides of the body, and mixed with these are dirty yellow hairs, which are, however, by no means abundant: upper parts of the body brown, pencilled with yellow; sides of the body, as well as the feet, yellow-brown; lips, throat, and abdomen, dirty yellow-white; moustaches brown.

The above short description is taken by myself from the specimen in the Paris collection; the original of all the descriptions under the names above quoted. It is evidently a young animal, and, M. Is. Geoffroy informs us, was withdrawn from spirit, and hence its colouring has, no doubt, been somewhat changed by this mode of preservation.

In having the root of the tail well clothed with hairs, the *L. didelphoides* approaches the *L. armata*, but M. Isidore Geoffroy, judging from Lichtenstein's description of the latter animal, concludes that *L. didelphoides* differs in having the hairs continued for a greater extent on the basal portion of the organ; in having the small scattered hairs which are observable in the remaining portion of the tail dark, instead of white; in the slenderness of the spines[1], and, lastly, in

[1] The spines of *L. armata* are said to be $\frac{3}{4}$ of a line wide, and 9 lines in length (Prussian measure)—having the length only twelve times greater than the width; whilst in *L. didelphoides* the spines on the middle of the back are, according to M. Geoffroy, $9\frac{1}{2}$ lines in length, and only $\frac{1}{3}$d of a line in width (French measure), the width being $\frac{1}{28}$th part of the length.

some peculiarities in the general colouring. I must confess these differences do not appear to me to furnish sufficient grounds for our regarding the two animals as distinct, since the colouring of the only known specimen of *L. didelphoides*, for reasons before stated, cannot be depended upon, and since the animal is admitted to be young, a state in which the spines differ from those of the adult animal. The admeasurements of the *L. didelphoides* are—

	Inches.	Lines.
From the nose to the root of the tail ...	7	4
Length of the tail	6	10
" of the hind foot and nails	1	3
" of ear, about		3½
From nose to ear	1	3

The animal is probably from Brazil; it was removed from the Museum at Lisbon to Paris, during the occupation of Portugal by the French.

LONCHERES OBSCURA.

Dusky Loncheres.

Loncheres obscura. WAGNER, Abh. der Bayr. Akad. der Wissensch. iii. p. 196, Tab. 2, fig. 5—12; Schreb. Säugth. Suppl. iii. p. 336.

Brown, freckled with yellow; belly yellowish; tail equal to the head and body in length, and provided with very short, scattered hairs.

Inhabits Brazil.

Two specimens of this species were brought by Spix from Brazil. According to Dr. Wagner, the animal is of a robust form, and has the short feet and thick tail which characterize most of the species of *Loncheres*. The upper parts of the body, and the shoulders and thighs, are clothed for the most part with spines, mixed with which are some scattered coarse hairs;

on the other parts, coarse hairs only are found. The incisor teeth are white, and the claws are brown.

The specimens were preserved in spirits.

	Inches.	Lines.
Length from nose to root of tail	6	6
" of the tail	6	5
" of ear		7½
Width of ditto		6
Length of fore foot, without the nails ...		9
" of middle finger, without the nail ...		4
" of the hind foot to the base of the middle toe	1	1
" of the middle toe, without the nail ...		5
Greatest breadth of the foot, which is in the middle		6

Of a larger specimen the head and trunk of the skeleton (measured in a straight line) was 7″ 8‴; and the series of caudal vertebræ, also, 7″ 8.‴

Loncheres semivillosa.

Nelomys semivillosus. Is. Geoff., Ann. des Sci. Nat. x. p. 125 (1838); Magas. de Zool. 1840, Mammifères, pp. 42 and 50, Pl. 23 (animal); and Pl. 28, figs. 7, 8, and 9, (the teeth).

Loncheres semivillosa. Wagn. Schreb. Suppl. iii. p. 338.

Tail equal to the head and body in length, moderately well clothed with fawn-coloured hairs, but these not hiding the scaly skin: body rufous-brown, freckled with yellow; under parts white, slightly suffused with rufous: moderately strong spines on the upper parts and sides of the animal.

Inhabits New Granada.

The flattened spines, which are found mixed with the hairs to form the clothing of this animal, are very numerous on the back, and are nearly an inch in length, and somewhat less than $\frac{1}{25}$th of an inch in width on this part: many of the spines (especially on the hinder parts of the body) are black towards the point, yellowish at the point, and horn-coloured at the root; others again (chiefly on the fore parts of the body) have the terminal half entirely of a blackish or dusky hue. On the upper surface of the head, as well as on the thighs, the spines are narrower than those on the back, and on the under parts of the body, hairs only are found. A very small portion of the tail, and that at the base, is covered with coarse hairs, being some few spines intermixed, and the remaining portion of the organ is moderately well furnished with fawn-coloured hairs, which, however, do not entirely hide the scaly skin beneath.

	Inches.	Lines.
Length from tip of nose to root of tail, very nearly	7	9
" of tail	7	9
" of hind foot and nails	1	4

From *Loncheres armata* and *L. didelphoides*, M. Geoffroy distinguishes this species by its tail being relatively longer; and, moreover, that author notices differences in the spines: thus the spines of *L. semivillosus* are said to be much narrower than those of *L. armata;* and, whilst in *L. didelphoides* the spines on the back are but a trifle narrower than those in the same part of *L. semivillosus,* there is a marked difference in the spines on the head, since in the former animal they do not surpass $\frac{1}{5}$th or $\frac{1}{6}$th of a *millemètre* in width, whilst in the latter they are nearly half a *millemètre*[1].

The above differences, however, M. Geoffroy observes, he should not have considered sufficient to establish his species, were it not that they are combined with certain modifications in the structure of the molar teeth.

[1] Close upon $\frac{1}{50}$th of an English inch.

Loncheres grandis.

Loncheres grandis. (NATTERER) WAGNER, Wiegm. Archiv für Naturgeschichte 1845, Pt. 2, p. 146.

Upper parts of the animal of a golden-fulvous hue, freckled with black; under parts yellowish; head black, slightly sprinkled with fulvous; feet brown: spines flexible. Length of head and body 11 inches (WAGNER).

With regard to this species, Dr. Wagner states that he has been induced to arrange it in the genus *Loncheres*, on account of a near affinity which it seems to evince to the *Loncheres cristata,* but that he had not been able to examine the skull and dentition of the only specimen which had come under his notice. The specimen was brought from the neighbourhood of the Amazon River by Dr. Natterer.

Loncheres macrura.

Loncheres macrura. (NATTERER), WAGNER, in Wiegm. Archiv für Naturgesch. 1842, vol. i. p. 360.

Tail nearly equal to the head and body in length, and provided with very short, scattered, white hairs: general colour of the animal inclining to fulvous, but palish on the sides; under parts of a hoary yellowish cast. Length of head and body together, 10¾ inches; of tail, 10 inches.

Inhabits Borba, Brazil.

Loncheres nigrispina.

Loncheres nigrispina. (NATTERER), WAGNER, l. c. p. 361.

Tail, excepting at its thickened base, densely and equally clothed with brown hairs, and without tuft at the extremity: upper parts of the animal of a glossy fulvous-brown, but with a sprinkling of black; feet and under parts yellowish white; sides of the snout yellowish hoary. Length of the head and body taken together, 9¼ inches; of the tail, 6 inches.

Inhabits Ypanema, Brazil.

Loncheres unicolor.

Loncheres unicolor. (Rüppell) WAGNER, l. c. p. 361.

General hue of an uniform palish ferruginous brown on the upper parts of the animal; the under parts, as well as the feet, still paler: tail furnished with long hairs: fur of a crisp and harsh nature; that on the back with an admixture of narrow, flattened, and grooved spines. Length of head and body, 10″ 9‴; tail, 7″ 9‴.

Inhabits Brazil.

Sub-genus, *Isothrix.*

Echimyina with the fur soft, and destitute of intermixed, flattened spines: dentition and general aspect as in the species of *Loncheres*[1].

LONCHERES (*Isothrix*) PICTA.

Bicoloured Loncheres.

(*Plate* 19, fig. 1.)

Nelomys pictus. PICTET, Notices sur les Animaux Nouveaux ou peu connus du Musée de Genève, 2de livr. p. 29, Pl. 7 (animal); and Pl. 8, the skull and other parts.

[1] The dentition of the species of Isothrix, according to Dr. Wagner, approaches that of the *Echimys*, and yet presents essential modifications, in which it perfectly agrees with the *Nelomys pictus* of M. Pictet, but upon referring to the last-mentioned author's description and figures of the dentition of *Nelomys pictus*, I cannot perceive any points of distinction between its dentition and that of other species of *Nelomys* (or *Loncheres*, as the group is designated in this work). The mere presence or absence of spines,

Brownish white ; a large oblong brown patch on the upper surface of the head ; back and shoulders, as well as the tail, brown ; but the latter, with the terminal portion, and likewise the basal portion beneath, dirty white : tail rather longer than the head and body taken together.

Inhabits Bahia—Brazil.

According to Prof. Pictet's description, nearly the whole of the head of the *L. picta*, as well as the neck, a great portion of the sides of the body, and the whole of the under parts, are dirty white, but irregularly and indistinctly suffused with brown, arising from the hairs on these parts being brown at the base, and white on the external half only[1]. The back of the animal, and the upper surface of the head, are for the most part clothed with hairs, which are pale brown at the root, and dark brown externally The brown on the head assumes the form of an oblong patch, which is pointed before and behind, and expanded between the ears, which,

or flat, stiff hairs, is certainly of very trivial importance, since animals agreeing otherwise in all essential particulars, differ much with respect to the size and strength of the spiny hairs which are mixed with those of the ordinary fur : they present, in fact, all the intermediate conditions, not only in the *Loncheres* section, but likewise in the genus *Echimys*.

[1] The perusal of Prof. Pictet's description of the colouring of the fur of the *Nel. pictus* strongly reminds me of certain abnormal conditions of colouring which I have noticed in several species of Rodents. Thus, in the British Museum, is a specimen of the *Oxymycterus hispidus* of Pictet, in which there are two large dirty white patches covering nearly the whole hinder parts of the animal, but separated in the mesial line of the back by a longitudinal band of the ordinary brown colour. The hairs on these patches are grey at the root like those on other parts of the animal, and mixed with them are a few hairs which are coloured entirely like those of other parts. This specimen is from the same locality as the *Nel. pictus*, and was accompanied by others presenting the usual colouring of the species. Again, M. Gervais, in the Voyage de la Bonite, figures and describes a specimen of the *Phælomys Cumingii*, in which the muzzle, ears, and tail, are the only parts which present the dark brown colouring usual in the species, the remaining parts being dirty white, but, as in the *Nel. pictus*, the hairs are white at the points only ; and, a condition between this and the normal one is exhibited by a specimen in the Museum of the Zoological Society ; in this, the colours are disposed nearly as in the *Nelomys pictus*

however, it does not reach. The line of separation of the brown and white on the sides of the body, commences at the elbow, and as it runs back, gradually ascends, until it reaches the root of the tail. On the fore parts of the body the line of separation of the dark and pale colours forms a curve, running backwards and upwards on either side from the shoulders, across which the brown descends in the form of a moderately broad band, to terminate on the middle of the fore-arm. The feet are covered with stiff, whitish hairs. The tail is rufous-brown above, at the base, and whitish beneath at the same part, the white being extended about two inches lengthwise, and the brown about an inch and a half; a space of three inches, or rather more, of the tail—being the terminal portion—is white above, and golden yellow beneath, and the intermediate portion of the organ is of a very dark brown: it is thick, and has a scaly skin; the scales, however, are hidden by the long and dense hairs.

	Inches.	Lines.
Length from tip of nose to root of tail ...	10	8
" of tail	12	10
" from nose to ear	2	0
" of ear		7½
Width of ditto		7½
Length of fore foot	1	1
" of hind foot	1	10⅓

The following species of the *Isothrix* section are only known to us through the brief descriptions published by Dr. Wagner, in the Archiv für Naturgeschichte, Part 2 for 1845, p. 146. They constitute part of the great collection formed by Dr. Natterer in Brazil.

Isothrix bistriata (Natterer) Wagner.

Above, varied with black and yellow, beneath yellowish; head with two black bands; tail long and hairy, black, with the base yellow. Length of head and body together, 11 inches; of tail, 10¼ inches.

Inhabits Rio Guapore.

Isothrix pagurus (Natt.) WAGNER.

Fore parts of the animal variegated with straw-yellow and black, the hinder parts with rust colour and black; under parts yellowish; snout fulvous on the upper surface. Head and body, 8″ 10‴ in length.

The tail is wanting in the specimen described by Dr. Wagner, but he thinks its absence is accidental. The skull and dentition of this species are said to agree in their structure with the same parts in the *Isothrix bistriata*. It inhabits Borba.

Isothrix pachyura (Natt.) WAGNER.

Above variegated with black and dirty yellow; beneath pale yellowish; tail densely clothed with hairs; above, brown, beneath yellow-white. Head and body, $10\frac{1}{8}$ inches; tail (perhaps not perfect), $6\frac{3}{8}$ inches.

Dr. Wagner states that the skull of this animal is unknown to him, but he considers the species as identical with the *Nelomys antricola* of Dr. Lund[1].

[1] It may here be observed, that Dr. Lund, in his papers on the Fossil Mammalian Remains discovered in the caverns of Brazil, divides the old genus *Echimys* into four genera; viz. *Phyllomys*, *Echimys*, *Loncheres*, and *Nelomys*—see Det Kongelige Danske Videnskabernes Selskabs Naturvidenskabelige og Mathematiske Afhandlinger, vol. viii. p. 243 (1843). Of these, it is only necessary to remark that—

Phyllomys is in all essentials the same as *Loncheres* of Illiger.
Echimys, Lund, is *Dactylomys*, Is. Geoffroy.
Loncheres, Lund, is *Echimys*, Is. Geoff. and others.

In the genus *Nelomys*, Dr. Lund includes two species—*Nelomys antricola*, a living species, and *N. sulcidens*, hitherto found only in a fossil state: on their dentition, and other characters, some remarks will be found under the head *Echimys antricola*.

The genus *Phyllomys* is established by Dr. Lund upon some upper jaws found in the Brazilian caverns, belonging to species still living, but which the Danish naturalist does not appear to have met with. The dentition of the animal (to which the specific name *Braziliensis* is given) departs slightly from

Genus, *Mesomys.*

Mesomys. WAGNER, Archiv für Naturgesch. 1845, p. 2, p. 145.

In general aspect resembling the species of Loncheres: teeth like those of an Echimys; body armed with strong spines; tail wanting.

Whilst the form of the head, ears, and feet, are entirely as in *Loncheres*, the dentition fully agrees with that of *Echimys fuliginosus.* This animal is hence intermediate in its character between the genera *Loncheres* and *Echimys*: from both, however, it is distinguished by the want of a tail, which, according to Dr. Natterer, is not an accidental peculiarity[1].

Mesomys ecaudatus. WAGNER, in Archiv für Naturgesch. 1845, Pt. 2, p. 145.

Upper parts of the animal fulvous, sprinkled with black; under parts ochraceous. Length of body, 6″ 8‴.

Inhabits Borba, Brazil.

The above account, which is from Dr. Wagner, leaves further details desirable.

the typical species of *Loncheres*, inasmuch as the crown of each molar tooth is divided by grooves into four narrow transverse plates. A fossil species, also placed by Dr. Lund in the genus *Phyllomys*, presents an intermediate condition in the structure of its molars, some of the transverse plates being united in pairs. It may be necessary to explain that the floors of the caverns alluded to are strewed with an abundance of bones of small Rodents, of species still inhabiting the district; these having been conveyed there, Dr. Lund informs us, by certain animals of prey, and more especially by a species of owl.

[1] It would have been well had Dr. Wagner stated whether more than one specimen of this species has been found, for it is difficult to conceive that the absence of a tail should not have been accompanied by modifications in the structure of the feet and other parts, unless that absence were accidental.

Genus, *Echimys*[1].

Echimys. (Geoffroy) DESM., in Nouv. Dict. d'Hist. Nat. x. p. 54 (1817) —Mammalogie, p. 290.

Echinomys. WAGN., Schreb. Säugth. Suppl. iii. p. 339.

Echimyina with the feet narrow and moderately long; upper molar teeth with one internal indenting fold of enamel, and one or two folds on the outer side.

Representatives of the genus Echimys are found from the northern parts of South America, southwards as far as Paraguay: they extend in the western direction, into Peru and Bolivia, but do not occur on the west side of the Andes.

The species of Echimys usually have the ears larger than those of the *Loncheres* division; their feet are longer and narrower, and have the toes, as I have already observed, more unequal in length; the muzzle, moreover, is longer, and pointed; whilst in *Loncheres* it is short, and somewhat obtuse. To these differences we have to add certain modifications observable in the structure of the molar teeth, and in the form of the palate. With regard to the latter, there is a tolerably well-marked point of distinction in the skulls of the species of the two genera: thus, in *Loncheres,* that portion of the palate which lies between the molar teeth, is long and narrow, whilst in *Echimys* the same part is short, and relatively broad: in both, the series of teeth on opposite sides of the jaw converge slightly in front. The broader and shorter palate of the Echimys arises from the proportionately small size of its molars[2].

[1] From ἐχῖνος, the sea-urchin, or the hedgehog; and μῦς, a mouse or rat: *Echinomys,* as Prof. Wagner observes, would have been a more correct compound.

[2] Upon comparing the molar teeth of the *Loncheres Blainvillii* with those

The structure of the molar teeth in the Echimys (Pl. 16, figs. 1 and 1A) is more simple than in the Loncheres: each upper molar has a deep internal fold, which divides the tooth into two unequal parts; the foremost, which is narrow and transverse, is the smaller; the hinder portion approaches somewhat to a triangular figure; broad externally, and almost pointed on the inner side; within it are seen two small transverse areas, surrounded by enamel, in the first and second molars, and but one transverse area in the third, as well as in the corresponding division of the lower molars (Pl. 16, fig. 1 B), which resemble the upper, excepting that they are (as usual) reversed in position, and that the foremost lower molar has a small extra lobe in front, partially separated by a small transverse area. In these molar teeth of the lower jaw, the primary fold of enamel in traversing the body of the tooth runs obliquely upwards, whilst in Loncheres it takes an opposite direction[1].

of the *Echimys Cayennensis*, I find that the series of these teeth in the upper jaw of the latter animal occupy one-third less space in the longitudinal direction, and in relation to the size of the skull, than they do in the *Loncheres*.

[1] The letters added to the figures of the molar teeth represented on Plate 16, are intended to indicate the corresponding folds of enamel in each. The primary internal fold is marked *a*, the principal external fold *a* 1, and the secondary folds, *b* 1 and *b* 2. Of these folds the primary internal one is the deepest, and consequently the last to be obliterated by the wearing of the crowns: it is the most constant in Rodents. When the folds *a* and *a* 1 both penetrate deeply into the body of the tooth, the latter almost invariably passes behind the former, and this circumstance would lead one to believe that in the molar of the Echimys, the foremost of the two small areas in the hinder division of the tooth represents the fold *a* 1; in which case, the second small area would represent the fold *b* 2. An examination of the teeth of several species of Echimys, both in the adult and young states, will no doubt enable us to determine these points, without which we cannot properly appreciate the distinctions which separate the molars of Echimys from Loncheres.

ECHIMYS CAYENNENSIS.

Cayenne Echimys.

(*Plate* 19, fig. 2).

Echimys Cayennensis.	(GEOFF.) DESM., Nouv. Dict. d'Hist. Nat. x. p. 59 (1817); Mammalogie, p. 292 (1822).
" "	IS. GEOFF., Magas. de Zool. année 1840, pp. 13 and 52.
" "	PICTET, Mém. de la Soc. de Phys. et d'Hist. Nat. de Genève, ix. p. 145. Pls. 1, 2, and 3, and Pl. 4, skeleton (1841-2).
" *setosus.*	(GEOFF.) DESM. Nouv. Dict. x. p. 59.
" *myosuros.*	LICHT., Abhand. der Königl. Akad. der Wissensch. in Berlin, 1818-19, p. 192, Pl. 1, fig. 2.
" *leptosoma.*	(LICHT.) BRANTS, Het Geslacht der Muizen, p. 150, (1827).
Mus cinnamomeus.	LICHT., Darstell. Part 7, Tab. 36, fig. 2 (1832).
Loncheres anomala.	KUHL, Beiträge zur Zoologie, p. 72 (1820).
" *Cayennensis.*	FISCHER, Synopsis Mammalium, p. 307 (1829).
? *Echimys longicaudatus.*	RENGGER, Naturgesch. der Säugth. von Paraguay, p. 236 (1830).
Echinomys leptosoma.	WAGN., Schreb. Säugth., Suppl. iii. p. 341 (1843).

Tail nearly as long as the head and body together; ears large; tarsi long; upper parts of the body chiefly covered with spines; sides likewise with spines, but with a considerable admixture of hairs: upper parts of the animal of a dusky brown colour; sides of the body, and more especially of the head, suffused with rufous; under parts white.

Inhabits Guayana and Brazil.

M. Pictet, who has furnished the most perfect account of the characters of this animal, calls attention to the absence of spines, as characteristic of the young animal up to the time that three molar teeth only are developed in each series, and that the animal measures about eleven inches in length, without including the tail. Its colouring, then, differs only from that of the adult in having an admixture of grey with the brown of the upper parts, and in having the feet of a less pure white. In this stage the animal, M. Pictet states, corresponds very nearly with the specimen upon which M. Geoffroy founded the *Echimys setosus*.

In a still more advanced stage, when the fourth molar begins to make its appearance, the animal still exhibits the greyish hue on the upper parts of the body, as in the first stage mentioned, but a slight rufous tint is visible on the cheeks, and the white on the under parts is more pure, and more distinctly separated from the brown hue of the upper parts. Spines also make their appearance, but as yet confined to the back of the animal.

In 1837, I examined a specimen of the *Echimys Cayennensis* preserved in the Paris Museum, which, agreeing very closely with Desmarest's description, and being deficient in a portion of its tail, as mentioned by that author in his account, is no doubt the original of his description[1]. It is about equal in bulk to the Black Rat (*Mus Rattus*); of a brownish yellow colour; dusky on the back, and white on the whole of the under parts of the head and body: the lower part of the muzzle, the fore feet, and the inner side of the hind feet, as well as the toes, are also white: the outer surface of the tarsus is brownish; the moustaches brown; the tail brown above, pale below, and clothed with short hairs, excepting at the apex, where they are about a quarter of an inch in length. Bristly, flattened hairs cover the shoulders and haunches, and strong, grooved spines, with a few intermixed hairs, cover the back. The brown-yellow colour of the sides of the body forms a well-marked line at its junction with the white of the under parts. Length from tip of nose to root of tail, 7″ 7‴; of the ear, 7‴; of the tarsus, 1″ 9‴. The imperfect tail

[1] The length of the animal, measured from the tip of the nose to the root of the tail, I found to be about an inch more than in the admeasurement given by Desmarest, but such a difference would occur were the animal measured in a straight line, in one case, and along the curve of the back (as it is my practice to do), in the other.

measures 5″ 6‴ : a very small portion only appears to have been lost.

Of the adult animal, the British Museum collection contains several specimens; they vary but little in their colouring: the general hue of the upper parts of the body is dusky brown, and sometimes the brown assumes a slight vinous or purple tint: on the sides of the body, and on the head, a rufous tint prevails—sometimes very bright, and inclining to yellow; and the under parts of the body, the lower half of the muzzle and cheeks, as well as the inner side of the fore legs, the fore part of the hind legs, and the feet, are white; generally very pure, and the hairs on these parts are uniform to the root. The white of the under parts of the animal, and the rufous tint on the side of the body, form a well-marked line where they meet, not gradually blending, as is often the case. The upper surface of the head is pencilled with black; the region of the eye is dusky, and the hairs of the moustaches brown; behind each ear is a rufous white, or pale rufous patch: the ears are tolerably large, and distinctly emarginated behind; of a whitish colour, but broadly margined with dusky; they are rather sparingly clothed with brown hairs, which are for the most part very short, but on the anterior margin, and on the inner surface immediately behind the opening, they are long. A slight pencilling of brown is observable on the sides of the body: the hind legs are dusky externally, especially near the heel; the tail is dusky above, and dirty white beneath, and the small hairs, with which it is for the most part scantily clothed, are of the same colour as the part from which they spring; towards the apex of the tail the hairs become longer, and at the point they are about half an inch in length; here they are entirely of a dirty white colour. The spines on the back of the animal average about ten or twelve lines in length, and two-thirds of a line in width; they are nearly white at the root, and shaded through grey into dusky-brown externally; the extreme point, however, is often of a paler tint, inclining to rufous. On the sides of the body the spines are less strong, and have the exposed end of a paler brown colour, and those on the haunches are pale yellow-brown at the apex—sometimes almost white. Spines, with a few hairs intermixed, cover the upper parts of the body; on the flanks and haunches the hairs are fully as numerous as the spines, and on other parts the spines are wanting. This description applies very closely to all the stuffed specimens in the British Museum, excepting that in one or two there is a faint yellow tint on the under parts. In the same Museum are two specimens (a male and female) preserved in spirits, and which were procured in the province of Minas Geraes; they appear to me to be without doubt identical in species, not only with the *Echimys Cayennensis*, but likewise with the animal described by Dr. Lund, from the same part of Brazil, under the name *Loncheres elegans*. In the female specimen I could discover but two mammæ; these were very distinct, placed one on each side of the body, about midway between the fore and hind legs, and exactly on the line which marks the separation of the yellow colour of the flanks, from the white of the under parts. The soles of the feet are of a dusky purple hue; the tarsus is remarkably smooth beneath, and the under surface of the toes presents numerous deeply-indented transverse incisions. The flesh of the ears is whitish, but these organs are very broadly margined with dusky.

The examination of several specimens (clearly of the present species) contained in the Leyden Museum, convinced me that the absence of spines in some parts of the body which are usually covered by them, is not always a mere indication of immaturity, but that, at certain seasons of the year, they are cast to be again replaced; for in one of the largest specimens, I found the spines almost entirely confined to the middle of the back: in a second, they covered the hinder half of the back, and extended to the root of the tail; on the remaining upper parts of the body were stiff hairs. The latter specimen is of a bright brownish rust colour, pencilled with brown-black on the back; the feet and under parts white; fore legs also white, excepting on the outer side, where they are coloured like the sides of the body, and in the region of the wrist, where they are dusky, as are also the hind legs, immediately above the heel; the thighs brown externally, white in front, and pale rufous internally[1]. The first of the two specimens alluded to, differed from the second in having the feet, and lower parts of the legs, both externally and internally, of an uniform pale rufous-brown; and in having the abdomen yellow-white[2]. These specimens are from Surinam.

[1] Its admeasurements are given in the fifth column.

[2] Admeasurements in the sixth column; its tail measured 6″ 2‴, but a small portion was wanting.

	Bahia. British Museum.		Bahia. Zoological Soc.		Minas Geraes. MALE. British Museum.		Minas Geraes. FEMALE. British Museum.		Surinam. Leyden Museum.		Surinam. Leyden Museum.	
	Ins.	Lines.	Ins.	Lines.	Ins.	Lines.	Ins.	Lines	Ins.	Lines.	Ins.	Lines.
Length from tip of nose to root of tail	9	3	8	6	8	4	8	0	8	8	10	6
" of tail	8	6	8	3	8	0	7	7	7	7		
From nose to ear	1	10	1	9½	1	9	1	9				
Length of ear		8		8		9		9½		8		7¾
Width of ditto at the base		10		9½		10		10				
Length of fore foot and nails		9½		9¾		10		10				
" of hind foot and nails	1	11	1	11⅓	1	10½	1	10½	1	10½	1	10
" of skull (to hinder part of auditory bullæ) ...	1	11⅓	1	10½								
Width of ditto	1	0	1	0								
" between orbits		5½		5¼								
Length of nasal bones		9		8½								
Width of ditto behind		2		2								
" " in front		2½		2⅔								
From incisors to molar teeth		5½		5								
Length of four upper molars, taken together		3¾		3¾								
" of lower jaw	1	1	1	1								
Height of ditto, from the condyle		5¾		5½								

Echimys setosus. (Geoff.) Desm.

About equal in size to the Common Rat (*Mus decumanus*); general hue bright rusty yellow, the back suffused with brown;

under parts, as well as the feet, white, with a very indistinct yellow cast: tail with scattered brownish white hairs, which are rather longer than in the Common Rat, and, near the apex, about half an inch in length. Spines are mixed with the hairs on the back of this animal, but are almost entirely hidden by the hairs; they are of a very pale grey-brown (almost white) at the root.

	Inches.	Lines.
Length from tip of nose to root of tail ...	7	10
" of tail (a small portion is wanting) ..	6	9
" of hind foot and nails	1	10
" of ear, about		6¾
" from nose to ear		6½

The above description is from the original specimen described by Desmarest, in the Paris Museum. The author just mentioned has undoubtedly (as pointed out by M. Isidore Geoffroy) committed an error in giving 5″ 6‴ (French measure), as the length of the animal, exclusive of the tail.

According to M. Isidore Geoffroy, the spines of this animal are less numerous than in the *E. Cayennensis*, and differ, moreover, in being narrower, longer, and less stiff, and hidden by the hairs forming the ordinary fur. The habitat of the specimen is not known, but M. Geoffroy states that the Paris Museum has received two specimens perfectly agreeing with it, excepting in being rather smaller, both of which are from Brazil—one was brought from Rio Janeiro by MM. Quoy and Gaimard. I am quite inclined to agree with M. Pictet's opinion with regard to the *E. setosus*—that it merely presents an immature condition of the *E. Cayennensis*. In proportion as the spines are more developed in this animal, the hairs with which they are mixed become more scanty—the growth of the former must, in fact, be at the expense of the latter.

Echimys longicaudatus. Rengger.

The *Echimys longicaudatus* of Dr. Rengger is regarded by several mammalogists as identical with the *E. Cayennensis*, but whether this view be correct, can scarcely be determined from the very brief account given in the "Saugethiere von Paraguay." Dr. Rengger states that the *E. longicaudatus* (which was procured in the northern part of Paraguay) agrees perfectly with his *E. spinosus*, excepting that its tail is longer, that the cheeks are reddish brown, the head, neck, back, and outer side of the limbs brown, and the points of the spines pale reddish brown. Head and body, 8 inches; tail, 6″ 4‴.

Echimys fuliginosus.

Echimys fuliginosus. Wagner, Schreb. Säugth. Suppl. iii. p. 343.

General hue of the upper parts of the animal blackish brown; sides of the body pale brown; feet and under parts white; narrow bristles on the back, and still narrower bristles on the sides of the body: tail rather shorter than the head and body together.

Inhabits Brazil.

This is most probably, as Dr. Wagner now supposes, a variety of the *E. Cayennensis*, differing from the ordinary conditions of that animal in the absence of the rufous colouring, and in having the spines narrower, these being about ten lines in length, and half a line in width.

	Inches.	Lines.
Length from tip of nose to root of tail ...	8	10
" of tail	7	8
" of ear		10
" of hind foot and nails	1	9

I only know this animal through Dr. Wagner's description.

ECHIMYS ALBISPINOSUS.

White-spined Echimys.

Echimys albispinosus. Is. Geoff., Guér. Mag. de Zool. année 1840, pp. 33 and 56, Plate 26,—molar teeth, Pl. 29, figs. 1, 2, and 3.

Tail scaly, and with short hairs; head, upper parts and sides of body, with spines, those on the upper parts black at the extremity, and those on the flanks white: general hue of upper parts brown, and of the flanks, rufous; the whole under parts are pure white.

Inhabits Bahia, Brazil.

With a general resemblance to the *Echimys Cayennensis*, this species, according to M. Geoffroy, is distinguished from that animal by the presence of spines on the head, as well as on all the parts of the body, excepting those which are white; the white colour of the spines on the flanks, moreover, forms one of its distinguishing characters—the one which suggested the specific name. The spines are as strong as those of the *E. Cayennensis*, but much less so than in the *E. hispidus*; those on the upper parts are greyish at the root, and black at the point, and it is the mixture of the black with the rufous interspersed hairs which produces the brown hue on these parts; whilst on the sides of the body, as well as the thighs and haunches, the rufous ground-colour has on it a white pencilling, arising from the protruded points of the spines; some of the spines, however, on these parts are of a very pale grey at the tip. The ears are naked, and tolerably large; the limbs are of a brownish grey colour; the fore foot, the toes of the hind foot, and part of the tarsus, are white. The tail is brown above, whitish below, where the hairs, which spring from between the scales, are also whitish; those on the upper surface of the tail are black.

	Inches.	Lines.
Length from tip of nose to root of tail ...	7	3
" of tail	5	11
" of hind foot	1	9

ECHIMYS HISPIDUS.

Strong-spined Echimys.

Echimys hispidus. (Geoff.) Desm., Nouv. Dict. d'Hist. Nat. x. p. 58; Mammalogie, Pt. 2, p. 292.
" " Is. Geoff., in Guér. Mag. de Zool. année 1840—Mammifères, pp. 9, 35, and 54, Pl. 27—molar teeth, Pl. 29, figs. 4, 5, and 6.
" " Pictet, Mém. de la Soc. de Phys. et d'Hist. Nat. de Genève, ix. p. 156, Pl. 5.

About equal to the Black-rat (*Mus Rattus*) in size; tail very nearly equal to the head and body in length, and well covered with longish hairs, excepting at the base; ears small: broad and strong spines cover the whole upper parts and sides of the animal, commencing on the back of the head; upper parts rusty-brown; feet, sides, and under parts of the body, rust-coloured.

Inhabits Bahia, Brazil.

The above, which are the leading characters of the *E. hispidus*, are extracted from my notes made upon the original specimen contained in the Paris Museum, from which notes, I may further add, that the hairs on the head are stiff, bristly, and grooved along the upper surface, and that the broad spines, which are pale at the root, and of a rusty-brown colour beyond, commence behind the ears, and cover not only the whole of the upper parts of the body, to the root of the tail, but extend *full* half way down the sides of the body. The hairs of the tail are of a yellowish brown colour; at the base of the organ they are short, and do not hide the scaly skin, but they gradually become longer and more numerous to the point, at which part they are upwards of half an inch in length.

The Paris specimen has been in that museum for many years, having originally been brought from the museum at

Lisbon, and is no doubt somewhat changed in its colouring; I will therefore extract some portions of M. Pictet's detailed description of an adult male specimen, contained in the Geneva collection, and which has recently been procured from Bahia. The spines on the upper parts of the body, in M. Pictet's specimen, are white on the lower half, then inclining to violet-grey, and at the apex they are of a bright fawn-colour[1]; the general hue of the upper parts is of a deep chestnut-brown. The head is covered above with narrow spines, as in the Paris specimen, and the broad spines are found covering the same parts as in that specimen, excepting that the neck is said to be armed with narrow spines like those on the head; the upper surface of the head, and the upper lip, are brown; the cheeks fawn-colour; the ears rather short, rounded, and with long hairs on the margin; the moustaches black. The outer surface of the hind legs presents a mixture of flexible spines and hairs, but on the lower half the legs have hairs only. The fore legs are whitish internally, and covered externally with hairs, which are brown at the base, but passing into yellow at the opposite extremity; the toes are grey-white: the hind feet are similarly coloured. In both fore and hind feet the middle and fourth toes are equal, whilst the second and fifth are shorter than these by the length of a joint. The under parts of the body are of a pale fawn-colour, or even whitish, this colour being gradually blended with the colour of the flanks; the lips, under surface of the neck, and the lower jaw, are nearly white. The tail is clothed with deep brown hairs, excepting on the under surface of the basal portion, where they are pale.

[1] M. Pictet gives the width of these spines as somewhat more than a *millemètre;* in this I fear there is some mistake; I feel certain, at least, that the spines in the Paris specimen are very much broader than this measurement indicates: I regret I am not able to give the actual size.

	Paris specimen. Inches.	Lines.	M. Pictet's specimen. Inches.	Lines.
Length from tip of nose to root of tail	7	6	7	11
" of tail	7	3	7	10½
" of head			2	0
" of ears		3½ [1]		7
" of fore foot				8
" of hind foot	1	2	1	3¾

ECHIMYS BRACHYURUS[2]

Short-tailed Echimys.

Echimys spinosus.	DESM., Nouv. Dict. d'Hist. Nat. x. p. 57 (1817); Mammalogie, p. 291 (1822).
? " "	RENGGER, Naturgeschichte der Säugth. von Paraguay, p. 234 (1830).
Mus spinosus.	LICHT., Darstell. Pt. 7, Pl. 36, f. 1 (1832).
Loncheres rufa.	LICHT., Abhandl. der K. Akad. der Wissenschaften in Berlin, 1818—1819, p. 192, (1820).
Echinomys brachyurus.	WAGNER, Schreb. Säugth. Suppl. iii. p. 346, (1843).
Rat premier, ou *Rat épineux.*	AZARA, Essais sur l'Hist. Nat. des Quad. de Paraguay ii. p. 73 (1801).
Échimys roux.	CUV. Règne Animal, i. p. 195.

Spines narrow, and hidden by the fur; ears moderate; tail about as long as the body; tarsus long: general colour rich brown (pencilled black and rufous); under parts dirty white; feet dark brown.

Inhabits Paraguay and Bolivia.

[1] According to my notes, the ears are "about" 3½'''. I presume their condition did not admit of accurate measurement.

[2] The name *brachyurus* applied to this species, is referred, by Dr. Rengger, to Illiger, but I cannot discover where this last mentioned author has published it; I, however, follow Dr. Wagner's example in adopting this name, not only because it is more appropriate than that of *setosus*, but because the latter too nearly resembles the names given to other species of *Echimys*, a circumstance which has been found very perplexing to others as well as myself.

Two specimens of a Spiny-rat, or Echimys, brought by Mr. Bridges from Bolivia, agree so closely in all essential points with Azara's description upon which the *Echimys spinosus* of authors has been founded, that I do not hesitate to introduce their characters under this head. The larger of the two specimens is scarcely equal to the Black Rat (*Mus Rattus*) in size, and both specimens agree in having a comparatively short tail, and in being, to all appearance, destitute of spines,—characters which at once separate them from all the preceding species here described. The fur on the sides of the body is tolerably long, glossy, and moderately soft, and each hair, below the exposed portion, is of a brownish ash colour, but very pale, and, indeed, almost white towards the root; near the point the hairs are of a rich rufous tint, and at the point they are dusky; the general hue produced by the mixture being rufous-brown: on the back, the general hue is of a deeper brown, owing to the hairs being black at the point, instead of brownish or dusky; the rufous pencilling is, however, still very distinct on this part: here there is, mixed with the ordinary fur, an abundance of stiff bristly hairs, averaging about 9 lines in length, and one-quarter of a line (or $\frac{1}{48}$th of an inch) in width; they are grooved along the upper side, as usual in the Spiny Rats, almost white at the root, and gradually shaded through brown into black, which is the colour of the exposed ends. That the animal appears to be destitute of spines, arises from the circumstance that each of the spines terminates in a long and extremely slender hair; the slender end, in many cases, being equal to one-fourth of the entire length of the spine, and in all is the only portion visible, until the fur is parted. The head is dark brown, with but a slight trace of the rufous pencilling on the upper surface; the chin, throat, and abdomen, are white, with a faint yellowish cast; on these parts the hairs are somewhat scanty and short, and they are uniform in colour throughout their length; the limbs are brown externally, and the feet are dusky brown; the tail is dusky above, and pale beneath, and clothed throughout with very small hairs, which are not sufficiently numerous to hide the scaly skin. The ears are sparingly clothed with brownish hairs; the moustaches are brown black, or some few of the hairs pale brown; the incisor teeth pale orange colour in front.

					From Azara.	
	Ins.	Lines.	Ins.	Lines.	Ins.	Lines.
Length from tip of nose to root of tail ...	5	10	6	6	7	8
" of tail	3	0	3	10	3	0
" of hind foot and nails	1	5	1	7	1	6
" of ear		5		5½		4½

Mr. Bridges obtained the two specimens alluded to, from the neighbourhood of the village of Chinoré, on the north side of the Andes of Cochabamba.

The four specimens which came under Azara's notice, were procured at Atira, which is situated about eight leagues to the east of Assumption, in Paraguay.

The burrows of the animal are excavated in a sandy soil, and are often found closely approximated: a burrow of this Echimys, opened by Azara, was four feet in length.

Echimys spinosus. Rengger.

The Echimys described by Dr. Rengger as the "*Rat épineux*" of Azara, must either be a distinct species, or we must conclude that all the specimens examined by Azara, as well as the Bolivian specimens described by myself[1], were imma-

[1] And I may further add, that Dr. Lichtenstein's description of a specimen, said to be from Brazil, agrees closely, in all essential points, with the individuals which I have examined.

ture, and that the form of the spines alters with age, since Dr. Rengger states of his *E. spinosus*, that he has never observed the spines terminate in a long hair-like point in that animal, as noticed by Azara, and as I find to be the case in the Bolivian specimens. With regard to these latter, I must observe, moreover, that in them the spiny hairs are strictly confined to the back: from Azara's description we do not learn whether such was the case in his animal. I will here add an abstract of Dr. Rengger's description.

Upper parts and sides of the body brown, or where the spines exist, brown mixed with brownish red, as the bristles have a brownish red point: throat, breast, belly, and inner surface of the limbs grey-white; tail brownish black. On the whole of the upper parts of the body, as well as on the flanks, the hinder part of the head, and the upper half of the limbs externally, are numerous spiny hairs mixed with those of the ordinary fur, and which project beyond, and almost entirely hide the latter. Length of head, 2″; of trunk, 5″ 6‴; of tail, 2″ 9‴.

Dr. Rengger states that he found this animal in the southern parts of Paraguay only, and that it was there very rare. They live often in great communities on the gentle declivities of sandy hills, where they excavate serpentine burrows of five or six feet in length, and scarcely more than six inches below the surface of the soil. The burrows sometimes have one, and sometimes several outlets; and, at their termination, is a bed or nest of dried grass. In one of the burrows opened, Dr. Rengger found two young spiny Rats: they were blind, and the bristles on the back were quite soft. The food of the Echimys, Dr. Rengger further observes, appears to consist of the roots of grasses, and the seeds and fruits of shrubs, for, in the districts in which it is found, no other vegetable substances suitable for it exist.

ECHIMYS INERMIS.

The Spineless Echimys.

Echimys inermis. PICTET, Notices sur les Animaux Nouv., &c. in Mém. de la Soc. Phys. et d'Hist. Nat. Genève, x. p. 33, Pl. ix. and Pl. x. figs. 1—8.

Feet moderately long; tail rather shorter than the head and body taken together, and clothed with long hairs, which entirely hide the scales; fur soft, and without any mixture of soft hairs; general hue brown-yellow; under parts yellowish white.

Inhabits Bahia, Brazil.

This species, which is readily distinguished from others of its genus by its fur being uniformly soft, has the tarsi considerably shorter than in the *Echimys Cayennensis*, and the outer toe of the fore feet relatively longer: its tail is provided with longer hairs than in other species of Echimys, and they are so abundant as almost totally to hide the scaly skin, especially on the basal part of the organ. Although the fur is soft, that on the back is composed of hairs of two kinds, the one presenting the usual appearance, and the other flattened and grooved like the spines of other species of Echimys: all these hairs have the basal half grey; above this they are brown; at the point they are deep brown, and they have a subterminal pale yellow ring. The muzzle is clothed with hairs which are annulated with brown and white, and the orbit of the eye is white above. The hairs of the moustaches are some of them white, and some black: the region of the mouth, and the under surface of the head, is white, slightly inclining to yellow; the throat has a mixture of grey and yellow colours; the limbs are clothed with hairs which are grey, with a yellow point, and the toes are uniformly grey. The under parts of the body and inner surface of the limbs are yellowish white. The hairs of the tail are black, excepting those on the under side of the basal half, which are whitish.

A second specimen, examined by M. Pictet, had the fur still softer than the first, and its general tint was more grey. It was killed in the crevice of a rock at Jacobina, in the province of Bahia.

	Inches.	Lines.
Length from tip of nose to root of tail ...	8	0
" of the tail	7	0
" from tip of nose to ear	1	9½
" of ear		9⅓
Width of ditto		8½
Length of the fore foot		9⅓
" of the hind foot	1	9¼

ECHIMYS ANTRICOLA.
The Cave Echimys.

Nelomys antricola. LUND, Det Kongel. Dankse Videnskab. &c. viii. p. 246, Pls. 22 and 23.
Echimys apereoides. LUND, l. c. viii. p. 98.

Body stout; the muzzle thick; ears short; tail long, and well clothed with hairs: upper parts of the animal grey-brown,

[1] Whether Dr. Wagner is correct in identifying his *Isothrix pachyura* with this animal it is impossible to determine, for the accounts given of both animals are far too brief for the decision. Should the *Is. pachyura* prove to be the same as the *Nelomys antricola* of Lund, then must it be placed in the genus *Echimys*, for I must insist upon the structure of the teeth as of higher importance than the more or less hairy condition of the tail, and the other external characters which induce Dr. Wagner to place his animal in the same section as the *Loncheres picta:* that species most decidedly belongs to the section of the Spiny-Rats (the Loncheres section), in which the upper molars have *three* indentations, or folds, of the enamel entering from the outer side. Whether the middle fold of these three joins the principal internal fold or not, is a matter of little importance, when we are simply seeking for the primary type of structure in the part in question. *Echimys antricola*, as I have elsewhere stated, has but *two* folds on the outer side of the upper molars.

with an admixture of rusty yellow ; under parts white: the hairs of the fur harsh, and adpressed.

Inhabits the interior of Brazil.

The *Echimys antricola*, briefly described by Dr. Lund in his papers in the Transactions of the Danish Society of Naturalists, &c., is evidently a distinct species, though closely allied to the *Echimys inermis*[1]. Like that animal, its fur is destitute of true spines, and its molar teeth have each but three folds of enamel, two on the outer side, and one on the inner, in the upper molars, and with the folds reversed in the lower molars: but, from the *E. inermis*, it must differ not only in being of a larger size[1], but in having a coarse fur, and a relatively shorter tail.

	Inches.	Lines.
Length from tip of nose to root of tail ...	10	0
" of tail	7	6

Genus, *Carterodon* (fossil).

Echimys et *Aulacodus*. LUND, in the Danish Transactions,

Echimyina with a short and broad skull; the incisor teeth large; those of the upper jaw each with a longitudinal furrow near the outer edge, and a longitudinal mesial ridge, and those of the lower jaw flat in front: molar teeth of the upper jaw with one internal fold of enamel, and two external folds; those of the lower jaw with the same number of enamel folds, but these reversed in position.

Carterodon sulcidens.

Echimys sulcidens. LUND, Det Kongel. Danske Videnskab. &c. viii. pp. 99 and 242, Pl. 25 (lower jaw).

[1] The skull of *E. inermis*, according to Prof. Pictet's figures, measures 2″ 0½‴ in length, and 1″ in width; whilst Dr. Lund's figures of the skull of the *E. antricola* represent it as 2″ 4‴ in length, and 1″ 2‴ in width. The animal is said to bear a close resemblance to the *Cavia Aperea*, both in size and colouring.

Certain remains of a small Rodent found by Dr. Lund in the caverns in the interior of Brazil, are referred by that author to the genus *Echimys*, and subsequently to that of *Nelomys*, but most recently to the genus *Aulacodus* of Temminck. They consist of portions of skulls and lower jaws, with the teeth; and, from a character furnished by the latter, the specific name *sulcidens* was suggested. Amongst numerous Mammalian remains, from the caverns in the same district, contained in the collection of the British Museum, are several crania of the same species, and from an examination of these, I have satisfied myself that the characters they present do not strictly belong to either of the genera above mentioned. The molar teeth very nearly agree in their structure with those of the Echimys, and can scarcely be said to differ from those of Aulacodus; and, consequently, are unlike the molars of Nelomys (or *Loncheres*). They are relatively larger than in Echimys, broader than long, and remarkable for the thickness of the enamel which covers them; the inner fold of enamel of the upper molars runs obliquely forwards and inwards from the posterior angle of the tooth, and at its termination touches the foremost of the two deep external folds. The upper incisor teeth, in being grooved, differ from those of *Nelomys* and *Echimys*, and they differ, moreover, in being relatively much larger: nor will they bear a very close comparison with the incisors of Aulacodus, for in that animal each upper incisor has three grooves, and these placed on the inner half of the tooth, whereas in the *Carterodon* there is but one groove, and that runs close to the outer edge; this groove is broad, and tolerably deep, and bounded on the inner side by a strongly convex ridge, which runs along the middle of the tooth; the inner half of the tooth is flat; and, what is rather singular, this portion of the anterior surface of the tooth, in a skull before me, is white, whilst the outer half of the tooth is of a deep orange colour[1] The fossil skull under examination differs from the skulls of the species of *Echimys* and *Nelomys*, not only by its short and broad form, but in having the zygoma deeper, and the malar bone destitute of a post-orbital process; the infra-orbital opening relatively small, and the incisive openings much larger. From the skull of the Aulacodus it differs, amongst other points, in wanting a separate groove for the infra-orbital nerve, and in being destitute of the strong sagittal and occipital crests, which are so remarkable in that skull. The condition of the fossil skulls alluded to is such, that I am inclined to believe the species may still be living in South America; they have scarcely lost their animal matter, and owe their very perfect state of preservation to a coating of stalagmite, which in some cases is so thin that it is removed without difficulty. I see that Dr. Lund also supposes the animal is still living, for he arranges it amongst the existing species, though he has not yet, as he states, met with it. Skulls and lower jaws of the *C. sulcidens* are abundant in nearly all the caverns visited by Dr. Lund. The following dimensions are taken from one of the most perfect of three crania[2]. It is figured on Plate 16.

	Inches.	Lines.
Total length of skull	1	7
Width of ditto		11¼
" between orbits		4
Length of zygomatic arch		8½
Greatest depth of ditto		2⅛
Length of nasal bones		6
Width of ditto		2⅓
" of incisor teeth		2¼
Distances between incisors and molars ...		3¾

[1] A second specimen differs only in having the orange colour replaced by yellow: the colour in this has evidently faded.

[2] Some imperfect skulls are a trifle larger than the specimen measured, and the lower jaw, of which the dimensions are given, apparently belongs to one of these.

	Inches.	Lines.
Length of four upper molars, taken together		4
Height of ant-orbital opening		$2\frac{1}{3}$
Width of ditto		$1\frac{1}{5}$
Length of lower jaw	1	$1\frac{3}{4}$
Height of ditto, from the condyle		$5\frac{1}{2}$

Genus, *Aulacodus*[1].

Aulacodus. TEMMINCK, Monogr. de Mammalogie, i. p. 245.

Echimyina with but four toes to the hind feet; tail short: incisor teeth very broad, those of the upper jaw with three longitudinal grooves, those of the lower, flat in front; molar teeth of the upper jaw with one internal fold, and two external folds of enamel: skull short and broad, and with the occipital crest much elevated; a vertical plate on the upper part of the anterior root of the zygomatic arch forming the outer boundary of a groove for the infra-orbital opening.

The genus Aulacodus presents all the essential characters of the Echimys group, and therefore furnishes a remarkable exception in the geographical distribution of that section, which, with this exception, and a second furnished by the *Petromys typicus*, is strictly confined to South America. The dentition of Aulacodus can scarcely be said to differ from that of the genus *Carterodon*, excepting in the structure of the incisors, and of the foremost molar of the lower jaw; this latter tooth having three indenting folds on the inner side, whereas in *Carterodon* there are but two. The incisor teeth are still broader and stronger than in the last mentioned genus, and, indeed, they are relatively more powerful than in any other known Rodent: the upper incisors are strongly arched; broader than deep, flattened at the sides, and slightly convex in front, where they each present three deep longitudinal grooves; one of these grooves is placed very nearly

[1] From αὖλαξ, a furrow; and ὀδοὺς, a tooth.

in the mesial line of the tooth, and the other two grooves are disposed between this and the inner edge of the incisor, the innermost being separated from the edge only by a narrow ridge ; the interspaces between the grooves are strongly convex in the transverse direction, and but little broader than the grooves ; that which separates the first mentioned groove from the second is broadest. The molar teeth are very nearly equal in size, and the crowns of these teeth are nearly square. The enamel covering the molar teeth is unusually thick, and the indenting folds which it forms penetrate from either side to the mesial line of the tooth, or very nearly so.

Of the five toes to the fore feet, the outer toe on either side is very small ; that which represents the thumb is the smallest, being a mere tubercle, with a short truncated nail ; the nail of the outer toe is moderately large, and curved ; the nails of the remaining three toes are large and strong, and fitted for burrowing ; they are solid, longer than the toes, and but little arched above. To the hind foot are three well-developed toes, furnished with large and powerful nails, and a rudimentary outer toe provided with a moderate sized nail. The ears are short, but very broad ; the tail short, being scarcely equal to half the body in length : it is rather sparingly clothed with stiff hairs.

But one species of this genus is known.

AULACODUS SWINDERIANUS

The Aulacodus, or Ground Rat.

Plate 16, fig. 2.

Aulacodus Swinderianus. Temm., Monogr. de Mamm. i. p. 248, Pl. 25.
" " Bennett, Proc. of the Zool. Soc. for 1830-31, p. 111.

Body covered with stiff and bristly hairs on all parts: general hue brown, pencilled with black and yellow, or, sometimes, with dirty yellowish white; tail dusky above, whitish below.

Inhabits Sierra Leone, the neighbourhood of the Gambia, and South Africa.

The Aulacodus was first described by M. Temminck, from a very young specimen preserved in spirit of wine, and its native country and adult characters were not known to us until a specimen was procured at Sierra Leone by Mr. Boyle, who having presented it to the Zoological Society, enabled Mr. Bennett to publish a detailed account of its peculiarities in the Proceedings of that Society A second specimen has since been received by the Society, and an adult specimen exists in the British Museum: the former was procured at the Gambia, and the latter is said to be from Port Natal, South Africa. At Sierra Leone the animal is known, according to Mr. Boyle, by the names *Ground-Pig* and Ground-Rat; and that gentleman states that it feeds upon ground nuts, cassada, and other roots. It burrows in the ground, and I have been informed that large holes formed by the Aulacodus are very numerous on some parts of the coast.

When full grown, the Aulacodus is but little short of two feet in length; it has a short head, a broad and obtuse muzzle, and small eyes: the ears project but little beyond the coarse hairs of the head, but are very broad; the limbs short and stout; the toes short, and provided with large and solid nails; the tail short, and clothed with short stiff hairs, which do not completely hide the scaly skin. The covering of the animal consists entirely of stiff bristly hairs, which are flattened, and grooved along the upper surface; they are of moderate length, and closely applied to the skin; those on the upper parts of the animal are of a pale ashy grey colour at the root, dusky in the middle, and black at the point; most of them have a subterminal brownish yellow ring. The general hue of the animal is brown; less rich on the sides of the body than on the back, owing to the yellow rings on the hairs of these parts being paler, and the black pencilling less distinct. The chin and upper lip are whitish; the throat is of a dirty yellowish hue, and the abdomen of a very pale brownish yellow tint, pencilled with grey-brown. The feet are pencilled with black and yellow. The ears are moderately well clothed with hairs on the inner surface, but chiefly towards and at the margin, and these hairs are for the most part of a yellowish white colour, and about a quarter of an inch in length. The hairs of the moustaches are some of them white, but the greater portion are black, and there are some few long and coarse hairs mixed with those of the ordinary kind, springing from the upper surface of the muzzle. The hairs on the tail average about a quarter of an inch in length; those on the upper surface are chiefly black, but many of them are in part rufous, and those on the under surface are of a rusty white tint.

The above description is taken from the specimen before referred to, which was received from the Gambia: the specimen in the British Museum is a trifle less rich in the colouring of the upper parts of its body, and has a shorter tail, but in other respects agrees very closely. The admeasurements of these specimens are as follows:—

	Specimen from the Gambia. Inches.	Lines.	Specimen from Port Natal. Inches.	Lines.
Length from tip of nose to root of tail	20	0	21	6
" of tail	8	6	5	6
" of ear		6		6
Width of ditto	1	4	1	6
Length of fore foot and nails ...	1	8	1	10
" of the nail of the middle toe		7		7½
" of hind foot and nails ...	3	5	4	0
" of the nail of the middle toe		8		9
Width of incisor teeth		5⅔		6

The specimen procured at Sierra Leone by Mr. Boyle is preserved in spirit of wine, and, according to Mr. Bennett, its length, measured in a straight line, from the nose to the root of the tail, is 17 inches; or, measured along the curve of the back, 20″; tail, 9″; head, 2½; fore leg, 3½; fore foot and toes, 1½; femur, 4½; tibia, 4½; tarsus and toes, 3½; ear, 1⅙ long, and 1″ broad.

The skull having been removed from this specimen, I have had an opportunity of comparing it with the crania of other Rodents, and the comparison has left no doubt on my mind as to the true affinities of the animal; it, in fact, *essentially* agrees with the skull of the Echimys, but at the same time presents certain peculiarities worthy of notice. It is short, and exhibits great strength in its parts. One of the most remarkable features of this skull is the great development of the occipital and lambdoidal crests; the former is raised so as to increase the height of the occiput by about a quarter of an inch; the lambdoid crest is rather less elevated, and springs chiefly from the large triangular interparietal bone. The frontal bones are separated by a groove, and a broad depression runs along the nasal bones, which, when viewed in front, present a double arch: these bones are of equal width at each extremity, and about one-third longer than broad. The lachrymal bone is of moderate size; the antorbital opening very large, and a second channel is formed (as in

Petromys, and the Octodons), for the infra-orbital nerve. The zygomatic arch is remarkably strong, having relation to the large size of the incisors, as has likewise the strongly developed crests on the hinder portion of the skull—the masticating muscles being attached to these parts. The malar bone is very deep in front, has a slightly raised post-orbital process, and an angular projection on the under side. The portion of the palate which lies between the molar teeth is slightly contracted, and has the sides parallel; it descends far below the level of the anterior half of the palate; in this latter are two large incisive foramina. The palatine bones run forwards to terminate opposite the second molar tooth, where there are two approximated palatal foramina opening forwards. The auditory bullæ are of moderate size. The occipital opening is nearly square, but presents a small notch in the upper margin. A very small portion of the mastoid bone is visible on the occipital surface of the cranium, which latter is remarkably flat, and nearly vertical.

	Inches.	Lines.
Total length of skull	3	6
Width of ditto	2	5
" between orbits	1	4½
" of the occiput	1	7
Height of the occiput	1	3
Length of nasal bones	1	2
Width of ditto behind		8
" in front		7¾
Length of zygomatic arch	1	8
Greatest depth of zygoma (which is beneath the anterior angle of the eye)		7½
Height of ant-orbital opening		8¼
Width of ditto		6
From incisor to molar teeth		10
Width of incisors		6
Length of four upper molars taken together		9
Width of palate between them		3¾
Length of incisive openings		6
Width of the two openings taken together ...		3½
Length of lower jaw	2	11
Height of ditto, from the condyle	1	3¼
Length of angular portion	1	5

SUB-FAMILY, DASYPROCTINA.

Dasyproctina (part). GRAY, Annals of Philosophy, vol. x. (New Series p. 341 (1825); List of the Mammalia in the Brit Museum (1843), p. 124.

Hystricidæ with semi-rooted molars arranged in parallel series, the cranium elongated; nasal bones rather short and broad; malar bone destitute of a descending process on the lower edge; a tolerably distinct post-orbital process, formed chiefly by the frontal bone, but in part likewise by the squamosal; scapula with the emargination in the fore part of the spine but moderately deep; clavicles wanting; feet formed for running, with the toes 5—3, or 5—5, terminated by sub-solid nails, which are but little arched; tail rudimentary; body clothed with hair only—no admixture of spines.

Habitat, South America—ranging from the northernmost parts, southwards to Paraguay and Bolivia: occur, likewise, in some of the West India Islands.

Excepting in those characters which are common to the *Hystricidæ*, the Rodents of the present section differ very widely from the Cavies with which they have always been associated, and the differences, for the most part, indicate that the Dasyproctina are higher in the scale of organization than the Caviina. I allude more especially to those modifications which are exhibited in that most characteristic part of the animal, the skull. We now find a tolerably broad and perfect palate, which is neither contracted nor suddenly raised between the anterior molars, as in nearly all the species of the preceding sections, nor broken up by large incisive openings, as in the lowest of them: here the small incisive openings are confined to the intermaxillary bones. With this more perfect condition of the palate, may be noticed a corresponding change in the condition of the sphenoid. Hitherto the openings in the sphenoid alæ have, from their large size, reduced the interspaces into mere narrow branches, but in the Dasyproctina the foramina in question are relatively small, and widely separated: the process corresponding to the external pterygoids is but little extended in the longitudinal direction, and the canal which runs forwards, between the pterygoids into the orbit, is very short. The internal pterygoid is proportionately large, being much more extended in the antero-posterior direction than in the preceding sub-families. The lachrymal bone is large, and perforated by the lachrymal canal[1].

Genus, *Cœlogenys*[2].

Cœlogenus.	F. Cuvier, Annales du Muséum, x. p. 203 (1807).
Cœlogenys.	Illiger, Prodromus Systematis Mammalium, &c., p. 92 (1811).
Osteopera.	Harlan, Fauna Americana, p. 126 (1825).
Cavia.	Linn., Erxl., &c.

Dasyproctina with the zygomatic arch enormously developed, and of great depth; the portion which is due to the superior maxillary bone, presenting a large hollow beneath; the malar bone deeper than long; a separate groove for the infra-orbital nerve; cheek-pouches within the cavity formed by the zygomatic arch; molar teeth rather longer than broad, the crown of each of these teeth presenting four or five deeply indenting folds of enamel; incisors slender; toes 5—5.

The genus Cœlogenys contains but one well-established species — the Paca. This animal is arranged by mammalogists with the Agoutis, and, notwithstanding that it

[1] In the two preceding sub-families, *Echimyina* and *Octodontina*, the lachrymal bone is remarkably small, and in several of the species in these sections it only in part encloses the lachrymal canal.

[2] From κοῖλος, hollow; and γένυς, cheek.

differs widely from those Rodents in some respects, I am inclined to follow the usual course, regarding the differences in question as of minor importance. The enormous development of the zygoma in the Paca would appear to be a purely adaptive peculiarity, and hence cannot be regarded as indicative of a distinctness of type. To the Agoutis the Paca approaches most nearly in the general structure of the skeleton of the trunk; and even in the skull there is much resemblance, if we do not consider the zygomatic arches. It has the same elongated form, and the length is due in great part to the great longitudinal extent of the frontal bones, for, as in the Agoutis, the nasal bones are relatively short and broad. The pterygoid canal is short, and the external pterygoid is perforated in the antero-posterior direction, as in the animals just mentioned.

The molar teeth in the Paca are relatively larger than in the true Agoutis, and they differ moreover in being longer than broad; those of the lower jaw resemble the corresponding teeth in the Coypu (*Myopotamus*) in having one external, and three internal folds of enamel, but, viewing the outer surface of the lower molar in the Coypu, we find the two principal lobes of the tooth produced into more or less prominent angles, whereas, in the Paca, the same lobes are rounded externally. The foremost lower molar in the Paca, as in the Coypu, has an increase in the number of indentations of the enamel. In the immature animal (where but three molars are developed) this is more strikingly the case, for then the foremost molar has four internal folds of enamel, and I find a horse-shoe shaped area, surrounded by enamel, in front of them. This is most probably a milk tooth.

The upper molar teeth in the adult Paca have each two internal folds of enamel, and three transverse areas enclosed by enamel, with the exception of the hindermost tooth, in which there are two external, and three internal enamel folds[1].

[1] If we regard a complicated Rodent's molar as normally having a primary external and internal indentation of enamel (marked *a* in the figures) dividing the body of the tooth into two equal, or nearly equal, parts, and as having two secondary indentations (marked *b* and *c*), each of which subdivides one of these two parts—the secondary folds, in the upper molars, entering from the outer side of the tooth, and, in the lower, entering from the inner side—then we may simply describe the lower molars of the Coypu and Paca as presenting the normal condition. The foremost upper molar of the Coypu is also normal, but the remaining upper molars depart from the type in having the hindermost of the secondary lobes internal instead of external. The upper molars of the Paca depart from the type in the same manner, and differ, moreover, in having an extra indentation on the hinder part of the tooth; and, in the hindermost molar, two extra enamel folds are visible, the additional lobes which they separate being joined to the back of the tooth, as we shall find also to be the case in the teeth of other mammals, in which the grinders are highly complicated.

Further, I have to remark, with respect to the enamel folds in the teeth of the Paca, that the principal internal indentation of the upper molars is often confluent with the foremost of the secondary folds, and that most of these latter, in the adult animal, become isolated from the enamel which surrounds the body of the tooth; the grooves, which separate the enamel plates, being deeper in the inner part of the tooth than near its margin.

CŒLOGENYS PACA.

The Paca.

Mus Paca.	LINN., Syst. Nat. 12, i. p. 81.
Cavia Paca.	SCHREB., Säugth. iv. p. 609, Pl. 171.
Cœlogenys Paca.	RENGGER, Säugth. von Paraguay, p. 251.
" "	WAGNER, Screb. Säugth. Suppl. iv. p. 52.
" *fulvus.*	F. CUVIER, Ann. du Mus. x. p. 207; skull, Pl. 9, figs. 1 and 2.
" "	PR. MAXIM., Beiträge, ii. p. 454.
" *subniger.*	F. CUV., Ann. du Mus. x. p. 206; skull, Pl. 9, figs. 3 and 4.
Osteopera platycephala.	HARLAN, Fauna Americana, p. 126.
Paca.	MARCGR., Bras. p. 224.
"	BUFFON, Hist. Nat. x. p. 269, Pl. 43; Suppl. iii. p. 203, Pl. 35.
Le Pay.	AZARA, Essais sur l'Hist. Nat. de Paraguay, ii. p. 20.
Spotted Cavy.	PENNANT, Synopsis of Quadr. p. 244.

Hair rather short, coarse, and by no means dense; limbs and upper parts of the body of a brown colour, varying in shade; under parts white; flanks with three, four, or five longitudinal white bands, which are more or less interrupted and broken up into spots.

Inhabits South America, extending from Cayenne southwards to Paraguay; and found likewise in Peru, but rarely; and in some of the West India islands.

The Paca is a stout-bodied animal about two feet in length; the limbs rather short; the head large, broad, and terminating in an obtuse muzzle, which is clothed throughout with fine hairs; the eyes, which are placed somewhat nearer to the ears than to the end of the muzzle, are large, directed obliquely upwards, and have a round pupil. The ears are rather below the medium size, and sparingly clothed with hairs. The feet are naked beneath, and provided with five toes, but the inner toe, both of the fore and hind feet, is very small, and furnished with a correspondingly small nail; the nails of the other toes are rather broad, but little arched, and nearly solid; they are moderately long. The tail, which is a mere fleshy tubercle, is naked. The fur of the animal is composed of one kind of hair only; the hairs are coarse, and closely applied to the skin.

In some specimens of the Paca but three white bands are visible on the sides of the body, a few white spots, above these, being all that remain in such specimens of a fourth band, which in other individuals is distinct; sometimes there are five lateral bands. The female Paca has two pairs of teats, one pair being pectoral, and the other inguinal.

Not only does our animal vary in the number of bands on the flanks, but likewise in the general tint of the upper parts of the body and of the limbs; and M. F. Cuvier, perceiving that differences of colouring in certain individuals were accompanied by peculiarities in the cranium, concluded that there existed more than one species of Paca; and, in fact, that two kinds might be distinguished—the one by a blackish umber colour, and by the outer suface of its skull being smooth; and the other by the general hue of the body being fawn-coloured, and the skull having the surface rough. To the first he gave the specific name *subniger*, and to the second that of *fulvus*.

The following description is from a specimen in the British Museum, which corresponds with M. Cuvier's

Cœlogenys subniger.

The prevailing hue is black-brown; the whole of the under parts are white, faintly tinted with yellow, as well as the inner surface of the limbs. The cheeks are whitish; the hairs of the moustaches are most of them white, but some few are blackish. Three narrow, longitudinal, white bands adorn each side of the body; the uppermost of these, which is the shortest, commences behind the shoulder, and terminates on the haunches; the remaining two extend both over the shoulders and haunches. All these bands are interrupted in parts, and they are more completely broken up into spots at their extremities.

	Inches.	Lines.
Length from tip of nose to root of tail ...	23	0
" from ditto to ear	4	0
" of ear		10
Width of ditto	1	1
Length of fore foot and nails	3	11
" of nail of middle toe		5¾
" of hind foot and nails	3	11
" of the nail of the middle toe		8

This specimen is from Granada—St. Fé de Bogota; its skull shows it to be an immature animal, since there are but three molar teeth developed on each side of the jaw: the outer surface of the zygoma, and of the frontal bones, is very nearly smooth, there being a very slight roughness on the fore part of the zygomatic process of the superior maxillary bone. I possess a cranium from the Province of Minas Geraes precisely corresponding with this.

Cœlogenys fulvus. F. Cuvier.

An animal agreeing with M. Cuvier's description of *C. fulvus* also forms part of the British Museum collection. It is rather less than the dark Paca just described (being perhaps younger), and differs in being of a bright rufous-brown colour on the whole upper parts of the body, as well as on the flanks and external surface of the limbs, if we except the white bands and spots on the latter parts. The under parts are white—not quite pure. On the sides of the body are three longitudinal, interrupted, white bands, and these extend on to the haunches, on which part are two rows of spots, one above, and the other below the three principal bands. Its skull cannot be examined. This specimen is said to be from tropical America.

That the different shades of colouring observed in the Paca do not indicate a distinction of species, is now satisfactorily proved by the observations made upon these animals by Dr. Lund in Brazil, and by those of Dr. Rengger, who studied them in Paraguay. Both these authors agree in stating that the Paca is subject to considerable variety of tint, and the former observes that specimens occur presenting all the intermediate shades between the dark and light varieties; whilst the latter author states that he has met with Pacas of different shades of colour, and these paired, and living in the same burrows[1].

The skull of the Paca, as I have before stated, if viewed without the zygomata, is of the same elongated form as that of the Agoutis, but the enormous size of the zygomatic arches, and these being thrown boldly outwards—most so at their hinder extremity—gives to the whole structure a width which only falls short of the length by about one-fourth part. The upper surface of the cranium is very little arched in the longitudinal direction, but a transverse section would be nearly semicircular. The frontal bones are very long; the

[1] According to Dr. Rengger, however, the usual tint of the Pacas in Paraguay is bright yellow-brown on the upper parts and sides of the body; the extremities dark yellow-brown; the rows of spots, or interrupted stripes, on the sides of the body (of which there are said to be five), and the under parts yellow-white. The dimensions of a very large female Paca, Dr. Rengger found to be as follows:—

	Inches.	Lines.
Head and body	25	6
Tail		4
Medium height	13	0
Length of skull	5	0
Breadth of ditto	3	3

nasal bones short and broad. The infra-orbital opening is proportionately smaller than in most other *Hystricidæ*, the space being contracted by the unusual development of the zygomatic portion of the superior maxillary, in which a long, narrow channel, opening above in the infra-orbital outlet, marks the course of the infra-orbital nerve. The zygomatic arch, in a skull in which but three molar teeth are developed on either side of the jaw, measures in the longitudinal direction about half the entire length of the cranium, and its depth is equal to half its length; but, in the adult animal, the length and depth are proportionately greater, the former being sometimes equal to two-thirds of the entire length of the skull. The hinder third of the arch (or rather less in the young animal), is formed by the malar bone, which has the greatest vertical diameter, and the remaining portion (which gradually decreases in width towards the fore part) is formed by the superior maxillary bone. This latter portion encloses a large oblong cavity beneath, which, in the living animal, is lined, as it were, by a continuation of the skin of the face, and beneath this is a pouch formed of the skin of the cheek: the pouch has a small outlet, which opens in front of the molar teeth. The uses of this pouch have not yet been determined. Two strongly elevated ridges, separated by a deep groove, run along the fore part of the palate, from the molar teeth almost to the incisors; the portion of the palate which lies between the molars is slightly contracted in front, and deeply emarginated behind. The squamous portion of the temporal bone extends backwards to the occipital plane.

Besides these leading peculiarities in the singular cranium under consideration, we have to notice the extreme roughness of the zygoma, and of the frontal bones, which is observable in the skulls of some Pacas. In such, the whole outer surface of the parts in question is studded with small wart-like excrescences, more or less confluent, and in some parts (especially on the malar bone) joining so as to form coarse and very irregular wrinkles[1].

[1] This roughness of the surface increases with age, as is clearly seen by the different conditions presented by a series of crania in the British Museum. In these crania the frontal bones and zygomatic arches are more rough in proportion as the molar teeth are more worn. But though this roughness occurs in some Pacas' skulls, there are others which remain smooth, even in the adult animal; and this fact leads us to take into consideration the views of M. F. Cuvier, which have already been alluded to—viz. that the rough and smooth crania belonged to different species of Paca — species distinguished not only, it is said, by the peculiarities just referred to, but by colour, and by a difference in the degree to which the zygomata are extended in the lateral direction. The difference of colour, it has already been shown, cannot be regarded as specific, and a comparison of the following dimensions will demonstrate that the lateral expansion of the zygomata in the smooth skull is not always less than in the rough skull, as has been stated.

	Very old. Rough.		Adult. Rough.		Adult. Rough.		Adult. Zygoma rough; frontal bones smooth.		Adult. Smooth.		Young, with three molars. Smooth.	
	Ins.	Lines.	Ins.	Lines.	Ins.	Lines.	Ins.	Lines.	Ins.	Lines.	Ins.	Lines.
Total length of skull ...	5	7	5	1	5	6	6	0	5	2	4	7
Greatest breadth ...	3	8	3	3	3	5	4	1	3	3	2	9½
Width between orbits .	1	7½	1	5	1	7	1	11	1	7	1	3
Length of nasal bones .	1	8	1	7	1	9	2	1	1	8	1	4
Width of ditto in front		10		9		11	1	0⅔		10		7¾
Length of zygoma ...	3	6	2	10	3	2	3	6½	2	10	2	7
Depth of a malar bone	2	1	1	7	1	9	2	0	1	7½	1	2
Of zygoma, in a vertical line, dropped from the anterior angle of orbit	1	7	1	3½	1	6	1	8	1	4½	1	2½

Not having been able, then, to discover any points of distinction in the skulls of the Paca, excepting that noticed in the condition of the surface of certain bones, I am induced to regard this distinction (when it presents itself in adult skulls) as a sexual one. This view has been suggested by Dr. Lund, and, as Dr. Wagner truly observes, requires confirmation.

Analogical considerations would lead us to suppose that the rugose cranium distinguished the male Paca; and, connected with this view, I must notice a

With respect to the habits of the Paca, we have only to observe, that it is generally seen, singly or in pairs, on the borders of the forests, or near the banks of rivers, but only after sunset, since during the day it lies concealed in some cavity beneath the root of a tree, or in a burrow of its own excavation; its burrow usually being about four or five feet in depth. Its food, Dr. Rengger states, consists of the leaves of different plants, and of fruits and flowers. Occasionally it visits the plantations of sugar-cane and melons, and does considerable mischief. The female has but one young at a birth, and this is said to follow the parent for two or three months. The range of the Paca is very extended, the animal being found nearly throughout South America, from its northernmost parts down to Paraguay. In Brazil it is tolerably plentiful; in Peru it is rare; and in Paraguay it is by no means common. It occurs in Bolivia, in the department of Sierra de la Cruz. The Paca is likewise found in Tobago and Trinidad. On the west side of the Andes it does not occur.

The accompanying woodcut represents the dark variety of the Paca, and a specimen in which the lateral lines are more broken up into spots than is usual.

peculiarity which is striking in the five adult rough skulls which lie before me. Not only are the more aged skulls more rugose than the others, but there is a considerable amount of variation in the pattern (if I may so term it), which the warty excrescences and rugæ assume. Parts which are rough in a cranium in which the molars are moderately worn, are smooth in a second cranium having indications of greater age in the condition of the teeth, whilst in this second skull other parts are more distinctly rugose. Can it be that this deposition of bony matter is periodical, and that it bears some analogy to the periodic deposition of bony matter which forms the horns of the male deer?

Cœlogenys laticeps (Fossil).

Cœlogenys laticeps. LUND, Det Kongel. Danske Vidensk. Selsk. Naturvidensk. og Mathem. Afhandl. viii. Pl. 20, fig. 1.

Founded upon a fossil cranium (from the caverns of Brazil), having the surface of the frontal bone and zygomatic arch smooth; and which differs, according to Dr. Lund, from the skull of the *Cœl. Paca* in the infra-orbital opening being larger, the upper part of the bony ring, which encloses it, narrower, and the zygoma thrown more boldly outwards[1].

Dr. Lund distinguishes a second fossil species of *Cœlogenys* from the same caverns, to which he gives the name

[1] Differences of the same kind, and almost equal in degree (judging from Dr. Lund's figures), are presented in the crania of the recent species of Paca.

Cœl. major. It appears to be chiefly separated from the recent species on account of its superior size.

From the caverns above alluded to, the British Museum contains numerous remains referrible to the genus Cœlogenys. The most characteristic specimen among these, is a considerable portion of a skull, in which I can find no points which would lead me to suppose it belonged to a species distinct from the Common Paca.

Genus, *Dasyprocta.*

Dasyprocta.	ILLIGER, Prodrom. Syst. Mamm. &c. p. 93 (1811).
Platypyga.	ILLIGER, l. c.
Chloromys.	F. CUVIER, Ann. du Mus. xix. p. (1812).
"	LESSON, Manuel de Mamm. p. 300 (1827).
Cavia.	ERXL., SCHREB., GMEL.
Mus.	LINN., Syst. Nat.

Dasyproctina with but three toes to the hind feet; the limbs long and slender; crowns of the molar teeth of a rounded form, with a single indenting fold of enamel, and four or five transverse, isolated grooves surrounded by enamel: the hairs covering the hinder parts of the back coarse, and very long.

Inhabits some of the West India Islands, and nearly all parts of South America (east of the Andes), down to Paraguay.

To this genus belong the Rodents commonly known by the name Agoutis; animals which, being common in many parts of South America, are easily captured, and, feeding upon almost any vegetable substance, are frequently brought to Europe to enrich our menageries. Most of the species are equal to the Common Hare in size, and they may all be distinguished by their slender and graceful form, combined with a very small, or rudimentary tail; an elongated head terminating in a broad and obtuse muzzle; moderate sized ears, which are of somewhat rounded form, but truncated, or slightly emarginated on the hinder part; large and moderately prominent eyes; five toes to the fore feet, of which the innermost is a mere tubercle, though supporting a nail; and three toes to the hind feet, which are long, and, like the fore feet, entirely naked beneath. The nails of the toes are but little arched, and nearly equal in depth and width, and by no means sharply pointed. This form of nail is found in burrowing and in running animals[1], but in the former it is combined with short and powerful limbs, a condition in these parts very unlike that presented by the Agoutis. These animals, then, we should infer, from their general structure, are fitted for running; and that they neither climb nor burrow, but are very swift runners, we are informed by those who have observed their habits in their native country. Though, however, the Agoutis do not burrow, they will scratch holes to obtain the roots of certain plants which are palatable to them.

The fur of the Agoutis consists of coarse hairs, which are for the most part of moderate length, but on the hinder third of the back are remarkably long—a peculiarity which suggested their generic name. The dentition in these animals more nearly resembles that of the true Porcupines than any other Rodents; the molar teeth being semi-rooted, their crowns of a rounded form, and the enamel folds being most of them isolated when the tooth is but little worn. The enamel which breaks up the surface of the tooth, here, is not continuous with that which surrounds the tooth, but merely encloses small grooves, which for the most part are arranged transversely. Of such grooves four or five are seen in each molar. One fold or loop of enamel, however, is seen in each tooth; it is that which is most constant in the molars

[1] In climbing Rodents, such as the Squirrels, the toes are provided with very deep and compressed nails, and these much curved, and sharply pointed.

of the Rodents, namely, the middle inner fold in the upper molar, or the principal outer fold in the lower molar. Not only in the structure do the molars of the Agoutis nearly resemble those of the true Porcupine, but these two sections of the Hystricidæ agree in having the molars of opposite sides of the jaw arranged in parallel series. The incisor teeth are somewhat slender, rather deeper than broad, and their anterior surface is convex.

Upon comparing the skeleton of the Agouti with the same part in any of the *Echimyina,* the most important differences are those which present themselves in the cranium: some of these have already been noticed, and with reference to this part of the structure, I will only further add, that the zygomatic arch is short, and the malar bone is proportionately slender, but suddenly deepens near its anterior extremity, where it sends out a post-orbital process: in a perpendicular line, above the point of this process, lies the centre of the orbit, which is bounded anteriorly in a great measure by the lachrymal bone, but partly likewise by the frontal, the maxillary, and the malar bones. Of the hinder division of the eye-cavity, the upper half only is enclosed by bone: the supra-orbital ridge is very prominent. The depression above the hinder root of the zygoma, which marks the course of the temporal muscle, is very contracted[1]. The squamosal or squamous element of the temporal, here does not extend to the plane of the occiput, as in the Paca's skull, the petro-mastoid intervening. There is scarcely any trace of the descending, lateral process of the supra-occipital bone, which is so characteristic of the skulls of the *Echimyina* and Octodons. A large opening in the superior maxillary is seen immediately above the anterior root of the zygoma,

[1] A side view of the cranium of an Agouti, showing the disposition of the masticating muscles, will be found on Pl. 6 *a*, fig. 1; the lettering of the figures is explained at p. 151.

which communicates posteriorly with the lachrymal canal, and, anteriorly, with the nasal cavity[1].

In the remaining parts of the skeleton there are but few points to arrest our attention. A skeleton of the Common Agouti in the British Museum, has 13 dorsal, 6 lumbar, 5 sacral, and 10 caudal vertebræ. The spines of the cervical vertebræ are very little elevated, if we except the vertebra dentata, which has the spine considerably elongated, and much extended in the antero-posterior direction. The spine of the foremost dorsal vertebra is short, but the second, and several following dorsal vertebræ, have the spine elongated. The scapula resembles that of the Hare in having a branch thrown out laterally from the anterior termination of the spine, but differs in having the supra-spinal fossa much larger: here the portion of the scapula which lies above the spine is strongly dilated in the middle[2]. The humerus has

[1] An opening, corresponding to this, is found in many rodent skulls, but being very conspicuous in the cranium of the Agouti, I here notice it. It is large in the skull of the Guinea-Pig, and of moderate size in that of the Chinchilla, where a very narrow strip of bone separates it from the lachrymal opening; immediately below the last-mentioned opening is seen a slender branch (being a portion of the superior maxillary) which runs upwards and backwards to join the frontal, and forms a partition between the orbital outlet of the spheno-palatinal canal (which is very large) and the opening connected with the lachrymal canal, before alluded to. The same remarks will apply to the skull of the Habrocoma, but here the small lachrymal bone does not enclose the lachrymal canal. In the Octodons, and the Echimys group, the opening into the lachrymal canal is either very small, or entirely wanting; it is wanting in the Coypu, and either can scarcely be traced, or is absent, in the skulls of Nelomys and Echimys: in these sections, moreover, the spheno-palatine opening is very small, and the small lachrymal bone rather leads to, than forms part of, the lachrymal canal.

[2] Such is also the case in the Coypu, Capromys, and Echimys, but in these animals the supra-spinal fossa is relatively smaller than in the Agoutis; and, whilst the outline of the upper, or cervical margin, of the scapula is rounded in the *Dasyproctina*, it generally forms a more or less distinct angle in the *Echimyina*. The lower margin forms a slight sigmoid curve in the *Dasyproctina*, and is straight in the animals belonging to the two preceding subfamilies.

the deltoid crest but moderately raised, and in this respect differs from the humerus in the species of the two preceding sections[1]. In the structure of the wrist, the Agouti and Paca present this difference, that, while the former has a small supernumerary bone (commonly found in the Rodents[2]) wedged in between the trapezoides and os magnum, the latter wants this bone. All the five toes of the fore foot are perfect, but the bones of the inner toe are small and slender, and notwithstanding that the clavicles exist only in a rudimentary condition in the Agouti, the fore feet are used by the animal to hold its food and convey it to the mouth. In the hind foot of the Agouti not only are the phalanges of the outer and inner toes wanting, but the metatarsal bones of these toes are likewise absent.

DASYPROCTA AGUTI.

The Yellow-rumped Aguti.

Mus Aguti.	LINN., Syst. Nat. 12, i. p. 81.
Cavia "	SCHREB., Säugth. iv. p. 613, Pl. 172 A.
Dasyprocta Acuti.	DESM., Mamm. Part 2, p. 357.
" *Aguti.*	WAGLER, in Isis, 1831, p. 618.
" "	(Long-nosed Agouti). BENNETT, Gardens and Menagerie of the Zool. Soc. i. p. 293.
" "	PR. MAXIM., Beiträge, p. 458.
" "	WAGNER, Schreb. Säugth. Suppl. iv. p. 42.
" "	F. CUVIER et GEOFF., Mamm. 1re livr.

[1] The deltoid crest is marked * on the bone *h* of Plate 16, fig. 1: in the *Echimyina* and *Octodontina* it is less prolonged in the direction of the shaft of the humerus, but is produced into a more or less distinct process—a process which is much developed in the Ctenomys and Coypu.

[2] Here the supernumerary bone in question articulates with the metatarsal of the second toe, and its position so far differs from that of the extra bone marked * in figure 2 of Plate 14. In the foot of the Hare the second metatarsal also articulates with the supernumerary bone, but to do so, the metatarsal runs up between the os magnum and os trapezoides, whilst in the Agouti the small extra wrist-bone descends to the level of the lower edge of the os trapezoides.

Agouti.	BUFF., Hist. Nat. viii. p. 375, Pl. 50.
"	SHAW, General Zoology, ii. Part 1, p. 24, Pl. 126.
Long-nosed Cavy.	PENNANT, Synopsis of Quadr. p. 245.

General tint olive-brown (the fur distinctly pencilled with black and yellow); mesial line of abdomen yellow or whitish; the long hairs covering the hinder portion of the back for the most part of a bright orange colour.

Inhabits Guayana, and the northern portions of Brazil.

The Yellow-rumped, or Common Agouti, though abundant in most of the northern portions of South America, does not occur, according to the observations of Dr. Natterer, in the southern parts of Brazil. It is subject to some slight variation in its colouring, arising from the yellow portions of the hairs of its fur being more or less rich—sometimes pale, and sometimes with a considerable admixture of red, but it may be distinguished by the long hairs which cover the hinder part of the back being chiefly of a rich yellow, or sometimes yellow-red hue. These hairs are most of them four inches, or more, in length; white, or nearly so, at the root, and sometimes entirely yellow beyond that part, but usually they present four or five narrow black rings. On other parts of the body the somewhat coarse hairs of the fur are of moderate length; black, and commonly with two yellow rings: on the abdomen the black is replaced by dusky, and the yellow is less rich; and along the mesial line the hairs are often entirely yellow, and sometimes white. The feet are pencilled with black and yellow, and the toes are dusky. In some specimens, which do not otherwise differ from this description, the feet are entirely black. The ears are large, and of a pink colour, excepting at the margins, where they are dusky. The chin is whitish, and the hairs of the moustaches black. The small fleshy tail is naked, and of a dusky hue.

	Zoological Society. Inches.	Lines.	British Museum. Inches.	Lines.
Length from tip of nose to root of tail	20	0	18	6
" from ditto to ear	3	6	3	7
" of ear	1	0	1	0
Width of ditto	1	4	1	4½
Length of fore foot, not including the nails	1	4		
" of the nail of the middle toe ...		4½		
" of the hind foot, without the nails	3	10	3	10½
" of the nail of the middle toe ...		7		5½
" of tail, about		7		8
" of skull			3	11
Width of ditto			1	11¾
" between orbits			1	2
Length of nasal bones			1	4
Width of ditto in front				7⅔
" " behind				8
Length of frontal bones			1	7½
From incisor to molar teeth			1	0
Length of four upper molar teeth, taken together				8
Width of palate between foremost molars				4½
Length of lower jaw			2	3½
Height of ditto, measuring from the condyle			1	0
Length of angular portion			1	1

DASYPROCTA CROCONOTA.

White-toothed Agouti.

Dasyprocta croconota. WAGLER, in Isis, 1831, p. 618.
" " SCHREB., Säugth. Pl. 172 B.
" " WAGNER, Schreb. Säugth. Suppl. iv. p. 44.

Incisor teeth white: general hue of the fur, rich olive-brown; the greater portion of the back, as well as the rump and upper part of the thighs, of a bright yellow-red colour; mesial line of abdomen white; nails of toes short.

Inhabits Brazil.

The *D. croconota* is founded, by Wagler, upon an Agouti brought from the Amazon River by Spix; it is said to agree very nearly in size with the Common Agouti, but to differ from that and other species of the genus, by its incisor teeth being entirely white, the fore part of these teeth being usually of an orange colour in the Agoutis. It differs, moreover, from the Common Agouti, according to the description of Wagler, in having the tarsi shorter, and the nails shorter and more compressed, as well as in the general hue of its fur being much richer. The head, neck, and fore part of the back, are pencilled with black and orange, and the sides of the body with black and yellow: the whole hinder half of the back, as well as the rump and upper part of the thighs, are of an uniform bright yellowish red colour: the hairs on these parts are, however, yellow at the roots; the mesial line of the abdomen is white, and the white is continued forwards to the chest; the ears flesh colour, broadly margined with dusky; the feet dusky brown.

The following dimensions, furnished by Dr. Wagler, indicate the differences of proportions which this species presents when compared with the Common Agouti.

	D. croconota.		*D. Aguti.*	
	Inches.	Lines.	Inches.	Lines.
Length from tip of nose to root of tail ...	17	9	18	0
" of head	3	11	3	7
" of tarsus (not including the nails)	3	5	3	10

DASYPROCTA PRYMNOLOPHA.

Dasyprocta prymnolopha. WAGLER, in Isis, 1831, p. 619; Schreb. Säugth. Tab. 172 c.
" " WAGNER, Schreb. Säugth. Suppl. iv. p. 46.

Fur pencilled with black and rich yellow; loins and sides of the rump golden yellow, more or less inclining to rufous; a tuft of long black hairs on the back of the head, and a broad black band extending from the middle of the back to the tail; abdomen golden yellow, with a mesial white line.

Inhabits Guayana.

Although I have seen several specimens of an Agouti agreeing with Dr. Wagler's *D. prymnolopha*, I am not aware from what part of South America they were procured. The specimen described by the author just mentioned is said to be from Guayana.

This is certainly one of the most beautifully coloured species among the Agoutis, and is readily distinguished by the broad black band which runs along the hinder half of the back, and is continued to the tail. This band is rendered the more conspicuous by the parts immediately adjoining it, on either side, being of a rich golden yellow, or sometimes of a bright yellow-rust, or yellowish chestnut colour. The hairs covering the hinder part of the back are very long, and the exposed end only of each of these hairs is black; at the root they are yellow. Another peculiarity which this species presents is the large tuft of black hairs on the hinder part of the head. The chin, angle of the mouth, and lower jaw, are white; the fore part of the neck, and the upper part of the chest, are pencilled with dusky and yellow, and the lower or hinder part of the chest, and the abdomen, are golden yellow, excepting in the mesial line, where the hairs are white. The ears are large, and of a pink colour, with the margin dusky.

	Inches.	Lines.	Inches.	Lines.
Length from nose to root of tail ...	19	6	18	9
" from nose to ear	3	6	3	8
" of tail, about				8
" of ear	1	0	1	0
Width of ditto	1	4	1	5
Length of hind foot, not including nails	3	9	3	8
" of nail of middle toe		6		6

The feet are black in the specimen from which the foregoing description is taken, which is in the museum of the Zoological Society, but in that described by Dr. Wagler they are freckled with black and yellow, as is also the case in one of two specimens in the British Museum. The second specimen in the National Collection differs from others which have come under my notice, in having very little black pencilling, excepting on the back of the neck and shoulders; its prevailing hue being bright golden yellow; the upper parts orange yellow, freckled with black; the tuft on the back of the head, as well as the broad band on the hinder half of the back, are, as in other specimens, perfectly black; the feet are golden yellow, slightly freckled with black, and the toes are black. The dimensions of this specimen will be found in the second column; those of its skull are as follows:—

	Inches.	Lines.
Length of skull	3	10½
Width of ditto	1	10
" between orbits	1	2
Length of nasal bones	1	3⅔
Width of ditto in front		6⅔
" " behind		7
Length of frontal bones	1	6⅔
From incisor to molar teeth		11½
Length of four upper molars, taken together		8⅓
Width of palate between foremost molars ...		3½
Length of lower jaw	2	3
Height of ditto, measuring from the condyle	1	0
Length of angular portion	1	0

The cranium from which the above dimensions are taken differs from that of the *D. Aguti,* in being narrower and longer, in having the muzzle more attenuated, the palate narrower, and in the form of the malar bone. This bone in the *D. prymnolopha* throws up a distinct post-orbital process, the point of which is removed from the malo-maxillary suture by a distance of about one line, whilst in the *D. Aguti,* where the post-orbital process is less produced, and forms an obtuse angle, the apex of that angle is divided by the malo maxillary suture; one half, then, of the post-orbital process is here formed by the zygomatic process of the superior maxillary. The lower margin of the angular portion of the lower jaw is much less thickened in the *D. prymnolopha,* and presents no trace of an angular process on the inner edge, whilst such a process is distinct in the lower jaw of the *D. Aguti.*

DASYPROCTA CRISTATA.

Crested Agouti.

Cavia cristata.	(GEOFF.) DESM., Nouv. Dict. d'Hist, Nat. i. p. 213.
Dasyprocta cristata.	DESM., Mammal. Pt. 2, p. 358.
" *variegata.*	TSCHUDI, Fauna Peruana, Pt. 5, p. 190 (1845).
Agouti à crête.	F. CUVIER et GEOFF., Mammif. livr. 52, tom. iii.

Fur long and harsh, pencilled with black and dirty brownish yellow, or rufous yellow, the black prevailing ; limbs dusky ; the long hairs on the hinder part of the back black, obscurely pencilled with rufous: a large tuft of long black hairs on the back of the head.

Inhabits Surinam (and Peru?)

In the Museum of the Zoological Society are three specimens of Agouti, which, agreeing very closely with M. Desmarest's description of the *Dasyprocta cristata*, I shall proceed to notice under this head. Compared with other Agoutis (the

Dasyprocta fuliginosa excepted) they are distinguished by the general dark hue of the fur; and, from the *D. Aguti* they further differ in being larger, in having the muzzle more elongated, the ears relatively smaller, and in possessing a distinct tuft, or crest, of long black hairs, springing partly from the back of the head, and partly from the adjoining portion of the neck. In two of these specimens the fur is black, pencilled with brownish yellow, the black prevailing, especially on the back; the legs are almost entirely black, and the feet are black. A dusky patch is observable immediately behind the base of the fore leg; the throat is whitish, and the abdomen is yellowish along the mesial line, with a slight admixture of dusky. The long hairs covering the hinder part of the back are for the most part black, being but obscurely annulated with brownish yellow. In the third specimen the hairs of the fur are annulated with reddish chestnut instead of the brown-yellow[1].

The dimensions of one of these specimens is given in column No. 1; in columns 2 and 3, are the admeasurements of two specimens of *Dasyprocta* in the British Museum, which Mr. Gray regards as specifically identical with the *D. cristata*—an opinion in which I agree. These latter specimens, however, differ considerably in their colouring from the animal described by Desmarest, and perhaps form a permanent variety. Their peculiarities I will point out under the head—

Dasyprocta cristata, variety.

Fur pencilled with black and dirty pale yellow (on the abdomen with the black replaced by dusky brown); the hinder half of the back of a bright rust colour; a distinct crest on the back of the head.

[1] The female of this species, according to Desmarest, has six mammæ, whilst the *D. Aguti* is said to have twelve. In the *D. punctata* I found four pairs of teats; the foremost pair being pectoral, and the hindermost inguinal.

The black is less pure in these animals than in the Zoological Society's specimens, and has a slight brownish cast. In one specimen the crest on the back of the head is black, and the long hairs on the hinder part of the back are of an uniform bright rust colour; whilst in the other, the crest is dusky brown, and the long hairs of which it is formed are annulated with dirty yellow; the hairs on the hinder part of the back are rather obscurely annulated with black; the rust colour prevailing. These hairs, in both specimens, are of a palish golden yellow colour at the root. The abdomen can scarcely be said to differ in tint from the sides of the body, and in this respect differs from the same part in the *D. Aguti;* there is, however, a whitish line on the lower part, in the mesial line.

	No. 1.		No. 2.		No. 3.	
	Ins.	Lines.	Ins.	Lines.	Ins.	Lines.
Length from tip of nose to root of tail ...	20	6	21	0	21	6
" from ditto to ear	4	0	4	2	4	4
" of ear		9		10		9
Width of ditto	1	3	1	4	1	4
Length of hind foot, without the nails ...	3	9	4	8½	4	8
" of nail of middle toe				7½		8
" of tail, about				6		6

Upon looking through Dr. Tschudi's description of the *Dasypr. variegata,* that animal appears to agree in all respects with the *D. cristata* of Desmarest; which thus, it would seem, extends into Peru.

DASYPROCTA FULIGINOSA.

The Sooty Agouti.

Dasyprocta fuliginosa. WAGLER, in Isis for 1832, p. 1220.
" *nigricans.* (NATTERER) WAGNER, in Archiv für Naturgesch. 1842, Pt. 1, p. 362; Schreb. Säugth. Supp. iv. p. 46.

Dasyprocta nigra. GRAY, Ann. and Mag. of Nat. Hist. vol. x. p. 264, Dec. 1842; Voy. of the Sulphur—Mammalia, Pt. 2, Pl. 16.

Fur sooty black, somewhat obscurely pencilled with white (or sometimes with pale yellow); feet black; ears tolerably well clothed with hairs.

Inhabits the northern provinces of Brazil.

Dr. Wagner has satisfied himself, by a comparison of the specimens, that the Agouti described by Wagler under the name *fuliginosa,* is the young of the *D. nigricans* of Natterer, and this latter is clearly the same as the *D. nigra* of Mr. Gray.

The various names given to this species all have allusion to its dark hue; and, if we add that it is larger than other species of Agoutis, and has the ears more perfectly clothed with hairs, we have noticed its leading characteristics.

The specimen in the British Museum, upon which Mr. Gray founded the *D. nigra,* is rather larger than the *D. cristata.* Its fur is of a sooty black hue, and somewhat obscurely pencilled with white. The white is most distinct on the flanks and under parts of the body, whilst it is almost wanting on the crown of the head, shoulders, and outer surface of the thighs. The hairs on the rump (which are four inches, and some of them more, in length) are white at the root, and indeed for the greater part, but their exposed ends are black: many of them, however, have the extreme point white, as well as the root. On the fore part of the back many of the hairs are perfectly black. The muzzle and chin are brownish; the hairs of the moustaches black; the throat dirty white; the fore legs are black in front and on the outer side, and dusky, with much white pencilling, on other parts; the feet are black. The ears, which are rather large, are well clothed with hairs; those on the outer surface are dusky brown, and those on the inner side are yellowish (slightly pencilled with dusky), excepting at the margin, where the hairs are coloured like those on the outer surface.

The admeasurements of this animal (which is said to be from tropical America) are as follows :—

	Inches.	Lines.
Length from tip of nose to tail	23	0
" from nose to ear	4	4
" of ear	1	1
Width of ear	1	6
Length of tarsus, not including the nails ...	4	9
" of nail of middle toe	0	9
" of tail, about		10

Of two specimens brought home by Dr. Natterer, one is from Borba, and the other from Rio Negro. The specimen described by Wagler was procured by Spix on the Amazon.

One of Dr. Natterer's specimens, according to the detailed description furnished by Dr. Wagner, has the hairs of its fur ringed with pale yellow instead of white—the black prevailing on the hinder part of the back; they are at the root whitish, and this is followed by a broad blackish brown space; a narrow, sub-terminal, pale yellow ring is seen on each hair, and the point is black. In the second specimen the hairs are black, with a single pale ring near the point. The white being much extended on the long hairs covering the hinder part of the back, prevails over the black in this part; on other parts of the back the black prevails. On the back of the neck is a narrow black tuft.

DASYPROCTA AZARÆ.

Azara's Agouti.

Dasyprocta Azaræ.	LICHT., Verzeichniss der Doubletten des Zoolog. 1823, p. 3.
" "	WAGNER, Schreb. Säugth. Suppl. iv. p. 39.

Dasyprocta punctata.	GRAY, Annals of Nat. Hist. vol. x. p. 264 (Dec. 1842); Zool. of the Voy. of the Sulphur—Mammalia, Pt. 2, p. 36, Pl. 15.
Chloromys Acuti.	RENGGER, Naturgesch. der Säugth. von Paraguay, p. 259 (1830).
L'Acouti.	AZARA, Essais sur l'Hist. Nat. des Quad. de Paraguay, ii. p. 26.

Fur on all parts distinctly pencilled with black and rich yellow, excepting on the abdomen, which is yellow, and often has a white mark running along the middle line.

Inhabits Paraguay, Bolivia, and the southern parts of Brazil.

Of this species I have examined many specimens, brought from Bolivia by Mr. Bridges, who informs me that these Agoutis inhabit the woods in the neighbourhood of Santa Cruz de la Sierra, and are known, in that part, by the name "Hoche colorado." Dr. Wagner states that Dr. Natterer found this animal in great numbers at Ypanema, in the Province of St. Paulo, Brazil, where it appears to take the place of the *D. Agouti,* which only occurs further north.

The specimens which have come under my notice varied but little in their colouring, and were almost equal in size to the *D. Agouti,* from which animal they differed in having the long hairs on the hinder part of the back as distinctly annulated with black and yellow as those on other parts of the back. The yellow was richer (inclining to rufous) in some specimens than others; and, in some, the bright yellow colour of the abdomen was not interrupted in the mesial line by the white mark which is usually seen in that part[1]. The feet are always darker than in other parts, having a greater admixture of black, which increases towards their extremities.

The animal described by Mr. Gray, under the name *D. punctata,* appears to me to be clearly identical in species with

[1] In one specimen there was but a small spot of white on the middle of the abdomen.

this Bolivian animal, differing only in having the feet black. Its fur is pencilled with black and yellow in about equal proportions; the yellow assuming a richer, and somewhat rufous tint, on the back. The throat and belly are yellow. The long hairs on the hinder part of the back have about five yellow, and the same number of black rings; the yellow assuming a paler hue towards the roots of the hairs. On the fore part of the back each hair has two yellow rings. The tip of the muzzle is dusky; the ears apparently flesh-coloured, with a dusky margin. No crest on the back of the head. Its dimensions are given in column No. 1; those in columns 2 and 3 are from Bolivian specimens.

	No. 1.		No. 2.		No. 3.	
	Ins.	Lines.	Ins.	Lines.	Ins.	Lines.
Length from tip of nose to root of tail ...	19	0	20	0	18	0
" from ditto to ear	3	10	3	9	3	7
" of ear		11	1	0	1	0
" of tarsus, including the nails ...	4	10	5	0	4	9
" of tail, about				9		

Dasyprocta caudata.

Dasyprocta caudata. LUND, Det Kongel. Danske Videnskab. &c. viii. p. 297 (1841).
" *pallida*, of the Leyden Museum.

Fur harsh, pencilled with black and bright yellow, the latter colour slightly prevailing: abdomen yellow; the long hairs covering the hinder part of the back annulated with black white, producing a grey tint.

Inhabits Minas Geraes.

Dr. Lund distinguishes a species of Agouti inhabiting Minas Geraes from the *D. Azaræ* by its being of larger size, by the greater length of its tail, and the grey colour of the hinder part of the back. The length of the Paraguay animal (*D. Azaræ*), Dr. Lund states, is 18, or sometimes 19 inches, whilst the Minas Geraes Agouti (named *D. caudata*) has an average length of 20 inches, and specimens are met with which exceed that length. The tail in the *D. Azaræ* is half an inch, whilst that in the *D. caudata* is an inch in length. Of two adult skulls of the last mentioned animal, one measured 4″ 1‴, and the other 4″ 4‴: this second skull belonged to a male, whilst the first was that of a female. The length of the skull of the adult *D. Azaræ* is given at 3″ 2‴[1].

The hairs of the fur of the *D. caudata* are said to be annulated with black and yellow, the general hue being olive-grey; the yellow becoming more distinct towards the abdomen, which is of a sulphur yellow: the extremities black, and the rump grey; a tint produced by the long hairs on this part being annulated with black and white.

Two Agoutis in the Leyden Museum are clearly referrible to this species: they are about equal in size to the *D. cristata*, and their ears, as in that animal, are smaller than in the *D. Aguti*. The hairs of their fur are harsh, and pencilled with black and yellow, the bright colour prevailing somewhat over the dark. Each hair on the back is of a bright yellow colour; nearly white at the root, and with three or four black rings. The long hairs over the rump are white, and have five or six black rings; the black and white occupying about equal portions of each hair. The chin and throat whitish; the abdomen bright yellow. The ears are clothed with hairs, which are of a rusty yellow colour, except near the apex, where the hairs are dusky, as are also those on the fore part of the outer surface. In one specimen the feet are black; and in the other they are black, pencilled with yellow[2].

[1] The dimensions of the *D. Azaræ* are taken by Lund from Dr. Rengger's account of that animal.

[2] Dr. Wagner will not admit that the *D. caudata* is distinct from the *D. Azaræ*, since (as he states in his Report on Zoology for 1842) among the

	Inches.	Lines.	Inches.	Lines.
From tip of nose to root of tail ...	21	0	22	0
" ditto to ear, about			4	0
Length of ear, about		9		9⅓
Width of ear	1	4	1	4
Length of hind foot and nails	4	6	4	9
" of fore foot and nails ...	2	2	2	3

DASYPROCTA ACOUCHY.

The Acouchy, or Olive Agouti.

Cavia Acouchy.	ERXLEBEN, Systema Regni Animalis, p. 354 (1777).
" *Acuschy.*	GMELIN, Linn. Syst. Nat. i. p. 121 (1788).
" *Acuchy.*	SCHREBER, Säugth. iv. p. 612, Pl. 171 B (1792).
" *Acuschy.*	SHAW, General Zoology, ii. Pt. 1, p. 27, Pl. 126, (1801).
" *Akuchi.*	DESM., Nouv. Dict. d'Hist Nat. i. p. 212 (1816).
Dasyprocta Acuschy.	DESM., Mamm. ii. p. 358 (1822).
" "	WAGNER, Schreb. Säugth. Suppl. iv. p. 48 (1844).
" *leptura.*	(NATTERER) WAGNER, l. c. p. 49.
" *exilis.*	WAGLER, in Isis, 1831, p. 619.
" *leporina*[1].	GRAY, List of the Mammalia in Brit. Mus. (184 3),

numerous Brazilian specimens contained in the Vienna collection, all of which he regards as the *D. Azaræ*, there are some in which the rump passes into grey; and, as he moreover affirms, there is no difference in size between these animals and the *D. caudata*. As, however, my own observations confirm Dr. Lund's, with regard to the superior size of the *D. caudata*, as compared with the *D. Azaræ*, and as Dr. Wagner has omitted to notice the most important point dwelt upon by Dr. Lund, viz. the great difference of size in the skulls of the two animals, it may be questioned whether the Vienna specimens referred to, do all actually belong to one species. They were collected in the province of San Paulo, which lies between Minas Geraes and Paraguay, and upon the supposition that the Agouti of one of the two latter districts is specifically distinct from that of the other, it is highly probable that the intervening country would be inhabited by both species. The *D. Azaræ*, however, would appear to be the more common species in the province of San Paulo, for Dr. Wagner, in his continuation of Schreber's great work, only notices one amongst the numerous specimens brought by Dr. Natterer from that province—"a full-grown male"—in which the hinder part of the back is grey.

[1] The *Dasyprocta leporina* of authors is founded upon a figure of a Rodent (called the Java Hare) given in Catesby's Carolina (Appendix, Pl. 18), and is

Akouchy. BARRERE, Essai sur l'Hist. Nat. de la France Equinoxiale, p. 153 (1741).
Akouchi. BUFFON, Hist. Nat. xv. p. 158 (1767); Suppl. iii. p. 211, Tab. 36 (1776)
Olive Cavy. PENNANT, Synopsis of Quadr. p. 246 (1771).

Tail slender, about two inches in length, and covered with white hairs on the apical half: general hue of the fur chestnut brown (the hairs pencilled with brown-black, and rich yellow-brown); hinder part of back dusky; under parts of the body of a deep golden yellow; feet rufous-yellow, obscurely pencilled with black; a bright rufous, or golden yellow spot behind each ear.

Inhabits some of the West India Islands (St. Lucia and Grenada), Guayana, and the northern parts of Brazil.

The Acouchy is a delicately made little animal, much smaller than either of the preceding species of Agouti, and readily distinguished from that circumstance, combined with the greater length as well as the slenderness of its tail. This organ is not naked as in other species of Agouti, but is tolerably well clothed (on the apical half at least) with silvery white hairs, of from one to two lines in length. Another character which will help to distinguish this species is the conspicuous bright rust-coloured, or golden yellow patch, which is situated immediately behind the ear. Its general colouring varies somewhat, but in most specimens

the *Cuniculus Javensis* of Brisson, *Mus leporinus* of Linnæus, and *Cavia leporina* of Erxleben. That it is an Agouti there can be no doubt, and that no such animal inhabits Java and Sumatra, as was supposed, is equally certain. According to the text of Catesby's work, it was of a reddish brown colour, had a very short tail (represented in the figure as a mere tubercle, like the tail of *Dasyprocta Aguti*), and was equal in size to a Hare. These last two points show that it could not be the Acouchi, that animal having a comparatively long tail, and being of much smaller size; and hence I cannot see the propriety of changing the specific name which has been attached to that anima so many years which have come under my notice, it may be described as rich chestnut brown — shaded almost into black on the hinder half of the back—a tint produced by the mixture of brown-black, and rufous or bright rust colour; the ground colour of each hair being dark, and the brighter colour forming two or three rings on each hair. The whole under parts are of a bright rufous, or sometimes golden yellow hue; and the fore legs, and fore part of the hind legs, are nearly of the same colour, but usually somewhat richer, and may be described as of a deep orange red tint; the feet are of this latter colour, but rather obscurely freckled with black. The ears are of moderate size, flesh-coloured, and very sparingly clothed with hairs. The hairs of the moustaches are black: the angle of the mouth is whitish. There is usually a slight pencilling of black on the chest.

In some specimens the brown-black of the fur is replaced by pure black, the paler annulations of the hairs incline more to yellow, the abdomen is of a less deep golden yellow, and the hinder half of the back is black. The bright spot behind the ear is always distinct.

In a female specimen in the British Museum I found eight mammæ: it agrees in its colouring with my general description[1]. A specimen in the same collection, agreeing with the second description, is apparently a male[2].

[1] The dimensions of this animal are given in column No. 1.

[2] See its anmeasurements (in which those of the skull are included) in column No. 2. The dimensions in columns 3 and 4 are taken from specimens in the museum of the Zoological Society.

	No. 1.		No. 2.		No. 5.		No. 4.	
	Ins.	Lines.	Ins.	Lines.	Ins.	Lines.	Ins.	Lines.
Length from tip of nose to root of tail	14	6	13	6	13	6	14	0
" from nose to ear	2	9	2	9	2	8	2	8
" of ear		9½		9		8		8½
" of tail	2	0	2	2			1	10
" of hind foot, without the nails	3	3	3	0	2	3	3	1
" of nail of middle toe ...		5½		6½				
" of skull			3	4	3	2		
Width of ditto			1	7½	1	6		
" " between orbits ...				11		10¾		
Length of nasal bones			1	1⅓	1	0⅓		
Width of ditto				6		6		
From incisor to molar teeth ...				10½		9⅓		
Length of four upper molars, together				6¼		6⅓		
Width of upper incisors ...				2⅓		2⅓		
" of palate, between the molars				3⅓		3¼		
Length of zygoma			1	2¾		2		
" of lower jaw			1	11		10		
Height of ditto				9⅔		10		
Length of angular portion ...				11		10½		

The skull of the Acouchy has the post-orbital process of the frontal and malar bones more developed than in the crania of any other Agoutis which have come under my notice. The post-orbital process of the malar bone is removed from the malo-maxillary suture by about one line. The sagittal crest is more distinct, and more elongated, than usual. The posterior emargination of the palate terminates in a line with the interspace of the last and penultimate molars. The ant-orbital opening is enclosed in front, as usual, by a portion of the superior maxillary. I have not seen a cranium of any Agouti in which the lachrymal bone forms any part of the bony ring which encloses the ant-orbital opening, as noticed by Cuvier in the Agouti skull described in the *Ossemens Fossiles*.

For a detailed account of the anatomy of this animal, we must refer to Professor Owen's papers in the Proceedings of the Zoological Society[1].

With regard to the habits of the Acouchy, we are not aware that it differs in any important point from the other species of Agouti. The following short notice from the pen of Prof. Bell[2] well depicts the habits as noticed in two specimens which that gentleman received alive from Guayana. After calling attention to the chief points of distinction between the Acouchy and Agouti, and especially noticing the greater length of the tail in the former, it being upwards of two inches in length in the living animal, and of equal thickness throughout, Prof. Bell observes, that "the animal frequently agitates this organ with a quick tremulous motion. Both the individuals are mild and gentle in their dispositions, but somewhat timid; they are, however, familiar with their master, and run to him whenever he enters the room in which they are kept, and about which they are allowed to range during the day. Their food is entirely vegetable; they are especially partial to nuts and almonds: they drink but little. They are extremely cleanly, and take great pains to keep their fur in order, in cleansing which they mutually assist each other. They leap occasionally in play to a considerable height, and frequently on springing from the ground to an elevation of two feet, descend on the spot from whence they rose. Their voice is a short, rather sharp, plaintive pur. The individuals, male and female, show great attachment to each other."

[1] See p. 75 of Proceedings for 1831, for an account of the viscera; and p. 100 of Proc. for 1832, for the peculiarities of the skeleton. In two specimens examined, Prof. Owen found small clavicles, "about the thickness of a small pin, and eight lines in length."

[2] Communicated to the Zoological Society at the meeting for Nov. 1830.

Dasyprocta leptura. WAGNER.

Judging from Dr. Wagner's description, I feel almost certain that the *D. leptura* is a mere variety of the Acouchy, in which the colouring is somewhat richer than usual. The specimen described was obtained by Dr. Natterer at the Rio Negro, the chief northern tributary of the Amazon. It is said to be freckled with rusty red, and black ; dark on the outer surface of the limbs, and very dark on the hinder portion of the back ; whilst on the sides of the body the rusty red prevails, the black gradually disappearing towards the under parts, which are of an uniform rusty yellow colour. The tail is slender, and has scattered reddish hairs, except at the point, where the hairs are whitish. Length, 12″ $2\frac{1}{4}$‴ ; tail, 2″ 0‴ ; ear, 1″ $2\frac{1}{2}$‴

Dasyprocta exilis. WAGLER.

That this supposed species is founded upon a young specimen of the Acouchy, in which the tail has been lost, I feel but little doubt. According to the description, it has the general colouring of the Acouchy, and the bright patch behind each ear, which is always seen in that animal. Its length is said to be $9\frac{1}{2}$ inches : the upper parts of the body brownish chestnut, freckled or pencilled with black, and the under parts reddish fulvous, but with a white stripe in the mesial line. This white stripe I have not noticed in the specimens of Acouchy which I have described, but I have observed that a white stripe, on the abdomen, is either present or absent in different individuals of several species of Agoutis. The specimen from which Dr. Wagler drew up his description was brought by Spix from the Amazon district.

Dasyprocta albida.

Dasyprocta albida. GRAY, Annals and Magaz. of Nat. Hist. for Dec. 1842, vol: x. p. 264.

The specimen described by Mr. Gray under the above name, is in such a bad condition—a skin deficient of the fore legs, and otherwise imperfect—that all that can be said about it is, that the animal must have been about equal in size to the Acouchi: its fur is long and coarse, and of a lead-grey colour, but very pale—inclining to white; all the hairs are white at the root, and there is a whitish patch above, and another below the eye. The hind foot (including the nails) is 3″ 5‴ in length; and the naked under surface, instead of being black as in other Agoutis, is of a flesh colour. It is from I. St. Vincent, West Indies, and possibly may be an accidental variety of the Acouchi, which is said to inhabit the neighbouring Islands.

SUB-FAMILY, HYSTRICINA.

Hystrices. BRANDT.

Hystricidæ with rooted, or semi-rooted molars; skull with the malar bone destitute of angular process on the lower margin; frontal bones very broad; feet short, the toes 5—5, 4—5, or 4—4; body more or less armed with spines.

Inhabit all the four quarters of the globe.

The *Hystricina,* or true Porcupines, form the subject of an excellent monograph by M. Brandt[1], who divides the group into two minor sections. One is composed of Porcupines which live upon the ground, and seek shelter in burrows which they themselves form. They have five toes both to the fore and hind feet; the soles of the feet naked and

[1] Mammalium Exoticorum Novorum vel minus Cognitorum, &c., 4to. Petrop. 1835, being a collection of papers upon Mammalia, originally published by the Author in the Transactions of the Imperial Academy of Sciences of St. Petersburgh, 6th Series, vols. ii. and iii.

smooth: the skull more or less elongated, and provided with a distinct lachrymal bone, which partly encloses the lachrymal foramen. The molar teeth semi-rooted, and arranged in parallel series; those of the upper jaw have one internal fold of enamel, and three or four folds entering from the opposite side of the tooth, but which soon assume the form of small isolated areas, disconnected with the margin of the tooth. The lower molars are like the upper, but with the enamel folds reversed in position. The species of this division are confined to the Old World, whilst those of the next are peculiar to the New. The latter are grouped by Brandt under the head *Philodendræ*, from their habit of climbing trees—indeed, they almost entirely live in trees. Their feet are usually provided with but four toes, and these are nearly equal in length, armed with long, compressed, and curved claws; sometimes however, the hind foot has five toes. The soles of the feet are thickly studded with minute, depressed warts. Their skull is short and broad, and is provided with a minute lachrymal bone, which forms no part of the lachrymal canal; the portion of the palate which lies between the molar teeth is distinctly on a lower level than the anterior portion, and the bony partition which separates the incisive opening (being part of the intermaxillaries) runs back above the palatal portion of the superior maxillary bones; whilst in the Old World Porcupines the incisive septum joins the superior maxillaries by a serrated suture, and is continuous with the plane of the palate, which latter is throughout on the same level[1]. The series of molar teeth of opposite sides of the jaw converge in front: these teeth are distinctly rooted. Each molar has a distinct fold of enamel on either side, and the crown, when but little worn, presents a deep

[1] Distinct anterior and posterior clinoid processes bound the pituitary depression in the skulls of the Old World Porcupines, but are wanting in the New World species.

transverse cavity, surrounded by enamel on each of the two lobes which are separated by the enamel folds. Incisor teeth small. To this division belong the genera *Erethezon*, *Cercolabes*, and *Chætomys*. The last named genus differs considerably in the structure of its molar teeth, and in some other points, from the typical *Philodendræ*. The section *Philogææ* contains the genera *Hystrix* and *Atherura*. We will commence with the American Porcupines.

Genus, *Chætomys*.

Chætomys. GRAY, Proceedings of the Zool. Soc. for Feb. 1843, p. 21; Voyage of the Sulphur—Mamm. p. 36.
Plectrochærus. PICTET, Revue Zool. 1844, p. 225.

Hystricina with the molar teeth longer than broad; each upper molar divided by transverse incisions into three lobes, of which the middle lobe is narrow and transverse, and the remaining two, which are broader, have each a deeply indenting fold of enamel, which enters from the outer side of the tooth: lower molars with one internal, and two external folds of enamel. Skull with the malar bone very deep in front, and throwing out a post-orbital process which nearly meets a corresponding process of the frontal bone; the palate contracted: head and body covered, nearly throughout, with slender, cylindrical, waved spines: tail clothed with short stiff hairs.

The genus *Chætomys* contains but one known species, an animal which has usually been associated with the *H. prehensilis* and *H. insidiosa* of authors, but which, as Mr. Gray has pointed out, differs considerably from other species of Porcupines, in the structure of its skull, as well as that of the teeth. M. Pictet has also called attention to these differences, but not being aware that the animal in which he noticed them had been previously described, applied to it a new specific, as well as generic name. With regard to the structure of the cranium, the most striking peculiarity consists in the orbital fossa being almost perfectly separated from the temporal, by the large post-orbital processes of the malar and frontal bones: the apices of these processes are only separated, in one skull before me, by a space of about one line, and in a second cranium they are still more nearly united. The malar bone is deeper than in any other Rodent, if we except the Paca. Compared with the crania of the Prehensile-tailed Porcupines, the skull of the Chætomys, moreover, presents these differences, that the portion of the palate which lies between the molar teeth is more elongated, and more contracted, and that the bony tube of the ear is elongated. The temporal ridges are prominent, and as they are continued backwards from the post-orbital process they converge but little. Another ridge, marking the points of insertion of the temporal muscles, runs obliquely downwards and backwards from beneath the post-orbital process: on this ridge is seen the suture which separates the frontal from the squamosal bone. I notice this, not as a peculiarity in the present skull, for this ridge is very characteristic of the crania of the New World Porcupines—differing here only in being oblique in its position instead of vertical, as in other species. The nasal bones are of moderate size, and their length exceeds their combined width by about one-third.

The incisor teeth are rather small, about equal in depth and width, and but slightly convex in front. The molar teeth are relatively longer and narrower than those of other Porcupines, and they are less distinctly rooted than in the following species of the New World section. Their length exceeds their width by about one-third. Each upper molar has the grinding surface divided into three lobes by two transverse incisions, which enter deeply into the body of the tooth: the middle lobe is narrow and transverse; the remaining two lobes are nearly oval, with the long diameter in the transverse direction of the tooth; and each of these lobes has a deep fold of enamel entering from the outer side[1]. When considerably worn this fold is isolated from the margin of the tooth. The lower molars have each one external fold, and two internal folds of enamel; the hindermost of the two internal folds, after penetrating pretty deeply into the body of the tooth, bifurcates[2]. The tail is moderately long, and nearly resembles that of the Common Rat, being distinctly covered with scales, and rather scantily clothed throughout with hairs, which, though short, are, however, relatively longer than in the Rat's tail. This organ has the hairs worn off from the upper surface of the apical portion, and is no doubt prehensile; but we are led to conclude that the power of grasping objects is less developed in the tail of the Chætomys than in that of the species of the genus next to be described (*Cercolabes*), since it is not distinctly thickened at the root as in those animals. The feet present the same structure as in the Cercolabes; the naked surface beneath is studded with minute tubercles; the toes are sub-equal, with long, much arched, and compressed claws; the fore part of the foot (both of the fore and hind leg) is narrow, and the hind part broad, having a large and nearly semicircular lobe on the inner side; the inner toe and the thumb are represented by a small nailless tubercle only.

CHÆTOMYS SUBSPINOSUS.

The Chætomys.

(*Plate* 21, fig. 1).

Hystrix subspinosa.	(LICHTENST.) KUHL, Beiträge, p. 71 (1820).
" "	PR. MAXIM., Beiträge, ii. p. 440 (1826).
Chætomys subspinosus.	GRAY, Proceedings of the Zool. Soc. for Feb. 1843, p. 21; Voyage of the Sulphur—Mamm. Pt. 2, p. 36, Pl. 18, figs. 1—6 (skull), July, 1843.

[1] Plate 18, fig. 1 *a*. [2] Plate 18, fig. 1 *b*.

Plectrocœbrus Moricandi. PICTET, Révue Zoolog. Aug. 1843, p. 225—7.
Cercolabes (*Sphiggurus*) *subspinosus.* WAGNER, Schreb. Säugth. Suppl. iv. p. 35 (1844).
Hystrix tortilis. ILLIG. OLFERS, Neue Bibl. der Reisebeschr. xv. p. 211[1].

General hue dull brown (sometimes varied with grey-white); under parts of the animal inclining to rufous-brown: tail and feet brown-black.

Inhabits the middle and northern parts of Brazil.

Specimens of this animal have been brought from Bahia, and the individual described is said to be from Cometa, in North Brazil. Of its habits we have no information. Its clothing consists of modified hairs, which may be described as presenting a condition intermediate between spines and bristles—stouter than the latter, and more flexible than the former. On the head, shoulders, and fore parts of the back, being rather stouter than elsewhere, and not yielding to the touch, the term spine would not be inaccurately applied to these modified hairs. On other parts of the body there is but little difference in the character of the hairs, excepting in their length[2]. They are waved like the strong bristles which spring from the sides of the muzzle in the Seal: those on the hinder part of the back, which are the longest, measure, most of them, about two and a half inches in length, whilst those on the head scarcely exceed half an inch in length. The general colour of the animal is dull brown; sometimes rufous-brown, richer on the under parts than the upper, and somewhat variegated with grey-white on the latter parts. This general tint arises from the visible half of the hairs being of a dark brown colour below the points, and pale brown at the point; or, in some specimens, there is a broad subterminal grey-white ring. The basal half of each hair is

[1] Quoted from Dr. Wagner's continuation of Schreber's great work on Mammalia; Olfers' work I have not been able to consult.

[2] A spiny hair of the back is represented in Pl. 21, fig. 1 *a*.

often pure white, and sometimes yellow-white. The tip of the muzzle is clothed with very short and slender hairs. The feet are also clothed with short hairs, and these are of a brownish black colour: the nails are pale brown. About two inches of the basal portion of the tail is clothed (on the upper surface at least), with long hairs like those on the back; the lower surface of the tail at the root is densely clothed with short stiff black hairs; and moderately stiff black hairs, about two lines in length, are seen on the remaining portions of the tail; though tolerably abundant, they are not sufficiently so to hide the scaly skin beneath. The ear is very small, and hidden, but the long dusky hairs, with which they are clothed, are conspicuous. The hairs of the moustaches are black. Three specimens of this species of Porcupine are contained in the British Museum collection. I add the dimensions of two of these, together with those of their skulls. They are adult animals.

	Inches.	Lines.	Inches.	Lines.
Length from tip of nose to root of tail	18	0	17	6
" of tail	11	0	10	6
Nose to ear	2	7	2	7
Length of fore foot and nails ...	1	11	1	10
" of the nails of the middle pair of toes		6		5
" of hind foot and nails ...	2	8	2	7
" of nail of second toe		7		6
Length of skull[1]	3	1		
" " to hinder root of zygoma	2	3	2	3
Width of ditto	1	9⅔	1	8
" between orbits	1	0	1	0¼
" immediately behind post-orbital process	1	5	1	4½
" from the apices of the post-orbital processes	1	8	1	7
Length of nasal bones				11¼
Width of ditto behind		6⅓		7
" in front				6½

[1] Pl. 18, fig. 1—⅔ds. nat. size.

Length of frontal bones, measured in the mesial line			1	3
Width of terminal portion of nasal process of intermaxillary bone .		1⅓		1⅔
Length of zygomatic arch	1	4	1	4½
" of orbital fossa		7		6¾
" of temporal fossa		7½		7
" of malar bone	1	4	1	1½
Height of malar bone in front ...		6⅔		7
" " beneath orbit		5¼		5½
" " from the apex of the post-orbital process ...		6⅔		7½
" immediately in front of zygomatic process of temporal bone		2½		2¾
Width of incisors		2¼		2⅓
From incisors to molar teeth		11		10⅔
Length of four upper molars, taken together		9¼		9¼
Width of palate between the foremost molars		2⅕		2⅕
" " between the hindermost molars		3		3
Length of lower jaw			2	1
Height of ditto, measuring from the condyle				11
Length of angular portion				10½

Genus, *Cercolabes*[1].

Cercolabes. Brant, Mém. de l'Acad. de St. Pétersb. 1835, VIme. Sér. tom. iii., Sci. Nat. tom. i. p. 391 ; Mamm. Exoticorum novorum vel minus rite cognitorum, &c. p. 55. 1835.

" Wagner, Schreb. Säugth. Suppl. iv. p. 29. 1844.

Coendu. Lacep. Tabl. 1803.

Coendous. Desm., Mamm. Pt. 2, p. 345. 1822.

Synetheres et *Sphiggurus.* F. Cuvier, Mém. du Mus. d'Hist. Nat. ix. p. 427. 1822.

Hystricina with the body (excepting on the under parts) covered with spines, or with spines and hairs ; the under parts sometimes with spiny hairs ; tail long and prehensile ; feet with four toes, armed with long and curved nails ; the hind foot, moreover, with a rudimentary inner toe, represented by a

[1] From χέρχος, a tail ; and λαμβάνω, to grasp.

small nailless tubercle; the palm of this foot much expanded: incisor teeth narrow, and sub-compressed; molar teeth nearly equal in size, of a subquadrate form, and the series of opposite sides of the jaw converging in front; each molar divided into two principal lobes by a deeply-indenting fold of enamel on either side, and each lobe presenting a transverse area surrounded by enamel.

Inhabit America, extending from Brazil, in the northern direction through Peru into Mexico, and in the southern direction into Paraguay and Bolivia.

The animals of this section living upon trees, are expert climbers, but their movements are slow. In many climbing mammals the hind foot is provided with a large thumb opposeable to the toes, and thus capable of grasping objects, but in the Tree-Porcupines the same end is accomplished by different means. The naked soles of the feet are expanded, and present a rough surface; the nails of the toes are very long, compressed, and curved; the hind foot is articulated with the leg in such a manner that the sole is directed inwards, and, to compensate for the want of a thumb, a large, and nearly semicircular lobe may be observed projecting from the inner side of the foot, which not only extends the surface of the sole, but which the animal has the power of bending inwards to a certain extent: this lobe is supported internally by several bones; the fore part, by the bones of the inner toe, of which there are two, a broad and tolerably long metatarsal, and one phalanx; the hinder part, by two other bones which have no analogue in the tarsus of many mammals[1]; one of these—a large bone[2]—is attached to the outer side of the navicular end of the astragalus, with which it articulates by a concave surface: to its lower side is joined the

[1] In several Rodents, however, supernumerary bones have been noticed in the same part.

[2] Marked with an asterisk in fig. 4 of Pl. 18, which represents the tarsus of one of these animals.

cuneiform of the thumb, and to its outer side is attached the second supernumerary bone[1], which is hatchet-shaped, narrow internally, where it articulates with the first-mentioned supernumerary bone, and expanded externally. The articular surface of the astragalus for the tibia is directed inwards, whilst that for the fibula is directed upwards, and hence the oblique position of the foot. In the fore foot, which also has a lobe on the inner side, though of smaller size than that of the hind foot, there is also a hatchet-shaped bone; this articulating with the large bone formed by the anchylosis of the scaphoid and lunare; and, as in most other Rodents, there is a small extra bone wedged in between the scapholunare, the trapezium, and the os magnum: the thumb has but one phalanx. The tibia and fibula, as in other climbing mammals, are widely separated, excepting at the extremities. In addition to these modifications, adapted to the climbing habits of the animal, we have to notice, that it has a long tail, which is thick and muscular at the base, and somewhat slender at the opposite extremity, where the absence of hair, on the supper surface, shows that that part is used in grasping. In ascending the trunk of a tree it is stated that the Tree-Porcupine uses its feet only, but when in the tops of the trees, which it seldom quits, it derives assistance and security from its tail, the end of which it coils round the slender branches. Here it is the upper surface of the tail which is applied to the branch, and the organ is therefore coiled in the reverse direction to that of the Prehensile-tailed Monkeys which inhabit the same country. Judging from the peculiar covering of the under surface of the basal half of the tail—which, as I have before said, is very powerful—one may suppose that when the animal is on a thick vertical branch, where the organ could not be used for prehension, it may still be of considerable assistance in supporting the

[1] Marked with a double asterisk in the figure.

body, and in such a manner as to leave the fore limbs free, in case they should be required for holding the food, which, in nearly all Rodents, is taken in the fore paws. The covering of the part of the tail in question, consisting of a very dense mass of short and stiff bristles, the points of which are directed backwards, would, when closely applied to the branch, afford a strong resistance; and, combined with the lateral pressure of the hind feet, suffice to sustain the body of the animal securely.

It needs only further to be added, with respect to the structure of the feet, that of the toes of the fore feet, the middle two are nearly equal in length, and the remaining two are nearly equal to each other; the inner toe, which is slightly the shorter of the two, has its nail terminating in a line with the middle portion of the nail of the adjoining toe. The toes of the hind feet are very nearly equal in length.

The muzzle of the Tree-Porcupines is thick, and obliquely truncated, so that the upper portion projects slightly over the lower part: the nostril openings are large, nearly round, and approximated, and are scarcely visible when the muzzle is viewed laterally. The whole of the muzzle[1] is clothed — rather sparingly—with minute velvet-like hairs. The upper lip is slightly notched, or emarginated immediately above the incisors, but presents no vertical groove. The eyes are small, but prominent: the ears are likewise small, and sparingly clothed with fine hairs, excepting on the inner surface of the lobe, where there is a tuft of long and stiff hairs.

The skull of these animals[2] is short and broad; the zygoma is somewhat slender; there is scarcely any trace of a post-orbital process either to the malar bone or the frontal: the ant-orbital opening is large, and enclosed externally by a

[1] See Plate 6 *a*, figs. 6 and 6 *a*.

[2] Pl. 81, figs. 2 and 3.

process of the superior maxillary only. The palate is rather deeply emarginated behind, and contracted between the foremost pair of molar teeth, where its width is usually about equal to the width of one of the molars: the auditory bullæ are moderately large, and the ear opens directly into the chamber without any prolonged tube. The molar teeth[1] are perfectly rooted, nearly equal in size, and their crowns are of a sub-quadrate form. Each of the upper molars has a moderately deep fold of enamel on the inner side, which runs obliquely inwards and forwards from the posterior angle of the tooth, and a still deeper enamel-fold on the outer side, which runs directly inwards, and meets that of the opposite side. Before the crown of the tooth is worn, each half presents a concavity surrounded by enamel, which is in contact with that which surrounds the lobe externally; but, the concavity being shallow, its area becomes contracted in proportion as the tooth wears, and in a moderately aged animal we find it represented only by a narrow transverse groove, surrounded by enamel, which is distinctly separated from that of the outer surface of each half of the tooth. Besides the principal enamel folds, there is a very small fold entering from the back of the molar, and which is often very indistinct. The hindermost upper molar is a trifle smaller than the others. The lower molars resemble the upper, excepting that the enamel-folds are reversed in position—thus, the oblique fold is on the outer side of the tooth—and, excepting that the foremost molar is longer and narrower than the rest, and its foremost lobe is distinctly smaller than the hinder lobe.

In a female Cercolabes described by Azara, that author found four mammæ, and I have found the same number in the female of a different species of the same genus—the *C. Novæ-Hispaniæ*[2]. This animal, which I examined immediately

[1] Pl. 18, fig. 3 *b* (1st and 2d right upper molars).

[2] On each side of the body I found a nipple situated at a distance of about

after death, as well as two specimens of the *C. prehensilis*, emitted the strong and disagreeable odour (somewhat resembling that of garlic) which has been noticed in other species.

The food of the Tree-Porcupines is said to consist of the fruits and leaves, as well as the tender bark of trees. These animals are usually seen singly; they pass the hot parts of the day in sleep, and seek their food in the morning and evening. They are very inoffensive, and soon become reconciled to captivity, though they show very little intelligence. The spines, or quills, with which they are armed, are barbed, that is, they have at the pointed end numerous minute spines (visible only with a lens), the points of which are directed backwards; and, as the quills are but loosely attached to the skin, it may occasionally happen that their points having pierced some object, the quills themselves may become detached from the animal and remain sticking in that object: and this circumstance perhaps has given rise to the erroneous statement made by some of the older authors, who have attributed to these animals the power of shooting their quills.

The Prehensile-tailed Porcupines are divided by M. F. Cuvier[1] into two genera—*Synetheres* and *Sphiggurus* (or, more correctly, *Sphingurus*), these being distinguished chiefly by certain peculiarities in the structure of the cranium. M. Brandt, regarding the peculiarities in question as of

two inches behind the base of the fore leg, and a second about an inch and a half behind this. The foremost pair were somewhat approximated, being only an inch and a half apart: the hinder pair were rather more widely separated.

[1] To M. F. Cuvier is due the credit of having first divided into sections the various very dissimilar animals previously associated under the name *Hystrix*, and for having carefully defined most of those sections.—See M. Cuvier's memoir entiled "Examen des espèces du genre Porc-épic, et formation des Genres ou Sous-Genres *Acan'hion*, *Eréthizon*, *Sinéthère* et *Sphiggure*," published in the ninth volume of the Mémoires du Muséum d'Histoire Naturelle.

minor importance, links the two above-mentioned sections together under the generic title *Cercolabes,* but adopts M. Cuvier's sections as sub-genera. The essential distinction of the species of the sub-genus *Synetheres,* according to M. Brandt, is, that they have the under part of the neck, and the chest and abdomen, as well as the inner surface of the limbs, covered with bristles of a more or less spiny nature; whereas those of the sub-genus *Sphingurus* have the same parts clothed with hairs only. Upon this distinction I may remark, that if we take a typical species of each section, viz. *C. prehensilis* for the section *Synetheres*, and *C. villosus* for the section *Sphingurus,* we might add that the *Synetheres* type is further distinguished by the upper surface of the cranium being very broad and convex, and the spines of the body being in no way hidden by fur; whilst in the *Sphingurus* type the spines are almost entirely hidden by a fur composed of very long hairs, and the skull is less broad between the orbits, and nearly flat in the same part. Now, an animal unknown to M. Brandt (the *C. Novæ-Hispaniæ*) presents all these characters noticed in the *C. villosus,* but has tufts of spiny bristles on the under parts of the body interspersed with a soft fur.

Regarding the distinctions just noticed as specific rather than as sectional, I will describe these animals under Brandt's generic name, and will first characterize those species in which the body is almost destitute of hairs, being covered with spines, and, in part, with spiny bristles.

CERCOLABES PREHENSILIS.

Brazilian Tree-Porcupine.

Hystrix prehensilis. **LINN., Syst. Nat. ed. 12, p. 76.**
" " SCHREB., Säugth. iv. p. 603, Pl. 168.
" " (Prehensile Porcupine). SHAW, Gen. Zool. ii. Pt. 1, p. 7, Pl. 123.

Hystrix Cuandu.	DESM., Nouv. Dict. d'Hist. Nat. vi. p. 44 ; Mammalogie, p. 346.
Synetheres prehensilis.	F. CUV., Mém. du Mus. ix. pp. 427 and 433, Pl. 20 ter. figs. 3 and 4 (skull) ; Dict. des Sci. Nat. xlii. p. 533.
" "	IS. GEOFF., Dict. Class. d'Hist. Nat. xiv. p. 215.
Coendu prehensilis.	LESSON, Manuel de Mamm. p. 291.
Cercolabes (Synetheres) prehensilis.	BRANDT, Mamm. Exoticorum, &c. p. 60, Pl. 9, figs. 5—9 (skull).
" " "	WAGNER, Schreb. Säugth. Suppl. iv. p. 30.
Brazilian Porcupine.	PENNANT, Synopsis of Quadr. p. 264, Pl. 24, fig. 1.
Le Coendou à longue queue.	BUFF, Hist. Nat. Suppl. vii. p. 305, Pl. 78.
Le Coendou.	F. CUV. et GEOFF., Hist. Nat. des Mammif. Fasc. 46.

Body covered with spines having a circular section, and which are white at the base and apex, and black in the middle—when in their ordinary position the black and white show in about equal proportions : under parts with spiny bristles : tail nearly equal to the head and body in length, the basal third armed with spines above ; the terminal third (or less) naked above, and the remaining portions clothed with stiff hairs, which are for the most part black—sometimes dusky and dirty white.

Inhabits Guiana, Brazil, and Bolivia.

This animal is frequently met with in Brazil and Guiana, and it occurs likewise in Santa Cruz de la Sierra, a district of Bolivia in which nearly all the Mammalia are identical in species with those of Brazil. Mr. Bridges, who brought to England specimens from this locality, informs me that it is most frequently seen on the Palm trees, and that it feeds upon the fruits of those trees.

On more than one occasion I have had an opportunity of seeing the Brazilian Porcupine alive, in the menagerie of the

Zoological Society, but I noticed no peculiarities of habits worthy of remark: excepting when tempted by food, it seldom descended from a perch which was placed in its cage, and on which it sat, resting entirely upon its hind feet, the long tail hanging down, and serving no doubt as a balance to the anterior part of the body, which was curved forwards and downwards, so as to bring the head nearly in contact with the feet: at such times the head would barely be distinguished from the round ball of spines which the animal resembled, were it not for the protrusion of the large fleshy muzzle; a part which, when the animal was the least excited, was almost constantly in motion.

The *C. prehensilis* is usually from about sixteen to eighteen inches in length, without including the tail, which measures but a trifle less. The upper surface of its head is much arched; the muzzle is very prominent, slightly recurved, rounded at the apex (with a slight depression, however, in the mesial line), and nearly flat in front, where, about midway between the upper surface of the nose, and the edge of the upper lip, are placed the nostril openings, which are moderately large, nearly round, but slightly pointed above, and are separated by a space of about one-sixth of an inch. The whole surface of the muzzle (with the exception of a narrow naked space round each nostril) is somewhat scantily clothed with small velvet-like hairs, mixed with which are some longer and coarser hairs, and, as usual, the long coarse hairs of the moustaches which take their origin on the sides of the muzzle. The chin is clothed with fine hairs like the muzzle, but all other parts of the head, as well as the body, are covered with spines, or spiny bristles. The spines commence at the distance of about half an inch in front of the eye, and are at first short and small, but rapidly increase in size as they spring from parts more remote from the muzzle; and on the middle of the back they attain their greatest length, which is about three and a half inches: here they have a diameter of one line, or a trifle more. The spiny bristles on the under parts of the body are short, white at the root, dusky in the middle, and white again at the point. Upon separating them, a very scanty growth of short and thin hairs is visible. The outer surface of the limbs is clothed with spines like the sides of the body, but towards the feet they become smaller, and are mixed with hairs. Besides the spines, a few very long bristly hairs spring from the legs, and sides of the body.

The feet are clothed with coarsish, party-coloured, dusky and whitish hairs: the toes are scantily clothed. The tail is covered above for about one-third of its length (or sometimes for half its length), with spines like those on the upper parts of the body, but the area thus covered gradually becomes narrower from the base of the tail, until it terminates in a point: about one-third (or rather less) of the apical portion of the tail is naked above, and the remaining portions are clothed with coarse hairs averaging about half an inch in length; on the under surface of the basal portion they are extremely dense; dusky, with a dirty yellowish tip; towards the end of the tail the hairs become scanty, and do not hide the scaly skin. The naked soles of the feet are brownish, and the naked part of the tail dusky brown. The ear is small, and almost entirely hidden by the spines of the head; it has a few long and very fine hairs scattered over its surface, a tuft of stiff hairs springing from the lobe.

	Inches.	Lines.	Inches.	Lines.
Length from tip of nose to root of tail (in a straight line)	16	0	18	0
" of tail	15	0	17	0
" from nose to ear	2	7		
Width of muzzle	1	4		
Height of ditto, measured from edge of upper lip	1	5½		

	Inches.	Lines.	Inches.	Lines.
Length of fore foot, not including the claws	1	11		
Length of the longest claw		7		
Width of fore foot		8		
Length of hind foot, without the nails	2	9	2	11
" of longest nail		$6\frac{1}{3}$		
Greatest width of foot	1	$3\frac{1}{2}$	1	$4\frac{1}{2}$

The above dimensions are taken from specimens immediately after death, the specimens having lived in the menagerie of the Zoological Society. A specimen in the Leyden Museum, and which is labelled as coming from Surinam, measures 20 inches in length, following the curve of the back; its tail is about 23 inches in length, and therefore proportionately longer than in the animals belonging to the Zoological Society; in other respects it does not differ, excepting that the hairs on the tail are dusky brown.

A specimen, in the British Museum collection, brought from Bolivia by Mr. Bridges, differs from others which have come under my notice, in being of a much paler hue, owing to the white portions of its spines being much more extended in proportion to the dark part; and the dark ring on each spine here assumes a brown-black hue. The dark ring on the spines of the head, as well as on those of the upper surface of the tail (excepting quite at the base) scarcely show, and hence these parts appear to be almost wholly white. The prevailing hue of the hairy portions of the tail is pale ash-colour; the stiff bristly hairs being dirty white, with a narrowish ashy brown ring above the middle. The spiny bristles on the belly are similarly coloured. The hairs of the moustaches are black, but a considerable portion (being the basal portion) of each hair is white; and the few very long bristly hairs which spring from the limbs and sides of the body are party-coloured, black and white. The specimen is a very large female, and exhibits four mammæ; one on each side of the body, about midway between the fore and hind legs, and a second about midway between this and the base of the fore leg. The animal is about 26 inches in length from the tip of the nose to the root of the tail, and its tail is apparently about 22 inches in length: the hind foot and nails is 4″ 2‴ in length, and the greatest width of this foot is 1″ 10″.

Another specimen in the British Museum is remarkable for its general dark hue, the visible portions of the spines (which are white at the root) being black, with the exception of a short white point, the white being extended not more than two or three lines from the tip of the spines. In several of the spines the white of the point is replaced by pale brownish yellow. The under parts of the body are dusky, but rather finely pencilled with whitish. The hairy portion of the tail is dusky, excepting at the base beneath, where the hairs have brown-white points. Length 21″ 6‴; tail about 19″ 3‴.

The skull of the *Cercolabes prehensilis*[1] is short and broad (the frontal bones very broad) and remarkable for the great convexity of its upper surface. The anterior half of the nasal bones is depressed, but their hinder half rises abruptly to form a continuous curve with the frontals, which curve is most elevated at the fore part of the last-mentioned bones, and gradually descends from that point to the occipital ridge. The series of molar teeth of opposite sides of the upper jaw are separated in front by a space about equal in width to one of these teeth, and they diverge very little behind, in which respect they differ from the molars of the *C. villosus*, where the two rows distinctly diverge. A muscular impression immediately in front of the foremost molar, which is very strongly marked in the skull of *C. villosus*, and some

[1] Plate 18, fig. 2.

nearly allied species, is in the *C. prehensilis* somewhat indistinct. The auditory bullæ are very large and prominent.

	Leyden Museum. Surinam.		British Museum.	
	Inches.	Lines.	Inches.	Lines.
Length of skull	3	3	3	7½
Width of ditto	2	1	2	2¼
" " between orbits ...	1	6	1	6⅔
Length of nasal bones	1	1		0*
Width of ditto, behind	1	0½		
" " in front		11¾		11⅔
From incisor to molar teeth		10½		11
Length of four upper molars, taken together		9		10
Longitudinal diameter of auditory bullæ		9		11
Transverse diameter of ditto		6		6
Length of lower jaw	2	2½	2	4
Height of ditto		11½	1	1
Width of incisors, together		3		3

Cercolabes platycentrotus.

Cercolabes platycentrotus. BRANDT, Mém. de l'Acad. de St. Pétersb. l. c. Mammalium Exoticorum, &c. p. 63, Pl. 2.

Body covered above by spines; and beneath by bristles, or spiny-bristles; the spines many of them grooved on the upper surface, the groove either extending from the base to the apex of the spine, or confined to the basal portion.

Habitat ————?

This species is founded on a single specimen in the Museum at St. Petersburg, an animal which we are informed greatly resembles the *C. prehensilis*, but which differs in having most of the spines grooved along the upper or anterior surface, either from the base to the middle, or as far as the point. The spines which are grooved in the whole length are for the most part dilated and broader than the rest. Some there are which are flattened, and which are compared to a narrow corneous

* All the sutures on the upper surface of this skull are obliterated.

lamina, contracted to a point at each extremity. Mixed with the grooved and flattened spines are others which are round, as in the *C. prehensilis*. All the spines are white at the base and apex, and black in the middle, as in the animal just mentioned.

	Inches.	Lines.
Total length of the animal	28	0
Of which the tail measures	14	9
From nose to ear	2	6

M. Brandt observes, with regard to the *C. platycentrotus*, that, although it bears a great resemblance to the *C. prehensilis*, and that individuals of that species are found in which there are a few grooved spines mixed with the ordinary cylindrical spines[1], nevertheless it can hardly be a variety of that animal, since the grooved spines prevail in the *C. platycentrotus*, and, moreover, the flat spines which have been alluded to are tolerably abundant in his animal, and are not found in the *C. prehensilis*. Lastly, in the region of the spiny portions of the tail and limbs there are flat and channelled bristles, which do not occur in the *C. prehensilis*.

CERCOLABES NYCTHEMERA.

Peruvian Tree-Porcupine.

Hystrix nycthemera. (Licht.) Kuhl., Beiträge zur Zoologie, p. 71 (1820).
Sphingurus bicolor. Tschudi, Fauna Peruana, Pt. 4, p. 186, Pl. 15, (1845).

Upper parts of the animal entirely black; the spines very abun-

[1] In the Leyden Museum I saw a specimen which appeared to me to be the *C. prehensilis*, but in which many of the spines were broadly channelled from the base nearly to the apex; its spines were rather longer than those of a specimen of the *C. prehensilis* from Surinam, in the same collection, and differed, moreover, in having the broad dark ring of a dusky brown hue instead of black. It is labelled as coming from Brazil.

dant, with the base white, and the apical third black, very acute, and rather larger than those of *Hystrix insidiosa*.—KUHL.

In addition to the above, we learn from Kuhl that his *H. nycthemera* belongs to the same section as the *H. prehensilis*, has a long tail, and is distinguished from *H. insidiosa* by the absence of long hairs mixed with the spines.

A small Cercolabes in the collection of the British Museum I have little doubt is the young of this species. Its characters are as follows:—

Upper parts and sides of the animal covered with well-developed spines, which hide a clothing of woolly hair; the woolly hair scanty on the back, but moderately dense over the rump, where it is not entirely hidden by the spines. Under parts of the head and body, inner surface of limbs, and fore parts of hind legs, clothed with soft hairs: feet covered with somewhat coarse hairs; tail, for about two-thirds of its length, beneath, covered by dense and stiff hairs; the apical portion rather sparingly clothed with coarse hairs; the upper, basal half, with spines.

***Colouring*.—The spiny parts of the animal pencilled with black and white, in about equal proportions; the spines being white, with long black points; some of them (more especially those on the haunches) of a pale sulphur-yellow at the root. Under parts palish dusky brown; the hairs being grey-white at the root, and dusky brown, or, some of them, black at the point. Tail black beneath, brown-black at the apex, and pencilled with black and rufous on the sides. Feet dusky brown, with a mixture of rufous-brown: the toes chiefly rufous-brown. The woolly fur hidden by the spines on the upper parts of the body is grey-white. Length of head and body 9½ inches; tail, 9¼ inches.**

Habitat, Brazil.

Sphingurus bicolor. TSCHUDI.

The following description of the *S. bicolor*, extracted from

Dr. Tschudi's account of that animal, seems to me to be likewise referrible to the species too briefly described by Kuhl.

Spines of a sulphur yellow colour, with dusky black points, and not intermixed with long hairs: under parts of the body covered by a pale brown fur, mixed with which are some very slender spines.

Inhabits Peru.

Dr. Tschudi informs us that he met with but a single specimen of this animal, and that as the Indians, to whom it was shown, were unacquainted with it, the species must be very rare in Peru. It was procured in the tropical forest on the east side of the Andes.

Judging from Dr. Tschudi's description, the *C. bicolor* must bear considerable resemblance to the *C. prehensilis*, but would be readily distinguished by its spines being black at the point. The same parts are covered by spines as in the animal just mentioned, viz. the upper parts and sides of the head and body, the outer surface of the limbs, and the upper surface of the tail; here leaving the terminal third of the organ unprotected by these weapons. They attain their greatest length on the middle of the back, where they measure $3\frac{1}{3}$ inches from the base to the point; and, on the hinder part of the back, they are said to decrease in length rapidly as they approach the tail. It is only on the hinder part of the body, and the outer surface of the hind legs, that hair is found mixed with the spines, and this fur is short, and of a coarse, but woolly nature: its colour is yellow-brown. The spines are of a pale sulphur yellow colour at the base, and black, or brown-black, at the point. The spines being in their ordinary position, the black prevails on the fore parts of the body, whilst the yellow is evident on the hinder parts, and this arises from the degree to which either the dark or pale hue is extended on each spine: thus, on the fore parts of the body, about one-third of each spine is yellow, the point is brown-black, and the intermediate space is pale brown, the dark and light colours gradually blending: on the hinder parts of the body, sulphur yellow prevails, only about one-fourth or one-fifth part of each spine being black. The spines on the upper surface of the tail are similarly coloured. The throat, breast, belly, and inner surface of the limbs, are somewhat scantily furnished with hairs, which are about a quarter of an inch in length, and mixed with these is a woolly under fur, besides some slender spines. These parts are described as of a very pale, almost whitish brown, colour; the terminal portion of the tail as reddish black, and the feet as dusky. The entire length of the animal is 26 inches, of which the tail is 13 inches.

Cercolabes spinosus.

Sphiggurus spinosus. F. Cuv., Mém du Mus. d'Hist. Nat. ix. p. 433, Pl. 20, ter. figs. 5 and 6 (skull); fig. 7 (feet); Dict. des Sci. Nat. xlii. p. 534.
" " Lesson, Manuel de Mammalogie, p. 291.
" " Is. Geoff., Dict. Class. d'Hist. Nat. xiv. p. 215.

Upper parts covered with spines, which are for the most part tri-coloured, white at the base, black towards the point, and orange-red at the point, and in no way hidden by hairs; under parts clothed with woolly hairs, amongst which are intermingled some coarse hairs: spines on the rump, and most of those on the tail, white, with a black point: tail black, except on the spiny portion.

Inhabits Brazil.

The following notes were made upon the specimen preserved in the Paris Museum, there labelled as the *Sphiggurus spinosus* of F. Cuvier, and which I presume is the original of that author's description.

Upper surface of the head, the cheeks, the upper parts of the body, and the flanks, covered with spines; those on the body being moderately stout. Mingled with these spines are a few inconspicuous slender hairs. Under parts clothed with a woolly fur, but with coarse hairs intermixed, and near the flanks the fur is almost entirely replaced by slender spines. Similar spines, mixed with hairs, cover the outer surface of the limbs. Nearly half the upper surface of the tail is covered with spines, amongst which are fine hairs like those on the upper parts of the body, but more numerous; the remaining portions of the tail, being the under parts, and the whole apical half (with the exception of about two inches of naked space on the upper side) covered with stiff hairs; scanty on the terminal portion, and very dense on the under surface of the basal portion.

Colouring.—Spiny parts of the body, with the exception of the terminal part of the back, strongly pencilled with black, white, and orange-red: on the fore parts and sides of the body (the spines being in their ordinary position) the red prevails slightly over the black, and the white is but little apparent. The hinder part of the back, and the spiny portion of the tail, are pencilled with black and white. The spines, taken separately, are white at the base, black towards the point, and orange-red at the point, the white usually occupying about half of the entire length of the spine, and the remaining portion is about equally divided between the black and orange colours. The spines of the rump are white with a black point, and so are those on the basal portion of the tail, if we except a few which are tipped with orange. The muzzle is clothed with brown hairs; the moustaches are black; the fur on the under parts of the body is brownish next the skin, and ashy grey externally; the coarser intermixed hairs are black. The tail, with the exception of the spiny portion, is dusky black.

	Inches.	Lines.
Length from tip of nose to root of tail ...	13	0
" of tail	10	0
" from nose to ear	2	5
" of fore foot and nails	1	10
" of nail of middle toe		6
" of hind foot and nails	2	7
" of nail of middle toe		$6\frac{1}{4}$

M. Isidore Geoffroy St. Hilaire doubts that this animal is distinct from the *C. villosus,* and accounts for the absence of the plentiful long hairs, which in that animal hide the spines, by relating a statement made by an Indian to M. D'Orbigny, to the effect, that the *C. villosus* loses its hair in the summer. I can readily believe that, like many (if not all) other mammals, the *C. villosus* sheds its fur from time to time, but I cannot believe that all, or nearly all, the hairs are shed at one time; in fact, I think it may well be doubted whether the *C. villosus* ever presents the same condition as the animal under consideration. If such were the case, it is very remarkable that of all the naturalists who have observed the *C. villosus* in its native country, none have noticed the fact.

Besides the conclusions I am inclined to draw from the foregoing considerations, I am further led to doubt the specific identity of the two animals under notice, from the circumstance that the spines of the *C. spinosus* are stouter, and relatively shorter, than in *C. villosus,* and that its tail does not present the bright rufous hue on the under surface, which appears to be a constant character in the last-named animal.

CERCOLABES NOVÆ-HISPANIÆ.

Mexican Tree-Porcupine.

Hystrix Novæ-Hispaniæ.	Briss., Règne An. p. 127 (1756).
" *prehensilis,* var. γ.	Schreb., Säugth. iv. p. 604 (1792).
" "	Gmel. Linn. Syst. i. p. 119 (1788).

Hystrix Mexicana. SHAW, Gen. Zool. ii. Pt. 1, p. 8 (1801).
Cercolabes Liebmanni. REINHARDT, in Archiv für Naturgesch. 1844, Pt. 1, p. 241.
Mexican Porcupine. PENNANT, Hist. of Quadr. ii. p. 393 (1781).
Hoitztlacuatzin seu Tlacuatzin spinoso. HERNANDEZ, Rerum Medicarum Novæ-Hispaniæ, &c., p. 322 (1651).

General hue black; the spines (which are nearly all of them hidden by the fur) sulphur yellow (or, some of them, white), with the points black: tufts of spiny bristles are mixed with the fur of the under parts of the body, and are conspicuous from being, for the most part, white.

Inhabits the east coast of Mexico.

We are indebted to M. Reinhardt for a careful description of this Mexican species of Tree-Porcupine; a description founded upon two specimens procured by Prof. Liebmann, after whom the animal is named. One of these specimens was found near Mirador, about twenty-five leagues from Vera Cruz, and the other at St. Francisco, a little town but a few leagues from the same spot. The animal, however, according to Prof. Liebmann, occurs generally in the eastern portions of Mexico, from the 22d degree of N. latitude down to the 16th, inhabiting the temperate mountain regions at elevations of between two and four thousand feet above the sea. It is still known to the natives by the name of Hoitztlacuatzin, a name under which the Mexican Porcupine is noticed by Hernandez nearly two hundred years back[1].

The *Cercolabes Liebmanni* is readily distinguished from other species of Tree-Porcupines, in which the spines are hidden by a long fur, not only by its almost uniformly black

[1] There can be no doubt that this is the Porcupine described by Hernandez, and consequently that it is the *Hystrix Novæ-Hispaniæ* of Brisson, and the *H. Mexicana* of Shaw, the characters of which are derived from the account given by Hernandez; but, as the old descriptions of the Mexican Porcupine have been erroneously supposed to refer to the *H. prehensilis*, M. Reinhardt thinks it desirable to rename it.

hue, but by the under parts of its body being provided with numerous spiny bristles besides the ordinary fur. These bristles are tolerably abundant, and spring from small tubercles on the skin, each tubercle giving origin to from three to five bristles; they are about three-quarters of an inch in length, and black at the point, but being white below that part, they form a strong contrast to the dusky colour of the fur: on the throat they are very numerous, on the under side of the neck they become stouter, and towards the sides of the neck they are replaced by true spines. Spines, as usual in the animals of this section, cover the upper part of the head, the cheeks, the upper parts and sides of the body, the outer surface of the limbs, and the upper surface of the basal half of the tail. The longest spines, which are found on the back of the animal, measure about an inch and a half in length; these are moderately stout, but the spines on the extremities become more slender in proportion as the parts from which they spring are more remote from the body: those near the feet might almost be described as bristles; the spines on the cheeks are stronger than in the *C. melanurus*, and much stouter than in *C villosus*. Most of the spines are of a pale sulphur yellow below the point, but some are white at the base; and all are black at the point. In a specimen now before me, the spines on the flanks, as well as those on the hinder part of the back, are not perfectly hidden by the fur, and those on the head are distinct, the fur being scanty on this part: elsewhere the spines are hidden by the long and soft hairs of the fur, which have the exposed ends of a sooty black hue, but are brown on the basal half. On the under parts of the body the hairs are somewhat short and scanty, and of a dusky brownish tint. Excepting on the basal half above, and a small portion at the apex, the tail is clothed with stiff black hairs.

Two specimens of the Mexican Porcupine are contained in

our London collections, and one of these, which is in the British Museum, has recently been examined by M. Reinhardt, who pronounced it to be the animal he had described. The other is in the museum of the Zoological Society, having died in the menagerie. This latter specimen agrees most closely with M. Reinhardt's description, whilst the former differs only in having the long fur of the upper parts of the body whitish next the skin, instead of brown. The dimensions are :—

	Brit. Museum.		Zool. Society.	
	Inches.	Lines.	Inches.	Lines.
Length from tip of nose to root of tail	19	6	17	0
" of tail	14	0	12	6

The length of the head and body of one of the specimens described by M. Reinhardt is 22″ 10‴, and that of the tail 12″ 5‴.

CERCOLABES MELANURUS.

Black-tailed Tree-Porcupine.

Sphiggurus melanurus. WAGNER, in Archiv für Naturgesch. for 1842, Bd. i. p. 360, ; Schreb. Säugth. Suppl. iv. p. 34.
" " GRAY, in the Annals and Magazine of Nat. Hist. for Dec. 1842, vol. x. p. 262.

General hue sooty black, pencilled with dirty white, the hairs of the fur (which are rather coarse) being sooty black, dirty white next the skin, and many of them with long pale points ; spines sulphur yellow (or white), with brownish black points : tail black.

Inhabits Surinam and the northern portions of Brazil.

The specimen of *C. melanurus* originally described by Mr. Gray is 15 inches in length from the nose to the root of the tail, and its tail measures about 11½ inches. The spines of its back, as well as those on the upper portion of the basal third of the tail, are perfectly hidden by the long dark fur, but those on the flanks are but partially concealed, and on the head there are but few hairs mixed with the spines. On this latter part the spines are white, with black points; those on the cheeks are somewhat stronger than in the *C. villosus*. On the body the spines are for the most part sulphur yellow at the root, and the brownish black at their points is but little extended. Spines are continued about half way down the outer surface of the fore legs, and of the hind legs the outer surface of the thigh only is armed; below the thigh the spines are replaced by long and coarse hairs, which are very numerous, and for the most part dirty white, but ringed with dusky below the point. The abdomen is rather scantily clothed with short, and by no means soft, hairs, which are dusky at the point, and most of them white at the root. The hairs of the moustaches are black. About one-third of the upper surface of the tail is clothed like the body, the remaining portions are covered with very stiff black hairs, and these are dense, except on the tip of the organ. The feet are covered with short and coarse black hairs, excepting on the upper surface of the tarsus, where they assume a dusky brown tint. The hairs on the back average about 2½ inches in length, but many are considerably longer: the greater portion of the hairs are fine and soft, but mixed with these are numerous longer and somewhat coarse hairs; the finer hairs are for the most part black, excepting at the base, where they are dirty white; whilst the longer hairs are some of them entirely dirty white, and others differing from these in having the middle portion dusky: the spines on the same part are very many of them about 13 lines in length, and the longest are scarcely 1¼ inches.

In this specimen the long hairs on the body are of a dirty greyish white colour at the point, and such is the case in a specimen in the Leyden Museum; but three specimens examined by Dr. Wagner are described as having the hairs of a delicate yellow at the point.

The *C. melanurus* may be distinguished from the *C. villosus*, *C. affinis*, and *C. nigricans*, by its tail being uniformly black, except on the basal portion of the upper surface, where it is clothed with spines and bicoloured hairs like the body. The stiff black hairs which cover other parts of the tail (if we except about two and a half inches of the apical portion) are very abundant—decidedly more dense than in the *C. villosus*; its fur is coarser than in the last-mentioned animal, and none of its spines are terminated by an orange-coloured point as in that species; its tail, moreover, is decidedly longer.

The dimensions of the specimens referred to by Dr. Wagner, which are in the Vienna Museum, are—

	No. 1.		No. 2.		No. 3.	
	Ins.	Lines.	Ins.	Lines.	Ins.	Lines.
From tip of nose to root of tail	17	8	15	6	15	4
Length of the tail	16	0	17	6	15	0

The specimen in the Leyden Museum is 13½ inches in length, without including the tail, which is 11½ inches. It is from Surinam, whilst two of the specimens (Nos. 2 and 3) in the Vienna Museum, were procured by Dr. Natterer at the Barra do Rio Negro. Possibly this species does not extend south of the Amazon.

CERCOLABES VILLOSUS.

Tri-coloured Tree-Porcupine.

(*Plate* 21, fig. 2).

Sphiggurus villosus. F. Cuvier, Mém. du Mus. ix. p. 434 (1822); Dict. des Sci. Nat. xlii. p. 534 (1826).

" " Lesson, Manuel de Mammalogie, p. 292 (1827).

Sphiggurus villosus.	Is. Geoff., Dict. Class. de l'Hist. Nat. xiv. p. 216, (1828).
Cercolabes insidiosus.	Brandt, Mamm. Exot. p. 71, Pl. 4 (animal); and Pl. 10, figs. 5—7 (skull). 1835.
" "	Wagner, Schreb. Säugth. Suppl. iv. p. 31, Pl. 148 a, (1844).
Sphiggurus variegatus.	Gray, in the Brit. Mus. collection.
? *Hystrix insidiosa.*	(Licht.) Kuhl, Beitr. z. Zool. p. 71 (1820).
? " "	Pr. Maximil., Beitr. z. Nat. ii. p. 434 (1826); Fig. in Abhandl. zur Naturg.
? *Sphiggurus spinosus.*	Rengger, Nat. der Säugth. von Paraguay, p. 240, (1830).
? *Hystrix Couiy.*	Desm., Mammalogie, Pt. 2. p. 345 (1822).
? *Le Couiy.*	Azara, Essais sur l'Hist. Nat. de Paraguay, ii. p. 105 (1801).

Spines of the head and upper surface of the basal portion of the tail, as well as those of the body, tri-coloured, viz. white, or pale sulphur yellow on the basal half, orange-coloured at the point, and ringed with black below the point; spines on the rump yellow-white at the base, and black at the point: fur very long, for the most part hiding the spines, and of a brown-black colour, but each hair with a long pale point—usually pale yellow—or rufous-brown: tail rusty yellow, or rust coloured, beneath at the base.

Inhabits Brazil.

Of this species I have examined six specimens,—two in the Paris Museum, labelled as the *S. villosus* of F. Cuvier, two in the Leyden, and two in the British Museum. All these specimens agree in having most of the spines (with the exception of those on the hinder part of the back) of a sulphur yellow colour, or white, at the root, black or dusky above the middle, and orange coloured at the point; the orange colour being more or less bright, and sometimes with a slight admixture of brown. The spines on the hinder part of the back differ from those of other parts in wanting the orange point; they are sulphur yellow, or white, with a black point[1].

[1] See *a*, in fig. 2 of Pl. 21; *b* and *c* represent spines from the back of the animal; spine *a* is from the hinder part of the back.

Here and there, mixed with the orange-pointed spines, may be seen shorter spines, in which the point is dusky. In young individuals the colouring of the spines is less intense; the black is replaced by dusky or brown, and the orange colour at the point is very pale—sometimes replaced by dirty yellow. Those spines which are most completely hidden by the fur have the sulphur yellow most distinct, whilst in those which are exposed, such as the spines of the head, this colour is replaced by white, or is only seen near the root of the spines. The long fur with which the animal is clothed is dusky black, or brownish black, but each hair is terminated by a long pale point, which is sometimes of an extremely pale brown-grey hue, sometimes yellowish, sometimes tinted with rufous, and always glossy. On the back, these hairs are from $3\frac{1}{2}$ to 4 inches in length: on the head they are scanty, and by no means hide the spines, whilst on other parts they are so dense that the spines are nearly all of them hidden by the fur. Occasionally specimens are seen in which the fur is less dense on the flanks and hinder part of the back, in which parts the spines consequently show themselves. The spines occur on the upper surface of the head, on the cheeks, the whole upper parts of the animal, as well as on the flanks, the outer surface of the upper half of the limbs, and the upper surface of the basal half (or rather less) of the tail. The longest spines (which are found on the hinder part of the crown of the head, and on the back) average about 2 inches in length, and have a diameter of $\frac{1}{24}$th of an inch. The under parts of the head and body, and the inner surface of the limbs, are clothed with shortish hairs of a dusky hue, but with pale points. The coarser hairs which cover the lower half of the limbs externally, and the feet, are similarly coloured. The lower surface of the basal half of the tail is densely covered with stiff bristly hairs, the visible ends of which are rusty yellow, or bright rust coloured, but the hidden portions are black; the terminal half of the tail is clothed with less stiff hairs, which become gradually more scanty towards the point, and are not sufficiently numerous to hide the scaly skin; they are black near the middle of the tail, and brown at the end: from two to three inches of the apical portion of the tail, above, is almost destitute of hairs.

	Paris Museum.		Paris Museum.		British Museum.		Leyden Museum.	
	Ins.	Lines.	Ins.	Lines.	Ins.	Lines.	Ins.	Lines.
Length from tip of nose to root of tail	13	0	15	6	18	0	18	0
" of tail	11	6	9	9	13	6	10	0
" of fore foot and nails ...	1	7	1	10	1	10		
" of nail of middle toe ...		6		6		7		
" of hind foot and nails ...	2	5	2	8	2	8		
" of nail of middle toe of ditto		6		7		7½		

The animal above described is, I think, without doubt identical with the *S. insidiosus* of Brandt, and of Wagner. It has been identified with the Paraguay Tree-Porcupine, but the following particulars, gleaned from Dr. Rengger's description of that animal, indicate differences which leave some doubt as to the accuracy of that identification.

Sphiggurus spinosus. RENGGER.
Hystrix Couiy. DESM.
Le Couiy. AZARA.

According to Rengger, this Paraguay animal is 22 inches in total length, of which the tail measures 9½ inches. Its spines are for the most part of a sulphur yellow on the lower half of their length, and chestnut brown at the apical half; some are yellow, with the point, only, brown; others, again, are yellow at the base and point, and brown towards the middle; whilst the spines on the forehead differ from the last-mentioned in being reddish yellow at the point. The spines on the neck and back are almost entirely hidden by hairs which are more than two inches in length, of a greyish brown colour, but with yellow-red, or reddish yellow, points. The spines are said to vary from 6 to 16 lines in length, and they occur in the same parts as I have noticed in the *H. villosus*.

Hystrix insidiosa. Pr. Maximilian.

This is also supposed, by most authors, to be the same species as the *S. villosus* of F. Cuvier, and my only reason for placing a note of doubt to the *Hystr. insidiosa* of Prince Maximilian, where that name occurs amongst the probable synonyms of F. Cuvier's species, is, that the Prince describes the spines of his animal as having brown-black points; the only exception noticed being in the spines on the head, which, as in *S. villosus,* have yellowish red points. The long hairs of the fur of the back are said to be of a grey-brown colour, but with bright red points. In having the tail yellow-red beneath, and, indeed, in all other respects, the description of Pr. Maximilian agrees with the *S. villosus*.

Hystrix insidiosa. (Licht.) Kuhl.

The original description of the *Hystrix insidiosa* of Lichtenstein is that given by Kuhl in his *Beiträge*. All we learn respecting the animal from this, or, I believe, from any other source, is, that it is a long-tailed Porcupine, and that the upper parts of its body are covered with hairs and spines, in contradistinction to the *H. prehensilis,* in which these parts are covered with spines only—in fact, that it belongs to the section *Spingurus*. Then for specific characters Kuhl informs us that it is of the size of the *Arctomys Marmota ;* its spines less than an inch in length, very sharp at the point, *of a straw colour at the base, and brown at the apex :* that they are most plentiful on the basal portion of the tail, on the neck, and above the eyes; that long and soft hairs, *of a pale ashy grey colour*, extend beyond the spines, and that the tail is clothed with bristly hairs.

Amongst the numerous specimens which Dr. Wagner informs us he has examined, of the animal which he describes under the name *insidiosus*, none are noticed as having the spines bi-coloured (as wanting the orange colour at the point), or as having the fur of a pale ashy grey colour. Neither does M. Brandt notice such characters in the two specimens he describes under the same name. On the contrary, the specimens examined by these authors have the dark fur and tri-coloured spines, as in the six specimens noticed by myself, and described under the head *C. villosus*; and hence I think it highly probable Kuhl's animal is a distinct species from the *C. villosus*.

Cercolabes nigricans. BRANDT.

Cercolabes nigricans. BRANDT, Mamm. Exot. p. 67, Pl. 3 (animal), and Pl. 10, figs. 2 and 3 (skull).

Spines but imperfectly concealed by the fur, for the most part of a pale sulphur yellow, or white colour, with a subterminal dusky or black ring, and a dirty orange-coloured point; the spines on the rump, and many of those on the tail, white, with a black point. Hairs of the fur black, or dusky, with the points yellowish brown, or yellowish. Tail with black or brown hairs (excepting on the upper surface of basal portion, where there are spines and hairs), on the apical portion, and with those on the under surface of the basal half, black, with the points of a palish brown-straw colour. Nasal bones rather larger than in *C. villosus*. From nose to root of tail 14½ inches; tail, 9 inches.

I see no points in M. Brandt's detailed description of his *C. nigricans* which appear to me to indicate that that animal differs specifically from the *C. villosus*. The species is founded upon a single specimen procured by Langsdorff in Brazil, and shot, as I suspect, whilst it was shedding its fur.

Cercolabes affinis.

Cercolabes affinis. BRANDT, Mamm. Exot. p. 76, Pl. 5 (animal), and Pl. 10, figs. 8 and 9 (skull).

Brown, inclining to yellowish grey; abdomen and extremities pale greyish brown; tail, on the under side of basal portion, with a mixture of brown and yellowish colours: spines on the forehead of a pale dirty orange colour at the point, those on the back for the most part brown or dusky at the point, the remaining portion being white, or tinted with yellow.

Inhabits Brazil.

The hairs of this animal are said to be rather less rigid, and its spines stronger than in the *C. insidiosus.* There are, moreover, but few of the spines on the back which are tinted with orange colour at the point, most of them being dusky or brownish at that part. The spines on the arms, as well as those on the rump, are but little hidden by the fur. The fur of the upper parts, which is less dense than in *C. insidiosus,* is of a somewhat hoary blackish brown hue, with the points of the hairs greyish yellow. The stiff hairs on the under surface of the basal portion of the tail are brown-black, with yellowish points; sometimes brown-yellow immediately below the point.

	Inches.	Lines.
Length from nose to root of tail	13	0
" of tail	10	6

But one specimen is noticed as presenting the above characters. The upper surface of its skull is more flat than in the animals described as *C. insidiosus* by Brandt (*C. villosus* of F. Cuvier), the incisive foramen narrower; and some other distinctions are noticed, which are certainly by no means important. According to the figures given by Brandt, the most striking difference is observable in the form of the zygomatic arch, the malar bone being considerably narrower behind than in *C. nigricans* and *C. insidiosus*, and having an obtuse post-orbital process, which is placed more forward than the less distinct corresponding angle noticed in the malar bone of the two animals last named. I am by no means satisfied that the *C. affinis* is distinct from the *C. villosus*.

Cercolabes pallidus.

General hue of an extremely pale fulvous brown; spines pale sulphur yellow, or white, with a short black point, and almost entirely hidden by the long fur; muzzle dusky brown; feet suffused with brown; tail brown-black beneath.

The fur of the animal is soft, has a silk-like gloss, and is remarkable for its pale colour; the hairs are almost uniform in tint from the root to the point, and that tint may be described as pale fulvous, or rufous-yellow, with a slight admixture of brown; the limbs, lower parts of the flanks, and the under parts of the body, are of a more decided brown hue, and the feet and muzzle are suffused with dusky brown. The spines on the upper surface of the head, and on the cheeks, are short, and almost pure white, with short black points, and those on other parts of the body differ only in having a sulphur yellow colour at the root, which is more or less distinct—very evident on the spines of the back, and scarcely to be traced on those covering the rump, and on many of those on the tail. The longest hairs on the back measure about an inch and a half in length. The limbs (excepting quite at the base externally), and the whole of the under parts of the animal, are destitute of spines, but on the outer surface of the legs (especially of the fore legs) there are numerous spiny hairs, mixed with the soft hairs of the ordinary fur, and which are coloured like the true spines. The hairs of the moustaches are black, and there are a few long, stiff, black hairs, springing from the limbs. The tail is short: on the upper surface of the basal half it is clothed with spines and long fine hairs, like those on the body, but the spines are very slender, and towards the sides of the tail they are replaced by bristly hairs, which are white, but have a longish black point. The remaining portions of the tail are clothed with coarse and somewhat short hairs, with the exception of the apical portion above, which is naked; the naked space extending about two inches from the point. The hairs of the tail are nearly all of them of a brownish black colour, the only exceptions being those which spring from the terminal portion of the organ, which are brown.

	Inches.	Lines.
Length from tip of nose to root of tail ...	12	6
" of tail, about	7	6
" of hind foot and nails	2	1

In the British Museum are two specimens of this pale-coloured Tree-Porcupine, one of which is said to be from the West Indies; they are both immature animals, the largest (of which the dimensions are given) having but three molars on each side of the jaw. Possibly this may be the *H. insidiosa* of Lichtenstein, but I think that Kuhl, who described that animal, would scarcely have omitted to notice the black hue of the under surface of the tail, which forms a strong contrast with the pale tint of other parts. The tail in this animal is decidedly shorter than in the *C. melanurus*.

Fossil *Cercolabina*.

Synetheres magna et *dubia*. Lund.

Dr. Lund has found the remains of two species of Tree-Porcupines in the caverns of Brazil, which he regards as distinct from the recent species; one of them, to which he gives the specific name *magna,* is supposed to have been as large as the Peccari.

Cercolabes fossilis.

Remains of a species of Cercolabes, or of an animal of a closely allied genus, contained in the British Museum, indicate the former existence of a very large species of this group, but a species which, however, must have been inferior in bulk to the *C. magna* of Lund. These remains consist of portions of the right and left branches of the lower jaw, one ramus containing two molar teeth, and the other, three; and this latter, moreover, shows the socket of a fourth molar. The three molars *in situ* are the three hindermost, and these measure together 11 lines in the longitudinal direction. Judging of the size of the anterior molar by its socket, the four molars must have measured, together, full fifteen lines, whilst in the *Cercolabes prehensilis* the four corresponding teeth measure but 10½ lines. The transverse diameter of one of the molars in the fossil jaw is fully 3½ lines: the depth of the jaw, below the penultimate molar, is 7½ lines. In the structure of the teeth there is no difference between the fossil and the recent species. The specimens referred to are from the caves of Minas Geraes, in Brazil.

Genus, *Erethizon*[1].

Erethizon. **F. Cuvier, Mém. du Mus. ix. pp. 426 and 432.**
Hystrix. **Linn.**

Tail short, thick, and depressed, covered above, at the base, with hairs and spines; at the apex, and on the under surface,

[1] From ἐρεθίζω, to irritate.

with stiff bristles; nostrils closely approximated; feet short and broad; toes 4—5, all armed with long and curved claws.

Inhabit North America, between the 37th and 67th parallels of latitude.

In the genus Erethizon the skull and dentition are essentially the same as in the Cercolabes group, and, indeed, the Canada Porcupine, the only known representative of this group, clearly is a mere modification of that South American type, the main differences being, that in the North American Porcupine the tail is shorter (being about equal to one-third of the length of the animal), much stouter, and somewhat depressed: this organ, moreover, wants the prehensile power which is found in the tail of the South American species, and is well covered throughout either by hairs or spines. The nostril-openings are more approximated, being separated—at their lower part especially—by an extremely narrow partition.

The hind feet present a well-marked difference, in being provided with a distinct inner toe, and that armed with a large claw, which is but little inferior in size to the claws of the other toes; these feet, moreover, want the projecting semicircular lobe on the inner side. All the claws are large, much arched, and but little compressed. The naked soles of the feet, as in *Cercolabes*, are covered by minute tubercles. The upper lip is slightly notched above the incisor teeth, but there is no dividing groove—no naked mesial line—such as we shall have to notice in describing the Old World Porcupines; even the narrow septum between the nostrils is not destitute of hairs. The body is stout, and covered on the upper parts with a long and dense fur, which hides the spines. The limbs are short and strong.

ERETHIZON DORSATUS.

Canada Porcupine.

Hystrix dorsata.	Linn., Syst. Nat. 12 ed. p. 76.
" "	Schreb., iv. p. 605, Tab. 169.
" "	Desm., Mamm. Pt. 2, p. 345.
" "	Sabine, Franklin's Journal, p. 664.
" "	Cozzens, in Annals of the Lyceum of Nat. Hist. of New York, i. Pt. 1, p. 190.
" "	Harlan, Fauna Amer. p. 190.
" "	Audubon and Bachm., Quadr. of N. Amer. i. p. 277.
Hystrix pilosus, Americanus.	Catesby, Carol. App. p. 30.
" *Hudsonius.*	Brisson, Règn. Anim. p. 182.
" *pilosus.*	Richards., Fauna Boreali Americana, p. 214.
Erethizon dorsatum.	F. Cuvier, Mém. du Mus. ix. p. 432, Tab. 20.
" "	Brandt, Mamm. Exot. p. 51.
" "	Wagn. Schreb. Säugth. Supp. iv. p. 27.
Canada Porcupine.	Forster, Phil. Trans. lxii. p. 374.
" "	Pennant, Quadr. ii. p. 126; Arct. Zool. i. p. 109.
L'Urson.	Buffon, Hist. Nat. xii. p. 426, Tab. 55.

Fur dark brown, the large and coarse hairs usually with long dirty white points; spines white, with the point dusky, or brown; stiff hairs on the under surface of the tail, dusky, or brown.

The Canada Porcupine is about two feet in length, has short legs and tail, and a stout body; the stout appearance of the body, however, in a great measure arises from the long and dense fur with which it is clothed. The fur is composed of hairs of two kinds; one, which is much more plentiful than the other, is of a soft and somewhat woolly nature; this is long; but the other kind of hair, which is coarse, and almost bristly in parts, is very long, measuring often six or seven inches in length on the back of the animal. The shorter hairs are sometimes brown, sometimes dark brown, and I have seen specimens in which it was almost black, but in all cases it is greyish next the skin. The longer hairs are also dark, but they usually terminate in long dirty white points; and, being tolerably abundant, these pale points give to the general hue as light greyish tint. In some specimens which have come under my notice, the long hairs were most of them of an uniform dark hue; those near the head, and on the hinder part of the back, only, having pale points. The spines, which are hidden by the long fur, are found on the whole of the upper parts of the animal in great abundance; commencing on the crown of the head; they extend downwards towards the cheeks, behind the eyes, and they cover likewise the sides of the body, and part of the outer surface of the hind legs, as well as the upper surface of the tail. Those on the head and fore parts of the body are slender; on the middle of the back they are about two and half inches in length, and about $\frac{1}{24}$th of an inch in diameter. On the hinder part of the back they are stouter, many of them one line in diameter. All the spines are white at the base, and nearly all of them are dark at the point; on the fore parts of the body they have the terminal half (or often less) brown —frequently yellow-brown—especially in the spines which spring from the region of the shoulders: on the hinder part of the back they are dusky or black at the point. In many of the spines of this part the dark hue is considerably extended, but in some it is confined to the extreme point. The muzzle is clothed with short hairs, if we except the usual long bristly hairs which spring from the sides, and which are here black. On the cheeks there is an admixture of spiny hairs. The fur on the under parts of the body is rather short; brown, without any admixture of white. On the limbs there is usually an abundance of coarse, white-pointed hairs, especially on the hinder and outer part of the hind legs. The feet are clothed for the most part with long and coarse dark hairs. The tail is covered on the upper surface, at the base, with spines and hairs like those on the body; on the sides with long and bristly dirty white hairs, and the under surface is densely covered with stiff bristles, which are short near the root of the organ, but gradually increase in length towards the opposite extremity; often these bristles are black, but sometimes they are brown. The ears are well clothed with dark hairs, and are hidden by the long fur of the surrounding parts.

Usually the spines on the head are but imperfectly hidden by the fur, and generally the spines on the hinder part of the back are but partially concealed, the fur being comparatively scanty at this part.

A specimen which lived for some time in the Gardens of the Zoological Society, and which is now in that Society's Museum, lost nearly all the fur from the spiny parts of the body.

A young Canada Porcupine in the British Museum differs from others which have come under my notice, in having the muzzle, the under parts of the body, inner surface of the limbs, and the feet, rufous-brown, the other parts being brown-black: a few white pointed hairs are seen near the head and on the hinder part of the back.

The specimens contained in the British Museum, and in the museum of the Zoological Society, vary from thirteen to twenty-five inches in length, without including the tail, which is equal to about one-third of the length of the head and body. I subjoin the dimensions of the Canada Porcupine given by Dr. Richardson and Dr. Bachman.

	From Dr. B.		From Dr. R.	
	Inches.	Lines.	Inches.	Lines.
Length of head and body	29	0	30	0
" of tail (vertebræ)	7	0	8	0
" of ditto to end of hair ...	8	6		
From heel to longest nail	3	6		

According to Dr. Richardson, the Canada Porcupine has been met with on the banks of the Mackenzie as far north as lat. 67°. Dr. Bachman informs us that the animal is rare in the northern and eastern parts of the United States; but is found, however, in the northern and western parts of New York, and is said to be increasing in numbers in some of the western counties of that state. Dr. Leonard obtained specimens from the mountains of Vermont. It occurs sparingly in the northern parts of Pennsylvania, and in a few localities in Ohio; and specimens have been obtained by Messrs. Audubon and Bachman from the Upper Missouri.

The Canada Porcupine climbs trees with facility, but, whether on a tree or upon the ground, it is slow in its movements. Although from this circumstance it cannot escape from its enemies by flight, it is not attacked with impunity. It is stated by Messrs. Audubon and Bachman, that a Canada Lynx was found in the woods in a dying state, owing to its mouth being filled with the quills of this Porcupine, and that they have heard of many Dogs, some Wolves, and at least one Panther, which had been found dead in consequence of the inflammation produced by these quills. The quills are loosely attached to the skin of the animal, and being barbed at the point, they readily fix themselves to whatever they penetrate, and if not removed, they, by the slightest motion, become more and more deeply inserted.

Dr. Bachman, who kept one of these animals in confinement for some time, relates, that upon one occasion it made its escape into his garden, and when there, was attacked by a large mastiff belonging to a neighbour. At the moment of attack, the Porcupine, by raising its spines and long fur, appeared suddenly to have become twice its usual bulk: it gave the dog a lash with its tail, by a sharp lateral movement of that organ; the dog immediately retracted, and commenced howling; its nose, mouth, and tongue, so beset with the Porcupine's spines, that it could not close its jaws. These spines were immediately extracted, but the dog's head was very much swollen for some weeks afterwards, and it was two months before it fully recovered.

The food of the Canada Porcupine consists of various vegetable substances, and, in the winter months, this animal subsists chiefly, if not entirely, upon the bark of various kinds of trees. According to several accounts, a single Porcupine will, by stripping them of their bark, kill an immense number of trees in the course of the winter. It seldom quits a tree until all the bark is eaten, and hence remains in the same tree for a long time—sometimes for a week; at the end of which time it will most probably commence its attack upon a second tree near the first. This fact is so well known, that a person desirous of obtaining one of these animals which he may have seen in a certain spot, will return to that spot after a considerable lapse of time with the expectation of finding the animal not far distant. As the animal is usually found on the branches, it is often necessary to cut down the trees to procure it: when approached, it is said to make a crying noise, like that of a child. The female Porcupine brings forth generally two young at a birth; but three, and even four young, have been found in the nest: this is placed in the hollow of a tree.

Erethizon epixanthus.

Erethizon epixanthus. BRANDT, Mamm. Exot. p. 55, Pl. 1 (animal); and Pl. 9, figs. 1—3 (skull).

Fur brown or brown-black, whitish next the skin; the longer and coarser hairs brownish yellow at the point: spines white, or yellowish at the base, and most of them brown-black, or dusky at the apex.

Inhabits the west coast of North America—California, Unalaschka, and Sitka.

Five specimens of an Erethizon from the west coast of North America, in the museum at St. Petersburgh, having the exposed ends of the longer hairs of the fur of a brownish yellow colour instead of white, as the same hairs are stated to be in the *E. dorsatus*, M. Brandt is inclined to suppose there are two species of *Erethizon*, but not having specimens of the Canada animal for comparison, he is not able to satisfy himself upon this point. The specimens examined by M. Brandt are from California and Unalaschka, and I may add that a similar animal is found at Sitka, a specimen from that part, which I saw in the Leyden Museum, agreeing with M. Brandt's description: its spines were, most of them, of a delicate yellow below the dark points.

PHILOGÆÆ, Brandt; or HYSTRICINA proper.

The more I examine the members of the Porcupine group, the more satisfied I am of the distinctness of the sections into which they have been subdivided: in fact, that we should follow Brandt in regarding these sections as sub-families: and I think it would be well to confine the name *Hystricina* to the Old World species, and (following the ordinary practice of taking the sub-family names from the typical genus of the sub-family) to adopt Mr. Gray's name, *Cercolabina*, for the New World section.

The principal distinctions of these two sections have already been noticed, and I have only to add, that the true *Hystricina* are further distinguished from the *Cercolabina*, by the upper lip being divided by a vertical groove which runs up to join a transverse fissure which unites the nostril-openings, as seen in fig. 7 of Plate 6 *a*[1].

[1] The difference will be seen upon comparing this figure with the fig. 6 *a*, on the same plate, which represents the terminal portion of the muzzle in one of the *Cercolabina*. Ordinarily, in the Common Porcupine and its allies, we

M. F. Cuvier divides the Old World Porcupines into two genera,—*Hystrix*, and *Acanthion*[1]; the former genus having for type the Italian Porcupine, and the latter being founded upon the *H. Javanica*. The genus *Acanthion* is not adopted by the Baron Cuvier, but in his second edition of the *Règne Animal* we find the sectional name *Atherura* (les *Atherures*) proposed for the long-tailed species named *Hystrix fasciculata* by Linnæus. Mr. Gray, in a paper upon these animals recently communicated to the Zoological Society[2], adopts the three divisions formed by the Cuviers, and adds a fourth sectional name to the Old World Porcupines, separating the *Histrix Javanica* (but only as a sub-genus) from other species of the genus *Acanthion*, and applying to this section the name *Acantherium*.

Setting aside the section *Atherura*, which I adopt, I must remark, with respect to the species included in the remaining three sub-divisions, that one of them—the *Histrix cristata* of Schreber—is pre-eminently distinguished amongst Rodents by the enormous size of the nasal cavity of its skull; and, as this cavity is enclosed above by the nasal bones, these bones are proportionately large; whilst the other species show a gradual decrease in the size of the cavity in question, and consequently of the nasal bones. This will be seen upon referring to Plate 20[3], in which the skulls of the animals under consideration are represented. It will further

perceive but a narrow vertical fissure in the upper lip, and a curved groove, marking the termination of the nose, but these grooves are formed by folds of the integument, which the animal has the power of separating to a certain extent (as represented in fig. 7), and it is only when thus separated that we can see the boundaries of the nostril-opening. The skin in these folds is hairless.

[1] Mémoires du Muséum, ix. p. 424.

[2] Proceedings for June, 1847; and Annals and Magazine of Natural History for November, 1847, vol. xx. p. 349.

[3] The nasal bones are marked *n* on the figures, which were drawn by myself assisted with the camera lucida (as are the greater portion of the skulls figured in this work), and are reduced to half the natural size.

be observed, upon inspecting this plate, that the intermaxillary bones (*int. m.* on the figures) vary considerably in form in the different crania, and especially in being more or less broad, or more or less contracted at their hinder extremity; and, lastly, it will be seen that there are considerable differences in the relative size of the parietal bones (*p.*) Of these skulls M. F. Cuvier was acquainted only with the two marked figs. 1 and 4 on the Plate. Fig. 1 exhibits the characters of the genus *Hystrix* according to this author, and fig. 2 those of his genus *Acanthion*, which he distinguishes as having the cranium less arched, the nasal bones much smaller, and the parietal bones more extended, as compared with fig. 1. Mr. Gray, giving most importance to the form of the intermaxillary bone, arranges together under the sectional name *Acanthion*, the species in which that bone is contracted behind, including the animals of which the crania are represented by figs. 1, 3, and 4, on Pl. 20; and places in the genus *Hystrix*, the *H. leucurus* (skull, fig. 2), distinguishing this genus by the intermaxillaries being truncated behind, and as broad at that part as in front. The two sub-divisions of the genus *Acanthion* are thus characterized—

* "The nasal very long, broad to the middle of the orbit"—ACANTHION proper.

** "Skull narrower in front. Nasal bone moderate to the front edge of the orbits. Malar bone with an obtuse post-orbital process"—ACANTHERIUM.

For further details respecting Mr. Gray's views upon the subdivisions of the Old World Porcupines, I must refer the reader to the paper already quoted. Mr. Gray notices a difference in the form of the molars, and certain differences in the form of the quills or spines, together with the absence or presence of a crest on the back of the neck, as helping to distinguish the genera *Acanthion* and *Hystrix;* but these differences do not exist in all cases, since Mr. Gray's *Hystrix leucurus*, and his *Acanthion Cuvieri*, do not present such points of distinction.

With respect to the distinctions derived from the structure of the cranium, I cannot but regard them as simply specific. Viewing them with reference to M. F. Cuvier's definition, we find the skulls, figs. 2 and 3, presenting intermediate conditions; and, in adopting Mr. Gray's views, we must associate animals which, although they agree in having the hinder portion of the intermaxillary bone contracted, present the greatest extremes in other cranial characters, and which, in fact, present the types of the two divisions previously founded by M. F. Cuvier. Thus—

Gen.—Acanthion, *Gray*, is = Hystrix, and Acanthion, *F. Cuvier*.
Sub-gen.—Acantherium, *Gray*, is = Acanthion, *F. Cuvier*.
Gen.—Hystrix, *Gray*, is founded upon an animal unknown to *F. Cuvier*.

Genus, *Hystrix*.

Hystrix. Linn., Syst. Nat. 12 ed. p. 76.

Hystricina with a short tail: the hinder part of the back armed with long cylindrical spines or quills.

Geogr. range: the southern portions of Europe, the middle and southern portions of Asia, and perhaps the whole of Africa.

The molar teeth in the species of *Hystrix* (Pl. 20, fig. 2 *b*) are arranged in parallel series, and their crowns are of an oval or round form; they are deeply implanted in the jaws, and divide at the base into very short roots, in most of them three in number. Each upper molar has a strongly indenting fold of enamel on the inner side, and when moderately worn, the crown presents three or four areas surrounded by enamel; the hindermost is very small and nearly round; the remaining three are narrow, placed transversely, and more or less curved, having the convex side of the curve directed to the front of the tooth: of these three, the hindermost is the longest and most curved; they all entered originally from the outer side of the tooth. In a much worn molar the inner fold becomes isolated from the inner side of the tooth, and the crown then presents five narrow areas surrounded by enamel. When but little worn, the internal enamel-fold is united with a fold from the outer side, so as to form a separate lobe of the anterior portion of the tooth, as in the uppermost of the four teeth figured[1]. The lower molars resemble the upper, excepting that the corresponding enamel-folds are reversed in position. The incisor teeth are moderately large, rather deeper than broad, and nearly flat in front; the upper and lower pairs are of equal width.

Both fore and hind feet are provided with five toes, but the inner toe of the fore foot is extremely short, and armed with a small blunt nail; the remaining toes have strong and moderately curved claws. The soles of the feet are naked and smooth beneath; there are three large fleshy pads at the base of the toes of the fore foot, and two others (wrist pads) behind these. On the sole of the hind foot there are four pads at the base of the toes, and two on the hinder half of the foot, one of which (on the outer side of the sole) is very long, extending from the middle of the sole to the heel.

I am acquainted with four well-marked species of Hystrix, two of which have a crest composed of very long bristles, and springing from the crown of the head and from the back of the neck; the remaining two have no crest.

Sect. 1. *Head and back of neck with a crest of long bristles.*

[1] The upper tooth in the figure 2 *b*, is the front upper molar, and is the last tooth in the jaw which is developed, its place being occupied until a late period by a deciduous molar, which, when present, is easily distinguished by its comparatively long and distinct roots: these may be seen without removing the tooth from the socket.

HYSTRIX CRISTATA.

The Crested, or Common Porcupine.

Hystrix cristata.	LINN., Syst. Nat. 12th ed. p. 76. 1766.
" "	SCHREB., Säugth. iv. p. 599, Pl. 167 (animal), and Pl. 166 (skull). 1792.
" "	F. CUVIER, Mém. du Mus. ix. pp. 424 and 430, Tab. 20 bis, figs. 1 and 2 (skull); Dict. des Sci. Nat. xlii. p. 528 (1826); Mammif. livr. 34.
" "	Is. GEOFF., Dict. Class. d'Hist. Nat. xiv. p. 213. 1828.
" "	BRANDT, Mém. de l'Acad. de St.-Petérsb. (1835), p. 371. Tab. 8, figs. 1 and 2 (skull), and Tab. 6, figs. 1, 4, 5, 6, and 7 (ear, feet, &c.) 1835; Mamm. Exot. p. 35, (same Plates).
" "	WAGNER, Archiv für Naturgesch. (1842), Pt. 1, p. 29; Schreb. Säugth. Suppl. iv. p. 17. 1844.
Acanthion Cuvieri.	GRAY, Proc. of the Zool. Soc. for June, 1847; Ann. and Mag. of Nat. Hist. for Nov. 1847, vol. 20, p. 353; and vol. 1 of Second Series, p. 246.

Muzzle rather sparingly clothed with small hairs; the longer quills of the back usually with about one-fifth of their entire

length, being the terminal portion, entirely white: bristles of the crest most of them dusky, with long white points, and some of them entirely dirty white.

Skull **(Pl. 20, figs. 1 and 1 *a*) with the upper surface strongly arched, the nasal bones enormously large, terminating posteriorly nearly in a line with the posterior root of the zygoma; together, their outline approaches to an oval form; intermaxillary bones very broad in front, and much contracted behind, where they are about half as wide as in front.**

Inhabits Italy and the northern and western portions of Africa.

The following description is taken from a specimen recently added to the British Museum collection, and which is from the Gambia. Its skull[1], which has been removed from the skin, although of very large size, does not display the fully adult condition of the teeth, the last molar in each series being but imperfectly developed.

The length of the animal, measured in a straight line from the tip of the nose to the root of the tail, is 28 inches. Its large and obtuse muzzle is but sparingly clothed with small hairs, and these of a dusky hue; on the upper lip, are numerous scattered, longer, and coarser hairs. The fore parts of the animal, as well as the limbs and under parts, are covered by spines, mingled with which—more especially on the under parts of the body—are some coarse hairs. The spines on the shoulders are most of them rather less than two inches in length; they are inserted into the skin by a slender stalk, above which they rather rapidly increase in size, so as to attain their greatest diameter at about the termination of the basal third of their length; from this part they become gradually narrower towards the point, but at a short distance from the point they are more suddenly contracted, so that the point itself (for the length of about two lines) is very slender, and somewhat flexible. The surface of these spines presents

[1] **It is the skull of this specimen which I have figured in Plate 20, fig. 1.**

several delicate longitudinal channels, and when cut transversely, their section approaches to a square figure, but is rather broader than deep. Excepting some slight variations in size, this description will apply tolerably well to all the spines found on different parts of the animal.

The crest is composed of immensely long bristles, very stout at the base, but tapering somewhat towards the opposite extremity; they are very abundant, and curved gently backwards: the crest commences on the crown of the head, extends along the back of the neck, and for a short distance on the back. Many of the bristles are as much as sixteen inches in length. The whole of the hinder half of the body (if we except the under parts), as well as the tail, is covered by quills. Of the quills there are three principal varieties: one kind is distinguished by their great length (from 12 to 16 inches), being comparatively slender, somewhat flexible, and varying scarcely in diameter for the greater portion of their length. These are most forward in position, and, excepting when erected, they almost entirely hide the next kind of quills to be noticed. This second kind of quills is much stouter and shorter than the first; they taper from near the base to the point, which is very sharp, and they are very stiff; the longest of these are about 6 inches in length, and 3 lines in diameter, at the thickest part. They cover the hinder third of the back, and the basal portion of the tail, but along the mesial line of the rump, it must be observed that, though the spines are stiff, and sharply pointed, they are much smaller than on other parts. Besides the stout quills in the region of the tail, there are a few very long and slender quills in these parts. The third kind of quill is very unlike the other two varieties; they spring from the tip of the tail, are hollow, most of them open and truncated at the end, and each one is supported upon a very slender stalk, usually about half an inch in length; their average diameter is about a quarter of an inch, and most of them present an oval section; some are pointed, and no doubt all are so at first. Usually they are about twenty in number. The feet are covered with very long and coarse hairs. The bristles springing from the sides of the muzzle are very long, being many of them seven or eight inches in length.

The prevailing colour of the animal is brown-black; a white band crosses the fore part of the neck, and extends about half way up the sides, becoming gradually narrower from the middle. The stout bristles forming the crest are brownish, but have very long white points; some of them are entirely white: the spines covering the shoulders, and most other parts of the animal not protected by the quills, are grey-brown at the base, and dusky or brown-black at the point: the feet are nearly pure black. The quills are most of them broadly annulated with black and white—white at the root and point. The long and slender quills, which cover the stouter kind, have moderately long white points. The quills on the rump—more especially in the mesial line—are chiefly black; those at the end of the tail, as well as nearly all the other quills springing from this organ, are entirely white; some few of the quills near the root of the tail are partly black and partly white.

	Inches.	Lines.	Inches.	Lines.
From tip of nose to root of tail ...	28	0	26	0
From ditto to ear	5	0	5	6
Height of free upper portion of ear		9		10
Total height of ear	1	8	1	10
Length of fore foot and nails ...	3	2	3	4
" of the longest nail		8		
" of hind foot and nails ...	4	3	4	2
" of longest nail		8		
" of tail, not including the quills	6	0		
" of one of the hollow quills at the end of ditto ...	2	6	2	6
" of which the slender stalk is		9		6
Diameter of the expanded hollow portion		3		5

The dimensions in the second column are from a specimen in the museum of the Zoological Society, and which is from Tunis: it differs from the Gambia specimen, in having most of the short quills on the mesial line of the rump, near the root of the tail, entirely white; and in having the hollow quills at the end of the tail much broader.

The leading peculiarities of the skull of the Common Porcupine have already been noticed: it is readily distinguished from the crania of other species of the genus, by the enormous size of the nasal bones[1], and by certain other differences which are noticed in the account of the next species. Subjoined are the dimensions of the skull of the specimen from which the foregoing description is taken (see column No. 1); those of a skull, in the museum of the Zoological Society, upon which Mr. Gray founded his *Acanthion Cuvieri*[2] (col. No. 2); in column No. 3 are the

[1] Although always to be distinguished by the great development of the nasal cavity, the large nasal bones, and the highly arched upper surface, the cranium of the Common Porcupine is subject to considerable variations in these parts, as will be seen upon comparing the dimensions. As a general rule, where any species is characterized by a maximum of development of certain parts, those parts are more subject to variation in the different individuals of the species than are parts which approach more nearly to the normal conditions.

[2] Mr. Gray admits that the animal described by me under the name *Hystrix cristata* is specifically identical with his *Acanthion Cuvieri*, but he gives a new specific name to this animal, because he supposes another species to be the true *H. cristata* of Linnæus. We have not the means of determining to which of the crested Porcupines Linnæus applied the name *cristata*: he says his animal is found in Africa and Asia, and it is more probable that he had the Common North African animal in view than the Asiatic species, which is now known to be distinct: be this as it may, when two species are confounded together, as have been the African and Asiatic, crested Porcupines, we should, to avoid confusion, adopt the names for the two species, as applied by the person who first clearly points out their distinctions. The most striking differences in the two animals are found in the structure of the skull; and, judging from the figure of the skull given by Schreber (more than fifty years back,—see his Säugth. Pl. 166), we cannot doubt that his *Hystrix cristata* is identical with the animal so named by F. Cuvier and Brandt (who have also figured the skull),

admeasurements of a skull from Tunis; it has but three molars, and belongs, therefore, to a young animal; and, lastly, the admeasurements of a skull of the *H. cristata*, described by Dr. Wagner (col. No. 4).

	No. 1.		No. 2.		No. 3.		No. 4.	
	Ins.	Lines.	Ins.	Lines.	Ins.	Lines	Ins.	Lines.
Total length of skull	6	4	5	7	5	2	5	3
Width of ditto from the outer surface of the zygomatic arches...	3	4	3	2	2	10		
" " between orbits ...	3	0	2	5	2	5	2	6
" of palate between second pair of molars		6				5⅓		
Length of nasal bones	3	9	3	0	2	11½	3	7
Greatest width of ditto, behind...	2	9	2	1½	1	0⅓		*
" " in front...	1	7	1	3	1	3		
Length of frontal bones, measured in the middle line of the skull	1	2½			1	0	1	0
" of parietal bones, ditto, to occipital crest ...	1	1			1	2	1	1
Width of intermaxillary in front (from the posterior outer edge of the alveolus of incisor, back to the intermaxillary suture, opposite anterior root of zygoma)	1	4½	1	1½	1	0⅔		
" of do. measured behind ...		3½		4⅓		3¾		4½
Vertical diameter of ant-orbital opening		5½		6		5		
Length of lower jaw	4	2	3	11			3	0
Height of ditto, measured from the condyle	1	5	1	3				

Wagner, and myself. Brandt first clearly defined the two species, and Wagner subsequently confirmed all the essential points of distinction noticed by him in a paper (evidently embodying the results of a careful examination) published in the Archiv für Naturgeschichte.

* Dr. Wagner gives as the width of one of the nasal bones on the hinder part, 1″ 5‴; in the skulls measured by myself, the same part gives 1″ 8‴ in No. 1; 1″ 1½‴ in No. 2; and 1″ 1⅓‴ in No. 3.

The Common Porcupine inhabits Italy (in the Appenines) Sicily, and Spain, to which parts it is supposed not to be indigenous, its true habitat being the northern parts of Africa. Dr. Wagner informs us that he has received a specimen of this animal from the Upper Nile; and on the western side of the continent we are certain that it extends as far as the Gambia. Mr. Fraser has recently brought to England living specimens from the neighbourhood of Tunis, and a skull, which accompanied them (now in the British Museum), I find agrees very closely with that figured in Pl. 20. This animal seeks its retreat in holes in rocks, or it burrows in the ground for shelter. Its food consists of various kinds of vegetable substances, and its flesh is well flavoured. When irritated, to defend itself it erects its formidable quills; and the long crest on the back of the neck is at the same time elevated. It emits a succession of short grunts, and, moreover, it makes a rattling noise with the tuft of hollow quills at the end of its tail, by rapidly vibrating that organ. Further, it will strike the ground with its feet, as do the Hares.

HYSTRIX HIRSUTIROSTRIS.

The Asiatic Porcupine.

Hystrix hirsutirostris. BRANDT, Mamm. Exot., &c. p. 39, Pl. 8, figs. 3—6 (skull and lower jaw). 1835.

" " WAGNER, in Archiv für Naturgesch. 1842, p. 29; Schreb. Säugth. Suppl. iv. p. 17. 1844.

" *cristata* et *leucurus.* GRAY, Proc. Zool. Soc., June 1847; Annals and Magaz. of Nat. Hist. for Nov. 1847, vol. 20, pp. 350 and 351.

" *leucurus.* SYKES, Proc. Zool. Soc., July 1831, p. 103.

Bristles of the crest black-brown, with but little admixture of white; muzzle densely clothed with hair; short quills along the mesial line of the rump, white; longest quills of the back with the white much extended on the terminal portion hollow quills of the tail very broad.

Skull with the nasal bones of moderate width, the lateral margins parallel, or nearly so, the hinder margin terminating in a line with the anterior boundary of the orbit; intermaxillary bones truncated behind, as broad at that part as in front.—See Pl. 20, figs. 2 and 2 *a*.

Inhabits continental India, and extends westwards to Syria and Lycia.

Although, when the crania of the African and Asiatic species of Crested Porcupines are compared, many well-marked differences present themselves, yet in the external characters of the two animals—in the colouring, in the quills and spines, or in size and proportions—the differences are by no means striking. Colonel Sykes, many years back, expressed an opinion that they were distinct, and pointed out certain characters which he imagined might serve to separate them, but the characters referred to are none of them distinctive[1]. Judging from the cranium, the head of the *H. hirsutirostris* must be less arched, and the muzzle less obtuse, than in the *H. cristata;* but these differences would not be obvious in stuffed specimens: the chief distinguishing character pointed out by Brandt is, that the muzzle in the

[1] The distinction on which Col. Sykes lays most stress is that the quills of the tail of the Indian animal are all of them white; whilst in the *H. cristata* such, he says, is not the case: and hence he proposed the specific name *leucurus* for the Indian species. According to my observations, all the quills on the terminal and under parts of the tail are white in both species; at the basal portion of the tail in the Indian Porcupine, nearly all the quills are white, but some few are party-coloured, whilst in *H. cristata* sometimes but few of the quills at the same part are white, the remainder being party-coloured, and sometimes many are party-coloured. Had Col. Sykes discovered any good points of distinction between the two animals, it would perhaps have been desirable, on the score of priority, to have adopted his specific name, though not very appropriate; but as he failed to do so, I adopt the *appropriate* name given by Brandt, who has clearly defined the species.

present animal is densely clothed with short hairs, whilst in the *H. cristata* the same part is but sparingly clothed. Beyond this, Brandt observes that the quills in the lumbar region are entirely white in *H. hirsutirostris*, and chiefly dusky in the *H. cristata*, and that the hollow quills at the end of the tail are broader in the former animal than in the latter. Dr. Wagner has noticed similar differences in other specimens of the two species; and he adds that the bristles of the crest, in the *H. cristata*, have very long white points, whilst in *H. hirsutirostris* they are almost entirely brown, and that the long quills of the back have the white more extended at the point in the latter than in the former animal.

All the above differences distinguish a specimen of the Indian Porcupine (from the Deccan) presented to the British Museum by Col. Sykes, from the Gambia specimen, which I have described as the *H. cristata*. A specimen from Nepal, in the same Museum, has the tail imperfect, wanting the terminal portion, and hence the terminal quills cannot be compared; but, except in being smaller (probably young) than Col. Sykes's specimen, it agrees in all essential points. The Deccan specimen has the foremost quills of the back rather more elongated, and more slender than the corresponding quills in the Gambia animal[1]. The hollow quills at the end of the tail in the latter are, most of them, about 3‴ in diameter, whilst those in the Indian animal are 4½‴. Two of the supposed peculiarities of the *H. hirsutirostris*, however, I have reason to believe will not always serve to distinguish that animal from the *H. cristata*; for in the Zoological Society is a specimen from Tunis, which agrees with *H. cristata* in other respects, but in which the hollow quills

[1] The terminal half, or nearly as much, of each of these quills in the Deccan specimen is white, and their average length is about 15 inches; the average length of the longer quills in the specimen of *H. cristata* described, is about 13 inches, and the white at the end occupies about ⅓rd of the entire length.

at the end of the tail are even broader than in the specimens of *H. hirsutirostris* which have been noticed, and this same Tunis animal has the short quills on the mesial line of the rump entirely white—at least those near the root of the tail; at a short distance from this point they are party-coloured, being about half black and half white. The skull of this specimen cannot be examined, but this can scarcely allow us to doubt the species, since I have the skull of a second specimen from Tunis before me, and that presents all the characteristics of the skull of the *H. cristata.*

	From the Deccan.		*Hystrix hirsutir.* Brandt.		*Hystrix hirsutir.* Wagner.	
	Ins.	Lines.	Ins.	Lines.	Ins.	Lines.
Total length of the animal	28	0	29	0	31	0
Of which the tail and quills are about ...	5	0				

Subjoined are the admeasurements of various crania, which appear to me to be referrible to the present species. They are at once distinguished from the crania of other species of *Hystrix*, by the broad and truncated frontal, or nasal, process of the intermaxillary bone, This bone is here nearly as wide behind as in front; whilst in all other species the fore part of the intermaxillary (measuring from the edge of the alveolus of the incisor, back to the intermaxillary suture, opposite the anterior root of the zygoma) is more than twice as broad as the hinder part, and sometimes equals four times the width of that part. Compared with the skull of *H. cristata*, the skull of *H. hirsutirostris* is somewhat narrower, and distinctly less arched above; the nasal bones are smaller, and differ, moreover, in having the sides nearly parallel, the hinder margin produced but little beyond the apex of the intermaxillary, and terminating nearly in a line with the anterior edge of the orbit: the frontal bones are much more extended in the longitudinal direction of the cranium, and they join the parietals behind by a distinctly serrated suture; in *H. cristata* there is an indentation between the frontal and parietal bones, and the suture is scarcely serrated: the palate is narrower, and the sphenoid alæ less widely separated. The lower jaw is distinctly deeper[1].

[1] All the above peculiarities apply equally well to two skulls which Mr. Gray regards as belonging to a distinct species, and which that author supposes is the true *H. cristata* of Linnæus. I cannot think, however, that certain differences noticed in the skulls in question, when compared with the skulls of *H. hirsutirostris*, are sufficiently important to indicate a distinction of species. One of these skulls belonged to an adult animal, which died in the Surrey Zoological Gardens, and its habitat is uncertain; the other belongs to a very young animal, in which but two molar teeth of each series are developed: it was brought from Xanthus. The adult skull differs from the skulls of *H. hirsutirostris* brought from India, in being rather narrower; but the chief distinction consists in the frontal process being considerably broader behind than the nasal bone. The young skull from Xanthus, on the other hand, is distinctly broader than the Indian skulls, and the intermaxillary is but little broader than the nasal bone. A cranium described by Dr. Wagner, which was removed from a Porcupine from Jerusalem, has the nasal bone and the intermaxillary equal in width. In the skulls of *H. hirsutirostris*, I may observe, the hinder part of the intermaxillary bone is usually about equal in width to the same part of one of the nasal bones, but sometimes the intermaxillary is somewhat distinctly broader than the nasal bone. The following admeasurements of the two parts in question in several crania will give an accurate idea of the amount of difference which exists between Mr. Gray's *H. cristata* and the *H. hirsutirostris*:—

	Intermaxillary.	Nasal bone.	
	Lines.	Lines.	
Specimen from the Surrey Gardens ...	$14\frac{1}{3}$	9	*H. cristata.* GRAY.
" from Xanthus	$9\frac{1}{3}$	8	
" from Jerusalem	8	8	*H. hirsutir.* WAGNER.
" in the Zoological Society's museum	10	8	
" from Continental India ...	11	9	
" from ditto			
" from Nepal	10	10	

In a skeleton from Nepal, presented to the British Museum by Mr. Hodgson, I find one more dorsal vertebra (and consequently an extra pair of ribs) than exists in the skeleton of the Gambia specimen of *H. cristata*, there being fifteen dorsal vertebræ in the former, and fourteen in the latter. This Nepal skeleton is remarkable for having eight cervical vertebræ; the eighth, having the transverse process perforated like the preceding vertebræ. There are four lumbar vertebræ in both the skeletons referred to.

We are certain that the *Hystrix hirsutirostris* is found in the central parts of Hindoostan; that it occurs in the upper, central, and lower parts of Nepal, we are informed by Mr. Hodgson; and that it is a native of Syria (in the neighbourhood of Jerusalem) we learn from Dr. Wagner. As travellers speak of "the Porcupine" as being found in Persia, and Dr. Griffith states that it occurs in the lower parts of Afghanistan[1], it is highly probable that it is the *H. hirsutirostris* which inhabits these intervening countries. Whether it be this species or the *H. cristata* which inhabits the Cape of Good Hope, I cannot determine; the only specimen from the Cape which I have seen is in so very bad a condition, that I will not venture to give an opinion.

[1] See Journal of the Asiatic Society, vol. x.

	Continental India.[1]						*H. cristata.* Gray.		Nepal.		Nepal.		Specimen in Zoological Soc.'s Museum.		*H. hirsutirostris.* Wagner. Jerusalem.	
	Ins.	Lines.	Ins.	Lines.	Ins.	Lines.	Ins.	Lines.	Ins.	Lines.	Ins.	Lines.	Ins.	Lines.	Ins.	Lines.
Total length of skull	5	9	5	10	6	1½	5	11	5	5	5	3	4	11	4	11
Width of ditto	3	5	3	4	3	5	3	0	2	11	3	0	2	8½	2	9
" between orbits	2	3½	2	6½	2	11½	2	4	2	2	2	1½	2	2	2	1½
Length of nasal bones	2	9½	2	11	3	2	2	8	2	6	2	3	2	2	2	3
Width of ditto, behind...	1	6⅓	1	9	1	7½	1	7½	1	6¾	1	6	1	3		
" in front	1	3¼	1	6	1	4	1	4½	1	4¼	1	2¾	1	1¼		
Length of frontal bones	1	8	1	9½	1	8½	1	10			1	8½	1	8	1	9
Width of intermaxillary bone behind		11		11½		11½	1	2½		9¾		10		10		8
" " in front		11½	1	1	1	0½	1	0⅓	1	0¼		11½		10		
" of palate between the foremost molars		4⅓		4½		4⅓		3⅔		4		4⅓				
" " between the hindermost		6¼		6¼		6		5¾		6		6⅔				
Vertical diameter of ant-orbital opening		7½		6¾		7		6		5		5½		5		
Length of lower jaw	4	1			4	3	4	0	3	9	3	10			3	5
Height of ditto	1	8			1	9½	1	9½	1	6½	1	8				

[1] One of these skulls has five molar teeth on one side of the upper jaw.

SECT. 2. *Head and back of neck destitute of a crest.*

HYSTRIX HODGSONI.

Crestless Nepal Porcupine.

Hystrix Hodgsonii. GRAY, Proc. Zool. Soc. for June, 1847; Annals and Mag. of Nat. Hist. for Nov. 1847, vol. xx p. 352.
" *alophus.* HODGS., Journ. of the Asiatic Soc. of Calcutta for Aug. 1847, p. 772, Pl. 32.

General hue brown-black; head brown; in the visible portions of the quills the black prevails over the white, but numerous very long and slender quills are almost entirely white: the spiny bristles, which form the chief covering of the animal, are rather long and slender, and have long hair-like points. —Skull, Pl. 20, fig. 3.

Inhabits Nepal.

This animal is easily distinguished from the *H. cristata* and *H. hirsutirostris* by the total absence of a crest on the back of the neck and head: excepting the hinder portion of the back, and the tail, almost every part of the animal is covered with spiny bristles, and these are relatively longer than in the two preceding species of Porcupines: the longest (which are found immediately in front of the quills) measure, many of them, as much as three inches in length; they all terminate in an elongated hair-like point, are rather distinctly grooved on the outer surface, and have a nearly square section. Mixed with these spiny bristles are a few hairs, and on the muzzle and feet they are replaced by hairs; on the latter, the hairs are very coarse. Quills cover the whole hinder half of the back, and extend down the sides of the body to the base of the leg: the tail is also covered with quills. The quills are proportionately shorter than in the Common

Porcupine, and, as in that animal, there are shorter, and stouter, sharply-pointed quills, and others which are slender and flexible. The longest of the latter measure about ten inches in length; they are for the most part white; some are entirely white, and others are white with a black ring about midway between their extremities. Of the shorter spines, some have the basal half white, and the apical half black; others are white, but with a very broad black ring towards the point. The quills on the tail are most of them partly black and partly white, but those on the under surface are entirely white, as are also the hollow truncated quills at the end of the tail. The prevailing colour of the animal is brown-black; the head is brown; the white collar on the neck is very indistinct. The spiny bristles are white at the root. The muzzle is well clothed with hairs, and the ears are rather small. The whole length of the animal is 20 inches, of which the tail, with its quills, measures about 4 inches; the hind foot and nails is 3″ 1‴.

The above description is taken from a specimen presented to the British Museum by Mr. Hodgson, which specimen is likewise the original of Mr. Gray's description. It is not quite full-grown, but we are indebted to Mr. Hodgson for a detailed account of the peculiarities of this species, drawn up from an adult male specimen. According to this author, "the Crestless Porcupine" of the sub-Himalayas, measures from 22 to 24 inches, from the tip of the nose to the root of the tail; its tail (without including the quills) measures 4 inches; and its weight is from 16 to 21 lbs.

"The head, neck, fore half of the body, and entire belly and limbs, are covered with spinous bristles, which have a pretty uniform length of from 2 to 3 inches, but are shortest and feeblest on the head and limbs. The hinder part of the body, or croup, and tail, only, are armed with true quills, of which the longest thick ones are about 7 inches, and the longest thin ones, about 12 inches. The tail is conico-depressed, thick, and about one-sixth of the animal's length. Its longest thick quills are 3 to 4 inches; and its longest thin ones, 5 to 6 inches. The rattle at the end of the tail consists of from 35 to 40 hollow cylinders, of about an inch in length, some of which are closed, and some open, at the distant end. The skin of the body is pure white. The iris brown. The nudish lips, and soles of the feet, fleshy-brown. The spinous bristles black, save on the head, where they are less deep-hued, passing to brown. The white collar is very narrow and vague. The quills white, with *but one* sub-central black ring; those of the inferior surface (only) of the tail, being all white."

	Inches.	Lines.
From nose to root of tail	23	6
Length of tail	4	0
" of ditto, including the hollow quills ...	5	6
Mean height	8	0
Girth of chest	18	0
Length of fore foot and nails	2	7½
" of hind foot and nails	3	4½
Ear, from the lobe	1	3

"Porcupines (says Mr. Hodgson) are very numerous, and very mischievous, in the sub-Himalayas, where they depredate greatly among the potatoe, and other tuberous or edible rooted crops. They are most numerous in the central region, but are common to all three regions. They breed in spring, and usually produce two young about the time the crops begin to ripen. They are monogamous, the pair dwelling together in burrows of their own formation. Their flesh is delicious, like pork, but much more delicately flavoured, and they are easily tamed, so as to breed in confinement. All tribes and classes, even high-caste Hindoos, eat them, and it is deemed lucky to keep one or two alive in stables, where they are encouraged to breed. Royal stables are seldom without at least one of them. The Partbattiahs call them *Dumsi;* the crested one[1], *Chotia Dumsi;* the uncrested, *Anchotia*[2] *Dumsi.* The Lepchas and Lambus of Sikim do not distinguish the two species, but call them both *Sathung* and O′—é, in their respective languages."

The valuable series of crania and skeletons which accompanied the extensive collection of Nepal animals presented by Mr. Hodgson to the British Museum, contains two skulls of the *Hystrix Hodgsoni,* one of which is figured in Plate 20 (fig. 3) of half the natural size: they are readily distinguished from the second species found in Nepal—the *H. hirsutirostris*—by the nasal bone being rounded behind, and much more prolonged beyond the fronto-intermaxillary suture: here they terminate in a line with the middle of the orbit. Another very striking distinction is seen in the form of the frontal process of the intermaxillary, this process being exceedingly contracted, and almost terminating in a point in *H. Hodgsoni,* and broad and truncated in *H. hirsutirostris;* the frontal bones are much less extended in the longitudinal direction, and the parietal bones are more extended than in the last mentioned animal. The nasal cavity is much smaller than in the skull of *Hystrix cristata;* the skull is much narrower, and very little arched above; the parietal bones are considerably more extended in the longitudinal direction; the ant-orbital opening is larger; the palate narrower between the molar teeth, and the crowns of these teeth are very nearly round. Both skulls present the above distinctions; one of them is somewhat fractured; the admeasurements of the other are as follows:—

	Inches.	Lines.
Total length of skull	4	10
Width of ditto	2	6
" " between orbits	1	8

[1] *Hystrix hirsutirostris.*

[2] Achotia, meaning uncrested; and Chotia, crested.

Length of nasal bones	2	6⅓
Width of ditto, behind	1	4⅔
" " in front	1	0
Length of frontal bones, in the mesial line ...		10¼
Width of intermaxillary bone, behind ...		1⅔
" " in front ...		10
" of palate, between the second pair of molars		3
Vertical diameter of ant-orbital opening ...		5½
Length of lower jaw	3	3
Height of ditto, from condyle	1	3

HYSTRIX JAVANICA.

Java Porcupine.

Acanthion Javanicum. F. CUVIER, Mém. du Mus. ix. pp. 425 and 431, Tab. 20 bis. figs. 3 and 4 (skull), 1822; Dict. des Sci. Nat. xlii. p. 530 (1826).

" " VAN DER HOEVEN, Nov. Act. Bonn. xix. i. p. 182, Tab. 19, fig. 4 (lower jaw). 1839.

Hystrix torquata, and *Acanthion Javanicum.* VAN DER HOEVEN, Tijdschr. iii. p. 110. 1836.

" *fasciculata.* MÜLLER, Verhandl. over de Naturlijke Gesch. der Nederl. Overzeeschte Bezittingen, Zool. i. p. 36. 1839.

" *brevispinosa.* WAGNER, Schreb. Säugth. Suppl. iv. p. 20. 1844.

Acanthion Javanicum. GRAY. Proc. Zool. Soc. for June 1847; Ann. and Mag. of Nat. Hist. for Nov. 1847, vol. xx. p. 353.

Body for the most part covered with stiff and somewhat depressed spines, which are distinctly grooved on the outer surface, and have prickly points; hinder half of the back, and the tail, covered with quills; the quills rather short: those at the end of the tail hollow, and pedunculated: general tint brown (or, sometimes, black), freckled with dirty yellow; the stouter quills brownish yellow, with a sub-terminal, broad, dusky ring; the longer and more slender quills, with the exposed ends dirty yellow.—Skull Pl. 20, fig. 4.

Inhabits Java, Sumatra, and Borneo.

The Java Porcupine is considerably smaller than the

H. cristata and the *H. Hodgsonii,* and is not only distinguished from those species by the absence of the crest, but by the spines, which form the chief covering of the body, being stouter, and by their terminating in a prickly point. This peculiarity in the spines also distinguishes the present animal from the crestless Porcupine of Nepal. In the three preceding species the spines terminate in a flexible hair-like point, but in the *H. Javanica,* though the spines terminate in an acute point, that point does not yield to the touch. The spines, moreover, are more distinctly grooved on the outer surface than in the other species of Hystrix, the broad longitudinal furrow on that part being very evident to the naked eye: their average length is about $2\frac{1}{4}$ inches, and their greatest width is about $\frac{1}{14}$th of an inch. On the head, and under parts of the body, the points of the spines are thin, elongated, and flexible. The quills are shorter even than in the *H. Hodgsonii;* the average length of the stouter quills being about 3 inches, and that of the more slender quills varying from 6 to 8 inches: these latter spring from near the middle of the back, and some from the sides of the body. The hollow quills at the end of the tail, which, as in the preceding species, are supported upon a long and slender stalk, are many of them about an inch and a quarter in length, and about $1\frac{1}{2}$ lines in width. The muzzle and chin are clothed with hairs; the hairs of the moustaches are very long, some of them brown, and some black. The feet are covered with long black bristly hairs. Usually the prevailing hue of the animal is brown, but freckled with dirty yellow. The spines are of a dirty yellow-white at the root, and brown beyond; many of them, however, have a short pale point. The stouter quills are of a dirty pale yellow, but have a broad, subterminal, dusky ring, often about an inch in width, and leaving a pale point of about an inch in length. The whole of the visible end of the longer and more slender quills is dirty yellow. The hollow quills at the end of the tail are dirty white. A distinct yellowish collar encircles the fore part of the neck, and is sometimes very broad at the lower part.

Five out of six specimens which I have examined agree with the above description; four of them are from Java, and form part of the collection in the Leyden Museum; a fifth specimen, in the same collection, differs in having the general hue black; in the dirty yellow of the quills and spines being replaced by an impure white, and in having a broad transverse white patch on the under side of the neck near the chest. This is perhaps from Sumatra, since Müller informs us that the specimens from that island are darker even than those from Borneo, which latter are of a deeper hue than the Java specimens. A specimen in the British Museum has a large dirty white patch on the neck, immediately in front of the chest.

	Inches.	Lines.
Length from tip of nose to root of tail ...	20	0
" from nose to ear	4	4
From upper portion of ear, about		7
Length of fore foot and nails	2	4
" of hind foot and nails	3	1
" of tail, including the hollow quills, about	3	6

The skull of the *Hystrix Javanica* has the same elongated and narrow form (as compared with that of *H. cristata*) which we observe in the skull of the *H. Hodgsonii*, but it differs much in other respects from the skull of the last mentioned animal. Viewing it above, two striking differences present themselves,—the comparatively small size of the nasal bones, and the greater extent of the frontals. Here the nasal bones are relatively narrower, and they are truncated behind, where they terminate in the same transverse line as the anterior margin of the bony ring, which encloses the ant-orbital opening. The frontal process of the intermaxillary is contracted, and shorter. The palate is very narrow—distinctly narrower than in *H. Hodgsonii*.

	H. Javanica. Leyden Museum.		*H. Javanica.* British Museum.		Hybrid. (*A. Flemingii.*[1])		Hybrid.	
	Ins.	Lines.	Ins.	Lines.	Ins.	Lines.	Ins.	Lines.
Total length of skull	4	7	4	4	4	4	4	6
Width of ditto	2	4½	2	4½	2	4¼	2	6¼
" between orbits	1	7½	1	8	1	6½	1	8½
Length of nasal bones	1	8½	1	10	2	0	1	10
Width of ditto, behind		11		11¼	1	0¼	1	0½
" " in front ...		9½		9		9¼		9¼
Length of frontal bones, in the mesial line	1	6	1	5½	1	2½	1	5
Width of intermaxillary bone, behind				3½		3½		3½
" of ditto, in front				9		9		9½
" of palate between second pair of molars ...				2¼		2		3⅓
Vertical diameter of ant-orbital opening				5		5		5
Length of lower jaw	3	1	2	11½	3	0	3	1
Height of ditto from the condyle	1	4¼	1	3½	1	5	1	4⅔

The habits of this animal, as recorded by Müller, do not differ from those of the *H. Hodgsonii*. They are said to live in pairs, in burrows in the ground, to seek their food by

[1] *Acanthion Flemingii.*

Acanthion Flemingii. Gray, Proc. Zool. Soc. for June 1847; Ann. and Mag. Nat. Hist. for Nov. 1847, vol. xx. p. 354.

The *A. Flemingii* is founded upon a skeleton of a Porcupine which died in the Surrey Zoological Gardens. The skeleton was purchased for the British Museum from Mr. Bartlett, who informed Mr. Gray that the animal to which night, and to be very destructive to the maise crops, and the potatoes. Their flesh is highly esteemed.

it belonged had square spines, and was destitute of a crest. Since the publication of Mr. Gray's account of this skeleton, Mr. Bartlett has learnt from the persons connected with the Gardens referred to, that this animal was a hybrid between the Common Porcupine and the Java species, and that the latter was the male. The skull differs very little from that of the *H. Javanica:* it has the same extent of interparietal bones; the frontals nearly agree, being but a trifle shorter; the same form of intermaxillary bones, the same contracted palate, and an obtuse post-orbital process in the middle of the upper edge of the malar bone, as in *H. Javanica.* The only differences in the two skulls, which are worthy of notice, are, that the occipital condyles are broader (3¼‴ in the hybrid, and 2¾‴ in the *H. Javanica*), and that the nasal bones are rather longer; are rounded behind instead of being truncated, and terminate in the same transverse line as the posterior edge of the bony ring of the ant-orbital opening, instead of the anterior edge. The skeleton of the trunk differs from that of *A. Javanicum* in having the ribs broader, and in this respect is intermediate between the skeletons of the two parents. The fourth rib, which is the broadest, measures 3¾‴ in width at about an inch from the articular end, whilst the same rib in *H. Javanica* measures 3‴. The fourteenth, or last pair of ribs, are well developed in the hybrid, whilst they are very small in the *H. Javanica;* but this, I think, is an accidental peculiarity in the latter animal, since the small ribs do not correspond on opposite sides of the body, one being considerably smaller than the other—the longest is but 15 lines in length. The spinous process of the *vertebra dentata* has a smaller antero-posterior diameter than in *H. Javanica,* and is more elongated, and the spinous process of the first dorsal vertebra is about one-fifth shorter than that of the second dorsal, whilst in *H. Javanica* the two processes in question are equally long. The lumbar vertebræ are four in number, and so are the sacral in both species. Of caudal vertebræ there are 17 in the hybrid. They are wanting in the skeleton of the *H. Javanica.*

A skeleton of a second hybrid between the same pair of Porcupines, agrees more nearly in some respects with that of the female parent, and in others with the male, but, on the whole, decidedly approaches most closely to that of the male (*H. Javanica*). The ribs are narrow, as in the last mentioned animal. Of vertebræ, there are 14 dorsal, 4 lumbar, 5 sacral, and 15 caudal. The skull differs from that of *H. Javanica* in being decidedly broader, and a trifle shorter; in having the palate broader, and likewise the nasal bones, but these bones terminate in the same line as in the *H. Javanica.* In the extent of the frontal and parietal bones there is scarcely any difference. The admeasurements of this skull are given in the fourth column of dimensions.

The animal (which, as well as the skeleton, are in the British Museum) differs very little from the *A. Javanica;* it has the same stiff and prickly spines on the upper parts of the body, and they are strongly grooved in the same

Fossil species of Hystrix.

Cuvier refers to a species of *Hystrix,* a molar tooth found in the Val d'Arno; and Messrs. Falconer and Cautley have noticed remains of Porcupines amongst their Sewalik Hill fossils.

Genus, *Atherura.*

Atherura. CUVIER, Règne Animal, 2d ed. (1829, i. p. 215).

Hystricina proper, with the tail nearly as long as the body, cylindrical, of moderate thickness, scaly, with scattered, small stiff bristles in the middle, and with a large tuft of long flat bristles on the apex; ears sub-ovate, clothed with small hairs: skull with the nasal cavity of moderate size.—See Pl. 18, figs. 5 and 6.

Habitat, Africa and Southern Asia.

The dentition of the species of this genus can be scarcely said to differ from those of the genus *Hystrix:* when but moderately worn, the upper molars in the *Atherura macroura* have three deep folds of enamel entering from the outer side of the tooth, and one internal fold, as represented in fig. 6 *b* of Pl. 18.

ATHERURA FASCICULATA.

Indian Brush-tailed Porcupine.

Hystrix fasciculata.	SHAW, General Zoology, ii. Pt. 1, p. 124 (1801).
" *macroura.*	GERVAIS, Voyage de la Bonite, Mammif. p. 60, Pl. 11, figs. 4 to 6 (skull and lower jaw).
Le Porc-épic de Malacca.	BUFF., Suppl. vii. p. 303, Tab. 77.

manner; but, above the shoulders, in the mesial line of the back, we can perceive a slight trace of the crest, which distinguishes the female parent; this crest is composed of very stiff, white-pointed bristles, which are from three to five inches in length. In colouring the animal agrees very closely with the dark variety of the *H. Javanica;* its prevailing hue being black.

Body for the most part covered with depressed spines, which are strongly channelled on the outer surface; tail, at the root, covered with similar spines; in the middle it has very short bristles scattered on the scaly skin; and at the apex it is provided with a large tuft of long flattened bristles: general hue brown—the spines white at the root; the bristles at the end of the tail dirty white, and often dusky at the point.

Inhabits Siam, and the Malay Peninsula.

Of a Porcupine agreeing with the *A. fasciculata* of Shaw, I have examined four specimens, all of which are from Siam, and form part of the Leyden collection.

In general appearance the animal reminds one of a large Rat, but the head is rather shorter than usual in the Rat group. The ears are semi-ovate, somewhat emarginated behind, and have thin, scattered dusky brown hairs (many of them rather long) springing from their surface. The hairs of the moustaches are very long and stout, and of a brown hue. The extremity of the muzzle is entirely clothed with hairs. The feet are short; and each foot is provided with five toes, but the inner toe of the fore foot is extremely short, and it has a short rounded nail; the inner toe of the hinder foot is less short, and has a larger nail, which is pointed like the nails of the other toes; these are rather short, moderately arched, and not very deep, neither are they much compressed. The upper surface of the feet is clothed with coarse brown hairs; the under surface is naked. The covering of the animal consists of depressed spines, which are flat on the under surface, and concave in the transverse direction, on the upper surface; they are sharply pointed, and broadest at a short distance from the root: their average length is about one inch, and their width, one line. The spines on the under parts, however, are smaller, and relatively narrower. Mixed with the spines on the back are some long and stout bristles, projecting, some of them, as much as three inches beyond the spines. Upon parting the spines a very scanty growth of fine, pale coloured hairs is visible. The tail is considerably shorter than the body, and moderately thick; about one inch, being the basal portion, is covered with spines like those on the body; beyond this, the organ is covered with rhombus-shaped scales, and presents numerous longitudinal ridges, each scale being ridged along the middle. The distal extremity of the tail is provided with a large tuft of flattened bristles (somewhat resembling thin and narrow strips of whalebone); they are from three to four inches in length, of a dirty white colour, but with the tip sometimes dusky.

The general tint of the animal is yellowish brown, freckled with dusky brown, especially on the back: the spines, taken separately, are brown-white at the root, and become gradually darker to the point: the points of the spines on the back are very dark, being of a blackish brown colour. The long and stout bristles which are mixed with the spines on the back are similarly coloured.

	Inches.	Lines.
Length from tip of nose to root of tail ...	18	0
" from nose to ear	2	3
" of ear		6
" of tail, not including the tuft ...	7	6
" of fore foot	1	7
" of hind foot	2	2

Atherura macroura.

Hystrix macroura.	LINN., Syst. Nat. 12th ed. p. 77 (1766).
" "	SCHREB., Säugth. iv. p. 607, Pl. 170 (1792).
" "	SHAW, Gen. Zool. ii. Pt. 1, p. 9, Pl. 124 (1801).
" *fasciculata.*	GRAY, fig. in Illustr. of Indian Zoology.
Porcus aculeatus, &c.	SEBA, Thesaurus, i. p. 84, Pl. 52, fig. 1 (animal), and fig. 2 (end of the tail).
Long-tailed Porcupine.	PENNANT, Syn. of Quadr. p. 263 (1771).

Body for the most part covered with flattened spines, which are grooved along the outer surface; those on the hinder part of the back long, being about four inches in length, and having nearly a square section: mixed with the spines of the hinder part of the back are some very long spiny bristles, the exposed ends of which are of a yellowish white colour: scales on the tail nearly square, and behind each scale springs a small stiff spine, and several very short bristles; apex of the tail provided with a large tuft of flat bristles, which are spirally twisted, and alternately contracted and expanded: upper parts of the animal deep brown; sides of the body freckled with brown and dirty yellow; abdomen dirty yellow; feet dark brown; tuft at the end of the tail dirty white.

Inhabits Sumatra?

The above are the principal peculiarities of an animal which I saw in the Leyden Museum. According to my notes it was larger than either of four specimens of the *A. fasciculata* from Siam, and differed in having the spines much longer, especially those on the hinder part of the back, some of which were five or six inches in length. It differed, moreover, in having the spiny bristles—which spring from the same parts (some of which were seven or eight inches in length)—of a yellowish white colour—at least their exposed ends; in having the tail proportionately shorter[1], and in the form of the flattened bristles at the end of the tail, these being alternately contracted and expanded[2]: these bristles measured from three to four inches in length, and they were,

[1] The notes referred to, state that all four of the Siam specimens alluded to agreed in having the tail relatively longer and more slender than in this animal; that they all had the flat bristles at the end of the tail of uniform width—not alternately contracted and expanded; and the long spiny bristles of the back were brown.

[2] This peculiarity is particularly noticed by Seba, who, moreover, figures the terminal portion of the tail to show the curious modification of the hairs or bristles forming the tuft on that part. Seba attributes irridescent colours to the animal—perhaps his specimen was preserved in alcohol, and was wet when examined.

most of them, three quarters of a line in width in the expanded parts. The hairs of the moustaches were immensely long, some of them as much as seven inches in length. Its dimensions are—

	Leyden Museum.		Brit. Mus.	
	Inches.	Lines.	Inches.	Lines.
From tip of nose to root of tail ...	20	0	16	3
" ditto to ear	2	11		
" of ear		8½		8
" of tail, not including the tuft ...	7	9	7	6
" of fore foot	1	10	1	8
" of hind ditto	2	11	2	5

I add the dimensions of a second specimen, clearly referrible to the same species, which is contained in the British Museum collection. It is an immature animal, having but three molar teeth developed on each side of its jaws; but agrees with the Leyden specimen in having long spines, with nearly a square section, on the hinder part of the back, in having the interspersed very long bristles almost entirely whitish, and in having those forming the tuft at the end of the tail flattened, and alternately contracted and expanded, but the expanded parts are narrower, being not more than two-thirds of a line in width, and generally less; and there are many interspersed simple bristles. The ordinary spines are white at the base, and dusky brown at the apex; or, many of them, especially on the sides of the body, have the basal half white, a short white point, and the intermediate space dusky brown. The longer spines on the hinder part of the back are from three to four inches in length, nearly square, but rather broader than deep, and strongly grooved on the upper surface. The still longer stout bristles on the same part, and on the sides of the body, are some of them entirely dirty white, and some of the same pale hue, but with a brownish space, of no great extent, near the middle. The under parts of the animal are of a dirty yellowish white colour, if we except the chest, which is somewhat suffused with brown. Very stout, but short, pointed bristles spring from between the square scales on the tail, and these are for the most part dusky; the tuft at the end of the organ is white. The very long hairs of the moustaches are some of them entirely dusky brown, and some dirty white, but most of them with the basal half (excepting quite at the root, where they are pale) brown, and the apical half, dirty white: long bristles spring from the cheeks. About one-third of the tail is covered with spines like those on the body. The native country of this specimen is not known[1].

The skull of the specimen last described differs much from a cranium figured in the *Voyage de la Bonite* (Pl. 11, figs. 4 and 5) under the name *Hystrix macroura*. Although belonging to an immature animal, the length of the skull in the British Museum, measured from the posterior root of the zygoma, to the front of the incisor, is equal to that of the adult skull represented in the figures above referred to, and hence we must conclude that the British Museum cranium would have been longer than the adult one, had the animal to which it belonged, lived to attain to maturity. The immature skull is much narrower (viz. 3 lines) than the mature one; the distance between the anterior edge of the ring, which encloses the ant-orbital opening, is greater by $1\frac{3}{4}$ lines, whilst the zygomatic arch is shorter. The most striking differences, however, are noticeable in the zygoma. The process of the intermaxilllary which encloses the ant-orbital opening, is represented in the figures in the *Voyage de la Bonite*, as being compressed from before backwards, whilst in the British Museum cranium the same process presents a broad outer

[1] Dr. Müller states that the *Hystrix macroura* (but under this name I suspect he includes the animals which have received the names *macroura* and *fasciculata*) generally inhabits the main land, that it appears to be tolerably abundant in Siam, and is found likewise at Malacca. In the Archipelago it is apparently confined to the eastern coast of Sumatra, most of the specimens which came under his notice having been from the district of Palembang.

surface (full two lines in width), and is narrower in the opposite direction. Here the malar bone is three times as deep under the orbit as it is behind, where it joins the zygomatic process of the temporal bone; whilst in the "Bonite" specimen the malar bone is a trifle narrower in front than behind; there is no trace of a post-orbital process to the frontal bone, which is so distinct in the figures; neither is the frontal bone contracted behind. The palate is deeply emarginated behind.—See Pl. 18, fig. 6 *a*.

I strongly suspect the skull figured in the "Bonite" will be found to belong to the *Atherura fasciculata*. The dimensions of the cranium removed from the specimen of *A. macroura* in the British Museum, are as folllows:—

	Inches.	Lines.
From the front incisors to the hinder root of the zygoma	2	3
Width from outer surface of zygomatic arches	1	6½
" between orbits	1	0⅔
Length of nasal bones		9¾
Width of ditto, behind		5½
" " in front		5½
Length of frontal bones	1	3
" of zygomatic arch	1	3⅓
From front of incisor to first molar ...	1	0¾
Width of palate between the second pair of molars		2¾

ATHERURA AFRICANA.

African Brush-tailed Porcupine.

Atherura Africana. GRAY, Annals and Mag. of Nat. Hist. for Dec. 1842, vol. x. p. 261; and for Nov. 1847, vol. xx. p. 355.

" *fasciculata.* BENNETT, Gardens and Menagerie of the Zooological Society, i. p. 175.

Body covered with depressed spines; long spines interspersed with the ordinary spines on the hinder part of the back,

having nearly a round section, besides very long spiny bristles: tail spiny at the base, scaly in the middle, and with a tuft of long flat bristles, which are alternately expanded and contracted.—Skull, Pl. 18, fig. 5.

Inhabits the west coast of Africa (Sierra Leone and Fernando Po.)

A specimen in the Museum of the Zoological Society, which is the original of Mr. Bennett's description in his work above quoted, has the ordinary spines on the body about an inch and a half in length, or some an inch and eight lines, and very nearly a line and a half in width: they are flat beneath, and concave in the transverse direction, above, and very sharply pointed. The under surface of their apical end is thickly studded with exceedingly minute prickles, the points of which are directed forwards; these prickles make their appearance at a short distance from the point, and may be traced nearly half way down the spine; the lower half is smooth. On the hinder part of the back there are many large spines, measuring about four inches in length, and nearly two lines in diameter at the thickest part, and these have nearly a round section. Besides these, there are some flexible spines, or spiny bristles, which are still longer: they have a round section, and are about a line in diameter at the thickest part, which is immediately above the root. The general colour is brown; on the back, dusky brown: the spines are dirty white at the root, and shaded through brown to brown-black, which is the colour of the point. On the sides of the body many of the spines are tipped with white; brown below the point, and white at the root. The under parts are brownish white, but there is a brown patch on the chest. The very long hairs of the moustaches are most of them whitish at the root, and brown beyond. The tuft at the end of the tail is white.

	Zool. Soc.		Zool. Soc.		Brit. Mus.	
	Ins.	Lines.	Ins.	Lines.	Ins.	Lines.
Length from tip of nose to root of tail ...	14	3	9	6	16	0
" of tail, not including the tuft ...	5	6	3	2	7	0
" of ditto, including the tuft	7	5	6	6		
" of fore foot			1	6	1	10
" of hind ditto			2	3	2	8
" from nose to ear			2	3		
" of ear				9		

The admeasurements in the second column are from a smaller specimen, which is preserved in spirit, the larger one being stuffed. In this small specimen (which no doubt is immature), the longest spines on the back are not more than 1¾ inches in length, and 1 line in width; they are depressed, and have a broad and deep channel along the upper surface. On the back are interspersed some stout bristles, which are some of them upwards of four inches in length. The curiously modified hairs, or bristles, forming the tuft at the end of the tail, are about 2½ inches in length: they are flat, and the apical third part of each bristle is dilated to a width of about 1 line, or sometimes rather more, but the width gradually decreases to the point; below this part the bristle is twice dilated, but less so than on the apical third; the intermediate parts are very slender. This specimen is from Sierra Leone.

I also add the admeasurements of a specimen in the British Museum, which is from Fernando Po.

The skull of the *A. Africana* presents many well-marked points of distinction when compared with that of *H. fasciculata*, and with that of *A. macroura*, as represented in the figure in the *Voyage de la Bonite*. The facial part is less attenuated, and the hinder portion of the nasal cavity is more expanded both laterally and superiorly, the palate is extended considerably beyond the posterior edge of the last molar teeth, and it is truncated behind.—See Pl. 18, fig. 5 *a*. Compared with the skull of *A. fasciculata*, there are other differences worthy of note: thus, the orbital portion of the lachrymal bone is twice as much extended in the vertical direction, whilst the facial surface is much smaller; the upper root of the zygomatic process of the superior maxillary is more expanded, and convex (instead of being flat) externally; and the malar runs up so as to form a considerable portion of the anterior boundary of the orbit, which is not the case in *A. fasciculata*: the distance between the anterior edge of the malar bone and the front edge of the ant-orbital arch is much less in the larger skull of *A. Africana*, than in the smaller skull of *A. fasciculata*, being 2¼ lines in the former, and 3¼ in the latter. The cranium is figured of two-thirds of the natural size—see Pl. 18, fig. 5: it is fully adult.

	Inches.	Lines.
Length of skull	4	0
Width of ditto	2	0
" between orbits	1	4
Length of nasal bones	1	0½
Width of ditto behind		7¾
" " in front		6⅓
Length of frontal bones, measured in the mesial line	1	7
Width of terminal portion of nasal process of intermaxillary		3
Length of zygomatic arch	1	7½
" of malar bone	1	2
Height of ditto, in front		6¼
" " immediately in front of the zygomatic process of the temporal bone...		1
Width of incisors		3½
From incisor to molar teeth	1	0¼
Length of four upper molars, taken together		9
Width of palate between the molar teeth ...		3½

Geographical Distribution of the Hystricidæ.

In the following tables asterisks are introduced opposite the names of the species in those columns which are headed by the names of the districts which such species are known to inhabit: thus, the first mentioned species (*Dolichotis Patachonica*) is found in La Plata and Patagonia; and, so far as we are aware, in no other parts. Upon inspecting these tables it will be seen that the great metropolis of the *Hystricidæ* is Brazil; that those which I have given reasons for regarding as the lowest species of the group, are either found at great elevations on the Andes, or in the more southern portions of the continent; that the species of the typical division, the *Echimyina,* almost entirely occur in the central parts of South America; but that this division has representatives from Cuba in the north, down to the southern portions of Chile[1]. The *Cercolabina* are also chiefly central, but extend as far north as Mexico; and a somewhat aberrant form (*Erethizon*) is found even in the northern parts of North America. The *Dasyproctina* have a tendency to increase in number in the northern parts of South America, and extend into the West Indian Islands. The two aquatic species *Myopotamus Coypus* and *Hydrochœrus Capybara* have each a very wide range.

The Old World Porcupines, though belonging to the same great family type (*Hystricidæ*), decidedly exhibit a very aberrant modification of that type.

[1] It is to this section that belong the two most singular exceptions in the distribution of the family—I allude to the *Petromys typicus* and *Aulacodus Swinderianus*, both of which are found in Africa.

	West Indies.	Venezuela, Guayana.	Peru—west of the Andes.	Peru—east of the Andes.	Brazil.	Bolivia - east of the Andes.	Paraguay.	La Plata.	Patagonia.	Chile.	Andes of Peru.	Andes of Chile and Bolivia.
Caviina.												
Dolichotis Patachonica	...	...	...	...	...	...	...	*	*	...	...	...
Cavia (Cerodon) rupestris	...	...	...	...	*	...	...	...	..	...	...	...
Cavia A. 1 (Cerodon) flavidens	...	...	...	...	*	...	...	...	...	...	...	...
" " *nigricans*	...	...	...	...	*	...	...	...	...	...	...	...
" " Spixii	...	...	...	...	*	...	...	...	...	...	...	...
" " Boliviensis	...	...	...	...	...	*	...	..	...	...	...	...
" " Australis	...	...	...	...	...	...	...	...	*	...	...	...
Cavia Aperea	...	...	...	...	*	*	*	...	...	...	...	...
" *Cutleri*, Tschudi	...	...	*	...	...	...	...	...	...	...	...	...
" fulgida	...	...	...	...	*	...	...	...	...	...	...	...
" leucopyga	...	...	...	...	*	...	...	...	...	...	...	...
Hydrochœrus Capybara	...	*	...	*	*	*	*	*	...	...	...	...
Chinchillina.												
Lagostomus trichodactylus	...	...	...	...	...	...	...	*	...	...	...	...
Lagidium Cuvieri	...	...	...	...	...	...	...	...	...	...	*	*
" pallipes	...	...	...	...	...	...	...	...	...	...	*	*
Chinchilla lanigera	...	...	...	...	...	...	...	...	...	...	*	*
" brevicaudata	...	...	...	...	...	...	...	...	...	...	*	...
Octodontina.												
Habrocoma Bennettii	...	...	...	...	...	...	...	...	...	*	...	...
" Cuvieri	...	...	...	...	...	...	...	...	...	*	...	...
Octodon Degus	...	...	...	...	...	...	...	...	...	*	...	...
" Peruana, *Tschudi*	...	...	...	...	...	...	...	...	...	...	...	*
" Bridgesii	...	...	...	...	...	...	...	...	...	*	...	...
" gliroides	...	...	...	...	...	...	...	...	...	...	...	*

	West Indies.	Venezuela, Guayana.	Peru—west of the Andes.	Peru—east of the Andes.	Brazil.	Bolivia—west of the Andes.	Paraguay.	La Plata.	Patagonia.	Chile.	Andes of Peru.	Andes of Chile and Bolivia.
OCTODONTINA—*continued.*												
Schizodon fuscus	...	...	...	...	...	...	...	...	...	...	...	*
Spalacopus Poeppigii	...	...	...	...	...	...	...	...	...	*	...	...
Ctenomys Braziliensis	...	...	...	...	*	*	*	*	...	...	...	...
" Boliviensis	...	...	...	...	...	*	...	...	...	...	...	...
" leucodon	...	...	...	...	...	*	...	...	...	...	...	...
" Magellanicus	...	...	...	...	...	...	...	...	*	...	...	...
ECHIMYINA.												
Capromys pilorides	*	...	...	...	...	...	...	...	...	...	...	...
" prehensilis	*	...	...	...	...	...	...	...	...	...	...	...
Plagiodontia ædium	*	...	...	...	...	...	...	...	...	...	...	...
Myopotamus Coypus	...	...	...	*	...	...	*	*	...	*	...	...
Cercomys cunicularius	...	...	...	...	*	...	...	...	...	...	...	...
Petromys typicus[1]	...	...	...	...	...	...	...	...	...	...	...	...
Dactylomys typus	...	...	...	...	?*	...	...	...	...	...	...	...
" amblyonyx	...	...	...	...	*	...	...	...	...	...	...	...
Loncheres cristata	...	*	...	...	*	...	...	...	...	...	...	...
" Blainvillii	...	...	...	...	*	...	...	...	...	...	...	...
" armata	...	*	...	...	*	...	...	...	...	...	...	...
" didelphoides	...	...	...	...	?*	...	...	...	...	...	...	...
" obscura	...	...	...	...	*	...	...	...	...	...	...	...
" semivillosa[2]	...	*	...	...	...	...	...	...	...	...	...	...
" grandis	...	...	...	...	*	...	...	...	...	...	...	...
" macrura	...	...	...	...	*	...	...	...	...	...	...	...
" nigrispina	...	...	...	...	*	...	...	...	...	...	...	...

[1] *Petromys typicus* inhabits South Africa.

[2] *Loncheres semivillosa* is from New Granada.

	West Indies.	Venezuela, Guayana.	Peru—west of the Andes.	Peru—east of the Andes.	Brazil.	Bolivia—east of the Andes.	Paraguay.	La Plata.	Patagonia.	Chile.	Andes of Peru.	Andes of Chile and Bolivia.
ECHIMYINA—*continued*.												
Loncheres unicolor	...	...	...	...	*	...	...	...	...	...	...	...
" picta	...	...	...	...	*	...	...	...	...	...	...	...
" bistriata	...	...	...	...	*	...	...	...	...	...	...	...
" pagurus	...	...	...	...	*	...	...	...	...	...	...	...
" pachyura	...	...	...	...	*	...	...	...	...	...	...	...
Mesomys ecaudatus	...	...	...	...	*	...	...	...	...	...	...	...
Echimys Cayennensis	...	*	...	...	*	...	...	...	...	...	...	...
" albispinosus	...	...	...	...	*	...	...	...	...	...	...	...
" hispidus	...	...	...	...	*	...	...	...	...	...	...	...
" brachyurus	...	...	...	...	...	*	*	...	...	...	...	...
" spinosus	...	...	...	...	...	...	*	...	...	...	...	...
" inermis	...	...	...	...	*	...	...	...	...	...	...	...
" antricola	...	...	...	...	*	...	...	...	...	...	...	...
Aulacodus Swinderianus[1]	...	...	...	...	...	...	...	...	...	...	...	...
DASYPROCTINA.												
Cœlogenys Paca	*	...	...	*	*	*	*	...	...	...	...	...
Dasyprocta Aguti	...	*	...	...	*	...	...	...	...	...	...	...
" croconota	...	...	...	...	*	...	...	...	...	...	...	...
" prymnolopha	...	*	...	...	...	...	...	...	...	...	...	...
" cristata	...	*	?*	...	...	...	...	...	...	...	...	...
" fuliginosa	...	...	...	...	*	...	...	...	...	...	...	...
" Azaræ	...	...	...	...	*	*	*	...	...	...	...	...

[1] *Aulacodus Swinderianus* is from West Africa.

	West Indies.	Venezuela, Guayana.	Peru—west of the Andes.	Peru—east of the Andes.	Brazil.	Bolivia—east of the Andes.	Paraguay.	La Plata.	Patagonia.	Chile.	Andes of Peru.	Andes of Chile and Bolivia.
Dasyproctina—*continued.*												
Dasyprocta caudata	...	...	...	...	*	...	...	...	...	...	...	...
" Acouchi	*	*	...	...	...	...	...	...	...	...	...	...
" albida	*	...	...	...	...	...	...	...	...	...	...	...
Cercolabina.												
Chætomys subspinosus	...	...	...	...	*	...	...	...	...	...	...	...
Cercolabes prehensilis	...	*	...	...	*	...	...	...	...	...	...	...
" platycentrotus (?)	...	...	...	...	...	...	...	...	...	...	...	...
" nycthemera	...	...	...	*	*	...	...	...	...	...	...	...
" spinosus	...	...	...	...	*	...	...	...	...	...	...	...
" Novæ-Hispaniæ[1]	...	...	...	...	...	...	...	...	...	...	...	...
" melanurus	...	*	...	...	*	...	...	...	...	...	...	...
" villosus	...	...	...	...	*	...	?*	...	...	...	...	...
" insidiosus	...	...	...	...	*	...	...	...	...	...	...	...
" nigricans	...	...	...	...	*	...	...	...	...	...	...	...
" affinis	...	...	...	...	*	...	...	...	...	...	...	...
" pallidus	*	...	...	...	...	...	...	...	...	...	...	...
Erethizon[2] dorsatus	...	...	...	...	...	...	...	...	...	...	...	...
" epixanthus	...	...	...	...	...	...	...	...	...	...	...	...
Hystricina[3].												

[1] Cercolabes Novæ-Hispaniæ inhabits Mexico.
[2] The two species of Erethizon are from the northern parts of North America.
[3] The six or seven species forming the true Hystricina belong entirely to the Old World, being found in Asia and Africa.

INDEX TO THE TECHNICAL NAMES.

INDEX TO THE ENGLISH AND OTHER LOCAL NAMES.

EXPLANATION OF THE PLATES.

Plate 2. Skull and lower jaw of *Lepus* and *Lagomys,* natural size; named on the plate.

Plate 4, Fig. 1. Skull of *Dolichotis Patachonica,* half the natural size; fig. 1*a,* lower jaw of ditto; fig. 1*b,* first and third lower molars, the former magnified.[1]

Fig. 2*b*. First and second right upper molars of *Cavia* (*Cerodon*) *rupestris*.

Fig. 3*b*. Left upper molars of *Hydrochœrus Capybara,* natural size.

Fig. 3*c*. Left upper molars of lower jaw.

Fig. 4. First and second right upper molars of *Cavia Cobaya*.

Plate 6, Fig. 1. First and second lower molars of *Cavia* (*Cerodon*) *rupestris*.

Fig. 2. Ditto of *Cavia* (*Cerodon*), A. 1; most probably the *Cavia flavidens* of Brandt.

Fig. 3. Second upper molar of *Cavia* (*Cerodon*) *Spixii*.

Fig. 4. First and second lower molars of ditto.

Fig. 5. Second upper molar of *Cavia* (*Cerodon*) *Boliviensis*.

Fig. 6. First and second lower molars of ditto.

Fig. 7. Second upper molar of *Cavia* (*Cerodon*) *Australis*.

[1] By mistake marked Pl. 16 on the earlier impressions struck off. Plate 16 (correctly so numbered) contains the skull of Myopotamus, Cercomys, &c.

Plate 6, Fig. 8. First and second lower molars of ditto.

Fig. 9. Ditto of *Cavia Aperea*.

Reference to the above teeth is made in the text at pp. 163 and 184.

Fig. 10. Palatal view of skull of *Cavia* (*Cerodon*), A. 1.

Fig. 11. Ditto, and upper view of skull of *Cavia* (*Cerodon*) *Spixii*.

Fig. 12. Palatal view of skull of *Cavia* (*Cerodon*) *Boliviensis*.

Fig. 13. Ditto of *Cavia* (*Cerodon*) *Australis*.

All the above figures are of the natural size: side views of most of these skulls will be found in Plate 10.

Plate 6*a*, Fig. 1. Side view of the head of an Agouti (*Dasyprocta*), showing the arrangement of the muscles (explained at p. 151). *a a*, is the masseter muscle; *b*, the temporal muscle; *d*, a portion of the digastric; *e*, the infra-orbital nerve.

Fig. 2. Side view of the head of the Viscacha (*Lagostomus trichodactylus*).

Fig. 3. Ditto, with portions of the masseter muscle removed to show the inferior layers of this muscle, and to expose the comparatively small temporal muscle, *b*.

Fig. 4. View of the under side of the lower jaw of the same animal. *a*, the masseter muscle; *c*, *c*, the pterygoid muscles; *d*, the digastric.

*a**, in the above figures, is a portion of the masseter muscle which passes through the sub-orbital (or here, from its position, antorbital) opening.

Fig. 5. Side view of the head of *Capromys pilorides*, half the natural size.

Fig. 5*a*. Front view of the muzzle of ditto; natural size.

Fig. 6. Side view of the head of *Cercolabes prehensilis*; half natural size.

Plate 6*a*, Fig. 6*a*. Front view of muzzle of ditto.

Fig. 7. Front view of muzzle of *Atherura Africana.*

Fig. 8. Ditto of *Lagostomus trichodactylus.*

In the above-named animals, the terminal portion of the muzzle is entirely clothed with small velvet-like hairs.

Fig. 9. Front view of muzzle of *Cricetomys Gambianus.*

Here the part left white on the figure is naked.

Fig. 9*a*. Nostril of the same animal, enlarged.

Fig. 10. Ear of *Lagostomus trichodactylus,* half the natural size.

Plate 8, Fig. 1. Skull and lower jaw of *Habrocoma Bennettii.*

Fig. 2 and 2*a*. ,, *Octodon Degus.*

Fig. 3 and 2*b*. ,, *Octodon Bridgesii.*

Fig. 4. ,, *Schizodon fuscus.*

Fig. 5. ,, *Ctenomys Magellanicus* (from Mr. Bennett's fig.)

Fig. 6. ,, *Ctenomys Braziliensis.*

Fig. 7. ,, *Spalacopus Poeppigii.*

All the above skulls are of the natural size.

Plate 10, Fig. 8. Skull and dentition of *Plagiodontia ædium,* copied (half the natural size) from F. Cuvier's figure.

Plate 10*, Fig. 1*a*. First and second right upper molars of *Chinchilla lanigera.*

Fig. 3. Skull and lower jaw of *Lagostomus trichodactylus,* two-thirds the natural size. The remaining figures are of the natural size.

Plate 12. Skull and lower jaw of *Capromys pilorides,* three-fourths the natural size.

Fig. 6. First and second right lower molars magnified.

Plate 14, Fig. 1. Skeleton of *Capromys pilorides,* reduced in size. See p. 291.

cr. the cranium.	*r.* radius.
cv. cervical vertebræ.	*cp.* carpus.
dv. dorsal vertebræ.	*mc.* metacarpus.
rb. ribs.	*pl.* phalanges.
lv. lumbar vertebræ.	*pv.* pelvis.
sv. sacral vertebræ.	*f.* femur.
cdv. caudal vertebræ.	*p.* patella.
s. scapula.	*t.* tibia.
cl. clavicle.	*fb.* fibula.
h. humerus.	*tr.* tarsus.
u. ulna.	*mt.* metatarsus.

Fig. 2. Carpus, and part of metacarpus, of *Myopotamus Coypus.*

sc. l. scaphoid and lunare joined.	†. supernumerary bone.
c. cuneiform.	*mc.* 1. metacarpal of the thumb.
ps. pisiform.	
tm. trapezium.	*mc.* 2. do. fore-finger.
ts. trapezoides.	*mc.* 3. do. middle finger.
m. magnum.	*mc.* 4. do. fourth finger.
un. unciforme.	*mc.* 5. do. fifth finger.
*. supernumerary bone.	

Fig. 3. Tarsus, and part of metatarsus, of *Myopotamus Coypus.*—See page 303.

cm. calcaneum.	*c.* 2. middle or second cuneiform.
a. astragalus.	
cb. cuboides.	*c.* 3. external or third cuneiform.
n. naviculare.	
*. supernumerary bone, or perhaps a dismemberment of the *os naviculare.*	*mt.* 1 to 5. metatarsal bones of the first (or inner) to the fifth toes.
c. 1. internal or first cuneiform.	

Plate 16, Fig. 1. Skull of *Myopotamus Coypus,* half the natural size.

Fig. 1*a*. Occiput of the same.

Fig. 1*b*. Third right upper molar of ditto.

Fig. 1*c*. Second right lower molar of ditto.

Fig. 2. Skull of *Cercomys cunicularius.* From M. F. Cuvier's figure.

Fig. 2*a*. Left upper molar teeth of ditto.

Fig. 3. Skull of *Dactylomys typus.* From M. F. Cuvier's figure.

Fig. 3*a*. Upper molars of the left side of the jaw of ditto.

Fig. 3*b*. Lower molars of the left side of ditto.

Fig. 4. Skull of *Loncheres cristata.*

Fig. 4*a*. Second right upper molar of ditto.

Fig. 4*b*. Second right lower molar of ditto.

Fig. 5. Skull of a species of *Echimys,* from Bolivia,[1] natural size.

Fig. 6. Left upper molar teeth of *Echimys Cayennensis.*

[1] This skull agrees so closely in size and proportions, and the animal itself agrees so well with M. Pictet's description of the *Echimys inermis,* that I think it must be of the same species, although the Bolivian animal has no tail—it has probably been lost through some accident.

In general tint, the Bolivian Echimys is rather brighter than the common rat (*Mus decumanus*), there being a greater admixture of yellow in the colouring; and the yellow is somewhat richer. The fur is rather more harsh to the touch; it is of moderate length, and composed of soft hairs, and others which are somewhat bristly. The latter are confined to the upper parts and sides of the body, average about an inch in length, are depressed, distinctly grooved on the upper surface, and about a quarter of a line in width; they terminate in long slender points. The fur on the upper parts and sides of the animal is pencilled with rich yellow and dusky black; on the sides of the body the yellow prevails. The hairs are grey-white at the root. The whole of the under-parts are white, the hairs being uniformly white to the root. The fore-legs are of an extremely pale ashy-brown externally; the fore-feet white, not quite pure. The hind-feet are ashy-white; immediately above the heel the hairs are somewhat dusky. The ears are very sparingly clothed with fine hairs, and rather smaller than in the E. Cayennensis; the fore-feet are larger, the hinder stouter, but shorter. The hairs of the moustacheos brown, or, some of them, white. Nails white. The specimen is a male, and measures, from the nose to the rump, 9″ 6‴; height of ear, 8‴; length of fore-foot and nails, 11‴; of hind-foot and nails, 1″ 9¼‴; from heel to end of nail of inner toe, 1″ 1½‴.

Plate 16, Fig. 6*a*. Left lower molars of the same.

Fig. 7 and 7*a*. Skull and lower jaw of *Carterodon sulcidens*, natural size.

Fig. 7*b*. Second right upper molar of ditto.

Fig. 7*c*. Incisor tooth of ditto, 1½ natural size.

Fig. 8. Skull of *Petromys typicus*, natural size.

Fig. 8*a*. Second right upper molar of ditto.

Fig. 9. Skull of *Aulacodus Swinderianus*, two-thirds the natural size.

Fig. 9*b*. Second left upper molar of ditto.

Fig. 9*c*. Incisor tooth of ditto.

Plate 18, Fig. 1. Skull of *Chætomys subspinosus*, two-thirds the natural size.

Fig. 1*a*. Second right upper molar of ditto.

Fig. 1*b*. Second right lower of ditto.

Plate 18, Fig. 2. Skull of *Cercolabes prehensilis* (below the natural size). Copied from Brandt's figure.

Fig. 3. Skull of *Cercolabes insidiosus*, ditto ditto.

Fig. 3*a*. First and second right upper molars of *Cercolabes pallida*.

Fig. 4. Right fore-foot of *Cercolabes Novæ-Hispaniæ*, slightly enlarged. The corresponding bones are marked with the same letters as in the foot represented on Plate 14, Fig. 3. The bones marked * and ** are supernumerary bones.

Fig. 5. Skull of *Atherura Africana*, two-thirds the natural size.

Fig. 5*a*. Posterior portion of the palate of ditto, natural size.

Fig. 6. Skull of *Atherura fasciculata*, two-thirds the natural size.

Fig. 6. Posterior portion of palate of ditto, natural size.

Fig. 6*b*. Second right upper molar.

Plate 18, Fig. 7. Skull of *Dasyprocta Acouchi*, two-thirds the natural size.

Fig. 7*a*. Second right upper molar of ditto.

Fig. 7*b*. Ditto ditto, much worn.

Fig. 8. Skull of *Cœlogenys Paca,* three-sevenths the natural size.

Fig. 8*a*. Third and fourth upper molars, natural size.

Fig. 8*b*. Third and fourth lower molars, natural size.

Plate 20, Fig. 1. Skull of *Hystrix cristata.*

Fig. 2. Skull of *Hystrix hirsutirostris,* vel *leucurus.*

Fig. 3. Skull of *Hystrix Hodgsonii.*

Fig. 4. Skull of *Hystrix Javanica.*

The above four skulls are represented of half the natural size.

ADDITIONS AND CORRECTIONS.

Page 155, line 3, *for* "depth," *read* "width."

" 163, " 9 from bottom, *after* "marked *a* in figs. 1, 2, 4, &c." *add* "of Plate 6."

" 177, " 1, under the dimensions, *for* "Pl. 6, fig. 3," *read* "Pl. 6, fig. 12."

" 265, " 6 from bottom, *for* "S. lat. about 75," *read* "S. lat. about 35."

" 281, The dimensions in the first column are those of the skull (Pl. 10*, fig. 4) of "*Ctenomys Boliviensis,*" not "*C. Braziliensis.*"

" 333, The letters referred to in the foot-note were accidentally omitted when lettering the plate.

" — " 2, *for* "figs. 1 and 1A," *read* "figs. 6 and 6A."—The plate in some copies is numbered 16; *see* Explanation of Plates.

" — " 11, *for* "Pl. 16, fig. 1*b*," *read* "Pl. 16, fig. 6*a*."

" 334, " 4, *instead of* "without including the tail," *it should no doubt be* "including the tail."

" 349, " 3 of diagnosis, *for* "without any admixture of soft hairs," *read* "without any admixture of stiff hairs."

" 350, The reference-mark for the foot-note *should be added* after the name "*antricola,*" which occurs four lines above.

" 361, " 5, *for* "pterygoids," *read* "pterygoid."

END OF VOL. II.

WILSON and OGILVY, 57, Skinner Street, Snowhill, London.

Fig. 1.

Fig. 2

Fig. 1. LEPUS SAXATILIS. Fig. 2. LAGOMYS PUSILLUS.

Engraved by H. S. Beckwith. *London, Published by H. Baillière, 219, Regent St. 1847.* *Printed by Brooker & Harrison.*

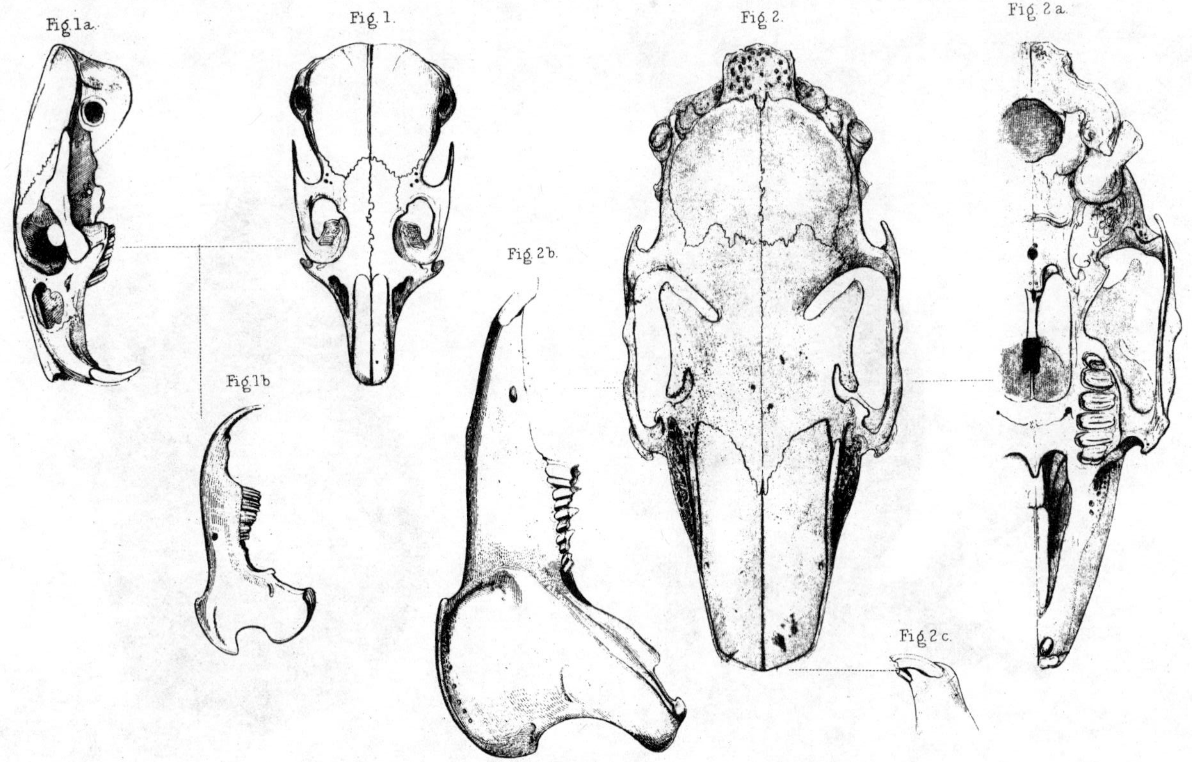

Fig. 1. SKULL AND LOWER JAW OF LAGOMYS ALPINUS; Fig. 2. SKULL AND LOWER JAW OF LEPUS TIMIDUS.

Printed by Brooker & Harrison

Fig. 1.

Fig. 2.

Fig. 1. DOLICHOTIS PATICHONICA. Fig. 2. CAVIA KINGII.

Engraved by H. S. Beckwith. *London Published by H. Baillière, 219, Regent St. 1847.* *Printed by Brooker & Harrison.*

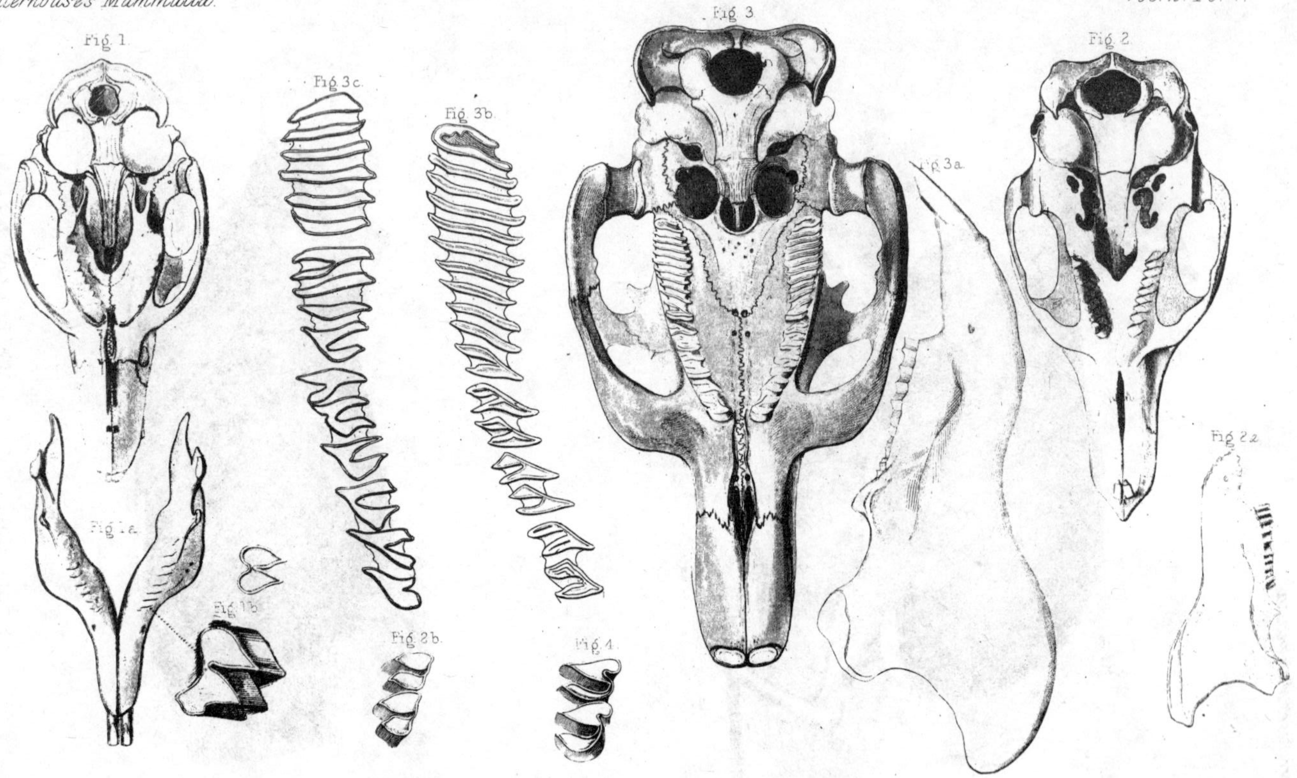

Fig. 1. DOLICHOTIS PATACHONICA. Fig. 2. CAVIA RUPESTRIS. Fig. 3. HYDROCHŒRUS CAPYBARA. Fig. 4. CAVIA COBAYA.

Engraved by H. S. Beckwith. London, Published by H. Baillière 219. Regent S.t 1847. Printed by Brooker & Harrison

Fig. 1.

Fig. 2 a.

Fig. 2.

Fig. 1. LAGOSTOMUS TRICHODACTYLUS. Fig. 2 & 2 a. HYDROCHŒRUS CAPYBARA & ITS FORE FEE

Engraved by H. S. Beckwith. *London. Published by H. Baillière 219, Regent S.t 1847.* *Printed by Brooker & Harrison.*

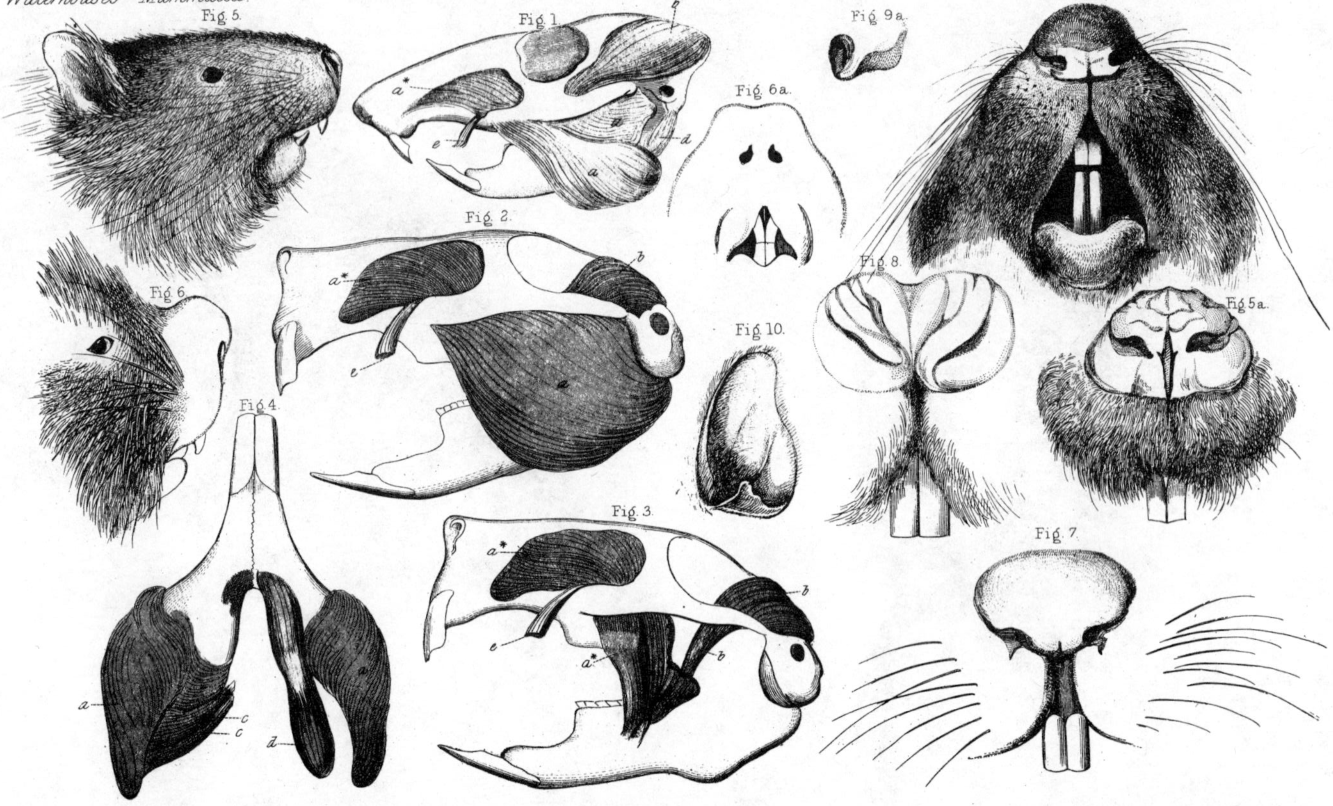

Fig. 1. DASYPROCTA. Figs. 2.4. LAGOSTOMUS. Figs. 5..8. FORMS OF MUZZLE IN HYSTRICIDÆ. Fig. 9. MUZZLE OF ONE OF THE MURIDÆ. Fig. 10. EAR OF LAGOSTOMUS TRICHODACTYLUS.

Engraved by H.S. Beckwith.

London: Published by H. Baillière, 219, Regent St. 1849.

Printed by Brooker & Harrison.

Fig. 1

Fig. 2

Fig. 1. LAGIDIUM PERUANUM. Fig. 2. HABROCOMA BENNETTII.

Engraved by H. S. Beckwith. London Published by H. Baillière, 219, Regent St. 1847. Printed by Brooker & Harrison

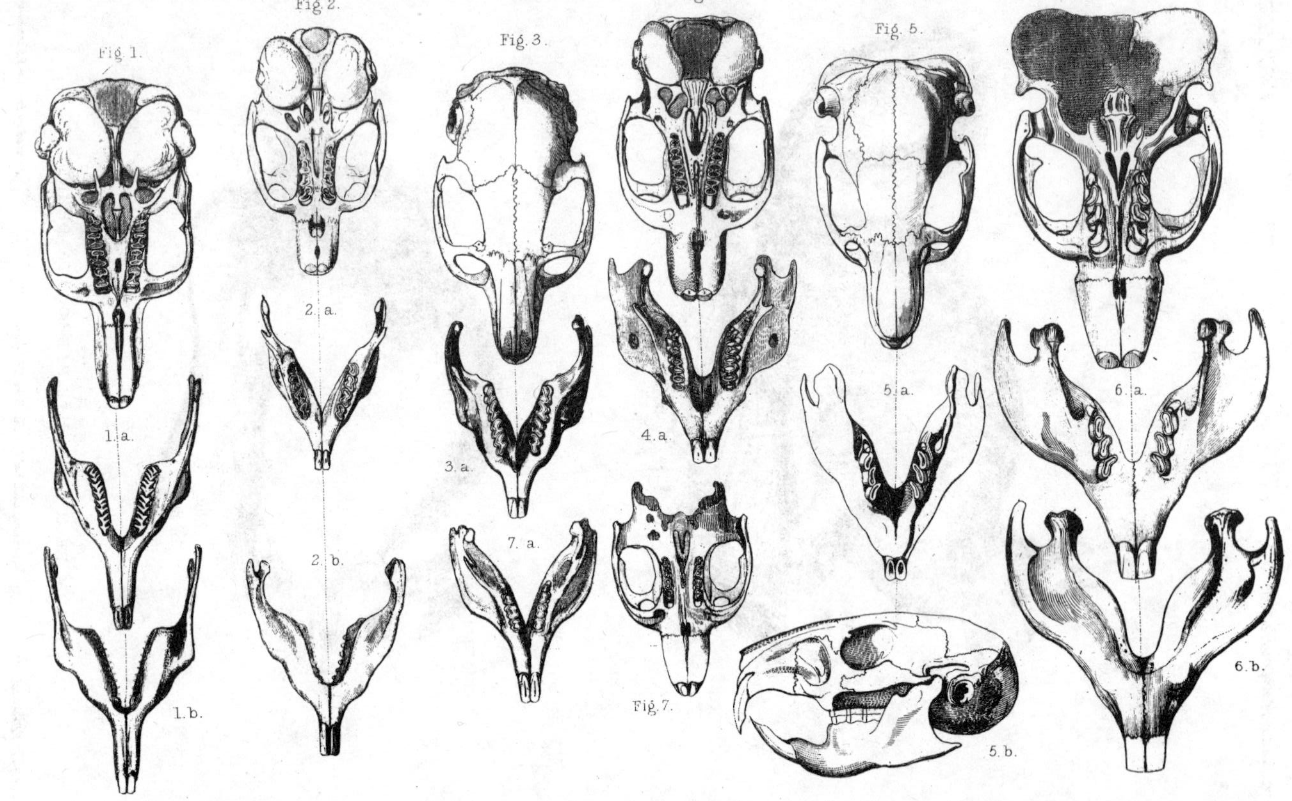

FIG. 1, HABROCOMA; _ FIG. 2, OCTODON DEGUS; _ FIG. 3, OCT. BRIDGESII; _ FIG. 4, SCHIZODON; _ FIG. 5, CTENOMYS MAGELLANICUS; _ FIG. 6, CT. BRAZILIENSIS; _ FIG. 7, SPALACOPUS.

Engraved by H. S. Beckwith. London, Published by H. Baillière, 219 Regent St. 1847. Printed by Brooker & Harrison.

Fig. 1

Fig. 2

Fig. 1 SPALACOPUS POEPPIGII. Fig. 2. CTENOMYS MAGELLANICUS.

Engraved by H. S. Beckwith London. Published by H. Baillière, 219, Regent St. 1847 Printed by Brooker & Harrison

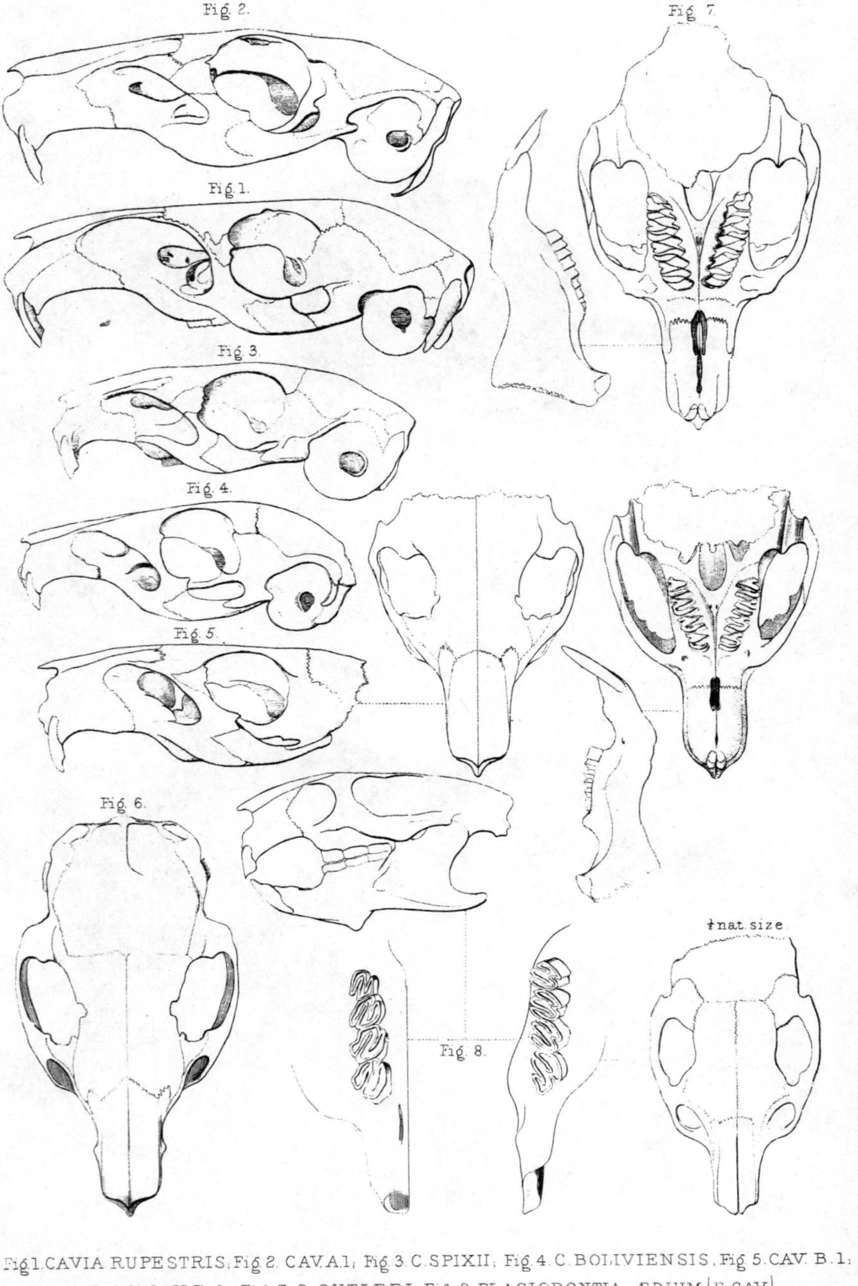

Fig. 1. CAVIA RUPESTRIS; Fig. 2. CAV. A. 1; Fig. 3. C. SPIXII; Fig. 4. C. BOLIVIENSIS, Fig. 5. CAV. B. 1;
Fig. 6. CAV. B. 2; Fig. 7. C. CUTLERI; Fig. 8. PLAGIODONTIA ÆDIUM (F. CAV.)

Engraved by H. S. Beckwith. London. Published by H. Baillière 219 Regent St. 1848. Printed by Brooker & Harrison.

Fig. 1.

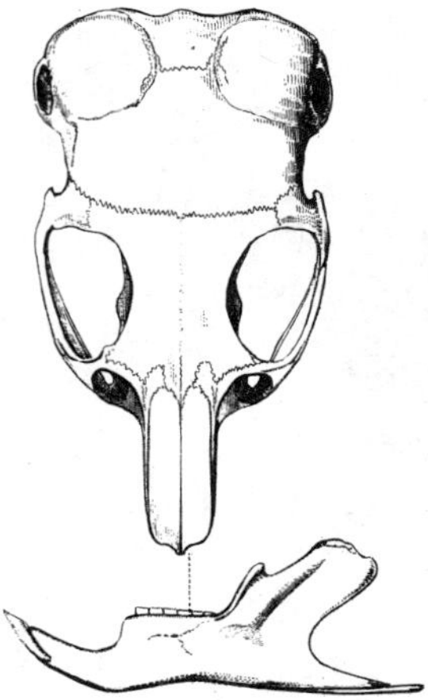

Fig. 1a.

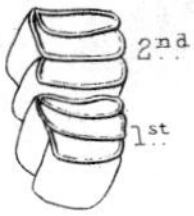

Fig. 2.

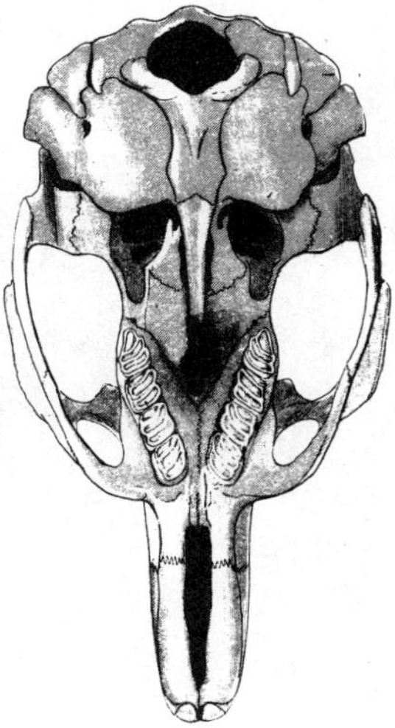

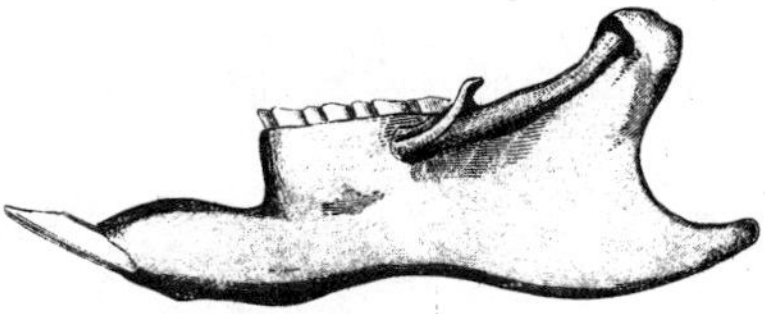

Fig. 3.

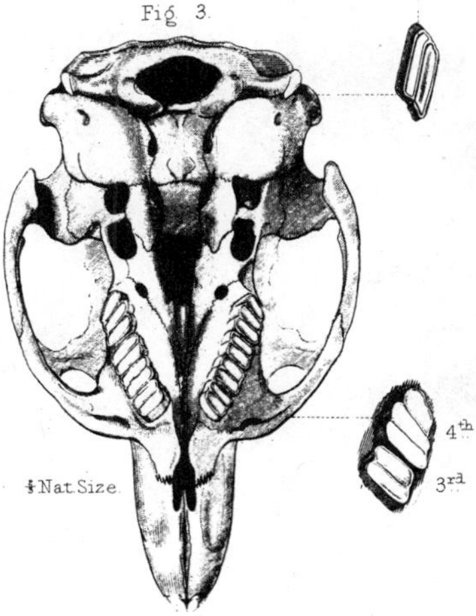

⅓ Nat. Size.

Fig. 4.

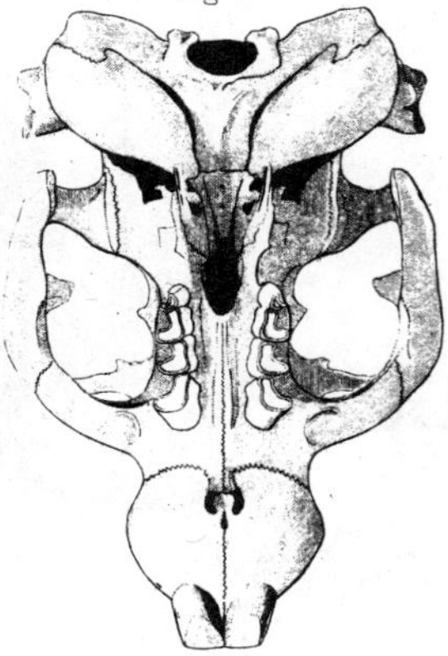

Fig. 1. CHINCHILLA LANIGERA; Fig. 2. LAGIDIUM CUVIERI;
Fig. 3. LAGOSTOMUS TRICHODACTYLUS; Fig. 4. CTENOMYS BOLIVIENSIS.

Engraved by H. S. Beckwith. *London: Published by H. Baillière, 219, Regent St. 1848.* *Printed by Brooker & Harrison.*

Fig. 1.

Fig. 2

Fig 1. SCHIZODON FUSCUS. Fig 2. OCTODON DEGUS.

Engraved by H. S. Bradwith London. Published by H. Baillière, 219 Regent St. 1847 Printed by Brooker & Harrison.

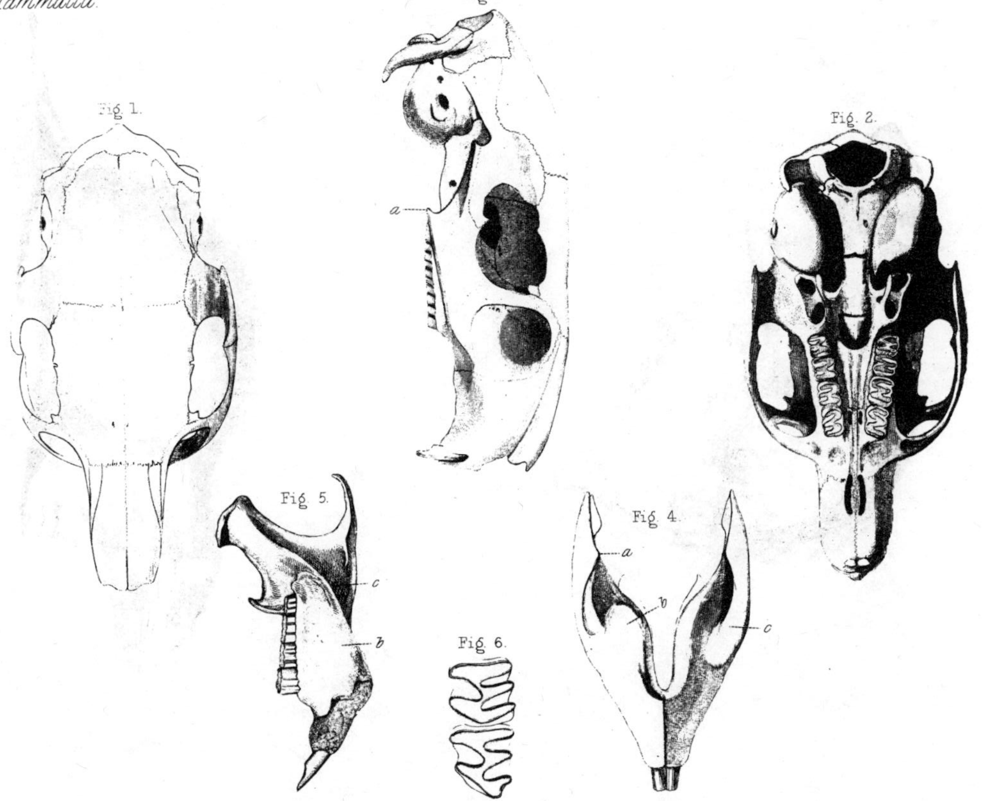

CAPROMYS.

Fig. 2

Fig. 1. CAPROMYS PILORIDES. Fig. 2. PLAGIODONTIA ÆDIUM.

Engraved by H. S. Beckwith *London. Published by H. Baillière 219 Regent St. 184[illegible]* *Printed by Brooker & Harrison.*

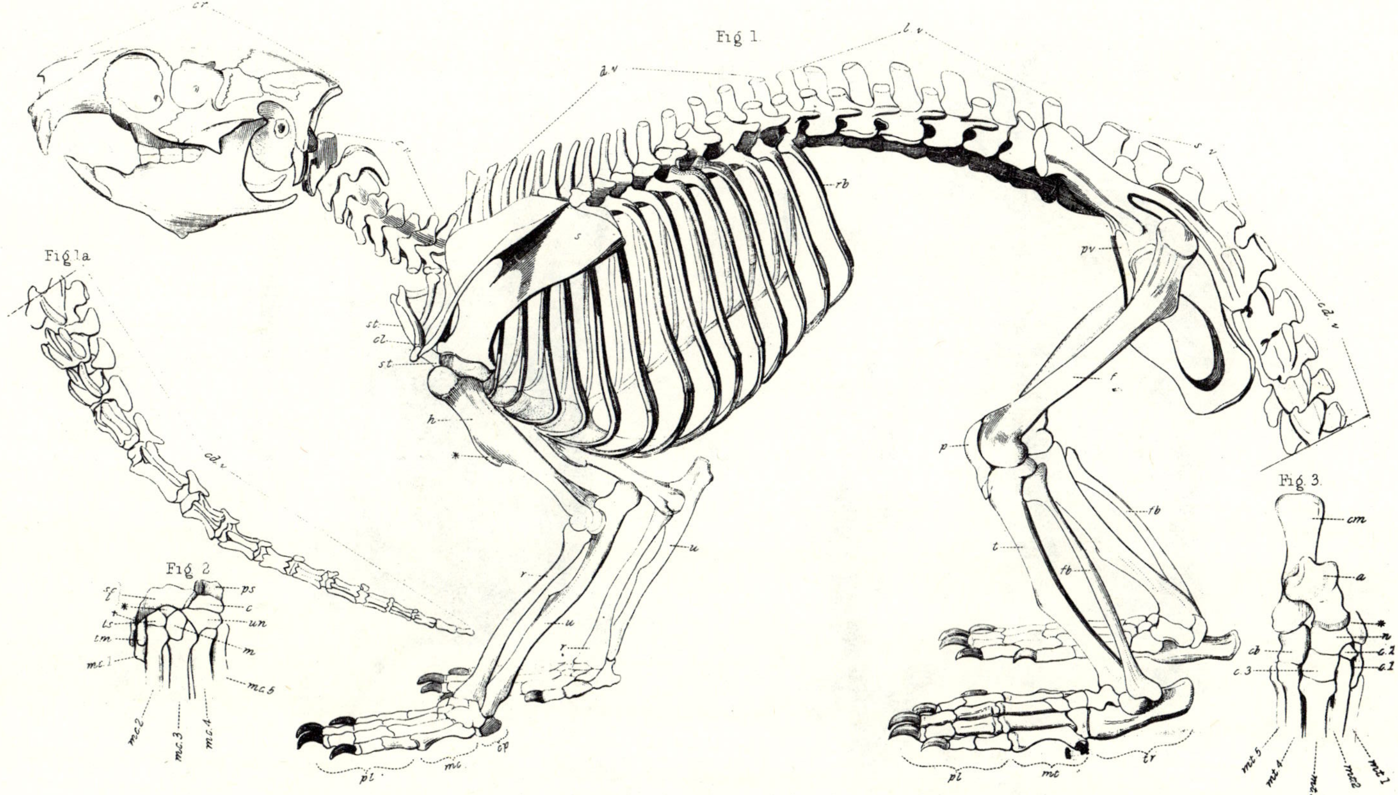

Fig 1. SKELETON OF CAPROMYS PILORIDES (R. DE LA SAGRA); Fig. 2. WRIST, & Fig. 3. TARSUS OF THE COYPU

Engraved by H. S. Beckwith. London. Published by H. Baillière, 219 Regent St. 1847. Printed by Brooker & Harrison.

Fig. 1.

Fig. 2.

Fig 1. MYOPOTAMUS COYPUS; Fig 2. AULACODUS SWINDERIANUS

Engraved by H. S. Beckwith. London. Published by H. Baillière 219, Regent St. 1848. Printed by Brooker & Harrison.

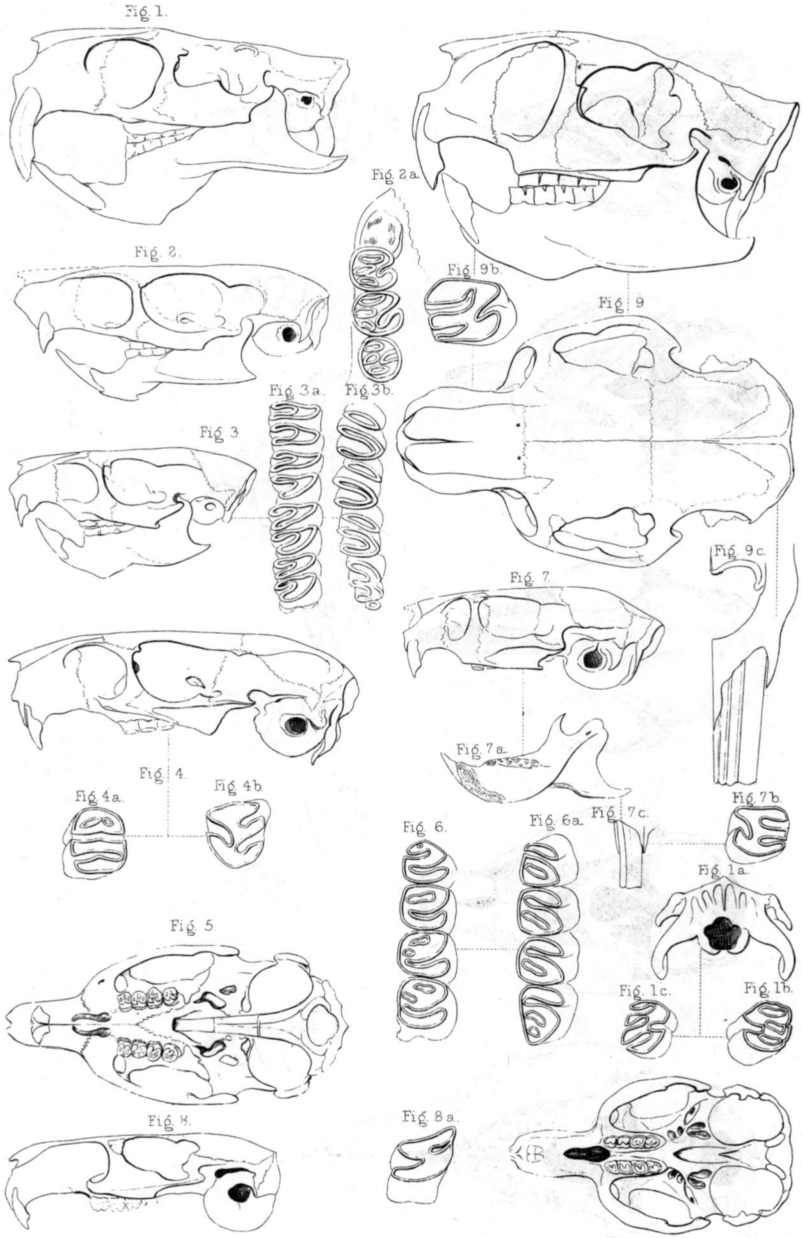

Fig. 1. MYOPOTAMUS, Fig. 2. CERCOMYS; Fig. 3. DACTYLOMYS; Fig. 4. LONCHERES; Fig. 5 and 6. ECHIMYS;
Fig. 7. CARTERODON; Fig. 8. PETROMYS; Fig. 9. AULACODUS.

Engraved by R. S. Beckwith. *London: Published by H. Baillière, 219, Regent St. 1848.* *Printed by Brooker & Harrison.*

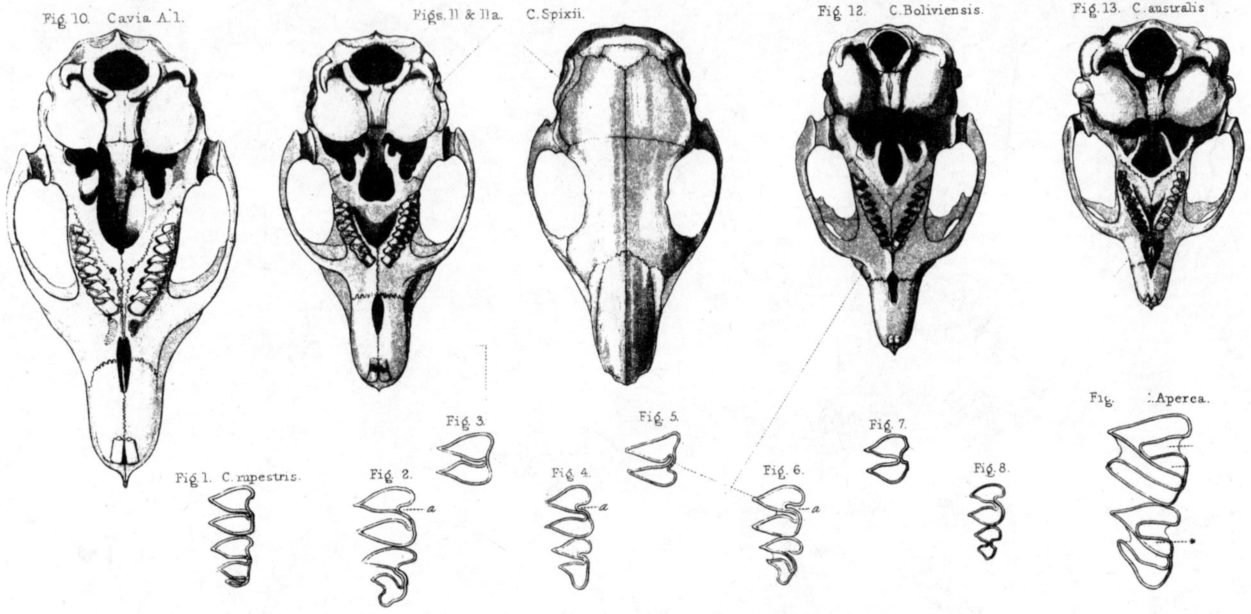

GENUS CAVIA

Figs. 3.5 & 7. SECOND MOLAR; UPPER JAW: Figs. 1.2.4.6.8 & 9. FIRST & SECOND MOLARS OF LOWER JAW.

Engraved by H. S. Beckwith. London: Published by H. Baillière, 219, Regent S^t. 1847. Printed by Brooker & Harrison.

Fig. 1.

Fig. 2.

Fig. 1. PETROMYS TYPICUS; Fig. 2. & 2 a. LONCHERES CRISTATA & ITS SPINE.

Engraved by H. S. Beckwith. London, Published by H. Baillière, 219, Regent S^t 1848. Printed by Brooker & Harrison

Fig. 1. CHÆTOMYS; Fig. 2. CERCOLABES PREHENSILIS; CERC. VILLOSUS; Fig. 4. FOOT OF CERCOLABES;
Fig. 5. ATHERURA AFRICANA; Fig. 6. A. MACROURA; Fig. 7. DASYPROCTA; Fig. 8. CŒLOGENYS.

Engraved by H. S. Beckwith. *London: Published by H. Baillière, 219 Regent St. 1848.*

Fig. 1.

Fig. 2.

Fig. 1. LONCHERES PICTA. Fig. 2. ECHIMYS CAYENNENSIS.

Engraved by H. S. Beckwith. *London: Published by H. Baillière, 219, Regent St. 1847.* *Printed by Brooker & Harrison.*

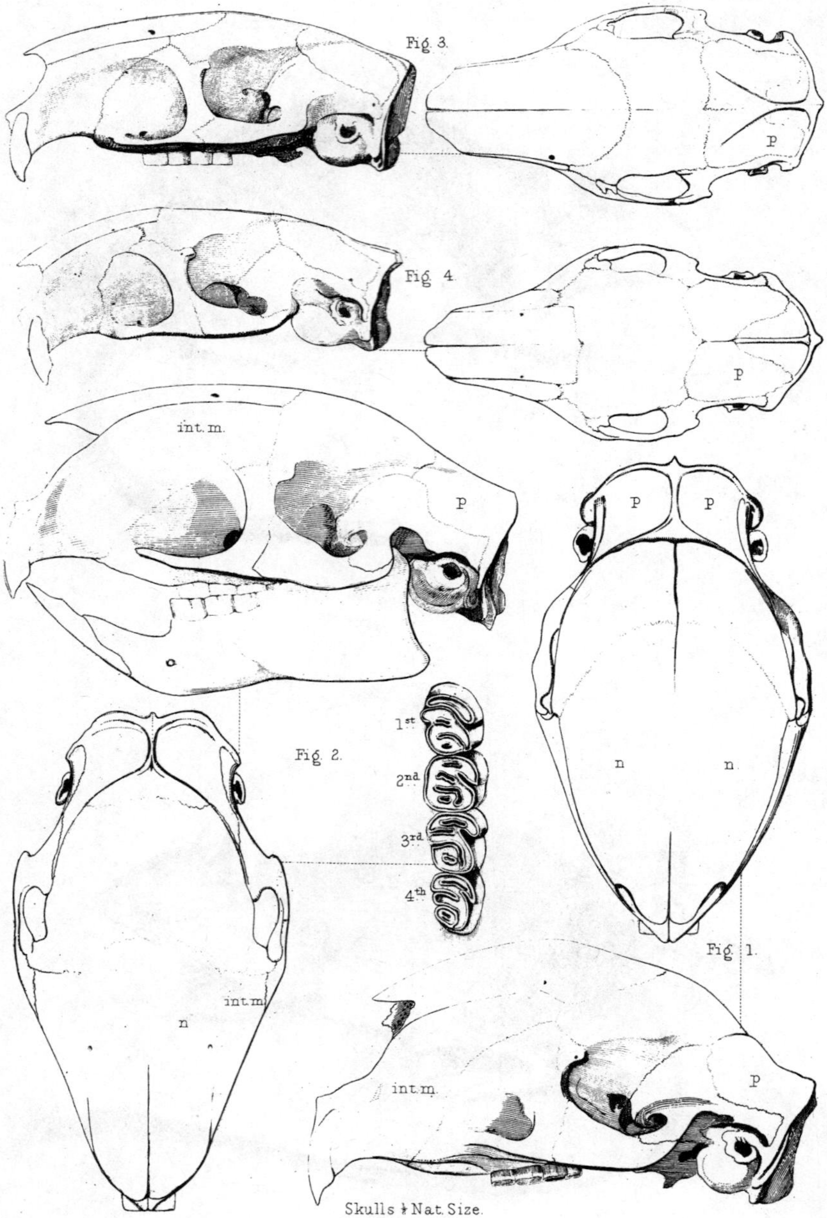

Skulls ½ Nat. Size.

Fig. 1. HYSTRIX CRISTATA, Fig. 2. H. LEUCURUS; Fig. 3. H. HODGSONII; Fig. 4. H. JAVANICA.

Engraved by H. S. Beckwith. *London: Published by H. Baillière, 219, Regent St. 1848.* *Printed by Brooker & Harrison.*

Fig. 1.

Fig. 2.

Fig 1. CHÆTOMYS SUBSPINOSUS; Fig 2. CERCOLABES VILLOSUS.

Engraved by H. S. Beckwith. London: Published by H. Baillière, 219 Regent St. 1848. Printed by Brooker & Harrison.

BIOLOGISTS AND THEIR WORLD

An Arno Press Collection

Adler, Kraig, editor. **Early Herpetological Studies and Surveys in the Eastern United States.** 1978

Adler, Kraig, editor. **Herpetological Explorations of the Great American West.** Two vols. 1978

Agassiz, [Jean] Louis [Rodolphe]. **Contributions to the Natural History of the United States of America.** Four vols. in two. 1857/1860/1862

Allard, Dean Conrad, Jr. **Spencer Fullerton Baird and the U.S. Fish Commission.** 1978

Altum, Bernard. **Der Vogel Und Sein Leben.** 1868

Azara, Don Felix de. **Apuntamientos Para La Historia Natural De Los Quadrúpedos Del Paragüay Y Rio De La Plata.** Two vols. in one. 1802

Baer, Karl Ernst v[on]. **Reden Gehalten in Wissenschaftlichen Versammlungen und Kleinere Aufsaetze Vermischten Inhalts.** Three vols. in two. 1864/1876/1873

Barrett-Hamilton, Gerald E. H. and Martin A. C. Hinton. **A History of British Mammals.** Two vols. 1910-1921

Boulenger, G[eorge] A[lbert]. **The Tailless Batrachians of Europe.** Two parts in one. 1897/1898

Brocchi, [Paul]. **Mission Scientifique Au Mexique Et Dans L'Amérique Centrale,...** Étude Des Batraciens De L'Amérique Centrale. 1882

Buffon, [Georges L. L.]. **The History of Singing Birds.** Translated from the French. 1791

Buffon, [Georges L. L.]. **The Natural History of Oviparous Quadrupeds and Serpents.** Translated by Robert Kerr. Four vols. in two. 1802

The Cabinet of Natural History and American Rural Sports. Three vols. in one. 1830/1832/1833

Candolle, A[ugustin] P[yramus] de and K[urt] Sprengel. **Elements of the Philosophy of Plants.** Translated from the German. 1821

Cassin, John. **United States Exploring Expedition During the Years 1838, 1839, 1840, 1841, 1842 Under the Command of Charles Wilkes, U. S. N.:** Mammalogy and Ornithology. Two vols. 1858

Chapman, Frank M. **Essays in South American Ornithogeography.** Edited by Keir B. Sterling. 1978

Cope, Edward D[rinker]. **The Vertebrata of the Tertiary Formations of the West.** One vol. in two. 1883

Cuvier, [Georges]. **The Class Mammalia.** Five vols. 1827

Donovan, E[dward]. **The Natural History of British Fishes.** Five vols. in two. 1802/1803/1804/1806/1808

Duméril, Auguste [Henri André], [Marie-Firmin] Bocourt and [François] Mocquard. **Mission Scientifique Au Mexique Et Dans L'Amérique Centrale,...** Étude Sur Les Reptiles. Two vols. 1870-1909

Flower, William Henry and Richard Lydekker. **An Introduction to the Study of Mammals, Living and Extinct.** 1891

Forbush, Edward Howe. **Birds of Massachusetts and Other New England States.** Three vols. 1925/1927/1929

Girard, Charles. **United States Exploring Expedition During the Years 1838, 1839, 1840, 1841, 1842 Under the Command of Charles Wilkes, U. S. N.:** Herpetology. Two vols. 1858

Grinnell, Joseph. An Account of the Mammals and Birds of the Lower Colorado Valley. 1914

Howard, H[enry] Eliot. **Territory in Bird Life.** 1920

Hume, Edgar Erskine. **Ornithologists of the United States Army Medical Corps.** 1942

Huxley, Leonard. **Life and Letters of Sir Joseph Dalton Hooker.** Two vols. 1918

LeConte, John L. and George H. Horn. **Classification of the Coleoptera of North America.** 1883

Linsley, E. Gorton, editor. **Beetles From the Early Russian Explorations of the West Coast of North America, 1815-1857.** 1978

Linsley, E. Gorton, editor. **The Principal Contributions of Henry Walter Bates to a Knowledge of the Butterflies and Longicorn Beetles of the Amazon Valley.** 1978

Murray, Andrew. **The Geographical Distribution of Mammals.** 1866

Osborn, Henry Fairfield. **Cope:** Master Naturalist. 1931

Peale, Titian R. **United States Exploring Expedition During the Years 1838, 1839, 1840, 1841, 1842 Under the Command of Charles Wilkes, U. S. N.:** Mammalogy and Ornithology. 1848

Phillips, John C. **American Game Mammals and Birds.** 1930

Rafinesque: Autobiography and Lives. 1978

Ray, John. **Synopsis Methodica Animalium Quadrupedum et Serpentini Generis.** 1693

Ray, John. **Synopsis Methodica Avium & Piscium.** [Edited by William Derham]. 1713

Richardson, John. **Fauna Boreali-Americana;** Or the Zoology of the Northern Parts of British America, Part Third: The Fish. 1836

Richardson, John, William Swainson and William Kirby. **Fauna Boreali-Americana;** Or the Zoology of the Northern Parts of British America, Part Four: Insecta, 1837

Riley, Charles V[alentine]. **[Nine] Annual Report[s] on the Noxious, Beneficial and Other Insects of the State of Missouri, 1869-1877** *and* **General Index and Supplement.** Ten vols. in three. 1869-1877/1881

Say, Thomas. **The Complete Writings of Thomas Say on the Entomology of North America.** Edited by John L. LeConte. Two vols. 1859

Schuchert, Charles and Clara Mae LeVene. **O. C. Marsh:** Pioneer in Paleontology. 1940

Sclater, William Lutley and Philip Lutley Sclater. **The Geography of Mammals.** 1899

Seton, Ernest Thompson. **Trail of an Artist-Naturalist:** The Autobiography of Ernest Thompson Seton. 1940

Smith, James Edward. **A Selection of the Correspondence of Linnaeus and Other Naturalists, From the Original Manuscripts.** Two vols. 1821

Stuckey, Ronald L., editor. **Development of Botany in Selected Regions of North America Before 1900.** 1978

Stuckey, Ronald L., editor. **Essays on North American Plant Geography From the Nineteenth Century.** 1978

Stuckey, Ronald L., editor. **Scientific Publications of Charles Wilkins Short.** 1978

Swammerdam, John. **The Book of Nature.** 1758

Wadland, John Henry. **Ernest Thompson Seton:** Man in Nature and the Progressive Era, 1880-1915. 1978

Waterhouse, G[eorge] R[obert]. **A Natural History of the Mammalia.** Two vols. 1846/1848

Weiss, Harry B. and Grace M. Ziegler. **Thomas Say:** Early American Naturalist. 1931

Wheeler, Geo[rge] M., [editor]. **[Reports Upon Insects Collected During Geographical and Geological Explorations and Surveys West of the One Hundredth Meridian During the Years 1872, 1873, and 1874].** 1875

Willughby, Francis. **De Historia Piscium** *and* **Icthyographia Ad Amplisimum Virum Dnum., Samuelem Pepys, Presidem Soc. Reg.** Two vols. in one. 1686/1685

Youmans, William Jay, editor. **Pioneers of Science in America.** Revised edition. 1896

A NATURAL HISTORY OF THE MAMMALIA

This is a volume in the Arno Press collection

BIOLOGISTS AND THEIR WORLD

See last pages of this volume
for a complete list of titles

A

NATURAL HISTORY

OF THE

MAMMALIA

G[eorge] R[obert] Waterhouse

Volume I

ARNO PRESS
A New York Times Company
New York • 1978

Editorial Supervision: MARIE STARECK

Reprint Edition 1978 by Arno Press Inc.

BIOLOGISTS AND THEIR WORLD
ISBN for complete set: 0-405-10641-6
See last pages of this volume for titles.

Manufactured in the United States of America

Library of Congress Cataloging in Publication Data

Waterhouse, George Robert, 1810-1888.
A natural history of the Mammalia.

(Biologists and their world)
Reprint of the 1846-1848 ed. published by H. Baillière, London.
Includes index.
1. Marsupialia. 2. Rodentia. I. Title. II. Series.
QL737.M3W37 1978 599.2 77-83843
ISBN 0-405-10739-0

A

NATURAL HISTORY

OF THE

MAMMALIA.

BY

G. R. WATERHOUSE,

CORRESPONDING MEMBER OF THE PHILOMATIC SOCIETY OF PARIS,

ASSISTANT IN THE DEPARTMENT OF MINERALOGY AND GEOLOGY OF THE BRITISH MUSEUM.

VOL. I.

CONTAINING THE ORDER

MARSUPIATA, OR POUCHED ANIMALS,

WITH

22 ILLUSTRATIONS ENGRAVED ON STEEL, AND 18 ENGRAVINGS ON WOOD.

LONDON:

HIPPOLYTE BAILLIÈRE, PUBLISHER, 219, REGENT STREET;

FOREIGN BOOKSELLER TO THE ROYAL COLLEGE OF SURGEONS,
AND TO THE ROYAL MEDICO-CHIRURGICAL SOCIETY.

PARIS: J.-B. BAILLIÈRE, LIBRAIRE DE L'ACADÉMIE ROYALE DE MÉDECINE.

LEIPSIC: T. O. WEIGEL.

1846.

ADDRESS.

The Natural History of Mammalia is intended to embrace an account of the structure and habits of all the known species of Quadrupeds, or Mammals; to which will be added observations upon their geographical distribution and classification. Since the fossil and recent species illustrate each other, it is also intended to include notices of the leading characters of the Extinct Species.

The genera, and many of the species, will be illustrated partly by engravings on steel, and partly by wood engravings; of the latter, between fifteen and sixteen hundred figures are executed. The modifications observable in the structure of the skulls, teeth, feet, and other parts, will be almost entirely illustrated by steel engravings.

For various reasons it has been thought desirable in this work to follow the course usually adopted by comparative anatomists; *viz.* to commence with the lowest species, and to trace through them to the highest, the various modifications of structure and functions. In the present case, one of the advantages arising from this mode of treating the subject is, that it will enable the author to lay before the public, at an early period, an account of some of the most interesting of modern discoveries amongst the Mammalian forms; and, in accordance with this plan, he has commenced with the Marsupial animals.

The work is issued in Monthly Parts, each containing three sheets of Letterpress and Engraved Plates, besides Woodcuts. Price, with the Plates coloured, 3s.; with the Plates plain, 2s. 6d. each Part.

The size of the work corresponds with Dr. Prichard's *Natural History of Man*, to which it will form a suitable companion.

ORDER—MARSUPIATA.

THE Marsupiata, or Marsupialia[1], are implacental mammalia, whose young are brought forth prematurely, and received, in most instances, into a peculiar pouch situated on the lower part of the abdomen of the female: both sexes are furnished with two bones, called marsupial bones, and which are attached to the anterior margin of the pelvis. The brain is deficient of both

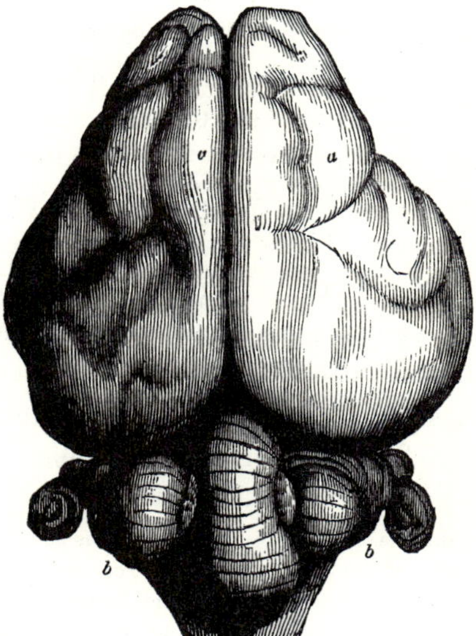

Brain of a Kangaroo—(Owen).

a. a. The Cerebrum.
b. b. The Cerebellum.

[1] From *marsupium*, a purse or bag —having allusion to the peculiar pouch with which the female marsupial animals are furnished.

corpus callosum and septum lucidum: the cerebrum is small in proportion to the animal, contracted in front, and its surface is smooth, or presents but few convolutions: the cerebellum is entirely exposed[1], and has the vermiform process large in proportion to the lateral lobes; the olfactory lobes are large. Two venæ cavæ enter the heart[2].

The order Marsupiata embraces a large assemblage of quadrupeds, amongst which are those animals familiarly known as Opossums and Kangaroos. At the present period the great metropolis of the order is Australia; certain species of the group, however, are found in the Molucca Islands, and one genus, containing many species, is peculiar to the New World, and although chiefly confined to the tropical portions, is met with as far north as the United States (where, however, one species[3] only is found), and extends in the southern direction to Buenos Ayres[4], on the east side of the Andes, and Valparaiso on the west[5]. The species found in the islands north of Australia (if we except New Guinea) all belong to one genus (Phalangista), which occurs likewise in Australia, but they agree amongst each other in having the fur short, dense, and crisp, and the tail with the apical portion naked, and studded with fleshy warts,—characters which, in the opinion of some naturalists, entitle them to the rank of a subgenus to which the name *Cuscus* is applied. No Phalan-

[1] In some marsupialia (such as the true Opossums and the Dasyuri), even the corpora quadrigemina, or optic lobes, are not covered by the cerebrum.

[2] We learn from Prof. Owen's Papers that in the Marsupialia the right auricle of the heart has no trace of a *fossa ovalis*, or an *annulus ovalis*. In the Ornithorhynchus, Meckel found a deep, but closed, *fossa ovalis*. "Ad characteres supra dictos addantur. Uterus bipartitus; vagina, vel in toto, vel partim septo divisa; canales, et intestinales, et genitalis in cloacam communem recepti; maribus scrotum ante penem positum."

[3] *Didelphys virginiana.*

[4] *Didelphys crassicaudata*, and *D. brachyura.*

[5] *Did. elegans.*

gers presenting these peculiarities have been found in Australia. In the following table the names of the various species of *Cuscus* are arranged under those of the islands in which they occur.

Celebes.	Amboyna.	Banda.	Waigiou.	Timor.	New Guinea.	New Ireland.
ursinus	...	...	...	...	...	...
...	maculatus	maculatus	maculatus	...	maculatus	...
...	chrysorrhos	...	...	...	...	...
...	cavifrons	cavifrons	...	cavifrons	...	cavifrons.

In New Guinea the Marsupialia increase in numbers, since, notwithstanding its coast line is not yet completed on our charts, and the interior is almost unknown, seven species of marsupial animals have been discovered in that island, and these appertain to six distinct genera, one of which only, it must be observed, is peculiar to New Guinea[1]. Of the species, five are peculiar; one is said to be identical with a species found in New South Wales[2], and the seventh is the *Cuscus maculatus*, included in the above table.

With respect to the Marsupialia inhabiting the continent of Australia, it will be seen, upon consulting the observations on the habits and ranges of the species given under their respective heads, that, *generally*, species which are very nearly allied, and have very nearly similar habits, are not associated together in the same limited district. Of four nearly allied species of Kangaroos, which, from their habits, are called

[1] The genus Dendrolagus.

[2] *Petaurus sciureus.*—Of this identification I cannot help feeling some doubt; the animal called *Petaurus sciureus* by M. Müller, in his table displaying the geographical distribution of the Mammalia of the Indian Islands, I think is more probably identical with the *Pet. ariel* of Gould, a nearly allied species found in North-Australia.

rock-kangaroos, and which constitute the little section *Petrogale*, one inhabits the hills of New South Wales, a second is found on the opposite side of the continent, a third has only been met with on the south coast, and the fourth on the north-west coast.

Many similar cases might be noticed. No doubt, in a large district, such as New South Wales, are found several species of the same section, as, for instance, the section *Halmaturus:* but of the several species of *Halmaturus*, or Brush-Kangaroos, inhabiting that portion of Australia, we learn from Mr. Gould that they each affect different minor districts. Some are fitted for one kind of scrub, and some for another; some prefer the swamps, and others the high table land; such a difference of habits is observed in the two species which are most nearly equal in size and power—the *H. ualabatus* and the *H. ruficollis* of Gould.

Viewing the Marsupialia as an order, it is one which presents a remarkable diversity of structure (and, consequently, habits), containing herbivorous, carnivorous, and insectivorous species: indeed, we find amongst the marsupial mammals analogous representations of most of the other orders of Mammalia. The *Quadrumana* are represented by the Phalangers, the *Carnivora* by the Dasyuri, the *Insectivora* by the small Phascogales, the *Ruminantia* by the Kangaroos, and the *Edentata* by the Monotremes. The *Cheiroptera* are not represented by any known marsupial animals, and the Rodents are represented by a single species only; the hiatus is filled up, however, in both cases, by placental species, for both Bats and Rodents are tolerably numerous in Australia, and, if we except the dog, which it is probable has been introduced by man, these are the only placental mammalia found in that continent.

The most striking peculiarity in the marsupial animals consists in the premature birth of their young, and consequently the imperfect state of development which they present at this period. The young of the great Kangaroo (Macropus major), which Professor Owen examined twelve hours after birth, "resembled an earth-worm in the colour and semi-transparency of its integument, adhered firmly to the points of the nipple, breathed strongly but slowly, and moved its fore legs when disturbed. The body was bent upon the abdomen, its short tail tucked in between the hind legs, which were one-third shorter than the fore legs; but with the three divisions of the toes more distinct. The whole length, from the nose to the end of the tail, did not exceed one inch and two lines[1]." The young of a species of Kangaroo (probably the *Halmaturus Derbianus*), described by Mr. Collie[2], that gentleman states was also perfectly naked; its size was nearly equal to that of the last and half of the middle joint of one's little finger.

Four days after the birth of the young Kangaroo, Professor Owen, being anxious to decide the nature of the connection between it and the nipple of the parent, and to ascertain whether so small a fœtus would manifest the powers of a voluntary agent in regaining the nipples, detached it, and, after two days, upon again examining the pouch, he found it empty; every portion of the litter was carefully examined, in hopes of finding the fœtus, but without success, and it is supposed that, owing to the young being disturbed, the mother had destroyed it. A similar experiment was tried by Mr. Morgan[3], on a fœtus about the size of a Norway rat, and which, after two hours' separation from the nipple, regained its hold, and sustained no injury from the interruption of the supply of nourishment. A young Kangaroo, born in the managerie of Sir Robert Heron, and which is described as being perfectly naked, having accidentally got out of the

[1] Philosophical Transactions, Part 2 for 1834.

[2] Zoological Journal, vol. v. p. 239.

[3] See Transactions of the Linnean Society, Part 2, for 1834.

pouch of the mother, and which was scarcely alive when pointed out to the keeper, was first fed with some milk, and afterwards restored to the pouch of the parent, where, five days after, it was found to be alive and apparently in a healthy condition[1].

An animal so little advanced at the time of its birth as the young marsupial, requiring a constant supply of food, and so ill fitted to bear the exposure which the more advanced young of other Mammalia are subject to, must, it would appear, perish, were not some peculiar provision for its safety provided, and in the pouch of the female marsupial animals we find such a provision. This pouch, when the animal is very young, has its orifice closed, and glued, as it were, to the body of the parent by a peculiar secretion. When the young animal is more advanced, this secretion disappears, and the young frequently leave the pouch to return at will: they do not entirely quit the pouch until they have attained a large size, as compared with the parent.

Closely connected with the pouch, and with the generation of the animals of the present group, are the marsupial bones which so peculiarly characterize it. These bones are even more constant than the pouch, being found in the Echidna and Ornithorhynchus, in which no traces of the pouch have been discovered; and in some of the Opossums, in which the pouch is only represented by two small folds[2].

The marsupial bones are elongated and more or less flattened, widely separated at their distal extremity, and converging as they approach the pubis to which they are joined:

[1] The particulars of this case will be found in a letter addressed to the Secretary of the Zoological Society, by Sir Robert Heron, in the Proceedings of that Society for July, 1840.

[2] The *Thylacinus cynocephalus* affords the only exception, hitherto found, of a marsupial animal in which the marsupial bones are wanting; at least they are here only represented by two cartilages.—See observations on this subject by Prof. Owen, in Proceedings of the Zoological Society, for Dec. 1843, p. 148.

they are found in both sexes of the marsupial animals; are relatively longest, straightest, and most slender in the Perameles; flattest, broadest, and most curved in the Koala,—sometimes, as in the Wombat, they are articulated to the pubis by two points. Around these bones the cremaster muscle winds, and they serve important purposes in relation to the generative economy of the Marsupiata. "In the female they assist in producing a compression of the mammary gland necessary to the alimentation of a peculiarly feeble offspring, and they defend the abdominal viscera from the pressure of the young, as these increase in size during their mammary or marsupial existence; and still more when they afterwards return to the pouch for temporary shelter[1]."

One of the most interesting features in the skull of the Marsupiata consists in the permanent separation of the bones; these do not anchylose in the adult and old animals as do many of the bones (especially those of the cranial portion of the skull) in the placental series: the temporal bone generally presents a permanent separation of the squamous, petrous, and tympanic elements. "I have observed," says Professor Owen, "this reptile-like condition of the bone in the mature skulls of an Ursine Dasyure, a Virginian Opossum, a Perameles, in different species of Potoroo (or Kangaroo-rats) and Kangaroo, in the Wombat and in the Koala." The palatine portion of the skull is generally very imperfect, presenting large openings. In all the species of the group, with the exception of the Echidna, Ornithorhynchus, and Tarsipes, the angle of the jaw is bent inwards, so that in viewing the underside of the jaw each ramus presents a more or less flattened and pointed process encroaching upon the interspace of the two branches behind, whilst in the

[1] See Professor Owen's paper on the *Osteology of the Marsupialia*, in the Proceedings of the Zoological Society for 1838.

placental mammalia the angle of the jaw is almost invariably on the same plane with the horizontal ramus.

In addition to these points of distinction between the placental and implacental or marsupial animals, there are others which are derivable from the number of the teeth. In the placental mammalia the normal number of incisors in both jaws is six, and of the true molars three on each side of each jaw, whilst in the present group no species has yet been found possessing six incisors in each jaw; the highest number found is in the Opossums of America, where there are ten in the upper jaw and eight in the lower: next follow the Dasyuri, in which they are eight above and six below; and in a great number of the species, as in the Kangaroo tribe, there are six above and two below. With the exception of the Wombat, which has two above and the same number below (as in the Rodents), there is no marsupial animal known possessing an equal number of incisors in the upper and lower jaws. With the exception of the edentate species of Marsupials, or those which are nearly edentate, like the Tarsipes; and also excepting the Myrmecobius, all marsupialia possess four true molars; that is, four molar teeth on each side of each jaw, which have never replaced other teeth in the vertical direction—a number not clearly made out to be found in any other group of quadrupeds, but possibly may prove to exist in the insectivorous quadrupeds forming the genus *Centetes*.

With regard to the structure of the molar teeth, they often correspond with the same teeth in the ordinary mammalia; when much complicated, as in the herbivorous, and especially the insectivorous species, they present the four principal cusps, as in the placental herbivora and insectivora. In the Kangaroo-rats, these four cusps or tubercles are distinct, but in the true Kangaroos the anterior and posterior pairs are joined, so as to present, before they are worn, two transverse sharp ridges. Besides the four principal cusps, complicated molar teeth have frequently a band or projecting ridge, more or less perfectly encircling the tooth at the base of the exposed portion, and this band throws up small tubercles in parts; frequently on the outer side of the tooth a small tubercle is thrown up from the anterior angle of the tooth, a second tubercle in the middle, and a third at the posterior angle. By the development or non-development of one or more of these parts, most of the varieties observed in complicated molar teeth may be explained; and in my descriptions I shall endeavour to trace and point out these modifications in the different genera.

Of the affinities of the various minor groups of which the order Marsupialia is composed, with the exception of one or two cases, there can be very little difference of opinion; the modifications of the extremities, combined with the characters furnished by the digestive system, very clearly indicate the natural divisions, and their relationship to each other.

In the structure of the stomach, Professor Owen points out three leading modifications[1]; it is simple in the genera *Didelphys, Myrmecobius,* and *Perameles,* and likewise in the *Dasyurus* and *Phalangista* groups; also simple in the Koala and Wombat, but in these two animals it is provided with a glandular apparatus situated to the left of the cardiac orifice; the third modification is exhibited in the Kangaroo section, where it is complicated by sacculi. The small intestine commences with a cœcum in all the groups with the exception of the *Dasyuridæ;* the cœcum, according to Professor Owen, is moderately long in the *Myrmecobius, Perameles,* and *Didelphys,* very long in the *Phalangistidæ* (including the Koala), long in the *Macropodidæ,* and short, wide, and with a vermiform appendage, in the Wombat. Setting aside the Monotremes, we find one type of dentition exhibited in the car-

[1] See the article "Marsupialia," in Todd's Cyclopædia of Anatomy and Physiology.

nivorous and insectivorous species, in which there are more than six incisors in the upper jaw, distinct canines, and at least two false molars on either side of each jaw: a second type is presented by the Kangaroo group, where the canines are wanting, or very small, and there exists but one false molar on each side of the jaw; from these latter herbivorous species should be separated the Wombat, in which the molar and incisor teeth differ in being rootless, and the latter are reduced in number as before stated. But we have yet to notice another large group, the *Phalangistidæ*, in which we find the teeth presenting intermediate conditions to the two great sections just mentioned; they differ from the first section in having but six incisors in the upper jaw, and in having the two foremost incisors of the lower jaw very large and nearly horizontal, and the others rudimentary or wanting; there is also considerable difference in the form of these teeth. From the second section they differ in the structure of the molar teeth, as will hereafter be pointed out, and (with one exception only —the Koala) in having more than one false molar to each side of each jaw.

In the various species composing the sections noticed in the following Table, wherever the inner toe of the hind foot is developed, it assumes the form of a thumb, and is opposeable to the other toes, and thus gives a prehensile power to the hind foot, adapting it to climbing: hence we find the thumb developed in all climbing species, whether carnivorous, insectivorous, or herbivorous; but in the ground-inhabiting species of each of these sections, the inner toe of the hind foot is reduced to a rudimentary condition, or wanting; in both the herbivorous and animal-feeding groups, however, we find a gradual transition from the thumbless to species possessing a well-developed thumb. In the *Didelphidæ*, for instance, the thumb is in many well developed; in others it becomes very small, as in some of the *Dasyuridæ;* and in this last mentioned group we find a tolerably distinct thumb in the Phascogali, a rudimentary thumb in the *Dasyurus macrourus,* no external thumb in *D. Maugei;* but here the internal metatarsal bone exists; and, lastly, this metatarsal bone is absent in the *Thylacinus.* In the herbivorous group, we find the intermediate stages in degradation of the hind foot, in passing from the *Phalangistidæ* through the *Peramelidæ*, to the Kangaroo group.

The principal modifications of the extremities are expressed in the following Table:—

I. Second and third toes (counting the toe corresponding to the great toe as the first) free.	
A. Thumb to the hind foot very small, or wanting	DASYURIDÆ. Mymecobius.
B. Thumb distinct	DIDELPHIDÆ.
II. Second and third toes of the hind feet slender, and united in a common integument.	
A. The four limbs of nearly equal length; the hind feet short, and furnished with a distinct opposeable thumb	PHALANGISTIDÆ. Phascolomys.
B. The fore limbs distinctly shorter than the hind, which are much developed—the foot long.	
a. Hind foot having an inner metastarsal[1], and sometimes one, and even two[2], of the phalanges of the thumb, or inner toe; fore feet with the outer toes rudimentary and nailless, or wanting . . .	PERAMELIDÆ.
b. No inner metatarsal bone; fore feet with all the toes well developed, and having claws	MACROPODIDÆ.

The mutual relations of the different genera of marsupials,

[1] Chœropus, no doubt, forms an exception.

[2] Professor Owen has observed a second phalanx to the inner toe of the hind foot, of a species of Perameles, the terminal phalanx being divided by a longitudinal fissure at the extremity, as in the nail-bearing phalanges of the other toes. This is remarkable, since such a structure would lead one to suppose the inner toe in this animal had a nail: in no marsupial have I yet met with a nail to the inner toe in question.

as indicated by the above structural characters, may be symbolically expressed by arranging them in the following manner:—

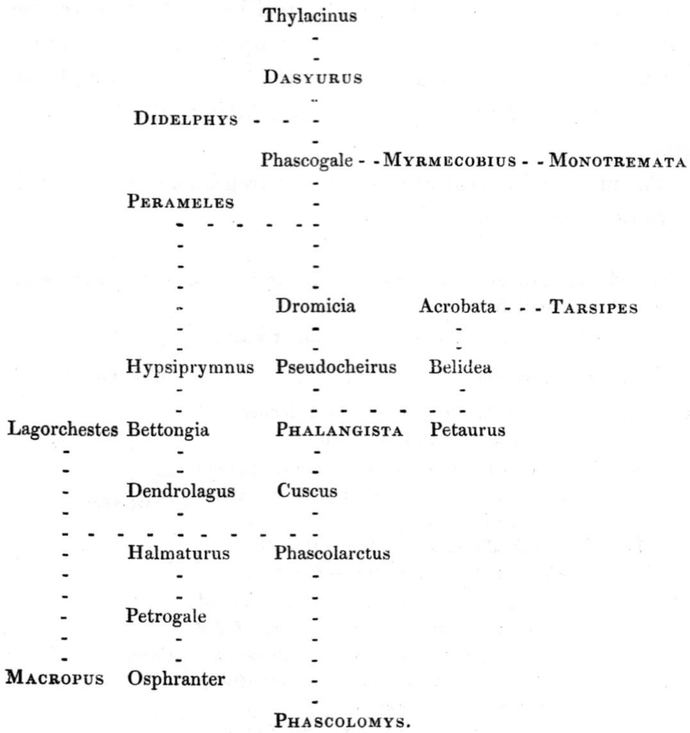

Remains of marsupial animals have been found in Europe, Australia, and South America. In the caverns which occur in the calcareous rocks in the interior highlands of Brazil, the bones of the small opossums are very numerous, but these are for the most part very nearly allied to, if not absolutely identical with, species still living in Brazil. In Europe, the most authentic remains of marsupial animals are those found in the gypsum quarries of Montmartre; the greater portion of a skeleton (fig. A), in which the marsupial bones remain in their natural position with regard to the pelvis (fig. B), in which the

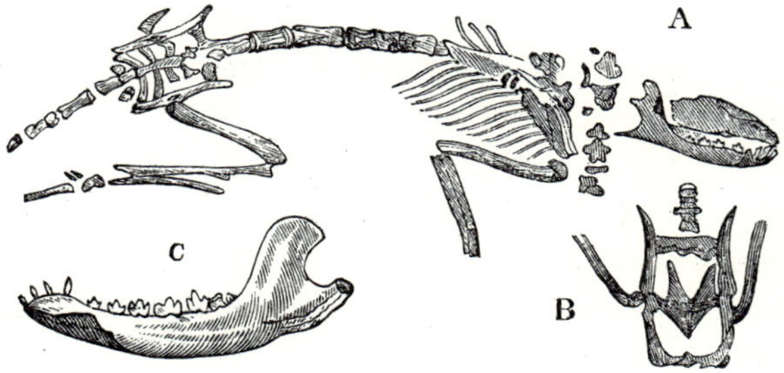

lower jaw presents the angular portion curved inwards, and in which the molar teeth correspond precisely with those of the genus Didelphys[1], leaves no doubt as to the existence of marsupial animals in Europe in the Eocene geological period: besides this Opossum-like animal, a lower jaw found in the same quarries has been referred to the marsupial group—it forms the genus Pterodon of De Blainville: a small fragment of a lower jaw, having a single false molar in situ, found in the London clay formation, near Woodbridge, in Suffolk, is supposed by Mr. Charlesworth[2] to have appertained to an animal of the marsupial group; I do not believe, however, that satisfactory evidence of the affinities of a quadruped can be deduced from such scanty materials. In formations considerably lower in the series than the Eocene—in the Stonesfield oolite—have been found several rami of lower jaws of small Insectivorous Mammalia, some of which, in their general form, and in the dentition which they present, approximate very closely to the Insectivorous Marsupialia; these, however, cannot, with propriety, be arranged in any known recent group of Marsupialia; they form the genera *Thylacotherium*, or *Phascolotherium*, of Professor Owen. To the latter genus belongs the

[1] See Cuvier's Ossemens Fossiles, 4to. ed. 1822, tom. iii. p. 284, plate 71.
[2] Magazine of Natural History for September 1839, p. 540.

Didelphys Bucklandi of Mr. Broderip, founded upon a very perfect ramus of a lower jaw (fig. c) presented by that gentleman to the British Museum.

The marsupial remains found in Australia are chiefly from the caverns of Wellington Valley, New South Wales. A considerable collection of these remains was formed by Major (now Sir T. L.) Mitchell, and are described in that gentleman's work[1] by Professor Owen. It appears, from the examination of these and other fossil mammalian remains from Australia, that they are referable, for the most part, to genera still presenting living species in the same country; there are some, however, which exhibit some very remarkable modifications of the marsupial structure, and form the types of new genera, which are noticed in their proper place.

Note on the rank of the section Marsupiata.

In the first edition of the *Règne Animal,* Cuvier treated the Marsupiata as a family of the order Carnivora; but, in the last edition, he forms an order of the marsupial animals, because, as he observes, they present so many singular features in their economy, and, above all, because they exhibit a kind of representation of three very different orders. But he immediately afterwards, apparently influenced chiefly by such considerations, states, that the group in question might be regarded as a *distinct class*, parallel to that of ordinary quadrupeds, and, in like manner, capable of being divided into orders.

In the most recently published classification of the Mammalia[2], the marsupial animals are regarded as a class, and are arranged parallel with the ordinary placental mammalia. By Prof. Owen, and some other naturalists, the present section is ranked as a subclass.

As considerable stress has been laid upon the correspondence of certain

[1] Three Expeditions into the interior of Eastern Australia, &c. by Major T. L. Mitchell, Surveyor-General.

[2] Mammifères: Classification parallèlique de M. Isidore Geoffroy Saint-Hilaire. 1845.

divisions of the Marsupiata to certain divisions of the placental mammalia, I will here show some of the comparisons which have been made, and I shall afterwards add a few remarks upon them.

The first parallel arrangement of divisions of Mammalia which follows, exhibits the views of Cuvier upon the correspondence of the groups of the placental and implacental series; next follows an illustration of those of M. Isidore Geoffroy, and lastly, of those of Prof. Owen.

Cuvier.

Marsupial.	*Placental.*
Didelphys . . Dasyuridæ . Peramelidæ .	Insectivora part.
Phalangista . Hypsiprymnus	Insectivora part.
Phascolomys .	= Rodentia.

Isid. Geoffroy.

Dasyuridæ . Didelphidæ . Peramelidæ . Myrmecobius . Tarsipes . .	Carnivora. Insectivora.
Phalangistidæ . Macropodidæ . Phascolomys .	Rodentia.

Owen.

Marsupial.	*Placental.*
Thylacinus . Dasyurus . .	Carnivora.
Phascogale . Myrmecobius . Perameles . .	Insectivora.
Didelphidæ . Phalangistidæ .	Quadrumana.
Phascolarctus .	American Sloths, or Arboreal Sun Bears.
Phascolomys .	= Rodentia.
Petaurus . .	= Pteromys.
Macropus . .	= Herbivora.

From an inspection of this list, it would *appear* that the groups of the placental mammalia are by no means strongly typified by those of the implacental series, or there would be a greater amount of correspondence in the views of these authors. Nevertheless, certain marsupial animals present a very striking resemblance in general appearance, and in the functions they perform, to certain sections of the ordinary Mammalia, and I perfectly agree, so far, in the justness of the comparison instituted by Prof. Owen between the carnivorous and insectivorous Marsupialia, and the carnivorous and insectivorous Placentalia. The herbivorous marsupials permit, in the same manner, of a tolerably close comparison with the higher herbivorous groups. Admitting these points of correspondence between the two so-called parallel classes of Mammalia, I cannot agree with Cuvier and Isidore Geoffrey St.-Hilaire if they educe from them grounds for raising the Marsupiata to the rank of a class parallel to the ordinary

Mammalia. Such an interpretation cannot be made of these analogies, for in so doing we must admit that those groups of Marsupialia, which typify the orders *Carnivora*, *Insectivora*, *Rodentia*, &c., are groups of equal *rank* to the orders they typify, and if we carry our analogies further, we shall find that such a view will be by no means borne out. For instance, in the structure of the brain, there is so much difference between the placental *Insectivora* and the *Carnivora*, that many naturalists do not admit of these two sections being arranged *near* to each other, and by almost all who have recently treated of the classification of the Mammalia, the Insectivora and Carnivora are admitted to be distinct orders. Now if the order *Carnivora* is represented by *Thylacinus* and *Dasyurus*, and the order *Insectivora* by *Phascogale*, *Perameles*, &c. almost the only point of distinction between the carnivorous and insectivorous marsupial groups consists in a slight difference in the structure of the teeth ! [1] Geoffroy, who established the genus Dasyurus, regarded the type of Temminck's genus *Phascogale* as a member of that genus; and Professor Owen, as well, I may say, as every mammalogist whose labours I am acquainted with, regard *Phascogale* and *Dasyurus* as members of the same *family*: how then can two species of the same family represent in rank two orders ? Again, upon a careful examination of the Wombat, I find so many points in common with the Phalangista group—that it is so intimately related to the Koala (which is more clearly an aberrant Phalangista), as indicated by the structure of the stomach, and the deficiency in the number of the false molars, and the total absence of tail—that I am inclined to regard the genus *Phascolomys* as presenting an aberrant form only of the *Phalangistidæ*: that the thumb should be reduced to a small size in this animal, which differs from others of its (supposed) family, in living upon the ground, I am prepared for, since in the *Dasyuridæ* the same thing takes place under similar circumstances. I am also prepared to find in an herbivorous group like the *Phalangistidæ* a difference in the structure of the molar teeth—in having them rooted in one case and rootless in another, for such happens in other herbivorous groups of Mammalia. Can I, then, regard the Wombat as constituting one order of Mammalia, and the Phalangers another, and at the same time suppose the one to represent the highest group of placental mammalia (Quadrumana) and the other the *Rodentia*, which is one of the lowest ?—might we not regard the last-mentioned group (*Rodentia*) as a class, because the *Sciuridæ* typified the *Quadrumana*, and the *Muridæ*, in their comparatively carnivorous habits, represented the *Carnivora*, the Agoutis the *Ruminantia*, and, lastly, because the Capibara quite as strongly typified the *Pachydermata* as does the Wombat the Rodentia. The analogies are here

[1] In stating this, it must be observed, I am comparing *Phascogale* with the *Dasyuri*, in which I include *Thylacinus*. *Perameles* differs *much* more from *Phascogale* than does that genus from *Dasyurus*.

sufficiently evident, as it appears to me, though not so striking as in the Marsupiata, but this may, I think, be accounted for.

Let us compare the Marsupiata, and the sections of which it is composed, with the Placentalia, in another point of view. In those groups which are universally admitted to be *orders*, there are most frequently very many species, and, what appears to me of more importance to notice, there are in each order many genera and several families. Sometimes, where an order does not contain many recent species, the comparatively few species present very striking modifications of structure, as is the case with the order *Edentata:* here, however, we know that a great portion of the species have become extinct, as well as in the order *Pachydermata*, which presents a similar condition, having but few recent species, and those exhibiting striking modifications of structure: then, again, we find that the groups admitted to be orders have representatives in widely separated portions of the globe, and not only each order is widely dispersed, but frequently the principal genera of which each is composed, and from which the other genera appear to radiate, have (or had) an almost universal distribution[1]. Some orders do not present all these conditions, not containing several genera of universal distribution, but all evince an *approximation* to them; and it is amongst those orders which fall short (if we may so speak) of these conditions, that we find the nearest approximation to parallelism to the *Marsupiata* as regards geographical distribution. The *Edentata*, like the *Marsupiata*, are found in both hemispheres; each of these divisions, however, has its chief metropolis in one hemisphere; both, low in grade of organization, are driven down, as it were, into the southern portions of the globe, but have existed in former times in the northern hemisphere. Both include species exhibiting very marked variations in structure and of habits. But the *Edentata* are correctly, as I think, regarded by most naturalists as forming an order.

When, on the one hand, we find the conditions presented by the principal divisions of the Marsupiata approximate most nearly to those of families of other orders, being separated from each other by comparatively trivial characters, and that the whole group presents the strongest anology to other groups which are regarded as orders, with respect to their geographical distribution, we are warranted, *cæteris paribus*, in regarding the Marsupiata as an order. But then it may be said, perhaps, that the amount of differential characters which serve to separate the Marsupialia from other orders is greater than that by which those orders are distinguished. If it be true that each minor group of

[1] In the order CARNIVORA we find (having a nearly universal range) the genera *Canis*, *Felis*, *Mustela*, *Ursus*, and *Phoca*; in the RODENTIA, *Sciurus*, *Mus*, *Hystrix* (Linn.) and *Lepus*; in the PACHYDERMATA, *Mastodon* (fossil), *Tapirus* (and its allies, found only in a fossil state), *Equus* (either recent or fossil), and *Sus* (Linn.)

Mammalia presents species of comparatively high and others of lower grade of organization, and I think it very certain, then should we be prepared to find, in the lowest *order* of the class, a great amount of difference, as compared with the higher orders, in those characters which, as they approximate to, or deviate from, a certain standard, are said to indicate a higher or lower grade in the scale of organization.

Now it is precisely in such characters that the most important distinctions of the section *Marsupiata* are manifested; and the raising that group to a rank above an order, is to admit that the amount of difference is greater than could be, *à priori*, anticipated in the lowest order of a class; and yet this most important branch of investigation has not, to my knowledge, been considered. It remains to be inquired whether there is not an increase in the ratio of the amount of differential characters as we descend in the scale, and whether there may not be an increase in the amount of variation exhibited in the species of the lowest division of any great group. There are not only grounds for believing such to be actually the case, but I think the embryologist would be prepared to account for some of these points—partly, perhaps, by the more rapid changes in the metamorphoses which a high animal undergoes in the earlier stages of its existence.

In the foregoing pages I have not alluded to the *Monotremata*, for although it is generally admitted that that group possesses a relationship to the *Marsupiata*, the nature of that relationship can only be determined by such investigations as I have above alluded to. If the views which are hinted at in this note[1] be well grounded, then is the *Monotremata* a family of the order Marsupiata.

SECTION I.—MONOTREMATA[2].

MAMMALIA possessing marsupial bones, wanting the corpus callosum to the brain, with the mass called *corpora quadrigemina* divided by a transverse fissure, and with the posterior part undivided: the sternum and shoulder bones joining, and encircling the fore part of the trunk: the sternum with the manubrium joined in front by an episternum, which is produced on each side at its anterior extremity into a long branch, attached to and

[1] I see no reason why similar considerations should not be brought to our aid with a view to determine the rank of the section Amphibia among Reptiles, about which there has been so much discussion.

[2] From μόνος, unicus; τρῆμα, foramen.

running along the upper surface of which is the clavicle; the space below the lateral branches on either side almost entirely occupied by a large flattened epicoracoid, which is bounded on the outer side by a coracoid bone: the jaws edentate, or provided with crushing teeth of a horny nature. The mammary glands abdominal, in the form of numerous elongated subcylindrical lobes, converging and opening into a small oval areola, which has not been seen to form a projecting nipple; cœcum small; facial bones produced and covered by a hairless skin; ear without any external concha; young (in Ornithorhynchus at least) naked.

The section Monotremata contains but two well-determined species, both of which are from Australia, and were originally described by Dr. Shaw towards the close of the last century,—the one under the name of *Myrmecophaga aculeata,* and the other under that of *Platypus anatinus.* The former, in many modern systematic works, appears under the name of *Echidna hystrix,* and the latter with the name *Ornithorhynchus paradoxus,*—a name first applied, in 1800, by Blumenbach.

An animal presenting such a remarkable combination of characters as the Ornithorhynchus, could not fail soon to attract the attention of the anatomist and physiologist, and since it was soon discovered to approach the oviparous classes of birds and reptiles in many of its characters, a question arose as to whether, like the Mammalia, it suckled its young, or, like the other two classes just mentioned, was oviparous.

Even after the Ornithorhynchus had been carefully dissected by Meckel[1], that able anatomist, who had discovered and well described the mammary glands, still doubted whether the animal might not lay eggs, inasmuch as the generation of the *Marsupialia* closely resembled that of the oviparous

[1] See his *Ornithorhynchi paradoxi descriptio anatomica,* folio, 1826; and *Beitrage zur Vergleichenden Anatomie,* 1808.

classes; and, on the other hand, as birds have been known to hatch the eggs within their body, and to give birth to a living chick, he thought it highly probable that the generation of the Ornithorhynchus, approaching still more nearly to birds and reptiles than the Marsupialia, might be oviparous. Geoffroy St.-Hilaire, Oken, De Blainville, and Prof. Owen, took active parts in the discussion of this question: the former supposing the mammary glands were wanting, removed the Ornithorhynchus from the class Mammalia, and arranged it, together with the Echidna, in a separate class, to which he affixed the name *Monotrema*[1]; and, subsequently, when Meckel had shown the existence of the glands in question, Geoffroy St.-Hilaire still firmly maintained his opinion[2]. These glands he conceived to be analogous to those situated along the flanks of the salamanders, or to the odoriferous glands observed on the sides of the abdomen in shrews. At an early period, De Blainville rightly conjectured that mammary glands would be found in the Ornithorhynchus, and that the animal would prove to be allied to the *Marsupiata*[3].

In two Memoirs published in the Philosophical Transactions[4], and a third paper published in the Transactions of the

[1] *Anatomie Philosophique*, tome i. 1818. Sir Everard Home, it must be observed, had previously pointed out the close relationship of affinity which existed between the Ornithorhynchus and Echidna, and called particular attention to various peculiarities in the sexual organs of these animals.—See his paper, On the Anatomy of the Ornithorhynchus, in the Philosophical Transactions for 1802, vol. 92, p. 67; and on the Ornithorhynchus hystrix, loc. cit. p. 348.

[2] See *Annales des Sciences Naturelles*, 1826, p. 457.

[3] An excellent summary of the interesting controversy relating to the mammary glands, &c., of which an outline only is here given, will be found in the article *Ornithorhynchus*, in the Penny Cyclopædia, from the pen of Mr. Broderip.

[4] On the Mammary Glands of the *Ornithorhynchus paradoxus*, Phil. Trans. 1832, part 2; and on the Ova of the *Ornithorhynchus paradoxus*, Phil. Trans. 1834, part 2.

Zoological Society[1], Professor Owen has thrown much light on this interesting subject:—the discovery of the mammary glands, by Meckel, has been confirmed, and a great amount of evidence is brought forward tending to prove that the Monotremata are allied to the *Marsupiata*, and are essentially ovo-viviparous mammals,—that they bring forth living young,—and that these are suckled by the parent.

Prof. Owen and some other authors agree in regarding the *Monotremata* as forming a distinct order of the class Mammalia and subclass Marsupiata,—an order which presents the lowest grade of organization among mammals, and which approaches most nearly to the oviparous classes of birds and reptiles.

The most essential characters of the group are strictly anatomical, and we must content ourselves here with a mere notice of some among the more striking points. The Monotremata, observes Prof. Owen, are allied to the Marsupiata by the absence of the *corpus callosum*, and by the presence of the marsupial bones, but differ in the absence of the abdominal pouch, and in not possessing teeth; in the simplicity of the bigeminal bodies, and in some remarkable modifications of the skeleton and generative organs.

The *corpus callosum*, it may be necessary to explain, is a portion of the brain which forms a kind of band connecting the two hemispheres of the brain, and that its presence was regarded as peculiarly characteristic of the mammalian order (it being absent in birds and reptiles, &c.), until Prof. Owen discovered that this band was wanting in the *Marsupiata*. The bigeminal bodies also form part of the brain, and come into view immediately upon separating the hinder part of the great hemispheres. In ordinary quadrupeds, the mass, here called bigeminal bodies, is divided by a transverse and a

[1] On the Young of *Ornithorhynchus Paradoxus*, Trans. Zool. Soc. vol. i. p. 221.

longitudinal groove into four parts, called *corpora quadrigemina;* in the Monotremata, the same mass is divided in the transverse direction; the anterior portion is also indistinctly divided by a longitudinal groove, but the hinder portion, which is much smaller, is undivided, and thus forms a single transverse band.

In the skeleton, one of the most striking peculiarities, as compared with other mammalia, is displayed in the structure of the chest and blade-bones: these together (in the Ornithorhynchus), form a ring encircling the fore part of the body; the upper part of this ring is formed by the two scapulæ, or blade-bones, and the fore, or under part, by the chest-bones. These latter, *sternal,* bones, are five in number: the foremost (the episternal) is a broad flat bone, considerably expanded at the lower extremity, and throws out, nearly at right angles, a long slender branch on each side at the anterior extremity. The second (the manubrium sterni) is also a broad flat bone, and of a triangular form; the three following, posterior, bones are very narrow. We have next to notice the clavicles, which are here long and slender bones, nearly meeting in the mesial line of the chest, and as they are extended outwards to the shoulder, are joined to, and run parallel with, the upper part of the T-shaped episternal bone. Lastly, in the chest, are four other bones; the two coracoid, and the two epicoracoid bones: the former are joined to the scapular bones, and as they run downwards, converge and meet the lower extremity of the episternal bone. The epicoracoids are flat, broad bones, which nearly occupy the whole of the interspace between the coracoids and the episternum; they partially overlap the inner side of the latter bone[1]. In

[1] In the above description of the shoulder and chest of the Ornithorhynchus, the author has followed Professor Owen's views, as regards the signification of the various bones.—See the article *Monotremata,* in Todd's Cyclopædia of Anatomy and Physiology.

this structure of sternum and shoulder, we perceive an approach to the Birds; and there is a still greater resemblance in these parts to the Lizards and Ichthyosauri, as is pointed out by Prof. Owen. The sternum and shoulder bones are *essentially* upon the same model in the Echidna[1].

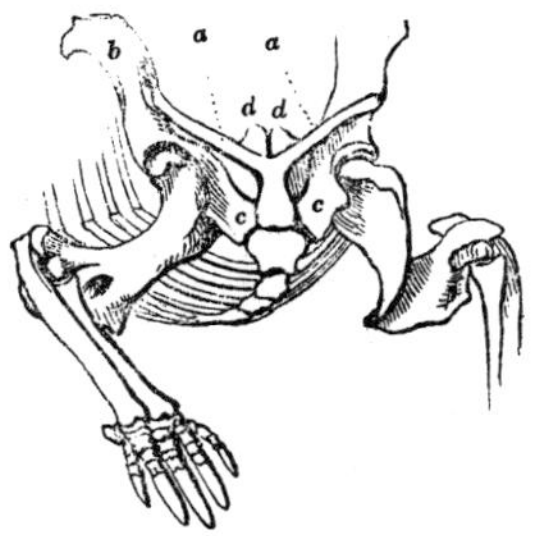

Sternum of the Echidna.

a. a. The two lateral processes of the episternum, along the upper surface of which are attached the clavicles.
b. The scapula, or blade-bone.
c. c. The coracoids.
d. d. The epicoracoids.

[1] For a detailed account of the anatomy of the Monotremata, we must refer our readers to the able article under that head, by Prof. Owen, in Todd's Cyclopædia of Anatomy and Physiology.

Genus ORNITHORHYNCHUS. Blumenbach.

Platypus,	SHAW, Naturalists' Miscellany, vol. x. Plates 385 and 386. Dated June, 1799.
Ornithorhynchus,	BLUMENBACH, Voigt's Magaz. ii. 1800.
Dermipus,	WIEDEMANN, Archiv. für Zoologie, &c. i. 1800[1].

BODY depressed, nearly oval, and clothed with a dense fur; head with the facial portion elongated, and forming a broad and depressed beak-like snout, covered by naked skin, which is produced into a lappet-like fold at the base of the snout: eye small: upper and under jaws furnished on each side, and towards the front, with a long narrow horny appendage; and towards the hinder, with a broad, nearly ovate, crushing tooth of the same horny substance: tongue short, and provided in parts with horny papillæ: legs short; the feet fitted for swimming; each foot with five well-developed toes, between which a membrane is extended—in the fore foot the membrane is produced considerably beyond the toes; claws of the fore foot large, solid, and depressed, and fitted for burrowing: tail rather short, broad, and depressed: stomach with the cardiac and pyloric orifices closely approximated; cœcum small. The male sex provided with a spur to hind foot.

[1] The name *Platypus* of Shaw, being the first published, would be here adopted, but that name had been previously applied to a genus of Insects. The name *Ornithorhynchus*, which is universally adopted, has reference to the bird-like snout of the animal; being from the Greek words ὄρνις, a bird; and ῥύγχος, a snout. The other two generic names were suggested by the peculiarities of the foot: *Platypus* is from πλατυς, broad; and ποῦς, the foot; and *Dermipus* is from δέρμα, skin; and ποῦς—in allusion to the feet being webbed.

THE ORNITHORHYNCHUS.

Ornithorhynchus anatinus.

Platypus anatinus, Shaw, Nat. Misc. vol. x. Plate 385—1799: General Zoology, vol. i. Part 1, p. 229, Plates 66 and 67.—Gray, in Cat. of Brit. Mus. Coll. p. 191; 1843.

Ornithorhynchus paradoxus, Blumenbach, Voigt's Magaz. (1800), ii. p. 305, Plate 41.—Home, Phil. Trans. 1800, p. 432; and for 1802, p. 67.—Cuvier, Règne Animal, Ed. 1829, tom. i. p. 235.—Meckel, Ornithorhynchi paradoxi descriptio anatom. Lips. 1826, fol.

Ornithorhynchus fuscus et *rufus*, Peron, Voy. de Découv. i. Tab. 34, Figs. 1 and 2.—Leach, Zool. Misc. ii. p. 136, Tab. 3.—Desmarest, Mammalogie, Part 2, p. 380.

Ornithorhynchus brevirostris, Ogilby, Proceedings of the Zool. Soc. for Nov. 1831, Part 1, p. 150.

Ornithorhynchus crispus et *lævis*, Macgillivray, Memoirs of the Wernerian Society, 1832, p. 127.

Length about 18 inches: fur rather short, dense, the under fur soft, and the outer fur rather crisp to the touch; general colour dusky-brown; on the upper parts of the body rather dark, on the under, paler.

Immature and young animals, with the fur soft to the touch,

of a bright-brown hue on the upper parts of the body, and whitish on the under parts. When from two to four inches in length only, the beak is proportionately much shorter than in the adult (being shorter than the head), and the body is destitute of hair.

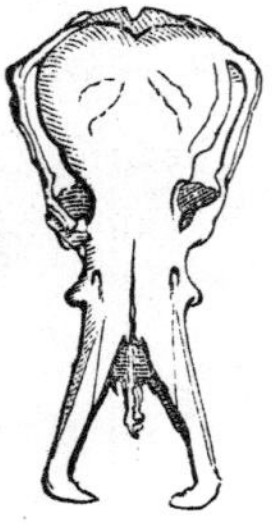

Skull of the Ornithorhynchus.

The skull of the Ornithorhynchus is much elongated, and has the facial portion remarkably produced and depressed, and, owing to the divergence of the superior maxillary and intermaxillary bones, is expanded in front. A stright line lying lengthways on the mesial portion of the upper surface of the skull, would touch very nearly at every part, from near the occiput to the anterior boundary of the facial bones. The temporal fossæ are narrow, and the orbits are of moderate size; the zygomatic arch is deep and moderately long, and sends up a post-orbital process; the zygoma appears to be formed entirely of meeting processes of the squamous portion of the temporal bone and superior maxillary. Professor Owen could find no malar bone either in the skull of a young Ornithorhynchus, or in that of an immature Echidna. The glenoid cavity for the lower jaw is concave in the transverse direction, and slightly convex from before backwards, and has no posterior descending process. The frontal bones are very small, and the parietals large. The sutures of the cranial bones soon become obliterated, and in this respect, as well as in their thinness and density, remind us of the skull of a bird. The nasal bones are large, and much extended in the longitudinal direction. The superior maxillaries are greatly expanded immediately under the anterior root of the zygoma, to give support to the large horny molar teeth, and are continued forwards, on the outer side of the intermaxillaries, to within a short distance of their apices. The intermaxillary bones are rather narrow, depressed, widely separated, and suddenly bent inwards (but do not meet) at their extremity. The palate is of great extent.

The rami of the lower jaw are long, nearly cylindrical in the middle, but expanded to form the alveolus of the great horny tooth. Each ramus, viewed from the outer side, presents a slight sigmoid curve, being slightly arched in the middle, and having the condyloid portion curved gently upwards. The condyle is broader than long, broad externally, and narrow internally, and moderately convex. On the inner side of the ramus is a tubercle, situated about two lines behind the molar tooth, and which gives attachment to the internal pterygoid muscle; and on the upper surface of the jaw, nearly in a line with this tubercle, is a depression indicating the point of insertion of the temporal muscle; immediately under this point, on the outer side of the jaw, is a very large cavity, which runs inwards, and is suddenly contracted beneath the molar tooth. The dental canal is a large opening, and may be seen on the inner side of the jaw, under the posterior margin of the tooth: it has two outlets; one on the outer side of the jaw, in a line with the front of the tooth, and the other on the under side, at a short distance from the extremity of the jaw. The rami meet, and join immediately behind the last mentioned opening, but are loosely attached, and diverge again at the apex. The apical portion of the ramus is depressed, and exhibits a long shallow groove along the outer edge, in which the anterior narrow horny tooth is situated.

	Inches.	Lines.
Total length of cranium	3	11
Greatest width (which is at the hinder root of the zygoma)	1	9
Width between orbits	0	6
Width at the tip of the intermaxillary bones	1	1
Length of zygomatic arch	1	3

In a skeleton in the British Museum collection, I find the skull considerably larger than indicated by the above dimensions; it being four inches and two lines in length, and two inches in width behind. The lower jaw of this skull is three inches and ten lines long.

The vertebræ in the skeleton are fifty in number, of which 7 are cervical, 17 dorsal, 2 lumbar, 2 sacral, and 22 caudal. The cervical vertebræ, as Prof. Owen observes, are remarkable for having the ribs separated for a longer period than usual from the body of the vertebræ, and in the axis vertebra the ribs are not anchylosed even in the adult animal[1]. In this character, as well as in the great number of dorsal vertebræ, and consequently ribs, we perceive one of the many points of approximation in the Ornithorhynchus to the Oviparous classes. The transverse processes to the caudal vertebræ are very broad, flat, and much produced: the tail is broadest near the middle, where its width in the skeleton before me is one inch and two lines, but the width gradually decreases from this point to the apex, in consequence of the gradual shortening of the transverse processes.

The humerus is a short, broad, and strong bone, and has the internal condyle perforated. The ulna and radius are in contact, and the former is remarkable for having the olecranon very suddenly expanded at the extremity. The hinder extremities are longer than the anterior, but the increased length is due almost entirely to the tibia and fibula, the femur being but little longer than the corresponding bone of the fore leg—the humerus. The fibula is remarkable for having the upper extremity continued considerably beyond the proximal end of the tibia, and with this free portion much expanded: a process is thrown out from the base of the expanded portion of the fibula to articulate with the tibia and femur.

The Ornithorhynchus inhabits New South Wales and Van Diemen's Land, and is called by the colonists the Water-mole—a name suggested by its aquatic habits, combined with some slight resemblance which it bears to the common European mole. Were we, however, to institute a comparison between the Ornithorhynchus and any animal from other parts of the world, it would be with the Mygale of the Pyrennees, or the allied species which inhabits Russia: here the resemblance, both as regards external appearance and habits, is very considerable. The native names of our animal,

[1] I find the ribs of the axis vertebra unanchylosed in the skeleton in the British Museum collection, which is considerably larger than those in the College of Surgeons' collection, where the same peculiarity has been noticed by Professor Owen.

Mr. Bennett informs us, is Mallagong and Tambreet; but the latter is the most common appellation with the Aborigines of New South Wales.

The most striking feature in the structure of the Ornithorhynchus is the beak-like face; a peculiarity which suggested the generic name given by Blumenbach, and which, in combination with the broad flat foot, likewise suggested to Shaw the name Duck-billed Platypus, or *Platypus anatinus*. The head is rather small, but the facial bones are much developed, and, together with those of the lower jaw, form a framework which is covered with a thick naked skin. The beak is much depressed, about one-third longer than broad, and broader in front than behind: the skin with which it is covered is in the living animal of a greyish-black colour, and covered with innumerable pores; but both at the front and sides of the beak is a free portion of the skin, which is uniformly smooth and flexible.

Fig. 2. Fig. 1.

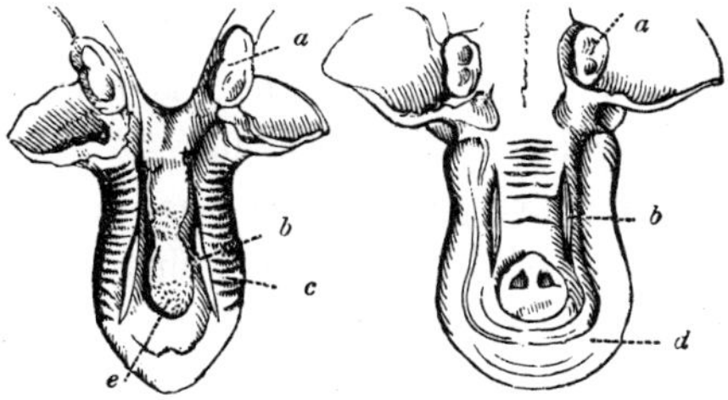

Jaws of the Ornithorhynchus.

Fore part of the upper and under jaws of the Ornithorhynchus, showing the inner side of the beak-like jaws.

Fig. 1. The upper jaw.

a. The horny molar teeth.

b. Long, narrow, anterior horny teeth.

d. Free portion of the integument projecting from the jaw.

Fig. 2. The lower jaw.

a. The horny molar teeth.

b. Narrow, anterior horny teeth.

c. Transverse ridges on the leather-like covering of the beak.

e. The tongue.

In front of the beak, the free membrane (*d*, fig. 2) measures half an inch in width, and at the sides rather less. At the base of the beak the skin forms a kind of fold or lappet, which averages about half an inch in width; at the angle of the mouth it is narrower, but immediately below the angle its width is nearly three quarters of an inch. This fold falls back over the fore part of the head and throat, and no doubt serves to prevent the mud, in which these animals are constantly grubbing in search of their insectivorous food, from soiling the fur: it may also serve to protect the eyes both from the mud and from the soil, when burrowing in the ground. The nostrils are moderately large, and situated near the extremity of the upper surface of the beak. The lower mandibles are covered, like the upper, with a naked skin, but here it is generally of a whitish colour; the free, marginal portion is narrower, and on the upper surface it presents numerous transverse ridges, which may be compared to the ridges observed in the beak of a duck, and no doubt answers a similar purpose — that of sifting or straining the water from the materials collected in the mouth, which is moreover provided with two tolerably large cheek-pouches for their reception. "An air-breathing, warm-blooded animal, which obtains its food by the capture of small aquatic animals (observes Prof. Owen) while submerged, must derive great advantage from the structure which enables it to transfer them quickly to a temporary receptacle, whence they may be extracted and masticated while the animal is floating on the surface, or at rest in its burrow."

The teeth in the Ornithorhynchus, unlike those of most mammalia, are of a horny substance: they are eight in number; four are situated (one on each side of each jaw) in the fore part of the jaws, and are in the form of long and narrow strips, being about three-quarters of an inch in length, and $\frac{1}{12}$th of an inch in width: they present a slightly trenchant edge, and are thus fitted for prehension: the other four teeth are of a somewhat irregular oblong oval form, about $\frac{8}{12}$ths of an inch in length, and $\frac{5}{12}$ths in width: the crown of these teeth is concave, but presents two transverse ridges; one near the middle of the tooth, and the other on the hinder part. These represent the grinder teeth, or true molars, of ordinary mammalia, and occupy the usual situation of such teeth; viz. immediately under the anterior root of the zygomatic arch. Besides these teeth, the Ornithorhynchus, in harmony with many other parts of its structure, presents a Reptilian character, in having horny teeth on its tongue; this is moderately long (terminating about half an inch from the end of the mouth), and has the upper surface of its anterior portion covered with minute papillæ; the hinder part is suddenly expanded, and at the same time is raised, and presents an acute angle in front, on which are observed a group of about six minute horny tubercles, and two others which are in the form of a depressed cone, and about $\frac{1}{12}$th of an inch in length. This bulb on the back part of the tongue would serve to prevent the passage of the materials collected in the mouth, together with the water, into the gullet, and to direct the former into those temporary receptacles, the cheek-pouches, which have their opening on each side, at the back part of the mouth[1]. The eyes are small, of a brown colour, directed somewhat upwards, and situated near the base of the beak. The external orifice of the ear is placed at a short distance behind the eye, and being hidden by the fur, is not easily found in the dead animal, though readily seen in the living, which has the faculty of closing or opening it at will. The legs are strong, and very short; the feet are provided each

[1] In two specimens of Ornithorhynchus obtained by Mr. Bennett, he found the cheek-pouches filled with mud and gravel; but in other cases their contents were found to consist of river insects, small shell-fish, &c.: these, however, were always mingled with mud or gravel.

with five well-developed toes; those of the fore feet are nearly equal in length; the inner toe, or thumb, is the shortest, and the outer one approaches this most nearly in length; they are furnished with long, broad, and somewhat depressed, solid claws, which are rounded at the extremity; not only are the interspaces of the toes webbed, but the web is extended about half an inch beyond the end of the claws, and thus gives great expanse to the foot when used for swimming; but when employed in burrowing, the free edge of the membrane is folded back. The toes of the hind feet are rather longer than those of the fore, and very nearly equal in length, if we except the inner toe, which is about one-third shorter than the others; the claws are long, curved, and pointed; the spaces between the toes are webbed, but here the web extends only to the base of the claws. On the heel of the male Ornithorhynchus, is a large and sharply-pointed moveable spur, which is curved upwards, and directed backwards: this spur is pierced by a minute tube, the outlet of which is near the point; and connected with this little tube is a large gland, which has been supposed by some to secrete a poisonous fluid, which is injected through the tube in question into any wound inflicted by the spur: but from all the evidence collected by Mr. Bennett, who tried various experiments upon himself, it does not appear that the supposition of the secretion being poisonous is well founded. A small rudimentary spur is found in the young female Ornithorhynchus; but this disappears when the animal becomes adult, and a small hollow marks its situation. The fur of the Ornithorhynchus is very dense, by no means long, and rather soft to the touch: it is composed of hairs of two kinds; the one is extremely fine and dense, and the hairs composing the outer fur, as it may be termed, are likewise fine, if we except that portion of each hair which projects beyond the under fur; these free points of the longer hairs are comparatively harsh, broad, and lanceolate, and are very glossy, and bent at an angle with the slender basal portions of the hairs. In this character of fur we can perceive a beautiful adaptation both to the burrowing and aquatic habits of the animal; for, when burrowing, were the longer hairs equally stout from the base to the point, and directed towards the tail, as usual, they would incommode the animal when moving backwards in its burrow; but being slender at the base, and expanded externally, the points readily accommodate themselves to any surface with which they may come in contact, and laying flat on each other, serve either to keep the water or the soil from penetrating to the under fur. The general colour of the fur is deep brown, but on the under parts of the body it is somewhat paler: in the young animal the fur is of a brighter brown tint, and the under parts of the body are whitish. The tail is short, depressed, and very broad, and covered with coarse hairs; these, however, are generally worn off on the under side of the tail in adult or aged individuals, probably by the friction of the ground, since the animal's legs are too short to elevate the body. The toes of the fore feet are naked; those of the hind are clothed, very nearly to the extremity, with short, adpressed, glossy brown hair, and there is a fringe of stiff pale-coloured hairs on the outer side of the foot. The three middle toes of the hind foot are so united by the skin as to be capable of very little lateral separation; the membrane which joins the little toe to the others is more ample, and so is that which joins the thumb to the second toe; and here, instead of merely filling up the interspace between the toes, there is a fleshy lobe (Pl. 2, fig. 3, *c.*) of full half an inch in length, continued from the end of the thumb beneath, and to the end of which the membrane is extended.

We have quoted Mr. Bennett in some of our preceding pages. This gentleman is the author of a very interesting account of the habits of the Ornithorhynchus[1], from which we shall proceed to make some extracts.

Mr. Bennett was so eager to obtain a view of the living Ornithorhynchus, after his arrival in Australia, that upon reaching a friend's estate at Mundoona, although after a long journey, he readily acceded to an offer to walk out to the banks of a neighbouring river frequented by the animal.—"We soon came to a tranquil part of the river," observes Mr. Bennett, "such as the colonists call a 'pond,' on the surface of which numerous aquatic plants grew. It is in places of this description that the *water-moles* are most commonly seen, seeking their food among the aquatic plants, whilst the steep and shaded banks afford them excellent situations for excavating their burrows. We remained stationary on the banks, waiting their appearance with some degree of impatience, and it was not long before my companion quietly directed my attention to one of these animals, paddling on the surface of the water, not far distant from the bank on which we were then standing. In such circumstances they may be readily recognized by their dark bodies just seen level with the surface, above which the head is slightly raised, and by the circles made in the water around them by their paddling action. On seeing them the spectator must remain perfectly stationary, as the slightest noise or movement of his body would cause their instant disappearance, so acute are they in sight or hearing, or perhaps both, and they seldom reappear when they have been frightened." On ordinary occasions, they do not remain more than a minute or two at a time on the surface of the water.

A burrow of an *Ornithorhynchus,* which Mr. Bennett opened, had its entrance on a steep part of a bank, situated

[1] Notes on the Natural History and Habits of the *Ornithorhynchus paradoxus,* by Mr. George Bennett.—Transactions of the Zoological Society, vol. i. p. 229.

about one foot from the water's edge, and concealed among the long grass and other plants. "This burrow ran up the bank in a serpentine course, approaching nearer to the surface of the earth towards its termination, at which part the nest is situated * * *. No nest had yet been made in the termination of the burrow, for that appears to be formed about the time of bringing forth the young, and consists merely of dried grass, weeds, &c. strewed over the floor of this part of the habitation." The expanded termination measured one foot in length and six inches in breadth, and the whole length of the burrow was twenty feet. Besides the entrance before alluded to, it appears there is usually a second opening into the burrows below the surface of the water, communicating with the interior just within the upper aperture. A burrow subsequently examined by Mr. Bennett terminated at a distance of thirty-five feet from the entrance; and that gentleman assures us that they have been found fifty feet in length.

From the burrow first opened by Mr. Bennett, a living female was taken, and placed in a cask, with grass, mud, water, &c., and in this situation it soon became tranquil, and apparently reconciled to its confinement. On his return home to Sydney, Mr. Bennett determined to indulge it with a bathe; and with this view, when he arrived in the vicinity of some ponds, he tied a long cord to its leg. "When placed on the bank it soon found its way into the water, and travelled up the stream apparently delighting in those places which most abounded in aquatic weeds. When diving in deep and clear water, its motions were distinctly seen: it sank speedily to the bottom, swam there for a short distance, and then rose again to the surface. It appeared, however, to prefer keeping close to the bank, occasionally thrusting its beak into the mud, from whence it evidently procured food, as, on raising the head after withdrawing the beak, the mandibles were seen in lateral motion, as is usual when the animal masticates. The motions of the mandibles were similar to those of a duck under the same circumstances. After feeding it would lie sometimes on the grassy bank, and at others partly in and partly out of the water, combing and cleaning its coat with the claws of the hind feet. This process occupied a considerable time, and greatly improved its sleek and glossy appearance."

The Water-moles are said to have a peculiarly fishy smell, more especially when wet, which probably proceeds from an oily secretion: they are used by the Aborigines for food; "but it is no particular recommendation of them," Mr. Bennett remarks, "to say they are eaten by the native Australian, as nothing in the shape of provender comes amiss to him, whether it be snakes, rats, frogs, grubs, or the more delicate opossum, Bandicoot and Flying Squirrel."

The Ornithorhynchus is captured by the natives when in its burrow: they first examine the interior of the burrow, to ascertain, by the presence of recent footmarks on the soil, whether it is inhabited; and if the examination proves satisfactory, they proceed to dig holes with pieces of sticks, from the surface of the ground into the burrow, at distances from each other, until they discover its termination, when the Australians consider themselves exceedingly fortunate should they find the young, since they are regarded as a great delicacy.

The young have been found in their nests, by Mr. Bennett, of about $1\frac{7}{8}$ in. in length, in the early part of December; and near the end of the same month he found young Water-moles of ten inches in length: these latter were kept alive for nearly five weeks, and their habits, whilst in captivity, are described in detail in his paper, which is illustrated by some admirable figures shewing their various attitudes, &c. The young were allowed to run about the room, but an old Ornithorhynchus, in the possession of our author, was so restless, and damaged the walls of the room so much, by her attempts at burrowing, that it was found necessary to confine her to the box. "During the day she would remain quiet, huddled up with her young ones; but at night she became very restless, and eager to escape. The little ones were as frolicsome as puppies, and apparently as fond of play; and many of their actions were not a little ludicrous. During the day they seemed to prefer a dark corner for repose, and generally resorted to the spot to which they had been accustomed, although they would change it on a sudden, apparently from mere caprice. They did not appear to like deep water, but enjoyed exceedingly a bathe in shallow water, with a turf of grass placed in one corner of the pan: they seldom remained longer than ten or fifteen minutes in the water at one time. Though apparently nocturnal, or at least preferring the cool and dusky evening to the glare and heat of noon, their movements in this respect were so irregular as to furnish no grounds for a definite conclusion. They slept much, and it frequently happened that one slept while the other was running about, and this occurred at almost all periods of the day. They climbed with great readiness to the summit of a bookcase, placing their backs against the wall and their feet against the bookcase; and thus, by means of their strong cutaneous muscles and of their claws, mounting with much expedition to the top. Their food consisted of bread soaked in water, chopped egg, and meat minced very small, and they did not seem to prefer milk to water."

The young are naked at first, and differ much from the adult in the form of the beak: this is very short in proportion, and has the margins smooth and fleshy; the tongue is proportionately large, reaching the extremity of the mouth, which is thus fitted for sucking. There is, moreover, according to Prof. Owen, a thin fold of integument surrounding the base of the mandibles, which extends the angle of the mouth from the base of the lower jaw to equal the breadth of the base of the upper one, which must increase the facility for receiving. the milk ejected from the mammary areola of the mother.

There is very little difference in the size of the male and female *Ornithorhynchus;* but the males, it would appear, are rather the larger. The average length, in fifteen specimens, Mr. Bennett found to be 1 foot 7 inches to 1 foot 8 inches in the males, and 1 foot 6 to 1 foot 7 inches in the females. One male specimen, shot near the Murrumbidgee River, measured 1 foot $11\frac{1}{4}$ inches. The following dimensions, taken from a recent specimen, will give an idea of the proportions of the parts:—

	Inches.
Length from the extremity of the mandible to the extremity of the tail	1 ft. $7\frac{7}{8}$
Length of upper mandible	$2\frac{7}{8}$
Breadth of ditto	$2\frac{1}{8}$
Length of lower mandible	$1\frac{6}{8}$
Breadth	$1\frac{3}{8}$
Length of fore leg	$3\frac{6}{8}$
Breadth of fore foot expanded	$4\frac{2}{8}$
Length of tail	$4\frac{5}{8}$
Breadth of ditto at the broadest part	$3\frac{2}{8}$
Hind leg to extremity of longest claw	4
Breadth of hind foot	$2\frac{4}{8}$

The other species of Ornithorhynchus enumerated in Zoological works, it appears to me, are founded upon individuals differing in age only. In the large and mature specimens, the fur is crisp, and of a dull brown colour; whilst in the smaller specimens it is of a bright brown hue, and soft to the touch, and, on the under parts of the body, is almost white. Those of smaller size and brighter colouring have received the specific names, *rufus*, *lævis*, and *brevirostris;* and I may further observe, that the specimen from which Dr. Shaw drew up his original account, and which is now in the

British Museum collection, presents a similar condition of the fur, and is of small size. The *O. fuscus* of Péron and Lesueur, and the *O. crispus* of Macgillivray, are large specimens, in which the fur is comparatively crisp, and of dull colouring. Mr. Macgillivray, it will be found[1], was himself subsequently of opinion, that his two species, *O. crispus* and *O. lævis*, were mere varieties of the *O. paradoxus*.

The head, feet, chest, bones, pelvis, &c. of the Ornithorhynchus are figured in Plate 2.

FIG. 1. The head viewed from above; half the natural size.
" 2. The same viewed from beneath.
(The remaining figures are of the natural size.)
" 3. The hind foot, viewed from beneath: *a*, the inner toe; *b*, the spur.
" 4. The fore foot, viewed from above: *a*, *a*, *a*, *a*, semi-cartilaginous appendages[2], springing from the under side of the apical portion of the toes, to support the membrane which fills up their interspaces.
" 5. The chest and shoulder bones: *a*, *a*, the blade bones (*scapulæ*); *b*, *b*, the coracoids; *c*, *c*, the epicoracoids; *d*, *d*, the clavicles; *e*, the episternum; *e**, *e**, lateral branches of the episternum; *f*, the foremost of the bones of the sternum, called the manubrium sterni; this is followed by three other sternal bones, *g*, which are small and narrow; *h*, the humerus.
" 6. The pelvis, shewing the position of the marsupial bones, *a*, *a*.

[1] See Memoirs of the Wernerian Society, vol. vi. p. 132.

[2] These appendages appear to be an extraordinary development of the fleshy pads which are observable on the under side of the apical portion of the toes in very many quadrupeds.

Genus, *Echidna*.

Echidna[1].	Cuvier, Tableau Élémentaire. 1797.
Tachyglossus[2].	Illiger, Prod. Syst. Mammalium et Avium, p. 14. 1811.

Rostrum naked, elongate, slender, and attenuated; mouth opening, small; tongue protractile, slender, cylindrical, and very long; palate, with horny papillæ. Body above, furnished with spines and hair intermixed: legs short and powerful; fore and hind feet, each with five well-developed toes, having large nails; the fore feet fitted for burrowing; the hind feet, in the male sex, furnished with a horny spur: tail very short: cœcum short. Habitat, Australia.

Two species only of the present genus are enumerated in systematic works, and one of these will perhaps prove to be a local variety: they are insect-feeding animals, and burrow in the ground. At a cursory glance they would somewhat resemble the Hedgehog, were it not for their long and slender snout.

[1] Ἔχιδνα, a viper. It is supposed the Echidna has the power of inflicting a poisoned wound with its spur; and this idea, it is probable, suggested the generic name given by Cuvier. It must be remarked, however, that there is no well-authenticated instance on record of such a wound having been inflicted by the animal.

[2] Ταχυς, quick; and γλωσσα, tongue.

ECHIDNA ACULEATA. The long-spined Echidna.

Myrmecophaga aculeata.	Porcupine Ant-eater, SHAW, Naturalists' Miscellany, vol. iii. pl. 109. 1792.
Aculeated Ant-eater.	SHAW, General Zoology, vol. i. pt. i. p. 175.
Ornithorhynchus Hystrix.	HOME, Philosoph. Transactions for 1802, p. 348.
Echidna Hystrix.	CUVIER, Règne Anim.
Tachyglossus aculeatus.	(ILLIGER), SCHREBER, Saugth. t. lxiii. B.
Echidna longiaculeata.	TIEDEMANN, Zoologie, i. 592.

About one foot in length: body stout, having the upper parts covered with strong spines, and the under parts, as well as the head and legs, clothed with brownish-black coarse hair; head with the facial portion produced into a slender and tapering snout, and covered with a naked skin; mouth opening small; tongue very long and flexible; legs short and strong, each foot furnished with five toes; those on the fore feet armed with large and strong claws; the inner toe of the hind foot provided with a broad and rounded nail, the other toes having long and curved claws; the claw of the second toe very long. Tail very small, and hidden by the spines and fur.

THE long-spined Echidna has long been known as a native of New South Wales, and the author is indebted to Mr. Gould for calling his attention to the fact of its also inhabiting the West Coast, that gentleman having recently received a specimen from the Swan River district.

The Echidna is a small animal, being about equal to the common Hedge-hog in size, but it has a powerful frame, fitted for burrowing habits.

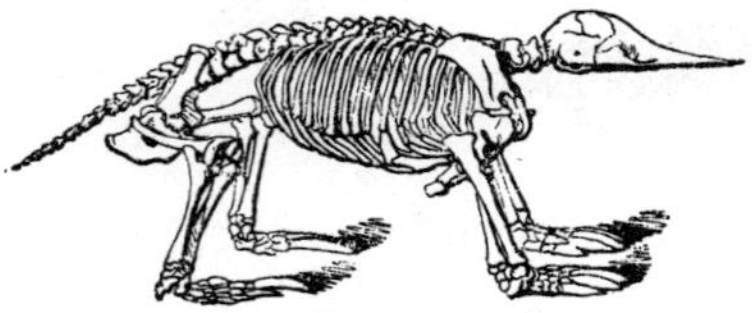

Skeleton of the Echidna.

Its food consists of ants, and probably other small insects, and these are captured in the same way as the Ant-eaters (*Myrmecophaga*) procure similar prey—by the tongue, which in both instances is protractile, very long[1], slender and flexible, and is constantly kept lubricated with a viscous matter, to which the ants adhere. To supply this secretion, the Echidna is provided with two enormous submaxillary glands, which extend from behind the ear to the fore part of the chest. There are no teeth to the jaws, but the palatal portion of the mouth is armed with several rows of strong

[1] In the figure given by Sir Everard Home, in the Philosophical Transactions (1802), the tongue is represented as upwards of six inches in length.

horny spines, the points of which are directed backwards, and on the upper surface of the tongue are numerous small horny warts, between which and the palatal spines the prey of the animal is, no doubt, crushed, before passing into the stomach.

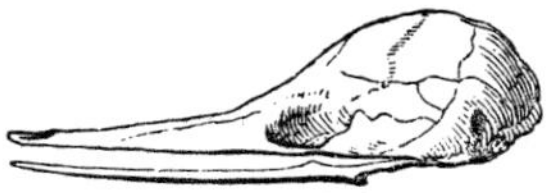

Side view of the Skull of the Echidna.

The skull might be compared to half a pear, cut lengthwise; the elongated facial bones, which are covered with a naked, blackish skin, representing the narrow end of the pear[1]. The muzzle is of a somewhat depressed form, and in an adult animal rather more than an inch and a half in length; its width at the base is three quarters of an inch, and near the tip, 4½ lines. The opening of the mouth is small, the angle being not more than four or five lines distant from the tip of the muzzle: the nostrils are situated on the upper surface, near the tip; the eyes are rather small, and of a black colour. The cavity of the ear is in the form of a long tube, and has its opening, which is large, and formed like an S, on the back part of the head;—it is compared, by Messrs. Quoy and Gaimard, to the larynx of a bird, being supported by cartilaginous hoops, in the same manner: there is no external auricle. The body is rather short and stout, and covered

[1] The following are the dimensions of the skull of an adult Echidna :—

	Inches.	Lines.
Total length	4	1
Greatest width (which is behind)	1	9¼
Width between orbits	0	7½
Length in front of orbit	1	11
Width immediately in front of orbit ...	0	10½
Width at the point of the muzzle	0	3
Length of zygomatic arch	0	11

with a thick skin, particularly on the back, where it has to support the strong spines: these are of a dirty-white colour, but more or less broadly tipped with black, sharply pointed, and average at about $1\frac{3}{4}$ inches in length; they commence on the back part of the head, and extend over the whole upper surface of the body. The points of these spines are directed backwards, and on the back, inwards, so that they cross each other in the mesial line: near the root of the tail they form a large tuft, radiating from two approximate centres, and hide the small rudimentary tail. The head (with the exception of the hinder part of the upper surface), and the lower half of the sides of the body, as well as the whole of the under parts and limbs, are covered with coarse brownish black hairs. The legs are short, and very strong; the fore feet are short and broad, and armed with large, solid, and nearly straight nails—that of the middle toe being usually about an inch in length, and a quarter of an inch in width; the shortest, which is found on the inner toe, is four or five lines in length: they are all rounded at the extremity. The hind feet are narrower, and less powerful, than the others, and have the inner toe very short, and apparently slightly opposeable; it is furnished with a broad and short nail, which is rounded at the extremity: the toe next the inner one is the longest, and is armed with an enormous claw, measuring sometimes an inch and a half in length; it is curved and nearly cylindrical, but concave beneath; the claws of the other toes are progressively shorter. The hind foot, when in its natural position, rests on its inner side, and perhaps in a great measure upon the thumb or great toe: by this arrangement the long claws are protected from wear when the animal is walking, and they have the concave surface presented outwards. The use of these claws, it would appear, is to cast away the earth which is loosened by the stronger fore feet and claws. Like the Ornithorhynchus, the present animal has the heel armed with a strong spur in the male sex;

and this spur is moveable, perforated, and supplied with a gland, and muscles capable of ejecting the secretion of the gland through the canal of the spur, as in the animal just mentioned. Messrs. Quoy and Gaimard tried, by irritating the animal, to induce it to inflict a wound upon themselves, that they might clear up the disputed point as to whether this apparatus was poisonous, but were as unsuccessful as Mr. Bennett, when he tried a similar experiment with the Ornithorhynchus; and they state that, after repeated inquiries, they could not learn that any accident had ever happened from a wound of the spur.

The French naturalists just mentioned procured a specimen of an Echidna (the *E. setosa*) at Van Diemen's Land, which they kept alive for some time. They describe it as an apathetic and stupid animal; and state, that for the first month after its capture it took no sustenance whatever, but at the end of that time it began to lap, and finally to eat some food prepared for it, consisting of a mixture of flour, water, and sugar. It avoided the light, and remained during the day partially rolled up, having its head bent forwards between its fore legs. The rapidity with which it burrowed was astonishing; being placed in a large case full of earth, containing plants, it worked its way to the bottom in less than two minutes. The naked snout, although highly sensitive, assists the feet in the labour.

Messrs. Bass and Flinders, when at Twofold Bay, state that their dogs found a Porcupine Ant-eater, but that the dogs could make no impression on the animal, which escaped by burrowing in the loose sand, not head foremost, but by sinking itself directly downwards, and thus presenting nothing but his prickly back to his adversaries.

Lieutenant Breton succeeded in keeping one of these animals alive for some time, feeding it at first upon ants' eggs and milk; but afterwards, when on shipboard, its food consisted of egg, liver, and meat, chopped very fine. It drank much water. Its mode of eating, Lieut. Breton observes, was very curious, the tongue being used sometimes in the same manner as that of the Chameleon, and sometimes it reminded one of the mode in which a mower uses his scythe, the tongue being curved laterally, and the food, as it were, swept into the mouth. The specimen died suddenly off Cape Horn, but the gentleman just mentioned expressed his opinion that the Echidna might be brought alive to England; and in the Proceedings of the Zoological Society for 1834, p. 23, will be found some hints on the mode of treatment of the animal in captivity.

In the spring of the present year a specimen of the Echidna arrived alive in England, and lived for a short time in the Zoological Society's menagerie. This specimen the author had not an opportunity of examining, but he has been kindly furnished with some notes[1] made by Professor Owen, who visited the Society's Gardens soon after its arrival, with a view to observe its habits. From these notes are the following extracts:—The animal was apparently in sound health, and active; it was placed in a large shallow box, having a wire-work top, and at the bottom a quantity of sand was deposited. In this sand it endeavoured to seek its natural shelter by burrowing; but finding it was too shallow, the Echidna commenced exploring its cage, thrusting its long slender nose into every fissure, and through the bars, to find some outlet through which it might effect its escape; and it was not until it had learnt that this was impossible, that it noticed its food, which consisted of bread and milk, in which some meal-worms were placed. Although it frequently had its nose in contact with the meal-worms it did not eat them. The milk was

[1] These notes were communicated to the Zoological Society in the month of July.

licked up by a rapid protrusion and retraction of the long cylindrical tongue. When seized, it offered but little resistance, and made no demonstration of defending itself by striking with its spur; its only action when irritated was to roll itself up into a ball, in which position the sharp points of the spines presented themselves in all directions. When asleep it likewise rolled itself up. Its temperature was 85° Fahr., being nearly 10° lower than that of a rabbit. The blood-discs, like those of the *Ornithorhynchus*, were flat and circular, and in fact resembled those of other mammalia in form, proportions, and colour: they were found to be rather larger than in the human species and the Apes.

ECHIDNA SETOSA. Short-spined Echidna.

Echidna setosa.	CUVIER, Règne Animal (1st edit. 1817), vol. i. p. 226; Nouvelle édit. i. p. 235.
Echidna breviaculeata.	TIEDEMANN, Zoologie, i. p. 592.

Fur harsh, long, and almost hiding the spines; head without spines; general colour, brown; head, and under parts of body, pale brown; eye encircled with dark brown. Total length, from 14 to 17 inches.

Inhabits Van Diemen's Land.

IN the *Echidna hystrix* the spines are long, and the fur short and scanty, and on the back of the animal the hair is not visible, whilst in the *Echidna setosa* the fur is so long as in a great measure to hide the spines.

The largest specimen of *E. setosa*, which has come under my notice, is contained in the Museum of the Zoological Society, having been presented by Lieut. Breton. This specimen, like all others having the short spines, is from Van Diemen's Land, and measures 17½ inches in total length;

its spur is small, being less than a quarter of an inch in length. In the same collection is a smaller specimen than the above, in which the spines and the spur are more developed.

The British Museum collection contains both sexes, and several specimens of each of the supposed species of *Echidna;* and there is, moreover, a half-grown specimen of the *E. hystrix.* The *E. setosa* is subject to some slight variation in tint, as well as in the texture of the fur: the spines also vary slightly, being somewhat longer in some specimens than in others; yet the differences observable in individuals are not such as to render it difficult to distinguish the *E. setosa* from the *E. hystrix;* nor can the differences existing between these two animals be those of age or sex, as some have supposed. The *E. hystrix* is confined to the continent of Australia, whilst the present animal is peculiar to Van Diemen's Land. M. Desmarest states, that the claws in the *E. setosa* are narrower than in *E. hystrix,* but I have not found any constant difference in this respect; indeed, I have not been able to discover any other differences, excepting those observable in the fur and spines; and this leads me to doubt there being two species, and to suspect that the comparatively humid climate of Van Diemen's Land may have had the effect of causing the fur to become longer and more dense; and if so, the increased development of the fur would, in all probability, affect the growth of the spines, by robbing them of their nutriment. The present species, or variety, was first described and figured by Sir Everard Home, in the Philosophical Transactions for 1802, p. 357, Pl. 13.

Since the above was in type, the author has had an opportunity of examining some specimens of the Short-spined Echidna preserved in spirits, and which are contained in the collection of the British Museum; one of these, a full-grown female, furnished the following dimensions:—

	In.	Lines.
Total length	15	0
Width of body, about	7	6
Length of the naked snout	1	$6\frac{1}{2}$
Width of ditto at the base	0	8
" near the apex	0	$3\frac{2}{3}$
Distance from the angle of the mouth to the apex of snout	0	$4\frac{2}{3}$
Length of the fore foot, including the nails ...	1	6
Width of ditto	1	9
Length of the nail of the middle finger	1	0
Width of ditto	0	$4\frac{1}{2}$
Length of hind foot, not including the nails ...	1	4
Width of ditto	1	$3\frac{1}{2}$
Length of longest nail (that on the innermost toe but one)	1	9
Length of nail of little toe	0	6
From the eye to the tip of the muzzle	2	1

The nostrils are situated on the upper surface of the muzzle, very near the tip, being about $\frac{1}{12}$ of an inch distant from that point, and are in the form of narrow openings of $\frac{2}{12}$ of an inch in length; they nearly meet posteriorly, and diverge slightly in front, being here separated by about $\frac{1}{8}$ of an inch, or rather more. The tongue, which is very slender, pointed, and nearly cylindrical (but slightly depressed), was protruded, and could be with ease drawn out of the mouth to the distance of more than three inches.

The head and feet of a female specimen of the *Echidna setosa* are represented, of the natural size, in Plate 2.

FIG. 7. The fore part of the head viewed from the side.
" 8. The fore foot viewed from above: *a*, the thumb or inner toe.
" 9. The hind foot viewed from beneath: *a*, the inner toe.

MACROPODIDÆ; or Kangaroo Family.

Dentition.—Incisors, $\frac{6}{2}$; canines, $\frac{0}{0}$, or $\frac{1-1}{0-0}$; premolars, $\frac{1-1}{1-1}$; molars, $\frac{4-4}{4-4}$ = 28 or 30[1].

Head elongated, the muzzle contracted; upper lip cleft; muffle[2] clothed with small hairs, or naked. Distinct eye-lashes springing from the eye-lid, as in man, in nearly all the species.

Clavicles slender and weak, especially in the large species of Macropus proper.

Fore limbs smaller than the hind—usually very small in proportion; the hands naked beneath, and having five well-developed fingers; each finger armed with a strong curved claw.

Hind legs large and powerful: the foot long; toes four in number, the inner, or first toe, being absent; the second and third toes long, but extremely slender, and united in one common integument, so as to have the appearance of a single slender toe with a double nail; the nails are distinct, oblong, and hollow

[1] The above is the usual and most simple method of expressing the number of teeth of different kinds: by incisors $\frac{6}{2}$, is meant there are six in the upper jaw, and two in the lower, and that they form a continuous series, or touch each other in either jaw; were they separated by a distinct interspace in the upper, and also in the lower jaw (as it sometimes happens), this would have been expressed thus, $\frac{3-3}{1-1}$, like the molar teeth, $\frac{4-4}{4-4}$; *i. e.* four on each side of each jaw. When we speak of "the three molars," or "the three premolars," it must be understood we mean the three molars, or the three premolars, on *either side* of the upper or under jaw, as the case may be; in all cases such expressions (where not otherwise mentioned) will refer to *one* series: that is, upon the supposition that we view the *whole* of the teeth of an animal as forming *four* series—a series on the right side of the upper jaw and another on the left, and the same in the lower.

[2] The French naturalists use the word "muffle" for that part at the end of the nose which is naked in the Ox, Dog, &c.; where the same part is covered by hairs, as in the Rabbit, the animal is said to have no muffle. The term will be used to designate the corresponding part of the nose, whether hairy or not, in this work; for there are intermediate conditions, and it will be convenient to have some definite term for the part in question.

beneath; the fourth toe much developed, and armed with a large solid claw, compressed above, broad and flat beneath, and more or less pointed; the fifth, or outer toe, well developed, but shorter and smaller than the enormous fourth toe, like which it is armed with a strong solid claw: fibula slender, and with its lower half closely applied to the tibia, though easily separable, excepting in *Hypsiprymnus*, where the lower portion is firmly joined to the tibia.

Tail long, and usually very powerful.

Pouch well developed, and opening forwards.

Mammæ four[1].

Stomach complex; cœcum long and simple.

THE incisor teeth of the upper jaw present a broad outer and inner surface, and are compressed from front to back; and so far resemble the incisors in man. The two foremost, though rather widely separated at the base, converge and meet at their apices: they are considerably arched in front in the longitudinal direction, and have the outer surface slightly convex in the transverse direction, and often presenting a shallow longitudinal depression. The second incisor is usually narrower than the first, dilated at the extremity, which is truncated, nearly flat externally, or slightly convex, and sometimes presents a vertical groove at this part; the third incisor is most frequently broader than either of the preceding pair, and has one or two external vertical folds; its apical portion is partly overlapped in front by the second incisor; the apices of these teeth are on the same plane, or nearly so, but the cutting edge of the foremost, on each side, descends obliquely from the outer to the inner side. The lower incisors are horizontal, long, compressed, and lanceolate, and have cutting external and internal margins; their outer surface is convex, and the inner surface is strongly convex in the transverse direction, in the middle, but concave near the margins; when

[1] I am not sure that this number of mammæ is constant.

the mouth is closed, the outer cutting edge of the lower incisors is brought in contact with the cutting edges of the two posterior incisors of the upper jaw on either side, and their points shut within the apex of the foremost pair of the upper jaw. In *Macropus major* (and perhaps in some nearly allied species) the rami of the lower jaw are loosely attached at the chin, and at the apex they are free, and the animal has the power of slightly separating the lower incisors, so that their outer cutting edges are brought more closely in contact with the upper incisors than they otherwise would be. The foremost of the molar series is a false molar, and differs in its form from the others, being laterally compressed; it sometimes has an indented fold of enamel both on the outer and inner side, in or near the middle of the tooth, the crown of which is thus divided into two parts, which are equal in length, but the posterior half has a greater transverse diameter. The true molars differ somewhat in size, there being a slight and gradual increase in size from the foremost to the last: the crown of each molar presents nearly a quadrate figure, but is longer than broad; it has two principal transverse ridges, which, before worn, are considerably elevated, and have trenchant edges; the foremost of these ridges is evidently formed by the junction of the anterior pair, and the hindermost by the union of the posterior pair of cusps, which are most commonly seen in complicated molar teeth, and which we find much less perfectly joined in the Kangaroo-rats. Besides these two principal eminences, are two other transverse ridges, which are smaller and less elevated; one is situated on the fore part of the tooth, and the other, which is less distinct, on the hinder part, but this latter is wanting in the molars of the lower jaw: these smaller ridges appear to represent what I have termed the band of the tooth, which in these animals is only developed in the parts mentioned. Besides the transverse ridges, is a longitudinal ridge, very nearly in the mesial line of the tooth, but which is interrupted in parts, merely serving to connect each pair of transverse cusps. All these ridges are covered externally with enamel, and when the molar is considerably worn we find it presenting two principal loops, or folds, of enamel, entering from opposite sides of the tooth, and meeting in the mesial line; two much narrower folds enter into the body of the tooth, in the same way, in the fore part of the tooth, but at the back of the tooth there is but one distinct narrow fold, and that enters from the outer side, and exists only in the upper molars.

The Kangaroos are vegetable-feeding animals, browsing upon herbage, like the Ruminants, and it appears that in some cases they chew the cud, like those animals[1]. Some are of great size, being nearly as tall as a man when in their common erect position; others are as small as the common hare, and indeed greatly resemble that animal in general appearance. They are remarkable, generally, for the small size of the anterior extremities, and for the slender proportions of the fore parts of the body, which are very flexible, and, on the other hand, for the great bulk of the hinder part of the body, and the large size of the hind legs and tail. When browsing they apply the fore feet to the ground, but usually they rest entirely on the hind feet and tail, and have the fore part of the body elevated, and inclining slightly forwards. The great and powerful tail serves as an extra limb, and is capable for a moment of supporting the whole weight of the body. In some of the smaller species of the present family the disproportion in size of the fore and hind legs is much less striking, and the tail is less powerful than in the large species of true Macropus, as now restricted; and in the

[1] "I have more than once observed the act of rumination in the Kangaroos kept in the Vivarium of the Zoological Society."—OWEN, article *Marsupialia*, in Todd's Cyclopædia of Anatomy and Physiology.

Tree Kangaroos of New Guinea we find the fore legs almost as long and powerful as the hind: the tail, which is clothed for the most part with short adpressed hairs in the ordinary Kangaroos, here is very bushy, as well as in the rock-inhabiting Kangaroos: these, and other modifications, however, observed in the various species, are pointed out in their proper places.

About the beginning of the present century but three species of the present group were known,—the *Macropus giganteus*, and *Macropus minor*, of Shaw, and the *Didelphys Brunii* of older authors. Illiger was the first to subdivide the genus *Macropus* of Shaw, instituting a new genus upon the smaller species of that author, to which he gave the name *Hypsiprymnus;* he at the same time unnecessarily substituted the new name *Halmaturus* for that of *Macropus*. In the *Dictionnaire des Sciences Naturelles*, article Kangaroo, M. F. Cuvier further subdivided the group, which had been augmented by several new species brought home by the French scientific expeditions, separating from the *Macropus giganteus*, and some nearly-allied species, certain Kangaroos, which differed in having the muffle naked. For his new section he retained Illiger's name, *Halmaturus*. The Kangaroo group is now very numerous in species, is regarded as forming a family, and has been subdivided into many genera. In some cases the authors of the new genera have not taken the trouble to define them, and in some the characters given are merely individual peculiarities.

As regards the section *Macropus proper* (as now restricted), characterised as having the muffle hairy, it is necessary to state, that the species which are arranged in that section differ as to the extent of the part of the muffle which is clothed with hairs; and in some cases, as in *M. rufus*, the muffle is quite as imperfectly clothed as in certain Halmaturi, in which the muffle is said to be naked. In the *Macropus leporoides*, the muffle is entirely covered with velvet-like hair;

this little animal forms the type of Mr. Gould's genus *Lagorchestes;* but even in this little section I find the muffle less perfectly clothed in a second and nearly allied species. Lastly, among the Kangaroo-rats, in which the muffle is described as "bald," is one species (*Bettongia rufescens*) in which that part is almost entirely covered with small hairs.

The presence or absence of canine teeth is dwelt upon as furnishing a distinguishing character of the Kangaroos and Kangaroo-rats; several species of Kangaroos, however, have canine teeth even when adult, and in all probability all possess them in a very rudimentary condition when young: they have been noticed by Professor Owen in the young *Macropus giganteus*. In two adult skulls of different species of *Halmaturus* in the British Museum (*H. Ualabatus* and *H. Thetidis*), I find the sockets of these teeth remaining. The canines, however, cannot be regarded as functional teeth in the Kangaroos, where they are always very small, whilst in the *Hypsiprymni* (or Kangaroo-rats) they are tolerably well developed.

In the structure of the incisor teeth of the upper jaw, the various species of *Macropodidæ* differ considerably; the *principal* modifications which I have noticed have presented themselves in the *Macropus giganteus, M. Bennettii, M. Thetidis, M. leporoides,* and in the genus *Hypsiprymnus,* and are pointed out in their proper places.

With regard to the premolars, and true molars, there are but two striking modifications of structure, and these are exhibited, on the one hand, in the Kangaroos proper, and, on the other, in the Kangaroo-rats; in these animals there are equally marked modifications in the structure of the cranium, and some less striking peculiarities in the extremities. I may add, that in the true Kangaroos, which are, I believe, the only Marsupials which are not of nocturnal habits, there are always distinct eyelashes to the eyes, springing from the eye-lids, as in the human subject: in no other Marsupials have I noticed true eyelashes: indeed, I believe the Macropi with eyelashes are the only Marsupials which roam about during the day, and this may account for the presence of these appendages.

Upon the whole, it appears to me, that the most important divisions which have been made of the group *Macropodidæ* are those which have received the names *Hypsiprymnus* and *Dendrolagus*. The third division, containing the great bulk of the species, can be subdivided into groups of minor value only; the minor divisions, however, will be convenient, though to a certain extent arbitrary, as may be inferred from the preceding observations.

It will be most convenient to notice here the

FOSSIL MACROPODIDÆ.

These are confined to Australia. The most important and earliest notice we have of the existence of fossil Kangaroos is that drawn up by Professor Owen upon a collection formed by Major (now Sir T. L.) Mitchell, and which is published in that gentleman's work entitled *Three Expeditions into the Interior of Eastern Australia*. The remains in question were found by Sir T. L. Mitchell, together with numerous others, appertaining entirely to Marsupial animals, in some caverns in the limestone rocks of Wellington Valley[1]. They were embedded in a fine red earth, more or less cemented together by stalagmite, and both bones and matrices closely resembled in their conditions those of the English caverns (of Torquay, for example) and of the caves in the

[1] An account of these caverns will be found in the second volume of the work quoted, and, from the pen of the same author, in the Proceedings of the Geological Society for April 1831.

interior islands of Brazil. Other Marsupial remains, including species of the present group, and also including some of the most distinct of the cavern species, have since been found by Sir T. L. Mitchell in the alluvial, or newer tertiary deposits, in the bed of the Condamine River, west of Moreton Bay; and we are indebted to Count Strzelecki for further discoveries of mammalian remains in the Wellington Valley caverns. Amongst the specimens contained in these collections, which have been presented partly to the Geological Society, and partly to the Museum of the Royal College of Surgeons[1], are remains appertaining to about six or seven distinct species of *Macropodidæ:* some approximate in size to the *Macropus major,* and with respect to these, as well as others of small size, Professor Owen remarks, that from want of skeletons of existing species of Kangaroos, he must leave doubtful the specific determination. The following three species are of very large size, and clearly differ from any of the hitherto-discovered large species.

Macropus Atlas (fossil): OWEN, in Mitchell's Journal, &c. Vol. II. (2d Ed.) p. 365, Pl. 47, fig. 1; and in Owen's Odontography, Pl. 101, figs. 3 and 5.

This species, Professor Owen observes, must have been at least one-third larger than the *Macropus major:* it is chiefly remarkable for the great size of the permanent spurious molar, in which respect it approaches the *Hypsiprymni;* and, inasmuch as this tooth wants the external vertical grooves, and the molar teeth have much elevated and sharp transverse ridges, the *M. Atlas* approximates most nearly to the

[1] The first of Sir T. L. Mitchell's collections was presented by that gentleman to the Geological Society; the other collection, as well as the collection formed by the enterprising traveller, Count Strzelecki, are presented to the College of Surgeons.

Hypsiprymnus Brunii. The molar teeth are proportionately shorter and broader than in *M. major*, and differ in having the longitudinal ridge in the central transverse valley almost obliterated; but, commencing at the bottom and middle of this valley, is a distinct sharp ridge, which runs upwards and inwards, and terminates at the inner salient angle of the anterior principal cusp.

	Lines.
Length of the second upper true molar teeth	$6\frac{1}{2}$
Width of ditto	$5\frac{2}{3}$
Length of the permanent spurious molar	$8\frac{1}{3}$

Several fragments of the cranium and lower jaw of the *M. Atlas* have been found both in the caves at Wellington Valley, and in the alluvial or newer tertiary deposits in the Condamine River, West of Moreton Bay. In the Museum of the College of Surgeons is a shaft of a right humerus (having a circumference of three inches below the deltoid ridge), a distal end of a femur, and a second phalanx of the longest toe of the hind foot, which Professor Owen thinks in all probability belong to the present species; they were contained in the same collections as the parts of the cranium referred to.

Macropus Titan (fossil): OWEN, in Mitchell's Journal, Vol. II. p. 365, Pl. 47, fig. 3; Odontography, Pl. 101, figs. 1 and 2.

Founded upon portions of the upper and lower jaw, containing molar teeth: these indicate an animal of equal size to the preceding, but which is readily distinguished from it by the comparatively small size of the first or spurious molar; in this respect more nearly corresponding with the *M. major*.

A fragment of the right side of the upper jaw, contained in the Museum of the College of Surgeons, possesses all the five molar teeth; the foremost of these are fractured. The whole series is 2 inches $4\frac{2}{3}$ lines in length; the length of the hindermost molar is $7\frac{3}{4}$ lines, and its width is $5\frac{3}{4}$ lines. The longitudinal ridge in the middle valley of the tooth is well developed

In the same museum is the distal half of a right humerus, a lower end of a left femur, and a corresponding part of the right femur, together with a small fragment of a shaft of a femur, which Professor Owen thinks probably belongs to the *M. Titan*. The above remains were found in the newer tertiary deposits in the bed of the Condamine River: the fragments upon which the author just mentioned founded the *M. Titan* were found in the caverns of Wellington Valley.

Macropus Goliah (fossil): OWEN'S MSS.

All that is known of this species is a fragment of the right side of the upper jaw, containing two molar teeth, which is from the newer tertiary deposits of the Darling Downs, Australia. Judging from the size of the teeth, this animal must have been even larger than either of the two preceding species, the two molars measuring together, in the longitudinal direction, one inch and a half, and the width of one of the molars being $7\frac{1}{2}$ lines; they are proportionately broader, therefore, than in *Macropus major*.

Besides these, Professor Owen characterises with a name a fourth fossil species—the

Macropus affinis (fossil) of the Descriptive Catalogue of the Fossil Organic Remains of Mammalia and Aves, contained in the Museum of the Royal College of Surgeons, p. 328.

It is founded upon a "portion of the left ramus of the lower jaw of a Kangaroo, with the penultimate and antepenultimate molars, showing the crowns much worn by mastication: the crown of the last molar has been broken off, and there are the remains of two molars anterior to the antepenultimate one; the extent of the four posterior molars is 1 inch 10 lines; the penultimate molar, besides its inferior size, differs from the corresponding tooth in *M. Atlas* in being narrower in proportion to its length, in having a relatively smaller talon, and no posterior one; it differs, *à fortiori*, from the antepenultimate molar of the *M. Titan*, inasmuch as this has a larger proportional anterior talon than in the *M. Atlas*. The teeth and the jaw of this specimen closely agree in size with those of the large male *Macropus laniger*, but the inner lobes of the penultimate molar are thicker in the fossil, and the jaw does not swell out so much on the outside of the alveolus of the last molar; there is also a longitudinal indentation on the outside of the alveolar process of the anterior molars. The present fossil, therefore, indicates either an extinct species of the size of the existing *M. laniger*, or it may have belonged to a female of a third gigantic extinct species." It is from the newer tertiary deposits in the bed of the Condamine River, west of Moreton Bay, Australia.

Amongst the specimens from the caverns of Wellington Valley, presented by Count Strzelecki to the College of Surgeons, the author noticed fragments containing molar teeth nearly corresponding in size to those of *M. Thetidis*, and others which must have belonged to Kangaroos as small as the *M. leporoides*.

There were also fragments which are clearly referrible to the genus *Hypsiprymnus*; upon these are founded the—

Hypsiprymnus spelæus, OWEN: Catalogue of the Fossil Organic Remains, &c., p. 332.

No. 1537 of the Catalogue mentioned is a fragment of the

right side of the upper jaw of the present species; it exhibits the characteristic premolar, and two of the true molar teeth; the three teeth together measure 9¾ lines, of which the premolar is 4¾ lines in length; it has about six vertical grooves on the outer side. In the breccia attached to this fragment is an incisor tooth of a Rodent, allied to the Rats.

Figs. 2, 3, 4, and 5, of Plate 3, will convey an idea of the general form of the skull, and of the relative position and form of the teeth, in the Kangaroo family.

FIG. 2. The skull viewed from beneath; *a a*, the posterior palatine openings; *i*, the incisor teeth; *p-m*, the premolar; *m*, the four molar teeth.

FIG. 2 *a*. The lower jaw viewed from the outer side; *i*, the incisor tooth: *p-m*, the premolar; *a*, the coronoid process; *b*, the condyle; *c*, the angle,—which is bent inwards.

FIG. 2 *b*. The left half of the lower jaw viewed from above.

FIG. 2 *c*. Shows the three incisors of the left side of the upper jaw.

FIG. 2 *d*. Represents the cutting surfaces of the same teeth.

FIG. 3. Molar tooth of the upper jaw, showing the grinding surface.

FIG. 4. Molar tooth of the lower jaw.

FIG. 5. An unworn molar tooth of the upper jaw, viewed from the inner side; *a*, the anterior principal transverse cusp; *b*, the posterior ditto; *c*, a small longitudinal ridge in the middle of the tooth; *d*, a portion of the *band*, which is prominent on the fore part of the tooth, wanting on the outer side, and slightly developed on the inner side, at *f*, and on the back of the molar, at *e*.—This tooth is magnified.

MACROPUS GIGANTEUS. The Great Kangaroo.

Didelphis gigantea.	SCHREBER, Saugeth. iii. p. 552, Tab. 154. 1778.
Macropus giganteus.	SHAW, Naturalists' Miscellany, Pl. 33. 1791.
Macropus major.	SHAW, General Zoology, vol. i. pt. 2, p. 505, Pl. 15. 1800.
Kangurus labiatus.	DESM., GEOFF. &c.
Halmaturus griseo-fuscus.	GOLDFUSS, in Isis, 1819, p. 266.
Macropus major.	Great Grey Kangaroo, GOULD, Monogr. Pl. 1.
Macropus ocydromus.	GOULD, Annals, and Magaz. of Nat. Hist. for 1842, vol. x. p. 1.
Macropus melanops.	GOULD, Proceedings of the Zool. Soc. for Jan. 1842, pt. 10, p. 10.

Fur moderate as to length and texture; general colour grey-brown; under parts of body paler than the upper, the hairs on these parts being mostly of them broadly tipped with whitish; upper surface of muzzle dusky-brown; around angle of mouth whitish; ears rather large, with long white hairs internally, and with dusky hairs on the outer surface: fore feet dusky-brown, the toes black; hind feet brown-white, the toes brown-black, and black at the extremity; tail black at the apex.

THE Great Kangaroo was discovered in 1770, during Cook's first voyage, whilst that celebrated navigator was stationed on the coast of New South Wales, to repair his vessel, which was in a very dangerous condition, having struck on a rock (and, indeed, was only saved by a portion of the rock, which broke off, and in a great measure filled the hole it had made). The Kangaroo was first seen by a party sent out to procure food for the sick. On the following day Cook himself and Sir Joseph (then Mr.) Banks had the pleasure of beholding this extraordinary animal, and soon afterwards a specimen was shot, from which, in all probability, the notes were made, and the figure drawn, which are published in Dr. Hawkesworth's Account of the Voyage (p. 561, fig. 20, and pp. 577-578). Skins of the animal, however, appear to have been brought home, from which some slight additional matter was added to the descriptions by Pennant, who gives the dimensions of the "largest skin" he examined; and it is probably from the same source that Dr. Shaw discovered the two, united, little inner toes, which had not previously been noticed.

The earliest technical name applied to the animal was that of *Yerbua gigantea*, given by Zimmerman[1], in 1777; and in the following year it received the name of *Didelphis gigantea*, from Schreber[2]. The accounts of both of these

[1] Zoologiæ Geographicæ Quadrupedum, p. 526.

[2] Saugethiere, vol. iii. p. 552, Pl. 154 (copied from the Plate in Hawkesworth).

authors, as well as those of Pennant[1] (in the first edition of his History of Quadrupeds), and Gmelin[2], which followed, are founded on the materials collected in Cook's Voyage. Governor Phillip[3], and White[4], in their Voyages, add further matter towards completing the history of this animal; and we find a summary (including some original observations, made upon living specimens, subsequently brought over to this country), given by Pennant, in the third edition of his general work upon Quadrupeds, which, so far as the habits and external characters of the Great Kangaroo are concerned, leave but little to be desired. Nor were appropriate generic and specific names wanting; Dr. Shaw having, in the meantime, supplied the deficiency of a proper generic title in the part of his Naturalists' Miscellany published in 1790, (Plate 33). We find, however, a number of new names given to the Kangaroo many years afterwards. There is, indeed, such a variety of names and different combinations of generic and specific appellations, that we have thought it desirable not to burden these pages with more than a few, which appeared necessary as a key, more especially, to the works of the principal authors on Mammalogy.

The Great Kangaroo inhabits New South Wales, Southern and Western Australia, and Van Diemen's Land, and is known to the colonists by the names "Boomer" and "Forrester." In the older works it is said to be called Kangaroo by the natives; and Mr. Gould informs us, it is the "Bundaary" of the Aborigines of the Liverpool range. It prefers low grassy hills and plains, and open parts of the

[1] Vol. ii. p. 306, Pl. 35: (the figure, like that given by Schreber, is copied from the Plate in Cook's Voyage).

[2] Linn. Syst. Nat. ed. Gmel. vol. i. p. 109.

[3] The Voyage of Governor Phillip to Botany Bay, p. 104, Pl. 10.

[4] Journal of a Voyage to New South Wales; appendix by John Hunter, p. 272, and Plate,

country, where it browses upon the herbage and low bushes. During the heat of the day it shelters itself among the bushes, tall grass, or ferns. In its more common position it rests on the hind legs and terminal half of the tail, with the anterior part of the body bent forwards, and but little elevated, but upon the slightest alarm the body is raised perpendicularly, and the animal is thus enabled to command a distant view. The senses of smell and sight in the Kangaroos are very acute, and as they are timid animals, they are very difficult to approach; they nevertheless not unfrequently fall a prey to the native dog, and are much hunted by the natives (their flesh being well-flavoured), who procure them by means of their unerring spears. Sometimes, discovering their retreat, the natives form a large circle, and gradually close upon them, and by shouting and yelling, Mr. Gould informs us, so terrify the animals, that they become confused, and are readily dispatched by means of the bommerengs, clubs, and spears. "Still, however formidable an enemy the Aborigines may have been," says this author, "the Great Kangaroo finds, at the present time, a far greater one in the white man, whose superior knowledge enables him to employ for its destruction much more efficient weapons than those of the more simple son of nature. Independently of the gun, he brings to his aid dogs of superior breed, and of so savage a nature, that the timid Kangaroo has but little chance when opposed to them. These dogs, which run entirely by sight, partake of the nature of the greyhound and deerhound, and, from their great strength and fleetness, are so well adapted for the duties to which they are trained, that the escape of the Kangaroo, when it occurs, is owing to peculiar and favourable circumstances; as, for example, the oppressive heat of the day, or the nature of the ground; the former incapacitating the dogs for a severe chase, and the hard ridges which the Kangaroo invariably endeavours to gain, giving him great advantage over his pursuers. On such grounds the females in particular will frequently outstrip the fleetest greyhound, while, on the contrary, heavy old males, on soft ground, are easily overtaken. Many of these fine Kangaroo-dogs are kept at the stock stations of the interior, for the sole purpose of running the Kangaroo and the Emu, the latter being killed solely for the supply of oil which it yields, and the former for mere sport, or for food for the dogs. Although I have killed the largest males with a single dog, it is not generally advisable to attempt this, as they possess great power, and frequently rip up the dogs, and sometimes even cut them to the heart with a single stroke of the hind leg. Three or four dogs are more generally laid on; one of superior fleetness to 'pull' the Kangaroo, while the others rush in upon and kill it. It sometimes adopts a singular mode of defending itself, by clasping its short, powerful fore-limbs around its antagonist, leaping away with it to the nearest water-hole, and there keeping it beneath the water until drowned: with dogs the old males will do this whenever they have an opportunity; and it is said, that they will also attempt to do the same with man. In Van Diemen's Land the *Macropus giganteus* also forms an object of chace, and, like the Deer and Fox of England, is hunted with hounds."

Like other animals whose forms differ considerably from those with which our eye is familiar, the Kangaroo, when first beheld, does not strike us as having agreeable proportions, and its movements appear awkward, especially when the animal is browsing, at which time it rests upon its four legs; requiring then to move but short distances, the body is outstretched, and the hinder parts, assisted by the tail, are suddenly brought close to the anterior extremities, and this movement is repeated so long as the animal continues to graze; but when it wishes to reach a distant spot, the fore legs are removed from the ground, and it attains its end by a succession of bounds, and with an ease which at once removes the impression of awkwardness. When in an open country, flying from its enemies, the Great Kangaroo is said to make leaps to the distance of fifteen feet and more: its body outstretched nearly horizontally, and the great tail in the same direction, the latter is not then applied to the ground, but serves as a balance, and to steady the course. "The prehensile faculty and unguiculate structure of the anterior extremities (as Professor Owen observes), appear to have been indispensable to animals requiring to perform various manipulations in relation to the economy of the marsupial pouch; and when such an animal is destined, like the Ruminant, to range the wilderness in quest of pasturage, the requisite powers of the anterior members are retained, and secured to it by an enormous development of the hinder extremities, to which the function of locomotion is almost restricted."

Without large canine teeth or horns, as weapons of defence, the Kangaroo, as is seen from the foregoing pages, is yet by no means to be attacked with impunity; its powerful hind legs, furnished with strong pointed nails, are formidable weapons, and the tail is so muscular, that it is capable of sustaining the whole weight of the body during the moment that they are used in striking. The fore claws are strong, and likewise used as weapons of defence.

In various accounts of the habits of the Kangaroo, the animal is said to be gregarious, living in flocks, which are generally headed by an old male; but Mr. Gunn, a good observer, who has published some interesting notes on the habits of various species of marsupial animals[1], attributes

[1] See Annals and Magazine of Natural History, vol. i. p. 104.

their being occasionally seen in flocks to the circumstance of their food being abundant in particular spots—as on recently burnt land, &c. Mr. Gould is also of opinion, that the Great Kangaroo is not, strictly speaking, a gregarious animal.

The ears in the Great Kangaroo are moderately large; the tail long and very thick, especially at the base; the fur moderate both as to length and texture, and the hairs of which it is composed are somewhat waved, giving it a slight woolly texture. The general colour is grey-brown, darkish on the back, and rather pale on the sides of the body; the under parts of the body, and inner sides of the limbs, are whitish; the hairs of the fur, both on the upper and under parts of the body, are brown-grey at the root; those on the back of the ear whitish-brown towards the point, and brown, or brown-black, at the point. The head is nearly of the same colour as the body; but the muzzle is somewhat dusky, and the lower parts of the cheeks are greyish; around the angle of the mouth the hairs are white, and there is sometimes an indistinct whitish mark running back on to the cheeks from that part; the chin is dusky. The ears are well clothed with white hairs internally; externally grey, but dusky towards the base. The prevailing hue of the fore-legs is whitish grey; sometimes a dusky mark is observed on the elbow: the fore-feet are freckled, or finely pencilled, with black and white, and the toes are black. The hind legs and base of the tarsi are very pale, and may be described as brown-white; the toes are brownish black, or black. The tail is clothed at the base with fur like that of the body, but passing onwards towards the opposite extremity, it becomes gradually harsher and adpressed, and at the apex, where the hairs are black, they are very harsh.

White, or brownish-white, varieties of this species (as well as of some others of the genus) sometimes occur: Mr. Gray regards his *M. albus* as one of these varieties. The posterior incisor tooth on each side of the upper jaw (Pl. 5, figs. 1 and 2,) is longer from front to back in this species than in others which I have had an opportunity of examining[1], and has two external vertical grooves, one situated rather in front of the middle, and the second placed between this and the anterior margin. The size and proportions of the cranium are expressed in the subjoined dimensions.

	No. 1.		No. 2.		No. 3.		No. 4.		No. 5.		No. 6.	
	Ins.	Lns.	Ins.	Lns.	Ins.	Lns.	Ins.	Lns.	Ins.	Lns.	Ins.	Lns.
Total length of skull . . .	7	6	7	0	6	7	8	2	6	6	6	11
Width	3	11	3	7	3	$5\frac{1}{2}$	4	4	3	$2\frac{1}{2}$	3	$9\frac{1}{3}$
Length of nasal bones . .	2	$11\frac{1}{2}$	2	9	2	10	3	6	2	$9\frac{1}{2}$	2	$11\frac{1}{2}$
Width of ditto at the base .	0	11	1	$1\frac{1}{2}$	1	0	1	$2\frac{1}{2}$	0	11	1	$0\frac{3}{4}$
" " near the apex	0	$8\frac{1}{2}$	0	8	0	9	0	$9\frac{1}{3}$	0	$7\frac{2}{3}$	0	$9\frac{1}{4}$
" between orbits . . .	...		1	3	1	$4\frac{1}{2}$	1	$9\frac{1}{2}$	1	6	1	$5\frac{2}{3}$
Length of palate	5	0	...		...		5	6	4	4	4	$7\frac{1}{2}$
Width of ditto opposite the third molar	1	2	...		...		1	$5\frac{1}{2}$	1	$2\frac{1}{4}$	1	$3\frac{2}{3}$
Length of the four posterior molar teeth, taken together, on either side of either jaw	1	11	...		...		1	11	...	?	...	
" of the three incisors of either side of the upper jaw	0	$9\frac{1}{2}$	0	10	...		0	$9\frac{1}{2}$	...	?	0	$9\frac{1}{2}$
" of the posterior incisor from front to back	0	5	...		...		0	$4\frac{3}{4}$	...	?	0	5
Distance between incisor and molar teeth of the upper jaw	2	8	...		...		2	8	2	4	2	1
Length of ramus of lower jaw	...		...		...		6	5	5	$0\frac{1}{2}$	5	6
Height in a vertical line dropped from the coronoid process	...		...		...		3	8	3	2	3	4

[1] Perhaps *M. fuliginosus* presents the same character: in *M. giganteus* the tooth in question varies somewhat as to length; the figures in Plate 5 give the extremes of variation I have met with.

	Adult Male. Brit. Mus.		Female. Brit. Mus.	
	Inches.	Lines.	Inches.	Lines.
Length from tip of the nose to the root of the tail . .	63	0	42	0
" of the tail	42	0	28	0
" from nose to ear	8	4	7	3
" of the ear	4	9	4	0
Width of ditto	3	8		
Length of the fore-arm and hand (without the claws) .	17	0	10	0
" of the tarsus (without the nails or claws) . . .	15	6	12	3

The first three of the columns in the Table in the preceding page give the dimensions of skulls of *M. giganteus,* but unfortunately the sexes of the specimens to which they belonged are not known; the remaining three columns of dimensions are from specimens of *M. ocydromus* in Mr. Gould's collection; the first (col. No. 4) gives the admeasurements of the skull of an adult male of very large size; it had the imperforate palate, and the general proportions, of *M. giganteus,* and it will be seen that, notwithstanding its greater size, as compared with the skull, the dimensions of which are given in column No. 1, the size of the molar teeth is the same The dimensions given in column 5 are from an adult female, weighing 50 lbs.: it had the three posterior molar teeth only remaining. The last column contains the admeasurements of an immature, though nearly adult male; it had four molar teeth on each side of each jaw, of which the foremost was a premolar; the last true molar was not developed.

I here subjoin descriptions of the *M. ocydromus,* and *M. melanops,* both of which appear to me specifically identical with the *M. giganteus.*

Macropus ocydromus. GOULD.

Fur short and somewhat woolly; general colour, brown-grey, slightly suffused with yellowish rufous, especially of the head and sides of the body; under parts, grey-white, tinted with yellowish rufous on the abdomen; muzzle brownish above, and at the sides, towards the tip; ears clothed at the base externally with sooty-brown fur, as well as the occiput, the remaining portions clothed with longish white hairs; limbs brownish white; fore-feet much pencilled with brown; the extremities of the toes black; tarsi slightly pencilled in front with brownish, the toes pencilled with blackish brown, and at the tip black; tail clothed, for a considerable distance from the base, with fur like that of the body both in texture and colour, but towards the apex the hairs gradually become harsher and adpressed; about eight or nine inches of the apical portion is brownish black; short hairs, like those on the tip of the muzzle, extend down on the muffle, terminating in a point in front, and leaving a naked space below of about 2½ lines in width; there is a naked space, of less width, also, bordering the nostrils. Height of a full-grown male, in its ordinary upright position, four feet six inches.

	Inches.	Lines.
Length from tip of the nose to root of tail ...	58	0
" of tail	36	0
Circumference of ditto at the base, about	13	0
Length of fore-arm (not including the hand) ...	13	0
" of hand (without claws)	4	6
" of central claw of ditto	1	10
" of tarsus	15	0
" from nose to ear	7	10
" of ear	4	10

A female, procured by Mr. Gould from the Sand Plains, near "Wongar Hills," Western Australia, and which is said to have been accompanied by a young one, and was therefore mature when killed, weighed about 40 lbs.: its general hue was almost uniform pale rufous brown; the toes and fingers suffused with black at the extremity; the occipital portion of the head was dusky; about five or six inches of the tip of the tail was also dusky, or blackish. Its dimensions, as nearly as they could be taken from an unstuffed skin, were—length from nose to root of tail, 3 feet; tail, 2 feet; tarsus and nails, 12 inches 6 lines; from nose to ear, 6½ inches; ear, 4 inches.

The largest cranium I have seen of a Kangaroo belonged to one of these animals; its dimensions are given in column No. 4, with those of some skulls of *M. giganteus;* it is in Mr. Gould's collection, and is labelled as belonging to an adult male weighing 160 lbs.

The only difference which I can perceive between Mr. Gould's specimens of *M. ocydromus* and the *M. giganteus*, consists in the fur of the former being shorter: they varied somewhat in tint, and a specimen of a Kangaroo living at the Zoological Society's Gardens, which Mr. Gould regarded as his *M. ocydromus*, differs in being of a darker colour than usual, and, in this respect, approaches the *M. melanops*.

Macropus melanops. Gould.

Fur moderate as to length, and inclining slightly to a woolly texture (resembling that of *Macropus giganteus*); general colour of the upper parts, sooty brown, the upper surface of the head, and the back of the ears at the base, rather darker than elsewhere—nearly black; sides of body, and outer side of limbs, yellowish brown; fore-legs grey, the feet black; tarsi brownish white, slightly freckled with blackish towards the toes, which have a much greater admixture of brownish black, and at the end are almost entirely of that hue; tail clothed at the base with fur like that of the body, but the hairs become gradually shorter, more harsh, and are closely applied to the skin on the apical third, where they are black; ears clothed with white hairs internally; externally, excepting at the base, the hairs are black and white in about equal proportions. The lower part of the cheeks, throat, and chest, have a whitish hue, the hairs on these parts being white at the point; but below the point they are grey, as are also the hairs on the abdomen, which are less distinctly tipped with white.

Length, from the tip of the nose to the root of the tail	37	0
" of tail	24	6
" of tarsi and claws	12	0
" from tip of nose to ear	6	4
" of ear	3	8
" of fore-arm and hand to end of nails, about	10	0

The only specimen I have seen of the *M. melanops* is that in the British Museum, which is a male, and is the original of Mr. Gould's description. It has no skull, and hence I have no opportunity of examining and comparing its teeth with those of *M. giganteus*, from which species it differs only (so far as I can perceive) in having its fur of a much darker hue, it being much suffused with sooty black, and the upper surface of the head being almost entirely black.

MACROPUS FULIGINOSUS. Sooty Kangaroo.

Kangurus fuliginosus.	Desm. Nouv. Dict. d'Hist. Nat. tom. xvii. p. 35, Pl. E. 22 (K. géant). 1817. Mammal. pt. 1, p. 263. 1820.
Kanguroo géant.	F. Cuvier, et Geoff. Hist. Nat. des Mamm. fasc. 2. 1819.
,, ,,	F. Cuvier, in Dict. des Sci. Nat. xxiv. p. 347. 1822.
Macropus fuliginosus.	Lesson, Manuel de Mamm. p. 225. 1827.
,, ,,	Gould's Monogr. of the Macrop. pt. 2, Pl. 1.
,, ,,	Waterh. Nat. Libr., Marsupialia, p. 200. 1841.
,, ,,	Gray, List of the Mamm. in the Collection of the Brit. Mus. p. 88. 1843.

About the same size as *Macr. giganteus:* general colour yellowish-brown; toes and apical third of the tail blackish; fur long, and inclining to a woolly texture: long fur, like that of the body, extends on to the base of the tail, covering rather more than one-third of its length; on the remaining portion the hairs are short and adpressed; a large dusky patch on elbows; on the fore feet and tarsi the hairs are much pencilled with black; under parts of the body rather paler than the sides, which are of a bright yellowish tint; but here, as on the upper parts, the hair is yellowish-brown at the root: ear with whitish hairs internally, but somewhat dusky at the apical margin, brown-black externally; head uniform in colour with the body; around the angle of the mouth, and on the chin, are long black hairs.

	Male.		Female.	
	Ins.	Lines.	Ins.	Lines.
Length from nose to root of tail	57	0	43	0
,, of tail	30	0	25	0
,, of tarsus (not including the nails)	12	0	10	6
,, of ear	4	4	4	0

Inhabits Kangaroo Island? [From the Paris Museum.

A male and female of this species form part of the Paris collection; and Mr. Gould informs us, there is a third specimen in the Museum at Leyden. It is from the former, which are the originals of M. Desmarest's accounts, that the above description is taken.

In Péron's "Voyage aux Terres Australes" (tom. ii. p. 75), mention is made of two large species of Kangaroos as occurring in the I. Decrés (Kangaroo Island), and it is supposed that the present animal and the *M. rufo-griseus*, are the two species referred to; the only grounds, however, for this supposition, appear to be, that specimens of *M. fuliginosus* and *M. rufo-griseus* formed part of the collection brought home by Peron's expedition.

In size and general proportions the *M. fuliginosus* so closely resembles the *M. giganteus*, that I think it will probably prove to be a variety of that animal. The name Sooty Kangaroo is most ill applied to the present animal, since its colouring is any thing but sooty, being, for the most part, of a brownish-yellow, rather bright on the sides of the body, and somewhat suffused with dusky on the middle of the back. Some allowance, however, must be made for the long effects of exposure to light, which has, no doubt, altered the colouring. Desmarest, whose first description was taken from these specimens about thirty years back, states the general colouring of the fur to be sooty-brown; darker on the back than on the flanks, and shaded into pale-grey on the under part of the neck, chest, and belly.

Sub-genus *Onychogalea.*

Genus *Onychogalea*. Gray, List of the Specimens of Mammalia in the Collection of the British Museum, p. 88. 1843.

Muffle clothed with hair; posterior upper incisor as narrow as the anterior one, or narrower, and with a single vertical groove; general form slender; fore legs small; tarsi long and slender; tail also long and slender, and furnished with a horny excrescence at the apex.

This little section contains some of the most graceful and prettily-coloured species of the Kangaroo tribe: they are of moderate or small size, and have short fur.

MACROPUS UNGUIFER. Nail-tailed Kangaroo.

Macropus unguifer. Gould, Proceedings of the Zoological Society for August, 1840, pt. 8, p. 93. Monograph of the Macropodidæ, pt. 1, Pl. 4.

Slender; tail very long; tarsi long; fore legs moderate; ears moderately long, and attenuated at the apex; fur very short, moderate as to texture; general tint, pale-reddish ochre, or fulvous; head, limbs, and tail, almost white; abdomen whitish; a palish-brown mark, commencing about the middle of the back, is continued over the rump, and extends along about four inches of the tail. This is clothed with small white adpressed hairs, but on the apical portion, commencing about eight or nine inches from the tip, is a slender black line on the upper surface, and this becomes gradually broader, and the dark hairs of which it is formed become also gradually longer, form a kind of crest, and, at the point of the tail, terminate in a long tuft, which hides a horny excrescence,

with which the tail is armed at this part; at the sides, the apical portion of the tail is brown, and beneath, where they are longer, they are brown-black. A whitish mark nearly crosses the haunches, running backwards from near the knee.

	Male. Inches.	Lines.
Length from tip of nose to root of tail ...	26	0
,, of tail	27	0
,, of tarsus, including the nail ...	7	6
,, from tip of muzzle to ear	4	0
,, of ear	2	7
Width of ditto	1	5
Length of fore arm and hand, including the nails, about	5	9
Height in ordinary erect position, about ...	22	0.

Inhabits the north-west coast of Australia.

I believe the only specimen which has reached Europe of the interesting species, is that deposited in the British Museum by Mr. Bynoe, of Her Majesty's ship, The Beagle, which was procured by that gentleman on the north-west coast of Australia. It is beautifully figured by Mr. Gould, in his Monograph of the *Macropodidæ*.

The *Macropus unguifer*, so called from the circumstance of its tail being furnished at the tip with a nail-like horny excrescence, together with the *Macropus frænatus*, which has likewise a horny termination to the tail, form a little section, to which Mr. Gray, in his "List of the Specimens of Mammalia in the Collection of the Museum," (1843), gives the name *Onychogalea*. *Macropus lunatus* is very closely allied to these animals, and, like them, has the tail terminated by a horny excrescence.

The muffle in *M. unguifer* is covered with hair, with the exception of a very narrow margin next the nostril-openings; the foremost of the three incisor teeth on either side of the upper jaw is distinctly the broadest, the other two are very nearly equal in width; the hindermost has a strong oblique external fold; these teeth are small, compared with the incisors of most Kangaroos. A canine tooth is present, but it is very small. The tarsi are very long and slender; the end of the nails of the double inner toe terminate $2\frac{1}{2}$ inches short of the end of the nail of the great central toe; and the tip of the nail of the outer toe is $1\frac{1}{2}$ inches short of the same point; the nails of the two larger toes are long, narrow, and much compressed above. The nails to the fingers are rather short and broad.

MACROPUS FRÆNATUS. Bridled Kangaroo.

Macropus frænatus. GOULD, Proceedings of the Zoological Society for August, 1840, pt. 8, p. 92. Monograph of the Macropodidæ, pt. 1, Pl. 3.

Form slender: fur short and soft, general tint gray, being finely pencilled with black and white; under parts of the body, and inner side of the limbs, white; a white mark on the cheeks, beneath which is a dusky line: ears of moderate size, pointed, grey externally, but edged with black at the apex; internally with white hairs: muzzle blackish in front of the eye; two conspicuous white marks run backwards from the occiput, and diverging, pass one on each side over the shoulder, and are recurved at a short distance behind the insertion of the fore leg; the space between these lines is black on the occiput, and brownish-black on the back of the neck: sides of the neck suffused with pale-ochreous yellow: tarsi and arms nearly white; hands and toes dusky, but most of the hairs round the nails of the former are white: tail long, and rather slender, coloured like the body at the base, but black along the upper surface of the apical third, and at the point, where the hairs are longer than elsewhere, and hide a small horny tubercle, with which the tail is terminated; the under surface of the tail is of a dirty yellowish-white hue.

		Inches.	Lines.
Length from the tip of the nose to the root of the tail		18	0
,,	of tail	14	6
,,	of ear	2	4
,,	of tarsus, including the nails	5	6
,,	from nose to ear	3	0

Amongst the numerous new species of the Kangaroo tribe discovered by Mr. Gould, this is certainly one of the most elegant. Its form is slender, and the two white marks which run backwards from behind the ear, and terminate on each side of the body a little behind the base of the fore leg, add to its beauty: these marks are rendered the more conspicuous by the intervening space on the back of the neck being almost black. Mr. Gould states that its weight varies from ten to fifteen pounds. The nearest point to the colony of New South Wales at which he met with it, was Brezi, on the River Mokai, whence it extended into the interior. "It inhabits," says Mr. Gould[1], "all the low mountain ranges similar to those of Brezi, whose elevation varies from one to five or six hundred feet, and which are of a sterile character—hot, dry, stony, and thinly covered with shrub-like stunted trees * * *. When started from its seat, which is formed like that of a hare, and sheltered by a tuft of grass, or a small bush, it bounds away with remarkable fleetness, generally giving the best dogs a sharp run, and frequently effects its escape by gaining the thick part of the trunk, or the hole of a decayed tree; and I recollect, on one occasion, that on being sharply pressed, the animal mounted the inside of the tree to an opening nearly fifteen feet from the ground, whence it leaped down before the dogs, and succeeded in reaching the hollow trunk of a fallen tree, from which it was finally taken by the hand."

The female, as is usual in this group, is considerably

[1] Monograph of the Macropodidæ.

smaller than the male. The specimen from which the description here given was taken, forms part of the Zoological Society's collection, having been presented, together with a second, by Mr. Gould. This second specimen differs in having the back part of the head and neck grey. Specimens are also contained in the British Museum collection. The dimensions given by Mr. Gould of the male and female are as follows:—

	Male.		Female.	
	Ins.	Lines.	Ins.	Lines.
Length from nose to root of tail	24	6	17	6
" of tail	1	7	1	3
" of tarsus and toes (including the nails) ...	6	6	5	0
" of arm and hand (including the nails) ...	5	0	4	0
" from the tip of the nose to the base of the ear	3	9	3	6
" of ear	3	6	3	3

MACROPUS LUNATUS. Crescent-marked Kangaroo.

Macropus lunatus. Gould, Proc. Zool. Soc. Aug. 1840, p. 93.

Fur soft and short; general tint, ashy grey; neck rufous; body beneath, grey-white; a distinct curved white mark behind the insertion of the fore-leg; feet whitish; ears rather long and attenuated; tail moderate, with a short crest of blackish hairs on the apical portion, the tip furnished with a small conical horny excrescence.

Inhabits the Swan River District, Western Australia.

About the size of a rabbit; fur very soft, and by no means long; general tint ashy grey, finely pencilled with dusky and yellowish-white; back of neck and shoulders, vinous rust colour; at a short distance behind the base of the fore-leg is a distinct curved white mark. Fur on under parts of the body pale grey, the hairs tipped with dirty-white; on the sides of the body a faint rusty tint is observed (sometimes the rufous tint on these parts, as well as the fore-part of the hind legs, is tolerably distinct); around the eye is a ring of a very pale rust colour, and the muzzle is suffused with the same tint. The ears are rather long and attenuated at the apex, have long white hairs within, and very minute hairs externally, which are dusky and very finely freckled with yellowish white; on the hinder half, the hairs are longer and almost white; the apex of the ear has a delicate fringe of blackish hairs. Fore-feet small, sometimes brown, and sometimes of a dirty-white colour. Tarsi also small, slender, and chiefly of a dirty-white colour; but the sides of the toes are suffused with palish brown. The tail is clothed for the most part with short adpressed hairs, having a general greyish tint; the upper surface is clothed throughout its length with somewhat longer hairs; on the apical portion they form a slight crest, which is usually blackish, and the under surface at the end of the tail is dusky; the tip of the tail is provided with a small conical horny appendage, like a nail, which is about an eighth of an inch in length.

			Male.—Brit. Mus.	
	Ins.	Lines.	Ins.	Lines.
Length from nose to root of tail	14	6	19	0
" " ear ...	2	2	3	0
" of tail	8	6	13	3
" of tarsus and claw ...	4	0	5	0
" of ear	1	8	2	7
" of fore-arm and hand to end of claws, about ...	1	11	3	10

The cutting edges of the upper incisor teeth, in the present species, terminate nearly all on the same plane; the three incisors on either side of the upper jaw are small, and of equal width, or perhaps the foremost rather exceeds the other two; this tooth has a longitudinal groove: the second incisor has no groove, and is somewhat rounded at the point: the third, or hindermost incisor, has an oblique external fold, and sometimes exhibits a mere notch rather behind the middle. The lower incisors are rather short and pointed. The fore feet are small, the central toe the longest, and the outer ones are very nearly equal in length. The hind feet are slender, and the nails of the toes are pointed, and rather small and slender; the nails of the double toe terminate about in a line with the middle of the nail of the outer toe, which is also the case in *Lagorchestes albipilis*.

Sub-genus *Lagorchestes*.

Lagorchestes. GOULD, Monograph of the Macropodidæ, Part 1.

Muffle clothed with velvet-like hairs; posterior upper incisor teeth small, the hindermost with a single vertical groove: tarsi and claws slender; fore legs small, the hands also small, and with smallish sharply pointed nails.

Mr. Gould separates from the Kangaroos which have the muffle clothed with hair, a group of small species, some of which bear considerable general resemblance to our common hare[1] (*Lepus timidus*), a resemblance, however, which is due chiefly to the texture and colouring of the fur, combined with the size of the animals. If compared with the Great Kangaroo, they differ considerably in the structure of their upper incisor teeth, the foremost pair being the broadest, and the hindermost incisor on either side being small, and about equal in width to the second, and this tooth has but one external vertical fold: these differences, however, are likewise observed in some other species of Macropus (as in *M. lunatus*), though perhaps in a rather less marked degree.

In separating as a distinct genus the present group, Mr. Gould was impressed with the idea, that the species of which it was composed were closely allied to the Kangaroo-rats, observing, with reference to the *Lagorchestes leporoïdes*, "that although belonging to that subdivision of the family

[1] The sectional name is derived from λαγος, a hare, and ὀρχηστης, a dancer.

which includes the Rat- and Jerboa-Kangaroos, it differs from them in inhabiting a different character of country (that is, the open plains), in having a hairy muzzle, and in the hands and nails being smaller, more slender, and more delicately formed than any other known species; points indicating that it is not a burrowing animal; and it is apparent, from the sharpness and spiny form of the fore-nails, that they are never used for the purpose of obtaining roots, as is decidedly the case with the Rat- and Jerboa-Kangaroos." But why assume that *Lagorchestes leporoïdes* is allied to the Kangaroo-rats, there being so much difference in the structure of the limbs, and the muffle being hairy? In these characters *L. leporoïdes* agrees with *M. lunatus* and its allies; and, in addition to this, I find the structure of the skull and teeth to be very different to that of the Kangaroo-rats[1], and, in fact, distinctly upon the same type as the true Kangaroos, which also, like *Lagorchestes*, inhabit the open plains. That *Lagorchestes* forms a natural group I admit, but I cannot regard it as entitled to the rank of a genus.

MACROPUS (*Lagorchestes*) LEPOROÏDES.

The Hare-Kangaroo.

Macropus leporoïdes.	**Gould, Proceedings of the Zoological Society for August 1840, p. 93.**
" "	**Waterh. Marsupialia, p. 204.**
Lagorchestes leporoïdes.	**Gould, Monogr. Part 1, Plate 12.**

Fur long and soft; on the upper parts of the body variegated with black, rust colour, and rusty white, the white most conspicuous, and the rust colour but little seen; the back of the neck and shoulders, and a considerable space round each eye, tinted with palish rust colour, sometimes inclining to

[1] See the account of the Kangaroo-rats, or Hypsiprymni.

buff yellow ; sides of the body and haunches suffused with rust colour; under parts grey-white, tinted with rust colour, but nearly pure white between the hind legs. Fore legs with a more or less strongly marked black patch at the base externally, but the hairs on this part pencilled with white; fore arm and hand with short brown hairs, pencilled with very pale brown: on the middle of the tibia is a dusky patch: tarsi impure palish rust colour, finely freckled with brown; toes brownish: tail clothed throughout with small adpressed hairs, which are partly black and partly white; beneath brown-white.

Inhabits South Australia.

This pretty little Kangaroo is nearly equal in size to the common Hare (*Lepus timidus*), and very much resembles that animal in the texture and colouring of its fur, circumstances which suggested the specific name. The head is rather short; the ears moderately long, attenuated, and somewhat pointed at the apex; internally well clothed with long white hairs, externally clothed with small black and white hairs: the upper lip is white; the tip of the muzzle completely clothed with minute brown hairs. The tarsi are long; the nails of the toes long, slender, and sharply pointed; the fore legs very small and delicate; the toes armed with slender nails. The tail is slender, and of moderate length. The fur of the back is nearly black next the skin; each hair has this black portion, which is of considerable extent, followed by reddish-brown; then a long space, which is rusty-white, and the tip is black: on the belly and chest the hairs are grey at the base, and rusty grey-white externally; between the hind legs they are white throughout. The foremost incisor tooth of the upper jaw is the largest, and the hindermost one, which has a distinct notch, is the smallest of the three on either side of the jaw. At a short distance behind the incisors is a small slender canine tooth.

	Inches.	Lines.
Length from nose to root of tail	19	6
" of tail	12	0
" of tarsus	5	7
" from nose to ear	3	5
" of ear	2	2
" from elbow to end of nails, about	3	4
" of skull	3	2
Width	1	10
Length of nasal bones	1	2½
Width at base	0	7
" near apex	0	3½
Length of frontals	1	1
Interorbital space	0	8
Length of palate	1	8½
" of three incisors	0	4
" from incisors to molar	0	8¼
" of premolar and three molars, taken together	0	10
Height of lower jaw	1	3¼
Width from back of condyle to front of coronoid process	0	8

The following observations relating to the Hare-Kangaroo are from Mr. Gould:—

"This singular little Kangaroo, which I have made the type of a genus, ranges widely over the interior; it is tolerably abundant in all the plains of South Australia, particularly those situated between the Belts of the Murray and the mountain ranges: I also hunted it successfully on the Lower Namoi; and skins were presented to me by the Messrs. Coxen, who, I believe, had obtained them on the Liverpool plains. Judging from what information I could gather respecting it, I believe it to be peculiar to the interior, and never to frequent that portion of New South Wales which lies between the ranges and the coast. * * *.

"The name of Hare-Kangaroo has been given to this species as much from the similarity of its form, its size, and the colour and texture of its fur, as from its habits assimilating in many particulars to those of the hare. I usually found it solitary, and sitting close, in a well-formed seat, under the shelter of a tuft of grass on the open plains: for a short distance its fleetness is beyond that of all others of its group that I have had an opportunity of coursing. Its powers of leaping are also equally extraordinary. I may mention an incident connected with the chase of the animal, which occurred to myself. While out on the plains in South Australia, I started a Hare-Kangaroo before two fleet dogs; after running to the distance of a quarter of a mile, it suddenly doubled, and came back upon me, the dogs following close at its heels: I stood perfectly still, and the animal had arrived within twenty feet before it observed me, when, to my astonishment, instead of branching off to the right or to the left, it bounded clear over my head, and on descending to the ground, I was enabled to make a successful shot, by which it was procured.

" Considerable diversity of colour is observable in different specimens, some being much redder than others; but the sexes are scarcely distinguishable by size."

Fig. 17, of Plate 5, represents the skull of the natural size; and fig. 17*a* shows the form and relative positions of the superior incisors, the canine, and three of the molar teeth.

MACROPUS (*Lagorchestes*) CONSPICILLATUS.

Spectacled Hare-Kangaroo.

Lagorchestes conspicillatus. **Gould, Proceedings of the Zoological Society for October, 1841, Part ix. p. 82. Monograph of the Macropodidæ, Part 2, Plate 13.**

In size, and in the colouring and texture of the fur, greatly resembling the Common Hare (*Lepus timidus*). Fur long and loose; on the upper parts of the body the hairs are black, but the curved and exposed ends are rusty-white, shaded through rusty-brown to black, which is the colour of the point; the whitish part is most conspicuous; sides of the body pale brownish rust colour; under parts brownish-white; the hairs here almost uniform to the root: an indistinct whitish mark crosses the haunches; fore legs and tarsi brown-white, but the hairs on these parts are blackish at the root; hands brown-white, slightly pencilled with blackish: tail rather sparingly clothed with small pale hairs; ear very short, narrow at the apex, clothed internally with pale hairs; externally, with hairs which are partly dusky and partly white: around each eye is a broad space, of a rich and bright rusty red colour: sides of the muzzle whitish, the tip above nearly black.

	Female.—Brit. Mus. Inches.	Lines.
Length from tip of nose to root of tail ...	18	6
" of tail	13	9
" of tarsus and claw	5	6
" from nose to ear	3	4
" of ear	1	3
" of fore arm and hand, to end of claw, about	3	6
Height, in ordinary erect position, about ...	14	0

From Barrow Island, north-west coast of Australia.

This species, so remarkable for its general resemblance to the Common Hare, may be distinguished from the *M. leporoïdes* by its ears being considerably shorter, the more brilliant rusty-red colouring round the eye, and the want of the black patch at the base of the fore leg: the muzzle, likewise, is more obtuse. In *leporoïdes* the muffle is entirely covered with velvet-like hairs, but in the present species there is a distinct naked margin next the nostrils, and a small naked space in front. Lastly, may be noticed, the structure of the fore legs and hands, which are larger and stronger than in *leporoïdes:* in this latter animal the hand is about 5½ lines in width, whilst in *M. conspicillatus* it is about 8 lines. The foremost incisor tooth of the upper jaw is the broadest, and the last is rather broader than the second, and has an oblique external fold. These teeth are small, as in *M. unguifer*, and indeed in most of the species having the muffle hairy, if we except the *M. giganteus*.

But two specimens of this species have been brought to

Europe, and it is to Captain Wickham and Mr. Bynoe, of H. M. S. The Beagle, that its discovery is due. The female, from which the above description was drawn up, was presented by the former of these gentlemen to the British Museum. Both the specimens were found on Barrow Island, which lies off the north-western coast of Australia, about thirty miles from the main land.

MACROPUS (*Lagorchestes*) FASCIATUS.

Banded Hare-Kangaroo.

(*Plate* 4, *Fig.* 2.)

Kangurus fasciatus.	Péron et Lesueur, Voyage aux Terres Australes, tom. i. p. 114, Plate 27.
" "	Desmarest, Mammalogie, Part 1, p. 274.
Halmaturus elegans.	Cuvier, Règne Animal, tom. i. p. 187.
Bettongia fasciata.	Gould, Monograph of the Macropodidæ, Part 2, Plate —.
Lagorchestes albipilis.	Gould, Annals and Magazine of Natural History for Sept. 1842, Vol. x. p. 2.

About the size of the Common Hare (*Lepus timidus*): ears moderate, attenuated at the apex; fore feet very small; tail about as long as the body; fur very long and soft, brown-grey, variegated with rust-colour, black and white; around the eye of a brightish rust-colour: numerous transverse narrow dark bands adorn the back, and are most conspicuous on the hinder part: on the whole of the upper parts and sides of the body, are very long interspersed white hairs; under parts of body dirty-white: the hairs springing from the sides of the two larger toes of the hind feet are very long (being many of them nearly an inch in length), rather harsh, and of a brownish-white colour.

Inhabits Western Australia.

The specimens of *Macropus fasciatus* in the Paris Museum being very old, have had the hair worn off from the tip of the muzzle, from which circumstance Mr. Gould (supposing the muzzle to be naturally naked at that part in the specimens in question) was deceived as to the section to which they belonged, and placed them in the genus *Bettongia,* and founded a new species, under the name *Lagorchestes albipilis,* upon some skins received from Western Australia agreeing with *M. fasciatus,* but which he imagined differed in having the muffle hairy. Upon a recent careful examination of the Paris specimens, I found that some few of the small hairs still remained, and indeed were distinct in one of the specimens, where, being in a hollow, they had not been exposed to friction.

In a skeleton of the *M. elegans,* contained in the Museum of Comparative Anatomy at Paris, I noticed that the molar teeth had distinct, and considerably elevated, transverse ridges, as in *Macropus;* that the foremost incisor tooth of the upper jaw was rather the narrowest of the three; the second and third incisors were very nearly equal, the latter had a distinct external notch; the first molar was shorter than in Hypsiprymnus, and had three external grooves; there were no canine teeth, and the nasal portion of the cranium was short. Canine teeth are absent in all Mr. Gould's specimens of *Lagorchestes albipilis* which I had an opportunity of examining, and the structure of the incisor teeth is as above described; they are represented in Plate 5, fig. 4. The fur of this animal is very long and soft; the ears of moderate size, and somewhat pointed; the tail about equal to the body in length; fore legs and feet very small, and the claws slender; tarsi slender, and of moderate length; the nails of the toes also slender, compressed, and concave beneath. The general colour of the fur is greyish, but it is variegated with black, white, and rust colour; the last-mentioned colour is most conspicuous around the eyes: on the back are numerous transverse, narrow, black bands; these are somewhat irregular, and not well defined; the spaces between the bands are partly of a rust colour and partly whitish; the white joins the dark band, and is gradually shaded into rust colour, to be followed by the next dark band; over the shoulders the bands are wanting. On the whole of the upper parts of the back, as well as on the sides of the body and the cheeks, are numerous very long interspersed hairs, which have the exposed portion white, but which, like the hairs which form the ordinary fur, are nearly black at the root. The under parts of the body are dirty white, having a considerable admixture of grey, the hairs being of the latter colour below the points. The ears have longish white hairs on the inner side; on the outer side the hairs are short, and finely freckled with brownish black, and white, the dark colour prevailing on the fore part, and the hinder part being entirely pale. The fore feet are of a dirty rust colour; the tarsi rusty white (or pale rust colour), and pencilled with blackish; the sides, both of the tarsus and toes, are chiefly of a very pale brown colour. The tail is rather slender, tolerably well covered with short adpressed hairs, the prevailing tint of which is brownish-grey; on the under side of the tail the hairs are somewhat longer, and of a brownish-white colour; on the upper surface is a narrow blackish streak, and on the apical third the hairs are longer than elsewhere, and form a small dark crest; at the point of the tail they are sometimes an inch in length. The muffle is entirely clothed with hair, with the exception of a small space in front, and a narrow line next the nostril openings.

The above description is taken from one of Mr. Gould's specimens of *Lagorchestes albipilis*, now in the British Museum, and agrees so perfectly with my notes made upon the specimens of *Kangurus fasciatus* in the Paris Museum, that it is unnecessary to make any extracts from those notes. I shall here subjoin their dimensions, together with those of the specimen above described:—

	Male. In Paris Mus.		Female. In Paris Mus.		British Museum.	
	Ins.	Lns.	Ins.	Lns.	Ins.	Lns.
Length from tip of nose to root of tail	17	6	17	6	17	0
" of tail	12	0	11	0	12	0
" of tarsus, including the nails	4	$3\frac{1}{3}$	4	7	4	3
" of fore arm, hand, and nails, about	3	4	3	4	3	2
" of ear	1	6	1	7	1	8
Width of ditto at the base	1	4	1	5	1	6
Length from nose to ear	3	6	2	11	3	0
" of the three upper incisors, taken together ...	0	$3\frac{1}{3}$	0	$3\frac{1}{2}$	0	$3\frac{1}{2}$

This pretty little animal is noticed in Dampier's Voyage, but was first carefully described and figured by Péron and Lesueur, in their account of the "Voyage aux Terres Australes;" who, moreover, brought home the specimens which are contained in the Paris collection. These were procured in the islands in Shark's Bay, on the west coast of Australia. It is said by these authors to inhabit the impenetrable low thickets, formed of a species of *Mimosa*, which are found in those islands; from these bushes they cut away the lower branches and spines, so as to form galleries communicating one with another, and where they take refuge in time of danger. The female brings forth but one young at a time.

Although abundant in the islands of Shark's Bay, Péron states that none were to be found on the main land. These little Kangaroos (as is the case with all those feeble animals which have neither the power of attack nor of defence) are, like the Hares, extremely timid. The slightest noise caused them to take flight to the thick brushwood in which their galleries are formed, and where it is impossible to pursue them: hence, although very common, they are difficult to procure.

The flesh of these animals is said to resemble that of the Rabbit, but has a slight aromatic flavour, arising, probably, from the nature of the plants on which they feed, nearly all of which are fragrant.

At the time that Péron visited the islands, all the females carried young in their pouch, and the courage with which they sought to save their offspring was truly admirable: although wounded, they flew with the young in the pouch, and never left them until, overcome with fatigue and loss of blood, they could no longer carry them; they then stopped, and squatting themselves on the hind legs, helped the young to get out of the pouch by means of the fore feet, and sought to place them in a situation favourable for retreat.

MM. Péron and Lesueur being unsuccessful in finding the Fasciated Kangaroo on the main land, imagined it was confined to the islands; such, however, is not the case, since M. Priess, an assiduous collector, has recently found it on the main land of Western Australia. The following notes, by Mr. Gilbert, upon the present species, are from the Proceedings of the Zoological Society for February, 1844.

"With respect to the Kangaroos, I have heard of the little silver-haired Lagorchestes (*Lagorchestes fasciatus*), and have tried hard to procure a specimen. It is a species well known to the natives of Moore's River, by whom it is called 'Nar-nine,' and is only to be found in densely thick scrubs on flats, and on the edges of swamps, where the small brush Melaleuca grows so thickly that it is almost impossible for a man to force his way through; its runs being under this, the animal escapes even the quick eye of a native. The only possible means of obtaining it is by having a number of natives to clear the spot, and two or three with dogs and guns to watch for it.

"This beautiful little animal makes no nest, but squats precisely like a Hare, as I have been assured by Mr. Johnson Drummond."

MACROPUS (*Lagorchestes*) HIRSUTUS.

Rufous-haired Hare-Kangaroo.

Lagorchestes hirsutus. GOULD, Proceedings of the Zoological Society for February 1844, p. 32.

Fur long; general tint grey-brown, much suffused with rufous on the hinder parts of the body; legs of a bright rust colour; very long rust-coloured hairs are abundantly interspersed with those forming the ordinary fur, more especially on the hinder parts of the body; under parts rusty white; eye broadly encircled with rust colour; ears moderately large, and rounded at the extremity; tail moderate, clothed with small stiff hairs, which scarcely hide the scaly skin beneath; those on the upper surface of a brownish black hue; on the under, paler.

Inhabits Western Australia.

This species is nearly equal to the common Hare in size; its fur is long and moderately soft; the upper parts of the body are grey, much tinted with rufous brown, and freely pencilled with white; the sides of the body, rump, hind and fore legs, of a bright rust colour, darkest on the hind legs, and least distinct on the fore legs; the throat, chest, and mesial line of belly, rusty white; the crown of the head, grey; a broad space around the eye is of a bright, but palish rust colour, and this tint extends on to the muzzle; a whitish line on the upper lip runs back past the angle of the mouth. The ears are moderately long, rather broad, rounded at the tip, and have longish white hairs internally; externally, they are pencilled with rusty yellow, and dusky; the former tint, however, prevails; the hinder half is almost entirely clothed with small white hairs. The fore feet are clothed with glistering yellowish white hairs; the tarsus is almost entirely of a pale-rust colour, but rusty white towards the hinder part; and the toes are obscurely suffused with brownish rust colour. The tail is clothed throughout with short, stiff, adpressed hairs, scarcely hiding the scaly skin; they are finely pencilled with black and rust colour at the base of the tail, but on the upper surface they soon assume an uniform brown black tint, which is continued to the point; on the under surface they are of a dirty pale rust colour, and towards the apex (about one inch from the point) is a naked scaly space of about an inch in length. The fur on the back is very nearly black next the skin, but a considerable portion of each hair is of a brownish rust colour; near the point the hairs are broadly annulated with white; and at the point, they are dusky or black. On the belly the fur is ashy grey next the skin.

	Inches.	Lines.
Length from nose to root of tail	17	0
" " to ear	3	6
" of tail	11	6
" of tarsus and claws	5	2
" of fore leg to end of claws	3	0
" of ear	1	9

The front incisor in *L. hirsutus* is very large (very much larger than the others); as broad as the one next to it and half the third taken together. The second is a larger tooth than the third, being longer and broader at the base; but the third, which is very short in the vertical direction, is more dilated at the apex, and presents a strong oblique groove on the outer side. The second tooth also has a groove, but this is scarcely perceptible on the outer side of the tooth; it traverses the crown of the tooth in the longitudinal direction, and is deeply indented on the under part of the tooth. I did not see this groove in the second incisor in a female specimen, which was perhaps older, and had the crown of the tooth more worn. This animal has a minute canine, which, when it was alive, could scarcely have penetrated the gum; it shows, however, in the skin, and is situated at the distance of about $1\frac{1}{2}$ lines from the incisors. The length of the three upper incisors taken together is 4 lines. (See Plate 5, fig. 5).

Mr. Gould says of this animal, that it is known to the natives of the York district, Western Australia, by the name "Woo-rup." It is distinguished from others of its genus by the long reddish hairs which are abundantly mingled with those of the ordinary fur on the hinder parts of back, and especially near the base of the tail.

The above description is drawn up from one of Mr. Gould's specimens now in the British Museum collection.

Sub-genus *Halmaturus*.[1]

***Halmaturus*. F. Cuvier.**

Kangaroos in which the muffle is naked in front; inhabitants, generally, of districts which are well clothed with shrubs.

The species of the present section agree in all *essential* characters with Macropus proper, differing only in having the muffle but partially clothed with hair. If a line drawn across the muzzle from the posterior angles of the nostrils be regarded as marking the hinder boundary of the muffle, the hair may be said to cover, usually, about half of the part in question, being continued forwards from the muzzle so as to cover a triangular area, the apex of which is situated in front. In the greater portion of the species, the point of termination is in a line with the middle of the nostril openings; sometimes it is rather in advance of that line, as in *M. agilis;* and sometimes the area which is clothed is in the form of an obtuse angled triangle, and then terminates somewhat behind the line drawn across from the middle of the nostril opening, as in *M. antilopinus* and *M. robustus*.

[1] From Ἅλμα-ατος, leaping; and οὐρὰ, tail.

I do not find any constant modification either in the structure of the skull, teeth, or extremities, combined with the differences of the muffle observed in *Macropus* (as restricted) and *Halmaturus*.

Some of the species of the present convenient, but, as it appears to me, arbitrary division, are of very large size, being in this respect equal, or but little inferior to, the *M. giganteus*, and, like that animal, have no large openings to the hinder part of the palatine portion of the skull; such are the *M. antilopinus*, *M. robustus*, and *M. rufus*.

MACROPUS (*Halmaturus*) ANTILOPINUS.

Antilopine Kangaroo.

Osphranter Antilopinus. Gould, Proceedings of the Zoological Society for October 1841, Part 9, p. 80.—Monogr. Part 2, Plate 7.

Male.—Fur short; the hairs stiff and closely applied to the body; the sides and under parts of the body, as well as the inner parts of the limbs, of a very pale rusty yellow, and the upper parts of a bright rusty red. This tint is extended on the outer side of the legs, and on the feet, but the toes are black: tail slightly suffused with rust colour, and dusky at the apex above; ears, with pale hairs internally: height, in ordinary position, about four feet.

Female.—Considerably smaller, and with the fur longer, less harsh, and less closely applied to the body, than in the male. General tint brown; greyish over the back of the neck and shoulders, suffused with rust colour, and pencilled with blackish on the back; under parts and limbs rusty white; fore feet brown, but with black hairs near the claws; hind feet pencilled with blackish in front, and shaded through brown into black on the toes; cheeks, upper lip, and chin, dirty yellow white; upper surface of muzzle brownish: height, about three feet.

Inhabits North Australia.

The only two perfect specimens of this fine animal at present in Europe are those which form the subject of one of the plates in Mr. Gould's Monograph. They now grace the collection of the British Museum, and their discovery is due to Mr. Gould's zealous assistant, Mr. Gilbert, who procured them, together with many other novelties, at Port Essington.

Captain Chambers, who placed several imperfect skins of the Osphranter or Antilopine Kangaroo in Mr. Gould's hands to assist him in his study of the group, informed him that he had seen individuals which were one hundred and seventy pounds in weight, indicating that the present species attains a size nearly one-third larger than the male specimen in the British Museum, that having weighed about one hundred and twenty pounds.

The same gentleman informs Mr. Gould that when hardly pressed in the chase, the Osphranter becomes exceedingly fierce and bold, and, while among the rocks, is a most dangerous animal to encounter, one of his finest dogs being tumbled over a precipice and killed by an old male.

Unlike most Kangaroos, where the fur is long, and the under fur, being abundant, is loosely applied to the body, the Antilopine Kangaroo is clothed with short stiff hairs, and these lie close to the skin, as in many of the Antelope tribe; a peculiarity which suggested the specific name. The present species is remarkable, moreover, for the great expanse of the nasal cavity of the skull; a character which induced Mr. Gould to regard it as the type of a new genus or subgenus, for which he proposed the name *Osphranter*. The character in question, however, I can but regard as a specific peculiarity. The muffle is very broad; the fore legs large and powerful, and the fore feet are furnished with very strong claws; the hind legs and feet are comparatively rather shorter than in the Macropus *giganteus;* the central toe is very large, whilst the two united toes, as well as the outer toe, are unusually small: the former terminate in a line with the latter, or very nearly so

The following dimensions are from the male and female Antilopine Kangaroos, figured by Mr. Gould in his Monograph, and which are in the British Museum:—

	Male. Inches.	Male. Lines.	Female. Inches.	Female. Lines.
Length from tip of nose to root of tail	53	0	39	0
" of tail	36	0	30	0
" of tarsus and nails ...	13	3	11	6
" from nose to ear ...	7	9	5	9
" of ear	4	0	4	0
" of fore arm and hands, including the claws ...	15	0	10	0

The skull of *Macropus antilopinus* is about equal in size to that of *M. giganteus*, but is shorter in proportion to its width; and the most striking difference is in the form of the muzzle, which, instead of becoming gradually narrower towards the apex, is quite as broad near the end as at the base, and is most broad rather in front of the middle; the nasal bones are shorter than in *M. giganteus;* the superior breadth of the muzzle is caused by the swelling outwards of the superior maxillary and nasal process of the intermaxillary bones, which thus gives a great expanse to the nasal cavity, which encloses very largely developed turbinated bones, and hence we may conclude the Antilopine Kangaroo possessed unusual acuteness of smell. The frontal bones are deeply concave above, between the orbits, and are much contracted immediately behind them; the temporal ridges meet to form a well-marked sagittal crest. The zygomatic arch is thrown more boldly outwards than in the *M. giganteus*, and is remarkable for its great depth; the lower posterior branch of the malar bone descends slightly below the level of the glenoid cavity; the palate is strongly concave in the longitudinal direction; the ordinary posterior palatine openings, situated in the palato-maxillary suture, are here represented by two very small perforations of about one line in length, and behind these are some other very small perforations; in fact, the palate is as entire as in *M. giganteus.* The foremost incisor tooth (see Pl. 5, fig. 15,) is but little broader than the second; the third is about double the width of the second, has a strong oblique external groove rather in front of the middle of the tooth, and there is a second groove in front of this, but this latter, which represents the foremost of the two grooves observed on the incisor of *M. giganteus,* is much less distinct than in that animal. In an adult skull I have seen the full number of molar teeth; but the cranium of a very aged individual presented but thirteen molar teeth in both jaws. The subjoined dimensions are from two skulls in the collection of Mr. Gould; the specimens were procured near Victoria, in North Australia, and the larger skull is said to have belonged to an animal weighing one hundred and twenty pounds, the largest which had been seen :—

	Male.		Female.	
	Ins.	Lines.	Ins.	Lines.
Total length of skull	7	1	6	0
Width	4	2	3	5½
" between orbits	1	2½	1	2
Length of zygomatic arch	3	1	2	7½
Depth of ditto behind	1	3½	1	0
Length from anterior root of zygoma to apex of intermaxillaries ...	3	6	3	1½
Width of muzzle in the middle ...	1	11	1	9
Length of nasal bones	2	11	2	6
Width of ditto behind	1	1½	1	1¾
" near the apex	0	11	0	9⅓
Length of palate	4	8	4	0
" of three anterior incisors taken together	0	9	0	8¼
—of which the posterior incisor is ...	0	4½	0	4½
Distance between incisors and premolar	1	10	(?)	
Length of the five molar teeth taken together	2	2½	(?)	
" of the lower jaw	5	7	4	8
Height of ditto, from the apex of the coronoid process	3	2	2	9

MACROPUS (*Halmaturus?*) ISABELLINUS.

Isabelline Kangaroo.

Osphranter (?) isabellius. GOULD, Proceedings of the Zoolog. Society for October, 1841, Pt. 9, p. 81.

The Yellow Wallaroo. GRAY, Catal. of the Mammalia in the Collection of the British Museum, p. 92.

Fur somewhat short and soft, and of a bright fulvous, or sandy red colour ; throat, under parts of the body, and limbs, whitish, somewhat tinted with pale yellow in parts ; fore feet and toes brown, the sides of the latter yellowish : tail of a somewhat paler hue than the body.

Inhabits Western Australia.

This species is founded by Mr. Gould upon a flat and imperfect skin, procured at Barrow Island, on the west coast of Australia (latitude about 21°), by Capt. Stokes, of H. M. S. Beagle, and which is now deposited in the British Museum. The characters of the animal, so far as they can be ascertained from such imperfect materials, are thus given in the Proceedings quoted :—The general colour of the skin is bright fulvous, or sand-red ; the fur is rather short, and soft to the touch ; the hairs are uniform in tint to the base : the throat and under parts of the body are white, faintly tinted with yellowish in parts. The fur on the belly is long, and very soft : the white or whitish colouring of the under parts, and the uniform fulvous colouring of the upper parts and sides of the body, do not blend gradually : the colour of the tail is nearly the same as that of the body, but is rather paler, and is nearly uniform. The fore-feet and toes above, are covered with brown hairs, but on the sides of the toes the hair is yellowish. The size of the animal is probably about equal to that of *Halmaturus Bennettii.*

MACROPUS (*Halmaturus*) ROBUSTUS.

Great Rock Kangaroo.

(*Plate* 7.)

Macropus (Petrogale) robustus. GOULD, Proceedings Zool. Soc. for August, 1840, Pt. 8, p. 92.

Petrogale robusta. GOULD, Monograph of the Macropodidæ, Pt. 1, 5th Plate.

Macropus robustus. WATERHOUSE, Marsup. p. 241.

Black Wallaroo of the colonists.

Male.—Fore legs powerful; tarsi short: fur rather short and harsh, general colour very deep slate grey, slightly suffused with brownish on the upper parts of the body; the under parts rather paler than the upper; chin with a black patch; ears white internally, and brown externally; feet black; hind feet sooty black, but paler on the inner side near the heel. Height about 3 feet 6 inches.

Female.—Much smaller than the male: general colour silvery-grey, slightly tinted with purple on the back; the under parts nearly white; fore-feet brown, toes blackish; hind feet pale, but with the toes brown-black.

Inhabits the mountain ranges in the interior of New South Wales.

According to the observations of Mr. Gould, the Great Rock Kangaroo inhabits the summits of sterile and rocky mountains, from which it seldom descends to the coverts at their sides, and never to their base. As yet it has only been found in the south-eastern portion of the continent, and is said to be tolerably abundant in the Liverpool range. Mr. Gould ascertained that it also inhabited many of the hills which branch off on either side of this great mountain chain, both towards the interior as well as towards the coast.

This species is extremely agile among the rocks, and its retreats are so well chosen among the crags and overhanging ledges, that it is nearly useless to attempt its pursuit and capture with dogs. It is a formidable, and even dangerous, animal to approach, for if so closely pressed that it has no other chance of escape, it will rush at and force the invader over the edge of the rocks, as the Ibex is said to do under similar circumstances. Independently of its great muscular power, this animal is rendered still more formidable by the manner in which it makes use of its teeth, biting its antagonist with great severity.

The ***Macropus robustus,*** Mr. Gould further observes, " may be regarded as a gregarious animal, four, six, and even more, individuals being frequently seen in company. On one of the mountains near Turi, to the eastward of the Liverpool Plains, it was very numerous; and from the nature of this and the other localities in which I observed it, must possess the power of existing for long periods without water, that element being rarely met with in such situations. The summits of the hills to which this species resorts soon become intersected by numerous roads and well-trodden tracks, caused by its repeatedly traversing from one part to the other: its food consists of grasses, and the shoots and leaves of the low scrubby trees which clothe the hills it frequents.

" Although much shorter in stature, and consequently less elegant in form, the fully adult male *M. robustus* equals in weight the largest specimens of *Macropus giganteus;* and so remarkable is the difference in the colour and size of the sexes, that had I not seen them together in a state of nature, I should have considered them to be different species, the black and powerful male offering so great a contrast to the small and delicate female."

The *Macropus robustus*, as compared with the Great Kangaroo (*Macropus giganteus*), differs in having the tarsi shorter, and the fore legs larger and more powerful; the fur is rather short, and has a somewhat shaggy appearance, especially about the head; its general colour is of a very deep slate-grey, obscurely suffused with brownish; the tarsi are brown behind, and gradually shaded into black on the fore part; the toes black, and the fore feet and wrists are also black; the upper side of the fore arm is brownish; the hind legs somewhat sooty, but pale on the inner side, especially near the heel; the head is nearly uniform in colour with the body, but slightly suffused with blackish at the tip of the muzzle on each side; a narrow white line is observed round the angle of the mouth; the lower lip is white, and there is a black patch on the chin; the throat and fore part of the neck are whitish; the under parts of the body are a trifle paler than the upper parts: the tail is brown above, and pale-brown beneath. The hairs on the back are of a palish slate-grey next the skin, and tinted with brownish externally, and the ends of the hairs on the outer side of the thighs are slightly suffused with a purplish rust tint.

The female is much smaller than the male, and has the limbs less powerfully formed; its colour is much paler, being silvery-grey; on the back is a faint purplish-brown hue; the abdomen is nearly white; the cheeks are greyish-white; on the chin is a dusky patch; the tail is dirty white, slightly tinted with brownish on the upper side; the legs paler than the body; the fore feet brown, but with the toes nearly black; the hind feet are pale, and the toes are brown-black.

	Male.		Female.	
	In.	Lines.	In.	Lines.
Length from tip of nose to root of tail	49	6	40	0
" of tail	34	6	30	0
" from nose to ear	8	0	7	0
" of ear	3	7	3	0
" of fore arm, hands, and claws	13	6	10	6
" of tarsus and toes, including the nails	12	0	10	2

Mr. Gould first described this animal as a member of the

section *Petrogale*, but he is now of opinion that it is very nearly related to the *M. antilopinus*, an opinion in which I agree. In the general form of the skull, *M. robustus* approaches the *M. giganteus*, and in those parts in which it deviates from the skull of the animal just mentioned, we can perceive strong indications of affinity to the *M. antilopinus*. The close relationship of the present animal to the one last mentioned is also displayed in the structure of the muffle, the large size and great power of the fore-legs, and the structure of the tarsi. As compared with the skull of *M. giganteus*, that of *M. robustus* differs in having the muzzle broader and the zygomatic arch deeper. The sagittal crest is well marked. The incisor teeth scarcely differ from those of *M. antilopinus;* the external grooves on the posterior incisor are indistinct, but the two can be traced, and they occupy the same position as in the incisor of the animal just mentioned. The skull differs from that of *M. antilopinus* in being more elongated, in having the muzzle narrower, and broadest behind; the frontal bones are less concave between the orbits. The palate is almost destitute of posterior palatine openings.—These notes are drawn up from a skull in Mr. Gould's collection, which furnishes the following dimensions :—

	Inches.	Lines.
Length of skull	7	7
Width	3	10
" of interorbital space	1	2
Length of nasal bones	3	2
Width of ditto at the base	1	3½
" " near the apex	0	10
Depth of zygomatic arch behind	1	1½
Length of palate	4	7
" of three upper incisors	0	9½
—of which the posterior incisor is	0	4½
Distance between incisors and molars ...	1	6½
Length of the five molar teeth taken together	2	2
Length of lower jaw	5	4½
Height from apex of coronoid process	3	1

MACROPUS (*Halmaturus*) RUFUS. Red Kangaroo.

Kangurus rufus. DESMAREST, Mammalogie, Supplément, p. 541, 1822.
" *laniger.* GAIMARD, Bulletin des Sciences par la Société Philomatique, Année 1822, p. 138.
" " QUOY et GAIMARD, Voyage de l'Uranie, p. 65, Plate 9, 1824.
Macropus laniger. GOULD, Monogr. Plate 2, Partt 1.

Male.—Fur short, moderately soft to the touch, and of a woolly, or rather, cotton-like texture: prevailing hue bright, but rather pale, rust colour; head greyish at the sides; chin and region of the mouth white, with a few small black spots, and one larger patch of the same colour above the angle of the mouth; cheeks with an indistinct whitish mark; ears rather large, white internally, externally greyish, but with some black hairs at the tip; limbs and tail nearly white; toes blackish, both large and powerful.

Female.—Limbs more slender; the fore legs considerably smaller in proportion: prevailing hue palish-grey, with an obscure vinous tint, and in parts, especially about the loins and haunches, of a bright, but rather pale rust colour; a distinct broad white mark on the cheeks.

Of this, which is one of the largest and most beautiful of the Kangaroo tribe, the National Collection contains four specimens, exhibiting the differences of sex and age; these were brought home by Mr. Gould, by whom the species is most beautifully figured in the first part of his Monograph of this interesting group. Until the arrival of these specimens in England, a single individual only existed in the European museums—that contained in the collection at Paris, which was procured during M. Freycinet's Voyage of the Urania, having been presented to the officers of that expedition by Mr. Fraser, the botanist, during their stay at Sydney, and was said to have been procured at Port

Macquarrie. This specimen is in bad condition, and Mr. Gould informs us, that the defective parts of its fur have been made up by finely-cut sheep's wool; whence has arisen, he supposes, the specific name *laniger*, most commonly used for the animal. This name, however, is not altogether inapplicable to the species, for its natural fur is of a somewhat woolly nature; or it might, perhaps, be more justly compared to a coarse kind of cotton, and the softness of texture arises from an almost total absence of the longer and coarser hairs, which in most Kangaroos hide the soft and dense under fur—a character which at once catches the eye, and assists in distinguishing it from its congeners. M. Desmarest's specific name of *rufus*, however, has the priority of that of *laniger* (Gaimard), and hence should be used in preference. Mr. Gould procured two of his specimens in South Australia, and the others on the plains bordering the Namoi. "From the works of Oxley and Sturt," Mr. Gould remarks, "we find that this species frequents the banks of the Murrumbidgee and Darling; we may consequently infer that it is very generally dispersed over the great basin of the interior of Australia, and it certainly is over the eastern portions of that continent. * * * The female is particularly attractive, from her graceful, slender, and elegant form, and from the snowy whiteness of her legs, and of the under part of the body, contrasted with the blue-grey tint of her sides and back. The male, especially when adult, has the red and white more blended into each other, the blue-grey, which distinguishes the female, being rarely, if ever, perceptible; hence has arisen the names of *red buck* and *blue doe* for the two sexes repectively: the female is also called the *flying doe*, from her extreme fleetness, for which her whole structure is so admirably adapted, that I have little hesitation in saying, that, under favourable circumstances, she would outstrip the fastest dogs: occasionally, however, both sexes are successfully chased, either from the chase being over soft muddy soil, or, in the case of the female, from her being encumbered by a large and heavy young one, which she has not been able to disengage from her pouch, and which she will always do, if possible, when hardly pressed." Two of Mr. Gould's specimens, male and female, were each procured by a single dog; the former was held at bay until the party came up and dispatched him, but not before he had made a fearful resistance.

The Red, and the Great Grey Kangaroo, sometimes inhabit the same districts; still, Mr. Gould states, they more commonly frequent localities of a different description; the latter resorting to grassy valleys, and brush growing on the dark soil, whilst the former apparently prefers the hard red stony ridges, clothed with box, and open plains, in the midst of which it may frequently be seen basking in the sun. The large male, which Mr. Gould estimated as having a weight of above 200 lbs., was killed whilst he and his party were making a forced march, between the Murray and Adelaide, at a time when all their provisions were exhausted; and its flesh, which is said to be excellent, lasted them four days. The dimensions of an adult male and female, in the British Museum, are as follows:—

	Male.		Female.	
	In.	Lines.	In.	Lines.
Length from tip of nose to root of tail (measuring over the curve of the back) ...	60	0	44	0
" of tail	38	0	2	9
" of hind foot and claws	14	6	12	0
" of fore leg to end of claws	17	0	12	0
" from ear to tip of muzzle	7	9	6	6
" of ear	5	0	5	0

	Female.		Male ?	
Total length of skull	6	6	?	
" width	3	4½	?	
Length from the orbit to the tip of the intermaxillary bones	3	3	3	7
Length of nasal bones	2	8	3	2
Width of ditto at the base	0	9½	1	2
" of ditto near the apex, immediately behind the free points	0	6⅔	0	7¾
" between the orbts	1	1	1	5

	Male. In.	Male. Lines.	Female. In.	Female. Lines.
Length of palate	3	10	4	3
Greatest width of ditto between molar teeth ...	1	$0\frac{1}{2}$		
Length of five molar teeth taken together ...	2	1		
" of three last molars taken together ...	1	6	1	$5\frac{1}{4}$
" of three incisors taken together ...	0	$8\frac{1}{3}$	0	$7\frac{1}{2}$
Width of the posterior incisor	0	$3\frac{1}{2}$	0	$2\frac{3}{4}$
Distance between incisor and molar teeth ...	1	7		
Width of muzzle in the middle	1	3	1	5

The dimensions in the first column are from a skull in the British Museum, which, compared with that of *Macropus giganteus*, does not differ much in its proportions; the zygomatic arch is thrown rather less boldly outwards; the interorbital space is narrower, and this part, which encloses the turbinated bones, is less inflated; the posterior upper surface of the frontals is more concave, and the outer surface of the lachrymal bone is much larger. The incisor teeth are considerably smaller; the foremost incisor is broader than the second, and is indeed but little narrower than the third, which has but one shallow vertical groove, situated rather in front of the middle—see Plate 5, fig. 3. The fifth, or hindermost molar, had penetrated the gum in this skull, but had not yet come into use; it exists with the other four molars, but the foremost of them is partially absorbed at the root, and no doubt would shortly have been cast. In the second column are some dimensions taken from an imperfect skull, but evidently belonging to an old individual of *M. rufus*. Upon comparing this skull with one of *M. giganteus*, of about equal size, I noticed, among other differences, that the nasal bones were longer, broader at the base, and narrower at the opposite extremity, and the space between the molars and incisors was distinctly less. The palate is without perforations, or with very small openings only, as in *M. giganteus*. The lower jaw differed from that of *M. giganteus* very considerably in the height of the ascending ramus. The height of the jaw, in a line dropped vertically from the apex of the coronoid process, was 3 inches 2 lines in *M. giganteus*, and but 2 inches 5 lines in *M. rufus*. The fore legs in *M. rufus* are much longer and more powerful than *M. giganteus;* the tarsus is long, but the great central toe is shorter in proportion than in the animal last mentioned. The hinder part of the muffle is clothed with hairs, and these terminate so as to form a point in front, which is situated about in a line with the middle of the nostril openings; laterally the hairs extend to the posterior angles of the nostrils, which are also fringed with hairs internally.

MACROPUS (*Halmaturus*) AGILIS. The Agile Kangaroo.

Halmaturus agilis. GOULD, Proceedings of the Zool. Soc. for October, 1842, Pt. 9, p. 81. Monogr. Pt. 2. Pl. 5.
" " HAMBRON ET JACQUINOT, Voyage au Pole Sud. Pl. 19.
" *Binoe.* GOULD, Proceedings of the Zool. Soc. for May, 1842, Pt. 10, p. 58.

Fur short, harsh, and closely applied to the body; general colour sandy or brownish yellow; on the back pencilled with black; under parts nearly white; feet brownish white; tail long, clothed with small adpressed whitish hairs; on the mesial part of the upper surface yellowish, but near, and at the apex, the hairs are dusky; a whitish mark crosses the haunches.
Inhabits North Australia.

The Agile Kangaroo, it would appear, extends along the whole of the north coast of Australia, specimens having been sent to Mr. Gould from Cobourg Peninsula, Port Essington, Raffles Bay, and from the shores of Torres Straits. " It is stated to be a most agile species, readily eluding the dogs employed in hunting it, by its extreme activity in leaping over the high grass; when chased, it frequently seeks shelter in the thick beds of mangroves, passing over the muddy flats in such a manner as almost to baffle pursuit."[1]

In very many quadrupeds, the clothing consists of hairs of two kinds; the one which composes what I have termed the under fur, is very soft, often woolly, and generally more or less hidden by longer interspersed hairs, which are of a harsher nature, and which may be said to constitute an upper or outer fur. Sometimes the longer hairs assume the form of bristles, and even spines, (indeed all the different stages may be seen in the different species of *Echimys*, or spiny rats of South America, as well as in the Old World Rats.) In cases where the longer hair assumes the form of spines, the under fur is *generally* but little developed, and this under fur is almost always abundant in Mammals inhabiting cold climates, and in those which live on the summits of lofty mountains. It is generally abundant also in nocturnal species, and those which inhabit damp and swampy situations, or that live in the water, even though they are denizens of a hot climate. On the other hand, the upper fur is often much developed in quadrupeds living in tropical climates: in the former cases the under fur is to protect the body from cold or wet, and in the latter, the long hair would appear to serve as a protection from the heat. Finally, I may remark, that the same animal exhibits great changes in the character of its fur when transported to different climates.

Having observed that the character of the fur varies with the climate, we are prepared to find Kangaroos, which, like the present species, or the *M. antilopinus*, inhabit the most tropical portions of Australia, with the fur composed almost entirely of the coarser kind of hairs,—the under fur being very little developed. In these two species the fur is short,

[1] Gould's Monograph.

and in the *M. antilopinus* it resembles that of the tropical antelopes, but in the Tree Kangaroos of New Guinea, (*Dendrolagus*), the outer fur is long, and may be compared to that of the monkeys.

The present species would appear to occupy an intermediate station between the true Kangaroos and the Wallaby Kangaroos (Halmaturi), the hair extending on to the muffle more than in the typical Halmaturi, being produced so far forwards as to be in a line rather below the centre of the nostril openings: there is, however, a naked space on each side next the openings in question of about $1\frac{1}{2}$ lines in width: the part covered with hair is gradually diminished in width towards the front, where it has a small angular emargination, or encroachment of the naked part.

The prevailing colour of the fur in *M. agilis* is sandy-yellow, but the back is somewhat pencilled with black, the hairs, many of them, being tipped with that colour,—below the point they are very nearly uniform to the root, being but indistinctly tinted with ashy-grey next the skin; the greyish tint is followed by a yellowish hue, which becomes paler as it approaches the point of the hair. On the under parts, the hairs are white, but slightly tinted with yellowish, and this latter tint is most evident on the abdomen, on the sides of which they are distinctly yellowish below the point, which is white; in some parts they are slightly suffused with pale grey at the root. The head has a whitish yellow patch above each eye; the muzzle is somewhat tinted with brownish; the crown of the head is also brownish, and this tint is extended backwards on to the neck; on each side of the occiput, behind the ears, is a whitish patch; the hairs on the occiput are directed forwards, and meet those of the head, which have the usual direction, in a line with the front margin of the ears. The ears are clothed with yellowish hairs externally, but are black at the apex, and along the anterior margin; on the inner side the hairs are white: a whitish mark crosses the haunches, running backwards from the knee, where it is most distinct. The fore legs are rather large, and the fore feet are also large, and armed with long and powerful claws; both fore and hind feet are of a brownish-white colour; the latter are of moderate size. The tail is long, rather slender, and clothed with short adpressed hairs,—it is chiefly white, but suffused with yellowish above at the base, and a mark of the same colour may be traced (though not very distinct) along the whole upper surface, excepting about two or three inches at the apex, where the hairs are a trifle longer and of a dusky brown colour.

	Inches.	Lines.
Length from the tip of the nose to the root of the tail	36	0
" of tail	34	6
" of tarsus	10	3
" from nose to ear	5	7
" of ear	2	7
" of arm and hand, including the nails, about ...	10	0

A skull of *Macropus agilis* now before me, as viewed from above, could scarcely be distinguished from one of *M. Bennettii;* the size and proportions are nearly the same; the facial portion of the cranium of *M. agilis,* however, is longer and a trifle narrower; the nasal bones are longer, broader, and less contracted in the middle than in *M. Bennettii:* viewed from beneath there is likewise but little difference, excepting in the size of the posterior palatal openings, these being very large in *M. agilis,* (*i. e.* nearly 9 lines in length), and small in *M. Bennettii,* where they are not more than 3 lines long. In the teeth of the two animals there are evident differences, both the molar and incisor teeth in *M. agilis* being larger. The skull is not that of an adult animal, having the fifth or hindermost molar tooth still in the socket; the four which are developed measure 1 inch 5¼ lines in length, whilst in

M. Bennettii, the corresponding teeth, taken together in the same manner, measure 1 inch 3¼ lines : other differences will be perceived upon comparing the following dimensions with those of the skull of *M. Bennettii.*

	Inches.	Lines.
Total length of skull	5	6
" width	3	0
Length of nasal bones, rather more than[1]	2	3
Width of ditto at the base	0	10
" at the apex, immediately behind the contracted free points	0	6½
Length of palate	3	4½
Width of ditto opposite the third molar tooth ...	1	0½
Length of the anterior four molar teeth taken together	1	5¼
" of the three incisors together	0	9
Distance between incisors and molars	1	3½
Length of hindermost incisor	0	4⅔

The hindermost incisor tooth of the upper jaw, which is considerably broader than the corresponding tooth in *M. Bennettii,* has a distinct external vertical fold situated rather in front of the middle of the tooth (see Plate 5, fig. 13). The foremost molar is broad, viewed externally, and presents two vertical narrow ridges on its outer side.

The lower jaw indicates more strength of the masticating muscles than in *Bennettii,* and this is likewise made evident by there being a distinct sagittal crest to the cranium, whilst in the older skull of *Bennettii,* the ridges formed by the temporal muscles are separated and less developed. The height of the jaw, measuring from the tip of the coronoid process, is 2 inches 3¼ lines; in *Bennettii* 2 inches 0½ lines. The incisor teeth are broader and rather shorter than in the last mentioned animal.—From specimens in the British Museum collection.

Mr. Gould describes a small Kangaroo from Port Essington,

[1] Their apices are broken off.

North Australia, which so perfectly resembles his *Halm. agilis* in all respects excepting size, that I cannot regard it as a distinct species: it is the

Halmaturus Binoe.

In size, this animal nearly agrees with the *Halm. Derbianus*, its dimensions being as follows:—

	Inches.	Lines.
Length from nose to base of tail	21	0
" of tail	20	0
" of tarsi	8	0
" of ears	2	3
" of hands and claws	2	4

MACROPUS (*Halmaturus*) PARRYI.

Parry's Kangaroo.

Macropus Parryi.	BENNETT, Proceedings of the Zool. Soc. for December 1834, p. 151; Trans. of the Zool. Soc. vol. i. p. 295, Pl. 37.
Macropus (Halmaturus) Parryi.	WATERH. Plate 18, and p. 206 of Vol. on Marsupialia in Naturalists' Library.
Halmaturus Parryi.	GOULD, Monogr. Part 2.

Silvery grey above, white beneath; upper surface of muzzle sooty black, but becoming paler and brownish on the posterior part; a distinct white line on the cheeks; fore and hind feet with the toes black; the tarsus and fore part of hind leg white; tail very long, and somewhat compressed, grey white, with the apex black: height about two feet six inches.

Inhabits New South Wales.

A delicately coloured species, and at the same time one of the most elegantly formed of the Kangaroo tribe. The specimen from which Mr. Bennett drew up his account, and from which the present description is taken, was brought alive to this country, and presented to the Zoological Society, by

Captain Sir Edward W. Parry, R.N., who states that it was procured at Stroud, near Port Stephens, in the latitude of about 30° south. It was caught by the natives, by whom it is called *Wōllăroo*, having been thrown out of its mother's pouch when the latter was hunted. At that time it was somewhat less than a rabbit; but, having continued in the possession of Sir Edward Parry for more than two years in New South Wales, besides six months on the passage home, it was no doubt full grown when it reached England. It was never kept in confinement until it was embarked for England, but lived in the kitchen, and ran about the house and grounds like a dog, going out every night after dark into the "bush" to feed, and usually returning about two o'clock in the morning to its friend, the man cook, in whose bed it slept. Besides what it might obtain in these excursions, it ate meat, bread, vegetables—in short, every thing given to it by the cook, with whom it was extremely tame; but it would allow nobody else to take liberties with it. It expressed its anger, when very closely approached by others, by a sort of half-grunting, half-hissing, very discordant sound, which appeared to come from the throat, and was not accompanied by any alteration in the expression of the countenance. In the day-time it would occasionally, but not often, venture out to a considerable distance from home, in which case it would sometimes be chased back by strange dogs, especially those belonging to the natives. From these, however, it had no difficulty in escaping, through its extreme swiftness; and it was curious to see it bounding up a hill and over the garden fence, until it had placed itself under the protection of the dogs belonging to the house, especially two of the Newfoundland breed, to which it was attached, and which never failed to afford it their assistance by sallying forth in pursuit of its adversaries.

A second specimen was presented to the Zoological Society

by James Macarthur, Esq., and lived in the menagerie for a short time.

Parry's Kangaroo is of a more slender form than most of the Halmaturi, and has a remarkably long tail. The ears are moderately large; the fur moderately long, and not very soft to the touch; on the upper parts of the body it is of a silvery-grey tint, but with a delicate brownish hue in parts, especially on the hinder portion of the back: the hairs on the under parts of the body, and inner side of the limbs, are of a pure white colour to the root, but those on the chest are slightly tinted with greyish at the point. The sooty brown colour of the upper surface of the muzzle is separated from a grey tint which prevails on the lower and hinder parts of the cheeks, by a distinct white band running from the upper lip backwards, and terminating nearly in a line with the hinder portion of the eye. The back of the head is also white, and so is the fore arm; but this latter is faintly tinted with grey externally. The ears have white hairs on the inner and outer surfaces; but they are dusky at the point externally, and clothed with long sooty black hairs at the base. The tail is furnished with short harsh hairs, which are nearly white on the upper surface, being but slightly tinted with grey; about six inches of the under surface of the apical portion, however, is covered with long black hairs, which form a kind of fringe; the tip of the tail is also black above.

	Inches.	Lines.
Length from tip of the nose to the root of the tail	34	0
" of tail	31	0
" from nose to ear	5	4
" of tarsus	9	6
" of ear	3	7
" of fore-arm and feet, claws included ...	8	4

Mr. Lambert notices a species of Kangaroo, in the Linnæan Transactions (Vol. VIII. p. 381, Plate 16,) under the name

Macropus elegans; but his description is so short that it will always be a matter of doubt as to the species to which his name should be attached. That it is of a silvery grey colour, and has the following dimensions, is all that we can glean from his account:—

	Inches.
Length from tip of nose to end of tail	$62\frac{7}{10}$
" " to eye	$2\frac{8}{10}$
" " to back part of skull ...	$5\frac{7}{10}$
" of tail	$26\frac{2}{10}$
" of ears	$3\frac{8}{10}$
" of hind leg, from the claws to the " knee," (by which the *heel* is no doubt meant)	10

These dimensions, combined with a silvery grey colour, are more applicable to the *Macropus Parryi* than to any other known species; and I find Mr. Gould agrees with an opinion I formerly expressed, that the *elegans* of Lambert was specifically identical with Parry's Kangaroo. Mr. Gray, however, associates *Lambert's* account with a very differently coloured animal, the *M. ruficollis,* and has been led to do so from an inspection of a coloured plate in a copy of the Linnæan Transactions, which formerly belonged to Sir Joseph Banks, and which is now in the British Museum Library. This plate represents the animal as of a very pale grey colour, with the under parts of the body, and apical half of the tail, white, the crown of the head, hands, and fore half of the foot dusky, or blackish; there is no trace of red on the neck or elsewhere. I think it must be intended to represent the *Macropus Parryi.*

The skull of *M. Parryi* greatly resembles that of *M. Bennettii,* but may be distinguished by the facial portion being more produced, the nasal bones longer, and less contracted in the middle, the sides being nearly parallel. The upper incisor teeth are *very* nearly the same; the groove in the hindermost of these teeth is in the middle. The foremost molar tooth, or premolar, is distinctly smaller in proportion, being $2\frac{3}{4}$ lines in length, whilst, in *M. Bennettii*, it varies from $3\frac{1}{2}$ to $3\frac{2}{3}$ lines: in this respect *M. Parryi* shows a nearer approximation to the typical Kangaroos, where this tooth is very small: on the other hand, we find the same tooth in *M. Ualabatus* still larger than in *M. Bennettii;* and it is most developed (combined with the normal Macropus structure of true molars) in the *M. Brunii* and the extinct *M. Atlas*.

	Inches.	Lines.
Length of skull	5	3
Width of ditto	2	$10\frac{1}{4}$
" between orbits	1	0
Length of nasal bones	2	1
Width of ditto at the base	0	$9\frac{2}{3}$
" near the apex	0	0
Length of palate	3	2
" of posterior palatine openings	0	$6\frac{1}{4}$
Depth of zygomatic arch behind	0	$8\frac{1}{3}$
Length of the five upper molar teeth taken together	1	7
" of the three upper incisors taken together	0	$7\frac{1}{4}$
" of the posterior incisor	0	$3\frac{2}{3}$
Distance between the upper incisor teeth and the molars	1	$5\frac{3}{4}$
Length of foremost molar of the upper jaw ...	0	$2\frac{3}{4}$

MACROPUS (*Halmaturus*) IRMA.

The Black-gloved Kangaroo.

Halmaturus Irma. JOURDAN, Annales des Sciences Naturelles for December, 1837, Vol. 8, p. 371; Comptes Rendus des Séances de l'Acad. des Sciences for October 9, 1837, p. 523.

Macropus (Halmaturus) manicatus. GOULD, Proceedings of the Zoological Society for October 1840, p. 127.

Halmaturus manicatus (Black-gloved Wallaby). GOULD'S Monogr. Part 1, 9th Plate.

Fur moderate; grey (distinctly pencilled with black and white), but exhibiting a yellowish cast, owing to the hairs being of that tint below the exposed ends; under parts pale grey, slightly suffused with yellowish; upper surface of head and muzzle sooty brown, nearly black on occiput and back of ears, which have, however, yellowish hairs at the base; internally they are yellow, but margined with black at the apex; cheek-mark yellow-white; chin with a black spot; legs and feet yellow, excepting the fore half of the latter, which is black; tail long, grey, with the tip black, and having long black hairs (forming a kind of crest), both on the upper and under surface of the apical half, which is somewhat compressed.

Inhabits Western Australia.

I feel no doubt that the *Macropus Irma* and the *M. manicatus* are varieties of the same species; since, however, they differ in some respects, it is necessary to state that the above diagnosis is drawn up from Mr. Gould's original specimen of the *M. manicatus*, a very elegant and beautiful Kangaroo. Its fur is moderately long and moderately soft; the visible parts of the hairs on the back are black and white, each hair having a tolerably broad white ring below the black point; the middle portion is of a palish rusty yelow, and the basal portion, or root, grey; on the neck and sides of the body the black and white parts are less distinct, and the yellow is more evident; and this is still more the case on the under parts of the body; indeed, the lower part of the abdomen is almost entirely yellowish. The upper parts of the head and muzzle are sooty brown, and a faint rufous hue is just traceable on the crown of the head; the occiput is almost black, and a broad dusky band runs downwards from the occiput along the back of the neck; rather more than the apical half of the ear externally is covered almost entirely with sooty black hairs, and a broadish black band runs down the anterior margin; the remaining external portion is clothed with yellowish white hairs; internally the ears are provided with long yellow hairs, but at the apex they are distinctly margined with black; and this hue descends about half an inch from the point of the ear. The legs, excepting at the base externally, and the feet, are ochreous yellow, the fore half of the latter excepted, which is black, and the black and yellow parts of the feet are separated by a well-defined line, as if the animal had thrust its feet into some black dye. The hairs covering the inner double toe of the hind foot are partly yellow and partly black. The tail is long, rather slender, and well clothed with adpressed hairs, which are partly black and partly white, in nearly equal proportions: on about the middle of the tail, both above and below, the hairs are considerably lengthened, and continue, in the form of a crest, to the apex, where they are fully an inch and a half in length: these crests, or fringes, are almost entirely black, but, at some distance from the point, some narrow white rings are visible on the hairs which form the upper crest, and these become more conspicuous as we proceed towards the middle of the tail: on the sides of the tail the hairs are adpressed, and but little longer at the tip than elsewhere; they are annulated with black and white almost in equal proportions, but a space of about an inch and a half from the tip of the tail is clothed with hairs which are almost totally black. The height of this animal, in its ordinary erect position, is about twenty-six inches.

	Inches.	Lines.
Length from tip of muzzle to the root of the tail	31	0
" of tail	27	0
" of tarsus and claws	8	10
" from nose to the ear	5	0
" of ear	3	6

The Black-gloved Kangaroo is said by Mr. Gilbert to be found equally abundant in all parts of the colony of Swan River, West Australia. It generally inhabits scrubby places, though occasionally seen feeding upon the open plains, but it always takes to the scrub when hunted : it runs very fast.

M. Jourdan's account is as follows:—Form very elegant; the limbs very delicate; head above and upper parts of body grey; chin with a black spot; cheeks and lips yellowish white; outer side of ears with the fore part brown, and posteriorly white; inner side yellow; the apex black; between the ears is a brown spot, which is extended somewhat on to the neck; chest, neck, and flanks, as well as the outer side of the legs, yellow; wrist and tarsus also yellow; the fingers and toes black; tail grey, blackish near the tip, terminated with white hairs, and having a double crest of hairs, of which the longer is on the upper side.

	Inches.	Lines.
Length from nose to tail	28	5
" of tail	24	10
" of fore leg	4	4
" of hind leg	17	9
" of ear	3	2

Inhabits the Swan River district.

Two Kangaroos in the British Museum collection agree so closely with M. Jourdan's description as to leave no doubt as to their specific identity with the *H. Irma*, and I am thus enabled to make a comparison between the animals named *Irma* and *Manicatus*, and to notice certain differences which, I repeat, I do not regard as specific. In the first place, both these specimens (as well as the specimen noticed by M. Jourdan) are smaller than Mr. Gould's specimen of *H. manicatus;* their general colouring is paler, the black and white pencilling of the hairs on the upper parts of the body being less distinct, and the hairs on these parts are reddish, rather than yellow, in the middle, and pale grey at the root: the reddish hue, which is very pale and delicate, shows but little on the upper part of the body, but is distinct on the neck and flanks; the under parts are grey-white, slightly suffused with reddish yellow; the upper lip is yellow, and the cheek-mark of a cream colour; the top of the head and muzzle are grey, and the crown of the head is slightly tinted with rust colour; the tip of the muzzle is sooty black, and so are the occiput and back of the ears, but the latter are brownish at the root, and have a large white patch behind at the base; on the inner side thé ears are yellow, and at the apex they are broadly margined with black; the fore arm and foot, and the outer side of the hind legs and tarsi, are yellow, but the anterior half of the feet is black; the black of the latter, however, is pencilled with brownish, and in parts with white, excepting on the toes; the double inner toe is pencilled with yellow and black. The hairs of the tail are pencilled with black and white, and are shortish and adpressed, excepting on the dorsal line of the apical half, where they are long and chiefly black, and on the under side, where longish black hairs extend from the tip to within about six inches of the root; near the apex of the tail these long hairs assume a brownish hue, and at the tip is a large tuft of yellowish white, or white, hairs; this pencil, or tuft, of pale hairs is about two inches long.

	Inches.	Lines.
Length from nose to tail	27	0
" of tail	22	0
" from nose to ear	4	4
" of ear	3	0
" of tarsus	8	3

The specimen furnishing the above characters is a male; the second specimen is a female, and differs only in being rather smaller, and in having the back of the ears brownish, and grey at the base; they have the same white posterior patch.

A skull of *M. manicatus*, for the loan of which I am indebted to Mr. Gould, presents the following dimensions :—

	Male*. Inches.	Lines.
Total length	4	$4\frac{1}{2}$
Width	2	$3\frac{2}{3}$
" of interorbital space	0	$9\frac{1}{3}$
Length of nasal bones	1	$10\frac{1}{2}$
Width of ditto at the base	0	8
Width of nasal bones near the apex ...	0	$4\frac{2}{3}$
Length of posterior palatine openings ...	0	2
" of three upper incisors	0	6
" of the upper premolar	0	$2\frac{1}{2}$
" of lower jaw	2	$11\frac{1}{2}$
Height of ditto from apex of coronoid process	0	7

The foremost incisor tooth is very broad, being equal in width to the third; the second is much narrower; the third has a deep vertical groove very near the middle of the outer surface.

MACROPUS (*Halmaturus*) GREYI.

Grey's Kangaroo.

Halmaturus Greyii. Gray, List of the Mamm. in the British Museum, (1843), p. 90.

Fur moderate; general colour pale ashy-brown, slightly tinted with yellowish; under parts of the body, as well as the legs and feet, of a pale buff yellow; toes black; chest greyish; head grey, with a broad blackish mark on the cheek, extending from the angle of the mouth to the eye; a broad yellowish mark below this, and beneath this again is a brownish band: extreme tip of the upper surface of muzzle black; crown of head very faintly tinted with rufous, and the back of the neck (if we except a dusky dorsal streak), and the ears externally, also pale rufous; the apical portion of the ear, however, is black; on the inner side of the ear the hairs are yellow, but at the apex they are black. Tail

* Not quite adult.

well clothed with hairs (rather longer, softer, and less adpressed than usual) of a very pale grey colour, washed, as it were, with yellow on the upper parts, and brown-white beneath; a considerable space at the apex covered with long dirty yellowish hairs.

Inhabits South Australia.

Two specimens of this beautiful species have recently been presented to the British Museum by the Hon. Capt. G. Grey, in honour of whom it has been named.

This species is allied to the *Halmaturus manicatus* of Gould, but differs not only in colouring, the general tint of the upper parts of the body being brownish, but pale and with an admixture of yellow, whilst in *H. manicatus*, or *Irma*, it is grey, but may be distinguished from either of the two animals mentioned by its having no crests to the apical portion of the tail: this organ is clothed at the sides, as well as on the upper and under surfaces, with long pale hairs at the apex. The palish tint of the upper parts of the body is produced by the mixture of white with pale rust colour and black, the visible portion of each hair exhibiting these colours; the hairs are, in fact, of a very pale grey at the root, rusty yellow in the middle, and this is followed by a white space, and the remaining portion is black, but between the white space and the black point is a narrow ring of a rusty-red colour. On the under parts of the body the hairs are of a pale buff-yellow colour externally, and pale-grey at the root. The head is grey above, obscurely tinted with rufous, and this latter tint is also observable on the back of the ears, as well as on the neck: immediately behind the naked tip, the muzzle is dusky black above, but the black hue is almost immediately blended into the general grey tint: on the sides of the muzzle are three longitudinal bands, of which the middle one, representing the ordinary pale cheek-mark, is pale yellow; the upper one almost black, but slightly pencilled with whitish, and the lower one is somewhat suffused with brownish: the ears are well clothed internally with rich yellow hairs, but they are rather narrowly margined with black at the apex; externally, the black extends downwards from the point for about half an inch: behind the eye is a yellowish spot. The chin and throat are tinted with fulvous, and there is a greyish spot on the former; the chest is greyish; below the chest the fur has a pale rusty-grey hue. The fore legs are grey-white at the base, and of a very pale fulvous colour, or fulvous-white beyond, and the hands are of the same colour, but the fingers are black, and the black extends very slightly beyond the base of the fingers; the hind legs and feet are coloured in the same manner: the thighs are somewhat greyish externally at the base, and the toes are black, if we except the long hairs which cover the nails, these being brownish. The tail is densely clothed with longish hairs, which on the upper surface are softer, and not closely applied to the skin, as is often the case in the Kangaroos; they average at about three-quarters of an inch in length, but are about an inch and a half long on the apical portion of the tail; here they are dirty yellowish white, which colour extends about four inches from the tip; on other parts they are of a pale greyish tint, suffused with yellow; on the under side the hairs are still long, but harsh and adpressed. The foremost incisor tooth of the upper jaw is distinctly the broadest; the middle incisor on either side is the smallest, and has a faint vertical groove; in the hindermost the vertical groove is distinct: the length of the three together is 6½ lines.

	Inches.	Lines.
Length from tip of nose to root of tail ...	30	0
" of tail	26	0
" of tarsus and nails	9	9
(of which the nail of the central toe, is	1	6)

	Inches.	Lines.
Length of ear	2	11
" from nose to ear	5	2
" of fore leg to the ends of claws of fingers	6	7

MACROPUS (*Halmaturus*) RUFICOLLIS.

The Red-necked Kangaroo.

Kangurus ruficollis.	DESMAREST, Nouv. Dict. d'Hist. Nat. tom. xvii. p. 37, 1817: Mammalogie, p. 274.
Kanguroo à cou roux.	F. CUVIER, Dict. des Sci. Nat. tom. xxiv. p. 348, 1822.
Macropus ruficollis.	LESSON, Manuel de Mammal. p. 228, 1827.
Halmaturus elegans.	(LAMBERT) GRAY, Catalogue of the Mammalia in the British Museum, (1843), p. 89.
" *ruficollis.*	GOULD, Monogr. of the Macropodidæ, Part 2, Plate 2.
Kangurus rufogriseus.	DESMAREST, Nouv. Dict. d'Hist. Nat. t. xvii. p. 36.
Halmaturus griseo-rufus.	GOLDFUSS, in Isis, 1819, p. 267.
" *leptonyx.*	WAGNER ?

Fur moderate; general colour rusty grey—chiefly grey on the back, and chiefly pale rust colour on the neck, shoulders, and in the region of the eye; under parts grey-white, slightly suffused with pale rust colour; ears white internally, but with the apical margin black; externally blackish: fore feet brown or blackish, finely pencilled with white; the toes dark brown or blackish: tarsi dark brown, but with the mesial portion nearly white: tail grey-white, slightly suffused with black at the apex.

Inhabits New South Wales, and King's Island.

The Red-necked Kangaroo was discovered by MM. Péron and Lesueur, in King's Island, (one of the largest of the Islands in Bass's Straits), in which they also found the Wombat, Echidna, and a species of Dasyurus (*D. maculatus*). These two travellers, together with M. Leschenault, were left by their party on King's Island in order to explore its productions, and their vessel not returning so soon as expected, all their provisions became exhausted, and they would have perished with hunger had they not fortunately met with a party of Englishmen who had settled there for the purpose of procuring the seals which abound on the shores of the small islands in Bass's Straits. From these settlers they received both food and shelter, until the vessel arrived, and M. Péron speaks in high terms of their kindness and rude hospitality. He found these seal fishermen living almost entirely upon the flesh of the Red-necked Kangaroo, Wombat, and Emu. The Wombat was readily captured, owing to its sluggish habits, but to procure the swift Kangaroos and Emus, the settlers had trained dogs to go out by themselves and hunt, and it is said they seldom returned without success. Having killed a Kangaroo, the dogs abandoned their prey, and going to their masters, made them aware of the circumstance: one of the men was then dispatched to bring home the game, to which he was guided by the dogs.

M. Péron observes that his party subsequently obtained a dog similarly trained, and that, in Kangaroo Island, they procured by its assistance twenty-seven of the largest-sized Kangaroos in a few days.

The specimens of *Macropus ruficollis*, in the British Museum, are from New South Wales, and we are informed by Mr. Gould that this animal was formerly common near Sydney, but is now gradually retiring before the advance of civilized man; it is still, however, abundant in the thick *Daveysia* scrub on the table-land behind Illawarra, particularly on the fine estate at Bong Bong, belonging to Charles Throsby, Esq.

If the *M. Bennettii* be specifically identical with the *M. ruficollis*, (and I feel no doubt that it is) then must we add Van Diemen's Land to the habitats above given.

The fur of the *Macropus ruficollis* is moderate both as to length and texture, and its general hue is pale rufous grey: on the back, each hair is brownish grey at the root, and this tint is followed by pale rust colour, then a broad white ring, and the point is dusky; the region of the eye, the neck, shoulders, and fore legs, are chiefly of a pale rust colour, but pencilled with white: the under parts of the animal are whitish, but slightly suffused with pale rusty grey, the hairs on these parts being of a vinous-grey tint next the skin, slightly tinted with pale rust colour in the middle, and white at the point. The head is of the same general colour as the body, but on the crown it is somewhat dusky, and the muzzle is brownish; the upper lip is white, and there is a tolerably distinct whitish cheek-mark: on the chin is a brown patch. The ears are clothed with white hairs internally, excepting at the apical margin, where they are black, or nearly so; and on the outer side they are covered with brownish-black hairs at the apex, and with hairs like those of the head at the base. The fore legs are of a pale rust colour, but very much pencilled with white; the fore feet are brown, finely freckled with white, and the toes are dark brown. The hind legs are almost entirely of a pale rufous tint externally; the tarsi brown, but whitish in the middle: here the hairs, however, are dark brown at the root: the toes are dark brown, if we except the hairs covering the base of the nails, which are pale: on the fore part of the hind leg is a whitish mark running up from the tarsus. The tail is tolerably well clothed with hairs, which are partly white, and partly black; the general hue is nevertheless very pale, and may be described as grey-white; the apical portion is suffused with black.

The above description is from a male specimen in the Paris Museum, procured during Péron's expedition, and which is moreover the original of Desmarest's description. A second specimen in the same collection is said to be from New Holland.

In the British Museum are specimens of the same species, procured by Mr. Gould from New South Wales; they are labelled *Halmaturus elegans*.[1]

The admeasurements of these specimens are as follows:—

	Male. Original Specimen.		Male. Second Specimen in Paris Mus.		Male. Brit. Mus.		Female. Brit. Mus.	
	Ins.	Lines.	Ins.	Lines.	Ins.	Lines.	Ins.	Lines.
Length from tip of the nose to the root of the tail .	39	0	41	6	41	0	35	6
" of the tail	25	0	25	6	31	0	26	6
" of the tarsus and nails .	9	1	8	10	10	0	9	6
" of which the nail of the central toe is .	1	1						
" from nose to ear . .	5	6	6	0	6	0	5	3
" of ear	2	5	2	6	3	1	3	0
" of fore-arm, hands, and claws, about . .	9	3	9	6	10	6	8	0
" of the three upper incisors taken together	0	6¼			0	7		

Macropus (Halmaturus) Rufo-Griseus.

"M. Geoffroy distinguishes under this name, as belonging to a distinct species, a female Kangaroo of the Paris Museum, of a tolerably large size, since its length, measured from the tip of the muzzle to the base of the tail, is about 3 feet 10½ inches; the head 8¾ inches; the tail 2 feet 2 inches, and the ear

[1] Mr. Gray is of opinion that the *M. ruficollis* of Desmarest is specifically identical with the *Macropus elegans* of Lambert, briefly noticed in the "Linnæan Transactions," vol. viii. p. 381. The description given by Lambert, however, is more applicable to the *M. Parryi*.

nearly 4 inches and 4 lines: all the upper parts of the body are of a rusty grey colour, the grey, however, predominates, and the under parts differ only in being paler; the extremities of the feet and tail are shaded with brown, and the under side of the latter is of the same colour as the upper: the hairs on the back are rust coloured at the base, have then a whitish ring, and the point is brown; those of the belly and chest have the whitish portion less extended.

The *Macropus rufo-griseus* is vaguely indicated as inhabiting New Holland; its great size, however, leads us to suppose that it is the species referred to by the French travellers, MM. Péron and Lesueur, as inhabiting Kangaroo Island, where a second large species, supposed to be the *M. fuliginosus*, is also found."—From Desmarest, in the Nouveau Dictionnaire d'Histoire Naturelle.

Very recently, the author visited the Paris Museum with a view to examine and determine, if possible, the three species of Kangaroos described by Desmarest under the names *Kangurus ruficollis*, *K. rufo-griseus*, and *K. Eugenii*, but was disappointed to find that the originals of M. Desmarest's descriptions of the two last mentioned species no longer existed. An animal to which the specific name *rufo-griseus* is attached in the Paris Collection, certainly agrees very closely with Desmarest's description, excepting that its size is less than indicated by that author. This animal is a female, and is clearly of the same species as the specimens named *ruficollis* in the same collection. Its general tint is grey, much suffused with yellowish rust colour, and this last mentioned tint is almost uniform on the back of the neck, shoulders, and sides of the body: the under parts of the body are of an impure white; the fur on the back is much pencilled with white; the hairs are dusky at the point, greyish at the base, shaded into pale rusty-yellow near the middle, and white, or nearly white, towards the point: the tail is of a dirty greyish white hue above, and dirty yellowish white beneath; the fore feet brownish, pencilled with white, and have the toes dark-brown; the tarsi are whitish in the middle, but the hairs on this part are brown at the base; the toes dark brown, with the exception of the hairs covering the nails, which are pale; the head is of the same general hue as the body, and, as in *M. ruficollis,* the region of the eye is tinted with pale rufous; the ears are clothed with white hairs internally, and are dusky externally; the fore part of the hind leg, immediately above the tarsus, is white.

	Inches.	Lines.
Length from the tip of the nose to the root of the tail	34	0
" of tail	22	6
" from nose to ear	5	0
" of ear	2	5
" of tarsus, including the nails	8	6
" of fore-arm, hands, and claws	8	0

The specimen from which the above description is taken is evidently very old, and may have been in the Paris Museum at the time that Desmarest drew up his description, but the difference of size, as I have before stated, forbids our believing it to be the *individual* described as *M. rufo-griseus.*

Macropus ruficollis, var. *Bennettii.*

Macropus Bennettii.	WATERHOUSE, Proceedings of the Zoological Society for October, 1837, Pt. 5, p. 103; Marsupialia, p. 211, Pl. 19.
Macropus (Halmaturus) fruticus.	OGILBY, Annals of Nat. Hist. for May, 1838, Vol. i. p. 219.
Halmaturus Ualabatus.	GRAY, Mag. of Nat. Hist. for Nov. 1837, Vol. i. (New Series), p. 583.
" *Bennettii.*	GOULD, Monogr. Pt. 1, 7th Plate.
" "	WAGNER, in Schreber's Saugeth. Suppl. 111—112 Heft, p. 115.

Fur long; moderate as to texture: general colour very deep grey, slightly tinted with rufous on the back of the ears at the base, and on the neck and shoulders; under parts grey-white; tail pale grey, black at the apex: the anterior half of the hind feet, the hands, and the apical half of the ear, externally, black: an indistinct whitish cheek mark: top of muzzle blackish.

Inhabits Van Diemen's Land.

That the *Macropus Bennettii* is a local variety only of the *M. ruficollis* I now feel little doubt; it differs, however, in having the fur longer, of a much darker general hue, being considerably suffused with black in parts, and exhibiting very little of the red tint which is so conspicuous in *M. ruficollis*.

The "Brush Kangaroo," as this animal is called by the colonists, is a native of Van Diemen's Land, where it is very abundant; "its flesh," Mr. Gould informs us, "is generally eaten, and highly esteemed, and its skin forms a considerable article of commerce, being largely imported from Van Diemen's Land into England for the manufacture of boots and shoes, besides being extensively used for the same purpose in the colony. It is universally dispersed over Van Diemen's Land, whose dense and humid forests afford it a retreat so secure as to preclude all chance of its extermination for centuries to come, although many thousands are killed annually. Advertisements may be frequently seen in Hobart Town newspapers, stating that three thousand skins are immediately wanted, and they are quickly supplied by the settlers, servants, and shepherds at the out-stations. They are either captured by dogs, or by snares set in their runs; the skins are generally taken off on the spot, and are afterwards stretched on the ground to dry; they are then sold for fourpence or sixpence each, to persons who visit the stock-stations of the interior for the purpose of collecting them, and who retail them again in Hobart Town, or Launceston, to the advertiser or others, for colonial consumption or for exportation."

Mr. Gould, moreover, states, that the Bennett's Kangaroo is gregarious in its habits, and although truly a brush animal, does not confine itself so strictly to localities of that description as the smaller species of the genus.

For many years past, specimens have lived and bred in the Gardens of the Zoological Society, as well as in the managerie of the Earl of Derby. In a large piece of enclosed ground in his Lordship's park, I had the pleasure of seeing many individuals of the Brush Kangaroo in a state of comparative freedom, and where they appeared to thrive well. When I entered the paddock in which they were kept, being all of them concealed amongst some heath, I was not aware of their presence until, approaching towards their place of shelter, they suddenly elevated the fore part of their bodies, and then darted off to a distant spot with great swiftness. When at rest they frequently assume a singular position; the fore feet are applied to the ground, and they at the same time sit upon their haunches, having the hind legs stretched forwards, and perfectly straight, as well as the tail, which lies between them. The young animal does not finally quit the pouch of the mother, until it has attained the size of a rabbit; at this time they do not differ in colouring from the parent. The female, as usual, is smaller and more delicately made than the male. The following is a description of an adult male and female which died in the menagerie of the Zoological Society.

The fur is long; that on the upper parts of the body is of a blackish grey colour next the skin, and black externally; but each hair is annulated with white or pale yellowish rust colour towards the tip. The general tint may be described as very dark grey, much pencilled with black, and less distinctly so with whitish on the back, but slightly washed, as it were, with pale rust colour; on the sides of the body the general hue is paler, the black being much less conspicuous;

and on back of the ears at the base, the back of the neck, the shoulders, the cheeks, and above the eye, a slight rusty hue prevails; the under parts of the body are grey-white, the hairs on these parts being grey, tipped with white. The top of the muzzle is blackish, and so are the ears externally, excepting at the base, the hands and wrists, and the fore half of the tarsus; internally, the ears are clothed with white hairs, but the apex is always more or less broadly margined with black: the chin is blackish; the upper and lower lips are margined with white, and there is a somewhat undefined whitish cheek-mark. The tail is clothed at the root with fur like that of the body, but beyond, the hairs are harsher and shorter, and the general colour is hoary grey, excepting on about five or six inches of the apical portion, where they are black; the under side of the tail is clothed with harsh, dirty white hairs, excepting at the apex, where they are chiefly black. On the posterior half of the tarsi, the hairs are hoary grey.

	MALE.		FEMALE.	
	Inches.	Lines.	Inches.	Lines.
Length from tip of nose to root of tail	35	0	30	3
" of tail	31	6	26	9
" from nose to ear	5	10	5	0
" of ear	2	11	2	11
" of tarsus	9	3	8	2
" from elbow to base of claws, about	9	3	7	0
			Sex unknown.	
Total length of skull	4	$11\frac{1}{2}$	5	6
Width	2	$6\frac{1}{2}$	2	$11\frac{1}{4}$
" between orbits	0	$10\frac{1}{2}$	0	$11\frac{1}{3}$
Length of nasal bones	1	$9\frac{1}{2}$	2	0
Width of ditto behind	0	$9\frac{1}{3}$	0	$9\frac{1}{2}$
" of ditto near the apex ...				
Length of palate	2	$11\frac{1}{2}$	3	3
" of posterior palatal openings	0	$4\frac{1}{2}$	0	$4\frac{1}{4}$
" of the five molars of the upper jaw, taken together ...	1	$5\frac{1}{2}$	1	$7\frac{1}{3}$
" of foremost molar, upper jaw	0	$3\frac{1}{2}$	0	$3\frac{2}{3}$

	Inches.	Lines.	Inches.	Lines.
Length of the three upper incisors, taken together	0	7½	0	8
" of the hindermost incisor ...	0	3½	0	4
Depth of the zygomatic arch behind ...	0	7	0	8
Distance between upper incisors and the molar teeth	1	3¾	1	3⅔
Height of the lower jaw, measured in a vertical line dropped from the apex of the coronoid process	1	10	2	1

The upper incisor teeth are represented on Plate 5, fig. 9

Macropus (*Halmaturus*) *leptonyx*.

Halmaturus leptonyx. A. Wagner, in Schreber's Säugeth. Suppl. 111-112, Heft. p. 116, November 30, 1842.

Upper parts of the body variegated with brown and whitish; under parts whitish; the nails of the hind feet much compressed, and somewhat concave at the sides; tail shorter than the body, and of a dirty yellowish colour beneath.

Inhabits ———— ?

Dr. Wagner compares his *Halm. leptonyx* with the *H. Bennettii*, from which he says it differs in having a much shorter tail, and in the compressed form of the claw or nail of the great fourth toe of the hind foot: it is described as follows. "Ears moderately long; the tail, which is densely clothed beneath with bristly hairs, shorter than the body; the external claw, as well as the central one, compressed inwards from the outer side; the fur dense, long, and moderately soft; colour of the upper parts of a dingy yellow-brown, with pencilling of a lighter tint, and, on the back of the neck, the brown is blended with clean ferrugineous yellow, whilst, upon the sides of the body and the posterior extremities, the brown colour is pencilled with grey-white, the latter tint predominating: on the upper parts, the hairs are dingy brown, and have a white ring below the point; and this ring is broader on the hairs of the sides of the body, and thus produces a lighter general tint on these parts; the whole of the under parts, as well as the inner side of the legs, are grey-white, but the hairs are grey-brown at the roots. The woolly hairs are more distinctly coloured. The upper surface of the head is of the same colour as the back, and on the cheeks is a whitish mark, extending backwards from the angle of the mouth; the muzzle is blackish; the ears have whitish hairs towards the margins internally, and, externally as well as internally, the apical portion has black hairs; the fore feet, and toes of the hind feet, are black-brown, and the tail is black, pencilled with whitish upon the upper surface; on the under surface the hairs are yellow white."

	Inches.	Lines.
The length of the body is	23	6
" of the tail	15	0
" of the ear	2	6
" of the tarsus and claw	7	0
" of the claw alone	1	2

In page 36, No. 3-4, for 1842, of Wiegman's Archiv., Wagner says, he has not satisfied himself whether his *H. leptonyx* is identical with my *H. Bennettii*, for, notwithstanding the similarity in colouring, &c., the form of the claw is so remarkable in *H. leptonyx* that he should consider it as a good species, unless that character be found in *Bennettii*.

With regard to these observations, I may state that, though I have found the nail or claw of the great central toe in *Bennettii* to vary somewhat in its form, I have not met with any specimen which can be said to agree with Prof. Wagner's description. The fact is, the claw in *Bennettii* offers nothing peculiar: in an adult specimen it is about one inch in length, and half an inch in width at the base, is gradually attenuated from the root to the extremity, which is more or less sharply pointed, and its transverse section would present the form of a triangle, but with the angles somewhat rounded, and the base of the triangle would be formed by the under surface of the claw. As these claws, however, are often much worn by usage, they of course will differ in different individuals, and sometimes be short and obtuse, as described by Mr. Ogilby; and in specimens kept in confinement, where they are not worn away at the point in proportion to their growth at the base, they often attain a considerable length, and sometimes very singular forms. In a specimen of *Macropus giganteus* in the British Museum, the claws of the fore foot, from some such cause, have attained the length of from two, to three or four inches, and are spirally twisted I suspect the *M. leptonyx* will prove to be founded upon a young specimen of *M. ruficollis*.

MACROPUS (*Halmaturus*) UALABATUS.
Black-tailed Kangaroo.

Kangurus Ualabatus.	LESSON et GARNOT, Zool. de la Coquille, tom. I, p. 161, Plate 7. 1826.
Macropus Ualabatus.	LESSON, Manuel de Mamm. p. 227. 1827.
Macropus Ualabatus.	WATERH. Marsupialia, p. 219. 1841.
Kangurus Brunii.	DESM. Mammal. p. 275. 1820.
Halmaturus Lessonii.	GRAY, Mag. Nat. Hist. Vol. 1, (New Series), p. 583. 1837.
Halmaturus nemoralis.	WAGNER, in Schreber's Saug. Suppl. Part 111-112, p. 114, November 30, 1842.

Fur moderately long, somewhat harsh to the touch, and somewhat glossy; general colour deep brown, suffused with reddish on the hinder parts of the body; feet and tail almost entirely black; the hairs longish, and very harsh; under parts of the body of a rusty yellow colour; ears with deep yellow hairs on the inner side; externally dusky, but of a rust colour at the base: a large blackish patch is situated immediately behind the base of the fore leg.

Inhabits New South Wales.

The present species has received the English name of *Black Wallaby* from Mr. Gould, but although, compared with most other species, its general colouring is dark, it is far from black. The English name Black-tailed *Kangaroo* has appeared to me less likely to mislead. In size the *M. Ualabatus* is scarcely inferior to the *M. ruficollis;* it must rank, therefore, among the larger species of its section, and from these it may readily be distinguished by its general dark colouring, its coarsish and somewhat glossy fur, the dark patch immediately behind the base of the fore leg, the rufous tint at the base of the ear in front, the reddish or yellowish tint of the abdomen, and the almost uniformly black tail.

"This well-marked species," observes Mr. Gould, "inhabits, with but few exceptions, all the thick bushes of New South Wales, especially such as are wet or humid. I hunted it successfully at Illawarra, on the small islands at the mouth of the Hunter, and on the Liverpool ranges. In the former localities it was frequently found in the wettest places, either among the high grass, and other dense vegetation, or among the thick mangroves, whose roots are washed by each succeeding tide. The islands at the mouth of the Hunter, particularly Mosquito and Ash Islands, are not unfrequently flooded to a great extent, yet it leaps through the shallow parts with apparent enjoyment, and even crosses the river from one island to the other. On the Liverpool range it as strictly keeps to such parts as are most humid; often near the crowns of mountains, which are frequently enveloped in fogs and dews."

The following is a description of a fine male specimen in Mr. Gould's collection:—

Fur moderately long, somewhat stiff, and more glossy than

in most other species of *Macropus* : general colour very dark brown, the visible portions of the hairs being pencilled with rust colour and black, but next the skin (on the back) they are almost black; head greyish; the crown, occiput, and region of the ears, of a rich rusty red colour; cheek-mark distinct, and of a rich yellow hue; muzzle dusky above; towards the apex, and immediately round the eye, the hairs are of the same colour; chin white; throat rich golden yellow, shaded gradually behind into the rusty red of the under parts of the body, which is deepest between the hind legs; the hairs on the under parts are, however, grey at the roots; a broad black patch is situated immediately behind the fore leg; ears with yellow hairs internally; externally, of a deep rusty red, suffused with brownish, but with the apical third deep brown, inclining to black; the margin pale; shoulders and fore legs greyish; the hairs on these parts pencilled with black and yellowish white; hands and wrists black, slightly pencilled with whitish; tarsi black, obscurely freckled with yellowish above, distinctly pencilled with this colour on the outer side, and of a rusty hue on the inner side; tail black, slightly pencilled with whitish at the base and along the sides; the hairs long, stiff, and glossy; beneath brownish.

A female specimen, also in Mr. Gould's collection, had the forehead, cheeks, shoulders, and sides of the belly, grey: the abdomen is sometimes of a palish yellow hue.

	No. 1.		No. 2.		No. 3.		No. 4.	
	Ins.	Lines.	Ins.	Lines.	Ins.	Lines.	Ins.	Lines.
Length from nose to ear . . .	34	0	32	0	29	6	34	0
" of tail, about	26	0	24	0	24	0	26	0
" from nose to ear . . .	5	3	5	0	4	2		
" of ear	3	0	2	6	2	6	2	7
" of tarsus, toes and claws .	8	4	8	0	7	3	7	8

The dimensions in the first column are from the male specimen above described, those of the second column are from a specimen in the British Museum, and the other two are from specimens in the Paris Museum. Column No. 3 gives the admeasurements of an individual brought from Port Jackson, in the expedition of the Astrolobe, and is the original of M. Lesson's description. The dimensions in the last column are taken from a very old specimen, which is, I suspect, the original of Desmarest's description under the name *Kangurus Brunii.* On the bottom of its stand is written "*Kanguroo Filandre, K. d'Aræ, Didelphis brun.*" I perfectly agree with Lesson, that this is the same as his *M. Ualabatus.* The *Didelphys Brunii* is a very distinct animal.

The following dimensions are from a skull of *M. Ualabatus* in the collection of the British Museum, and which is labelled as having been procured in Mosquito Island :—

This skull is rather narrower, and more elongated than usual; the nasal bones are long, very convex, but little expanded behind, and somewhat contracted in the middle; the bony palate terminates very nearly in a line with the hinder margin of the last molar; the posterior palatine openings are rather small compared with many of the Halmaturi; the occipital opening is very large, and has a distinct notch in the upper margin; the premolar tooth is unusually large, its length being equal to the first true molar added to about one-third of the length of the second, whilst, in most of the Kangaroos, the corresponding tooth is about equal to the first true molar in length, as in *M. Thetidis,* and sometimes shorter, as in *M. giganteus.* In the more elongated form of the skull, and greater size of the premolar, the *M. Ualabatus* evinces an approach to the New Guinea Kangaroo, *M. Brunii.* The incisor teeth of the upper jaw are represented on Plate 5, fig. 12.

	Inches.	Lines.
Total length of skull	5	3
Width	2	8
Length of nasal bones	2	$3\frac{1}{3}$
Width of ditto at the base	0	$9\frac{2}{3}$
" " near the apex	0	$6\frac{1}{2}$
Length of frontal bones	1	9
Width between orbits	1	$0\frac{2}{3}$
Length of palate	3	$1\frac{1}{2}$
Width of ditto opposite third molar	0	$11\frac{1}{4}$
Length of posterior palatal openings	0	$5\frac{1}{2}$
" of the five molar teeth taken together ...	1	$7\frac{2}{3}$
" of the premolar	0	$4\frac{2}{3}$
Distance between incisors and premolar	1	2

MACROPUS (*Halmaturus*) EUGENII.
Eugene Island Kangaroo.

Kangurus Eugenii. Desm. Nouv. Dict. d'Hist. Nat. tom. xvii. p. 38; Mammalogie, p. 274.

Macropus Eugenii. Lesson, Manuel de Mammalogie, p. 227.

Halmaturus Dama, and *H. gracilis.* Gould, Proceedings of the Zoological Society; the former in the Part for February 1844, p. 32, and the latter in the Part for June 1844, p. 103.

"Total length, measured from the tip of the muzzle to the root of the tail, about twenty-one inches (French measure); head four inches, and tail rather more than one foot in length. Fur very soft, somewhat like that of *M. ruficollis;* its general tint grey brown, but mixed with rust colour near the shoulders, on the back of the neck, and upper surface of the head, as well as the fore legs; the whitish colour of the under parts pretty distinctly separated from the deeper colour of the upper parts; under surface of the tail white, slightly tinted with reddish; upper surface grey brown; each hair of the back is grey at the root, and annulated with brown, and with whitish externally, but the extreme point is brown; the hairs of the shoulders and back of the neck are grey at the root, then rust coloured, followed by white, and rust colour at the apex."

The above description is from Desmarest: unfortunately the specimen upon which it is founded no longer exists. In size and colouring the *M. Eugenii* must greatly have resembled the *M. Derbianus;* but there is no mention in the diagnosis of a black mark on the back, such as is noticed in that species. A small Kangaroo, described by Mr. Gould under the specific name *Dama,* wants this black mark, and in other respects agrees with Desmarest's description. One of Mr. Gould's specimens of *M. Dama* now in the British Museum collection is from the neighbourhood of Moore's River; and another is labelled as from Wangan Swamps, both in Western Australia, where it is by no means improbable we should find the *M. Eugenii,* supposing Desmarest to be correct with regard to the habitat of the original specimen: that specimen, he says, to his knowledge, once was labelled as being from St. Peter's Island, and subsequently the label was changed for one giving Eugene Island as the habitat; both islands, however, are in Nuyt's Archipelago. Following are descriptions of the *M. Dama* and *M. gracilis* of Mr. Gould, which I regard as the same species. The dimensions of *M. Eugenii* are reduced to English measure, and arranged with those of the *M. Dama.*

Halmaturus Dama. GOULD.

Male.—General colour above, rusty brown, but considerably pencilled with black; back of neck, shoulders, and fore legs bright rust colour; flanks and hind legs externally suffused with rust colour; under parts white; chest yellowish: head grey, obscurely tinted with rust on the crown and in the region of the eye; tip of muzzle brownish; cheek-mark moderately distinct; ears grey externally, becoming nearly black at the apex: feet rusty white; toes slightly freckled with brown: tail grey. Weight about 10 lbs.

Female. — General colour greyish, but little tinted with rust colour; neck, shoulders, and fore legs, of a pale rust colour, much pencilled with white; body beneath, grey-white: tail with a black mark along the upper surface of the apical portion.

This animal is said to be called "*Dama*" by the natives.—Inhabits Western Australia.

	M. Eugenii.		MALE. M. Dama.		FEMALE. M. Dama.	
	Inches.	Lines	Inches.	Lines	Inches.	Lines
Length from tip of nose to root of tail	22	8	23	0	22	0
" of tail, about	13	0	15	0	14	0
" of tarsus and nails			6	0	5	8
" of the great central toe-nail			0	9	0	7¾
" from nose to ear	4	4	3	9	3	9
" of ear			2	6	2	3
" of fore arm, hands, and nails			5	9	4	9
" of hand and nails			1	9	1	5½
of which the nail of central finger is			0	5	0	5½
" of the three upper incisors ...			0	5½		

M. Dama differs from *M. Derbianus,* not only in wanting the black mark on the back of the neck, but in having the fore legs much more slender, and the hands smaller, the tarsi longer, and the ears larger: the upper incisor teeth are rather larger—they are represented on Plate 5, fig. 8. The smaller size of this animal, its rufous fore legs, and the structure of the upper incisor teeth, serve to distinguish it from the *M. Thetidis* of the East coast.

Mr. Gould's *Halmaturus gracilis,* I feel no doubt, is specifically identical with the *H. Dama.* Two specimens in the British Museum collection, and which formerly belonged to Mr. Gould, present the following characters:—

Macropus gracilis. Gould.

Fur soft; general colour ashy grey, slightly suffused with rufous on the back of the neck, shoulders, fore legs, and on the outer and hinder parts of the hind legs and sides of the tarsi; under parts white, very faintly tinted with grey, and in parts with yellowish; ears externally, and tail above, grizzled with black and white, the latter dirty yellowish white beneath, and with a line along the upper surface of the apical portion, formed of blackish hairs; feet rusty white, the toes brownish.

Inhabits Western Australia.

	Female.		Male.		H. Dama.	
	Inches.	Lines	Inches.	Lines	Inches.	Lines
Length from tip of nose to root of tail	18	6	18	0	18	6
" of tail	12	6	12	0	12	9
" of tarsus and nails	5	1	5	4	5	3
" of ear	2	0	2	0	2	1
" of fore arm, hand, and nails	?		4	0	4	2
" of the three upper incisors	0	$5\frac{1}{3}$	0	$5\frac{1}{2}$	0	5

The two specimens of *H. gracilis* referred to are labelled as being from "Walyema Swamps, 40 miles N.E. of Northam, Western Australia:" they are young animals, as may be seen by the condition of their incisor teeth; the posterior incisor of the upper jaw, for instance, in one specimen, is a new tooth, the cutting edge of which had not attained the same level as the other two incisors, and is perfectly unworn; as is also the case with the other specimen, though the same tooth is here a little more advanced. Both the specimens were shot when casting their fur, and hence their general colour is paler, and more grey, than is that of the specimens of *M. Dama* described; for the darker hue in the latter case is due to the longer and harsher hairs of the fur, which hairs being almost entirely wanting in *M. gracilis*, the softer grey under fur is exposed. An immature specimen of *M. Dama*, sent with the male and female from which my descriptions are taken, and forming part of the British Museum collection, furnished the dimensions given in the third column. The structure of the incisor teeth is the same in *M. gracilis* and *M. Dama;* and it will be seen by the dimensions just referred to, that (making a slight allowance for measurements taken from skins) the proportions are the same.

MACROPUS (*Halmaturus*) THETIDIS.

Pademelon Kangaroo.

Halmaturus Thetidis, F. Cuvier et Geoff., Mammifères, Tab. 55.
" " (F. Cuvier,) Lesson, in the Zoological portion of M. Bougainville's "Journal de la Navigation autour du Monde de la Frégate Thétis, &c." tom. ii. p. 305, Pl. 37.
" (*Thylogale*) *Eugenii*, Gray, Magazine of Natural History, 1837, vol. i. (New Series,) p. 583.
Macropus Eugenii, Waterhouse, in Catal. of the Mammalia preserved in the Museum of the Zoological Society, 1838, p. 66, sp. 644, Marsupialia, p. 232.
Halmaturus nuchalis, Wagner, in Schieb. Saugeth., Suppl. 111—112, Heft. p. 128, Nov. 1842.
" *Thetidis* (*Pademelon Wallaby*,) Gould, Monogr. Pt. 2, and Pl. 6.

Fur moderately long and soft: general hue of upper parts brown-grey; under parts white, but sometimes tinted with pale rusty yellow on abdomen; shoulders, back of neck and flanks, bright rusty-red; feet uniform brown, by no means dark; fore legs grey: tail clothed with short hairs; above grey, beneath brownish-white: ears grey at the base externally.

Inhabits New South Wales.

The "*Pademelon Wallaby*" of the colonists, is a small species, about twenty inches in height when sitting up on its hind legs, and is an inhabitant of New South Wales, where it is very abundant. Mr. Gould states, that it is strictly a brush animal, and that he has met with it in all the districts, where the low shrubs abound, from Illawarra to the Hunter; he had moreover received specimens from Moreton Bay. As an article of food it is highly prized, its flesh being tender and well flavoured, and very like that of the Common Hare.

Specimens of this species were brought alive to Europe, by the officers of M. Bougainville's voyage of the French vessel Thetis, and these lived for some time, and brought forth young, in the Menagerie of the *Jardin des Plantes* of Paris. It is upon these specimens that M. F. Cuvier founded his *Halmaturus Thetidis,* naming the species after the vessel just mentioned. M. F. Cuvier, however, in his account, alludes to a specimen, as belonging to the same species, which he states had long existed in the Paris Museum, where it had been regarded as the young of the *Macropus ruficollis.* Now M. Desmarest had previously characterized an animal under the name of *Kangurus Eugenii,* founded as he observes upon a Kangaroo of the Paris Museum, which had been regarded as the young of the *Macropus ruficollis,*[1] and from this circumstance, (together with a tolerably close agreement in the descriptions given by the two authors mentioned,) I was induced to sink the name *Thetidis* as a synonym of *Eugenii.*[2] Mr. Gould is of opinion that the *M. Eugenii,* which is said to be from Eugene Island, on the West Coast, is distinct from the New South Wales animal, *M. Thetidis,* and that it is identical with the *M. Derbianus,*

[1] I found no Kangaroo with the name *Eugenii* in the Paris Museum. It is somewhat remarkable that M. F. Cuvier should have made no mention of *M. Eugenii,* when he described the *H. Thetidis.*

[2] In page 232 of the volume on Marsupialia in the "Naturalists' Library."

which also inhabits the West Coast. One phrase in Desmarest's description favours this opinion, for he states, that the fore feet are red like the neck, &c., a character found in *Derbianus*, but not in *M. Thetidis:* he makes no mention, however, of the broad black mark on the back of the neck, which is observed in *M. Derbianus*, nor can I find any points either in his description, or in that given by Péron and Lesueur (the original describers of *M. Eugenii*) which would satisfactorily settle the question. There are several small Kangaroos so closely allied to the species under consideration, that it will be necessary to give a somewhat detailed description of each.

The following notes were recently made by the author upon the original specimens of M. F. Cuvier's *Halmaturus Thetidis*.

Male.—Neck and shoulders of a bright rust colour; hinder parts of the back of a rusty brown hue, distinctly pencilled with black, and rather less distinctly pencilled with rusty white; sides of the body chiefly of a rust colour, but freely pencilled with white; a pale rusty white mark crosses the haunches; under parts of the body, as well as the fore part of the hind legs, white; the outer and hinder parts of the hind legs chiefly of a rust colour: fore-arms ashy grey, the hands almost uniform in tint with the fore-arm, but very indistinctly suffused with brown: tarsi brown, almost uniform in tint, and by no means dark: tail sparingly clothed with short stiff hairs; on the upper surface chiefly of a black colour, but slightly grizzled with white; the under parts dirty yellowish white: head pale rust colour, much pencilled with black, especially on the vertex; muzzle palish brown; upper lip white; the pale cheek mark indistinct; ears with white hairs internally, and grey externally, the hairs being partly black and partly white.

The fore legs in this male specimen are large and strong,

and the tarsi are short and stout: on the fore half of the body the hairs of the fur are short, stiff, and very closely applied to the skin; those on the neck, both above and below, are directed forwards, and towards the sides, slightly outwards; immediately behind the shoulders the hairs on either side of the body are directed inwards, and the points, meeting in the middle line of the back, form a little crest: on other parts the hairs have the usual direction—those on the back are of a very pale grey at the root; each hair is then brown, and this colour is followed by a ring of white, which is shaded through rust colour into black: on the chin, throat, and chest, the hairs are uniformly white to the root, and on the abdomen they are nearly so, being but slightly tinted with grey at the root.

The remaining two specimens in the Paris Museum are females, and are of a smaller size than the male just described; they, moreover, have the fore legs much smaller and more slender, and the tarsi less stout. The fur of the neck is soft, loose, and nearly erect; the white hairs of the throat and chest differ likewise in being tinted with pale grey at the root, like those of the abdomen. The body is rusty brown above, freely pencilled with black, and rusty white; the back of the neck and shoulders are of a bright rusty red colour, and so are the flanks, though here of a paler hue: a rusty white mark crosses the haunches: the ears are grey externally, as well as the occipital portion of the head; the muzzle is brownish; the fore-arms pale yellowish grey, the hands pale brown, and the tarsi of a deeper brown; the tail is sparingly clothed with short stiff hairs, which have a general grey tint; on the under parts the hairs are of a dirty yellow colour.

In this species the front of the muzzle is naked, but the small hairs extend on to the hinder part, so as nearly to form a simicircle, running downwards, or forwards, on each side from the posterior angle of the nostril opening: there is also a narrowish naked space beneath the nostril, but this does not extend back to the posterior angle. The second incisor tooth of the upper jaw is rather broader than the first, and the third is distinctly the broadest, and has a notch behind, marking the separation of the posterior lobe from the body of the tooth, and is so far back that but a very small portion of the lobe in question is visible when the tooth is viewed from the outer side.

	Male.		Female.		Female.	
	Inches.	Lines	Inches.	Lines	Inches.	Lines
Length from tip of nose to root of tail	26	0	24	0	23	0
" of tail	18	9	14	6	16	3
" from nose to ear	4	3	3	9	3	11
" of ear	2	3	2	0	1	10
" of fore-arm, hands, and nails	6	8	4	8	4	7
" of the nail of middle finger...	—	9	—	7		
" of the tarsus and nails ...	6	6	5	4	5	6
" of the nail of the central toe	1	2	—	10		
" from tip of central toe to the apex of that of outer toe ...	1	0	—	11		
" from ditto to the base of nails of the inner double toe ...	1	7	1	3½		
" of the three incisors of the upper jaw	—	?	—	6		

In general form, the skull of *M. Thetidis* nearly resembles that of the cranium of *M. Ualabatus;* the nasal bones, however, in the last mentioned animal are less produced in the mesial line, behind, and they are broadest at the base, whilst in *M. Thetidis* the nasal bones are nearly of the same width from the base to the intermaxillary suture, in front of which they are considerably contracted; the posterior palatine openings are much larger, and the premolars smaller. In the skulls of both these animals I have seen the alveoli for small canine teeth.

The following dimensions are from two skulls of *M. Thetidis*, one of which is contained in the collection of the British Museum, and is from Mosquito Island, at the mouth of the Hunter River. Its dimensions are given in the first column, and it is figured in Plate 2; the other skull is in the College of Surgeons, and was brought home by Mr. Gould.

	No. 1. Inches.	Lines.	No. 2. Inches.	Lines.
Total length of skull	4	4½	4	0
Width	2	3⅔	2	2¼
Length of nasal bones	1	10½	1	6¾
Width of ditto at the base ...	—	8	—	7
" " near the apex ...	—	4½	—	4
Length of frontal bones ...	1	7	1	6
Width between orbits	—	8½	—	8
Length of palate	2	8	2	4
Width of ditto opposite the third molar	—	10		
Length of posterior palatine openings	—	8		
" of the five molar teeth taken together ...	1	3¼	1	3⅓
" of the premolar	—	2¾		
" of the three incisor teeth taken together ...			—	6
" of lower jaw	3	4½		
Height of ditto in a vertical line dropped from the apex of the coronoid process	1	9½	1	8

FIG. 2.—Plate 2 represents the skull as viewed from beneath, half the natural size; 2, *a*, the lower jaw viewed from the outer side; 2, *b*, the lower jaw viewed from above; 2, *c*, the incisor teeth of the upper jaw represented of the natural size; 2, *d*, shews the cutting surface of the same teeth.

MACROPUS (*Halmaturus*) PARMA.

Parma Kangaroo.

Halmaturus Parma. GOULD, in Mr. Gray's List of the Mammalia in Brit. Mus. 1843, p. 91.

Fur moderate; general colour rich rufous brown, pencilled with whitish; much pencilled with black on the back; under parts dirty rusty white; neck and shoulders brownish rust colour, the former with a longitudinal black mark; throat and chest pure white: tail sparingly clothed with short, black hairs above; beneath dirty white; feet brown.

Inhabits New South Wales.

In the list of species of Mammalia contained in the British Museum collection, the name of *H. Parma*, occurs as a species described by Mr. Gould in the Proceedings of the Zoological Society; I have not, however, found any animal noticed under that name in the work mentioned. Mr. Gould informed me that he had doubts whether he had described it. The Parma Wallaby, I think, merits the distinction of a species. It is intermediate between the *H. dorsalis*, and the *H. Derbianus*, and may be distinguished from either by its deep reddish brown colour, and the distinct large white patch on the throat and chest; the hairs forming this patch are white to the root; in *H. Derbianus* they are distinctly grey next the skin, and in *H. dorsalis* they are very slightly tinted with grey at the root in the same parts, and this circumstance, combined with the general form and superior size of *H. Parma*, caused me at first sight to think it might be a variety of the latter animal; I soon perceived, however, that it differed much from *H. dorsalis* in the form and size of its incisor teeth (see Pl. 5, fig. 7), and in the proportion of the tarsus; the three incisors on each side of the upper jaw in *H. Parma* measuring together only five lines, whereas they are $7\frac{1}{4}$ lines in *H. dorsalis;* the tarsus of the present animal is much shorter, as will be seen by the dimensions. The fur is moderate, both as to length and texture; the general colour is deep reddish brown, pencilled with white, and much pencilled with black on the back; on the sides of the body the white is less distinct, and, as the black is wanting, or nearly so, the general hue is paler; the fur on these parts is of a very deep grey next the skin; on the under parts of the body each hair of the fur has the basal half grey, and the external half whitish, but tinted with rust colour; on the throat and fore part of the chest, however, the hairs are uniform white. The back of the neck and shoulders, as well as the fore legs, are brownish rust colour, and there is a narrowish longitudinal black mark, which extends from near the occiput along the back of the neck. The head is ashy grey, tinted with rufous, and finely tinted with whitish; the pale cheek-mark is indistinct; the chin brownish; the back of the ears are clothed with hairs like those of the head; on the inner side there are scarcely any hairs; the few which are seen are whitish. The feet are brown, finely pencilled behind with very pale brown. The tail is very sparingly clothed, and, excepting at the base, the scales are distinct: on the upper surface the small, stiff, scattered hairs are black; quite at the root, the tail is clothed with hairs like those of the body; on the under side the hairs are more numerous, and of a dirty white colour. The foremost and hindermost incisor teeth are about equal in width; the middle incisor is but little more than half the width of the other two; the posterior incisor has an indistinct notch on the hinder third part.

	Inches.	Lines.
Length from tip of nose to root of tail ...	26	0
" of tail	19	0
" of tarsus	6	3
" of ear	2	1
" from tip of nose to ear	4	1
" of fore-arm, including the hands and claws	5	3

MACROPUS (*Halmaturus*) DORSALIS.

Black-striped Kangaroo.

Halmaturus dorsalis.	GRAY, Magazine of Natural History, for November, 1837, Vol. i. (New Series), p. 583.
" "	GOULD, Monograph, Pt. 1, Pl. 8.
Macropus (*Halmaturus*) *dorsalis.*	WATERHOUSE, Naturalists' Library, Marsupialia, p. 230.

Fur moderate; general colour brownish grey; neck and shoulders rusty red; under parts of the body white; fore and hind feet with the toes brown-black; tarsi whitish behind; tail grey above: a distinct black mark, commencing rather low down on the back of the neck, extends about half way along the back.

Inhabits New South Wales.

The *M. dorsalis* is considerably larger than the *M. Derbianus*, which also has the black dorsal stripe; in this latter animal the stripe in question is almost confined to the neck, whilst in *M. dorsalis* (where it is more distinct), it is chiefly confined to the back; the under parts of the body are pure white in the present species, or very nearly so, whilst in *M. Derbianus* they are of a dirty greyish white, and not unfrequently somewhat suffused with reddish. The subjoined description is taken from the original specimen described by Mr. Gray, which is in the collection of the Zoological Society. Specimens brought home by Mr. Gould may now be seen in the National Collection.

The fur is of moderate length, and somewhat harsh to the touch; its predominant hue brownish grey, indistinctly suffused with rufous; the occiput, back of neck, shoulders, and fore legs, are of bright rusty red colour; the chin, throat, and the whole of the under parts of the body, are white, and the hairs on these parts are uniform to the root: a well defined black mark extends along the back, commencing in a point over the shoulders, and terminating about half way down the back: on the cheeks is a whitish mark, which has its origin on the upper lip, and terminates beneath the eye: the ears are clothed internally with whitish hairs, and externally, at the base, with hairs of the same rusty grey tint as those of the head; but, at the apical portion of the ear, the hairs are shorter, and of a dusky colour: the fore and hind feet have the toes brownish black; the tarsi are whitish behind, but freely pencilled with dark brown on the anterior half; the tail is as long as the body, grey above, and of a dirty yellowish tint beneath; rather sparingly clothed above with short stiff hairs, which do not perfectly hide the scales, these being usually developed on this organ in inverse proportion to the hairs; on the under surface the hairs are longer, and more dense; an indistinct whitish mark crosses the haunches externally. The foremost incisor tooth of the upper jaw is nearly twice as broad as the second, and the third is rather broader than the first; it has a vertical groove in the middle of the outer surface, or perhaps even slightly in advance of the middle line—see Plate 5, fig. 14.

	Inches.	Lines.
Length from tip of nose to root of tail	27	1
" of tail	21	0
" of tarsus	7	6
" from nose to ear	4	8
" of ear	2	5

Fine male specimens sometimes exceed these dimensions, being about 30 inches in length from the nose to the root of the tail; and the females usually furnish dimensions rather less than those above given. A skull of a male specimen of *M. dorsalis*, in Mr. Gould's collection, presents the following dimensions:—

	Inches.	Lines.
Total length	4	8
Width	2	4½
" between the orbits	0	9½
Length of nasal bones	1	9
Width of ditto behind	0	7¼
" " near the apex	0	5¾
Length of posterior palatine openings	0	7¾
" of the three incisors of upper jaw ...	0	6⅔
Distance between upper incisors and molars ...	1	1¾
Extent of molars	1	5
Length of premolar tooth, upper jaw	0	3
" of lower jaw	3	2½
Height of ditto, from the apex of the coronoid process	1	8½

Like others of the section of Kangaroos to which the Black-striped Kangaroo belongs, the present species frequents those districts which are well clothed with dwarf shrubs. It is apparently confined to the interior of New South Wales; and is said to be found in great numbers in the scrubs clothing the sides of the hills that run parallel to the rivers Mokai and Namoi; and Mr. Gould found it especially abundant at Brezi, to the northward of the Liverpool plains, and in the Brigaloe brush, on the Lower Namoi. He often shot these animals for food, their flesh being excellent. They are trapped by the natives, both for their flesh and for their skins, which are used for clothing.

MACROPUS (*Halmaturus*) DERBIANUS.

Derby's Kangaroo.

Halmaturus Derbianus. Gray, Magazine of Natural History, vol. i. (New Series), p. 583; Gould, Monogr. Part 1, Pl. 11.

Macropus (Halmaturus) Derbianus. Waterhouse, Marsupialia, Naturalists' Library, p. 234, Plate 21.

Halmaturus Houtmanni. Gould, Proceedings of the Zoological Society for Feb. 1844, p. 31.

Fur long and moderately soft; grey, indistinctly tinted with rust colour; back of the neck, hinder part of back near the root of the tail, the base of the tail, and the limbs, rusty red; under parts of the body dirty white, slightly washed, as it were, with rusty yellow; fore feet and tarsi pale rust colour, very finely pencilled with blackish in front: tail clothed with short hairs, grey above, and dirty white beneath; a broad, but not strongly defined, black mark, extends from the occiput along the back of the neck.

Inhabits Western and South-western Australia.

The original of Mr. Gray's description was said to be from Swan River; it lived for some time in the Earl of Derby's Menagerie, and when it died was presented by his lordship to the Museum of the Zoological Society. Subsequently, a living specimen was presented to the same Society, by its active corresponding member, the late T. B. Harvey, Esq., and specimens have been procured from islands on the West Coast of Australia by Mr. Gould. This gentleman, who has taken much pains in ascertaining the true habits of the various species of Marsupial animals, is of opinion that the *M. Derbianus* is indeed strictly confined to the islands of the West coast, and states that it is found in Kangaroo Island; he had moreover seen the skin of an animal apparently identical with the present species, which he was informed inhabited Rottness and Garden Island. These islands are often covered with dwarf Eucalypti, forming what are called "scrubs," which, as Mr. Gould observes, afford a secure asylum for the small Wallabys. In these scrubs they form runs; and such is the dense nature of the vegetation, that nothing larger than a dog can follow them; still the residents in these islands procure them in great abundance, principally by snares, a simple noose being placed at the entrance of the runs; they are sought both for their flesh, which is well flavoured, and for their skins.

The present species much resembles the *H. Thetidis*, but may be distinguished by the broad dusky mark on the back of the neck, and the red colour of its legs and feet; the fore legs in Thetidis being grey and the feet uniform brown: the prevailing tint of the upper parts of the body is moreover grey, very distinctly pencilled with black and white, whilst in *Thetidis* it is brownish; the fur is usually longer, and that on the under parts of the body is never of so pure a white, exhibiting a slight greyish tint, and being generally more or less suffused with pale rust colour: the hairs on these parts are more distinctly tinted with grey next the skin; the hinder part of the back near the root of the tail is generally of a rufous tint, and so is the base of the tail, but in the *H. Thetidis* these parts are of the same brown tint as the body. The hairs on the back in *H. Derbianus* are grey at the root, shaded then into a blackish hue, which is followed by rust colour, white, and black. The upper lip is whitish, and there is a pale (nearly white) mark on the cheeks. The ears are blackish externally, and the hairs on the inner side are white, or nearly so.

	Inches.	Lines.
Length from tip of nose to root of tail	22	0
" of tail	17	0
" from nose to ear	3	10½
" of ear	2	0
" of tarsus	6	0

Upon a little Kangaroo found in the small islands, called Houtman's Abrolhos, situated on the west coast of Australia, north of the Swan River district, is founded the

Halmaturus Houtmanni. GOULD.

Brownish grey, pencilled with black and white; sides of body and outer side of hind legs suffused with rust colour; back of neck, shoulders, and fore legs, of a bright rust colour; under parts grey-white, slightly tinted with yellowish on the chest: a black stripe runs from near the occiput along the back of the neck: the hairs of the hands freckled with black and rusty white; tarsi of a very pale rusty brown, freckled in front with black; the toes suffused with black: tail black above, freckled with whitish towards the base, and almost entirely black near and at the apex; beneath dirty white: head greyish; cheek-mark moderately distinct; ears grey at the base externally, but dusky beyond. Weight of the male about 11 lbs.; of the female about 8 lbs.

In the female the rusty tint on the neck and other parts is less distinct.

	Male.		Female.	
	Inches.	Lines.	Inches.	Lines.
Length from tip of nose to root of tail	26	0	24	0
" of tail	13	0	12	6
" of tarsus and nails	5	5	5	5
" of the great central toe nail...	0	9	0	10
" from nose to ear	3	10	3	8
" of ear	1	9	1	9
" of fore-arm	4	0	3	6
" of hand and nails	1	10	1	8
" (of which the nail of central finger is	0	7	0	6)
" of the three upper incisors ...	0	$4\frac{2}{3}$	0	5

The specimens above described originally belonged to Mr. Gould, and are now in the British Museum; other specimens from the same locality have received from Mr. Gray the specific name of *Emeliæ,* in the List of Mammalia in the Collection of the British Museum, p. 90. I can discover no difference between these several specimens and the *M. Derbianus,* excepting that generally their colour is somewhat less bright, having but little of the reddish hue, but this is by no means a constant difference. Numerous skulls contained in Mr. Gould's collection I have compared with the skull of a specimen of the *M. Derbianus* belonging to the Zoological Society, and the only difference I could perceive was, that the last mentioned cranium was a trifle larger, and had the nasal bones broader behind; but then I found the skulls of *H. Houtmanni* to vary somewhat in the width of the nasal bones; in some being only 6 lines wide at the base, and in others 7⅓ lines at the same part, and in other species I have found the skulls to vary fully as much as is indicated in the following dimensions.

	M. Houtmanni.				*M. Derbianus.*	
	MALE.		FEMALE.			
	Inches.	Lines.	Inches.	Lines.	Inches.	Lines.
Total length of skull	3	9	3	7	3	11½
" width	2	0½	1	11½	2	1⅔
Width of inter-orbital space . .	0	7¾	0	7½	0	9⅓
Length of nasal bones	1	5½	1	4¼	1	4¼
Width of ditto behind	0	6½	0	7¼	0	9
" " near the apex . .	0	4	0	4½	0	4½
Length of posterior palatine openings	0	6½	0	6⅓	0	5½
" of the three upper incisor teeth	0	4½	0	4⅓	(?)	
" from upper incisors to molars	0	10¾	0	10	0	9¼
" of upper premolar . . .	0	2⅓	0	2¼	0	2½
" of the five molar teeth taken together	1	1⅓	1	0¼	1	2⅓
" of lower jaw	2	7	2	6¾	2	9
Height of ditto, in a vertical line dropped from the coronoid process	1	6	1	5	1	7

For the form of the upper incisor teeth, see Plate 5, fig. 6.

HALMATURUS BILLARDIERI.

Red-bellied Kangaroo.

Kangurus Billardierii. DESMAREST, Mammalogia, p. 452, sp. 843.
Macropus (Halmaturus) Billardieri. WATERH. Marsupialia, Naturalists' Library, p. 227; Halmaturus, id. GOULD'S Monogr. Part 1.
" *rufiventer.* OGILBY, Proceedings of the Zoological Society for February 1838, p. 23; and in Annals of Nat. Hist. for May 1838, p. 220.
Halmaturus (Thylogale) Tasmanei. GRAY, Annals of Nat. Hist. for April 1838, p. 108.
" *brachytarsus.* WAGNER, Schreb. Saug. No. 111-112, p. 121, November 30, 1842.

Ears short and rounded; fur long and rather soft; upper parts of head and body deep brown; under parts chiefly of a rusty red, and sometimes yellowish tint; lips and chin yellowish; feet brown; toes dark brown; tail moderately long, with short and harsh hairs, brown above, dirty yellowish beneath, and somewhat suffused with the same tint at the sides: height, about 18 or 20 inches.

Inhabits Van Diemen's Land.

The Red-bellied Kangaroo is readily distinguished from other small species of its group by its short ears, long dark-coloured fur, and the rufous and sometimes yellow tint of the under parts of the body. It appears to be confined to Van Diemen's Land and some of the islands in Bass's Straits. Mr. Gould regards this animal as strictly a gregarious one; hundreds, he observes, generally inhabiting the same localities. It frequents gullies, and the more dense and humid parts of the forest, particularly those that are covered with rank high grass, through which it forms numerous well-beaten tracks. From these coverts it seldom emerges, and never even approaches the outskirts of the forest, except at night; hence it is seldom seen by ordinary observers. It is very easily taken with snares, in the form of a noose, which are placed in its run; and thousands are captured in this way solely for their skins. It is one of the most highly esteemed for the table, being one of the best flavoured of the small Kangaroos, and is very generally eaten in Van Diemen's Land. Its weight is usually about fifteen or twenty pounds.

The "Wallaby[1]," as our present animal is called by the colonists, was first described by M. Desmarest. The original specimen in the Paris Museum presents the following characters:—Tail shorter than the body; ears short and rounded; fur long and rather soft; general tint very dark; on head, upper parts of body, legs, and upper surface of tail, brown; under parts dirty yellow; lips and tip of chin yellowish white; fore and hind feet brown; toes dark brown; tail brownish yellow beneath; hairs of the fur brown-grey next the skin, each hair annulated with brownish white near the apex, and brown-black at the point; fur like that of the body covers the basal portion of the tail; on the remaining parts the hairs are harsh, shorter, and somewhat adpressed; ears clothed internally with yellowish white hairs; externally the hairs are of the same colour as those on the upper surface of the head.

	Paris Mus.		Zool. Soc.	
	Inches.	Lines.	Inches.	Lines.
Length from tip of nose to root of tail	21	0	25	0
" of tail	10	9	1	$2\frac{1}{2}$
" of tarsus (claws not included)	4	8	5	2
" of ear	1	9	1	8
" from tip of nose to ear ...	3	8		

In other specimens which I have examined, the colouring was found to vary somewhat, especially as regards the tint of

[1] This must not be confounded with the "Wallaby" of New South Wales, which is a distinct species.

the under parts of the body; in some it is dirty yellow[1], and in others of a rusty red tint: the upper parts of the body are sometimes very dark, being much pencilled with black. The posterior outer margin of the ear is of a rusty yellow colour; the remaining portions of the back, or outer surface of the ear, are black, or nearly so. The tarsi are short and stout, and so are the fore legs. The hair of the muzzle terminates in a line with the posterior angle of the nostrils. The second column of the dimensions gives the proportions of a male specimen, the original of Mr. Ogilby's *Macropus rufiventer*. The skull is rather large in proportion to the animal, and affords the following measurements:—

	Inches.	Lines.
Total length	4	1½
Width	2	1½
From front of anterior incisor to the posterior palatal opening	1	7
Width between anterior molars	0	6½
" between posterior ditto	0	9¼
Length of the five molars of the upper jaw taken together	1	1¾
" of the three incisors of either side of upper jaw taken together	0	5
" of lower jaw, from condyle to base of incisor	2	11
" from hinder part of condyle to fore part of coronoid process	1	0
Height in a vertical line dropped from apex of coronoid process	1	6

The upper incisor teeth are represented in Plate 5, fig. 11.

[1] A specimen which recently died in the Zoological Society's menagerie has the under parts of the body almost of a cream colour.

MACROPUS (*Halmaturus*) BRACHYURUS.

Short-tailed Kangaroo.

Kangurus brachyurus. QUOY et GAIMARD, Voyage del'Astrolabe, Zoologie, tom. i. p. 114, Plate 19.

Halmaturus (*Thylogale*[1]) *brevicaudatus.* GRAY, Catalogue of the Mammalia in the British Museum.

About equal in size to the Common Rabbit. Ears short and rounded, and rather densely clothed internally with rusty yellow hairs: fore feet proportionately rather large; hind feet short; both anterior and posterior extremities of a darkish brown colour: tail short and slender, and sparingly clothed with small hairs, which do not hide the scales: fur long, somewhat glossy, and soft; its general tint, on upper parts of body, brown, and on under parts, whitish, but suffused with yellow: on the head is a slight rusty tint, especially in the region of the ears.

Inhabits the region of King George's Sound.

This species is remarkable in its group for its short and comparatively slender tail, which is not more than about an inch and a half in circumference at the base, and resembles

[1] In the Magazine of Natural History, vol. i. (New Series), p. 583, Mr. Gray separates, as the type of a sub-genus of *Halmaturus*, the *M. Thetidis* (supposed at that time to be the *M. Eugenii*), and gives as the characters of the section, "*the hinder fold of the hinder upper cutting tooth scarcely larger than the front one, so that the tooth appears only notched behind: front incisor short, simple.*" To this sub-genus the name *Thylogale* is applied; and, in the List of the Mammalia in the British Museum, the present animal (*M. brachyurus*) is associated with *M. Thetidis*, as a second species of *Thylogale*. This section I cannot adopt, because it is founded upon characters which are too variable in nearly allied species of Kangaroos to be of importance; and the two species which are associated together are by no means nearly related, if we may judge from the structure of the skull and extremities; nor do they even agree in the structure of the incisor teeth—compare fig. 16 *a* of Plate 5 with fig. 2 *c* of Plate 3; and, beyond these objections, I may add, that the species selected as the type of the section *Thylogale*, must, in fact, be regarded as the type of the section *Halmaturus*, as defined by F. Cuvier.

that of a Rat, in being sparingly clothed with minute stiff hairs, between which, rings of small blackish scales are distinctly visible. The shortness of its ears also renders it conspicuous; these are rounded, and much hidden by the long fur of the head; they are well clothed with hairs, and those on the inner side are yellow, or rusty yellow; externally they are of the same reddish-brown tint which prevails on the head and back of the neck, but is somewhat brighter in the region of the ears. The hairs on the back are grey next the skin, broadly annulated with yellow towards the point, and black at the point, but there are numerous long interspersed hairs, which are almost entirely black, and as these are most plentiful on the middle of the back, they give to that part a deeper hue: on the sides of the body the hairs are coloured in the same manner, but the yellow portion is paler, and the tips are brownish. The hairs on the under parts of the body are grey next the skin, and of a very pale yellow tint externally. The feet are clothed with deep brown, short hairs. The head is proportionately rather large.

Until recently the *M. brachyurus* was unique in the Paris Museum, where the original specimen, brought home by the Astrolabe Expedition, is deposited. This specimen was found dead at King George's Sound, whence a specimen has been since procured by Mr. Gould, which is now deposited in the British Museum. From this specimen the above description was drawn up, and upon comparing it with my notes made upon the Paris individual, I find them to agree perfectly, with the exception of some slight differences in the dimensions. The measurements of the two specimens are as follows:—

	Brit. Mus. Inches.	Brit. Mus. Lines.	Paris Mus. Inches.	Paris Mus. Lines.
Length from nose to root of tail ...	15	6	18	0
" of tail, about	7	6	7	9
" of tarsus	3	2	4	1½
" of ear	1	3	1	2

The length of the skull, according to Messrs. Quoy and Gaimard's Plate, is 3 inches and 4 lines.

A skull of the *M. brachyurus*, in the British Museum collection (see Plate 5, fig. 16), is remarkable for the shortness of the muzzle, which, though narrow in front, is broader behind than usual; the zygomatic arches are thrown boldly outwards, and are very long, as compared with other Kangaroos: the greatest breadth of the skull is towards the posterior root of the zygomatic arches; the frontal bones are narrow; the nasal bones short, convex above, but little dilated behind, and with the sides very nearly straight. The foremost and hindermost of the upper incisor teeth are about equal in width; the second incisor is narrower; the third has a distinct notch at the apex, and situated rather behind the middle of the tooth (fig. 16*a* of Plate 5); the premolar tooth has a greater transverse diameter than usual, and, indeed, is as broad as the first true molar, and its length is nearly the same. The true molars have the anterior ridge, and the small mesial ridge, very little developed.

	Inches.	Lines.
Length of skull, about[1]	3	0
Width near the posterior root of the zygoma ...	1	$8\frac{3}{4}$
" beneath the orbits	1	5
" of interorbital space	0	$5\frac{1}{2}$
Length of nasal bones	1	$0\frac{1}{2}$
Width of ditto	0	$4\frac{1}{4}$
" near the apex	0	$2\frac{2}{3}$
Length of zygomatic arch	1	$7\frac{3}{4}$
" from the front of the zygomatic arch to the apex of the intermaxillaries	0	$11\frac{1}{2}$
" of the three upper incisors, taken together	0	4
Distance between upper incisors and premolar ...	0	5
Length of the five molar teeth, taken together ...	0	$10\frac{3}{4}$
" of the lower jaw	1	$10\frac{1}{3}$
Height in a vertical line dropped from the apex of the coronoid process	1	$1\frac{2}{3}$
Width of coronoid process, from front to back ...	0	$5\frac{3}{4}$

[1] The occipital bone is wanting, and allowed for in above dimension.

In the British Museum is a small Kangaroo, which appears to me to be specifically identical with the *M. brachyurus,* but which differs in being of a rich reddish brown colour. The fur on the back is pencilled with bright rust colour and black; on the under parts of the body it is grey, but suffused with very pale rust colour, the hairs being grey at the root, and tipped with the brighter colour; the ears are of a bright rust colour internally, and the head is almost entirely of the same hue, being but slightly pencilled with black: the fore and hind feet are brownish black, but rusty at the sides: the tail is brown-black above, somewhat reddish at the sides and at the base, and of a dirty rusty white beneath. The specimen is a female, and evidently far from adult. Its dimensions are—

	Inches.	Lines.
Length from nose to root of tail, about	15	0
" of tail, about	7	0
" of tarsus and nails	3	6
" from nose to ear, about	2	9
" of ear, about	1	1

From Augusta, Western Australia.

Sub-genus HETEROPUS.

Heteropus[1]. JOURDAN, "Comptes Rendus" for October 9, 1837, p. 522; and Annales des Sciences Naturelles, tom. vii. p. 368.

Petrogale[2]. GRAY, Magazine of Natural History for November, 1837, vol. i. (New Series), p. 583.

Kangaroos with muffle naked; the hind foot short and stout, and densely clothed with coarse hairs; the nails smallish: tail cylindrical, and clothed with long hairs, especially on the apical portion.

Inhabitants of rocky situations.

[1] From ἕτερος, altered; and ποῦς, the foot—having allusion to the somewhat abnormal structure of the hind foot, observed in the species of this section.

[2] From πέτρος, a stone; and γαλῆ, a weasel, or cat.

Nearly at the same time Mr. Gray and M. Jourdan characterised the present division as a distinct genus; but to Mr. Bennett is due the credit of having first drawn attention to the peculiarities presented by the *M. penicillatus*, which is the type of the section[1].

We have described the plain-inhabiting Kangaroos, and those which frequent the "scrubs;" the present section contains Kangaroos which are fitted for inhabiting the rocks. Whilst the Kangaroos of the plain have the fore part of the body slender and light, great strength in the hinder parts, combined with a long leg and foot, adapting them to fleetness, the tail powerful, and assisting in the support of the long body, we perceive certain modifications in the form and structure of these parts in the Rock-Kangaroos which adapt them to their particular habitats: the body, more compact in form, requires but little assistance from the tail for its support, the leverage being less, and the hind feet are, though powerful, comparatively short, and remarkably rough beneath, being thickly covered over this part with hard tubercles, which no doubt prevent the foot from slipping: the nails of the two larger toes are shorter than usual, and, indeed, in some of the species, scarcely project beyond the fleshy pads with which the toes are terminated, and on the upper surface of which the nails are placed. A long and slender foot, with long nails, as in the typical Kangaroos, it is obvious, would be ill adapted to an animal which has to leap to, and balance itself upon the small ledges of the rocks. The tail is large, but not thickened at the root, as in the plain Kangaroos; and, unlike the tail in those animals, it is clothed with long hairs, which, gradually increasing in length from the base of the tail, become very long and bushy at the opposite extremity: it serves to steady the animal in its leaps, and to balance the

[1] See the Proceedings of the Zoological Society for January, 1835.

body when perched in situations which require it, but is of little assistance in supporting the weight of the trunk. The muffle is naked, as in the scrub-inhabiting Kangaroos.

In the skull in the Rock-Kangaroos, the muzzle and nasal bones are narrow, and the zygomatic arches are more prominent than usual; but in these respects the skull in the Rock-Kangaroos greatly resembles that of *M. leporoïdes* figured in Plate 5, fig. 17; that skull, however, when viewed from beneath, presents a rare exception among the Macropus group, in having the auditory chamber expanded, and enclosed externally by a thin and nearly spherical bony plate; a character in which it approaches the skulls of the Rat-Kangaroo group, but which is not found in the *Heteropus* section. The angle of the lower jaw is less raised than in most other Kangaroos. The two posterior upper incisor teeth are rather small; the last presents a deep notch in the crown (see Plate 5, fig. 10).

MACROPUS (*Heteropus*) PENICILLATUS.

Brush-tailed Rock-Kangaroo.

(*Plate* 1.)

Macropus penicillatus. GRAY, in Griff. Ann. Kingdom, v. p. 527.
Petrogale penicillata. GRAY, Magazine of Nat. Hist. for November 1837, Vol. 1. (New Series), p. 583.
" " GOULD, Monogr. of the Macropodidæ, Part 2, eighth Plate.
(?) *Heteropus albogularis.* JOURDAN, Comptes Rendus, &c. for October 1837, p. 552; and Annales des Sciences Naturelles for December 1837, tom. viii. p. 368.

Fur long; general colour deep purplish grey; chin, mesial line of throat, and chest, white; sides of body sooty brown, almost black immediately behind the fore leg; abdomen brown or yellowish; muzzle and occiput dusky; cheek-mark greyish white; ears with pale yellow hairs internally; externally black, but broadly margined behind and at the apex, with yellow, and greyish at the base; tail long and very bushy, being clothed with long, coarse, black hairs, excepting at the base, where they resemble those of the body; feet black, or nearly so, the tarsi densely clothed with long coarse hairs; back of the neck with an indistinct blackish mark.

Inhabits New South Wales.

The first specimens of the Brush-tailed Kangaroo brought to Europe are probably those contained in the museum of the Linnæan Society; more recently a specimen was presented to the museum of the Zoological Society by Sir Edward W. Parry, and which being exhibited at one of the Society's scientific meetings[1], Mr. Bennett called attention to the peculiarities of the tail, and the difference in the structure of the upper incisor teeth, as compared with those of the Great Kangaroo. In a note by Sir Edward W. Parry, which accompanied the specimen, that gentleman states that it was shot among rocks near Liverpool Plains, New South Wales. "As several of the same kind were seen together on more than one occasion, they appear to be gregarious. They seemed to prefer the neighbourhood of rocky ground, in which they had holes, to which, when hunted, they retreated. The first intimation received of these animals, by a gentleman referred to in the note, was, that monkeys were to be seen in a particular situation; and the manner in which they jumped about when he first approached a number of them left the same impression on his mind." They were so wild that he experienced great difficulty in obtaining a specimen.

The specimens of the Brush-tailed or Rock Kangaroo in the British Museum were also procured by Mr. Gould from the Liverpool Range; and others were obtained on the sides of the mountains facing Yarrundi on the Dartbrook, a tributary

[1] Proceedings of the Zoological Society for January 1835, Part 3, p. 1.

of the Hunter. Mr. Gould has likewise ascertained that it is very abundant on the Turi, and the other mountains situated to the eastward of the Liverpool Plains. According to the gentleman last mentioned, the Rock-Kangaroo is strictly gregarious, assembling in such numbers as to form well-beaten paths along the sides of the mountains. " Their agility in leaping from rock to rock," Mr. Gould observes, " is truly surprising, often alighting upon ledges so slight and narrow that it appears almost impossible for them to retain their footing: this power tends greatly to their protection, as neither the wily aborigine, nor their still greater enemy the Dingo, can follow them to their retreats. When closely pursued, and during the heat of the day, they seek shelter in the crevices and caverns of the rocks: to the caverns they evince a great partiality, usually, however, selecting those which have more than one outlet; a precaution rendered necessary by the frequent intrusion of the Dingo, who also resorts to similar situations. The *Macropus penicillatus* is strictly nocturnal in its habits, and during the night frequently leaves its well-beaten tracks among the rocks for the grass beds on the crowns or at the base of the mountains, but never so far from its haunts as to be unable to retreat again on the slightest alarm. It also readily ascends the trunks of sloping trees, wearing a kind of track or path on the back."

" As an article of food, the flesh of this animal is most excellent."

A specimen of the Rock-Kangaroo which was living in the menagerie of the Zoological Society in 1836, displayed the goat-like propensity of perching itself upon some small ledge, and appeared to be remarkably fond of leaping on to a narrow shelf which was raised about three feet from the ground in its den: on this it would sit and balance itself, although it would have appeared almost impossible that it could do so; and, after a short time, it would descend to leap up again

In leaping and perching itself, the tail evidently assisted in balancing the body; on the other hand, owing to its want of strength at the root, it was of but little assistance in supporting the body, and in slow progression, as compared with the same organ in the typical Kangaroos.

The animal under consideration is subject to some variation in its colouring; sometimes there is only a small white spot on the chest; the tail and feet are brownish, and the abdomen of a pale dirty yellow colour. The tail is generally brownish beneath, excepting at the apex, and often has the under surface at the root of a bright fulvous colour. The sides of the body present scarcely a trace, in some individuals, of the dusky hue, excepting near the fore leg; and just above this axillary patch is a whitish mark, more or less distinct. The tarsi are usually paler behind than in front.

	Female.		Female.		Male.	
	Inches.	Lines.	Inches.	Lines.	Inches.	Lines.
Length from nose to root of tail .	24	0	25	0	28	6
" of tail (without including the hair, which is sometimes three inches long at the tip)	21	0	18	0	24	0
" of tarsus and claws . . .	5	6	5	8	6	10
" from nose to ear	4	4	3	11	4	6
" of ear	2	2	2	0	2	9
" of fore arm, hands, and claws, about	4	9	4	9	5	0

The above dimensions have been taken from specimens in the museum of the Zoological Society and in the British Museum. Those in the second column are from a female in the latter collection, and were taken with a view to afford a comparison with a female (apparently adult) of the *M. lateralis*.

	Inches.	Lines.
Length of skull	4	2
Width of ditto	2	2½
" of interorbital space	0	9
Length of nasal bones	1	10
Width of ditto behind	0	5⅔
" near the apex	0	3¼
Length of posterior palatine openings	0	7¼
" of the three upper incisors taken together	0	5¾
" from the posterior incisor to the premolar	0	10½
" of premolar	0	3½
" of the anterior three true molars taken together	0	9¾
" of lower jaw	2	9
Height of ditto from apex of coronoid process	1	5½

The foremost upper incisor tooth is equal in width to the third, and much arched in front; the second is much narrower; the third has a deep notch on the outer side at the apex, which is situated on the posterior third of the tooth; the premolar is longer than the foremost true molar (see Pl. 5, fig. 10)

Heteropus albogularis. JOURDAN.

This species is described as having a longitudinal brown line on the head; the cheeks whitish; ears black externally, yellow internally; throat white; chest and belly rust colour; neck, and upper parts of back, grey; lower part of abdomen rusty yellow; feet and tail deep brown, the latter terminated with white.

	Inches.	Lines.
Total length, including the tail	39	7
Fore legs	4	9
Hind legs	11	10½
Body	23	9
Tail	22	2
Tarsi	3	2
Head	4	4¼

Inhabits the mountains south-west of Sydney.

I feel very little doubt that the above description is drawn up from an animal which does not differ specifically from the *M. penicillatus* of Mr Gray; the white termination to the tail, which is the only difference worthy of notice, is very probably an individual peculiarity.

MACROPUS (*Heteropus*) LATERALIS.

Black-flanked Rock-Kangaroo.

Petrogale lateralis. GOULD, Monograph of the Macropodidæ, Pt. 2, 9th Pl.

Fur moderately long and soft ; general colour brownish grey ; head and fore parts of the body almost destitute of the brown hue, which, somewhat admixed with rusty yellow, is conspicuous on the hinder parts : a distinct white cheek-mark is followed above by a broad blackish band running through the eye : a black mark runs from the occiput along the back of the neck ; sides of the body with a broad brownish black band, and with a narrower white band immediately above this ; mesial line of abdomen of a pale buff-yellow colour : feet brown-black, more or less suffused with rich brown in parts : tail but little bushy at the basal half, which is nearly of the same general hue as the body ; the apical half bushy, having long black, or nearly black, hairs.

Inhabits Western Australia—Swan River district.

Mr. Gilbert states that this species is only to be met with amongst the rocks in the interior, which are intersected with caverns; that it is remarkably shy, seldom venturing out during the day, but feeds at night in the little open patches of grass. He had never known it to stray more than two or three hundred yards from its retreats. When alarmed, it leaps most extraordinary distances from rock to rock, and with the greatest rapidity.—*Gould's Monogr.*

The Black-flanked Rock-Kangaroo greatly resembles the Brush-tailed species of New South Wales, but differs in being rather smaller, in having the fore part of the body more grey, the black mark on the back of the neck more distinct, as well as the black and white flank bands; the tail, moreover, is much less bushy. The top of the head, back of the neck and shoulders, and the front of the fore legs, are grey; the hinder parts of the back are of a deeper hue, and are much suffused with yellowish brown; here the hairs are of a deep brown colour next the skin, and fulvous towards the point; this is followed, on each hair, by a broad white space, and the tips of the hairs are black. In front of the eye is an oblong black patch, and immediately behind the eye, the fur is also dark, but less so than in front; below this dark band thus formed, is a distinct white cheek-mark. The ears are black externally, excepting at the base, where the fur is of a brownish white, and sometimes of a yellowish tint; on the inner side of the ears, the hairs are dirty white. The occiput is nearly black, and a line of the same hue runs from this part along the back of the neck, and extends somewhat on to the back. The hinder part of the fore legs is sooty black, and a broad brownish black band runs from thence along the flanks (encroaching somewhat upon the abdomen) to the hind legs; above this band is a narrow pale mark, which is almost pure white at first, near the fore leg, but becomes less distinct, and suffused with brownish, as it approaches the hind legs. The fore feet and toes are black, more or less mottled with rich brown; sometimes almost entirely of the latter colour. The tarsi are densely clothed with long black-brown hairs; the upper surface of the foot, however, is usually suffused with brown in the middle. The nails of the central and outer toes are short, and nearly conical; they scarcely project beyond he fleshy pad of the toe, and are hidden by the long hairs of the toes. The basal half of the tail is well clothed, but by no means bushy, and the general colouring of this part is almost the same as that of the body; the apical half is bushy and black; sometimes more than half the tail is black, or nearly so. The throat and mesial line of the abdomen are of a buff-yellow colour, inclining to white; the fore part of the neck and chest are usually greyish, but the latter is sometimes yellowish in the middle, and blackish at the sides.

	Inches	Lines.	Inches.	Lines.
Length from tip of nose to root of tail	24	0	23	0
" of tail	17	0	16	0
" from nose to ear	4	0	3	8
" of ear	1	10½	2	0
" of tarsus	5	3	5	0
" from end of nail of outer toe to end of that of great central toe			1	3
" from end of nails of inner double toe to the same			1	4
" of fore arm, hands, and claws, about	4	9	4	0

The first of the two columns of dimensions gives the size and proportions of a male—from Mr. Gould; the second, those of a female in the British Museum collection.

The frontal bones are very concave between the orbits in a skull of the *M. lateralis*, in Mr. Gould's collection; the nasal bones are very narrow in front, and broad behind, and at this part there is a small opening on either side between the nasal bones and the nasal process of the superior maxillaries; the posterior palatine openings are large. This skull belonged to an animal which was not quite adult: its dimensions are as follows:—

	Inches.	Lines.
Length of skull	3	7
Width of ditto	2	0
" between the orbits	0	8½
Length of nasal bones	1	6
Width of ditto behind	0	6½
" " near the apex...	0	2⅔
Length of the three upper incisors, taken together	0	4½
Distance between the incisor and molar teeth	0	8½
Length of the lower jaw	2	6
Height of ditto from the apex of the coronoid process	1	4

MACROPUS (*Heteropus*) INORNATUS.

Unadorned Rock-Kangaroo.

Petrogale inornata, GOULD, Proceedings of the Zoological Society for January, 1842, Pt. 10, p. 5. Monograph of the Macropodidæ, Pt. 2, Pl. 10.

General colour sandy grey, grizzled over the shoulders, and becoming much lighter on the flanks; an indistinct line of a lighter hue runs along the face under the eye; a dusky red patch behind the elbow; under parts sandy white inclining to rufous on the lower part of the abdomen; arms and tarsi sandy grey passing into dark brown at the extreme tips of the toes; basal half of the tail sandy brown, the remainder black, the former colour extending along the sides of the tail for some distance towards the tip; ears sandy grey, bordered by a very narrow line of dark brown on the inner edge; a dark patch at the occiput passing into a dark line down the forehead.—GOULD.

	Inches.	Lines.
Length from nose to tail	22	9
" of tail	15	3
" of tarsus, toes and nails	5	3
" of arm, hand and nails	5	0
" from tip of nose to base of ear ...	4	6
" of ear	1	10½

Inhabits the north coast of Australia.

This animal is about equal in size to the *M. lateralis*, and is readily distinguished from that and other species, (with the exception of *M. concinnus*,) by the absence of markings on the sides of the body, and the pale colour of the back of the ears, as well as of the feet. It was discovered by Mr. Bynoe, of H.M.S. Beagle, on the north coast of Australia, and is deposited by that gentleman in the British Museum.

MACROPUS (*Heteropus*) BRACHIOTIS.

Short-eared Rock-Kangaroo.

Macropus (*Petrogale*) *brachyotis*. GOULD, Proc. Zool. Soc. for Oct. 1840, Pt. 8, p. 128. Monogr. Pt. 1, 6th Pl.

Fur short, and rather closely applied to the body; general tint of upper parts, ashy brown, suffused with vinous or purplish rust colour; sides of the body, pale vinous grey; under parts dirty yellowish white: head pale brown, exhibiting the usual whitish cheek mark; the cheeks almost white; occiput with an indistinct dusky line; ears short and pointed, with pale hairs internally; externally dusky: a rusty black patch on the body immediately behind the base of the fore leg; fore feet brown; nails of the toes very short, and scarcely projecting beyond the fleshy portion, which is extremely rough beneath: tail with the basal half grey; the apical half bushy, having stiff black hairs, averaging about one inch in length.

Inhabits the north-west coast of Australia.

The specimens upon which Mr. Gould founded the *Petrogale brachiotis* were presented to him by Capt. George Grey, Governor of South Australia, who procured them at Hanover Bay, on the north-west coast, while traversing that previously unexplored region. Capt. Grey observes, that it is a very wild and shy animal, frequenting, in the day time, the highest and most inaccessible rocks, and only coming down to the valleys to feed early in the morning, and late in the evening. When disturbed in the day time it bounds among the roughest and most precipitous rocks, apparently with the greatest facility, and is so watchful and wary that it is by no means easy to get a shot at it. The heat of the sand rocks, amongst which it is always found, is, it is stated, very great, amounting in the hottest part of the day sometimes to 136°

The short-eared Rock-Kangaroo is readily distinguished from the *M. penicillatus* and *M. lateralis*, by the absence of the black band on the sides of the body, the only remains of this dark hue being confined to a patch immediately behind the base of the fore leg; its general colouring is paler, and the fur is much shorter. The tail is less bushy; its bulk is moreover inferior; and the proportionately small size of the ears is an important distinguishing character. In its smaller size, and in the reddish hue of the upper parts of the body, it approaches the *M. concinnus*, but, besides other differences pointed out in the description, that animal does not possess any dark mark or spot on the sides of the body.

	Inches.	Lines.
Length from the tip of the muzzle to the root of the tail	21	0
" of tail	16	6
" from tip of the nose to the ear	3	8
" of ear	1	11
" of tarsi, without including the claws ..	5	0

MACROPUS (*Heteropus*) CONCINNUS.

Rufous Rock-Kangaroo.

Petrogale concinna. GOULD, Proc. of the Zool. Soc. for May, 1842, p. 57.

Fur moderately long and somewhat soft; general colour bright rusty red, but the fur of the back is pencilled with white and reddish black; under parts of body yellowish white; limbs and sides of body of a pale rust colour, the fore legs and hands inclining to white: tarsi brownish white, slightly pencilled with brown; sides of head rusty white: tail bushy, the apical third clothed with hairs, which are chiefly whitish, but tipped with black.

Inhabits the North-west coast of Australia.

I have seen but one specimen of this species of Rock-Kangaroo, which is so remarkable for its brilliant colouring and small size, being scarcely equal in bulk to most of the Kangaroo-rats: this specimen was brought to England by Lieut. Emery, of H. M. S. Beagle, and is now in the British Museum: no peculiarities in its habits are mentioned in the only published account which has as yet appeared.

The *Macropus concinnus* may be readily distinguished from its congeners, not only by its small size and bright colouring, but by the absence of any black spot behind the base of the fore leg, The head is of a palish ash colour above, slightly suffused with rust colour, this latter tint being most conspicuous above the eyes; the cheeks are rusty white, and have an indistinct greyish brown mark extending forwards from the front of the eye; the ears are of moderate length, narrow and somewhat pointed, brownish (but very pale) externally, and with a few white hairs internally. The fur on the back is grey next the skin, and this tint, at the root of each hair, is followed by brilliant rusty red, then a broad space which is white, and the tip is deep rusty brown; on the under parts of the body the fur is grey next the skin, and has the visible portion yellowish white. The fore legs are rusty white, and the hands a brown-white; the hind legs are of a pale rust colour externally; the tarsi are brownish white, but slightly pencilled with brown: on the back of the neck is an indistinct trace of a mesial darker mark. The tail is clothed at the base with fur like that of the body; beyond this the hairs are of a harsher nature, at first about half an inch in length, and on the apical third they are about an inch and a half in length, of a brownish white colour, but tipped with black.

	Inches.	Lines.
Length from nose to root of tail	13	9
" of tail	9	9
" of tarsus	4	0
" from tip of nose to ear	3	0
" of ear	1	3
" of fore legs and hands to end of nails .	3	6
" of skull	2	11
Width of ditto	1	$8\frac{1}{3}$
" between orbits	0	$7\frac{1}{4}$
Length of nasal bones	1	$1\frac{1}{4}$
Width of ditto behind	0	7
" " near the apex	0	$2\frac{1}{4}$
Length of palate	1	$6\frac{1}{3}$
" of posterior palatine openings ...	0	$4\frac{2}{3}$
" of three anterior incisors	0	$3\frac{2}{3}$

The animal to which the above skull belonged was not quite adult, since the crown of the last molar tooth is but little raised above the level of its alveolus, and what is rather remarkable, there are four teeth in front of this on one side of the upper jaw (on the opposite side it has been thrust out by the pressing forwards of the hinder teeth,) and the foremost of these has the form and structure of a true molar, and yet it does not appear to be a milk tooth. The foremost of the three upper incisors is the largest, and the hindermost has a deep groove behind the middle. The muzzle is short, and much contracted in front; the nasal bones are much dilated behind, and there is a small opening on either side at this part between the nasal bones and the superior maxillary, as in *M. lateralis;* in front, the nasal bones are very narrow: the zygomatic arch is thrown boldly outwards beneath the orbits, and the lateral ridges of the frontal bones are parallel; these bones are nearly flat between the orbits. The occipital opening is notched above.

As compared with the skull of *M. penicillatus,* that of *M. concinnus* differs in having the muzzle shorter in proportion, the nasal bones shorter, and much broader behind, the interorbital space broader: in *penicillatus* the frontal bones are slightly contracted behind the orbits, but this is not the case in *M. concinnus*.

MACROPUS BRUNII. Le Brun's Kangaroo.

Filander.	LE BRUN, Voyages par Moscovie, en Perse, et aux Indes Orientales, tom. ii. p. 347. 1718.
Didelphis Brunii.	SCHREBER, Saugeth. iii. p. 551, Pl. 153. 1778.
" *Asiatica.*	PALLAS, Act. Acad. Sci. Petrop. for 1777, Pt. 2, p. 229, Tab. 9, figs. 4 and 5. 1780.
Javan Opposum.	PENNANT, Hist. of Quad., p. 305. 1781.
Didelphis Brunii.	GMEL., Linn. Syst. i. p. 109. 1788.
" "	(Javan Opossum.) SHAW, Gen. Zool., i. Pt. 2, p. 480. 1800.
Macropus veterum.	LESSON, Manuel de Mamm. p. 227. 1827.
" *Bruinii.*	FISCHER, Syn. Mamm., p. 283. 1829.
Didelphis Brunii.	QUOY et GAIM., Voy. de l'Astrolabe, Zool. p. 116, Pl. 20. 1830.
Hypsiprymnus Brunii.	MÜLLER, Zoogd. der Indesch. Archipel. Pt. 4, Pl. 21, Head, Pl. 22, fig. 3; Skull, Pl. 23, figs. 7 and 8, and Pl. 24, fig. 7; Bones of the hind leg, figs. 8 and 9. 1841.
Halmaturus Asiaticus.	GRAY, in List of Mamm. in Brit. Mus. p. 91. 1843.

Head narrow, and very long; ears short; tail moderate; fore legs strong: fur very short, soft, and composed of hair almost entirely of one kind, radiating from a point on the mesial line of the back a little behind the shoulders; general colour greyish brown suffused with yellowish, especially on the sides of the body; under parts pale dirty yellow: ears blackish externally.

Inhabits New Guinea.

This singular animal is the first of the Kangaroo family with which naturalists became acquainted, being imperfectly described, but better figured, as early as the year 1711, by Le Brun; its characters were subsequently more carefully pointed out by Pallas, and it is upon the accounts of these two authors that all the various descriptions and notices in systematic works,[1] chiefly under the specific names *Filander* and *Brunii*, have been founded until a comparatively recent period. Several specimens of the Filander were seen, in a state of captivity, at Batavia, by Le Brun; these, however, must have been transported from New Guinea, whence it has since been procured during the French expedition of the Astrolabe, and still more recently by the naturalists of an expedition sent out by the Dutch Government, an expedition which has added much to our knowledge of the natural history productions of the islands in the Indian Archipelago. One of the specimens of this last mentioned expedition is now in the British Museum, and enables me to give an original description. The dimensions of this specimen are as follows:—

	Inches.	Lines.
Length from the tip of the nose to the root of the tail	29	3
" of tail	18	3
" of tarsus	6	0
" from nose to ear	5	1
" of ear	1	10
" of fore leg, from elbow to ends of fingers	7	6

The Filander, like the Tree-Kangaroos, has the fur radiating from a point rather behind the shoulders, and the hair on the neck directed forwards as in those animals. The fur is remarkably short, rather soft, and has very little gloss; on the crown of the head the hairs have their points directed inwards and backwards, and there meeting the hairs of the neck, which have the points directed forwards, a small tuft is formed at

[1] Desmarest must be excepted, he having described a very distinct species (*Macropus Ualabatus*) under the name *Macropus Brunii*, supposing that it was identical with the New Guinea animal.

their point of junction, which is on the back of the head. The general tint of the animal is brown, by no means dark, and slightly inclining to grey-brown on the back; the sides of the body are of a somewhat brighter colour, being slightly tinted with yellowish; the whole of the under parts, as well as the fore legs and feet, are of a dirty yellowish white; the hind legs are of the same tint externally as the sides of the body, but paler on the inner sides; the tarsi are of an uniform palish brown. The ears are rather small, and rounded at the tip, clothed externally with short and almost velvet-like black hairs; on the inner side of the ears are but few hairs, and these are greyish. The tail is well clothed with short and soft hairs, brown on the upper surface, and brown-white beneath; on the sides of the tail, the hairs, instead of pointing backwards as usual, are directed upwards; the tip of the tail is almost destitute of hair, (apparently worn off by friction,) and exhibits the scales very distinctly. The head is of a pale brown colour, and the muffle is naked.

In the large size of the premolar tooth, and in the possession of a distinct canine, *Macropus Brunii* approaches the *Hypsiprymnus* group, in which it has been arranged by Dr. Müller, and the elongated form of its skull would lead one to compare it with the *Hypsiprymnus minor*, but although the skull is elongated in both instances, there are many important differences observable in its structure when the two are compared. The zygomatic arch is deeper and longer than in *H. minor*, and the orbit is more advanced; the nasal bones terminate in a line with the anterior boundary of the orbit, whilst in *H. minor* the root of the nasal bones is situated considerably in advance of the same point. In the structure of the intermaxillary bones, of the true molar teeth, (judging from Dr. Müller's figures) and of the auditory bullæ, *M. Brunii* is conformable with the Macropus type, and to these points, which separate *M. Brunii* from *Hypsyprymnus*, we may add the structure of the anterior extremities. On the other hand, in the peculiar form of the nasal bones, which are contracted in the middle, and expanded almost equally at each extremity, and the broad interorbital space, the *M. Brunii* evinces an affinity with the Tree-Kangaroos (*Dendrolagus*) which inhabit the same country. This affinity is also indicated in the comparatively large size of the premolars (if the dotted lines in Dr. Müller's figures indicate the size and form of these teeth correctly), and in the presence of canine teeth. In *Dendrolagus,* moreover, we find the zygomatic arch narrow in the vertical direction, as compared with the true Kangaroos, and in this respect they agree with *M. Brunii,* as well as in the form of the lower jaw, which has the angle much less raised than usual; indeed the skull of a Tree-Kangaroo, if more elongated, would more nearly resemble that of *M. Brunii,* than any other species of the *Macropodidæ.* The upper incisor teeth are remarkably small, and the posterior incisor apparently has no external vertical groove in *M. Brunii,* and an approach to this animal is further evinced in the Tree-Kangaroos, in the structure of these teeth, and more particularly in the *Dendrolagus ursinus.* Lastly, I may notice that the fore legs are comparatively large in *M. Brunii,* and that the hairs radiate from a centre, in the back of the neck, in the same way as in *Dendrolagus.* The animal under consideration, however, I must observe, differs from the species of *Dendrolagus* in having the muffle naked, and in possessing tolerably long posterior palatine openings to the skull. The dimensions of the skull of *M. Brunii,* taken from Dr. Müller's figures, are as follows:—

	Inches.	Lines.
Length of skull	5	2
Width	2	2
Length of nasal bones	1	11
Width at base	0	$5\frac{1}{4}$
Width near apex	0	$4\frac{1}{2}$

	Inches.	Lines.
Width of interorbital space	0	11
Length of palate	2	9
" of three incisor teeth, of either side of upper jaw, taken together ...	0	4
" from incisors to canine	0	5
" of canine	0	1
" from canine to molars	0	8
" of premolar	0	6¼
" of the four true molars	1	0½
Width of palate between the foremost molars	0	9
Length of posterior palatine openings about	0	6½

Genus, *Dendrolagus*[1].

Dendrolagus. MÜLLER, Verhandelingen over de Nateurlijke geschiedenis der Nederlandische overzeesche bezittingen.

Kangaroos with the anterior extremities large and powerful, being but little inferior in size to the posterior extremities; the claws of the fore feet very large, curved, and pointed; those of the two principal toes of the hind feet somewhat compressed, and curved; the hind foot, tibia, and femur, nearly equal in length: muffle clothed with small hairs as far forwards as the anterior angles of the nostrils: tail long, cylindrical, and somewhat bushy. Skull with the interorbital space broad; the nasal bones short, contracted in the middle, and expanded at both extremities; the zygomatic arch but little prominent, and with but a small vertical diameter; the palate almost destitute of posterior openings; the lower jaw with the angle but little raised; the three upper incisor teeth of nearly equal size, and the hindermost of these teeth apparently without any external vertical groove; the premolar large; a canine tooth present in the upper jaw, but of small size[2].

[1] From δένδρον, a tree; and λαγος, a hare.

[2] The characters derived from the skull and dentition are gleaned from the drawings given in the great Dutch work upon the productions of the Islands of the Indian Archipelago, above quoted; the author has not had an opportunity of examining the skulls themselves.

Two species of this section are known, both of which inhabit New Guinea, and are said to ascend the trees, for which habit their strong fore legs, added to the curved and powerful claws, are adapted.

The *Phalangistidæ* in their skull and dentition approach the Kangaroo group in such a manner, that one would be inclined to regard the two families as furnishing arboreal and ground types of a large natural division, and to suppose, in fact, that the Phalangers might, without impropriety, be looked upon as tree Kangaroos; it is interesting, however, to observe that in the Dendrolagus section we have examples of species which, although their extremities are modified to enable them to ascend trees in quest of their food, are strictly conformable to the Kangaroo type in all essential parts of their structure, and do not evince any very direct affinity to the *Phalangistidæ*—any such approach, in fact, as would lead us to regard them as a link between the Phalangers and Kangaroos; it is, indeed, in certain species of the *Hypsiprymnus* group, that we find the nearest approximation to the *Phalangistidæ*, in the structure of the skull and teeth, and even in the possession of a prehensile power in the tail.

DENDROLAGUS URSINUS. Ursine Tree-Kangaroo.

(*Plate* 1.)

Dendrolagus ursinus. MÜLLER, Zoogdieren der Indischen Archipel. Pt. 4, Plate 19; head, Pl. 22, fig. 1; skull, Pl. 23, fig. 1—3, and Pl. 24, fig. 1; bones of hind legs, figs. 2 and 3.

" " GOULD, Monograph of the Macropodidæ, Pt. 2, Plate 11.

Head conical; ears short; fur long, coarse, and glossy black: the hairs of the neck directed forwards, and meeting those of the head, which have the usual direction, form a transverse crest between the ears, which are densely clothed with very long hairs, black, excepting at the point, where they are of a deep brown hue: head clothed with short hairs; yellowish brown; of a deeper brown hue around the eye, and on the muzzle; cheeks brown-yellow; under parts of body brown, also clothed with short hairs: tail very long, densely clothed with moderately long, and very harsh black hairs, but around the base they have a deep brown hue.

Inhabits New Guinea.

This animal has received the specific name *Ursinus* no doubt on account of a certain superficial resemblance which it has to a small bear, arising in a great measure from the nature of its fur, which differs much from that of the ordinary Kangaroos, not only in being harsh and glossy, but in being composed of one kind of hair only: it would appear that that kind of hair which forms the chief clothing in the ordinary Kangaroos is here entirely, or almost entirely, wanting, and that the hairs representing the longer interspersed hairs in the fur in those animals, here forms the entire coat. With all the essential characters of the true Kangaroos, we find, in these tree-climbing animals, the limbs modified for their different mode of life—the long hind legs of the Kangaroo proper, are replaced by comparatively short legs, and the fore legs are but little inferior in size to the posterior limbs; the strong fore feet are armed with stout and long claws, compressed, and much curved, and fitted for clinging to the inequalities of the bark of the trees. The enormously long tail no doubt helps to balance the animal whilst on the branches of the lofty trees which it ascends in quest of food.

The hairs on the back average about an inch and a half in length; those on the head, as far back as the ears, and on the whole of the under parts of the body, are short; on the tail, which is densely clothed throughout, they are very harsh, and by no means decumbent; they average perhaps at about half an inch in length, or rather less; the tail gradually tapers from the base to the apex, and is about two inches in diameter (including the hair) at the former part, and three-quarters of an inch at the distal extremity. On the upper parts and sides of the body, as well as the limbs (excepting at the base internally), the fur is black and glossy, and very nearly uniform to the skin, an indistinct brownish hue being only observable quite at the root of the hairs. The tail is black, but tinted with brownish at the root. The ears are densely clothed with very long hairs, which completely conceal them; the hairs springing from the tip of the ear are brown, but the rest are black. The head in front of the ears, and the whole of the under parts of the body, are brown, but varying in intensity in parts, being darker around the eye and on the muzzle, and yellowish on the cheeks; the belly is also yellowish, whilst the chest assumes a deeper hue. The muffle appears as if naked, but has, in fact, very minute hairs scattered throughout: the hair on the muzzle, above, is very short. The fore legs are very strong, and so are the hands and the claws with which they are provided;—the claw of the middle finger being at least three-quarters of an inch in length. The hind feet are strong, but comparatively short: the largest toe has the nail an inch in length, strong, somewhat compressed and curved. The toes are less unequal in size than in the typical Kangaroos; the end of the claw of the outer toe terminates in a line with about the middle of the great central toe, and the nails of the double inner toe extend about $\frac{1}{12}$ of an inch beyond the base of the central one.

The specimen from which the above description is taken is a female, and forms part of the British Museum collection; its dimensions are as follows:—

	Inches.	Lines.
Length from tip of nose to root of tail ...	20	6
" of tail	23	9
" of tarsus	4	10
" from tip of nose to ear	3	6
" of ears	1	3
" of fore leg (from elbow joint to the end of the nails)	7	3

Length of skull[1]	4	3
Width of ditto	2	$2\frac{1}{2}$
" " between orbits	0	$10\frac{1}{2}$
Length of nasal bones	1	4
Width of ditto behind	0	7
" " near the apex	0	$5\frac{1}{4}$
Length of zygoma	2	$5\frac{1}{2}$
" from anterior root of zygoma to apex of intermaxillaries	1	$4\frac{3}{4}$
" of the three upper incisors taken together	0	$4\frac{1}{6}$

DENDROLAGUS INUSTUS.

Brown Tree-Kangaroo.

Dendrolagus inustus. Müller, Zoogdieren van den Indischen Archipel. Pt. 4, Pl. 20 ; head, Pl. 22, fig. 2 ; skull, Pl. 23, figs. 4—6, and Pl. 24, fig. 4 ; bones of hind leg, figs. 5 and 6.

" " Gould, Monograph of the Macropodidæ, Pt. 2, Pl. 12.

Fur rather long, and somewhat harsh ; brown, pencilled with brown-white ; under parts of head and body impure white, fore and hind legs brown-white ; feet dusky brown.

Inhabits New Guinea.

This species is about the same size as *D. ursinus*, from which it differs not only in being of a brown colour, but in

[1] These dimensions are taken from the figures in the Plates of Dr. Müller's work, already quoted.

having the muzzle and tarsi rather more elongated, and the ears less densely clothed with fur: the hairs of the back do not so distinctly radiate from a point, rather behind the shoulders, as in *D. ursinus;* over the shoulders, however, the points of the hairs are directed outwards, and on the back part of the neck they are directed forwards, but are semi-erect, and those of the head are directed backwards. The fur is rather less harsh than in *D. ursinus;* its general hue is deepish brown on the upper parts of the body, but here each hair is brown at the base, shaded into brownish black externally, whilst at the point they are of a very pale brown colour, inclining to white; on the under parts of the body the exposed portions of the hairs are white, or very nearly so, but in the middle they are of a very pale brown colour, and at the base still paler, and nearly white in some parts. The sides of the head are of a pale brown colour, and the upper surface is dusky brown. The muzzle is clothed with very short hairs. The ears are tolerably well clothed with longish hairs, brown on the inner side, and dusky on the outer: the fore arms and hind legs are brownish white, but the hairs on these parts are brown at the root: the hinder part of the haunches is whitish, and so is the tail beneath at the base; this member is well clothed with longish, harsh hairs, partly brown-white, and partly pale brown; its general hue is paler than that of the body. The fore and hind feet are dusky brown, but pencilled with whitish on the hinder parts.—From the Paris Museum.

	Female.	
	Inches.	Lines.
Length from tip of nose to root of tail ...	27	0
" of tail	25	0
" from nose to ear	4	2
" of ear	1	6
" of fore-arm and hands without the nails	6	8
" of the nail of the middle finger ...	1	6
" of tarsus without the nails	4	7
" of the nail of the central toe ...	1	1

	Inches.	Lines.
Length of skull[1]	4	3⅓
Width of ditto	2	3⅔
" between orbits	0	11
Length of nasal bones	1	6½
Width at base	0	7
" near the apex...	0	5¼
Length of zygoma	2	2½
" from anterior angle of zygoma to apex of intermaxillaries	1	3¼
" of the three upper incisors taken together	0	5⅔

The skull, it would appear, in *Dendrolagus inustus*, is broader than that in *D. ursinus*, the zygomatic arch deeper, and the superior incisor teeth broader. According to the figures of the leg bones given in Plate 24 of Müller's work, the tibia is rather shorter than the femur in *D. ursinus*, whilst in *D. inustus* the femur is the shorter of the two bones.

The author is not aware that any detailed account of these two singular animals has yet appeared; the only notice he has met with is the portion of the great work on the Dutch possessions in the Indian Archipelago, which treats of the general features of New Guinea, and of the natives.—(See Bijdragen tot de Kennis Van Niew Guinea). Here will be found a short description of the two species of *Dendrolagus*, and with respect to their habits, Dr. Müller merely observes that they live in the trees.

Genus *Hypsiprymnus*.

Macropodidæ with a distinct canine tooth in the upper jaw; the anterior upper pair of incisors descending considerably below the level of the remaining two pairs; the premolar long

[1] According to the figures given by Dr. Müller.

(sometimes as long as the anterior two true molars taken together), compressed, and presenting a cutting edge, the outer and inner surfaces with small vertical grooves; the true molars successively decreasing in size from the foremost to the hindermost; their crowns nearly square, and divided by a transverse and a longitudinal groove into four blunt tubercles; auditory bullæ large, and as it were inflated; zygomatic arch with a small vertical diameter: toes of the fore feet armed with very long, compressed, and but slightly curved solid claws; the two lateral toes much shorter than the others, and with the claws small in proportion.

Whilst the œsophagus terminates in the middle division of the stomach in the true Kangaroos, in the Rat-Kangaroos (Prof. Owen states) it is removed from the commencement of the middle sacculated compartment to its termination[1]. The cœcum is much shorter than in the great Kangaroos.

The Rat-Kangaroo, or Potoroos, are all of small size, as compared with most other species of the Kangaroo family, being, for the most part, about equal in bulk to the Common Hare or Rabbit. Their body is of a more compact form, the fore parts being less elongated; and the ears being small and rounded, gives them a different aspect, when compared with the typical Kangaroos; but some of the smaller species of *Macropus* could not be distinguished from the Rat-Kangaroos in these respects. In external characters, one of the most striking points of distinction is in the structure of the fore foot, the toes being more unevenly developed in the Rat-

[1] Todd's Cyclopædia of Anatomy, &c. iii. p. 301. "The stomach is as singularly complicated as in the Kangaroos, and the complication is essentially the same in both, arising from the sacculation of the parietes of a very long canal, by a partial disposition of shorter bands of longitudinal fibres; but in the Hypsiprymni this sacculation is confined to that part of the stomach which lies to the left of the œsophagus, while the right division of the cavity has the ordinary form and structure of the pyloric moiety of a simple stomach. The left cardiac division is enormously developed; in relative proportion, indeed, it is surpassed only by the true ruminant stomachs."—Owen, loc. cit.

Kangaroos,—the three central toes proportionately longer than we find them in *Macropus*, and the lateral toes smaller; the nails, moreover, are of a different form: in *Macropus* they are broadest, and concave beneath, whilst in *Hypsiprymnus* they are much compressed, solid, and broadest above. These differences, observable in the structure of the fore foot and claws, are connected with certain differences in the habits of the species of the two divisions. Whilst the true Kangaroos browse upon the herbage, the Rat-Kangaroos, we learn, feed much upon the roots of certain plants, which they scratch up with their fore feet. In the hind feet no constant difference is perceptible. In the structure of the skull and teeth there are many differences worthy of remark:—The frontal bones are larger in *Hypsiprymnus* than in *Macropus;* the muzzle is more compressed, and more pointed at the extremity; the zygomatic arch is much more slender, its vertical diameter being considerably less; the pterygoid processes are less developed; the intermaxillary bones are less produced: a vertical line dropped from the tip of the nasal bones, in the animals under consideration, would, in most species, very nearly touch the front edge of the foremost incisor, but the same tooth in Macropus would be found advanced considerably beyond the corresponding line. The auditory bullæ, which, as Prof. Owen has pointed out, are, in almost all the Marsupialia, formed by the expansion of the great alæ of the sphenoid, and not by a portion of the temporal bone, as in the Rodents, here assume a hemispherical form. I have not met with this, as it were, inflated auditory chamber, in any of the species of the *Macropus* division, excepting in the *Lagorchestes leporoïdes* and the *L. conspicillatus*[1]. With

[1] Since my account of the species *Lagorchestes* was printed, I have had an opportunity of examining a skull of the *L. conspicillatus*, contained in Mr. Gould's collection: this differs much from the skull of *L. leporoïdes*. The muzzle is shorter and broader, the zygomatic arch does not project so suddenly

regard to the upper incisor teeth, the differential characters are, that they are proportionately smaller, less compressed from the outer to the inner side; the foremost pair descend much below the level of the other two pairs, and these latter pairs are not so widely separated from each other as in *Macropus*, owing to the contraction of the palate between them. The canine teeth are always present, and usually pretty well developed; they are very small, however, in

in front, and above all it differs in its great strength, the bones being thicker in proportion than in any of the species of Macropodidæ I have met with: the temporal ridges meet, and form a distinct sagittal crest: the palate is remarkable for presenting four posterior palatine openings, two of which are nearly round, and situated entirely in the palatine bones, and are rather more than 1½ lines in diameter; the other pair are oblong, of about the same size, and are placed at some little distance in advance of the palato-maxillary suture. The canine tooth is more developed than usual, and in this character, as well as in having the auditory bullæ prominent, the *L. conspicillatus* approaches *Hypsiprymnus;* the premolar is moreover large for a *Macropus*, though smaller than in *Hypsiprymnus.* The foremost upper pair of incisor teeth are very much larger than the other two pairs; the palate, however, is not contracted between the lateral pairs of incisors, as in *Hypsiprymnus;* and the structure of the true molar teeth, the great depth of the zygomatic arch, the small size of the frontal bones (which are considerably contracted between the orbits), combined with the structure of the fore feet in the animal, all tend to show that its more direct affinity is with *Macropus.* The dimensions of this skull are as follows:—

	Adult Male.	
	Inches.	Lines.
Total length of skull	3	4
Width	2	0½
" between orbits		7¼
Length of nasal bones	1	4
Width of ditto behind		7⅓
" " near the apex		4¾
Depth of zygomatic arch behind		5½
Length of the three upper incisors, taken together		4
Distance between posterior incisors and premolar		5
Length of the five molar teeth, taken together	1	1½

Hypsiprymnus campestris: the premolar tooth is larger, and almost always has numerous distinct vertical grooves both on the outer and inner sides. The true molar teeth differ, in being successively smaller from the foremost to the last, and the crowns of these teeth have the two principal transverse ridges much less developed. A more accurate idea of the structure of the unworn grinding surface of the molar tooth of a *Hypsiprymnus* (see Plate 10, figs. 5 and 5 *a*) would be conveyed by describing them as of a quadrate form, and presenting four equidistant blunt tubercles, which are joined in pairs by transverse ridges, but with these ridges less elevated than the points of the tubercles: there is a slight trace of the band of the tooth both on the front and back of each molar, as in *Macropus*. The hindermost molar is generally small, and sometimes almost round[1].

On Plates 6, 8, and 10, are represented the skulls of the various species of Hypsiprymnus, or Rat-Kangaroos. Since the last tooth to make its appearance in these animals is the permanent premolar, it very frequently happens that the skulls in collections do not show this tooth. At different ages previous to the animals having attained maturity, the skull will present two molars belonging to the milk, or first series, and from one to a perfect series of the true molars. When all the true molars are just developed, and there are six molars on each side of either jaw, it will be found that the foremost of each row corresponds almost perfectly in structure with the permanent premolar, excepting that it is of rather smaller size; and the second tooth, which resembles a true molar, excepting in being rather smaller, are teeth of the first series; for they will be replaced in the vertical direction by another tooth, and this is the permanent premolar. The skulls, figs. 2 and 4, of Plate 10, belonging to animals which were not quite adult, present the condition of

[1] Cases occur in which the last molar tooth is absent; and, what is more extraordinary, I have observed an extra molar tooth on each side of the upper jaw, in a species of *Hypsiprymnus*.

the teeth just described; and to show the size and form of the permanent premolar, it has been necessary to remove a part of the bone from the outer side of the jaw. Fig. 2 *a*, and fig. 4 *a*, represent the fore part of the upper jaw, in which is seen the three incisors, followed by the canine, and then by two milk teeth, above which milk teeth is the permanent premolar (*), which, as it grows, thrusts out the two milk teeth, and assumes their place, as in the adult skull, fig. 3 *a*.

The letters added to the figures in Plate 8 may here be explained:—

a, is the occipital bone.
b, the interparietal bone.
d, the temporal bones.
c c, the sphenoid.
d′, expanded portion of the great ala of the sphenoid, forming the auditory bulla.
e e, parietal bones.
f, palatine bones.
f′f′, posterior palatine openings.
g g, frontal bones.
h h, lachrymal bones.
i i, malar bones.
k k, nasal bones.
l l, maxillary bones.
l′ l′, palatal portion of the maxillary bones.
m m, intermaxillary bones.
in. incisor teeth.
ca. canine.
p-m. premolar.

Of the lower jaw (figs. 2 *a*, and 3 *b*.)

A, is the horizontal branch, or ramus.
B, the ascending ramus.
1, the coronoid process.
2, the condyloid process.
3, the angular process.
i, the incisor tooth.
p-m, the premolar.
m, the four true molar teeth.

The Rat-Kangaroo[1] may be subdivided into three minor groups, two of which have received names; the third contains the *Bettongia rufescens* of Mr. Gray, which I shall proceed to describe.

Section, or *Sub-Genus*, 1.

Muffle almost entirely clothed with hair; bony palate without posterior openings; tarsus long.

HYPSIPRYMNUS (Sect. 1) RUFESCENS.

The Rufous Rat-Kangaroo.

Bettongia rufescens. GRAY, Magazine of Natural History for Nov. 1837, vol. i. p. 584.
Hypsiprymnus melanotis. GOULD, Monogr. Part 2.
,, ,, OGILBY, Proceedings of the Zool. Soc. for May, 1838, p. 62.

Fur moderately long and soft, the interspersed harsher hairs long: general tint bright rusty red, but much pencilled with whitish; under parts impure white; ears with white hairs internally, excepting at the margin, where they are rusty red; externally, densely clothed with soft black hairs; fore legs and feet white, or nearly so; tarsi dusky brown: tail moderately well clothed, the hairs adpressed, and tolerably long on the apical portion; of a dirty white colour, but rather finely pencilled with dusky on the upper surface at and near the base: muffle clothed with small hairs, but with a narrow naked space next the nostril openings.—Skull, Plate 10, fig. 1.

Inhabits New South Wales.

[1] The animals in question are usually called Kangaroo-Rats; but as they belong to the Kangaroo group, and not to the Rat tribe, I think all will agree in the propriety of the above alteration of the relative position of the two terms.

Hypsiprymnus rufescens is one of the largest species of the Rat-Kangaroos; and, as the names which have been given to it imply, is remarkable for the red hue of its fur, and the black colour of its ears; the black, however, it must be observed, is confined to the back of the ear. It not only differs from its congeners in size, and the colouring of its fur, but may be distinguished by its having the muffle nearly covered with fine velvet-like hairs, these extending nearly as far forward as the anterior angle of the nostrils; there is, however, a naked space around each nostril opening.

Mr. Gould informs us that this species is very common in New South Wales, inhabiting the but little elevated stony ridges, especially in those parts in which shrubs and grasses abound. It appears, observes this gentleman, to be dispersed over the whole of the colony, from the coast to the interior highlands. Its nest is composed of grasses, and is frequently placed under the shelter of a fallen tree, or at the foot of some low shrub. During the day the little animal lies coiled up in its nest, but it occasionally reposes in a "seat," like the Hare-Kangaroo (*Lagorchestes*), but it never sits in the open plains. On being pursued, it runs with great swiftness for a short distance, but from the circumstance of its invariably seeking shelter in the hollow logs, it falls an easy prey to the natives, who seek it for food. Its food consists, like others of the genus, of various kinds of roots and grasses.

Messrs. Ogilby and Gray both drew up their original descriptions of the present species from a specimen contained in the Museum of the Zoological Society, which presents the following characters:—

Fur long, loosely applied to the body, and having numerous very long and coarser interspersed hairs, the visible portion of which is chiefly white; each of these hairs, however, is black at the point, and has a rusty red space immediately below the black portion: the shorter hairs have the visible portion bright rusty red, and all the hairs on the upper parts and sides of the body are grey at the root; on the under parts of the body the hairs are white (not very pure), and slightly tinted with grey at the root. The head presents but little of the red tint; on the cheeks, immediately below the eye, the hairs are very coarse, and tinted with white, red, and black; the upper part of the muzzle is brownish: the ears are clothed internally with long whitish hairs, excepting at and near the margin, where they assume a rust colour; on the outer side of the ear they are dense, long, and soft, and of a black colour, but at the margin they are shorter and whitish. The fore legs are whitish, rather strong, and the hands are furnished with long whitish claws. The tarsi are clothed with glossy brown hairs. The tail is chiefly of a dirty white colour; on the upper surface variegated with dusky, but chiefly at the base; the hairs are moderately long (averaging about half an inch), and sufficiently numerous to hide the skin; they are rather harsh to the touch on the upper part of the tail, and very coarse on the under surface: though towards the tip of the tail the hairs are rather longer than on the middle, they do not form any crest or brush, as in *Hypsiprymnus penicillatus*, and some other Rat-Kangaroos. Of the subjoined dimensions, those contained in the first column denote the size and proportions of the specimen above described; and those in the second column are from a specimen in the British Museum collection:—

			Male.	
	Inches.	Lines.	Inches.	Lines.
Length from tip of nose to root of tail	19	0	20	0
" of tail	14	9	16	0
" of tarsus and claws	5	8	5	9
" of ear	1	8	1	8
" of fore legs, hand, and claws, about			4	0

The skull of *B. rufescens* differs from the crania of other species of the genus, in wanting the posterior palatal openings, which are generally so large; here the palatine bone is of a square form, and the palato-maxillary suture crosses opposite the interspace of the second and third true molar teeth; at the anterior angles of the palatine bone is a minute perforation. The skull, moreover, is remarkable for its breadth, and for the shortness of the facial portion, and consequently of the nasal bones, and of the interspace between the premolar tooth and the incisors. The palatal interspace between the hinder pair of incisors is much contracted, being not more than three-quarters of a line in width, and this contraction is owing to the comparative great transverse diameter of the teeth in question. The palate is broad and concave; the zygomatic arch comparatively deep, measuring towards the hinder part about four lines from the upper to the lower edge.

Although the skull has not been removed from the skin of the specimen of *Hypsiprymnus rufescens* in the Zoological Society's museum, I have been enabled to make one or two admeasurements of the foremost teeth, and of the distances between them; and these I have added in the second column,—the first giving the dimensions of an imperfect cranium contained in the Royal College of Surgeons' museum.— See Plate 10, fig. 1.

	Inches.	Lines.	Inches.	Lines.
Length from the posterior root of the zygomatic arch to the apex of the intermaxillary bones	2	7½		
" of the zygomatic arch	1	6		
Width of skull	1	11		
Length of nasal bones	1	2		
Width of ditto at the base	0	7½		
" " near the apex	0	4		
Length of the three incisors of upper jaw, taken together	0	4¼	0	4
" from posterior incisor to canine ...	0	1⅓	0	1½
" from canine to premolar ..	0	1¾	0	2
" of premolar (the milk tooth) ...	0	3¼		

	Inches.	Lines.	Inches.	Lines.
Length of ditto and three molar teeth together (the foremost molar being a milk tooth)	0	11¼		
" of palate	1	8		
Width of ditto between the premolar teeth	0	8		
" of interorbital space	0	6¾		
Length of permanent premolar	0	4⅓	0	4½
Height of lower jaw, measuring from apex of coronoid process	1	1¾		
Distance between hinder part of condyle and anterior margin of coronoid process, measuring in a horizontal line			0	8⅔

Sub-genus 2. *Bettongia.*

Bettongia. GRAY, Magazine of Nat. Hist. Vol. i. (New Series), p. 584.

Muffle naked; tarsus long.

The tail is generally provided with longish hairs on the upper surface of the apical portion, forming a bushy crest on this part, in the species of this section, some of which (if not all) have a prehensile power in the tail; the prehensile power, however, is apparently very limited, for the organ in question is not capable of being used, as in the Phalangers, and some other Marsupials, for sustaining the weight of the body.

HYPSIPRYMNUS (*Bettongia*) CUNICULUS.

Tasmanian Rat-Kangaroo.

Hypsiprymnns cuniculus. OGILBY, Proceedings of the Zoological Society for May 1838, p. 63; WATERHOUSE, Marsupialia, p. 186.

Bettongia setosa. GRAY, Magazine of Nat. Hist. vol. i. (New Series), p. 584; List of Mammalia in the collection of the British Museum, (1843), p. 93.

Bettongia cuniculus, Tasmanian Jerboa Kangaroo. GOULD, Monogr. on the Macrop. Part 2, 14th Plate.

Fur moderate, of an ashy brown colour, pencilled with whitish; on the under parts of the body dirty white: ears with pale yellowish hairs internally, pale brown externally; feet brown-white: tail palish brown above, brown-black at the apex, where there are longer hairs, forming a bushy crest on the upper surface; under parts of tail brown-white. The extreme point of the tail sometimes white, if not always so.—Skull, Plate 10, fig. 2.

Inhabits Van Diemen's Land.

The original of Mr. Ogilby's description of *Hypsipr. cuniculus*, contained in the museum of the Zoological Society, is said to have come from New South Wales, but it is possible there is some mistake as regards this habitat, since Mr. Gould, in his visit to Australia, found the species in Van Diemen's Land only. That gentleman informs us, that it is generally distributed over the island, and prefers the open sandy, or stony forest land, rather than the thick and humid brushes. It is the largest of the Rat-Kangaroos, and in general colouring resembles the *H. penicillatus;* it is not only distinguished, however, by its superior size, but by its having a white tip to the tail. The proportions of the crania of the two animals differ considerably.

The *H. cuniculus* of the Zoological Society's collection has the fur moderate as to texture, and tolerably long; the general tint of the animal is grey-brown, but the hairs on the upper parts of the body are rather broadly annulated with dirty rusty white towards the point, and dusky at the point—next the skin, or at the root, they are grey: the under parts of the body are dirty white; the ears are clothed with very pale yellowish hairs internally, and more densely clothed with pale brown hairs externally: the feet are of a very pale brown colour, or might be described as brown-white: the tail is brown above, and brown-white beneath; on the upper surface of the apical portion the hairs are longish, averaging nearly three quarters of an inch, and of a brown-black hue; these bushy hairs extend for about two inches from the end of the tail.

The extreme point of the tail is perhaps wanting in this specimen; for in a specimen in the British Museum in which the tail is perfect, about one inch of the apical portion is white. A second specimen has the tail somewhat suffused with rust colour, and the apical third is clothed with bushy brown-black hairs above—the point is wanting. This latter specimen is the original of the *B. setosa* of Mr. Gray, alluded to in the Magazine of Natural History, having been regarded by that gentleman as the *Hyps. setosus* of Mr. Ogilby—a very distinct species. The dimensions of this specimen are given below in the second column, whilst those of the *B. cuniculus*, of the Zoological Society, are contained in the first column:—

	Inches.	Lines.	Inches.	Lines.
Length from tip of nose to root of tail	17	6	16	0
" of tail	13	0?	12	0?
" of tarsus and claws	5	0	5	0
" of ear	1	2	1	2

Fig. 2, of Plate 10, is taken from a skull of the *B. cuniculus* in the Museum of the Royal College of Surgeons. It was procured by Mr. Gould in Van Diemen's Land. The skull of the specimen originally described by Mr. Ogilby differs in having the nasal bones a trifle broader behind, and distinctly narrower in front, and the inter-orbital space rather broader. Its dimensions are given in the first column, whilst those of the specimen figured (which is fractured behind) are contained in the second column:—

	Zool. Soc.		Coll. Surg.	
	Inches.	Lines.	Inches.	Lines.
Total length of skull, about	?		3	4
" width			1	$10\frac{1}{3}$
Length from tip of nasal bones to hinder margin of frontals	2	6	2	6
" of nasal bones	1	$4\frac{1}{2}$	1	$4\frac{3}{4}$
Width of ditto at base	0	8	0	7
" " near the apex	0	$3\frac{1}{2}$	0	4

	Zool. Soc. Inches.	Zool. Soc. Lines.	Coll. Surg. Inches.	Coll. Surg. Lines.
Width of inter-orbital space	0	$9\frac{3}{4}$	0	$9\frac{1}{4}$
Length of three upper incisor teeth, taken together	0	4	0	$4\frac{1}{3}$
Distance between incisors and canine tooth	0	1	0	$\frac{3}{4}$
" canine and premolar ...	0	$3\frac{3}{4}$	0	4
Length of permanent premolar	0	$3\frac{9}{10}$	0	$3\frac{9}{10}$
" of the premolar and four true molars, taken together			1	$0\frac{1}{4}$

HYPSIPRYMNUS (*Bettongia*) GRAII.

Gray's Rat-Kangaroo.

Hypsiprymnus Graii. GOULD, Proceedings of the Zoological Society for December, 1840, Pt. 8, p. 178; WATERHOUSE, Marsup. p. 190.

Hypsiprymnus Lesueuri (?) QUOY et GAIMARD, Voyage de la Coquille.

Fur ashy grey, pencilled with white, or brown, pencilled with rusty white; under parts pale yellow: tail rusty brown above, assuming a deepish brown hue towards the apex; usually about two inches of the apical portion white; tarsi brown-white, or very pale rusty brown: head usually suffused with yellow, especially on the sides; foremost pair of incisor teeth rather broad, not compressed. Skull, Plate 10, fig. 3.

Inhabits Western and Southern Australia.

The present species of Rat-Kangaroo, which is named in honour of Mr. Gray, head of the Zoological Department of the British Museum, inhabits both the Western and Southern districts of Australia, and in both districts is associated with the *H. penicillatus* (or at least with the *H. Ogilbyi*, which I regard as a variety of that animal).

Many specimens of *H. Graii* have come under my notice, and although these exhibited considerable variation in their colouring, and sometimes approximated very closely to other species, yet, with the assistance of the skull, I have found no difficulty in distinguishing them. Some individuals might be confounded with the *H. Gaimardi,* and others again approximate very nearly to the *H. penicillatus:* the broader form of the anterior pair of incisors, however, will help, in combination with other characters about to be noticed, to distinguish the *H. Graii* from either of those two species. The skull is proportionately broader, the auditory bullæ larger, the premolar tooth larger, and placed nearer the canine, than in *H. penicillatus:* the auditory chambers are much larger, and the interorbital space is more contracted, than in *H. Gaimardi.* Other points of distinction will be perceived in the proportions of the skull and teeth upon comparing the dimensions, hereafter given, with those of the crania of the species just mentioned.

The following is a description of an adult male specimen in the British Museum, which formed part of Mr. Gould's collection, and which, I believe, is the original of that gentleman's description in the Proceedings of the Zoological Society; if so, it is from the Swan River district.

Fur moderately long, and soft: general tint brownish ash colour, but pencilled with white and blackish; under parts dirty white. Ears externally clothed with long hairs of a brownish white colour, but tipped with brown; internally with shorter hairs, of a palish yellow colour, but brown on the margin of the ear. Feet brown-white, or very pale brown. Tail clothed throughout with adpressed hairs, and these of a rufous brown tint on the upper surface of the tail, and dirty white on the under surface; the apical portion of this organ, however, is entirely white, the white occupying about one-fourth of the entire length. The hairs of the fur on the back are grey at the root, then brownish white; this is followed by a slight rusty red tint, and the points are black. On the under parts of the body the fur is of a pale grey hue next the skin, and impure white externally; on the throat, the hairs are uniformly white. The muffle is small, the hair terminating above in a line with the posterior angle of the nostrils.

	Inches.	Lines.
Length from nose to root of tail	18	0
" of tail	12	0
" of tarsi	4	8[1]
" of ear	1	0

The following dimensions are from skulls of *H. Graii*, for the loan of most of which I am indebted to Mr. Gould:—

	Old Male.		Male Adult.		Female, Old.		Specimen in Zool. Soc. Mus. Adult.		Lesueuri, Quoy and Gaim.	
	In.	Li.	In.	Li.	In.	Li.	In.	Li.	In.	Li.
Length of skull	3	0$\frac{1}{2}$	2	11$\frac{1}{4}$	2	10$\frac{2}{3}$	2	10	2	11$\frac{1}{2}$
Width of ditto	1	10$\frac{1}{3}$	1	9	1	9$\frac{1}{2}$	1	8$\frac{3}{4}$	1	9
Length of nasal bones	1	2$\frac{3}{4}$	1	1$\frac{2}{3}$	1	2$\frac{1}{4}$	1	0	? 1	3
Width of ditto at the base	0	6	0	6	0	6	0	6$\frac{1}{3}$	0	6$\frac{1}{3}$
" " near the apex ...	0	2$\frac{3}{4}$	0	3$\frac{1}{4}$	0	3	0	3		
Length of frontal bones	1	1$\frac{2}{3}$	1	1$\frac{1}{2}$	1	0			1	0
Width of ditto between orbits ...	0	7$\frac{3}{4}$	0	7$\frac{1}{2}$	0	7$\frac{1}{2}$			0	8$\frac{2}{3}$
Length of palate	1	0	1	0	1	0$\frac{1}{2}$		?	1	1
" of three incisors on either side of upper jaw ...	0	4	0	3$\frac{3}{4}$	0	4	0	3$\frac{3}{4}$		
Space between incisors and canine ...	0	0$\frac{7}{12}$	0	0$\frac{7}{12}$	0	0$\frac{3}{4}$	0	1	0	1
" between canine and premolar	0	2	0	2$\frac{1}{2}$	0	2$\frac{3}{4}$	0	1$\frac{3}{4}$	0	2$\frac{3}{4}$
Length of premolar	0	3$\frac{3}{4}$	0	4	0	4$\frac{1}{6}$	0	4	0	4$\frac{1}{4}$
Space occupied by premolar and molars on either side of upper jaw	0	11	0	11	0	11	0	11	0	11$\frac{1}{3}$
Length of auditory bullæ	0	7$\frac{3}{4}$	0	7$\frac{3}{4}$	0	7$\frac{3}{4}$	0	7$\frac{1}{2}$	0	7$\frac{1}{2}$
Width of ditto	0	6$\frac{1}{4}$	0	5$\frac{3}{4}$	0	6$\frac{1}{4}$	0	6	0	5$\frac{3}{4}$
Height of ascending ramus of lower jaw, in a vertical line dropped from apex of coronoid process ...	1	2	1	0$\frac{1}{4}$	1	1$\frac{1}{3}$				
Length from back part of condyle to anterior margin of coronoid process	0	8$\frac{1}{2}$	0	8	0	8				

[1] The dimension of the tarsus, as given in the volume on Marsupial Animals, in the Naturalist's Library, is incorrect.

The skull of *H. Graii* approaches most nearly to that of *H. Gaimardi*, but differs in being rather broader, in having the auditory bullæ much larger, and the palate shorter—characters pointed out by Messrs. Quoy and Gaimard, as distinguishing, from *H. Gaimardi*, a skull found by them in Dirck Hartog's Island, Western Australia, upon which they found the *Hypsiprymnus Lesueuri*. I may add, that the zygomatic arch is rather deeper, and the anterior pair of incisors are broader, than in *H. Gaimardi*[1]. In a drawing of the skull of *H. Lesueuri*, made at Paris, and kindly lent me by Mr. Owen, I find the anterior part of the cranium is somewhat mutilated, and wants the incisors: hence this latter character was not noticed. Judging from the drawing in question, I can feel scarcely a doubt that the *H. Lesueuri*, and the *H. Graii*, are specifically identical, and if this view be correct the former of these two names should be used. The dimensions in the last column are taken from the drawing referred to. The skull, of which the dimensions are given in the first column, is remarkable for possessing five true molar teeth, the last of which is very small—less than one line in diameter. In the skull in question, the permanent premolar is perfectly developed, and considerably worn, and so are the molars, and, moreover, the little extra molar does not correspond in form and size to the ordinary last molar, that is here represented by the penultimate tooth.

[1] The width of the two foremost incisors together, and in their natural position, varies from nearly 3 lines to nearly 3⅓ lines, measured at the base, in *H. Graii*; in *H. Gaimardi* they give about 2¼ lines, and in *H. penicillatus* about the same—sometimes 2½ lines; they are much more compressed in the two species last mentioned.

HYPSIPRYMNUS (*Bettongia*) GAIMARDI.

Gaimard's Rat-Kangaroo.

Kangurus Gaimardi.	DESMAREST, Mammalia, Supplement, p. 542, Sp. 842, 1822.
Hypsiprymnus White.	QUOY and GAIMARD, Voyage de l'Uranie, Zool. p. 62, Plate 10, 1824.
Kangurus lèpturus.	QUOY and GAIMARD, Bullet. des Sci. Nat. January, 1824, tom. i. p. 271.
Hypsiprymnus Phillippi.	OGILBY, Proceedings Zoological Society for May, 1838, p. 62.
" *formosus.*	OGILBY, *l. c.* p. 62.
" *minor.*	(Potoroo), Cuv. Règ. Animal, p. 185.

Fur long and soft; prevailing hue ashy brown, much suffused with rusty yellow; under parts white, suffused with yellow; ears internally clothed with yellow hairs; feet dirty white, the heel and sides of the hind foot yellowish rust coloured; tail of a brightish rust colour above, paler beneath; the apical third clothed on the upper surface with longish, bushy, brown-black hairs, but at the extreme point with a few white hairs. The fur both on upper and under parts of the body is grey next the skin; the hairs on the back are blackish at the point, and annulated with yellowish white below the point. From specimen in the Paris Museum. [One inch or more of the apical portion of the tail is generally white.]

Inhabits New South Wales and South Australia.

Being the only Kangaroo-Rat known to the French zoologists, and coming from the same part of Australia, it was natural that the present species should have been confounded with the Potoroo of White. This last mentioned animal (which is the *Macropus minor* of Shaw), however, is a distinct species, as is proved by the skull of White's original specimen still preserved in the College of Surgeons' Museum[1]. The "Potoroo" of the French naturalists was found by M. Gaimard in the neighbourhood of Port Jackson, and formed part of the collection added to the Paris Museum by Freycinet's voyage. It was first described in 1822, by Desmarest, under the name *Kangurus Gaimardi*, and in 1824 received the specific names *Lepturus* and *Whitei* by Messrs. Quoy and Gaimard. The specimen from which these authors drew up their accounts is likewise the original of the above description[2]. Its dimensions are as follows:—

	Lines.	Inches.
Length from nose to root of tail ...	13	6
" " to ear	2	9½
Length of tail	13	0
" of tarsus, to base of claws ...	4	1
" of fore-arm and hand to base of claws about	3	2
" of ear	1	0

I feel no doubt that this is the same species as the *Hypsiprymnus Phillippi* of Mr. Ogilby, the original specimens of which are contained in the Museum of the Linnean Society, and are from New South Wales. They present the following characters:—

About the size of a rabbit: the tail is long, and the tarsi long and slender; the ears of moderate size, and rounded form. Fur long, and moderately soft; general tint grey-brown, slightly washed with rusty yellow; under parts grey-white, very faintly tinted with yellow; feet very pale brown; ears well clothed with fur, externally of the same colour as

[1] See the account of *Hypsiprymnus minor* vel *murinus*.

[2] On the under side of the stand of this specimen I found written "Potoroo White; Hypsiprymnus White, Quoy and Gaimard; Macropus minor, Shaw; Kangaroo-Rat, Phillip's Voyage; Potoroo Rat, Desmarest; and Kangaroo de Gaimard."

that of the head, and internally, with yellow hairs. Tail furnished above with moderately long, and somewhat adpressed rich brown hairs; beneath, with hairs of a very pale brown colour: a crest of long dark brown hairs runs along the upper surface of the apical portion of the tail, occupying about one-third of its entire length; at the point is a tuft of long white hairs. The fur on the upper parts of the body is deep grey next the skin, and each hair is yellowish white externally, but dusky at the point; the longer interspersed hairs are black at the point, and annulated with whitish below the point. The fur on the belly is of a very pale grey hue next the skin.

	Inches.	Lines.
Length from nose to root of tail ...	16	0
" of tail	13	0
" of tarsus, without the claws ...	4	3
" of ear	1	2

I am indebted to Mr. Owen for the loan of a careful drawing of the skull of *Hypsiprymnus Whitei*, or *Kangurus Gaimardi*, which he had made whilst at Paris. This skull is intermediate in some respects between the skulls of *H. penicillatus* and *H. Grayi;* it differs, however, from both in having the auditory bullæ smaller, and the interorbital space broader. From *H. penicillatus* it may be distinguished by its great breadth, the zygomatic arches being thrown out somewhat more boldly, and the premolar tooth longer from front to back; the palate is likewise rather longer.

A skeleton in the College of Surgeons, bearing the name *Hypsiprymnus Hunteri*, has the skull agreeing so closely with that of the *H. Whitei*, as to leave no doubt on my mind of the specific identity. This skeleton measures from the tip of the skull to the end of the tail 26″-6″; the scapula is 1″-7‴; the humerus 1-5$\frac{1}{3}$; the ulna 2-1$\frac{2}{3}$; the radius 1-9$\frac{1}{2}$; the hand 1-4$\frac{1}{2}$; femur 3-7; tibia 4-7; tarsus 4-9$\frac{1}{2}$. The fibula has nearly two-thirds of its length (distal extremity) anchylosed to the tibia.[1] The vertebræ are—cervical, 7; dorsal, 13; lumbar, 6; sacral, 2; caudal, 24=52.

In Mr. Gould's collection is a skull labelled as belonging to the *Hypsipr. Whitei*, which also agrees in *the smaller size of the auditory bullæ, and the greater width of the inter-orbital space,* (as compared with that of *H. penicillatus*), but which has the nasal bones considerably broader than in the Paris skull above alluded to, or in the *H. Hunteri*. I am still inclined, however, to think the skull in question is correctly labelled, for in the College of Surgeons' Museum is a fourth skull, which presents an intermediate condition of the nasal bones. I subjoin the dimensions of these skulls.

	H. Whitei, Par. Mus.		H. Hunteri Coll. Surg.		H. Whitei, Gould's Collectn.*		Skull of Young Animal in Coll. Surg.	
	Ins.	Lines.	Ins.	Lines.	Ins.	Lines.	Ins.	Lines.
Total length of cranium	2	11	3	0	2	11	2	6
Width	1	8	1	8½	1	8	1	4⅔
Length of nasal bones	1	2½	1	2¾	1	2	1	0
Width of ditto, at the base	0	6¾	0	6⅓	0	7¾	0	6⅓
Ditto near the apex, at the root of the free portion	0	3	0	3⅔	0	3⅔		
Space between orbits	0	9⅓	0	9½	0	9¼	0	8
Length of frontal bones	0	11			1	0⅔		
" of palate, to the posterior emargination	1	3			1	2¼		
" of auditory bullæ ...					0	5¾	0	5½
" of the three incisors on either side of the upper jaw, taken together ...	0	4	0	4	0	4		
Width of the anterior pair of incisors, taken together			0	2¼	0	2¼		

[1] In the *H. murinus*, the tibia and fibula are distinct.

* See Pl. 10, fig. 4.

	H. Whitei, Par. Mus.		H. Hunteri Coll. Surg.		H. Whitei, Gould's Collection.		Skull of Young Animal in Coll. Surg.	
	Ins.	Lines.	Ins.	Lines.	Ins.	Lines.	Ins.	Lines.
Between incisors and canine ...	0	0¾	0	0¾	0	0¾		
" canine and premolar ...	0	3⅔	0	4	0	*3¾		
Length of the premolar	0	3¾	0	3½	†	{2½ 3¾		
Space occupied by the premolar and four molar teeth	0	11	0	11	0	‡9¼		
Length from front of anterior incisor, to front of premolar ...	0	9½	0	9¼	0	9½		
Height of lower jaw, measuring from the apex of the coronoid process downwards			0	9	0	9½		

I have to remark, that the dimensions given in the last column are those of the skull of a young animal having but two molar teeth, in addition to the premolar, on each side of either jaw, and of these teeth the hindermost molar only, is a permanent tooth. This is the young animal alluded to in "Marsupialia," p. 183, as presenting the external characters of Mr. Ogilby's *H. formosus*. The original of Mr. Ogilby's description is in the Museum of the Linnean Society, and may be described as of a greyish yellow hue, somewhat suffused with pale rust colour; under parts dirty white; tail brown above, very pale brown beneath; towards the apex, on the upper surface, the hairs are somewhat longer, and assume a deep brown hue; about one inch of the apical portion, how-

* In this case the premolar is a milk tooth.

† The smaller dimension gives the length of the milk tooth, and the larger that of the permanent one.

‡ Three of these molars only are permanent teeth.

ever, is white,—not the "*latter half*" white, as described by Mr. Ogilby. It measures from the tip of the nose to the root of the tail, 11 inches; the tail is 10 inches, and the tarsi, without including the claws, 3″ 11‴.

HYPSIPRYMNUS (*Bettongia*) PENICILLATUS.

Tufted-Tailed Rat-Kangaroo,

(*Plate* 9.)

Bettongia penicillata, GRAY, Mag. Nat. Hist. for November, 1837, vol. i. (New Series,) p. 584. WATERHOUSE, Marsup. p. 183. GOULD, Monogr. of the Macropodidæ, Pt. 1, Pl. 14.

Hypsiprymnus murinus, OGILBY, Proc. Zool. Soc. for May, 1838, Pt. 6, p. 63.

" *setosus*, WATERHOUSE, Catal. of the Mammalia of the Zool. Soc. (1838), p. 65.

" (*Bettongia*) *Ogilbyi*, GOULD'S MSS. WATERHOUSE, Marsupialia, p. 185.

Bettongia Gouldii, GRAY, List of the Mammalia in the British Museum Collection, 1843.

Fur long and moderately soft: general colour ashy brown; sides of head and throat slightly suffused with yellowish; under parts dirty white, very obscurely tinted with yellow; ears clothed with yellow hairs internally: tail brown, somewhat pencilled with pale rust colour: the apical third, clothed above with long, bushy, black hairs; the under surface brown-white, brown towards and at the apex: fore feet brown-white; tarsi pale brownish ash colour.—Skull, Pl. 6, fig. 3.

Inhabits New South Wales.

The following detailed description is drawn up from a specimen in the Museum of the Zoological Society, which is the original both of Mr. Gray's diagnosis of *Bettongia peni-*

cillata, and of that of Mr. Ogilby of his *Hypsiprymnus murinus*. The name *setosus* was attached to the specimen in question, by mistake; an error which also appeared in the Catalogue of the Zoological Society's Collection; the *H. setosus* of Mr. Ogilby being in fact specifically identical with the Potoroo of White—*H. minor*, or *H. murinus* of authors.

The *H. penicillatus* is about equal in size to the common Rabbit; its fur is tolerably long, and moderate as to texture: general colour, ashy brown, but freely pencilled with white, and on the back pencilled likewise with brownish black; the under parts of the body dirty white, and indistinctly suffused with yellow: a faint yellowish tint is observed on the cheeks and throat. The ears are clothed internally with yellow hairs; externally, the hairs are chiefly brownish white, but at the margin they are brown-yellow. The top of the muzzle is brown near the apex; the naked part, or muffle, terminates in a line with the posterior angle of the nostril opening. The tail is brown, pencilled with rusty white; the apical third, clothed with long, bushy, black hairs; these become gradually longer to the tip, where they are about three-quarters of an inch in length: the under parts of the tail are clothed with stiff adpressed hairs of a very pale brownish colour; a trifle darker towards the end of the tail. The tarsi are of a pale brown tint, and the fore feet are brown-white. The fur on the back is grey next the skin; each hair is brown-white towards the point, and brownish black at the point. On the abdomen and chest, the hairs are of a pale grey hue next the skin.

	Inches.	Lines.
Length from tip of nose to root of tail ...	13	6
" of tail	11	3
" of tarsus and claws	4	4
" from nose to ear	2	10
" of ear, about	0	10
" of fore-arm and hand, about ...	2	11

Hypsiprymnus (Bettongia) Ogilbyi.

Mr. Gould distinguishes, under the name *B. Ogilbyi,* a species of Hypsiprymnus, which is found in the Swan River district, and also in South Australia. This I cannot think is sufficiently established as a species, distinct from the *Hypsiprymnus penicillatus.* The first two or three specimens of *B. Ogilbyi* which came under my notice, having the feet of a deepish brown colour, the tail of a rusty brown hue, and the apical third black, below as well as above, I imagined they might, perhaps, prove distinct from the *H. penicillatus;* but recently I have examined many specimens, both from West and South Australia, and others from New South Wales, and found them to vary somewhat in each of those localities, as to their colouring. All that can be said is, that the specimens of the Tufted-tailed Hypsiprymni from the western and southern districts are generally somewhat darker in the colouring of the feet and tail than those from New South Wales; but it is certainly, in some cases, difficult to distinguish these, which I can but regard as local varieties, by this difference of the colouring.

Amongst a series of skulls, a great portion of which were lent me by Mr. Gould, and some of which are contained in the British Museum Collection, in the museum of the Zoological Society, and that of the Royal College of Surgeons, I could discover no points of distinction between those which were from specimens of *B. penicillata* and those named by Mr. Gould *B. Ogilbyi.* Of the skulls of *B. penicillata,* from New South Wales, but few specimens have come under my notice; but of *B. Ogilbyi,* I have examined many specimens of both sexes, and of young and adult individuals: the size and proportions of some of these skulls are expressed in the following table of measurements:—

	B. Ogilbyi. Female, Adult. Mr. Gould's Collection.		*B. Ogilbyi.* Female, Adult. Mr. Gould's Collection.		*B. Ogilbyi.* Male, Adult. Mr. Gould's Collection.		*B. Ogilbyi.* Male, Adult. College of Surgeons. From Mr. Gould's Collection.		*Bettongia penicillata?* Male. Mus. Zoolog. Society.		Skull from New South Wales. Coll. Surg. From Mr. Gould's Collection.		*Bettongia penicillata.* Specimen *b*, Brit. Mus. Catalogue. From Liverpool Plains.	
	Inches.	Lines.	Inches.	Lines.	Inches.	Lines.	Inches.	Lines.	Inches.	Lines.	Inches.	Lines.	Inches.	Lines.
Total length of skull	3	0	3	1	2	10	2	$10\frac{2}{3}$	3	0	2	11	3	0
Width from the outer side of zygomatic arches	1	$7\frac{1}{3}$	1	$7\frac{1}{3}$	1	7	1	$6\frac{1}{2}$	1	$7\frac{3}{4}$	1	$6\frac{3}{4}$	1	$7\frac{1}{2}$
Length of the nasal bones	1	2	1	2	1	1	1	1	1	$2\frac{1}{2}$	1	$1\frac{1}{2}$	1	$2\frac{2}{3}$
Width of ditto at the base	0	$5\frac{3}{4}$	0	6	0	6	0	$5\frac{1}{3}$	0	$6\frac{3}{4}$	0	$6\frac{3}{4}$	0	6
" " near the apex	0	$3\frac{1}{4}$	0	3	0	$3\frac{1}{3}$	0	3	0	$3\frac{1}{3}$	0	$3\frac{1}{2}$	0	$3\frac{2}{3}$
" of interorbital space	0	8	0	$8\frac{1}{4}$	0	$7\frac{3}{4}$	0	$8\frac{1}{4}$	0	$8\frac{2}{3}$	0	8	0	$8\frac{1}{4}$
Length of three incisors on one side of the upper jaw, taken together	0	$3\frac{2}{3}$	0	$3\frac{2}{3}$	0	4	0	$3\frac{2}{3}$	0	$3\frac{3}{4}$	0	$3\frac{10}{12}$	0	$3\frac{10}{12}$
Distance between incisors and canine	0	$1\frac{1}{6}$	0	1	0	1	0	$0\frac{3}{4}$			0	$0\frac{3}{4}$	0	$0\frac{3}{4}$
" between canine and premolar tooth ...	0	$4\frac{2}{3}$	0	5	0	$4\frac{1}{2}$	0	$4\frac{1}{2}$			0	$3\frac{3}{4}$	0	$4\frac{3}{4}$
Length of premolar	0	$3\frac{1}{6}$	0	$3\frac{1}{6}$	0	$3\frac{1}{3}$	0	$3\frac{1}{6}$	0	$3\frac{1}{2}$	0	$3\frac{2}{3}$	0	$3\frac{1}{4}$
" of molar teeth and premolar together ...	0	$9\frac{1}{2}$	0	$9\frac{1}{2}$			0	$9\frac{1}{2}$	0	10	0	$10\frac{1}{2}$	0	9
" of palate, to the foremost edge of posterior opening	1	2									1	$1\frac{1}{2}$	1	$1\frac{1}{2}$
Height of lower jaw, measured in a vertical line dropped from the apex of coronoid process	0	$10\frac{3}{4}$	0	10	0	$10\frac{1}{3}$			0	10	0	$9\frac{1}{2}$	0	$9\frac{3}{4}$
Distance between back part of condyle of lower jaw and fore part of coronoid process, measured in a horizontal line	0	$6\frac{3}{4}$	0	7	0	$6\frac{3}{4}$					0	7	0	7

These dimensions show the extreme of variation which I have observed in the skulls of numerous specimens of the Rat-Kangaroo under consideration: they also teach us that, unlike the true Kangaroos, there is not any marked difference in the size of the cranium of the two sexes. As regards the fifth and sixth columns, it will be perceived that the two skulls, the dimensions of which are there given, have the nasal bones rather broader behind than the other skulls, and the premolar tooth is a trifle longer from front to back. The animal from which the skull of the sixth column is taken was not sent home; that belonging to the skull of the fifth column is in the museum of the Zoological Society, and is the specimen figured in "Marsupialia" (Plate 16). Of this specimen (which is a male), the habitat is unfortunately not known: in its colouring it is, in some respects, intermediate between the *B. penicillata* and *B. Ogilbyi*, having the rich rusty brown tail of the latter, but the tarsi paler, though considerably suffused with rust colour, especially about the heel. Its dimensions are given below, in column A.

The dimensions of a skull given in the seventh column are those of a male specimen of *Hyps. penicillatus* in the British Museum, from Liverpool Plains, New South Wales, (specimen *b* of the Museum Catalogue). This individual agrees closely with the original specimen in the Zoological Society's Museum; but I do not perceive any yellow tint on the under parts of the body, which are here impure white; and on the chest is a patch of white hairs which are uniform throughout, having no grey at the root as usual. Its dimensions are given in column B.

In two specimens of *B. Ogilbyi*, in Mr. Gould's Collection, from the vicinity of York, Western Australia, the general colouring of the fur was brown, but freely pencilled with yellowish white; the cheeks and sides of the body distinctly suffused with yellow; the under parts yellow-white; ears with yellow hairs internally; externally with hairs of the same colour as those of the head, but brownish at the margin; the tarsi of a rusty brown hue, but with the toes of a deepish brown, especially at the sides; tail rusty brown, with the apical half, which is covered with bushy hairs, brown-black; the under side brown, assuming a deep brown tint at and near the tip. The dimensions are given in columns C and D: those of column C are from a male specimen, and those of column D are from an individual which is apparently a female.

A female specimen in the British Museum Collection (*b* of the Catalogue) differs from the above in having the fur somewhat softer, and the apical third of the tail perfectly black beneath as well as above. For its dimensions, see column E. This is no doubt an immature individual.

The British Museum collection contains likewise individuals from South Australia, which have, as in the one last mentioned, the apical portion of the tail black both above and below, and one in which it is brown beneath. They vary as to the intensity of the colouring of the feet. I may notice also a specimen in the Zoological Society's Museum, in which there is a small tuft of white hairs at the end of the tail. The skull presents all the usual characters of the species.

	A		B		C		D		E	
Length from nose	In.	Lines.	In.	Lines.	In.	Lines.	In.	Lines.	In.	Lines.
to root of tail	15	0	14	6	15	0	16	0	13	0
" of tail ..	12	9	12	0	13	0	12	0	10	6
" of tarsus and claws ...	4	2	4	4	4	4	4	3	4	2
" of ear	1	$1\frac{1}{2}$	1	1	1	0	1	0		11

As regards the skull in *H. penicillatus* (pl. 6, fig. 3), I have only to remark, that it is narrower than other species of the Bettongia section which have come under my observation; the zygomatic arches are less boldly thrown outwards than usual, and have the middle portion of the zygoma nearly straight; the face is somewhat elongated, and acutely pointed; the nasal bones narrow; the auditory bullæ large; the palate deeply emarginated posteriorly; the incisor teeth are much compressed; the canine moderate as to size; the premolar short from front to back, as compared with other species, and deeply sulcated in the vertical direction, both internally and externally; the number of grooves on either side is about seven. The foremost molar of each series is the largest, and the hindermost the smallest, and is about equal in width to two-thirds of that of the foremost molar; the upper outline of the skull, as viewed from the side, presents a gentle and nearly even convex curve, but descends somewhat suddenly towards the occipital crest, and the interorbital space is nearly flat.

Although there are some slight variations in the width of the nasal bones, and even in the antero-posterior extent of the premolar tooth, it does not appear that these are connected with differences in the animals, nor that such differences distinguish the *H. Ogilbyi* from the *H. penicillatus*.

A great portion of the Marsupialia possess a prehensile power in the tail, and in the tree-climbing Opossums and Phalangers this organ is used as a fifth hand, hanging by its means, as they do, from the branches of trees. In the great Kangaroos, the tail, though not prehensile, still assists in locomotion This, however, does not appear to be the case in the Kangaroo-rats. In *Hypsiprymnus minor*, and some nearly allied species, in which the tail resembles that of a rat, being sparingly clothed throughout with short hairs, that organ appears neither to have any prehensile power nor to be used to support part of the weight of the body, as in the Kangaroos; but in the Bettongias,—in which the apical portion of the tail is furnished above with a brush of long hairs, and clothed beneath with short hairs which are closely applied to the skin, the organ in question possesses the prehensile power—a power of encircling and holding objects; but here the tail is used, as Mr. Gould informs us, for carrying grasses, &c., with which these animals form their nests. The nest, Mr. Gould states, is placed in a hollow in the ground, excavated for its reception, and, its upper surface being on a level with the surrounding herbage, it requires the practised eye of the native to discover it. During the day the nest is generally tenanted by one, and sometimes by a pair, of these small Rat-Kangaroos, and these lie perfectly concealed from view, there being no apparent outlet. It would seem, that after they have crept in, they drag some grass over the entrance. In the evening they sally forth in quest of food, which consists of grasses and roots; the latter being procured by scratching and burrowing, for which the strong claws of their fore feet are well adapted.

Bettongia Gouldii.

The specific name *Gouldii* is applied by Mr. Gray to a very young animal contained in the British Museum Gallery, which it appears to me merely presents the immature condition of the *Hypsiprymnus penicillatus*. This specimen is about the size of a Common Rat *(Mus decumanus,)* being 7½ inches in length from the tip of the nose to the root of the tail. The fur is of a brownish tint, the hairs being pencilled with black and yellowish white: the under parts of the body are white, but in parts suffused with yellow. The tail is nearly 7 inches in length, of a rusty brown colour at the base, and the terminal half is black both above and below. The tarsi are brown, and, as is usual in young animals, are disproportionately large, being 3¼ inches in length, including the nail of the central toe. The animal in question is from

South Australia. From the same quarter the British Museum collection has received the skull and foot of a Rat-Kangaroo, the former of which is represented on Plate 6, fig. 1. This skull, as will be seen upon comparing the figure alluded to with the skull, figure 3, on the same plate, greatly resembles that of *H. penicillatus*, but differs in being considerably smaller, and in having the nasal bones shorter: the palate moreover, is rather more deeply emarginated behind, and the hindermost molar of each series is proportionately smaller, being about half a line only in diameter. The foot which was sent with it, as appertaining to the same animal, resembles that of *H. penicillatus*, but is more slender, and the claws, or nails, are narrower. As the permanent premolar and all the true molars are fully developed, the animal must have been adult, and there would seem to be some grounds for the belief that there exists in South Australia a species of Rat Kangaroo, nearly allied to, but distinct from the *H. penicillatus;* the general resemblance which exists, however, between the two skulls, figs. 1 and 3, Plate 6, is such, that I am rather inclined to believe that the smaller one belonged to an individual of *H. penicillatus*, which from some cause was stunted in its growth. Following are the dimensions of the small skull.

	Inches.	Lines.
Total length of skull	2	$6\frac{1}{2}$
Width of ditto	1	$5\frac{2}{3}$
Length of nasal bones		11
Width of ditto at the base		5
" of ditto near the apex		$2\frac{2}{3}$
Length of frontal bones	1	0
Width of interorbital space		$7\frac{1}{2}$
Length of palate		11
" of auditory bullæ		6
" of three incisors of upper jaw		$3\frac{7}{10}$
" between incisors and canine ...		$0\frac{1}{2}$
" between canine and carnassier		$3\frac{3}{4}$
" of carnassier		3
" of ditto, and four molars, taken together		$8\frac{1}{3}$

HYPSIPRYMNUS (*Bettongia*) CAMPESTRIS.

Plain Rat-Kangaroo.

Bettongia Campestris. Gould, Proceedings of the Zoological Society, for June, 1843, Part 11, p. 81.

Head broad and short; ears short and rounded, well clothed with yellow-white hairs internally; brown externally, excepting at the base behind, where they are white: fur long and soft, prevailing tint very pale ochreous yellow, but pencilled with black; under parts pale buff-yellow; legs and feet yellowish, the former slightly tinted with rust colour, and this hue is observed near the root of the tail; the toes of the fore feet slightly suffused with brownish. Tail rather sparingly clothed with very pale adpressed hairs. Skull, Plate 6, fig. 2.

Inhabits South Australia.

This is a very distinct species, remarkable for its short and broad head, and general pale yellowish colouring. The hairs of the back are grey at the root, yellow in the middle, then blackish, followed by a long yellow-white space, and black tip; on the chest and belly they are pale grey at the base, and yellowish externally, but on the lower part of the abdomen the grey is wanting. The upper lip is white: the muffle is naked and broad; the fore legs small, and the claws of the fore feet are long, and white: the tarsi are very long, and of a rusty white hue. The tail is moderately long and slender, but sparingly clothed with small pale hairs on the upper surface and at the sides, and scales are visible as in the rat's tail; on the under part, the hairs are more dense, harsher, and of a brown-white colour. The sides of the body, and outer surface of the hind legs, are of a more distinct yellowish hue than other parts.

	Inches.	Lines.
Length from nose to root of tail	15	6
" of tail	13	0
" from tip of nose to ear	2	8
" of ear	1	1
" of tarsus and claws	4	10
" of skull	2	8
Width of ditto	1	8
Length of nasal bones	1	4
Width of ditto, at the base		$10\frac{1}{3}$
" of ditto, near the apex		$4\frac{1}{2}$
Length of frontal bones		9
Width of ditto between the orbits		$8\frac{2}{3}$
Length of three incisors of upper jaw, taken together		$3\frac{3}{4}$
Space between incisor and canine		$1\frac{1}{4}$
" between canine and premolar		3
Length of premolar		3
" of series of molars on one side of upper jaw, including the premolar		$10\frac{1}{4}$
" of auditory bullæ		$5\frac{1}{2}$

The skull of *Bettongia campestris* differs considerably from the crania of other species of the genus; it is short, and most remarkable for the great width of the nasal bones; these, and the frontals, present nearly a flat surface, but the latter are slightly concave between the orbits: the auditory bullæ are rather small. The two foremost incisors are broad; the second incisor on either side is unusually small, the third moderate: the canine is a minute tooth, scarcely more than a quarter of a line in width. The premolar is shortish from front to back, and the outer surface is concave in the same direction, and exhibits but a faint indication of three or four vertical grooves; the little lobe on the inner side and back part of this tooth is rather more developed than usual: the molar teeth are rather large in proportion to the skull; the last molar is larger than in either *B. Graii* or *B. penicillata*. The palate is very deeply emarginated behind, and consequently very short, terminating in a line with the hinder part of the premolar tooth. The skull of which the dimensions are given is from an old male, and has the temporal ridges very distinct.

Sub-genus 3, *Potoroüs.*

Potoroüs. DESMAREST, Nouv. Dict. d'Hist. Nat. tom. xxviii. p. 79. 1819.
" " Mammalogie, Part 1, p. 271. 1820.
Hypsiprymnus. ILLIGER, Prod. Syst. Mamm. et Avium, p. 79. 1811.[1]

Head elongated; tarsi short; tail sparingly clothed with short stiff hairs, and exhibiting a scaly skin: muffle naked.

[1] Of the names *Potoroüs* and *Hypsiprymnus*, applied in a generic sense, the former by Desmarest, and the latter by Illiger, to the animal called the Kangaroo-Rat, in White's Journal, that given by Desmarest (which is formed from the native name Potoroo) was undoubtedly the first proposed, since it is quoted by Illiger in his Prodromus, where he first defines the genus *Hypsiprymnus.* I am not aware, however, whether the section was characterised in the first edition of the Nouveau Dictionnaire d'Histoire Naturelle, in which, according to Desmarest, the name Potoroüs first occurs, not having been able to obtain a sight of that edition. In the second edition of this work, where Desmarest points out the peculiarities of his genus, he complains of Illiger's having substituted a new name for the one he had previously proposed. It is clear that the definitions of both authors are founded upon the account in White's Journal, and apply to the animal hereafter described under the name *Hypsiprymnus murinus*, an animal which is apparently distinct from the Kangaroo-Rat of Governor Phillip's work, though regarded by Desmarest, and many other authors, as specifically identical with White's Kangaroo-Rat. Since Illiger's classical name has been very generally adopted for the whole Rat Kangaroo group, and "has well taken root," I have thought it desirable to retain the term *Hypsiprymnus* (it has reference to the animal having the hinder part of the back much raised, when the fore feet are applied to the ground, being compounded of the Greek words, ὕψος and πρυμνα,) in the same sense, that is, for the genus, and to use Desmarest's name *Potoroüs*, for the minor section or sub-genus, of which his *Potoroüs murinus* is the type. We cannot conveniently, it may be observed, use either of the names for the *whole*, and at the same time for *part* of the Rat-Kangaroo section.

The species of the present section have the body more compact, the hinder legs shorter, and the head much more elongated and pointed than the Bettongias; the molar teeth are proportionately smaller, and the anterior pair of incisors of the upper jaw are longer, descending much below the level of the other incisors. To these principal points of distinction between the Bettongias and the Potoroos, we may perhaps hereafter have to add the anchylosed and unanchylosed condition of the tibia and fibula: having seen but the skeleton of one species in each, I am not aware, however, whether such a difference is constant. In a skeleton of Bettongia, which I feel no doubt is referrible to the *B. Gaimardii*, contained in the Museum of the Royal College of Surgeons, the bones in question are anchylosed at their lower extremity, whilst in more than one skeleton of the *H. murinus*, I have found the tibia and fibula distinct throughout.

HYPSIPRYMNUS (*Potoroüs*) MURINUS,

Rat-Tailed Hypsiprymnus, or Rat-Kangaroo.

Poto Roo, or Kangaroo-Rat. WHITE's Journal of a Voyage to New South Wales, p. 286, and Pl. . 1790.

Macropus minor. SHAW, General Zoology, vol. i. Part 2, p. 513, Pl. 116. 1800.

Hypsiprymnus murinus. ILLIGER, Prod. Syst. Mamm. p. 79. 1811.

Potoroüs murinus. DESM. Nouv. Dict. d'Hist. Nat. tom. xxviii. p. 79—80, Mammalogie, Part 1, p. 271.

Hypsiprymnus setosus. OGILBY, Proceedings of the Zoological Society for November, 1831, Part 1, p. 149.

" *Peron*. QUOY et GAIMARD, Zool. de l'Uranie, p. 64.

" *Bettong* of the natives of New South Wales. Mus. Lin. Soc.

" ? *myosurus*. OGILBY, Proc. Zool. Soc. for May, 1835, Part 6, p. 62.

Fur long, loose, slightly glossy, and rather harsh to the touch: general colour, dusky brown; the upper parts of the body much pencilled with black, and less so with pale brownish yellow; under parts dirty yellowish white; feet dark brown; tail sparingly clothed with short stiff hairs, which do not hide the scaly skin beneath; the hairs black, excepting at the extreme point, where they are white. Skull, Pl. 8, fig. 3.

Inhabits New South Wales. Specimen in the British Museum.

The *Hypsiprymnus minor*, or *murinus*, is readily distinguished from others of the present group, described in the preceding pages, by its elongated head and short tarsus, to which may be added, its rat-like tail, which being furnished with short stiff hairs, and these not very abundant, the scaly skin beneath is but partially hidden. Its form, moreover, is less slender than usual, and its fur is long, and of a dark hue; on the upper parts of the body it may be described as dusky brown, a general tint produced by the admixture of black and pale brownish yellow, the visible portion of the longer and coarser hairs being black, and that of the shorter fur being chiefly of a pale yellow hue: the under parts of the body are of a dirty yellowish white tint, but the fur covering these parts, as well as that of the back, is of a deepish grey colour next the skin. The ears are short and rounded, clothed internally with dirty white hairs, and on the outer side with hairs of the same colour as those of the head; the feet are brown. The muffle is not only naked in front, but a narrow naked space is continued backwards for about a quarter of an inch on the upper surface of the muzzle.

	From N. S. Wales. FEMALE. Inches.	Lines.
Length from tip of nose to root of tail ...	15	6
" of tail	9	3
" from nose to ear, about	3	7
" of ear	1	1
" of tarsus, including the nail of the middle toe	3	2

The present species was first described by Hunter, under the name Potoroo, or Kangaroo-Rat, in the Appendix to White's Journal, and from the description, and somewhat rude figure there given, it would have been difficult to determine which of the numerous species of Rat-Kangaroos since discovered, the Potoroo of White should be referred to, were it not that the skull of that animal is still preserved in the Museum of the Royal College of Surgeons. By the aid of that skull we are enabled clearly to identify the Potoroo of White's Journal, (upon which Shaw founds his *Macropus minor*,)[1] with the *Hypsiprymnus murinus*, of Pander and D'Alton,[2] and with the *H. Peron* of Quoy and Gaimard, founded upon a skull contained in the Paris Museum, of which Professor Owen has been so kind as to lend me a drawing.

Mr. Ogilby states that the animal to which he has given the name of *H. setosus* is known in the colony of New South Wales by the native name "Bettong;" and this remark no doubt has reference to the Rat-Kangaroo, so labelled in the collection of the Linnean Society, which specimens not only agree with Mr. Ogilby's description, but also with the animal I identify with the *Macropus minor*, of Shaw.

The animal described by Mr. Ogilby under the name *Hypsiprymnus myosurus*, agrees so perfectly with the *H. murinus* in every respect excepting certain differences observable in the skull, that I cannot think those differences are specific. It is supposed to have been brought from

[1] It is evident that Shaw's account of the *Macropus minor* is chiefly taken from Hunter's description of the Potoroo, though he likewise refers to the Kangaroo-Rat of Phillip's Voyage in that account. The specific name *minor* is, now that the animal is placed in the genus *Hypsiprymnus*, particularly inappropriate, there being other *Hypsiprymni* of smaller size, and scarcely any which are of superior bulk to the *H. minor*. It would be well to adopt in its stead the next oldest specific name of *murinus*.

[2] Skelete der Beutelthiere.

New South Wales, and the same animal is certainly common in Van Diemen's Land. To describe its colouring would be to repeat the description already given of *H. murinus:* I will therefore merely add its dimensions, together with those of two similar animals from Van Diemen's Land.

	H. myosurus, Ogilby.		Van Diemen's Land Specimens.			
	Inches.	Lines.	Inches.	Lines.	Inches.	Lines.
Length from tip of nose to root of tail	15	0	13	6	19	0
" of tail	10	6	9	6	9	10
" from nose to ear, about ...	3	11	3	11	4	2
" of ear	1	2	1	2	1	3
" of tarsus, including the claws	3	4	3	2	3	3

The dimensions contained in the second and third columns are from specimens contained in the British Museum, both of which are males. The original of Mr. Ogilby's description of *H. myosurus* is also a male.

The animal figured by White was a female, and other specimens having the same proportions in the skull, which have come under my notice, have been females, whilst those in which the skull resembled fig. 2 of Plate 8 were males. The dimensions of the skull, fig. 2, (which forms part of the British Museum collection), are contained in the third column of measurements given on the next page. The dimensions in the fourth and fifth columns are from crania in the Royal College of Surgeons: they differ from the skull of Potoroo of White, (Pl. 8, fig. 3,) chiefly in being larger, in having the muzzle proportionately more elongated and narrower, and in the frontal bones being produced on either side into a small post-orbital process. With regard to this last-mentioned character, I must observe, however, that in the figure of the skull of *H. Peronii*, now before me, a small post-orbital projection is represented; this skull in other respects resembling White's Potoroo.

Following are the dimensions of the skull of the Potoroo of White, that of *H. Peronii*, and of three skulls of the *H myosurus* of Mr. Ogilby.

	Potoroo of White. Col. Sur.		H. Peron		H. myosurus. Brit. Mus.		H. myosurus. Coll. Surg.			
	Ins.	Lin.	Ins.	Lin.	Ins.	Lin.	Ins.	Lin.	Ins.	Lin.
Total length of skull ..	3	1½	3	2⅓	3	8	3	9	3	9½
Width	1	6⅔	1	6¼	1	9½	1	9⅓	1	8⅓
Length of nasal bones ...	1	4⅔	1	4½	1	6	1	7⅓	1	8¼
Width of ditto at base ...		6		5½		4¾		4		5½
" near the apex		2⅓		2⅔		3		2¾		2½
Interorbital space		7½		7		9½		9		
Length of three incisors taken together, on either side of the upper jaw		3½				4¼		4		
From posterior incisor to canine		1⅓				1½		1⅔		
Between canine and premolar		3¼				3		3		
Length of upper premolar ...		3⅙				3¾		3⅙		
" of premolar, and other molars taken together ...		10½				11¾		10⅔		
From front of foremost incisor to back part of last molar	1	7½			1	10	1	9		
From ditto to front of premolar		9¼				10½		10⅔		

HYPSIPRYMNUS (*Potoroüs*) GILBERTII.

Gilbert's Hypsiprymnus.

Hypsiprymnus Gilbertii. GOULD, Proceedings of the Zoological Society for February, 1841, Part ix. p. 14; Monograph of the Macropodidæ, Part i. Plate 15.

Hypsiprymnus micropus[1] (Gould's MSS.) WATERH. Naturalist's Library, Marsupialia, p. 180.

Fur long, loose, and rather harsh: general colour deep greyish brown; upper parts of body pencilled with black and rusty white, or yellowish white; under parts dirty white; feet brown: skull with the facial portion broad, the outer surface of the superior maxillary bones being very convex.—Skull, Plate 8, fig. 1.

Inhabits King George's Sound.

Skins, or stuffed specimens, of this animal, are difficult to distinguish from the *Hypsiprymnus murinus;* the form of the head, however, must be different in the living animal. The fur is perhaps rather more harsh, and more distinctly pencilled with yellowish or rusty white: on the back the hairs of the fur are of a slate grey colour, and rusty brown externally; the longer and harsher hairs, which are abundant, have the basal half slate grey, and the remaining portion first white, slightly tinted with yellowish rust colour, and then black: some of the interspersed longer hairs have the

[1] Mr. Gould first determined to give the name *micropus* to this species; and when he afterwards applied that of *Gilbertii*, I had no opportunity of altering the former name, which I had adopted in the little volume on Marsupialia, in the Naturalist's Library. There can be no question about the priority, however, since the part of the Proceedings containing Mr. Gould's description was published before the Naturalist's Library volume referred to, which did not appear till August, 1841.

exposed portion entirely black. The fur on the belly is pale grey next the skin, and dirty white externally; the fore and hind feet are brown, but greyish above, excepting on the toes; the ears are densely clothed with brown hairs, excepting at the margin, where they are black, or nearly so.

	Inches.	Lines.	Adult Male. Inches.	Lines.
Length from nose to root of tail ...	15	6	15	0
" of tail, about	7	0	8	6
" from nose to ear ...	3	2		
" of ear	1	1		
" of fore-arm, hand, and nails, about			2	8
" of tarsus	3	7	3	2
" of outer toe and nail ...				8
Outer toe short of central one ...				8
Length of central toe and nail ...			1	3
" of inner toes				6¼
Inner toe short of central one (*i.e.* from tip of nail of central toe to end of nails of the joined inner toes)				9½

The skull of *Hypsiprymnus Gilbertii* closely resembles that of *H. murinus*, as will be seen by comparing figures 1 and 3 of Plate 8; but the former differs in having the muzzle broader, caused by the greater convexity of the facial parts of the superior and intermaxillary bones, and the nasal cavity is therefore more expanded: the molar teeth are also smaller; the premolar in *H. Gilbertii* is but little longer than the foremost true molar, whilst in *H. murinus* it is equal in length to the first true molar tooth, added to that of the second.

	Female. Inches.	Lines.
Total length of skull	3	3
Width	1	6½
Length of nasal bones	1	5¼
Width of ditto at the base		6½
" near the apex		3

	Female. Inches.	Lines.
Width between orbits		8
" of muzzle in a line with the front of the premolar		$10\frac{2}{3}$*
Length from front of foremost incisors to premolar		$10\frac{1}{4}$
" of the five molar teeth, taken together		$9\frac{1}{2}$
" (of which, the premolar is a trifle more than		$2\frac{1}{2}$)

HYPSIPRYMNUS (*Potoroüs*) PLATYOPS.

Broad-faced Rat-Kangaroo.

Hypsiprymnus platyops. Gould, Proceedings of the Zoological Society for June, 1844, p. 103.

Head short and broad; muzzle naked at the extremity only; ears small; tarsi short: fur very long, and rather soft; general colour grey brown, with distinct pencillings of white; under parts brownish white; feet brownish white; the tarsi finely pencilled with brownish in front: tail with small hairs applied to the skin, and of a deepish brown colour, almost black on the apical half; beneath, brown-white.

Inhabits Western Australia.

This is a small and very distinct species, readily distinguished from *Hyps. minor* and *H. Gilbertii*, by its having the tip of the muzzle naked in front only; in the two species just named the naked part of the muffle being extended somewhat on the upper surface. The hairs constituting the fur, are, on the back, grey at the root, then yellowish brown, and this is followed by a long space in each hair which is white, and this again is followed by black, that being the colour

* The same part in a skull of *H. murinus*, of the same size as the above skull, measures $8\frac{1}{2}$ lines.

of the tips of the hairs: on the under parts of the body each hair is pale grey at the root, and dusky white externally: the feet are dirty white, indistinctly grizzled with brownish; this latter tint being most distinct on the sides of the toes. The ears are short and rounded; externally clothed with longish hairs, which are partly brown and partly white; the hairs on the inner side of the ears are dirty white. The upper incisors are very small; the posterior two on either side of the jaw project very little from the gum; the foremost two, however, are much longer, descending much below the level of the crowns of the hinder incisors: the canine is small, and situated scarcely one-twelfth of an inch distant from the incisors. The zygomatic arches (so far as one may judge from the skull as enclosed in the skin) must be thrown boldly out from the cranium, and thus give the breadth to the face which suggested the specific name.

	Inches.	Lines.
Length from nose to root of tail	12	0
" of tail (about)	7	6
" of tarsus, including the nails ...	2	8
" of ear		10
" from nose to ear, about	2	7

The specimen from which the above description is taken formerly belonged to Mr. Gould, and is now in the British Museum collection. It is labelled as coming from "Walyema Swamps, about forty miles north-east of Northam, Western Australia." It is the only individual of the species yet transmitted to England. The fur on the upper parts of its body is very distinctly pencilled with white.

As partaking, to a certain extent, of the characters of the Kangaroo group on the one hand, and of the Wombats on the other, I will here notice the extraordinary fossil remains described by Professor Owen under the names of *Diprotodon* and *Nototherium*.

Genus, *Diprotodon*,[1] (fossil.)

Diprotodon. OWEN, in the Appendix to Mitchell's Three Expeditions into the Interior of Australia, 8vo. 1830, vol. ii. p. 362. Report on the Extinct Mammals of Australia, &c. Report of the British Association for 1844, p. 223. Catal. of Fossil Organic Remains of Mammalia and Aves contained in the Museum of the Royal College of Surgeons.

The genus *Diprotodon* contains but one species, the *D. Australis*, an animal which Professor Owen judges must have been of superior bulk to the Rhinoceros. The species was originally founded upon a portion of a tusk, and a fragment of a jaw from Wellington Valley; other specimens have been since transmitted to Professor Owen from the alluvial deposits in the bed of the Condamine River,[2] westward of Moreton Bay, and the district of Melbourne near Port Phillip, has also yielded some interesting fragments of the Diprotodon.[3] Besides portions of the upper and lower jaws, which are clearly referable to the *D. Australis*, the College collection contains some vertebræ, for the most part imperfect,—fragments

[1] This name has reference to the two large tusks with which the fore part of the lower jaw is provided.

[2] The River Condamine is situated in lat. 28° S., long. 130° E.; and Sir Thomas Mitchell (who has presented to the College of Surgeons the specimens above referred to) remarks with regard to this river, that "it is remarkable for forming large basins at some places, and losing its course in swamps at others, and at other parts, again, cutting its course in a deep channel, through deep beds of alluvium, in which these bones are thus brought to light." Catal. of Foss. Mamm. &c. in the College of Surgeons.

[3] These fragments were found by Patrick Mayne, Esq. during the operation of sinking a well, and are now in the museum of the Royal College of Surgeons, having been presented by Dr. Hobson, of Melbourne.

of ribs, of a blade-bone, and of some of the long bones, together with a right heel-bone,—which there are grounds for believing belonged to the same animal.

The various fragments of the lower jaw show that the Diprotodon possessed one incisor tooth and five molars in each ramus. The incisor is remarkable for its great size: it is very long, being deeply implanted in the jaw, and its direction approaches the horizontal, but the extremity is slightly curved upwards. A section of this tooth presents a rounded outline, and gives a vertical diameter of one inch and a half, whilst the transverse measurement is nearly an inch. The apices of these incisors were bevelled off as in the incisors of a Rodent, and the resemblance to the cutting teeth of the Rodents is further manifested by the enamel being almost entirely confined to the outer surface of the tooth; the upper and the greater portion of the inner surface being destitute of enamel. The three anterior molar teeth exhibit a marked successive increase in size, whilst the posterior two are nearly equal, and all these teeth are rooted, and with the exception of the first, are longer than broad, and like the molars of the lower jaw of the Tapir, or those of the Kangaroo, the crown presents two elevated transverse ridges, but these ridges are still more elevated than in either of the animals just mentioned: these principal ridges, or cusps, have a slightly elevated ridge running downwards and inwards from their outer angle, which clearly represents the ridge similarly disposed in the Kangaroo's molar tooth, though it is less developed in the Diprotodon, and here there is scarcely a trace of the longitudinal ridge which joins the two principal cusps in the molar tooth of the Kangaroo. Besides the two principal cusps, the band of the tooth is developed into a strong ridge in front, and a still stronger one on the hinder part of the molar; in this respect differing from the molar tooth of the lower jaw of the Kangaroo, where the band forms a still more prominent ridge on the fore part of the tooth, and is almost always wanting on the hinder part, though it may be seen in some species, as for instance in the fossil species to which Professor Owen has given the name *Macropus atlas*. The length of the fourth molar tooth, in one fragment of a lower jaw is $1\frac{7}{12}$ inch, and its width $1\frac{1}{12}$ inch; the length of the fifth molar is $1\frac{3}{4}$ inch, and its width $1\frac{1}{4}$ inch. The lower jaw is remarkable for the great extent and depth of the symphysial portion, (its vertical diameter, anterior to the molar series, is 4 inches,) a structure which has relation to the great development and deep implantation of the incisors, and which leaves a long toothless interval between these teeth and the molars. The angle of the jaw is distinctly bent inwards, and the lower edge of the ramus at this part presents a broad flattened surface—characters in which we perceive a manifestation of the Marsupial affinities of the animal, and when we find the molar teeth approximating very closely in their structure to those of the Kangaroos, accompanied by dental formula of another herbivorous Marsupial division, the Wombats, where the incisors, as in Diprotodon, are but partially covered by enamel, we have strong grounds for regarding the Diprotodon as a Marsupial animal. But these are not the only grounds adduced by Professor Owen, in proof that the animal under consideration belonged to the Marsupialia; numerous fragments of other parts of the skeleton found together with the parts of the jaws noticed, which at the same time, from their correspondence in mineral condition, and in relative proportions with the jaws, it is highly probable belonged to the same animal, all more or less bear out the conclusion. I shall here only notice two of the most complete bones which were found in the bed of the Condamine River, with the portions of the lower jaw. The first of these is the body of a dorsal vertebra measuring $2\frac{1}{4}$ inches in antero-posterior diameter, 3 inches in vertical diameter, and $4\frac{3}{4}$ inches in the transverse direction. The most remarkable character presented by this body of a vertebra consists in a deep transversely oblong depression situated in the tract between the neurapophyses, a character very rarely found in mammalian vertebra, but which is found in the dorsal and lumbar vertebræ of the Wombat. The second bone is a right os calcis, or heel bone, which measures 6 inches in length and $5\frac{1}{2}$ in breadth; it presents two large articular surfaces at right angles to each other upon its upper and anterior part, has a short calcaneal process, which is broad, depressed, and bent upwards, and is perforated at the base; there is moreover a thick obtuse process, directed downwards from the internal and under part of the bone. Of this calcaneum, Professor Owen remarks, that it presents some features of that of the *Mylodon* and *Mastodon*, and others which are peculiar to itself, but its essential characters, that of having the articular surfaces two in number, are only found in the Wombat and Koala.

Figures of several of the specimens alluded to in the preceding pages will be found in the Catalogue of Organic Remains, &c. in the Museum of the College of Surgeons.

Genus, *Nototherium*,[1] (fossil.)

Nototherium. OWEN, in the Report of the British Association for 1844, p. 231; and, in Catalogue of the Fossil Organic Remains contained in the Museum of the Royal College of Surgeons, p. 314.

[1] From νότος, the south, and θηρίον, beast.

The genus *Nototherium*, founded upon some portions of lower jaws discovered in the bed of the Condamine River, and presented to the Royal College of Surgeons by Sir Thomas Mitchell, contains two species, both of great size, being it is supposed equal in bulk to the Rhinoceros. The remains in question demonstrate that these Mammals are distinguishable from species of other genera, by the angle of the jaw being curved inwards, and by the absence of lower incisive tusks, combined with a short *symphysis menti*, and, apparently, only four molar teeth, these teeth being rooted, and some of them at least having two transverse elevated cusps.

Nototherium inerme. Owen.

Of this species the College collection contains a right half of a lower jaw, of which the hinder part is imperfect, and a fragment of a second right ramus, added to which is an astragalus, or ankle-bone, which it is probable belonged to the same animal. The dentition in the most perfect portion of a jaw, Professor Owen observes, " consists of molar teeth exclusively, four in number, which increase in size as they approach the posterior part of the series: a small portion of the anterior end of the symphysis is broken away, but there is no trace in that part of the socket of any tooth, and it is too contracted to have supported any tusk or defensive incisor. The length of the jaw is 11 inches: the molar series, which commences one inch in advance of the posterior border of the symphysis, is 6 inches in extent; each tooth is implanted by two strong and long conical fangs, the hindermost being the largest, and both being longitudinally grooved upon the side turned to each other. The first tooth is wanting, and the crowns of the rest are broken away: the base of the third remains, which gives an indication of a middle transverse valley, which most probably separated two transverse eminences."

The second fragment of a jaw exhibits the concavity along the lower part of the inner surface of the ramus, formed by the bending in of its lower margin continued from the angle of the jaw.

The astragalus alluded to is described as a broad, subdepressed, and subtriangular bone, the angles being rounded off, especially the anterior one; the upper or tibial surface is quadrate, concave from side to side, in a less degree convex from before backwards: a ridge extending in this direction divides the tibial from the fibular surface, which slopes outwards at a very open angle, and maintains a nearly horizontal aspect, presenting an oblong trochlea for the support of the fibula, shallower, and one-third smaller than that of the tibia. The tibial articular surface is not continued on the inner side of the astragalus, but its anterior and internal angle, which becomes convex in every direction, is directly continued into the anterior scaphoidal convexity, which sweeps round a deep and rough depression, dividing the outer and anterior part of the tibial trochlea from the corresponding half of the scaphoidal convexity; this has the greatest vertical extent at its inner part, where it is separated by a narrow rough transverse channel from the part which rested upon the os calcis. The calcaneal surface is single, and covers almost the whole of the under part of the astragalus: the greatest proportion of it is flat and reniform; an angular tuberosity or process being continued from the concave margin, where the pelvis of the kidney, to pursue the comparison, would be situated. On the inner margin of this calcaneal surface, opposite the tuberosity, a small triangular flattened surface is continued upwards upon the inner and posterior side of the astragalus, and nearly touches the inner and posterior angle of the tibial trochlea. The length of this astragalus is 4 inches 8 lines; its breadth is 3 inches 5 lines; its depth (at the base of the scaphoidal convexity) is $2\frac{1}{2}$ inches.[1]

This great fossil astragalus from Australia, Professor Owen remarks, offers great and remarkable peculiarities, and these are exclusively, but most closely, repeated in certain Australian genera of *Marsupialia*, and especially in the bulkiest of the existing vegetable feeders (the Wombats) which are not saltatorial. "The inference can hardly be resisted, that the rest of the essential peculiarities of the Marsupial organization were likewise present in that still more bulky quadruped, of which the fossil under consideration once formed part.[2]

Nototherium Mitchellii. Owen.

With the molar teeth (at least the penultimate and last), of equal, or very nearly equal size, with those in the jaw of *Nototherium inerme*, the present species has a depth of jaw below the middle of the penultimate molar of $3\frac{1}{4}$ inches, whilst the more perfect portion of the jaw of *N. inerme*, measured at the same part, only gives a dimension of $2\frac{3}{4}$ inches. The thickest part of the jaw of *N. Mitchellii*, beneath the same molar, is $2\frac{1}{4}$ inches, but in that of *N. inerme* it is only 1 inch and 11 lines. A marked distinctive character of the *N. Mitchellii*, as compared with the *N. inerme*, consists in the

[1] Owen, Catal. of the Organic Remains, &c. p. 319-20.

[2] Of this singular astragalus, as well as of the half jaw of *Nototherium inerme*, a beautiful figure will be found in the part of the College Catalogue already referred to.

position of the last molar tooth, which is in advance of the origin or base of the coronoid process, instead of being internal to, and hidden by that part when the jaw is viewed from the outer side. These and some other differences observable in an imperfect posterior half of the left ramus of a lower jaw, found in the bed of the Condamine River, Professor Owen regards as indicating a second speeies of *Nototherium,* which the Professor has named in honour of Sir Thomas Mitchell, who has exerted himself so much in procuring materials for the elucidation of the organic remains of that interesting continent, Australia.

PHASCOLOMYIDÆ; or Wombat Family.

Phascolomina. Gray, Annals of Philosophy, xxvi. 1825.
Phascolomyidæ. Owen, Classification of the Marsupialia. Proceedings of the Zoological Society for January, 1839, Part 7, p. 19.

Marsupial Mammals with incisor teeth $\frac{2}{2}$; premolars $\frac{1-1}{1-1}$; true molars $\frac{4-4}{4-4} = 24$; all the teeth rootless.

It is in accordance with the views of Professor Owen that the author separates the Wombats from the other Marsupialia as a distinct family, but it may be questioned whether these animals should not be regarded as forming, as I have before stated, an aberrant section of the *Phalangistidæ*, to which they appear to bear the same kind of relationship as that which exists between the Voles (*Arvicolæ*) and the true Rats, among Rodents, the Wombats differing *chiefly* from the Phalangers in having rootless teeth; but also they differ in the reduced number of incisors of the upper jaw. The habits of the Wombats, as compared with those of the Phalangers, likewise involve other differential characters; such as those observable in the structure of the toes of the fore feet, with their short, broad, and solid nails fitted for burrowing, and the small development of the thumb of the inner toe of the hind foot, this organ not being required for prehension in an animal living upon the ground like the Wombat. The section *Phascolomyidæ* contains but two recent species; they form the

Genus, *Phascolomys*.[1]

Phascolomys. GEOFFROY, Notice sur une Nouvelle Espèce des Mammifères, &c. Annales du Mus. d'Hist. Nat. ii. p. 364. 1803.

Vombatus. GEOFFROY, Bulletin des Sciences par la Soc. Philom. iii. p. 185. 1803.

Phascolomys. ILLIGER, Prod. Syst. Mamm. p. 77. 1811.

Amblotis. ILLIGER, loc. cit. p. 78.

Body stout; head large and having the upper surface flattened; muzzle obtuse, the muffle naked, the naked part terminating behind in an angle; nostrils very widely separated behind, but converging in front; upper lip cleft; eyes small; ears small and pointed; limbs equal, short and stout; feet naked beneath; the fore feet with five short and stout toes, each furnished with a broad, solid, and but little curved nail; hind feet provided with five toes, of which the first, or innermost, is very small, nailless, and placed nearly at right angles with the foot; the second, third, and fourth toes joined, and having (as well as the fifth toe) long and curved nails, which are concave beneath: tail rudimentary.

Stomach simple; provided with a special gland situated to the left of the cardiac orifice.

Cæcum very short, wide, and with a vermiform appendage.

Till very recently, the genus *Phascolomys* contained but one known species, the Wombat; and this animal was originally described and figured in Collins' Account of the English Colony of New South Wales:[2] and it is upon this description that Geoffroy founded the genus *Vombatus* in 1803. The account of the dentition of the Wombat given in Collins

[1] From φάσκωλος and μῦς, pouched rat.

[2] 4to. London, 1802. See pages 153 to 158.

work, is, however, very obscure and erroneous :[1] hence when Geoffroy subsequently had an opportunity of examining the animal in question, specimens having been brought alive to Europe by Capt. Baudin, not finding it agree with his published description, he regarded it as constituting the type of a new genus, which he characterised under the name *Phascolomys*. Some years afterwards, Illiger, not aware of the errors in the definition of the genus *Vombatus*, adopted that genus as well as Geoffroy's genus *Phascolomys*, but he proposed the new name of *Amblotis* in lieu of that of *Wombatus*. The dentition of the Wombat is remarkable for its general resemblance to that of the Rodents, which animals it also resembles in its gnawing propensities. The incisor teeth, as in the Rodents, are two in number, in each jaw, and widely separated from the other teeth; they are rootless and very long, curved, and approach to a cylindrical form, but are somewhat compressed (see Pl. 3, fig. 1 *i*.) The molar teeth are also long, curved, and rootless; those of the upper jaw have the convex side of the curve on the inner side, whilst the molars of the lower jaw are reversed in this respect. The true molars are divided into two nearly equal parts by a fold of the enamel entering deeply into the body of the tooth on one side, and a slight indentation on the opposite side; in the molars of the upper jaw the deep indentation is on the inner side, but it is on the outer side of those of the lower jaw. The foremost

[1] The error in the number of the teeth in Collins' work arose probably from a misprint. I may here mention that a specimen, alluded to in the work just mentioned, and which was procured on Preservation Island, was sent by Governor Hunter to Newcastle-upon-Tyne, accompanied by a drawing and description, which latter having been transmitted to Bewick, were published by him in the sixth edition of his History of Quadrupeds. At the British Association Meeting, held at Newcastle, Mr. Gray called attention to the specimen in question, which still exists in the museum of the Natural History Society of that town, and pointed out the identity of the so-called genera *Amblotis* and *Phascolomys*.

molar of each series, which is a premolar, as is often the case with the premolar teeth, represents one half of a true molar; in the lower jaw it is nearly cylindrical, and in the upper jaw it differs in having a small anterior lobe. The two rows of molar teeth of either jaw converge in front, so that they are there separated but by a narrow space, especially those of the upper jaw. A more perfect idea of the structure of the teeth of the Wombat will be obtained upon referring to

PLATE 3.

i. i. FIGS. 1. Are the incisor teeth.
p—m. The premolars.
FIG. 1 *b.* Is the masticating surface of a true molar.
FIG. 1 *c.* Represents the same tooth, viewed from the inner side, and
FIG. 1 *d.* Is the same, viewed from behind.

Fossil remains of a Wombat have been found in the caves of Wellington Valley, approaching very near to the recent species; but after a minute comparison of a considerable portion of a cranium and lower jaw, procured by Sir Thomas Mitchell in the caves mentioned, with the corresponding parts of the recent Wombat, Professor Owen found some differences in the proportions of the teeth which induced him to regard the fossil species as distinct, and in Sir Thomas Mitchell's work,[1] the Professor proposes for it the name,

Phascolomys Mitchellii, (fossil.)

"In this species," Professor Owen observes, "the molar teeth have the antero-posterior diameter greater in proportion to the transverse; as compared with the molars of *Ph. Wombat;*

[1] Three Expeditions into the Interior of Australia, &c. p. 368. Pl. 48.

the first grinder is also relatively larger, and of a more prismatic form; the upper incisors are less compressed and more prismatic; this difference is so well marked, that once appreciated, any one might recognise the fossil by an incisor alone. There is a similar difference in the shape of the lower incisor. The fossil is also a little larger than the largest Wombat cranium in the Hunterian collection."

PHASCOLOMYS WOMBAT.

The Wombat.

Phascolomys	*Wombat.*	PERON et LESUEUR, Voyage aux Terres Australes.
"	*fossor.*	SEVASTIANOF, in Mém. de l'Acad. Imper de St. Petersb. i. p. 444. 1809.
"	*Wombatus.*	LEACH, Zool. Miscell. ii. p. 101, Pl. 96. 1815.
"	*fusca.*	DESM. Nouv. Dict. des Sci. Nat. xxv. p. 500, Tab. G. 44, f. 1. 1817.
"	*Wombat.*	DESM. Mammal. Part 1, p. 276. 1820.
"	*Bassii.*	LESSON, Manuel de Mammal. p. 229. 1827.
"	*ursinus.*	GRAY, List of the Mammalia in the Brit. Mus. p. 95. 1843.
Womback		BEWICK, Quadr. sixth Ed. p. 522. 1811.
Wombat.		COLLINS, Account of the English Colony in New South Wales, &c. p. 153, with Plate. 1802.
Badger.		Of the Colonists.

Length about 3 feet: fur coarse, and moderately long; its general tint grey-brown; on the under parts of the body paler than the upper: ears well clothed with whitish hairs internally, and externally with hairs of the same colour as those of the head, excepting at the tip, where they are blackish: feet black: tail a mere tubercle, of about half an inch in length.

The Wombat is found in New South Wales, South Australia, and Van Diemen's Land, as well as some of the Islands of Bass' Straits. It is a burrowing animal, and remains concealed in its underground retreat during the day, quitting its hole in the night to feed: its food consists chiefly of roots and grass. The flesh is said to resemble pork in its fatness and flavour, but not in colour or texture, being red and coarse. According to Mr. Bass' account, in Col. Collins work already quoted, the Wombat is by no means active, and has a hobbling or shuffling gait somewhat like that of a bear. When at Cape Barren Island, situated in the straits named after him, Mr. Bass chased one of these animals, and having overtaken it, " by placing his hands under its belly, he suddenly lifted it from the ground and laid it upon its back upon his arm as a child would be. It made no noise," the account proceeds, " nor any effort to escape, not even a struggle. The countenance was placid and undisturbed, and it seemed as contented as if it had been nursed by Mr. Bass from its infancy. He carried the beast upwards of a mile, and often shifted him from arm to arm, sometimes laying him upon his shoulder, all of which he took in good part; until being obliged to secure his legs while he went into the bush to cut a specimen of a new wood, the creature's anger arose with the pinching of the twine; he whizzed with all his might, kicked and scratched most furiously, and snapped off a piece from the elbow of Mr. Bass' jacket with his grass-cutting teeth. Their friendship was here at an end, and the creature remained implacable all the way to the boat, ceasing to kick only when he was exhausted."

Specimens of the Wombat which have been brought to England have generally evinced a gentle disposition. The specimen dissected by Sir Everard Home, and which was brought from one of the islands in Bass' Straits, by Mr. Brown, the eminent botanist attached to Flinders Voyage, lived as a domestic pet in the house of Mr. Clift for two years. This animal was a male, measured 2 feet 2 inches in length, and weighed about 20 lbs. The observations made by Sir Everard Home on the habits of this animal whilst in confinement, correspond pretty closely with those already given. " It burrowed in the ground," that author observes, " whenever it had an opportunity, and covered itself in the earth with surprising quickness; it was very quiet during the day, but constantly in motion in the night: was very sensible to cold; ate all kinds of vegetables, but was particularly fond of new hay, which it ate stalk by stalk, taking it into its mouth like a beaver, by small bits at a time. It was not wanting in intelligence, and appeared attached to those to whom it was accustomed, and who were kind to it. When it saw them, it would put up its fore paws on their knees, and when taken up would sleep in the lap. It allowed children to pull and carry it about, and when it bit them, it did not appear to do it in anger, or with violence."[1]

A female Wombat which lived in the Zoological Society upwards of five years, and an account of the anatomy of which is published by Professor Owen,[2] weighed 59½lbs. The fur of the Wombat is tolerably long and very coarse; its general hue is grey-brown: next the skin the hairs on the upper parts of the body are of a dusky brown colour; and each

[1] Philosophical Transactions for 1808, p. 304.

[2] Proceedings of the Zoological Society for May, 1836, Part 4, p. 49.

hair of the ordinary fur has the exposed portion chiefly of a dirty white colour; but the longer and coarser hairs are black at the point: on the under parts of the body the hairs are for the most part of a dirty white colour, but dusky at the root. The general tint of these parts is paler than that of the back. The naked muffle is black; the ears small, pointed, and well clothed with hairs. The legs are short and strong, and the feet broad, naked beneath, and covered with minute, round, fleshy tubercles; the claws are large; those of the fore-feet solid, (that is, not concave beneath), but slightly curved, and depressed; those of the hind feet are curved, slightly compressed, and concave beneath. The hairs of the moustaches are numerous, strong, and of a black colour, as are also some long bristly hairs which spring from the cheeks. The tail is a mere tubercle, and consequently hidden by the fur. The skeleton of the Wombat presents certain peculiarities which are well worthy of attention; for instance, the number of its ribs (and consequently of the dorsal vertebræ) is unusally large, being fifteen, whilst twelve or thirteen are usually found in the Marsupialia. The body of the atlas vertebra remains permanently cartilaginous[1]; the humerus, besides having the inner condyle perforated, has an opening between the condyles; the patella, or knee-bone, is wanting.

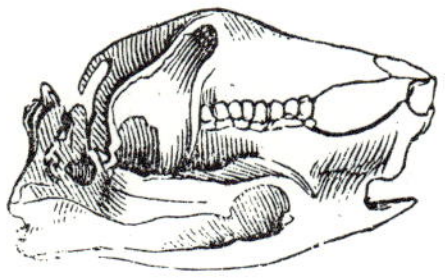

Side view of the Skull of the Wombat.

[1] This character, however, is not peculiar, since, according to Prof. Owen, the atlas vertebra of the Koala also retains the cartilaginous condition, and the Professor also calls attention to the general imperfect condition of the part in

In the general form of the skull, the Wombat approaches the Phalangista group more nearly than any other section of *Marsupiata*. The cranium is remarkable, not only for its large size, as compared with the bulk of the animal, (being larger than in the largest Kangaroos), but for its strength: the bones are thick and dense, and the large size of the temporal fossæ, deep and strong zygomatic arch, combined with the great development of hinder parts of the lower jaw, all indicate great power in the masticatory muscles. The general figure of the skull, as viewed from above, approaches to an oval; it is, however, broadest at the posterior root of the zygomatic arch, and thence becomes gradually contracted to the fore part of the zygoma, in front of which it is suddenly contracted; the muzzle, however, is broad, though very short. The upper surface of the skull is very nearly flat above—on the hind part it is quite flat, and the fore part is slightly convex. The nasal bones are nearly flat, short, broad, and much expanded posteriorly. The temporal ridges are strongly marked, though not much elevated; they converge, but do not meet posteriorly in the skull of an aged animal: the occipital crest is considerably produced. The zygomatic arch is stout, deep, and compressed, excepting beneath the eye, where it forms a large horizontal platform: the glenoid cavity or articular surface for the condyle of the lower jaw is of great extent, and forms a concave curve in the transverse direction, but in the longitudinal direction it is narrow and convex; it presents no anterior or posterior process to enclose the condyle of the lower jaw, which, therefore, can move backwards, forwards, and laterally. The palate is narrow, and much contracted between the anterior molars; the

question in Marsupial animals. In most of the species the neurapophyses of the atlas vertebra either remain permanently separated from the body, or are only anchylosed when the animal is aged.

incisive openings are narrow and rather small (half an inch in length); the posterior palatal openings are also small—half an inch long, and a quarter of an inch in width at the widest part, which is behind. The cavity corresponding to the sphenoidal *bulla ossea* in other Marsupials, is here (as pointed out by Prof. Owen) excavated in the lower part of the squamous element of the temporal bone. The condyle of the lower jaw is of great width, but narrow, and convex in the antero-posterior direction: the inflected angular portion is of great size, its width being very nearly equal to the width of the interspace between the two rami behind: the symphysis menti is of great extent: the coronoid process is large, and much elevated. The proportions will be illustrated by the following dimensions, which are taken from a cranium in the British Museum, which evidently belonged to an aged individual. Scarcely any of the sutures are obliterated in this skull—even the sutures indicating the boundaries of the elements of the occipital and temporal bones are distinct; the frontal sutures, and the suture between the nasal bones, are partially obliterated.

	Inches.	Lines.
Total length of skull	8	0
Width	5	11
" between orbits	2	6
Length of nasal bones	2	10½
Width of ditto in front		9⅔
" behind	2	2⅔
Greatest depth of zygomatic arch, behind ..		11½
Length of palate	4	7½
Width of ditto between the posterior molars ...	1	0½
" between the anterior molars ...		5
Longitudinal extent of the five molars of the upper jaw, on either side	2	2
Length of lower jaw	5	11
Height in a vertical line dropped from the coronoid process	3	0
Longitudinal extent of symphysis menti	3	2

	Inches.	Lines.
Width of condyle	1	7
Antero-posterior diameter of ditto		3
Distance between the outer edges of the two rami, behind	5	10
Width from the inner edge of the inflected angular portion to the outer edge of the ramus	2	2

The specimen of the Wombat already alluded to, as having lived in the Menagerie of the Zoological Society, was of a grey colour, and measured 3 feet in length: the following dimensions are from a male specimen, preserved in brine, in the British Museum Collection.

	Inches.	Lines.
Length from tip of nose to root of tail ...	28	0
" of ear	2	0
" from nose to ear	6	0
" of fore foot and nails	3	2
Width of ditto	1	11
Length of hind foot and nails	3	5
Width of ditto	1	6

The feet of this animal are figured in Plate 12.

FIG. 7. Represents the under side of the fore foot, half the natural size.
" 7 *a*. The hind foot, half the natural size.

PHASCOLOMYS LATIFRONS.

Broad-fronted Wombat.

Phascolomys latifrons. OWEN, Proceedings of the Zoological Society for 1845.

Frontal bones broad, and presenting a well-marked supra-orbital ridge, and post-orbital process; width of these bones in proportion to the length of the skull, as 33 to 75: width of skull in proportion to length, as 4 to 5: incisor teeth broadest in front, the upper pair distinctly broader than the lower.

Of the Broad-fronted Wombat, all that is known is a skull sent from South Australia to Professor Owen. This skull presents so many marked differences when compared with that of the *Phasc. Wombat,* that no doubt can be entertained of the existence of two distinct species of Wombats. I have sought in vain, however, amongst the specimens of Wombats contained in our museums, for an animal which might be identified with Professor Owen's new species. In none have I found the incisor teeth presenting the broadest surface in front, a peculiarity in which the *P. latifrons* differs from *P. Wombat,* where the broadest part of the incisor is at the side. The new species differs moreover in having the upper incisors distinctly broader than the lower, whilst in the Common Wombat the upper and lower incisors are very nearly equal in width when viewed in front. The following points of distinction presented by the skull of the *P. latifrons,* when compared with that of the *P. Wombat,* are for the most part pointed out in some notes from the pen of Professor Owen, who has kindly placed them at my disposal.

The skull of *Phasc. latifrons* is rather smaller and broader in proportion to its length; the upper incisors have a semi-oval transverse section; the convex enamelled surface directed more forward, and longitudinally substriated. The lower incisors narrower, trihedral, with the enamelled outer surface flat. The first lower molar tooth relatively larger; the last relatively smaller. The lower jaw is shorter, more suddenly curved behind, and has the symphysis deeper; the intermaxillary part of the skull is higher in proportion to the width, and less convex externally; the palate is less contracted between the foremost molars, and the palatine portion of the intermaxillaries is wider and very concave. The nasal bones are broader, forming the whole upper surface of the anterior third of the skull. The interorbital part of the cranium is much broader, and presents a well-marked supra-orbital ridge and post-orbital process, both of which are almost obsolete in *Phasc. Wombat*. The temporal fossæ are not bounded, as in the last mentioned animal, by two nearly parallel ridges, but are continued by a convex tract to the upper surface of the cranium : the supra-tympanic depression is much larger.

The subjoined dimensions of the skulls of the two animals under consideration will assist in conveying a more accurate idea of the differences of proportion they exhibit.

	P. latifrons.		*P. Wombat.*	
	Inches.	Lines.	Inches.	Lines.
Total length of cranium	6	3	8	1
Greatest width, (which is at the posterior part of the zygomatic arch)...	5	0	5	9
Length of nasal bones	2	0	2	11
Width of ditto behind	2	5	2	2
" " near the apex ...	1	2		10
Length of frontal bones	2	5	?	
Width of ditto between orbits ...	2	9	2	6
" of skull behind orbits, (where contracted by the temporal fossæ)	1	3	1	11
Length of palate	3	7	4	7
Width between the anterior molars ...		5½		3¾
" " posterior ditto ...		10½	1	0½
" of superior incisor teeth ...		9		9
Depth of ditto		3		5
Distance between incisor teeth (upper jaw) and molars	1	6½	1	11
Total extent of row of molar teeth ...	1	9	2	1½
Length of lower jaw	4	4	5	11
Height in a vertical line dropped from coronoid process	2	9	2	11
Width of lower incisors		6		8
Depth of ditto		3		3⅔

PHALANGISTIDÆ; or, PHALANGER FAMILY.

Dentition.—Incisors, $\frac{6}{2}$; canines, $\frac{1-1}{1-1}$, or $\frac{1-1}{0-0}$; constant premolars, $\frac{1-1}{1-1}$; true molars, $\frac{4-4}{4-4}$, or $\frac{3-3}{3-3}$

Head moderate; the facial portion rather short; upper lip cleft; muffle naked.

Limbs equal; fore feet with five well-developed toes, having compressed and curved claws; hind feet with five toes, of which the first, or innermost one, is large, nailless, placed at right angles with the rest, and opposeable to them; the second and third toes more slender and shorter than the others, united in a common integument very nearly to the extremity, and furnished with curved hollow nails; the fourth and fifth toes have curved and compressed claws.

Tail sometimes absent, but almost always long, and more or less prehensile; in some species wanting the prehensile power.

Pouch well developed.

Mammæ two or four.

Stomach usually simple, sometimes provided with a cardiac gland; cæcum present, and, in most species, very much developed.

The Phalangers, so called from their having the second and third toes of the hinder foot united in a common integument, form the fourth family of Vegetable-feeding Marsupial Mammals—that is, taking the groups in the ascending order, and viewing the Wombats as forming a family.

Although the diet of the species of different groups already described, no doubt varies to a certain extent, yet, on the whole, we might say the Kangaroos are more especially grass- and herb-feeders; the burrowing Wombats, root-feeders: the gigantic extinct Diprotodons and Nototheriums probably derived their sustenance from the twigs and leaves of the underwood or scrub, and the present family (*Phalangistidæ*) is composed of animals which are expert climbers, and living upon trees, feed upon their leaves, buds, and fruits.

The Phalangers are nocturnal in their habits, remaining concealed during the day on the branches, or in the hollows of trees; at twilight they quit their hiding places, and climb amongst the branches to seek their food. Generally speaking, they are not very active in their movements; but, among the smaller species, forming the section *Petaurus*, are some to which this remark will by no means apply, since they are extremely agile. The Phalangers possess six incisor teeth in the upper jaw, and two in the lower; a canine on either side of the upper jaw, and a premolar and four (or rarely three) true molars on each side of both jaws; but besides these, there are, in most of the species, some small teeth placed between the canine and principal premolar in the upper jaw, and between the great incisor and the molars, already noticed, in the lower jaw[1] These small teeth are never more than three in number on either side of each jaw: they vary in the different species, and not unfrequently there are fewer on one side of the jaw than on the other, in the same species. If we except these small premolars, which are sometimes entirely wanting, as in the Koala, the dentition of the Phalangers is numerically the

[1] Whether any of these small teeth, when present in the lower jaw, represent the canine in the carnivorous or insectivorous species of Marsupials, it is difficult to determine, but it is highly probable that that tooth is represented by the foremost of the small teeth in question. In the Phascogales, where the two foremost of the lower incisors are large, compared to the same teeth in the Dasyuri, their increased development is, as it were, at the expense of the posterior incisors, which are very small, and the canine which follows them is but moderately developed: in species presenting still larger lower front incisors, like the Phalangers, we should be prepared, therefore, in the first place, for the absence of the posterior incisors; and, in the next, for a reduced size of the canine.

same as in the Rat-Kangaroos; and, indeed, the structure of the teeth does not differ much.

The anterior upper pair of incisors (as in the Hypsiprymni) are larger and longer than the rest: the large lower incisors are nearly horizontal, or directed obliquely upwards. The canine, though by no means large, is usually rather more developed than in the Rat-Kangaroos; the premolar, which is contiguous with the true molars, is shorter and broader than in the animals just mentioned, and the true molars have the four principal cusps more developed.

The family may be divided into three principal sections or genera, readily distinguished by obvious external characters. As approaching in certain characters most nearly to the Wombats, we will commence with the genus *Phascolarctus*[1], the only known species of which (the Koala) is distinguished from other Phalangers by the absence of a tail: next follows

[1] With regard to the position of the Wombat, and the Koala (*Phascolarctus*) in a natural system, I may observe, in the first place, the Wombat, *cæteris paribus*, shows some affinity to the *Phalangistidæ* in the possession of a thumb, which, though short, is very broad, and sufficiently distinct; then beyond this we have to add, that the limbs are equal, the tibia and fibula are widely separated, excepting, of course, at the extremities; and the stomach is simple, as in the Phalanger group. On the other hand, we perceive in the Koala an animal possessing all the essential characters of *Phalangista*, but in which the stomach is provided with a peculiar glandular apparatus, and the tail is wanting, as in the Wombat. The two animals, moreover, agree very closely in the structure of the humerus; they agree in the non-possession of a patella, in the absence of ligamentum teres, and in the outermost of the articular surfaces of the upper extremity of the tibia being continuous with the articular surface of the fibula. The skull of the Koala, as compared with that of a typical *Phalangista*, differs in having the posterior palatine openings confined to the palatine bone, which is also the case in the Wombat: the lower jaw differs in the greater extent of the *symphysis menti*; and lastly, an approximation to that Rodent-like type of dentition which is exhibited by the Wombat, is perceptible in the Koala, in the smaller development of the posterior incisors and canines of the upper jaw, and the total absence of any of those premolars which, in the typical Phalangers, intervene between the canine and the five

Phalangista, in which the tail is long and prehensile; and, lastly, the genus *Petaurus*, in which the tail does not possess the power of grasping objects. The *Petauri*, moreover, may be recognised by their possessing a loose membrane extended from, and filling up, the space between the fore and hind legs. Other differences observable in the species of these sections are pointed out in their proper places.

PHASCOLARCTUS.

Phascolarctos. De Blainville, Prod. d'une nouv. Distrib. Systèm. du Règne Anim.; Bullet. de la Soc. Philom. de Paris, p. 108. 1816.

Lipurus. Goldfuss, in Isis, p. 271. 1819.

Teeth.—Incisors, $\frac{6}{2}$; canines, $\frac{1-1}{0-0}$; premolars, $\frac{1-1}{1-1}$; true molars, $\frac{4-4}{4-4} = 30$. Posterior upper incisors and canines small; true molars, each with the crown divided into four pyramidal tubercles.

Body, stout.

Head moderate, the facial portion very short; muffle naked; ears moderate, clothed with long hairs.

Fore feet with two inner toes, which are considerably shorter than the others, slightly opposeable to the remaining three, of which three, the central one is the longest: the nails of all the toes long, curved, very deep, and much compressed.

Hind feet with the first toe, or thumb, very far back, large, very broad, and nailless; the second and third, united toes, con-

molars of the upper jaw, and between the incisors and the corresponding teeth in the lower jaw. These characters, *combined*, I think indicate that I have properly located the Wombat in the tabular view of my arrangement of the Marsupial animals, given in p. 12, in which I have endeavoured so to place the several genera, that those which have the most characters in common, are most nearly approximated.

siderably smaller and shorter than the remaining two, of which the first is the largest.

Tail wanting.

Stomach with a cardiac glandular apparatus: *cæcum* greatly developed[1].

PHASCOLARCTUS CINEREUS.

The Koala.

(*Plate* 9, Fig. 2.)

Lipurus cinereus. GOLDFUSS, in Isis, p. 271. 1819.

Phascolarctos fuscus. DESMAREST, Mammalogie, p. 276. 1820.

" " " Dict. des Sci. Nat. t. xxxix. p. 448. 1826.

" *Flindersii.* LESSON, Manuel de Mammal. p. 221. 1827.

" *fuscus et cinereus.* FISCHER, Synopsis Mammal. p. 285. 1829.

" " " WAGNER, Schreb. Saug. 111-112 Heft, p. 92. 1842.

" *cinereus.* GRAY, List of the Species of Mammalia in the Coll. of the Brit. Mus. p. 87. 1843.

Koala Wombat (Patterson). HOME, in Phil. Trans. p. 304. 1808.

Le Koala ou Colak. DESMAREST, Nouv. Dict. t. xvii. p. 110, Tab. E. 22, fig. 4. 1817.

Wombat of Flinders. KNOX, in Edinburgh New Philos. Journal, p. 111. 1826.

Koala, and Wombat of the natives; Native Bear, and Native Sloth of the colonists.

Length of head and body about two feet: fur dense, woolly, and moderately soft: general colour ashy grey, slightly inclining to brown; hinder part of back dirty yellowish white; under parts dirty white; inner side of hind legs of a brownish

[1] The cœcum in the Koala, the only known species belonging to the present genus, is said to be more than three times the length of the animal. In a young male specimen, dissected by Mr. Martin, the cœcum measured 4 feet 2 inches in length, and was slightly puckered equidistantly, or nearly so, throughout its whole length, into sacculi, by a slight longitudinal (mesenteric) band of muscular fibres. See Proceedings of the Zool. Society for 1836, Part iv. p. 111.

rust colour: ears densely clothed with long hairs; white on the inner side, and greyish on the outer, but blackish on the anterior margin: feet whitish.

Inhabits New South Wales.—Specimen in the British Museum.

An excellent account of the habits of the Koala, or Native Bear, as it is frequently called by the colonists, appeared as early as 1808, in the "Philosophical Transactions[1]," from the pen of Colonel Patterson, formerly Governor of New South Wales.

I am informed by Mr. Gould that the Koala inhabits the range of country extending nearly from Moreton Bay to Port Phillip. It was known to Colonel Patterson as an inhabitant of the forests, about fifty or sixty miles to the south-west of Port Jackson, whence, it is stated, the first specimen was brought to that port[1]. "The New Hollanders," Colonel Patterson observes, "eat the flesh of this animal, and therefore readily join in the pursuit of it: they examine, with wonderful rapidity and minuteness, the branches of the loftiest gum-trees, and, upon discovering a Koala, they climb the tree in which it is seen with as much ease and expedition as an European would mount a tolerably high ladder. Having reached the branches, which are sometimes 40 or 50 feet from the ground, they follow the animal to the extremity of a bough, and either kill it with the tomahawk, or take it alive. The Koala feeds upon the tender shoots of the blue gum-tree, being more particularly fond of this than of any other food: it rests during the day on the tops of these trees,

[1] This account will be found in a memoir communicated by Sir Everard Home to the Royal Society, entitled, "An Account of some Peculiarities of the Anatomical Structure of the Wombat," &c., in which the author, unfortunately, confounds the Koala with the Wombat, being misled by the name "Wombat," which, it appears, is sometimes applied by the natives to the Koala as well as to the Wombat. See Phil. Trans. for 1808, p. 304.

feeding at ease, or sleeping. In the night it descends and prowls about, scratching up the ground in search of some particular roots; it seems to creep rather than walk: when incensed or angry, it utters a long shrill yell, and assumes a fierce and menacing look. They are found in pairs, and the young is carried by the mother on its shoulders. This animal appears soon to form an attachment to the person who feeds it."

I learn from Mr. Gould, that, unlike most quadrupeds, the Koala does not fly upon the approach of man; that it is very tenacious of life, and when even severely wounded it will not quit its hold of the branch upon which it may be at the time. It has been frequently compared to the bear in its movements and mode of climbing; and, indeed, in appearance the animal is not unlike a small bear.

The Koala is usually about two feet in length, and when on all fours stands 10 or 11 inches in height: the girth of the body is about 18 inches. Its limbs are of moderate length, and powerful; the hands and feet large, and admirably adapted by their structure to tree-climbing habits. The toes of the fore feet are so arranged, that the two innermost of the five are opposed to the other three; and all the toes, both of the fore and hind feet (if we except the innermost one of the latter), are provided with large, curved, very deep, and compressed claws. The innermost toe of the hind feet is large, nailless, assumes the form of a thumb, and is used as such, being opposed to the toes in grasping, as is the thumb of the human hand to the fingers. The head is rather large, the muzzle short, and nearly naked both at the sides and on the upper surface, these parts being merely clothed (and rather sparingly so) with small velvet-like hairs: the part thus sparingly clothed is most extended on the upper surface of the muzzle, here reaching back, about one inch and a half, from the tip of the nose, whilst at the sides only half an inch, or rather more, of the muzzle is destitute of the ordinary fur. The ears are of moderate size, and pointed, and entirely hidden by the very long hairs with which they are clothed, these latter being, for the most part, about two inches in length: on the inner side of the ears the hairs are white, and on the outer side they are of the same grey hue as those of the head, excepting those which spring from the anterior margin of the ear, which are chiefly black. The eye is rather large, and, like other Marsupial animals (with the exception of the Kangaroos), are not protected by eye-lashes; there are, however, a few long bristly hairs springing from immediately above the eye: the hairs of the moustaches are small and scanty. The fur of the animal is tolerably long, dense, of a wool-like quality, and rather soft to the touch; its general hue may be described as ashy grey, somewhat suffused with brown —a tint produced by the hairs being brown below the point, and whitish at the point. The hinder part of the back is of a dirty yellowish-white hue. The under parts of the head and body, as well as the inner side of the fore legs, and the posterior part of the hind legs, are white, but not very pure; the hairs covering the feet have the visible portions whitish, but they are dusky brown at the root, and a slight pencilling of this darker hue is generally observable on the toes. The inner side of the hind legs is a brownish rust colour. The muffle is naked, and, like the naked soles of the feet, appears to have been black in the living animal.

A fine female specimen of the Koala, in the British Museum collection, presents the following dimensions, to which are added (in a second column) some of the principal admeasurements taken from a specimen in the Museum of the Zoological Society:—

	Inches.	Lines.	Inches.	Lines.
Length of head and body, together	29	0	25	0
" from tip of nose to ear ...	5	3	4	6
" of ear, about	2	0	1	9
" fore foot, not including the claws	3	6	3	0
" hind foot, without the claws	3	7	3	1
" thumb of ditto	1	8		

A very young Koala in the Museum of the Zoological Society presents some features worthy of notice. This little animal, which measures about 11 inches in length, instead of having the woolly fur of the adult, is clothed with hairs which are moderately soft, short, and closely applied to the skin: on the mesial side of the back, a little behind the shoulders, the hairs radiate, and running forwards over the neck they meet those of the head having an opposite direction, and form a kind of crest at the line of junction: on the rump there is another of these centres from which the hairs radiate[1]. The ears, which are much pointed, and have the posterior edge emarginated, are clothed with hairs of about a quarter of an inch in length. Its colouring is the same as in the adult.

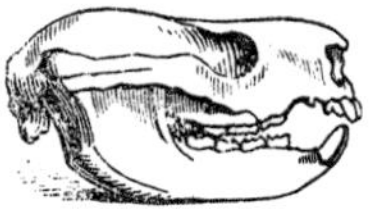

Skull of the Koala.

The skull of the Koala is remarkable for its oblong quadrate form, the shortness and great width of the nasal bones, which are truncated in front, the parallelism of the zygomatic arches, their great length, and the great protuberances formed by the auditory bullæ; these are nearly spherical, but slightly compressed in the lateral direction, in the young animal, and are much more compressed and considerably elongated in the

[1] I have noticed a similar arrangement of the hairs of the fur in a young Sloth.

vertical direction, in the adult. The pterygoid processes of the occipital bones are much elongated,—10 lines in length in an adult skull. The posterior palatine openings are by no means large, as compared with the Phalangers proper, and are confined to the palatine bones, which assume the form of a vertical plate, of ⅛ of an inch in depth, behind the palatine openings in question.

In the lower jaw, the points most worthy of notice are,—the great depth of the rami, the angle being less twisted inwards than in most other Marsupials—(in this respect resembling the extinct *Diprotodon* and *Nototherium*); the *symphysis menti* being more extended, and the rami more firmly united at this part, than in the typical Phalangers. The four hindermost incisors of the upper jaw and the canines are proportionately smaller than in the animals just mentioned, and the molar teeth are larger, and the four tubercles, which the crowns of the true molars present, are more angular, having each the form of a three-sided pyramid, or nearly so.

The following admeasurements are taken from a skull of an adult Koala, in the Hunterian Museum:—

	Inches.	Lines.
Length of skull	5	0
Width beneath orbits	2	8
Length of nasal bones	1	3¼
Width of ditto behind	1	2
" " at the apex	0	10
Length of zygoma	2	11½
Greatest depth of ditto	0	10
From anterior root of zygoma to apex of inter-maxillaries	1	6
Length of palate, including the palatine bones ...	2	4
Width of ditto between premolars	0	9
Three upper incisor teeth, taken together ...	0	4¾
From posterior incisor to canine	0	2
From canine to premolar	0	4
Length of the five molars, taken together	1	5
Length of lower jaw, measuring to apex of angle	3	10
Height, measuring from apex of coronoid process	2	6

Genus, *Phalangista.*

Phalanger.	STORR, Prodromus Methodi Mammalium, p. 33. 1780.
Phalangista.	CUVIER, Tableau Général des Classes des Animaux, in Leçons d'Anatomie Comparée, tom. i. 1799.
Balantia.	ILLIGER, Prod. Syst. Mamm., &c. p. 77. 1811.

Teeth.—Incisors, $\frac{6}{2}$; canines, $\frac{1-1}{}$; premolars, $\frac{2-2}{1-1}$; true molars, $\frac{4-4}{4-4}$:
—(these teeth are nearly constant).
Tail prehensile.

The number of the teeth in the different species of Phalanger varies, but the differences observable are not important, since they arise from the appearance or non-appearance of certain small teeth, situated between the molars and incisors, which are not always constant even in the same individual Phalanger; one of these teeth will sometimes be found, for instance, on one side of the jaw, and not on the other: all the species, however (with the exception of those belonging to the section *Dromicia*), possess four true molar teeth on each side of each jaw, and each of the series is preceded by a premolar, which is in contact with the true molars: in front of this, and more or less widely separated, is a second premolar in the upper jaw, and which, in most species, is a well-developed tooth, having the elongate conical form of a canine: in front of this, again, is a moderately well-developed canine tooth, which is usually situated more forward, with relation to the intermaxillary suture, than in most other quadrupeds; in most species the tooth is intersected by the line of the suture in question, and sometimes the whole canine is situated in front of the suture—that is to say, the visible portion of the tooth, for I suspect it will always be found to have its root in the superior maxillary bone. Lastly, there are six incisors in the upper jaw In the lower jaw there are usually two very small teeth, situated between the two long and nearly horizontal incisors, and the five constant molars.

The upper true molar teeth, in the species we shall first describe, are rather longer than broad, of nearly equal size, if we except the hindermost of each series, which is rather smaller than the others, and have four principal cusps, which are but moderately elevated and angular, and have a tendency to join in pairs, as in the Rat-Kangaroos, by a ridge running in the transverse direction from each of the two outer cusps to the inner tubercle. The hindermost molar of the upper jaw has but three principal tubercles. The principal premolar (or that which joins the true molars) is rather smaller than the true molars, of nearly equal length and breadth, and terminates either in a point, or in a cutting edge of small extent. The true molars of the lower jaw resemble those of the upper, excepting that their transverse diameter is less in proportion to their length.

The nasal portion of the skull is short; the remaining portion of the cranium, as viewed from above, presents a nearly oval outline; the interorbital space is moderately contracted; the zygomatic arches long and deep; the palate presents a large posterior palatine opening. The angular portion of the lower jaw is large, and bent inwards nearly at right angles with the plane of the ascending ramus; the condyle is transverse.

The genus *Phalangista* has been subdivided into four minor sections:—

Section 1. *Cuscus.*

Cuscus. LACÉPÈDE.
Ceonyx. TEMMINCK, Monogr. de Mamm., tom. i. p. 10.

Tail with the basal portion only covered with hair: ears short, almost hidden by the fur of the head; eyes with a vertical or nearly vertical pupil.

The animals belonging to this section are of moderate size (being for the most part about equal to the domestic cat, in this respect), have the fur dense, and of a more or less woolly nature: similar fur covers a portion of the tail, but usually about half the tail is destitute of hair, and presents numerous fleshy tubercles. The species of Cuscus are inhabitants of certain islands in the Indian Archipelago, and, indeed, are confined, so far as we know, to the islands of Celebes, Amboyna, Banda, Waigiou, Timor, New Guinea, and New Ireland.

PHALANGISTA (*Cuscus*) URSINA.

The Ursine Phalanger.

Phalangista ursina. TEMMINCK, Monographies de Mammalogie, t. i. p. 10. Skull of the adult, Pl. 1, figs. 1, 2, and 3; skull of a specimen of intermediate age, Pl. 2, figs. 1 to 5; skeleton, Pl. 4.

" " WAGNER, in Schreb. Saug. 109-110 Heft, p. 69.

Rather larger than the domestic cat: fur of moderate length, dense, and woolly: general colour black, freckled with yellow; head chiefly of the latter colour; under parts of body yellow: ears densely clothed, the hairs yellow, excepting at the base externally, where they are black, and more or less pencilled with yellow.

Messrs. Lesson and Garnot[1] make two minor divisions of the Cuscus group: the first, characterised by the ears being very short, almost hidden by the fur, and well clothed internally, includes the present species, together with the *Cuscus chrysorrhos*, *C. maculatus*, and *C. macrourus;* the second division includes but one species, *C. orientalis*, in which the

[1] See Voyage de la Coquille. Zoologie, p. 150.

ears are somewhat longer, not hidden by the fur, and naked within.

The Ursine Phalanger is readily distinguished from others of its section by its fur being chiefly of a black colour; most of the hairs on the upper parts of the body, however, are usually tipped with yellow, producing a freckled appearance; on the head the yellow colour predominates, and the muzzle, and lower part of the cheeks, scarcely exhibit a trace of the black hue. The under parts of the head and body, as well as the inner side of the limbs at the base, are of a dirty yellow colour; the fur of the abdomen is dark, however, next the skin. The ears are clothed at the base, externally, with black and yellow hairs, like those of the head; on other parts they are densely covered with longish yellow hairs. The naked parts of the feet, tail, and muzzle, appear to have been black, or at least of a dusky hue, in the living animal. Such is the colouring of two specimens in the British Museum collection. A specimen in the Zoological Society's Museum differs in having the general colouring of the fur brownish black, and the parts above described as yellow are of a brownish-white tint: on the upper and fore part of the back the black prevails, and on the hinder half of the back the pale colour is most conspicuous, whilst on the head, tail, and under parts of the body, scarcely any black is perceptible. The basal half of the tail is clothed with fur of the same texture as that of the body; the apical half is destitute of hair, and the naked part is more extended on the under than the upper surface by about two inches: the whole of the naked portion of the tail presents small scattered fleshy tubercles. The toes are but sparingly clothed with hairs; the claws are very large, and of a brownish colour.

Mr. Temminck informs us, that the young of this species differ from the old, in having the fur of a pale colour, and that the old specimens are perfectly black: those of moderate size brownish black; and those which he supposes to be the young of the year, have the fur of the upper parts of the body of a rusty brown tint; that on the cheeks, the flanks, and the limbs, yellowish brown, and on the under parts yellowish. Adult specimens are said to measure about 22 or 23 inches in length, from the nose to the root of the tail, and to have the tail about 20 or 21 inches in length; whilst the height of the adult animal varies from $10\frac{1}{4}$ inches to $10\frac{3}{4}$.

The dimensions of the specimen in the Zoological Society's Museum are as follows:—

	Inches.	Lines.
From tip of nose to root of tail	20	0
Tail	19	0
From nose to ear	3	5
Ear	0	8
Tarsus, without including the nails	3	3

An imperfect skull of *P. ursina*, in the British Museum, closely resembles the cranium figured in M. Temminck's second Plate, and evidently belonged to an animal which was not quite adult. Besides the $\frac{6}{2}$ incisors, and $\frac{5-5}{5-5}$ molars, which are evidently teeth of the permanent series, it has, in the upper jaw, a well-developed canine-like tooth about midway (or rather behind that point) between the incisors and molars, which has a double fang, and between this and the incisors is seen the point of the permanent canine, which is scarcely above the level of the bone: the root of this tooth is situated entirely in the superior maxillary bone, but, when fully developed, it would appear to belong to the intermaxillary teeth, being situated almost entirely in front of the intermaxillary suture. The canine is apparently in the same condition in the skull figured by M. Temminck, and hence this tooth, we may suppose, is the last one to make its appear ance in these animals. From the skull of a younger animal, also in the Museum, I learn that the principal premolar tooth is developed at a relatively earlier period than in the Hypsiprymni, it being more advanced, in the skull in question, than the third true molar. In the Rat-Kangaroos, as I have before noticed, it is the last tooth to make its appearance.

The bones of the skull of *P. ursina* are much thicker and stronger than in the *P. vulpina;* the muzzle is proportionately shorter and broader; the zygomatic arch deeper. The nasal bones are dilated at each extremity, where their width is the same, or very nearly so. The rami of the lower jaw are considerably deeper than in the *P. vulpina,* and the *symphysis menti* forms a more obtuse angle; and, in connection with this difference in the angle formed by the symphysis, we find the lower incisor teeth directed more upwards. All the teeth are larger in proportion than in *P. vulpina,* and in the relative proportions of certain teeth there are differences: thus the anterior upper two incisors are longer in proportion to the other two pairs, and the principal premolar is larger in proportion to the true molars, being equal in length, and nearly equal in width, to the foremost true molar. The anterior margin of the post-palatine opening is on a line with the middle of the first true molar.

The adult skull, according to M. Temminck's figure, is four inches and one line in length, and two and a half inches in width. The two skulls alluded to as being in the British Museum collection yield the subjoined dimensions: those in the first column are from a nearly adult skull; those in the second are from the skull of a young animal, in which only two of the true molars are perfectly developed: in both skulls there are two small teeth immediately behind the lower incisor, the foremost of which, I suspect, represents the lower canine:—

	Inches.	Lines.	Inches.	Lines.
Total length of skull	?*		3	1½
Width	2	2	1	10½
Length of nasal bones	1	3⅓	1	0¼
Width of ditto behind	0	6⅓	0	6
" " in front	0	6	0	4½
Length of palate to the posterior palatine opening	1	0½	0	11½
" principal premolar and four true molars, taken together	1	2⅓	?	
" lower jaw	2	7	2	2¾
Height of do. from apex of coronoid process	1	4	?	
Depth of ditto beneath the second true molar	0	6½	0	5

Ursine Phalangers, according to M. Temminck, are very numerous in the dense woods of the southern parts of the Island of Celebes. The inhabitants of that island are acquainted with no varieties of this animal: it is said to eat flesh. These Phalangers are seen but little during the day, at which time they squat on the branches of the trees, and are hidden by the foliage.

With regard to these Phalangers eating flesh, I must observe, that this is not likely to be their ordinary food. We know that other Phalangers feed chiefly upon vegetable substances, though they do not refuse flesh, at least when in confinement.

PHALANGISTA (*Cuscus*) CHRYSORRHOS.
Yellow-rumped Phalanger.

Phalangista chrysorrhos. TEMMINCK, Monographies de Mammalogie, t. i. p. 12.

" " WAGNER, in Schreb. Saug. 109-110 Heft, p. 70.

Fur short, dense, and of a cotton-like texture; general colour ashy grey, somewhat suffused with black on the back, and

* According to the proportions of other skulls of *P. ursina*, the length of this one should be 3″ 7‴: the back part is broken in this specimen.

paler on the head; ears well clothed with white hairs: rump and upper surface of tail golden yellow; under parts white: a blackish longitudinal band on the flanks.

Inhabits Amboyna.

The author has never seen this species, of which very few specimens have been brought to Europe. It is described by M. Temminck as being equal in size to the Wild Cat (*Felis Catus*), as having the muzzle short, and with the upper surface on a line with the forehead; the ears very short, hidden by the fur, and clothed throughout with hair; the tail about equal in length to the body (not including the head), covered with fur above, and on the sides, for about two-thirds of its length; the remaining portion naked, wrinkled, and of a yellowish colour; the fur short, close, and woolly; the longer silky hairs of the same colour as the ordinary fur: the fur of the head of a pale ashy grey colour, but that on the ears whitish; the upper parts of the body, the flanks, and the outer side of the limbs, ashy grey, more or less suffused with black; the rump and upper part of the tail, of a golden-yellow tint, brighter in the adult animal than in young individuals; the inner side of the limbs, and the under parts of the neck, white; the chest, and mesial line of abdomen, also white; the white of the abdomen separated from the grey of the flanks by a black band, in adult specimens, but with the band indistinct in the young; the region of the pouch of a rusty tint; the feet of a bright rust colour; the tip of the nose brown; the claws yellowish brown. Full-grown specimens measure, from the tip of the muzzle to the root of the tail, about 23 inches; tail, 13 inches.

The dimensions of an adult skull, according to the figures given by M. Temminck are—

	Inches.	Lines.
Length from posterior boundary of glenoid cavity to apex of intermaxillary bones	3	0*
Width	2	2
" between orbits	0	8
Length of nasal bones	1	3
Width of ditto at the base	0	6
" " at the apex	0	3¼
Length of palate, to anterior margin of post-palatine openings	1	1
From front of foremost incisor to principal premolar	0	9⅔
Length of principal premolar and four true molars, taken together	1	2

The principal premolar tooth of both jaws is represented in the skull figured by M. Temminck, as having the crown higher than in the corresponding tooth of *Phalangista ursina,* and as terminating in a conical point: the canine tooth more advanced, and the foremost canine-like premolar more distinctly separated from the other teeth: the corresponding tooth in *P. ursina* is placed close to the principal premolar, whilst in *P. chrysorrhos* it is separated by a space of about one line in width. The forehead of the skull is much arched. There are two small premolars in the lower jaw.

* The skull has the hinder part fractured; hence the total length cannot be given.

PHALANGISTA (*Cuscus*) MACULATA.

The Spotted Phalanger.

Phalanger mâle.	BUFFON, Hist. Nat. tom. xiii. Plate 11. 1765.
Phalangista maculata.	DESM. Nouv. Dict. d'Hist. Nat. tom. xxv. p. 472, Pl. M. 35 f. 3 (1817); Mammalogie, Part i. p. 266. 1820.
" *Papuensis.*	DESM. Mammal. Suppl. Part ii. p. 541. 1822.
" *Quoy.*	QUOY et GAIMARD, Voyage de l'Uranie, Zoologie, p. 58, Atl. Pl. 6. 1824.
" *maculata.*	QUOY et GAIMARD, l. c. p. 59, Pl. 7.
Cuscus maculatus.	LESSON et GARNOT, Voyage de la Coquille, Zoolog. p. 150, Atl. Pl. 4 (1826); Bullet. des Sci. Nat. tom. viii. p. 96. 1826.
? " *macrourus.*	LESSON et GARNOT, Voyage de la Coquille, Zool. i. p. 156, Atl. Pl. 5.
Phalangista maculata.	TEMMINCK, Monogr. de Mamm. tom. i. p. 14; Skull, Pl. 3, figs. 1-6. 1827.

Fur dense and woolly, of a dirty white colour, and irregularly clouded or spotted with brown or black; under parts of head and body white.

Inhabits the islands Amboyna, Waigiou, Banda, and New Guinea.

This species is subject to considerable variation in its colouring, and this is not only due to age, but M. Temminck, who has examined many specimens, informs us that he can perceive slight differences in the hue of the fur of individuals brought from different islands. The author just mentioned describes the *P. maculata* as being rather smaller than the two preceding species. Its tail is nearly equal to the head and body in length; about half is clothed with fur, the remaining (naked) portion is covered with ridges, and is of a yellowish colour. The ears are small, well clothed with fur, and hidden; the forehead is arched; the tip of the muzzle is naked and black. The fur is short, dense, and woolly, and intermixed with the ordinary fur are some longer hairs, which are longest and most abundant in the young animal. On the upper parts of the body the fur is usually of a dirty yellowish white tint, but exhibits numerous irregular blotches of dark brown: the face is covered with coarser hairs, and these are of a yellowish, and sometimes rusty yellow tint: on the head and sides of the neck, the hairs of the fur exhibit a mixture of whitish and grey, but the tuft which covers the ears is generally white: the chin, as well as the whole of the under parts of the body, and inner side of the limbs, are pure white; usually immaculate: the feet are of a bright rust colour. The hairy portion of the tail is white; sometimes exhibits a few dark spots, but is usually uniform in tint. The claws are yellowish. The eye is represented in the figure given by Messrs. Lesson and Garnot, as being of an orange colour, and with a vertical pupil. Sometimes the upper parts of the body and outer side of the limbs are covered with numerous irregular patches of dirty white and dark brown, and the lighter and darker parts are of about equal proportion.

In the young, the dark and light parts are less defined, owing to the longer interspersed hairs being very numerous, and of a dark colour on the light parts of the ordinary fur, and whitish on the dark patches. The ground colour, moreover, in the young, is sometimes of an ashy brown tint; and, finally, very young specimens are said to be entirely of an ashy grey hue.

In the adult animal the lower jaw presents two small teeth on either side, between the great incisors and the principal premolar, and in the young there are three of the small teeth, besides which there is a very small tooth immediately in front of the principal premolar, which is wanting in the adult.

The length of an adult animal, measuring from the tip of the nose to the root of the tail, is about 20 inches; the tail is about 17 inches. An adult specimen, preserved in spirits, measures very nearly three feet in total length, including the tail, which was nearly $18\frac{1}{2}$ inches; about $10\frac{3}{4}$ inches of which, on the upper surface, was covered with fur, and but $4\frac{3}{4}$ inches on the under surface was similarly clothed.

The skull of the adult, according to M. Temminck's figures, is—

	Inches.	Lines.
In length	3	10
Width	2	5
" between orbits	0	8
Length of the nasal bones	1	5
Width of ditto at the base	0	$6\frac{3}{4}$
" " at the apex	0	4
Total length of palate	1	11
Length of posterior palatine openings	0	9
From front of anterior incisor to the same part of principal premolar	0	$9\frac{1}{3}$
Length of premolar and four true molars, taken together	1	$2\frac{1}{2}$

Like others of the genus, the *P. maculata* lives in trees: its flesh is eaten by the natives of the islands in which it is found, and is said to be well-flavoured.

Messrs. Lesson and Garnot state that they were not successful in keeping alive some specimens which they purchased of the natives in the island of Waigiou. These animals were slow and dull, were constantly licking the naked parts of the feet and tail, and rubbing the face with the former; drank much; ate bread, which they held in their hands, but preferred meat. When two were placed together, they fought with fury: upon being in the slightest degree disturbed they growled like a cat, and when provoked they seized with their hands, and endeavoured to bite, those who irritated them. Their great red eyes, surmounted by a thick eyelid, gave them a stupid aspect. In a specimen dissected, the cœcum was found to be very ample, being about 18 or 19 inches in length. The stomach was filled with a green substance, apparently vegetable remains.

Cuscus macrourus. Lesson et Garnot.

General colour ashy grey, clouded with brownish; under parts white.

Messrs. Lesson and Garnot state that this species approaches very nearly to the *P. maculata*, but they consider themselves authorised in separating it as a distinct species, on account of its small size, the form of the head, and the size of the tail, as compared with the proportions of the body. It is about $13\frac{3}{4}$ inches in length from the tip of the nose to the root of the tail; the tail is about $18\frac{1}{2}$ inches in length, and densely clothed with fur for rather more than $10\frac{3}{4}$ inches of its length. The head is three inches in length. The ears are more prominent than in the *P. maculata*, and clothed with white hairs internally and externally; the forehead and upper surface of the muzzle are on the same plane; the muzzle is pointed and attenuated; like the forehead it is clothed with rust coloured hairs; the eyes are encircled with brown. The body and outer side of the limbs are of an ashy grey colour, clouded with brown; the throat and under side of the neck are white, and the abdomen and inner side of the thighs are impure white. The hairs covering the toes are black; the nails yellow; the fur of the tail is of an ashy reddish tint, and somewhat dusky towards the naked part.

Of the *Phalangista macroura* Messrs. Lesson and Garnot state that they had only seen one specimen. Now that specimen, excepting in having the tail longer in proportion to the body, it would appear resembles the young of *P. maculata*, and the describers should have informed us whether the dimensions were taken from a stuffed specimen or not, and whether the conditions of the skull and teeth indicated that the animal was adult.

PHALANGISTA (*Cuscus*) ORIENTALIS.

Valentyn's Phalanger.

Didelphis orientalis. PALLAS, Miscellanea Zoologica, p. 59. 1766.
" " SCHREBER, Saugethiere iii. p. 550, Tab. clii. 1778.
Phalangista rufa et *alba.* GEOFF. Catal. du Mus.
" *rufa.* DESM. Nouv. Dict. d'Hist. Nat. tom. xxv. p. 473. 1818.
Cuscus albus. LESSON et GARNOT, Voyage de la Coquille; Zoologie, tom. i. p. 158, Atlas, Pl. 6.
Phalangista cavifrons. TEMMINCK, Monogr. de Mammal. t. i. p. 17. 1827.
" " WAGNER, in Schreb. Saug. Suppl. 109-110 Heft, p. 73. 1842.
Cuscus orientalis. GRAY, List of the Mammalia in the British Museum, p. 84. 1843.

Ears naked internally; forehead concave; fur dense, and somewhat silky: male white; female of a pale reddish brown, and with a longitudinal dusky mark on the back.

Inhabits Amboyna, Banda, Timor, and New Ireland.

This species is equal to the Common Rabbit in size: its fur is rather short, dense, moderately soft, and has a slight silky lustre. The male specimens are perfectly white, even when young, but in aged individuals the fur is somewhat tinted with yellowish; the naked portions of the tail, and of the feet, ears, and muzzle, are of a livid white hue, and so are the nails. The females vary considerably in tint, but may be always distinguished by a dark band, which, commencing on the forehead, runs along the back, and terminates before reaching the tail. This band is generally of a chesnut brown colour. In adult females the fur varies, according to Temminck, from yellowish brown to nut brown, or to grey brown, more or less mixed with ash colour. In the young female, the fur is more or less tinted with rust colour, but is slightly suffused with ashy grey, and sometimes is silvery or whitish. The chin, and the whole of the under parts of the body, as well as the inner surface of the limbs, are white, slightly tinted with pale grey; the region of the pouch is rust coloured; the naked portion of the tail is yellowish in the adult, and whitish in the young.

The ears in the *P. orientalis* are not hidden by the fur of the head, as in other species of the *Cuscus* section, and they differ, moreover, in being naked on the inner side. The forehead is concave—rendered so partly by the prominence of the two temporal crests. The tail is about equal to the body in length. In full-grown animals the head and body together measure about 21½ inches in length, and the tail is about 15¼ inches. They rarely attain the total length of three feet three inches (tail included), according to M. Temminck. The females are rather smaller than the males, the largest being about 32 inches in length.

Besides the ordinary well-developed teeth, there is a very small tooth on each side of the upper jaw, situated immediately in front of the principal premolar, and the large inferior incisors are followed by three small teeth; thus differing from the adults of other species of *Cuscus*, where there are but two small teeth on either side of the lower jaw, behind the great canines.

The dimensions of the skull, figs. 7, 8, and 9, Plate 1 of Temminck's Monograph, are as follows:—

	Inches.	Lines.
Total length of skull	3	5
Width	2	2¼
" between orbits	0	6½
Length of nasal bones	1	3½
Width of ditto behind	0	5½
" " in front	0	3¾
Total length of palate	1	11⅓
Length of posterior palatine openings	0	9⅔
From apex of intermaxillary bones to principal premolar	0	10½
Length of principal premolar and four true molars, taken together	0	11⅔

In a specimen forming part of the Zoological Society's collection, the upper parts of the body are of a pale ashy brown colour, and the head rusty yellowish; the under parts dirty yellowish white: a deep brown line runs along the top of the head, and extends nearly the whole length of the back; woolly fur, like that of the body, covers about five inches of the upper surface of the tail, and terminates in a point; on the under surface it is less extended by about two inches; the rest of the tail is naked.

	Inches.	Lines.
Length from nose to tail	17	0
" of ear	0	6
" of tail	13	0

Of two stuffed specimens in the British Museum, one is perfectly white; the other, which is young, is of a pale rusty brown colour: neither have any mark on the back, and they are, therefore, both probably males.

Geoffroy, supposing the sexes of the present animal to be distinct species, gave to them the names *rufa* and *alba*, neither of which names is appropriate when applied to the opposite sex to that which received it; hence it is, we presume, that Temminck, in his "Monographies," proposed the new name *cavifrons* for our animal. We think, however, he might, without impropriety, have restored the old name of *orientalis*, which Pallas had applied to it as early as the year 1766.

The earliest notice of the *Phalangista orientalis* is that of Valentyn[1], who describes it under the name *Coes-coes;* the name by which it is known to the natives of Amboyna. It is described in great detail by Buffon, in the article headed "Phalanger femelle[2];" but this author committed an error in giving the *Phalangista maculata* as a variety of his animal —an error which is copied by Schreber, and several other authors. Messrs. Lesson and Garnot state, that the *P. orientalis* (*Cuscus albus* of their work) is abundant in New Ireland, where it is called *Kapoune* by the natives: that it is slow in its movements, and, like others of the genus, lives upon trees; and notwithstanding the care which the animals take to conceal themselves, they are discovered by a fœtid odour which they emit. Often, in traversing the forests of this island, the scent of the Cuscus was distinctly perceptible to the authors. It is stated by Cuvier, that when the Phalangers see any person, they suspend themselves by the tail, and, if looked at steadfastly, they will drop to the ground through fatigue[3]; and Messrs. Lesson and Garnot suppose

[1] Omstandig Verhaal van de Geschiedenissen en Zaaken, in Amboina, vol. iii. p. 272, (accompanied by a bad figure). Fol. 1726.

[2] Hist. Nat. xii. p. 108, Pl. 10.

[3] I find a similar statement in the work of Valentyn, already alluded to, whence Cuvier probably learned the fact. Possibly the *Cuscus* may have some kind of instinct allied to that of the North American Opossum, which induces it to feign death when in danger.

that it is by such means the numerous specimens were procured which were daily offered to them by the natives. The Phalangers were brought alive, but with the legs broken, and a piece of wood thrust into the mouth to prevent their biting[1]. The natives are extremely fond of the flesh of these animals, which is very fat, and they cook the Phalangers entire (removing only the intestines) upon peat coals. The teeth of the animal are strung together by the natives to form ornaments for their war implements, as well as to adorn their persons.

Section 2. *Trichosurus.*

Trichosurus. LESSON, Dict. Cl. d'Hist. Nat. t. xiii. p. 333. 1828.
Phalangista proper of TEMMINCK, LESSON, GRAY, and some other authors.

Tail densely clothed with fur, with the exception of a part of the under surface, commencing at the point, and more or less extended towards the root of the tail: ears distinct, usually long; eyes with the pupil round: fore feet normal (*i. e.* with no marked separation of the two inner from the three outer toes): cœcum very long[2].

The species of this section inhabit Australia: their fur is longer and less dense than that of the Phalangers of the islands north of Australia. In specimens of *P. vulpina*, living in the menagerie of the Zoological Society, I observed that the pupil of the eye was round, and not narrow, and nearly vertical, as in the species of the *Cuscus* section; but I

[1] In so hot a climate, were animals not brought alive to the market, they would too soon become unfit for food.

[2] In *Phalangista vulpina* the cœcum was found to be 25 inches in length, in an animal measuring 19 inches; Mr. Martin gives 16½ inches as the length of the cœcum of a specimen, the anatomy of which he describes in the Proceedings of the Zoological Society for 1836—the animal measuring 20½ inches from the tip of the nose to the root of the tail.

must observe, that the iris of the eye was of so dark a colour, that I could scarcely distinguish the pupil from that part. In Cuscus the iris is of an orange colour, or red.

PHALANGISTA (*Trichosurus*) VULPINA.

The Vulpine Phalanger.

(*Plate* 9, fig. 1.)

Didelphis vulpina.	SHAW, General Zoology, vol. i. Part 2, p. 503. 1800.
" *lemurina.*	SHAW, l. c. p. 487, Pl. 110.
Phalangista vulpina.	DESM. Nouv. Dict. d'Hist. Nat. tom. xxv. p. 475 (1817); Mammalogie, Part i. p. 267. 1820.
" "	TEMMINCK, Monographies de Mammalogie, tom. i. p. 5. 1827.
Phalangista melanura.	WAGNER, in Schreb. Saug. Suppl. 111-112 Heft, p. 81. 1842.
" *fuliginosa.*	OGILBY, Proceedings of the Zoological Society for September, 1831, p. 135.
" *Cuvieri* (Gray).	WATERHOUSE, in Naturalists' Library; Marsupialia, p. 268. 1841.
" *felina.*	WAGNER, in Schreb. Saug. Suppl. 109-110 Heft, p. 76. 1842.
Phalanger de Cook.	F. CUVIER et GEOFF. Mammifères, Plate 45. 1824.
Vulpine Opossum.	PHILLIP, Voyage to Botany Bay, p. 150, Pl. 16. 1789.
Wha Tapoa Roo.	WHITE, Journal of a Voyage to New South Wales, p. 278, and Plate. 1790.

Fur long; general colour grey: under parts of body pale yellow, or yellowish white: muzzle and chin blackish: ears nearly naked on the inner side, having but a few scattered pale hairs; externally well clothed, excepting at the point, with whitish hairs, but with a large black patch at the base: feet yellowish, more or less suffused with brown: tail bushy, black, excepting at the base; the apical third naked beneath: chest almost always with an oblong patch of deep rust-coloured hairs.

Inhabits New South Wales, Western Australia, and North Australia.

The Vulpine Phalanger is about equal in size to the Common Cat; its ears are long, and somewhat pointed; the tail is about equal to the body in length. The fur is long, loose, and moderately soft to the touch; its general colour is grey, the visible portions of the hairs being partly black and partly white; on the back the fur assumes a somewhat deeper hue than on the sides of the body, owing to there being very long black hairs rather plentifully interspersed with the ordinary fur on that part. The muzzle and chin are blackish, but the former is pale near the tip, and the naked muffle is of a whitish flesh colour. The eyes are encircled with blackish hairs; the ears are almost naked within, and of a brownish pink colour, but there are a few scattered pale hairs on this part; on the outer side, excepting near and at the point, and a narrow space along the anterior margin, they are clothed with a dense and moderately long fur, which is white at the posterior angle of the ear and towards the apex, but black elsewhere. The hairs of the moustaches are long, numerous, and black, and there are a few long, black, bristly hairs springing from above the eyes. The throat, as well as the whole under parts of the body and inner sides of the limbs, are of a pale buff-yellow tint, but on the chest is a large oblong patch of deep rust-coloured hairs. The feet are yellowish white, but more or less suffused with brown on the toes; the posterior part, and the outer side of the hind feet, are often dusky; the naked soles of the feet are flesh-coloured; the claws are dusky. The tail is clothed at the root with fur of the same texture and colour as that of the body; beyond, the fur is more bushy, of a harsher nature, and black; the extreme point of the tail, and the under surface of about the apical half, are naked. The region of the pouch of the female (as is very generally the case in the Marsupial animals) is clothed with rusty red hairs.

	Lines.	Inches.
Length from the tip of the nose to the root of the tail	18	0
" of tail	13	0
" from tip of nose to ear	3	2
" of ear	1	11
" of tarsus (claws not included) ...	2	7
" of fore-foot, without the claws	1	11

The foregoing description is taken from a female Vulpine Phalanger in the Museum of the Zoological Society, and will agree pretty closely with most of the specimens which I have seen from the continent of Australia; I have notes, however, of some variations in the colouring of specimens from the main land, which I shall briefly notice.

Var. 1.—This differs from the specimen just described in having the sides of the body chiefly of a bright rusty red hue, and the same reddish tint is observable on the back, though less distinct, the hairs of the fur on this part being reddish below the point, white near the point, and black at the apex: the cheeks and under parts of the animal are of a yellow colour; the feet yellowish white, but the outer side of the tarsus is blackish.—From South Australia. Mus. Zool. Soc.

Var. 2.—Resembles the last, and is from the same quarter, but differs in the rusty hue being paler, and in having the cheeks and feet suffused with a rich brown colour.—Mus. Zool. Soc.

In the British Museum are specimens from the neighbourhood of the River Nammoy, and from that of the Yaroundi, agreeing almost perfectly with my first description. The same might almost be said of a specimen from Perth, but in this specimen the under parts of the body are almost of a pure white, the ears are grey at the base externally, where usually black or dusky; and the rusty chest mark is absent. With regard to the chest mark I must observe, that it varies much in extent, and is more or less distinct. I have a description of a Vulpine Phalanger, from Port Essington, before me, but, upon looking it over, can perceive no points of the least importance in which this specimen differs from the individual first described. Its fur was rather more dense and crisp.

The dimensions of some of the specimens above alluded to are as follows:—

	PERTH. Female.		YAROUNDI. Male.		YAROUNDI. Female.		NAMOI.	
	Ins.	Lines.	Ins.	Lines.	Ins.	Lines.	Ins.	Lines.
Length from tip of nose to root of tail	19	0	20	0	21	0	17	0
" of tail	11	0	12	0	13	0	10	9
" of ear	1	10	2	1	2	1	1	8
" of fore-foot and nails ...	1	6	1	10	1	11	1	6
" of hind-foot and nails ...	2	4	2	9	2	10	2	3
Width of three upper incisors taken together		3½		3½		4		3½

Numerous specimens have from time to time formed part of the Zoological Society s living collection, and from my own observations they appeared to be by no means intelligent animals. During the day-time they were usually asleep, but towards the evening they became active, and on the alert for their food, consisting of bread and milk, and various vegetable substances, including fruits. Whatever eatable was given to them was taken by, and held between the hands in the same manner as a squirrel holds a nut. Occasionally a dead bird was given to these animals, which were evidently fond of such food, and most particularly the brain, which was the first part consumed.

When in their native haunts they inhabit the large trees (usually the Eucalypti), selecting such as have the heart of the branches, or trunk, decayed, since it is in the hollows of these trees that they secrete themselves during the day-time. At night they leave these hiding-places, and climb the branches to feed upon the buds, leaves, and fruits, and sometimes they descend to the ground, where they probably find herbs to their taste. Whilst climbing, their prehensile tail assists them to maintain a firm hold of the branches; in captivity I have noticed that in descending from one perch of their cage to another, or to the floor, the tail invariably encircled the perch they were quitting, until the animal was again securely lodged.

Phalangista melanura. WAGNER.

The *P. melanura* of Dr. Wagner is founded entirely upon the description and figure of the animal called *Phalanger de Cook,* in the great work of Geoffroy and F. Cuvier. These authors committed a singular error in identifying the animal they describe with the *Phalangista Cookii* of authors, which belongs to a different section of the Phalanger group. The figure alluded to is by no means accurately coloured, if we may judge from the text which accompanies it; and this, it appears to me, clearly refers to the *P. vulpina.*

Phalangista fuliginosa. OGILBY.

General colour brown-black; muzzle, chin, feet, and ears externally, black; throat, chest, and abdomen, of a fulvous brown colour, the last mentioned part of a deeper hue than the chest, &c.

Inhabits Van Diemen's Land.

The animal described by Mr. Ogilby, under the name *Phalangista fuliginosa*, died in the menagerie of the

Zoological Society, and was supposed to be from Sydney, but though numerous specimens of Phalangers of similar colouring and proportions have recently been forwarded to England from Van Diemen's Land, no such animal, to our knowledge, has been sent from Sydney, nor, indeed, from any part of the main land.

The following description is taken from the specimen just referred to.

Ears long, and pointed; tail very bushy, and equal to the body in length. Fur very long, loose, and moderately soft to the touch; its general colour almost black, especially on the back of the animal; but on the sides of the body a rich brown hue is perceptible; and the throat, chest, and under parts, are of a rich brownish fulvous hue, rather deeper on the abdomen than elsewhere. The fur on the back is pale brown next the skin; each hair assumes a deeper hue towards the point, and the visible portions of the hairs are black; on the sides of the body there is less of the black at the points of the hairs, the brown below the points is of a richer hue, and is visible; here, moreover, the hairs are many of them annulated, near the point, with rusty yellow, especially towards the shoulders: the chin and muzzle are almost entirely black, as well as the back of the ears, the feet, and the tail. The ears are naked internally, or nearly so; externally they are well clothed with fur, excepting near the point and along the anterior margin, which parts are covered only with small adpressed hairs, as in *P. vulpina*. About six inches of the apical portion of the tail beneath is naked.

The dimensions in the second column (p. 291) are from an animal, in the collection of the British Museum, which agrees with the above description; it is from Van Diemen's Land, as is also the animal whose dimensions are given in the third column. This latter specimen differs from the former in being of a light grey tint, and, indeed, its colouring corresponds perfectly with that of ***P. vulpina;*** I must observe, however, that the fur on the sides of the body is composed of hairs which are of a pale rufous hue, but annulated with white near the point, and black at the point, and that this rufous tint is slightly visible when the hairs are in their natural position. In some specimens from Van Diemen's Land the rufous tint is deeper, and very conspicuous on the sides of the body, and visible even on the back. Others I have seen which were intermediate in their colouring between the deep blackish brown specimens and the grey individuals. On the other hand, I may observe that, although the red hue on the sides of the body is not common in the New South Wales' specimens of ***P. vulpina***, yet it does sometimes make its appearance, as may be seen in individuals contained in the British Museum collection; and, from South Australia, I have seen several specimens having the fur of the upper parts of the body of a very distinct rufous grey hue. In the fourth column I have given the dimensions of a beautiful and uncommon variety of our animal, in which the whole of the fur is nearly of an uniform yellow-white, or cream colour; the cheeks and under parts of the body are, however, of a more decided yellow. The specimen, which is in the museum of the Zoological Society, is from Van Diemen's Land. A similarly coloured, but smaller specimen, from the same island, forms part of the British Museum collection.

As the almost black specimens are sent from Van Diemen's Land, accompanied with others, which are of very deep rufous brown tint, much suffused with black on the back; by others, which are of a rich rufous grey; by others, again, which are grey; and, lastly, by individuals which are of a cream colour, and as all these specimens agree very closely in size and proportions, we can but conclude that they form one and the same variable species.

Upon comparing the skins or stuffed specimens of this Van Diemen's Land Phalanger, with those of the *P. vulpina* of New South Wales, the only tolerably constant difference I could perceive was, that the Island animal was rather larger than that of the main land; and two skulls, contained in the Museum of the College of Surgeons, and known to be from Van Diemen's Land specimens, presented a corresponding difference of size when compared with the crania of the *P. vulpina*. I could find no sensible difference of proportion in these skulls, but the dimensions in the next page will speak for themselves.

	Zoological Society.		British Museum.		British Museum.		Zoological Society.	
	Ins.	Lines.	Ins.	Lines.	Ins.	Lines.	Ins.	Lines.
Length from the tip of the nose to the root of the tail ...	22	0	22	0	18	6	22	0
" of tail	13	0	14	0	12	0	13	0
" from nose to ear ...	3	11	3	8	3	4		
" of ear	2	1	2	0	2	0	2	2
" of hind foot and claws ...	3	5	3	6	2	8	3	4
" of fore foot and claws ...	2	7	2	7	2	0		
Width of three upper incisors taken together		4¼		4½		4		

	Van Diemen's Land.				Port Essington.				Habitat uncertain.					
									1		2		3	
	Ins.	Lines	Ins.	Lines	Ins.	Lines	Ins.	Lines	Ins.	Lines	Ins.	Lines	Ins.	Lines
Length of skull	3	9½	3	9	2	11	2	10½	3	6½	3	4	3	1½
Width	2	3	2	3	1	9	1	10	2	2¼	2	0¾	1	11
" between orbits		5½		5		5		5		5		6		5½
Length of nasal bones	1	5½	1	5⅓	1	1½	1	1¼	1	5⅔	1	4	1	3
Width of ditto behind		8½		8¼		6¼		6¼		8⅓		6¾		7
" " in front		3¾		3½		3¼		3		4		3¾		3
" of three upper incisors, taken together		4				3⅔		3⅔		4¼				
From front of foremost incisor to principal premolar		11½				10		10		11¼	1	0		11
Length of the five molar teeth, taken together	1	0½	1	0⅓		9⅔		10¼		11		11		11

(1) Dimensions of a skull in the Hunterian collection; (2 and 3) skulls in the museum of the Zoological Society, entered as from specimens of *Phalangista vulpina*, which died in the Menagerie.

A skull of *P. vulpina*, in Mr. Gould's collection, which is from the Swan River district, measures 3″ 3‴ in length, and 1″ 11½‴ in breadth; and in the Hunterian collection is a skull, the proportions of which are almost precisely the same, being 3″ 3‴ in length, and 2″ 0‴ in width. It is to be regretted that our collections do not contain well-authenticated skulls of *P. vulpina* from New South Wales.

A series of specimens of the Phalangers under consideration, accompanied with their skulls, both from New South Wales and from Van Diemen's Land, is required to enable us to determine, in a satisfactory manner, whether the animals called *vulpina* and *fuliginosa* are specifically identical. The evidence which I have been able to collect leads me to believe they are, and that we must attribute the difference of size (which is but little), and the great tendency to variation in colouring, on the part of the Van Diemen's Land animal, to local causes—such as food and climate.

Phalangista Cuvieri. GRAY.

I formerly thought, with Mr. Gray, that a certain specimen of Phalangista, to which that gentleman gave the name *Cuvieri*, was specifically identical with the *Phalanger de Cook* of Geoffroy and F. Cuvier, and distinct from other species; but upon re-examination of the specimen in question I have changed that opinion, so far as relates to its being a distinct species. It *may* be identical with the "Phalanger de Cook," but I suspect that that animal will prove to be the *P. vulpina* of the Continent of Australia, whilst the *P. Cuvieri*, I think, should decidedly be associated with the island variety, or species, whichever it may be—the *P. fuliginosa*, *P. Cuvieri*, differing from *P. vulpina* in having the feet larger, as well as the upper incisor teeth, and in these parts agreeing perfectly with the *P. fuliginosa*, from the grey specimens of which it differs only in being paler, and in having the tail less bushy; but, with respect to these differences, I must observe that the animal had lived in confinement for some time previous to its death, and when it died it had shed the greater portion of the longer and coarser hairs of the fur; and it is the dark points of these hairs which give the deeper general hue to specimens which are in better condition.

Phalangista felina. WAGNER.

This animal is described as being of the same size as *P. vulpina*, and of a rusty-red colour, but with the back suffused with black; the ears externally, as well as the lips and feet, blackish; the tail bushy, cylindrical, and perfectly black.

	Inches.	Lines.
Length from tip of nose to root of tail ...	23	0
" of tail	15	9
" ear	2	2

It is said to be from New Holland.

According to the dimensions given, the animal is rather larger than the *P. vulpina*, and agrees in size with *P. fuliginosa*, with some of the varieties of which it also agrees in colouring. I cannot think it is a distinct species.

PHALANGISTA XANTHOPUS.

Yellow-footed Phalanger.

Phalangista xanthopus. OGILBY, Proceedings of the Zoological Society for Sept. 1831, Part I. p. 135.

Fur loose and soft, of a deep grey colour above, having a considerable admixture of black, especially on the hinder parts of the back; under parts of body and feet yellowish white, the fingers suffused with brownish: muzzle, above dusky brown, at the sides sooty; chin dusky; ears naked within; externally, densely clothed with fur of a dirty white colour, but with a large dusky patch at the base: tail shortish, grey at base, black beyond, excepting about two inches of the apical portion, which is white: chest with a deep rust-coloured mark.

Inhabits Australia: the exact part not known.

The above description is taken from a specimen in the museum of the Zoological Society, which is the original of Mr. Ogilby's description in the Proceedings above referred to. Its dimensions are as follows:—

	Inches.	Lines.
Length from nose to root of tail	16	0
" of tail	10	6
" from nose to ear	2	9
" of ear	1	8
" of tarsus	2	0
" of fore-foot	1	1

I have never seen a second specimen of this species, which in most respects very closely resembles the *Phalangista vulpina*, but differs in having the apical portion of the tail white.

I cannot regard this as a well-established species; other specimens are required for examination, and above all, it would be desirable to examine its skull and dentition. With regard to the white tip to the tail, I must observe, that I have seen specimens of the *Petaurus breviceps* (an animal of the same family), in which the tip of the tail was also white, though it is usually black in that animal[1].

In the catalogue of the Sydney Museum, I find "the interior of Australia, from Iron-Bark Range, near the Glenelg River," given as the habitat of *Phalangista xanthopus*.

[1] In some packages containing about fifty or sixty skins of *Petaurus breviceps*, there were two or three specimens which differed in no way from the others, excepting in having the tail tipped with white.

PHALANGISTA CANINA.

Short-eared Phalanger.

Phalangista canina. **Ogilby, Proceedings of the Zoological Society for December, 1836, Part 3, p. 191.**

Above deep grey, beneath white, faintly tinted with yellowish; muzzle brown; sides of chin dusky; ears short, dusky at the base externally; tail bushy, black, excepting at the base, where it is coloured like the body; feet blackish.

Inhabits the interior of New South Wales.

This animal is about the same size as the *Phalangista vulpina,* to which it bears a considerable resemblance; it may, however, be readily distinguished by its comparatively short ears; these are rounded, nearly naked internally, and externally furnished at the base with fur of the same kind as that on the head, and of a blackish hue, but towards the posterior margin the hairs are whitish: the tail is very bushy, and nearly equal to the body in length: the fur of the animal is long, dense, and somewhat woolly, and its general hue is grey, it being rather finely pencilled with black and white; on the under parts of the body the fur is white, but each hair on these parts is indistinctly suffused with yellow externally, and is greyish next the skin; on the chest is a narrow rusty brown mark; the muzzle is dusky, and the eye is surrounded by the same dark hue; the feet are blackish; a small portion of the tail at the base is coloured like the body; the thick bushy hairs on the remaining portion are black; the apical third of the tail beneath, and the extreme point, are naked; the hairs of the moustaches are black, and the claws are pale horn colour.

	Inches.	Lines.
Length from nose to root of tail	22	0
" of tail	13	6
" from nose to ear	3	6
" of ear, about	1	2
" of tarsus	3	3
" of fore-foot	2	2

The specimen from which the above description is taken is in the museum of the Zoological Society, being also the original of Mr. Ogilby's description. It is said to be from the country beyond Hunter's River, about eighty miles north of Sydney. Mr. Gould informs me that, according to his observations, the Short-eared Phalanger was confined to the "scrub" districts, whilst the Vulpine Phalanger was found on the trees of the open plains.

Section 3. *Pseudochirus.*

Pseudochirus. OGILBY, Proceedings of the Zoological Society for March, 1836, Part 4, p. 26. Observations on the Opposable Power of the Thumb, Magazine of Nat. Hist. for Sept. 1837, vol. i. (New Series), p. 457.

Hepoona. GRAY.

Trichosurus. LESSON, Nouveau Tableau du Règne Animal, p. 189. 1842[1].

Phalangers with the two inner toes of the fore foot separated from, and partially opposable to, the other three; the tail clothed, excepting at the apex beneath, with short adpressed hairs; the ears short and rounded; and with six molar teeth, forming a continuous series, on either side of the upper jaw.

[1] M. Lesson, in the work above quoted, adds the date 1829 after the name *Trichosurus* (which is there applied to the group of which *Phalangista Cookii* is the type), and gives *Pseudochirus* of Mr. Ogilby as a synonym, with the date 1836! The fact is, as far I can ascertain, M. Lesson never, until 1842, separated the section under consideration from *Phalangista* proper, nor do I perceive that he was aware of the principal peculiarity which distinguishes *P. Cookii* from other Phalangers, and which was first pointed out by Mr.

The present section contains but two well determined species, and these are distinguishable from the species of the preceding section by their having the tail clothed with comparatively short hairs, the toes of the fore-foot being nearly equal in length, and divided, the inner two from the outer three, so as to be, to a certain extent, opposable to them. The second and third toes of the hind feet are not only united to the extremity (or very nearly so), but the latter toe is united to the fourth for about one-half its length. The inner toe, or thumb, is longer than in *P. vulpina*, extending, when directed forwards, very nearly to the extremities of the second and third toes, whilst in *P. vulpina* the thumb terminates on a line with the base of the same toes, and in this last mentioned animal the third toe is scarcely joined to the fourth. The very broad and short ears form another distinction for the present section, and lastly may be noticed the difference in the molar teeth. The dental formula in *P. Cookii* (the type of the section) is, incisors, $\frac{6}{2}$; canines, $\frac{1-1}{0-0}$; premolars, $\frac{3-3}{3-3}$; true molars, $\frac{4-4}{4-4}=38$. Of the premolars enumerated, there are $\frac{1-1}{2-2}$ which are small and unimportant, but the second of the three upper premolars is a well-developed tooth, and is contiguous with the third, which corresponds to the tooth which I have called the principal premolar. The incisor teeth are proportionately smaller than in *P. vulpina*, and the foremost pair of the upper series are but little larger than the others.

Ogilby. The name *Trichosurus* was orginally used by M. Lesson to distinguish the Australian Phalangers, having hairy tails, from the naked-tailed species of the Indian Islands, and consequently includes, besides the *P. Cookii*, the *P. vulpina* and *P. nana*.—See Dictionnaire Classique d'Hist. Nat. tom. xiii. p. 333, 1829. As *P. Cookii* and *P. nana* have since been made the types of new genera, or subgenera, we must restrict Lesson's sectional name to the Phalangers of which *P. vulpina* is the type.

PHALANGISTA (*Pseudochirus*) COOKII.

Cook's Phalanger.

Phalangista Cookii.	DESMEREST, Nouv. Dict. d'Hist. Nat. xxv. p. 478. 1817.
" "	TEMM. Monogr. de Mamm. tom. i. p.
" *viverrina.*	OGILBY, Proceedings of the Zoological Society for Nov. 1837, Part 5, p. 131.
" *Banksii.*	GRAY, Annals of Natural Hist. for April, 1838, vol. i. p. 107.
Phalanger de Cook.	CUVIER, Règne Animal, tom. i. p. 179, ed. 1817; tom. i. p. 183, ed. 1829.
? *Phalanger de Bougainville.*	CUVIER, Règne Animal (1829), p. 183.
New Holland Opossum.	PENNANT, Hist. of Quad. vol. ii. p. 301. 1781.
White-tailed Opossum.	SHAW, Gen. Zool. vol. i. Part 2, p. 504. 1800.
Ring-tailed Opossum of the Colonists.	

Fur long and soft: upper parts of body grey; flanks, and outer surface of limbs, suffused with bright rust-colour; under parts white, or yellowish white: ears well clothed externally, and the fur on this part dusky, excepting towards and at the posterior angle, where it is white, or yellowish: tail coloured like the body at the root, then black, and at the apex, white.

Inhabits New South Wales.

The *Phalangista Cookii* is about one foot, or rather more, from the tip of the nose to the root of the tail; its tail is about equal to the body in length. The ears are broad, but short and rounded, nearly naked on the inner side, and densely clothed on the outer surface, if we except a narrow space next the apical margin. The legs are short. The fur is rather long and dense, and moderately soft to the touch; on the upper parts of the body its general tint is pale grey, obscurely tinted with rust-colour; on the head, sides of the body, outer side of limbs, and under part of the tail at the base, a rusty yellow hue prevails: the chin, throat, and under parts of the body, as well as the inner surface of the limbs, are white, or yellowish white: the sides of the muzzle are blackish; the fur on the back of the ear is chiefly of the same blackish hue, but on the hinder part it is white (sometimes rusty white), and there is a small patch of white hairs on the side of the head immediately joining the posterior angle of the ear. The tail is clothed with fur like that of the body at the root, but receding from this part the hairs soon become harsher, shorter, and closely applied to the skin; these adpressed hairs are black on the middle of the tail and white at the tip: the longitudinal extent of the part of the tail which is clothed with white hairs varies considerably in different individuals. The apical third of the tail is naked beneath. The feet are of a pale rust colour, or sometimes brownish in the middle, and nearly white at the sides.

The present species of Phalanger was discovered in Cook's

first voyage by Sir Joseph Banks, and originally described by Pennant, from a specimen brought home by that gentleman[1]. During Cook's last voyage a second specimen was procured, and is figured in one of the plates illustrating the account of that voyage[2]. This second specimen was found at Adventure Bay, in Van Diemen's Land, whilst the first was from Endeavour River, on the north-east coast of the continent of Australia.

Subsequently specimens were brought to Europe by some of the French expeditions, and being recognized as the animal figured in Cook's voyage, the species was named after our celebrated navigator—I believe originally by Geoffroy. More recently, Mr. Ogilby, observing that a specimen sent to the Zoological Society from Van Diemen's Land differed in its size and colouring from others which he had examined, and which were known (some at least) to be from the main land, was induced to found a new species, under the name of *P. viverrina,* upon the Island specimen, retaining the name *Cookii* for the continental animal. Mr. Gray was also of opinion that the Van Diemen's Land and the continental Ring-tailed Phalangers were distinct, and proposed to name the one, originally discovered at Endeavour River, *P. Banksii,* and that from Van Diemen's Land, *P. Cookii*[3]. Mr. Gray has since changed his opinion, associating the Ring-tailed Phalangers under one specific name, and I find that Dr.

[1] See Hawkesworth's Voyages, vol. iii. p. 586.

[2] Cook's Third Voyage, first edition (1784), vol. i. p. 109, and Pl. 8.

[3] See Annals of Natural History, vol. i. p. 107. In the same volume will be found a discussion between Mr. Gray and Mr. Ogilby, relating to the nomenclature of these two supposed species; this discussion I do not think it necessary here to enter into, since I regard the animals as specifically identical. I will merely observe, that they were first separated by Mr. Ogilby, and that the specimens which first received the name *Cookii* by Geoffroy and Desmarest, agree in their colouring with the rufous-grey specimens of New South Wales.

Wagner, in his continuation of Schreber's great work upon Mammalia, likewise associates the animals named *P. Cookii* and *P. viverrina*.

It is certain that amongst the extensive series of these animals now in the British Museum collection, there are specimens from the continent of Australia which agree in their colouring with the *P. viverrina*, and on the other hand, amongst those from Van Diemen's Land, there are individuals which have the same pale grey hue as the *P. Cookii* of Desmarest.

As in the case of the Vulpine Phalangers of Van Diemen's Land, the Ring-tailed species in that island varies considerably in its colouring, often assumes a dark, sooty hue, and is sometimes white, or cream-coloured, whilst the specimens from New South Wales vary but little: from the opposite coast, however, I have seen several specimens which were fully as dark as any found in Van Diemen's Land. I will now notice some of the varieties of *Phalangista Cookii*, first observing that all the animals of this species which I have seen, have a conspicuous white, or yellowish white, fringe of hairs around the hinder angle of the ear, and the apical portion of the tail white: sometimes less than a quarter of the tail is white, and sometimes more than half. The tail is often black, or nearly so, on the part immediately preceding the white portion, and becomes paler, and coloured like the body, at the root.

With regard to the New South Wales specimens, I have only to remark, that they almost always agree very closely with the description heading this account. Sometimes the fur is red-grey, and the rusty red on the flanks and limbs is very distinct; in other specimens the upper parts are grey, and the legs and sides of the body are but little tinted with rust colour; and in the young the colouring is always rather darker than in the adult. The specimens from the Swan

River district are frequently very dark: three out of six, which are before me, are almost black on the upper parts of the body, but slightly inclining to grey: the under parts are white, and the muzzle and feet are sooty black. One specimen is brown, slightly suffused with reddish; the body beneath white, tinted with yellow in parts; the feet deep brown. The remaining two are of a pale grey hue on the upper parts of the body, and yellowish white beneath; in one of these specimens the feet are very pale, inclining to white; in the other they are brown.

As presenting the most general features of the Van Diemen's Land specimens, I will describe the specimen in the Zoological Society's collection, which has already been referred to, being the original of the

Phalangista viverrina. OGILBY.

General colour sooty-grey, the dark fur slightly pencilled with whitish on the head and fore parts of the body of the animal, and with brown on the hinder parts; chin, throat, and abdomen, yellowish white: muzzle dusky brown; cheeks paler than on the upper parts of the head, and with a whitish spot situated a little behind the eye: the dense fur on the back of the ear black, excepting at the posterior margin and hinder angle, where it is white: outer surface of the limbs rusty brown; feet black: tail brown-black, finely pencilled with pale brown, of a brownish rust colour at the root, and with the apical third white, excepting on the under surface, where it is naked.

A second specimen, in the same collection, is worthy of notice, differing from the one just described in having the general hue rather darker; the tail black, but with the apical third white; the abdomen grey; the chest white, and a broad white mark extending from the chest to the middle of the abdomen; the chin is dusky at the sides.

In the British Museum collection are five specimens from Van Diemen's Land, in which the general hue of the fur is pale rufous grey on the back, and bright rust colour on the sides of the body and limbs; of these, three have rusty white feet, and in two the feet are brown: on the cheek of several, is a dusky longitudinal mark, whilst in others the cheeks are almost entirely white. In all, the under parts of the head and body are white, or yellowish white. These specimens, therefore, resemble the individuals found in New South Wales.

The following are the dimensions of some of the specimens alluded to:—

	New South Wales.		*P. Cookii*, Desm., Paris Mus.		Swan River district. FEMALE.		*P. viverrina.* Mus. Zool. Society.		Van Diemen's Land. MALE.		Van Diemen's Land. FEMALE.		New South Wales. FEMALE.	
	Ins.	Lines.	Ins.	Lines.	Ins.	Lines.	Ins.	Lines.	Ins.	Lines.	Ins.	Lines.	Ins.	Lines.
Length from tip of nose to root of tail ...	13	6	14	0	15	0	15	0	15	0	14	0	15	0
" of tail ...	12	0	12	6	13	0	13	0	14	0	14	6	15	0
" from nose to ear	2	4	2	1	2	3½	2	5	2	2	2	3	2	2
" of hind foot ...	1	8			2	0	1	11	2	0	2	0	1	11
" of ear ...		9½		10½		11		11	1	1	1	1	1	0

The subjoined dimensions are from a skull of *Phalangista Cookii* in the museum of the Zoological Society.

	Inches.	Lines.
Total length of skull	2	6½
" width	1	4⅓
" " between orbits		3⅓
Length of nasal bones		10½
Width of ditto at the base		5
" near the apex		3
Length of palate to the anterior margin of the great palatine opening		9½
From front of foremost incisor to principal premolar		6¾
Length of the principal premolar and four true molars taken together		8¾

Phalangista Bougainvillii.

Phalanger de Bougainville. Cuvier, Règne Animal, ed. 1829, p. 183.
Phalangista Bougainvillii. Wagner, in Schreb. Saug. 111-112 Heft, p. 82. 1842.

Size of a squirrel: upper parts of the body ash-coloured, under parts white; the posterior half of the tail black, and the hinder half of the ear white.

The above short description is from Cuvier, who states that the specimen from which it is taken was brought home by the Baron Bougainville in his last voyage. I strongly suspect it will prove to be the *P. Cookii.*

PHALANGISTA (*Pseudochirus*) CANESCENS.

The Hoary Phalanger.

Phalanger grisonannt. Humbron et Jacquinot, Voyage au Pole Sud, &c. Zoologie—Mammifères, Pl. 16.

Ears very small: two inner toes of the fore foot distinctly opposable to the other three; thumb of hind foot directed backwards : general colour grey-brown ; under parts of body impure white ; sides of face fulvous ; upper surface of head with a broad, dusky, longitudinal mark.

Inhabits ?

A very distinct species of Phalanger is represented in one of the plates of the work above quoted, but since the letter-press of this great French work is not yet published, or at least has not reached us, I am unable to give any account of the animal beyond such as may be gleaned from the plate. According to the figures, the "Phalanger grisonnant[1]" is evidently allied to the *P. Cookii*, having the same *general* conformation of skull, teeth, feet, and tail ; the two inner toes of the fore feet, however, are still more distinctly opposed to the remaining three than in the animal just mentioned, and the thumb of the hind foot is directed completely backwards. The limbs and upper parts of the body, as well as the tail, are represented as of a grey-brown colour, the under parts of the body impure white, the sides of the face yellowish, or fulvous, the upper surface of the head with a broad, dusky, longitudinal mark, and the claws yellow. The ears are proportionately much smaller than in *P. Cookii.*

The young animal is depicted as having a dusky longitudinal stripe along the back.

The size and proportions of the parts should be very nearly as follows :—

	Inches.	Lines.
Length from tip of nose to root of tail ...	12	0
" of tail	11	0
" of ear		5
" of fore foot, including the nails ...		$7\frac{1}{3}$
" of hind feet, including the nails ...	1	1
" of skull	2	$1\frac{1}{2}$
Width of ditto	1	$3\frac{1}{2}$

[1] The authors give no technical name.

The skull is represented as having no posterior palatine openings. The teeth are—incisors, $\frac{6}{2}$; canines, $\frac{1-1}{0-0}$; molars, $\frac{7-7}{6-6}$: as in *P. Cookii*, there are six contiguous molars on each side of the upper jaw.

Section 4. *Dromicia.*

Dromicia. Gray, in Appendix to Grey's Journal of two Expeditions in Australia, p. 407. 1841.

Incisors, $\frac{6}{2}$; canines, $\frac{1-1}{0-0}$; premolars, $\frac{3-3}{4-4}$; true molars, $\frac{3-3}{3-3} = 36$.

Ears moderate, nearly naked, and folded; toes with the nails small; tail, excepting at the base, where it is covered with fur like that of the body, clothed with small adpressed hairs; naked beneath at the extremity.

The little Phalangers composing this section very much resemble the Dormice amongst Rodents, not only in general appearance, but also, to a considerable extent, in habits: they are readily distinguished by the reduced number of their molar teeth, affording a rare exception amongst the Marsupialia in having but three of these teeth on either side of each jaw. The foremost pair of incisors in the upper jaw are larger and longer, in proportion to the other two pairs, than in *Phalangista* proper; the latter teeth are very small; the canines are moderately developed, and situated distinctly behind the intermaxillary suture: these are followed by two minute teeth, separated from each other, and from the principal premolar which follows: this last mentioned tooth is compressed, pointed, and two-rooted. Of the three true molar teeth the first is rather the largest, and the last distinctly smaller than the others; this latter has but three pointed tubercles, whilst the foremost two true molars have each four pointed tubercles, of which the two on the outer side of the tooth are the largest. The incisors of the lower jaw are narrow, very long, and pointed; they are followed in each ramus of the jaw by three or four[1] small teeth, a principal premolar, and three true molars.

As is very generally the case in the smaller species of a natural group, the skull, to a certain extent, resembles that of the young individuals of the larger species; as, for instance, in the small development of muscular ridges, and proportionately larger size of the cerebral cavity. Such are the chief differences which distinguish the skull of the small Phalangers under consideration from those of the larger species of the group. The auditory bullæ are rather large, and partially divided by an oblique indentation: the palate has four posterior openings. The rami of the lower jaw are comparatively slender, and the inflected angular portion of the jaw is in the form of a slender pointed process.

In *Dromicia concinna* the nails of all the fingers, and of the fourth and fifth toes, are very small, and partially embedded in the upper surface of the fleshy pad which terminates these members[2]. The nails of the second and third, joined, toes of the hind foot are larger, and free: the first toe is nailless as usual. In the living animals the ears are considerably crumpled, and more or less pendant.

[1] In the skull before me, of *Dromicia concinna*, I find even five of these small teeth, but on one side of the jaw only, the other side of the jaw having four.

[2] I presume the same structure is found in *P. gliriformis*, but in dried skins the fleshy pad at the end of the toe necessarily shrinks, and the nails are then free.

PHALANGISTA (*Dromicia*) NANA.

Dormouse Phalanger.

Phalangista nana (Geoff.)	DESMAREST, in Nouv. Dict. d'Hist. Nat. tom. xxv. p. 470. Mammalogie, Pt. 1, p. 268.
" *gliriformis.*	BELL, Transactions of the Linnæan Society, vol. xvi. p. 121, Pl. 13.
" *nana* (Geoff.)	WATERHOUSE, Naturalists' Library, Marsupialia, p. 279.
Dromicia gliriformis.	GRAY, in List of the Mammalia in the British Museum, 1843, p. 85.
" "	GOULD, Mammals of Australia, Pl. 8.

Fur very soft, and moderately long; general colour ashy grey, suffused with pale reddish brown; under parts white, tinted with rusty yellow on the chest; hind feet white; fore feet grey; ears large; head rather paler than the body, dusky round the eye; tail very thick at the base, where it is clothed with fur like that of the body; the remaining portions very scantily clothed.

Inhabits Van Diemen's Land.

I had the pleasure, a short time since, of examining three or four living specimens of this interesting little animal in the menagerie of the Zoological Society. Very fat, and sluggish—during the day-time at least, I was struck with their resemblance to the Dormouse, but, as Mr. Bell correctly observes, they are broader, more depressed, and larger. That gentleman's accurate description, made from living specimens, I cannot improve, and shall therefore give in his own words—omitting only some few of the less important parts. "The head is broad across the ears, from whence it tapers to the nose, which is somewhat pointed; the nostrils are narrow, and of a semicircular form; the upper jaw, which is elongated, overhangs the under, and almost entirely conceals it; the eyes are very large, remarkably prominent, and of a jet black colour: the ears are of considerable size, erect[1], totally destitute of hair, and of an uniform mouse colour; the fur is very soft and thick, and its general hue is rufous grey; the under parts are more sparingly covered with fur, of a pale yellowish grey colour, the yellow predominating at the sides, and especially on the throat; the general colour of the face is also yellowish; the sides of the neck, as well as the throat, are buff; there is a blackish ring around the eye, and a darkish ring partially surrounds the ear at the anterior part, interrupted by a distinct white spot behind; the feet are almost entirely concealed by the fur, both when the animal is at rest, and when in an active state. The breadth of the body, combined with the length of the fur, and the extent to which the skin of the sides is attached to the legs, namely, as far as the carpi and tarsi, gives it very much the appearance of a *Petaurus*, to which this species evinces a remarkable approximation. The tail is nearly as long as the head and body together, and is remarkably broad and thick at the base, to more than half an inch from the origin, at which part it becomes contracted, and then gradually tapers to the extremity; it is hairy, being more thickly covered on the upper part, and especially at the base, where it partakes of the general colour of the upper parts of the body, becoming more scantily furnished towards the point, and there is on the under part, at the extremity, a space of about half an inch in length, which is entirely naked; the tail is more or less prehensile throughout its whole length, but especially towards its extremity."

	Inches.	Lines.
Length of head and body	3	10
" of tail	3	6
Breadth of tail at the root		6
Length of ear		5
Breadth of ditto		5

[1] The ears were always pendant in some specimens which lived in the menagerie of the Zoological Society, and they were not quite destitute of hair, though very nearly so.

The largest of the specimens living in the Zoological Society's Menagerie slightly exceeded in its dimensions the admeasurements above given, and differed from others somewhat in its colouring; its length was a trifle more than four inches measured in a straight line from the tip of the nose to the root of the tail, which was three inches and ten lines in length. The basal portion of the tail was very broad, being about three-quarters of an inch in width, and slightly depressed; at about an inch from the root it became rather suddenly contracted, and thence gradually decreased in diameter to the extremity; it was rather sparingly clothed with palish hairs, which were closely applied to, but did not hide the skin, which was flesh-coloured; the basal, incrassated portion of the tail, was clothed with fur like that of the body, but more closely applied to the skin; about half an inch of the apical portion was perfectly naked. All the naked parts, such as the under parts of the feet, tip of the muzzle, and ears, were of a delicate flesh-colour. The fur was extremely soft, on the upper parts of the body ash-coloured, but suffused with rusty brown, and obscurely tinted with yellowish about the muzzle; the under parts were of palish rusty brown, somewhat darker in hue on the chest. In another specimen the under parts of the body were white, slightly suffused with rusty yellow on the chest; the white of the under parts extended upwards on the sides of the body, where it was separated from the greyish colour of the upper parts by an indistinct dusky mark: the same dusky hue was visible on the outer, or rather towards the hinder parts of the fore legs. This specimen, when dead, gave the following dimensions—the tail was imperfect.

	Inches.	Lines.
From nose to tail, along the course of the back	4	2
From nose to ear		11
Length of tail		8
" of hind foot and nails		6½

The ears in all the specimens were partially folded, the apical portion lopping over. An excellent figure of the present animal, made from life, will be found in Mr. Gould's Mammals of Australia.

The habits of the *Phalangista nana*, Mr. Gould observes, "are extremely like those of the Dormouse, feeding on nuts and other similar food, which they hold in their fore paws, using them as hands. They are nocturnal, remaining asleep during the whole day, or, if disturbed, not easily roused to a state of activity; and coming forth late in the evening, and then assuming their natural rapid and vivacious habits, they run about a small tree, which is placed in their cage, using their paws to hold by the branches, and assisting themselves by their prehensile tail, which is always held in readiness to support them, especially when in a descending attitude. Sometimes the tail is thrown in the reverse direction, turned over the back, and at other times, when the weather is cold, it is rolled closely up towards the under part, and coiled almost between the thighs. When eating, they sit upon their hind quarters, holding the food in their fore paws, which, with the face, are the only parts apparently standing out from the ball of fur, of which the body seems at that time to be composed. They are perfectly harmless and tame, permitting any one to hold and caress them, without even attempting to bite, but do not evince the least attachment, either to persons about them, or even to each other."

Mr. Gould states that the *Dromicia gliriformis* is more particularly abundant in the northern portions of Van Diemen's Land; that of all trees it appears to prefer the Banksias, whose numerous blossoms supply it with a never-ceasing store of food, both of insects and sweets During the day it generally slumbers, coiled up, in some hollow branch or fissure in the trees. That gentleman has observed, that during the months of winter it is less active, undergoing, in fact, a kind of hybernation, somewhat similar, but not to the same extent, as in the Dormouse.

It would be interesting to learn whether, during this partial hybernation, the accumulation of fat at the root of the tail decreases. We have instances in two or three other species of Marsupials of an incrassated tail, as in the *Didelphys elegans* of Chili, and the *Phascogale crassicaudata;* but here the tail is not always equally thick, and in the *Didelphys elegans* I have seen specimens in which it was not incrassated, and others in which it was very thick.

As there may be a difference of opinion with regard to the specific identity of the *Phalangista nana* of Geoffroy and the *P. gliriformis* of Bell, I will add a description taken by myself from the original specimen of the *P. nana,* which is contained in the Paris Museum, first observing, that the only difference I could perceive was that of size, Geoffroy's specimen being considerably smaller than the individual described by Mr. Bell; and this difference I think is attributable partly perhaps to age, but more to a tendency which these animals have to attain extreme fatness when kept in confinement.

The *Phalangista nana* of the Paris Museum is about equal in bulk to the Common Mouse; its fur is soft, dense, and of a yellow-grey colour; the head pale dirty yellow; a brown mark runs through the eye: the ears have extremely minute hairs on the outer side, and are naked within: fur like that on the body extends for about three-quarters of an inch on to the tail, and the remaining portion of that organ is covered with very small adpressed hairs, brownish on the upper surface, and pale beneath: the whole of the under parts of the body, and the lips, are whitish.

	Inches.	Lines.
Length from tip of nose to root of tail ...	2	7
" of tail	2	6
" of ear		3½
" of hind foot		5⅓

The specimen was procured by M. Péron at Maria Island, situated close to Tasman Peninsula, Van Diemen's Land.

PHALANGISTA (*Dromicia*) CONCINNA.

Beautiful Pigmy-Phalanger.

Dromicia concinna. GOULD, Proceedings of the Zoological Society for January, 1845, Pt. 13.—Mammals of Australia, Pt. 1, Pl. 9.

Plate 11, Fig. 2.

Fur very soft; upper parts of the body rusty brown, or ashy grey suffused with rust colour; under parts white; a dusky patch in front of eye; feet white: tail with minute brownish hairs.

Inhabits the Swan River district, Western Australia, and South Australia.

This little animal is rather smaller than the common Dormouse (*Myoxus avellanarius*). From the *Dromicia gliriformis* it may be distinguished by its much smaller size, its tail not being incrassated at the base, and its colouring, which is usually of a delicate rusty brown hue on the upper parts of the body, but sometimes greyish, slightly suffused with rust colour only: the white of the under parts is strongly separated, on the sides of the body, from the darker colour of the upper parts; the dark patch in the region of the eye is nearly confined to the anterior angle, but is continued in the form of a narrow line across the upper part.

	Inches.	Lines.	Inches.	Lines.
Length from nose to root of tail ...	3	6	3	7
" of tail	2	10	3	0
" from nose to ear		7½		
" of tarsus		5½		5½
" of ear		6¾		6

In the British Museum collection are three specimens[1] agreeing with the above description; of one of these, the largest, I have given the dimension in the second column, those of the first column being from a specimen in Mr. Gould's collection. A skull removed from one of the Museum specimens is somewhat fractured, and in the first of the admeasurements given below, allowance has been made for the lost, occipital portion.

	Inches.	Lines.
Its length is about		9
The width is		$6\frac{3}{4}$
Total length of palate		5
From front of foremost upper incisor to back of last molar		$4\frac{1}{4}$
Length of lower jaw from condyle to base of incisor		$5\frac{1}{2}$
Height of ditto from apex of coronoid process		$2\frac{3}{4}$

According to Mr. Gould, the *D. concinna* is abundantly, and very generally distributed over the colony of Swan River. Its habits being strictly nocturnal, it secretes itself during the day in the hollows of trees, and at night leaves its retreat for the flowering branches of low shrubby trees. At that time it is said to be very active, and when kept in confinement will leap from side to side of its cage in chase of insects, of which it is exceedingly fond. The sexes present but little difference either in size or colouring; in some specimens the under parts of the body are slightly tinted with buff colour.

PHALANGISTA (*Dromicia*) NEILLII.

Neill's Phalanger.

Upper parts of the body grey, under parts white; a black patch in front of the eye; size less than the Common Mouse.

[1] Two of these are from Western Australia, and the third is from South Australia.

The author is indebted to Mr. Neill, Deputy-Assistant Commissary General at King George's Sound, for many interesting notes relating to the Marsupial animals found in the above mentioned part of Australia, and likewise for some specimens of the smaller species, preserved in spirits, which were forwarded with the notes. Amongst these specimens I find the present species of Phalanger, which I have taken the liberty of naming after its discoverer, since it decidedly differs from any of the hitherto described species of the group. One of its most remarkable features is its small size, it being less than either of the two pigmy species just described; and I should not have introduced it into the list of species, had I not satisfied myself, by an examination of its skull, that it was an adult animal, having all the true molar teeth developed. The specimen is a male.

Mr. Neill states that the animal is called Jaa-jat by the aborigines of King George's Sound, and that it is found under the dead bark of trees—Eucalypti, &c., and also in holes in trees which have been excavated by fires. "The specimen sent was taken alive by my own hand, and, by dint of care, lived in confinement for upwards of two months, during which time it was fed upon sugar mixed with bread-crumbs. The general colour, when the animal was first caught, was a light azure grey, inclining to steel colour over the whole of the back, the fore legs, and the thighs[1]. The breast and belly were pure white, and the fore legs were also edged with white.

"When in a state of repose the ears of the little animal were pendant, but when startled, or in activity, they were erect: their colour is brownish pink: the upper half of the ear is thinly covered with very fine hairs, the lower half (externally?)

[1] As seen in the clear spirits, the fur on the upper parts of the body appears to be grey, slightly suffused with brownish, and there is an evident dark patch in front of the eye.

of the same grey tint as the body." The natives of King George's Sound, Mr. Neill moreover states, never molest this pretty little creature, either because it is too insignificant for an article of food, or perhaps some superstitious feeling forbids them. They inform him that it feeds upon young shoots of grass, as well as upon the honey of the Banksias, Xanthorœa, &c.

	Inches.	Lines.
Length from tip of nose to root of tail ...	2	3
" of tail	2	7
" of ear		5
" of fore foot		3¼
" of hind foot		5½
" of thumb of ditto		2
" of skull		8½
Width of ditto		6
Length of nasal bones		3
" of palate		4
" from outer side of upper premolar to back of last true molar ...		3¾
" of auditory bullæ		2⅓
" of lower jaw, measured from the condyle to the base of the incisors		4¾
Height of ditto from the apex of the coronoid process		2½

The bones of the skull are thinner than in *Phalangista gliriformis*, and the temporal ridges are scarcely perceptible: the auditory bullæ are large, and the posterior root of the zygomatic arch is much inflated to increase the air-cells of the ear chamber; towards the hinder part of the palate are two oblong openings, each rather more than a line in length, and behind these are two minute round openings The fleshy pads of the hands and feet are distinctly striated in the transverse direction, and on the intermediate spaces are minute tubercles, and each toe is terminated by a fleshy pad having concentric striæ. The tips of the nails do not project beyond these pads.

Genus, *Petaurus*.

Petaurus.	SHAW, Naturalists' Miscellany, Pl. 60. 1791.
"	DESM., Nouv. Dict. d'Hist. Nat. tom. xxv. p. 400. 1818.
Petaurista.	DESM., Mammalogie, Part 1, p. 268. 1820.
Phalangista.	ILLIGER, Prod. Syst. Mamm., &c., p. 78. 1811.

Phalangistidæ having a membrane extended from the fore to the hind leg, and filling the interspace of these legs; the tail well clothed with hair throughout, and generally very long.

The Petauri, or Flying Phalangers, in general appearance greatly resemble the Flying-Squirrels, having, like those animals, a membrane extended between the fore and hind legs, and which serves, to a certain extent, to sustain the animal in the air, when descending from a height, after the manner of a parachute. In some *Petauri* the tail is bushy and cylindrical like the large Flying-Squirrels (*Pteromys*) of India and the Indian Islands; in others (the smaller species) the tail is distichous, and in this respect they resemble small fiying-squirrels of the sub-genus *Sciuropterus*.

The dentition of some of the Flying Phalangers greatly resembles that of certain species of *Phalangista*, and M. F. Cuvier, disregarding the external peculiarities presented by the *Petauri*, associates them with the true Phalangers, of which he forms two sections, in each of which are species possessing the flying membrane, and others destitute of this appendage: for these sections he retains the names *Petaurus* and *Phalangista*. In the latter group M. Cuvier associates the animals of the Cuscus section together with the *Phal. vulpina*, and the *Petaurus sciureus* of authors, and in the former group he arranges the *Phal. Cookii*, the *Petaurus taguanoides*, and the *P. macroura*. With regard to the

Petaurus taguanoides and the *Ph. Cookii*, there certainly exists a great similarity in the dentition, but the possession of the lateral membranes joining the anterior and posterior limbs, and a difference in the structure of the feet of the former animal, should not be entirely overlooked; and with regard to the group *Phalangista*, as constructed by F. Cuvier, I have to observe that he is quite incorrect in his statement that the teeth of the *Petaurus sciureus* resemble those of the Cuscus section, or of *Phalangista vulpina:* the teeth in *Pet. sciureus*, indeed, differ more from the Phalangers with which they are associated, than do those of *P. Cookii* from *P. vulpina*. On the whole, the more correct mode, as it appears to me, of expressing the relationship of the flying and non-flying Phalangers, would be to arrange the species of these sections in two parallel series, as in my table at p. 12, though that table would more closely express the parallelism of the group, had the section *Petaurus* been placed opposite *Pseudochirus*, and that of *Belideus* been shifted rather higher, for the animals of this latter section are intermediate in their characters between *Petaurus* and *Acrobata*, and are not represented by any known species of the *Phalangista* genus.

Sub-genus 1. *Petaurista*.

Petauri with broad, and rather short, rounded ears, which are densely clothed with long fur on the outer surface; the toes of the fore foot nearly equal in length; the flank membrane extending only to the elbow joint; with seven well developed molar teeth, forming a continuous series on either side of the upper jaw, and six in the lower; the true molars having pyramidal cusps.

M. Desmarest divides the Petauri into two sections, making

the *P. taguanoides* the type of his first section, to which he gives the name *Petaurista,* and *Pet. pygmæus* the type of his second section. From the first of Desmarest's subdivisions we shall have to separate certain species which differ from his type, and, in fact, it will be necessary to restrict his sectional name to the *P. taguanoides.* The dental formula in this animal is, incisors, $\frac{6}{2}$; canines, $\frac{1-1}{0-0}$; premolars, $\frac{3-3}{1-1}$; true molars, $\frac{4-4}{4-4}$. The incisors of the upper jaw are arranged laterally, and the three on either side are placed close together; the two foremost, belonging to opposites sides of the jaw, are separated from each other by a narrow space; they are narrow at the base, and wide and somewhat compressed, above the base. The next incisor, on either side, is larger than the posterior one, and about half the height of the first; narrow at the base, and wide and truncated at the apex. The third incisor is small, and but little expanded at the apex. The canine is small, being in size about equal to the posterior incisor; its tip is rounded, and it springs from the maxilla a little behind the intermaxillary suture. The first false molar is minute and conical, separated by a considerable space from the canine, and also from the second premolar, and has but one fang. The next two premolars have two fangs each; they are broad at the base, and compressed at the apex: the foremost presents an anterior larger, and a posterior small and compressed tubercle; the hindermost is divided at the tip into three compressed cusps. The true molars are nearly square, but rather longer than broad; the crown of each, with the exception of the last, presents four angular cusps. In the last there are but three of these cusps; two in front and one behind. The incisors of the lower jaw are large, nearly cylindrical at the base, and somewhat dilated, flattened and pointed beyond, and have two sharp edges. The space between the incisors and principal premolar, I have found edentate in two skulls of *P. taguanoides*, but in a third skull I noticed a minute tooth in the same part. The single premolar on each side is placed close to the true molars, compressed in front, and expanded behind; a small anterior tubercle is separated from the body of the tooth by a slight transverse incision. The true molars resemble those of the upper jaw, excepting that they are rather narrower, and the last tooth has four, instead of three, cusps. The skull of ***Pet. taguanoides*** differs from the skulls of the species of the next section, in being smaller in proportion to the bulk of the animal, in having the zygomatic arches stronger, and flattened at the sides, in being deeply concave, and much contracted between the orbits, instead of flat, and in having the palate deeply emarginated behind. The nasal bones are much shorter. The toes of the fore foot are nearly equal in length, if we except the inner toe, which is distinctly shorter than the rest; they are provided with large, curved, and compressed claws, as are the toes of the hind feet, with the exception of the thumb, which, as usual, is nailless. The second and third of the hind toes are joined to the extremity, and the third is joined to the fourth for about half its length. The hands and feet are naked beneath. The Petauri, like the true Phalangers, are nocturnal, and hide themselves during the day in the hollows of trees.

PETAURUS (*Petaurista*) TAGUANOÏDES.

Taguan Flying-Phalanger.

Petaurus taguanoïdes.	DESMAREST, Nouv. Dict. d'Hist. Nat., tom. xxv. p. 400. 1818.
Petaurista "	DESMAREST, Mammalogie, Pt. 1, p. 269. 1820.
" *Peronii.*	DESMAREST, Nouv. Dict. d'Hist. Nat., tom. xxv. p. 400.
Petaurus taguanoïdes.	WATERHOUSE, Marsupialia, p 283, Pl. 27.

Ears rather short, very broad, and densely clothed with fur on the outer side, like that of the head; tail cylindrical, longer than the head and body taken together: fur very long and soft; its general colour on the upper parts of the body brown-black; on the flank membranes pencilled with whitish: throat, as well as the under parts of the body of the animal, impure white: ears fringed with white behind: tail black, but more or less brownish at the root. Varieties occur in which the upper parts of the body are more or less of a dirty white colour.

Inhabits New South Wales.

The Taguan Flying-Phalanger has a short and small head, broad ears, which, with the exception of a narrow space at the apex, are clothed with long dense fur, like that on the head, and a very long, bushy, and cylindrical tail[1]. The flank membrane, which fills the interspace between the fore and hind legs, is attached to the fore leg as far forwards as the elbow-joint, and to the hind leg as far as the base of the inner toe or thumb. The fur of the animal is very long and loose, and soft to the touch: its general colour on the upper parts of the body is brown-black; on the head, and back of the ears, brownish, and on the flank membrane it is pencilled with whitish, though the ground-colour is the same as that of the body: the feet, muzzle, and chin, are almost black; the throat, chest, under parts of the body and flank membrane, as well as the inner side of the limbs, are impure white; the wrists and ankles, however, are black on the inner side as well as on the outer: the long hairs near and at the posterior margin of the ear, are white, or whitish, and project from the edge of the ear in the form of a fringe; the hairs covering the hind feet, and along the back of the leg to the root of the tail, are very long: the tail is of a black, or brown-black colour, but is almost always paler at the root, and along the under surface for some considerable distance from the root; here the hairs are sometimes yellowish brown, and sometimes brown-white.

	Inches.	Lines.	Inches.	Lines.
Length from tip of nose to root of tail	20	0	20	0
" of tail	22	0	21	0
" from tip of nose to ear ...	2	2		
" of ear	1	4	1	4
" of hind foot and nails ...	2	2	2	3
" of fore foot	1	11	2	1

[1] The average diameter of the tail, including the fur, is about two inches.

This animal varies considerably in its colouring, and is often of a greyish black hue on the upper parts, the dark hairs being more or less pencilled with grey; the flank membranes are generally more distinctly pencilled, and consequently of a paler general hue than the back, and the limbs are black externally. Specimens which are totally white, and others which are white and irregularly variegated with grey or dusky, are not rare.

	Inches.	Lines.
Length of skull	2	4½
Width of ditto	1	5½
" between orbits		4
Length of nasal bones		8¼
Depth of zygomatic arch behind		3½
Length of palate	1	2
" of the six upper, contiguous, molar teeth, taken together		9¼
" of the lower jaw	1	6
Height of ditto in a vertical line dropped from apex of coronoid process		11
Length of five contiguous molar teeth, taken together		9¼

Mr. Gould informs me that the Taguan Flying-Phalanger is chiefly confined to the "scrub" districts of New South Wales.

Petaurus Peronii.

Desmarest describes this animal as having the body brown above, and white beneath; the head brown, particularly around the eyes; the muzzle suffused with yellow; the ears very pointed, brown above, and whitish at their base internally, and this colour is somewhat extended on to the cheeks; chin, deep brown; flank membranes above, brown, variegated with grey; haunch brown, shaded into yellowish; thighs externally, as well as the hind feet, of a deep brown; tail cylindrical, rather longer than the body, brown, and having about half an inch at the extremity yellowish white; under surface of the neck, inner surface of the limbs and flank membranes, as well as the abdomen, yellowish white. Length of head and body, 8 inches 2 lines; tail, 9½ inches, French measure.

In the Paris Museum I noticed a Flying-Phalanger which agreed so precisely with the above description, both as to size and colouring, that I could scarcely doubt it was the original of Desmarest's description. This specimen, it appeared to me, was a young individual of the ***Petaurus taguanoïdes:*** it had the ear densely clothed with fur on the outer surface, as in that species, from which it differed only in having the tip of the tail white; a difference which cannot be regarded as important, since other species sometimes have the tail white at the point, although normally of a different colour at that part. The difference which Desmarest notices as existing between his ***P. Peronii*** and the ***P. taguanoïdes***, viz. that the flank membrane in the former extends to the elbow only, whilst in the latter it extends to the wrist, does not exist: I have always found the flank membrane terminate at the elbow in ***P. taguanoïdes:*** one other difference, it would appear from the description, separates the two species, since the ears of ***P. Peronii*** are described as "*très-pointues:*" possibly this is a misprint for *très-poilues*.

Subgenus 2. *Belideus*[1].

Petauri with long and nearly naked ears, a bushy tail, the lateral membrane extending to the outer finger; the outer two fingers of the hand long, and equal to each other, or very nearly so; the second and third fingers distinctly shorter than these;

[1] From βέλος, a dart or javelin.

the second, the shortest of the latter two, and the inner, or first finger, very short. The dental formula is, incisors, $\frac{6}{2}$; canines, $\frac{1-1}{0-0}$; premolars, $\frac{3-3}{4-4}$; true molars, $\frac{4-4}{4-4} = 40$.

There are three well determined species in this section, the *P. australis*, *P. sciureus*, and *P. breviceps*, to which Mr. Gould has added a fourth species, *P. ariel*. They have the anterior incisor teeth of the upper jaw large, somewhat suddenly dilated immediately below the fang, and assuming a triangular form. In *P. australis* they are broader than in either the *P. sciureus* or *P. breviceps*. The next incisor, on either side, is smaller than the posterior one, narrow at the base, and broad at the apex. The third incisor is broad, and has a sharp incurved cutting edge. The canine is tolerably large, separated by a narrow space on either side, from the false molars, or incisors; compressed and pointed, and its anterior and posterior edges are sharp; the exposed part of this tooth is more elongated than that of either of the molars. The first premolar on either side is large, broad, compressed, and pointed, has a slight indication of an anterior and posterior lobe, and two distinct fangs. The second premolar is small, short, and compressed, and has a minute anterior lobe: this tooth is separated by a considerable space from the first premolar, and by a narrow space from the third; the latter touches the first true molar, is narrow in front, and presents one triangular, and pointed cusp. The first true molar is considerably larger than the following molars, each of which is smaller than the preceding, so that the last is not equal in bulk to one-half of the first. With the exception of the last, all the true molars present four somewhat blunt and rounded tubercles, and in general appearance very much resemble the corresponding teeth in the squirrel. The last molar has but three tubercles—two in front, and one behind. The incisors of the lower jaw are long, compressed, and pointed, and have the upper and lower edge sharp; they are almost horizontal in their direction, being but slightly curved upwards. Next follows, on each side, a series of four small premolars, the hindermost of which has two fangs, whilst the others appear to have but one. The true molars nearly resemble those of the upper jaw, though they are narrower and larger. The first has a large irregular anterior lobe, which is more elevated than any other part of the tooth, and is divided into two tubercles: the three posterior molars have each four tubercles. The space occupied by the true molars is relatively much less in *Belideus* than in *Petaurista*.

PETAURUS (*Belideus*) AUSTRALIS.

The Yellow-bellied Flying-Phalanger.

Petaurus australis.	SHAW, Naturalists' Miscellany, Pl. 60. 1791.
Didelphis Petaurus.	SHAW, General Zoology, Vol. i. Part 2, p. 496, Pl. 112. 1800.
? ——— *macroura.*	SHAW, Zoology of New Holland, No. 3, p. 33, Pl. 12; General Zoology, Vol. i. Part 2, p. 500, Pl. 113.
Petaurus flaviventer.	DESM. in Nouv. Dict. d'Hist. Nat. tom. xxv. p. 403. 1818.
Petaurista ———	DESM. Mammalogy, Part 1, p. 269. 1820.
Belideus ———	GOULD, Mammals of Australia, Part 1, Pl. 3.
Hepoona Roo.	HUNTER, in White's Journal of a Voyage to New South Wales, p. 288.

Ear very long: fur soft; general colour on the upper parts of the body, greyish, suffused more or less strongly with fulvous; the upper and under surface of the wing membranes, a broad dorsal stripe; the chin, muzzle, and back of the ears, at the base, dusky: legs black, throat and abdomen yellow-white, or bright yellow.

Inhabits New South Wales.

The Yellow-bellied Flying-Phalanger is usually rather

more than a foot in length, without including the tail, which is cylindrical, bushy, and considerably longer than the head and body taken together. The ears are of an elongate ovate form, and but sparingly clothed with hairs; indeed, nearly naked, excepting on the outer side at the base, where they are covered with fur like that on the head. The fur of the animal is long and soft, and subject to some variation in colouring; the ground colour on the upper parts of the body, however, is most frequently greyish, but much suffused with brownish yellow; the sides of the muzzle are dusky,[1] and the eyes are encircled with the same dark hue; the ears are black, and so is the fur of their base externally, but at the posterior angle it is yellowish: a broad blackish mark runs along the back to the root of the tail, and the flank membranes are dusky above and beneath, but edged with a yellowish fringe of hairs. The limbs and feet are black externally, and the former are all black on the inner side in the region of the wrist and ankle. The chin is dusky, and the throat, abdomen, and inner side of the limbs at the base are yellow. The tail is dusky, but suffused with brownish yellow, excepting on the apical third, which is black. The sides of the neck are greyish, and there is usually a pale spot on the cheek behind the eye. Sometimes the upper surface of the head and body are chiefly of a brownish black hue, and the yellowish grey assumes the form of two longitudinal bands, separating, on either side, the broad dark band of the back from the dark colouring of the upper surface of the flank membrane. In other specimens the prevailing hue on the upper parts is fulvous brown, the dorsal black band being narrow, and the bright colour extending somewhat on to the flank membrane: the abdomen of a rich yellow hue.

[1] As I frequently use this term, I may as well explain that by it I mean *a dark hue approaching to black.*

The naked portions of the hand and feet, in the living animal, are of a yellowish flesh colour, and the naked tip to the nose is of a pale pink hue. In the third column I have given the dimensions of a specimen taken immediately after death. It lived for some time in the Gardens of the Zoological Society. The dimensions in the other two columns are taken from stuffed specimens:—

	Ins. Lns.	Ins. Lns.	Ins. Lns.
Length from tip of nose to root of tail	14 0	14 0	12 0
" of tail	19 0	18 0	15 6[1]
" of ear	1 9	1 9	1 11
" of fore foot, including the nails	1 6	1 7	1 6½
" of hind foot and nails ...	1 7½	1 7½	1 8

Mr. Gould furnishes us with the following interesting account of the habits of the present species. "The Yellow-bellied, or Long-tailed Flying-Phalanger," this gentleman observes, "is common in all the brushes of New South Wales, particularly those which stretch along the coast from Port Philip to Moreton Bay. In these vast forests, trees of one kind or another are perpetually flowering, and thus offer a never failing supply of blossoms, upon which the Yellow-bellied Flying-Phalanger feeds: the flowers of the various kinds of gums, some of which are of great magnitude, are the principal favourites. Like the rest of the genus, it is nocturnal in its habits, dwelling in holes of trees and in the hollows of the larger branches during the day, and displaying the greatest activity at night, while running over the small leafy branches, frequently even to their very extremities, in search of insects, and the honey of the newly-opened blossoms. Its structure being ill adapted for terrestrial habits, it seldom descends to the ground, except for the purpose of

[1] Without including the fur.

passing to a tree too distant to be attained by springing from the one it wishes to leave. The tops of the trees are traversed by this animal with as much ease as the most level ground is by such as are destined for terra firma. If chased, or forced to flight, it ascends to the highest branch, and performs the most enormous leaps, sweeping from tree to tree with wonderful address: a slight elevation gives its body an impetus, which, with the expansion of its membrane, enables it to pass to a considerable distance, always ascending a little at the extremity of the leap; by this ascent the animal is prevented from receiving the shock which it would otherwise sustain."

Petaurus macrourus.

The *Didelphis macroura* is described by Shaw as about equal in size to the Black Rat (*Mus rattus*), of a dark brown-grey colour above, and whitish beneath; the head and neck also whitish, but a dusky stripe runs along the top of the head almost to the nose; the ears whitish, moderately large, and slightly rounded; the upper parts of the fore feet whitish; and the lower half of the tail of a deeper black than the beginning. The above description was drawn up by Dr. Shaw, upon a specimen sent over by Mr. White. I strongly suspect it is the young of the animal described by the same author, also from White's collection, under the name of *Petaurus australis*.

PETAURUS (*Belideus*) SCIUREUS.

Squirrel Flying-Phalanger.

Didelphis sciurea. SHAW, Zoology of New Holland, No. 4, Pl. 11, p. 29. 1794. General Zoology, vol. i. Pt. 2, p. 498. 1800.
Petaurus sciureus. DESMAREST, Nouv. Dict. d'Hist. Nat., tom. xxv. p. 403. 1818.
Petaurista sciurea. DESMAREST, Mammalogie, Pt. 1, p. 270. 1820.
Belideus sciureus. GOULD, Mammals of Australia, Pt. 1, Pl. 4.
Norfolk Island Flying-Squirrel. PENNANT, History of Quadrupeds.—Phillip's Voy. to Botany Bay, p. 151, Pl. 17.
Sugar Squirrel of the colonists of New South Wales.

Ear moderately long; tail very bushy, especially at the base, where it is as broad as the body: fur extremely soft; general colour delicate ashy grey: a longitudinal black band commences near the tip of the nose, and terminates before reaching the root of the tail; eye encircled with black; ear black at the base externally, white at the posterior angle: a black patch on the cheek (which is white) immediately beneath the ear opening: flank membrane edged with white, but blackish towards the edge, both above and below: chin, throat, and under parts of body, white: tail black at the apex.

Inhabits New South Wales.

To the above description I need only add, that the naked tip to the nose, and naked palms of the feet, are of a delicate flesh-colour in the living animal; the ears are of a brownish flesh-colour, becoming darker towards the point. The black stripe on the upper parts of the animal is always rather suddenly expanded on the crown of the head; the fore legs are grey externally, but become blackish towards and at the wrist; the hands are greyish brown above: the hind legs are also greyish externally, but there is an undefined dusky mark on this part; the feet are greyish behind; in front, and on the toes, the hairs are white. The hairs on the throat, chest, and mesial line of the abdomen, are uniform white (sometimes yellowish) to the root; towards the sides of the abdomen they are faintly tinted with grey at some little distance below the point, and on the under surface of the flank membrane they are black, or nearly so, and for the most part tipped with white; this membrane is extended to the base of the outer finger, which has a fringe of white hairs on the outer side. A largish grey patch is generally observed on the sides of the throat, and, besides the black spot under the opening of the ear, there is a second black spot behind the white fringe which adorns the posterior angle of the ear.

	Inches.	Lines.	Inches.	Lines.
Length from tip of nose to root of tail	8	6	9	0
" of tail	8	6	10	0
" of ear		$9\frac{1}{2}$	0	11

	Inches.	Lines.	Inches.	ines.
Length of fore foot, including the nails	1	3	1	2
" of hind foot and nails ...	1	1	1	2
" of skull	1	10		
Width of ditto	1	2¼		
Length of nasal bones		7½		
" of frontal bones		8½		
" of palate		11⅓		

According to Mr. Gould, the Squirrel Flying-Phalanger is very generally dispersed over the whole of New South Wales, where, in common with other Phalangers, it inhabits the large and magnificent gum-trees. Nocturnal in its habits, Mr. Gould observes, "it conceals itself during the day in the hollows of the trees, where it easily falls a prey to the natives, who capture it both for the sake of its flesh and skin, which latter, in some parts of the colony, they dispose of to the colonists, who occasionally apply it to the same purposes as those to which the fur of the Chinchilla, and other animals, is applied in Europe. At night it becomes extremely active in its motions, but during the day it is sluggish. I observed that it prefers those forests which adorn the more open and grassy portions of the country, to the thick brushes near the coast. By expanding the membrane attached to the sides of its body, it has the power of performing enormous leaps, and of passing from tree to tree without descending to the ground."

That these animals have the power of changing their course to a certain extent when descending, parachute-like, from a height, is evident from the following circumstance related by Mr. Broderip[1]. On board a vessel sailing off the coast of New Holland was a Squirrel Petaurus, which was permitted to roam about the ship. On one occasion it reached the mast-head, and as the sailor who was despatched to bring it

[1] See Penny Cyclopædia, article Marsupialia, vol. xvi. p. 461.

down, approached, it made a spring from aloft to avoid him. At this moment the ship gave a heavy lurch, which, if the original direction of the little creature's course had been continued, must have plunged it into the sea. All who witnessed the scene were in pain for its safety; but it suddenly appeared to check itself, and so to modify its career that it alighted safely on the deck.

Our little animal is figured and described in Phillip's Voyage to Botany Bay, under the name of Norfolk Island Flying-Squirrel, but whether the animal is really found in that island, so remote from the coast of Australia, the author does not inform us. Possibly it may have been introduced there by the shipping.

PETAURUS (*Belideus*) BREVICEPS.

Short-headed Flying-Phalanger.

Petaurus (*Belideus*) *breviceps*. WATERHOUSE, Proceedings of the Zoological Society for Nov. 1838, Part 6, p. 152; Marsupialia (Vol. xi. in Naturalists' Library), p. 290, Pl. 29.

Tail long and cylindrical: general colour of the upper parts of the body ashy grey; a black stripe commencing near the tip of the muzzle, runs along the back; ears black externally at the base, and white at the posterior angles; flank membrane blackish above, but edged with white; under parts of head and body, white: tail black at the tip.

Inhabits New South Wales.

In colouring this animal greatly resembles the *Petaurus sciureus*; it is, however, of a smaller size, and always has a long cylindrical tail: the size and proportions of its skull also differ. The head is short, the ears moderate, almost naked, being very sparingly furnished with dusky hairs, excepting at the base externally, where they are covered with fur like that of the head, and of a dusky or blackish hue; at the hinder angle, however, the fur is white. The tail rather exceeds the head and body in length, is cylindrical, and but moderately bushy. The fur is very soft, and its general tint on the upper parts of the body is ashy grey: a dusky longi tudinal line extends from between the eyes along the back; on the back it is rather indistinct, and towards the hinder part it is obliterated. The hairs of the tail are nearly of an uniform ashy grey hue, but somewhat inclining to dusky; on rather more than two inches of the apical portion the hairs are black. The flank membrane is blackish above, but white at the edge; the white fringe which margins this membrane is continued along the hinder part of the arm, and terminates at the tip of the little finger. The upper surface of the fore foot is smoky black; the hind foot is of a deep grey colour above, inclining to black: a dusky mark is observable along the outer side of the hind legs, and the external margin of the hind foot is fringed with long hairs. The under parts of the head and body, as well as the inner side of the limbs, are white, or nearly so.

	Lines.	Inches.
Length from tip of nose to root of tail	6	6
" of tail	7	0
" of hind foot, not including the nails ..	1	1
" of ear		9
" of skull	1	3½
Width of ditto	1	0
Length of nasal bones		5⅓
" of frontal bones		6¼
" of palate		8

The short-headed Flying-Phalanger must be very local, since it has been rarely sent home amongst the skins of mammals from Australia; I once, however, had an opportunity of examining an immense number of skins of this animal which had been prepared for the furrier: they were readily distinguished from the few specimens of ***P. sciureus***, with which they were mixed, by their smaller size, and the comparatively slender and cylindrical tail. In some few specimens the extreme point of the tail was white.

Petaurus (*Belideus*) *ariel.*

Belidea ariel. Gould, Proceedings of the Zoological Society for January, 1842, Part 10, p. 11.

General colour of upper parts of head and body ashy grey; a black stripe, commencing on the muzzle, extends back nearly to the root of the tail; ear with a blackish patch at the base externally; flank membrane dusky above; the edge, as well as the whole of the under parts of the body, pale yellow: tail black at the apex.

Inhabits North Australia; Port Essington.

In size and proportions this species so closely resembles the ***Petaurus breviceps***, that I do not think we are justified in regarding it as distinct, on account of the slight differences which its colouring presents, when compared with the animal just mentioned. The differences alluded to consist in the upper parts of the body being rather paler, and the under parts of a delicate yellow. The tail is cylindrical, as in ***P. breviceps***. The fore and hind feet are of a pale yellowish tint: a narrow black mark, commencing between the eyes, runs along the back, and extends nearly to the root of the tail: the eyes are narrowly encircled with black, and a black ring surrounds the ear at the base, but is interrupted at the posterior angle, where the hairs are pale yellow. The upper surface of the head is rather paler than the back, which is of a pale ash colour, but indistinctly suffused with yellowish: the upper surface of the flank membrane is blackish, especially near the margin—this latter is pale yellow: the anterior part of the fore-arm and region of the wrist, as well as the posterior part of the hind leg, are dusky.

	Inches.	Lines.
Length from tip of nose to root of tail	6	0
" of tail	7	0
" of ear		8
" from tip of nose to ear	1	3½

Dr. Müller, in his list of the New Guinea Mammalia, includes the ***Petaurus sciureus.*** That author has not yet described the animal alluded to, and I cannot help thinking it is more probably the present species, since this is found on the north coast of Australia, where, I believe, the ***P. sciureus*** does not occur.

Subgenus 3. *Acrobata.*

Acrobata. DESMAREST, Mammalogie, Pt. 1, p. 270. 1820.

Tail moderately long, clothed above and beneath with short adpressed hairs, and fringed on either side with longer hairs: ears moderate, well clothed externally with fine hairs: feet with small claws; thumb of hind foot large; flank membrane scarcely extending to the wrist. Dental formula:—incisors, $\frac{6}{2}$; canines, $\frac{1-1}{0-0}$; premolars, $\frac{3-3}{4-4}$; true molars, $\frac{3-3}{3-3} = 36$.

The anterior upper pair of incisor teeth in *Acrobata* are considerably larger than the other two pairs, and are much expanded at the extremity; the posterior two incisors, on either side, are small, and the second is the smallest: the canine is a well developed tooth, and is placed close to the incisors, and distinctly springs from the superior maxillary bone: this tooth is followed, on either side, by three well developed premolars, which are compressed and pointed, and present nearly a triangular outline when viewed on the outer or inner side, but each of these teeth has a very small anterior and posterior cusp at the base; they are all two-rooted, and the middle tooth is a trifle larger than the others, which are about equal in size; the hindermost premolar is in contact with the true molars, the remaining two are separated from each other, or from the canine, or the last premolar, by narrow interspaces: the three true molars are small, and successively decrease in size from the first to the last; the two foremost have two outer, pointed cusps, and two smaller inner cusps; the last molar has but three cusps. The two incisors of the lower jaw are very long, pointed, and nearly horizontal, and are followed on either side by two small, simple premolars; then two other well developed premolars, having the crown much elevated and pointed (the foremost most so), and a small posterior cusp; they are two-rooted. The three true molars have each four cusps; the outer anterior cusp of the first of these teeth is most elevated. All these teeth of the lower jaw form an uninterrupted series.

This dentition, it will be perceived, agrees very closely with that found in *Dromicia*, the chief difference consisting in the greater development of the premolars in *Acrobata*. In both cases there is an approach to the insectivorous type of dentition (most marked in *Acrobata*), as evinced in the small size of the true molars, when compared with those of the large Phalangers, and the greater importance which the premolars assume: the peculiar triangular form, and the presence of two roots, in many of these teeth, remind us strongly of the premolars in the small insectivorous Marsupials, such as the Phascogales[1].

[1] I cannot help here calling attention to certain points which I have endeavoured to illustrate elsewhere, viz. that the affinities of species of one group to species of another, are more near, or more remote, according to the rank of

PETAURUS (*Acrobata*) PYGMÆUS.

Pigmy Flying-Opossum.

Didelphis pygmæa.	SHAW, Zoology of New Holland, No. 1, Pl. 2, p. 5, 1794.—General Zoology, vol. i. Pt. 2, p. 501.
Petaurus pygmæus.	DESMAREST, Nouv. Dict. d'Hist. Nat. tom. xxv. p. 405. 1818.
Petaurista (Acrobata) pygmæa.	DESM. Mammalogie, Pt. 1, p. 270. 1820.
Petaurus (Acrobata) pygmæus.	WATERHOUSE, Marsupialia, p. 293, Pl. 30.

Fur rather short and soft; on the upper parts of the body of an ashy grey brown colour, and on the under parts, white, or yellowish white: ear with a dusky patch in front externally, and whitish behind: tail ashy brown, paler below than above: eye encircled with black.

Inhabits New South Wales.

This pretty little animal, which is the "Opossum Mouse" of the colonists of New South Wales, is said to be exceedingly numerous in the vicinity of Port Jackson. It is about the size of the Common Mouse, and of an ashy brown colour, inclining to grey, on the upper parts of the body, flank mem-

the groups exhibiting those affinities. In the present case we have an instance, on the one hand, of two species of *Phalangistidæ* of different genera, approximating very closely; and, on the other hand, of a species of the family *Phalangistidæ* approaching to certain *Dasyuridæ*. Now, the approximation of the species of the two genera of the same family, is much closer than that evinced by the Acrobata to the Phascogale (these being of different families), inasmuch as the resemblance in the latter case is only in the dentition, and that but partial, since the structure of the molar and incisor teeth in Acrobata is in accordance with the Phalangista type, and unlike that of the Phascogales; and, in the conformation of the extremities, I can perceive no tendency on the part of *Acrobata* to assume the characteristics of the *Dasyuridæ*; the second and third toes of the hind foot are not less perfectly united in *Acrobata* than in other *Phalangistidæ*.

brane, and outer side of the limbs; and yellowish white on the under parts. The feet are rather paler than the legs externally: the lower part of the cheeks and the upper lip are yellowish white; the eye is encircled with blackish, and this dark hue is considerably extended in front of the eye. The ears are of moderate size, rounded, moderately well clothed with fur externally, where they are dusky towards the fore part, and whitish behind; on the inner side of the ear near the apex, and from the apical portion, spring numerous long and extremely fine hairs. The hairs of the moustaches are numerous, slender, and of a dusky brown colour. The tail is about equal to the body in length, tolerably well clothed with hairs both above and beneath, but the short hairs on these parts lie close to the skin; on the sides the tail is fringed with longer hairs, and, including the hairs, its horizontal diameter is about a quarter of an inch; the upper parts of the tail are of the same ashy brown hue as the body, and the colour of the under surface differs only in being paler. The fur of the animal is rather short, and very soft; on the back it is grey next the skin; on the throat and chest each hair is uniform yellowish white throughout its length, but on the abdomen the hairs are for the most part grey at the root.

	Inches.	Lines.	Inches.	Lines.
Length from tip of nose to ear	3	7	3	6
" of tail	2	9	2	10
" of ear		3½		3½
" of fore foot		4¼		
" of hind foot		5		

The dimensions in the first column are taken from a specimen in the British Museum, and those in the second give the size of a specimen in the museum of the Zoological Society.

The following are the admeasurements of a skull of *Acrobata pygmæa* in the British Museum :—

	Inches.	Lines.
Length of skull		$9\frac{3}{4}$
Width of ditto		$6\frac{1}{2}$
Total length of palate		5
From front of foremost incisor to back of last molar		$4\frac{1}{3}$
Length of nasal bones		$3\frac{1}{2}$
Width of interorbital space		$2\frac{1}{2}$
Length of lower jaw		$5\frac{3}{4}$
Height of ditto in a vertical line dropped from the apex of the coronoid process		$2\frac{1}{4}$

The bones of the skull of *Petaurus pygmæus*, like those of the little Phalangers forming the section *Dromicia,* are extremely thin and semi-transparent; the cerebral portion of the cranium is large in proportion to the facial part; the occipital opening is very large, and the zygomatic arches are slender: the angular portion of the lower jaw is in the form of a slender pointed process, directed inwards and backwards so as to form an obtuse angle with the plane of the horizontal ramus: the coronoid process is also slender.

As compared with the skull of a species of *Dromicia* now before me (the *D. Neillii*), that of *P. pygmæus* differs in certain points which are worthy of notice. In the Dromicia the posterior root of the zygoma is much expanded, or, as it were, inflated, to increase the air-cells of the auditory chamber, whilst in the Acrobata there is no such expansion of the zygomatic process of the temporal bone. The auditory bulla is much more prominent in the Dromicia than in the small Flying-Phalanger, and is almost entirely formed by an expansion of the petrous element of the temporal bone; in *Petaurus pygmæus* the auditory bulla is formed partly by the same portion of the temporal bone, but chiefly by the great ala of the sphenoid.

Genus, *Tarsipes*.

Tarsipes. GERVAIS and VERREAUX, Proceedings of the Zoological Society for January, 1842, Pt. 10, p. 1.

Teeth.—Incisors, $\frac{2-2}{2}$; canines, $\frac{1-1}{0-0}$; molars, $\frac{3-3}{3-3}$ (or more). All the teeth of the upper jaw minute and simple, and separated from each other; those of the lower jaw are also separated, and minute, if we except the two incisors, which are well developed, long, slender, and pointed, have the upper and lower edges sharp, and are horizontal in their direction.

Head.—With the muzzle elongated and slender; the muffle naked: mouth opening small: tongue long, slender, and sharply pointed.

Limbs.—Fore and hind legs nearly of equal length, the latter the longer: fore feet with five smallish toes, each somewhat thickened at the extremity, and with a minute scale-like nail impressed, as it were, into the flesh, on the upper surface of the toe; but this nail far from reaches either the side, or the extremity of the toe. Hind feet also with five toes, of which the innermost assumes the form and functions of a thumb, is slender, rather long, and nailless; the second and third toes very short, joined to the extremity, and furnished with small pointed nails, which are directed upwards almost at right angles to the plane of the toe: fourth toe twice as long as the second and third, joined, toes; and furnished, like the fifth, which is considerably shorter than the fourth, with a scale-like nail on the upper surface.

Tail long, slender, rather sparingly clothed with small stiff hairs, and having the skin scaly: a small space at the apex beneath is naked.

Skull with the bones very thin and semi-transparent: lower jaw consisting of two slender rami, without any distinct coronoid, or angular portions, and with an oblong perforation behind.

Stomach small and simple; the walls very thin and transparent at the cardiac extremity; the œsophagus terminating about midway between the extremities; the cardiac end nearly spherical: the mesial portion much contracted; and, from that part to the pylorus, the stomach becomes gradually narrower. Intestines exceeding the entire length of the animal by about one half, simple, slender, and destitute of cœcum[1].

Female with a distinct pouch: mammæ four in number.

It will be seen that the dental formula above given does not agree precisely with that given by Messrs. Gervais and Verreaux, and the difference, there are good reasons for believing, arises from a want of constancy in the number of the very minute teeth with which the jaws of our little animal are furnished. In a carefully preserved skull before me, I do not even find the same number on opposite sides of the jaw, there being three molars on one side of the upper jaw, and four on the other. The canines, which are rather more developed, and the lower incisors, are probably the only teeth which will be found constant in all individuals.

In two skulls of Tarsipes, I find four minute transparent incisors situated on the fore part of the intermaxillary bones; these are distinctly separated from each other, and, between

[1] The specimen dissected by me was a female, and not in the best of conditions for examination. I feel pretty certain, however, that the points above noticed will prove to be correct. I found the small intestines of a little Phalanger (the *Dromicia Neillii*), although the animal was of smaller size than the Tarsipes, not only proportionately longer, but having a greater diameter. The stomach was simple, had the longitudinal diameter about one-fourth greater than the vertical, the cardiac end large and spherical, the depth of the opposite end about one-third less than that of the cardiac, and the œsophagus entering near the middle. The cœcum was $8\frac{1}{2}$ lines in length; the small intestines about 5 inches, and the large about $1\frac{1}{4}$ inches in length; the length of the animal being $2\frac{1}{4}$ inches, from the tip of the nose to the root of the tail. The cœcum is therefore smaller in proportion in this little Phalanger than in the larger species of the group, and such is found to be the case in the small Petauri, in which the dentition approaches to that of the insectivorous Phascogales, where the cœcum is absent—in some at least, I may say, according to my own observations.

the hindermost on either side, and the canine, there is a considerable hiatus, and between the canine and the little molars there is a long vacant space. The number of molar teeth in each of these two skulls is, specimen No. 1, $\frac{3-4}{3-3}$; specimen No. 2, $\frac{3-?}{2-3}$*. The molars of the upper jaw are rather widely separated from each other; those of the lower jaw are separated only by narrow spaces, and are placed about midway between the extremities of the ramus.

In the structure of the feet Tarsipes agrees *essentially* with the Phalangistidæ, and is approached most nearly by the small species of that group forming the section *Dromicia,* in which the ends of the fingers and toes are expanded, and the nails small. Amongst the skeletons of Marsupial animals which I have had an opportunity of examining, I also find the general structure of that of Tarsipes most assimilated to the *Phalangistidæ.* It is indeed very similar to the skeleton of *Petaurus sciureus* in most respects (I will point out the differences hereafter), and I anticipate that we shall find a nearer approximation in the skeleton of the Dromicias. In the elongated muzzle and concave palate of Tarsipes, we perceive modifications of these parts suited to lodge the long and slender tongue, which, it appears, is thus formed for the purpose of collecting honey—a food requiring no mastication—and hence the rudimental condition of the teeth. The long and sharp, horizontal, lower incisors, we may readily suppose, would assist in the operation of piercing, and opening a passage for the tongue, into those parts of the flowers in which the honey is lodged.

But one species of the genus *Tarsipes* is known, and that inhabits the west coast of Australia.

* Part of the left side of the jaw is wanting in this specimen.

TARSIPES ROSTRATUS.

Long-snouted Tarsipes.

(Plate 11, Fig 1.)

Tarsipes rostratus. GERVAIS and VERREAUX, Proceedings of the Zoological Society for January, 1842, Pt. 10, p. 1; Guérin's Magasin de Zoologie, année 1842, Mammifères, Pls. 35—37.
" " GOULD, Mammals of Australia, Pt. 1, Pl. 1.
" *Spenseræ.* GRAY, in Annals and Mag. of Nat. Hist. for March, 1842, vol. 9, p. 40.

Fur short, adpressed, and rather harsh; general tint grey, more or less suffused with rust colour; sides of body distinctly tinted with rust colour, or rusty yellow; under parts rusty yellow, or rusty white; back with three longitudinal dark lines: tail dusky above, greyish beneath.

Inhabits Western Australia.

The Tarsipes, so named on account of its feet having a considerable resemblance to those of a little animal found in the Indian Islands, and called the Tarsier, is certainly one of the most interesting of Mammalian forms recently discovered. Let the reader imagine a little quadruped of the size and general form of a mouse, but with a long, slender, and pointed muzzle, and to this we will add, that its fur is shorter and coarser, and more closely applied to the skin, than in the mouse; that its colouring is richer, and that it has three dark longitudinal stripes on the back, of which the two outermost are rather indistinct. Such is the *general* aspect of the Tarsipes; but our animal requires to be more closely examined, and in so doing we shall find that its characters essentially differ much from those of the mouse, that they indicate a lower grade of organization, and a difference of habits, and food. The female Tarsipes possesses a distinct pouch, and the male presents the corresponding characteristics of the Marsupial Mammalia: in structure, the feet are adapted for climbing, and are more suited for grasping small twigs than for ascending large trees, since the toes are provided only with very small nails, which are for the most part embedded in the upper surface of the expanded fleshy pad with which each toe is terminated. The long and slender, opposable thumb of the hind foot is nailless, the second and third toes are united to the extremity, but the two nails of these toes are distinct, pointed, and hollow beneath, and are the only nails which have the points free: they are no doubt used (as I have observed in some other Marsupials) for cleansing the fur. The head is long, and the muzzle is very long, slender, and pointed; the ears rather small, rounded, and clothed with very small hairs. The eyes are small, of a black colour, and very prominent. The limbs are of nearly equal length, but the hinder legs are slightly longer than the anterior. The tail is longer than the head and body taken together, and clothed with small stiff hairs, like the tail of a mouse, and these do not completely hide the scaly skin. The colouring of the fur varies somewhat in different individuals, but may be described as generally grey, more or less suffused with rust colour on the upper parts of the body, and yellow on the under: a narrow black line, commencing at the back of the head, is continued along the back to the root of the tail, and on each side of this is a broadish grey space, followed by a rusty brown longitudinal band, usually not well defined, and blending externally into the rich rust colouring which is almost always observed on the sides of the body. The upper surface of the head is brown along the mesial portion; the sides of the face are tinted with rust colour: the hairs of the moustaches are numerous, moderately long, and black. The small hairs with which the ears are clothed are by no means abundant; on the inner surface of the ear they are yellowish, those on the outer side are darker, especially near the front margin, where they assume a dusky hue. There is a small lobe near the posterior angle of the ear internally. The feet are clothed above with small glistening pale hairs. The tail is dusky above, and greyish beneath. The hairs on the back are of a deep lead-grey colour next the skin, and those on the abdomen are of a paler grey at the root.

In some specimens the sides of the body are yellowish, and the under parts are nearly white. The male specimens are generally smaller than the females.

	Male.		Female.		Female.	
	Ins.	Lines.	Ins.	Lines.	Ins.	Lines.
Length from tip of nose to root of tail[1] ...	3	4	3	9	3	0
" of tail	3	6	4	0	3	7
" from nose to ear	1	0	1	2		11½
" from nose to eye		6		7¾		6
" of ear		4		4		4¼
" of fore foot		3¾		3¾		4
" of hind foot		7¼		7⅓		7½
" of thumb of ditto		2¼				2¼
" of the second and third, joined, toes		1⅙				1⅙
" of the fourth toe		3				3
" of the outer toe		2				2
" of middle finger of hand		1⅔				1¾

Mr. Neill having kindly sent me a female specimen pre served in spirit, I will avail myself of the opportunity to notice the leading points which the entire skeleton of this interesting animal presents, since only some few parts (the

[1] The dimensions in the two first columns are taken from skins, and those in the third column from a specimen preserved in spirit. All of these specimens are from King George's Sound.

skull, tibia and fibula, and the radius) have hitherto been noticed.

The skull is proportionately large, being about equal to half the body in length, and is remarkable for the thinness and semi-transparency of the bones of which it is composed, the long and tapering facial portion, and the short and slender zygomatic arch, and consequently the small size of the fossa, of which it forms the outer boundary. Viewing the under surface of the skull, the great extent of the auditory chamber, and the imperfect condition of the palate, are also striking.

The cranial portion of the skull occupies rather more than one third of the entire length, approaches to a spherical form, and presents a smooth surface. On the occiput scarcely any trace of a crest is perceptible, and the sutures which, in most Marsupials, mark the boundaries of the separate bones of which it is composed, are here obliterated, but all the other sutures of the cranium are distinct : the occipital opening is large. The interparietal bone is very large, and transverse ; the parietal bones are also large, whilst the frontal bones are rather small, and considerably contracted between the orbits. The lachrymal bones are tolerably large, and present the usual perforation. The zygomatic arch is extremely slender, and short, being in length equal to one-fifth of that of the skull; it is composed almost entirely of the malar bone, which sends up a small post-orbital process. The glenoid cavity is very small. The nasal bones are long, narrow in front, and much expanded behind. The large ear chamber presents an anterior and a posterior convex swelling, of which the foremost is the larger, and is formed partly by the sphenoid ala, and partly by the squamous portion of the temporal bone ; the hinder bulb, which is more convex, is due to the petrous and mastoid elements of the temporal bone. I cannot trace the tympanic bone. The palate is strongly concave in the transverse direction; presents two long and narrow incisive openings, and two posterior openings, which are also narrow. The channel on the back part of the skull, which is bounded on either side by the auditory bullæ, is continued forwards and joins the post-palatine openings: the anterior half of this channel, which is much contracted, is bounded by two long and thin plates of bone, which appertain chiefly to the anterior sphenoid; and from about the middle of these plates, a thin process appears to have been thrown across the valley, but it is partly broken in the skull before me; this process no doubt formed the posterior boundary of the palate, and is a portion of the palatine bone; in front of it, a longitudinal ridge forms the separation of the two post-palatine openings. The anterior root of the zygoma is perfectly hollow, and a small perforation leads into it from the under side. The lower jaw is composed of two very slender, and almost straight rami, and presents neither angular portion nor coronoid process: they converge and meet in front, but are by no means firmly attached to that part, and they terminate posteriorly in a small semicircular condyle: the hinder third of the ramus is divided by a long narrow perforation into an upper and a lower thin branch: these branches meet behind at a short distance from the condyle, and at the point where they join in front we can perceive a faint trace of the coronoid process, indicated by the ramus of the jaw being slightly expanded in the vertical direction at that part.

Of cervical vertebræ, I find the usual number: the atlas vertebra has the lower boundary of the ring formed by a thin cartilage, in the middle of which a minute transverse ossicle represents the body of the vertebra. The vertebra dentata has a moderately developed spine. The third, fourth, fifth, and sixth cervical vertebræ are totally destitute of spinous process; the seventh has a small spine. The dorsal vertebræ are thirteen in number, and are provided with slender, and but moderately elevated spinous processes; these processes, from the first to the eighth, are directed obliquely backwards; the remaining processes of the dorsal vertebræ, as well as those of the lumbar and sacral vertebræ, are upright. The ribs are compressed, and slender. The sternum is composed of six long and slender bones. There are five lumbar vertebræ, each of which has a moderately developed transverse, and a short spinous, process. Of sacral vertebræ there are three; their transverse processes join the ilium. The caudal vertebræ are very numerous, being thirty-four in number [the total number of vertebræ, therefore, is sixty-two]; they present little worthy of notice, if we except that the transverse processes of the basal three are much developed, flat, and expanded: the fourth caudal has but a very small transverse process, and the remaining vertebræ are simple, or nearly so: between the interspace of each pair, beneath, is a small bone having the outer surface flattened—it represents the hæmapophysis of the vertebra. The spine of the scapula is straight, or very nearly so, inclines over the infra-spinal part of the scapula, and is produced in front into a long, and extremely slender (almost hair-like) acromion process. The upper and lower edges of the scapula run nearly parallel at the anterior third portion, and there form a narrow neck; the lower edge is very nearly straight, and, passing backwards, gradually recedes from the spine, and the narrow infra-spinal fossa is truncated behind, so that its outline at that part forms a right angle with the line of the spine: the supra-spinal portion of the scapula is much larger, and, from the neck, is expanded so as to form nearly a semicircle. The clavicle is long and slender. The humerus is slender, has no perforation at the lower extremity: the deltoid ridge forms an obtuse angle, at the termination of the anterior third; the external and internal tubercles are small; the former rises scarcely as high as the head of the humerus; the outer condyloid ridge terminates in an obtuse angle at the commencement of the lower fourth of the humerus. The remaining bones of the fore leg present nothing to arrest our attention: the ulna and radius are in contact nearly for their entire length, being but slightly separated in the middle; the olecranon is well developed. The pelvis offers no peculiarities, excepting that it is more elongated than usual. The marsupial bones are straight, small, and slender. The femur is straight, has the head directed inwards and forwards, and joined to the shaft by a short neck; a depression for the ligamentum teres is visible; the two trochanters are small. The patella is present. The tibia and fibula are distinct, in contact at the lower extremity, but, commencing about the middle, they rather suddenly diverge, owing to the bending forwards of the shaft of the tibia, which, I may observe, is considerably compressed: at the proximal extremity, the fibula (which is unsually slender below) is suddenly expanded and flattened, sending out a square process posteriorly, and a smaller process in front to join the head of the tibia. With respect to the structure of the feet, it will be sufficient to remark, that they resemble those of the species of *Phalangista* in all essential points.

I have before said that the skeleton of Tarsipes bears a very great resemblance to that of *Petaurus sciureus*. In both animals nearly all the cervical vertebræ are destitute of spinous processes[1]: the scapula is very nearly the same in both, but the lower posterior angle is more produced in *P. sciureus*, and the acromion process is expanded. The principal differences are, that in Tarsipes the bones of the limbs are proportionately shorter than in the *Petaurus;* the lumbar vertebræ, as well as the bones of the pelvis, are likewise shorter: the humerus differs in not being perforated at the lower extremity, and in having the outer condyloid ridge less developed; the trochanters of the femur are smaller, and the lesser trochanter is placed nearer to the head of the femur; the tibia is more arched, and more compressed at its upper half, and the fibula is rather more expanded at its proximal extremity. In the general form of the skull *Tarsipes* is approached most nearly by the small species of *Perameles*.

"The Tarsipes is generally found in all situations suited to its existence, from Swan River to King George's Sound, but from its rarity, and the difficulty with which it is procured, notwithstanding the high rewards offered, the natives only

[1] The spinous processes of the cervical vertebræ are small in the Phalangers generally, and very small in the *Petaurus australis*, but I have seen no marsupial skeleton in which they are wanting excepting in *Petaurus sciureus* and Tarsipes. Probably in the minute species belonging to the sections *Acrobata* and *Dromicia* the same structure will be found.

brought me four specimens; one of these, a female, I kept alive for several months, and it became so tame as to allow itself to be caressed in the hand without evincing any fear, or making any attempt to escape. It is strictly nocturnal, sleeping during the greater part of the day, and becoming exceedingly active at night: when intent upon catching flies, it would sit quietly in one corner of its cage, eagerly watching their movements, as, attracted by the sugar, they flew around; and when a fly was fairly within its reach, it bounded as quick as lightning, and seized it with unerring aim, then retired to the bottom of the cage, and devoured it at leisure, sitting tolerably erect and holding the fly between its fore-paws, and always rejecting the head, wings, and legs. The artificial food given it was sopped bread, made very sweet with sugar, into which it inserted its long tongue, precisely in the way in which the Honeyeaters among birds do theirs into the flower-cups for honey; every morning the sop was completely honey-combed, as it were, from the moisture having been drained from it by the repeated insertion of the tongue: a little moistened sugar on the end of the finger would attract it from one part of the cage to the other; and by this means an opportunity may be readily obtained for observing the beautiful prehensile structure of the tongue, which I have frequently seen protruded for nearly an inch beyond the nose; the edges of the tongue near the tip are slightly serrated. The tail is prehensile, and is used when the animal is climbing, precisely like that of the *Hepoona* (*Phalangista*). The eyes, although small, are exceedingly prominent, and placed very near to each other; the ears are generally quite erect. When sleeping, the animal rests upon the lower part of the back, with its long nose bent down between its fore feet, and its tail brought over all, and turned down the back. Mr. Johnson Drummond shot a pair in the act of sucking the honey from the blossoms of the *Melaleuca*; he watched them closely, and distinctly saw them insert their long tongues into the flower precisely after the manner of the birds above mentioned[1]."

Mr. Neill, whose notes I have before quoted[2], informs me that the Tarsipes is known to the natives of King George's Sound by the names of "Tait," and "Noolbenger;" that it is very common in the district mentioned, and makes its nest in the overhanging leaves of the *Xanthorrhœæ*, and Kingias. In repeated dissections of these little animals, immediately after their capture, Mr. Neill only found in the stomach a transparent fluid, like honey; and this gentleman is on this account inclined to put faith in the statements of the natives, who inform him that the "Tait" lives upon honey, which it procures by thrusting its long and slender tongue into the cups of the flowers. As nearly all the flowering plants in Australia, Mr. Neill observes, produce honey, it is probable these animals can obtain this food throughout the year.

The specimen of Tarsipes described by Mr. Gray, was, like all other specimens which have come under my notice, from King George's Sound; and, according to the notes quoted by Mr. Gray, the animal is said to inhabit the low scrubby and heathy looking bushes of that district. The notes in question were forwarded to Mr. Gray, together with the Tarsipes and several other interesting animals, by the Governor of South Australia, Capt. George Grey.

Before passing on to the next family, I must not omit to notice that parts of a skull of a species of *Phalangista*, agreeing pretty nearly with *Phalangista vulpina*, were discovered by Prof. Owen amongst the Mammalian remains found by Sir Thomas Mitchell in caverns of Wellington Valley, which have already been alluded to.

[1] From Gould's Mammals of Australia. [2] See page 316.

PERAMELIDÆ, OR PERAMELES FAMILY.

Peramelidæ. WATERHOUSE, in Naturalists' Library—Marsupialia. 1841.
Peramelina. GRAY, Annals of Philosophy, xxvi. 1825.

Dentition :—Incisors, $\frac{10}{6}$; canines, $\frac{1-1}{1-1}$; premolars, $\frac{3-3}{3-3}$; true molars, $\frac{4-4}{4-4} = 48$. The teeth rooted; premolars compressed and pointed; true molars tubercular.

Head elongated, the facial part narrow, and pointed; muffle naked; nostrils lateral, and with a longitudinal groove between them; upper lip slightly cleft; ears moderate or very large, provided with two prominent fleshy lobes running backwards from the anterior angle, and a third lobe, also on the inner side, situated near the lower posterior angle.

Limbs unequal—the posterior legs being considerably longer than the anterior. Fore feet with the outer toes rudimentary. Hind feet with the inner toe rudimentary or absent; the second and third toes joined to the extremity, and having small nails; the fourth toe very large, and the outer toe moderate—sometimes rudimentary.

Tail usually short, and clothed with small hairs; sometimes long, and clothed, in parts at least, with very long hairs.

Pouch of the female with its entrance usually directed towards the tail.

Mammæ eight?[1]

Stomach simple. Cœcum of moderate size.

The animals of the present family are of small size, the largest known species being scarcely equal in bulk to the Common Hare. They occur in Van Diemen's Land, and, on the continent of Australia, they have been found in all the

[1] Prof. Owen found eight nipples in the *Perameles nasuta*, arranged in two slightly curved longitudinal rows: possibly this number may not be constant in all the species of *Peramelidæ*.

principal parts which have been visited, up to the north coast. One species (the *Perameles Doreyanus*) has been discovered in New Guinea.

In the structure of the hind feet, the *Peramelidæ* greatly resemble the Kangaroos, as will be seen upon comparing fig. 5 (right-hand figure) with fig. 6 (lower figure) of Plate 12. The Perameles' foot (fig. 5) here represented, however, is proportionately shorter, and the fourth, or principal toe, is less developed than in the Kangaroo's foot; such a distinction is general in the two groups: the *Peramelidæ*, indeed, have the fore and hind limbs less unequal than the Kangaroos, and I strongly suspect they do not progress by leaps like the species of the section last mentioned. Another difference observable in the two feet represented on the plate, consists in the presence of a small inner toe to the foot of the Perameles; but this, I must observe, is not a constant difference, the toe in question being absent in some of the *Peramelidæ*, as, for instance, in the *Perameles lagotis*, where the tarsus is proportionately longer than in most others of the group. We shall have to notice a singular modification of the feet in another species of the present family—the *Chœropus*. The fore feet differ from those of the Kangaroos in having the outer toes rudimentary. In the structure of the skull and teeth there exists much difference in the species of the two sections. The *Peramelidæ* may be readily distinguished from all other Marsupials by the number of their incisor teeth, of which there are ten in the upper, and six in the lower jaw. In no other Australian mammal are there more than eight incisors in the upper jaw: ten of these teeth are found in the Marsupials of America (the true Opossums), but those animals have eight incisors in the lower jaw. The true molar teeth in the animals under consideration are adapted to insect diet, and such we know to be the food of these animals, though, according to the accounts of good observers residing in

Australia, they will likewise eat vegetable substances[1], and we may add that a specimen of the *Perameles lagotis*, which lived in the menagerie of the Zoological Society, even refused meal-worms when offered, and was fed upon bread and milk, almonds, &c.[2]

Fossil remains appertaining to this family have been found by Sir Thomas Mitchell in the caverns of Wellington Valley.

Genus, *Perameles*[3].

Perameles. Geoffroy, Mémoire sur un nouveau genre de Mammifères à bourse, nommé Péramèles—Annales du Muséum, tom. iv. p. 56. 1804.

Isoodon. Desmarest, in Nouv. Dict. d'Hist. Nat. tom. xvi, p. 409. 1817.

Peramelidæ with five toes to the foot, of which the two outermost are rudimentary, and nailless, the remaining three well developed, and furnished with large, strong, and solid nails, covering ungual phalanges, which are cleft above, in the longitudinal direction, almost to the root: hind feet with five toes, of which the innermost is rudimentary, and nailless, and sometimes entirely hidden beneath the skin; the second and third toes extremely slender, and joined to the extremity, but with two distinct hollow nails; fourth toe very large, and provided like the fifth (which is well developed) with a solid nail, sheathed upon an ungual phalanx, which is divided like those of the principal toes of the fore foot.

[1] Some of the species of Perameles are said to feed upon the bulbous roots of plants imported into the colonies. It would be well to ascertain whether they do not destroy these bulbs to obtain insects which may have attacked them.

[2] It must be observed that the molar teeth in this animal are less tubercular than in others of its family, and the extremely worn condition of these teeth in the skulls of all the specimens which have come under my notice, would seem to indicate that they fed upon food which required more mastication than insects.

[3] From the Latin Pera, *a pouch*, and Meles, *a badger*.

The incisor teeth of the upper jaw are arranged so as to form a semi-oval figure; small, compressed, and when viewed from the outer or inner side, present nearly a square form; the two foremost are separated by a narrow space, the following three, on either side, form a continuous series with the first, but the fifth is usually more or less separated from the rest. The canine is moderate, or large, compressed, pointed, and somewhat recurved, and distinctly separated from the false molars, as well as from the incisors. The false molars are more or less separated by intervening spaces, compressed, and, when viewed externally, present a triangular figure, but at the base is a minute anterior and posterior cusp: the two foremost of these teeth are two-rooted; the third, which is in contact with the true molars, is sometimes three-rooted, and has a small tubercle on the inner side. The crowns of the true molars are nearly square, but broader on the outer side than the inner; the masticating surface of each tooth presents eight more or less sharply-pointed tubercles, of which four form a longitudinal series on the outer side of the tooth, two are situated in the mesial line of the tooth, are larger than the rest, and are joined by oblique ridges to the outer tubercles, so that they form the apices of two triangles, two of the smaller tubercles marking the angles at the base of each triangle; and, lastly, on the inner side of the tooth are two other tubercles situated opposite the two principal tubercles just described; their apices are much less elevated than those of the other tubercles; in some species they are of equal size, whilst in others the hindermost of these two eminences is very small[1]. The hindermost of the upper true molars, it must be observed, will not agree with the foregoing description, since the crown of this tooth is of

[1] This is important to notice, for we shall find that it is the absence of this posterior inner tubercle which gives the triangular form to the crowns of the true molars in the *Dasyuridæ*.

a more or less triangular form, and presents but five tubercles, or sometimes four tubercles, of which three form the outer side of the tooth, and the fourth, which is larger, is placed opposite the middle one of these, and there is an inner lobe which is not produced into a pointed eminence. The incisors of the lower jaw approach more or less to the longitudinal in their direction, and are in contact with each other; the four foremost are deeper from the outer to the inner side than in the opposite direction; the hinder one on either side is bilobed at its extremity. The canines are moderate. The false molars scarcely differ from those of the upper jaw, but sometimes want the small anterior tubercle. The true molars are rather longer and narrower than those of the upper jaw; they present two external and two internal principal cusps, and a very small anterior tubercle.

The genus *Isoodon* of Geoffroy and Desmarest is founded upon a supposed difference in the dental formula of *Perameles nasuta*, and *Perameles obesula*, which does not exist. That most excellent mammalogist, Desmarest, was but imperfectly acquainted with the *Perameles obesula*, and was misled (as was no doubt Geoffroy likewise) by the accounts given by others of the dentition of that animal.

Section 1. *Macrotis*.

Macrotis. REID, Proceedings of the Zoological Society for December, 1836, Pt. 4, p. 131.

Peragalea. GRAY, in Appendix to Gray's Journal of Two Expeditions in Australia, vol. 2, p. 401. 1841.

Ears very large; auditory bullæ in the form of a double bulb; tail long, and clothed with long hairs; tarsus long, the metatarsus clothed with hair beneath; innermost toe of the hind foot wanting; lower half of the fibula firmly joined to the tibia; pouch with its opening directed towards the head of the animal[1].

The above are the principal points of distinction which present themselves upon comparing the *Perameles lagotis* (the only known species of the present section) with other species of *Perameles:* the differences are, several of them, comparative. In other species the tarsus is partially covered with hair beneath, but in none is it so much clothed as in the animal just mentioned. Of the inner toe of the hind foot there is no trace externally, and in the skeleton we find only a rudimentary metatarsal bone, whilst in the animals of the next section (Perameles proper) there are always one or two phalanges to the inner toe, and a small nailless tubercle is visible before the flesh is removed. The fur of the Rabbit-eared Perameles is remarkable, as compared with that of its congeners, for being extremely soft, and having no admixture of coarse and spiny hairs.

[1] Such appeared to me to be clearly the case in a stuffed specimen of the *Perameles lagotis* preserved in the museum of the Zoological Society. This specimen, which is the original of Mr. Reid's description, was stated to have been procured in Van Diemen's Land, but I was informed by the gentleman who sent it to England that this was an error, he having obtained it in the Swan River district. Before this specimen was mounted, and in some respects could then be more carefully examined, Mr. Reid ascertained that it possessed nine mammæ, or nipples, eight of which were arranged at equal distances from each other, and around the ninth.

PERAMELES (*Macrotis*) LAGOTIS.

Rabbit-eared Perameles.

Perameles lagotis. REID, Proceedings of the Zoological Society for December, 1836, Pt. 4, p. 129.

" " WATERHOUSE, Naturalists' Library—Marsupialia, p. 153, Pl. 12.

Peragalea lagotis. GRAY, List of the Specimens of Mammalia in the Collection of the British Museum, 1841, p. 96.

" " GOULD, Mammals of Australia, Pt. 1, Pl. 12.

Native Rabbit of the colonists, and *Dalgheit* of the natives of Western Australia.

Plate 13, Fig. 1.

Ears very large, nearly as long as the head; fur very long and soft, of a pale grey on the upper parts of the animal, delicate vinous red on the sides, and white on the under parts: fore legs white, but dusky at the base externally; feet white, the hind feet clothed with black hairs on the under surface; the leg dusky immediately above the heel: tail at the root clothed with fur like that of the body, the remaining portion with coarse hairs, long on the upper surface (very long towards the end of the tail), and shortish on the under parts; on the basal half of the tail these hairs are black, and on the apical, white.

Inhabits Western Australia—Swan River district.

The very large and elongated ears of the present species no doubt suggested for it the name of "Native Rabbit," by which it is known to the colonists of the Swan River district, Western Australia—the only part of Australia, I believe, in which it has hitherto been found. According to the information furnished us by Mr. Gould, the Rabbit-eared Perameles is tolerably abundant over the whole extent of the grassy districts in the interior of the Swan River colony, and is usually seen in pairs. It commonly selects those spots where, the soil being loose, it is enabled to excavate its burrows with facility. Like the Rabbit it flies to its subterranean retreat for safety, when pursued, and as the burrows are both deep and long, frequently eludes the pursuer. Its flesh is sweet, and resembles that of the Rabbit. The food of the Rabbit-eared Perameles consists of insects, and the animal is said to be particularly fond of a certain large grub (probably the larva of a large species of Buprestes, found abundantly in the district), which is found at the roots of the Acaciæ, and which is in equal request with the natives, who also eat it.

The specimen already alluded to as having lived in the menagerie of the Zoological Society, was very active in the evening, but usually slept during the day-time, when, sitting upon its haunches, with its head thrust between its hind legs, it appeared like a large ball of fur. It was an exceedingly savage animal, bit very severely, and would not readily unfix its hold of any thing it happened to seize with its teeth. When walking, the hind legs only were used, and these were very widely separated. The tail assisted slightly in supporting the body, which was but little raised in front.

The *Perameles lagotis* is about equal in size to the Common Rabbit, and has a remarkably long and pointed muzzle, which is not only naked at the tip, but a naked area runs back from that part, on the upper surface of the muzzle, and terminates in a point, distant nearly an inch from the end of the snout. The ears are nearly of an elongate oval shape, but are somewhat contracted near the apex, and form a cylindrical tube at the base: they are clothed only with extremely minute hairs, and these are for the most part very sparingly distributed, and of a pale colour; near, and at the anterior margin, they are more numerous; those next the margin are whitish, but at a short distance from the margin they are brown. The eye is rather small, and is placed rather nearer to the base of the ear than the tip of the muzzle. The fur of the animal is very long, and extremely soft; on the upper parts of the body it is of a delicate grey hue; on the sides of the body of a pale vinous tint, or red inclining to purple; and on the under parts white: here the hairs are uniform to the root, but on the upper parts of the body they are pale grey next the skin, whitish above the middle, and have long dusky points. The muzzle is slightly suffused with black. The hairs of the moustaches are moderately long, and black. The fore legs are dusky at the base externally, but white below; the feet are also white, but the under surface of the hind foot is clothed almost entirely with longish black hairs; the black hairs extend along the outer side of the foot to the base of the outer toe, but on the opposite side the hairs assume a brownish hue on the palm of the foot, in the region of the joined inner toes. There is a smallish naked space on the under surface of the heel, a second still smaller on the protuberance formed by the projecting bases of the inner metatarsal bones; the under surface of the toes is also naked, and there is a large fleshy pad at the base of the toes, which is destitute of hair. The tail is rather shorter than the body of the animal; at the root it is covered with fur like that of the body; the remaining portion has the first half black, and the terminal half white: the hairs covering these parts are dense and coarse, shorter on the under surface than the upper, being here about half an inch in length, or less towards the extremity; on the upper surface they average towards the base (on the black portion) about one inch in length, but they gradually become longer as they approach the tip of the tail, where they are as much as 2½ inches, or more, in length in some individuals. The extreme point of the tail is naked.

The male specimens are rather larger than the females.

	Inches.	Lines.	Inches.	Lines.
Length from tip of nose to root of tail	18	0	16	4
" of tail	9	6	9	7
" from nose to eye			2	5
" " to ear	4	6	5	2
" of ear	3	9	4	0
Breadth of ditto at the base ...	2	1	2	0
Length of fore foot, not including the claws			1	6
" of hind foot, not including the claws	3	10½	4	1

The skull of the *Perameles lagotis* is much elongated, especially in the facial portion; its zygomatic arches are thrown more boldly outwards, and are proportionately stronger than in other species; the interorbital space is more contracted, and the temporal fossæ are therefore larger; thus indicating greater powers in the muscles used in mastication, as does also the development of a sagittal crest: this, however, is very little raised. The muzzle is rather suddenly contracted in width immediately in front of the orbits, and is nearly of the same diameter from the tip as far back as the second premolar. The nasal bones are very long and narrow, and very little expanded behind: they are joined at the apex, in a skull in the museum of the Royal College of Surgeons, by two small bones, being part of the nasal cartilage, which is ossified, as in the Pigs. The palate presents two long and narrow incisive foramina, a large opening in the middle, the anterior boundary of which is in a line with the second premolar, and the posterior boundary in a line with the hinder margin of the second true molar; and on the back part of the palate are six small round openings, four of which are arranged nearly in the same transverse line, and the remaining two are placed in advance of the middle pair; the two outermost of the posterior row are partly bounded by the palatine bone, and partly by the palatine portion of the superior maxillary bone. The palatine bone, and palatal portions of the superior maxillary, interlace very singularly in the skull before me, the palatine bone running forwards so as to form three lobes, a broad central lobe, and a narrower one on either side, the interspaces being filled by the maxillary bone. The auditory bullæ are chiefly formed by an expansion of the alæ of the sphenoid, but partly by the petrous bone; the sphenoidal portion is large, presents nearly an oval outline, is very convex, and runs, from within, obliquely forwards and outwards; the part due to the petrous element of the temporal bone is also convex, long, narrow, and runs parallel with the sphenoidal portion—it is comparatively small. The rami of the lower jaw are narrow in the vertical direction; the coronoid process is moderately elevated; the angular portion assumes the form of an isosceles triangle. The dimensions in the first of the two columns here subjoined are taken from a skeleton in the Royal College of Surgeons, and those in the second are from some parts of a skeleton in the British Museum: this latter must have belonged to a very large individual.

	Inches.	Lines.	Inches.	Lines.
Length of skull	4	1		
Width of ditto	1	9¾	2	0
" between orbits		5⅓		6
Length of nasal bones	1	9½	1	10½
Width of ditto behind		3⅔		4
Length from posterior root of zygoma to apex of intermaxillary bones	3	3½	3	6
" of palate	2	3½	2	4¾
Width of ditto between penultimate molars		7¼		8¼
Length of great palatal opening ...		8		7¾
" of the foremost of the two auditory bullæ		7½		9
Transverse diameter of the same ...		4½		5⅓
Length of zygomatic arch	1	3	1	5
Depth of ditto in the middle ...		2½		3
From orbit to apex of nasal bones ..	2	1½	2	2¼
Length of lower jaw			3	3
Height of ditto in a vertical line dropped from the coronoid process			1	2½
Length of scapula	2	2⅔		
Width of ditto in the middle ...		11½		
Length of humerus[1]	2	1½		

[1] This is perforated between the condyles as well as on the inner condyle.

	Inches.	Lines.	Inches.	Lines.
Length of ulna	2	11	3	3
" of radius	2	4	2	6
" of fore foot	1	8	1	9
" of femur	3	0½		
" of tibia	3	9½	4	1½
" of hind foot	4	2	4	4
The outermost toe of the hind foot terminates short of the principal toe				9½
The joined toes terminate short of the principal toe				12½

The vertebræ are—cervical, 7 ; dorsal, 13 ; lumbar, 6 ; sacral, 1* ; caudal, 23.

With regard to the dentition I need only remark, that the hindermost incisor, on either side of the upper jaw, is separated from the preceding one by a very narrow space only (a space about equal to its own diameter) ; the canine teeth are larger than usual, sometimes being nearly half an inch in length ; the molar teeth have the transverse diameter greater than in other species of *Perameles*, and these teeth are not expanded on the outer side : the last molar is of the same form as the others, but of smaller size.

Section 2. *Perameles proper.*

Feet, tail, and ears, proportionately short ; the hind foot with a rudimentary inner toe, naked beneath in front, and at the heel. Fur distinctly composed of hairs of two kinds, the one forming a soft under fur, the other hairs coarse, flattened, and longitudinally grooved. Tail clothed with very short adpressed hairs only. Pouch opening backwards.

* But one vertebra joins the os innominatum, and none of the vertebræ in the sacral region are anchylosed.

PERAMELES MACROURA.

Thick-tailed Perameles.

Perameles macroura. GOULD, Proceedings of the Zoological Society for February, 1842, Pt. 10, p. 41.

Thick-tailed Bandicoot. GRAY, List of the Specimens of Mammalia in the Collection of the British Museum, p. 96. 1843.

Fur moderate as to length, and harsh to the touch; on the upper parts of the body pencilled with black and yellow in about equal proportions; on the sides of the body the yellow prevails; under parts yellow, or yellow-white: tail about equal to half the body in length, sparingly clothed with small stiff hairs; above black, beneath brown-white; ears moderate.

Inhabits North Australia.

The specimens upon which Mr. Gould founded his *Perameles macroura* were procured at Port Essington. It may be desirable to notice, with respect to the specific name selected, that it has reference to a comparison between the present species and the *P. obesula* and its allies, the tail of the *Per. macroura* being proportionately shorter than in the *P. lagotis*. Besides the bristly hairs which cover this animal, there is a scanty under fur of fine hairs, which, on the upper parts of the body, is of a brown-grey hue; on the under parts all the hairs are sometimes of a delicate yellow, and sometimes of a yellow-white tint throughout. The feet are brown-white; those of the hind legs, however, are indistinctly tinted with yellow. The tail resembles that of a Rat in being sparingly clothed with small stiff hairs, which are not so numerous as to hide the scaly skin beneath. The ears are of moderate size, have the hinder margin straight, and are clothed with small hairs, of a yellow colour on the inner surface, and dusky on the outer, if we except the hinder part, where they are pale.

A male specimen in the British Museum collection presents the following dimensions :—

	Inches.	Lines.
Length from tip of nose to root of tail ..	16	0
" of tail	7	2
" from nose to ear	3	6
" of ear		10
Width of ditto		10
Length of fore foot and nails	1	7
(of which the nail of the largest toe is)		8
" of hind foot and nails ...	3	2
(of which the nail of the largest toes is)		7
Distance between posterior incisor and canine		1¼
" between canine and first molar ...		1½
" between first and second molars ...		0½
From front of foremost incisor to the first true molar	1	1

Perameles macroura greatly resembles the *P. obesula*, but besides having the tail longer, it presents other differences; attaining a larger size, and having the posterior upper incisor almost in contact with the others, and broader than in *P. obesula;* the hind feet, moreover, are very nearly uniform in tint—not pencilled with black, as in the animal last mentioned. The skull I have had no opportunity of examining.

PERAMELES OBESULA.

Short-nosed Perameles.

Didelphys obesula (Porculine Opossum). SHAW, Naturalists' Miscellany, vol. viii. Tab. 298; General Zool. vol. i. Pt. 2, p. 490.
Perameles obesula. GEOFFROY, Ann. du Mus. iv. p. 64, Pl. 45.
Isoodon obesula. (Geoffroy) DESMAREST, in Nouv. Dict. d'Hist. Nat. xvi. p. 409.
Perameles fusciventer. GRAY, in the Appendix to Gray's Journal, ii. p. 407.
" *affinis.* GRAY, List of the Mammalia in the British Museum, 1843, p. 96.

Head moderately long; fur very harsh to the touch; on the upper parts of the body pencilled with black and rich yellow in about equal proportions; on the under parts yellow-white: ears clothed with small hairs, yellow internally, and dusky brown externally, but paler towards the hinder margin; feet palish yellow, slightly pencilled with black. Tail short, clothed with small hairs, dusky above, and yellowish white beneath.

Inhabits New South Wales, South Australia, Western Australia, and Van Diemen's Land.

The hairs composing the fur of this animal are, as is most commonly the case, of two kinds; all that are visible (as the hairs lie in their ordinary position) are very harsh to the touch, flattened, pointed, and glossy; upon dividing these coarse hairs, a soft, but somewhat scanty, under fur is visible. On the upper parts of the body the coarse hairs, which are of moderate length, are greyish white at the root, black at the point, and broadly annulated with ochreous yellow at some little distance below the point; the under fur is grey: on the under parts of the body the coarse hairs are yellowish white, but almost pure white at the root, and the under fur is white, or nearly so. Towards the end of the muzzle the hairs assume an uniform dusky brown hue; the lips, chin, and throat, are whitish. The hairs of the moustaches are weak, and by no means numerous. The ears are rather small, and tolerably well clothed with small hairs; those on the inner side are yellow, but towards the margin brownish; on the outer side they are dusky, but paler on the posterior part, and there is a faint indication of a pale spot at the base near the anterior margin. The fore feet are whitish; the tarsi are dirty white, tinted with yellowish, and freckled with blackish on the upper surface; on the inner side they are of a delicate yellow. The tail is about one third of the length of the body, or rather less, and clothed with small hairs; about one inch of the basal portion has the hairs longer, more nearly like those of the body, and variegated with black and yellow; beyond this part the upper surface is dusky; on the under surface the hairs are of a dirty yellowish tint.

The short-nosed Perameles (Porcaline Opossum of Shaw[1]) has an unusually wide range, being found in New South Wales, South Australia, King George's Sound, the Swan River district, and Van Diemen's Land. I have examined specimens from each of these localities, and taken much pains to satisfy myself of their specific identity. The males I have found usually larger than the females; their fore feet are proportionately larger, and so are the canine teeth.

The colouring varies somewhat in different individuals of *P. obesula*, and is darker than that of other species, if we except the *P. macroura*.

[1] I have examined the original specimen described by Shaw, and feel no doubt that it is a young individual of the present species. I may add, that the skull figured by Geoffroy in the Annales du Muséum, and supposed to belong to the same species, is decidedly that of *P. obesula*.

	MALE. British Museum. Habitat?		FEMALE*. British Museum. Van Diemen's Land.		Museum Zoological Society. New South Wales.		FEMALE. New South Wales. Mr. Gould's Collection.		MALES. Darling Downs, New South Wales. Mr. Gould's Collection.		FEMALE? King George's Sound. British Museum.		MALE. Perth, Western Australia. Mr. Gould's Collection.		MALE. Perth. Adult, but not old. Weight 2¾ lbs.	
	Ins.	Lines	Ins.	Lines	Ins.	Lines	Ins.	Lines	Ins.	Lines	Ins.	Lines	Ins.	Lines	Ins.	Lines
Length from tip of nose to root of tail ...	12	6	13	6	13	6	15	0	13	0	12	0	15	6	14	0
" of tail	4	3	4	3	4	6	5	0	5	3	4	3?	5	6	5	6
" from nose to ear	3	0	3	1	3	0	3	3	3	0	2	9	3	2	3	2
" of ear		8½		9		9		9		8½		8½		9		9
" of fore foot and nails	1	5½	1	5	1	5	1	6	1	7	1	5	1	6	1	5
" of hind foot and nails	2	6	2	5	2	5	2	8	2	7½	2	6½	2	8	2	7½

* On the stand of this specimen I find a note to the effect, that four young, of about an inch in length, were found in its pouch.

Usually, the upper parts of the body of this species are distinctly pencilled with black and yellow, and the under parts are of an impure white—not unfrequently yellowish. The specimen from the Darling Downs, the dimensions of which are given in the fifth column, is of an unusually rich tint, the hairs on the upper parts of the body being pencilled with black and rich yellow-brown, inclining to orange; the under parts of the body are of a dirty yellow-white tint; the hind feet pale rusty brown; the inner surface of the ears is for the most part clothed with rich yellow hairs; externally the ears are blackish.

The skull of *Perameles obesula* differs from that of most other species of Perameles in having the muzzle less attenuated and less elongated, and the palate less imperfect—that is to say, the openings are proportionately smaller: the incisive openings are narrow, about three lines, or rather less, in length; a second pair of openings are situated opposite the first premolar; these are unusually small, varying from half a line to one line in length; the third or principal pair of palatine openings are sometimes divided and sometimes confluent; they commence opposite the principal premolar, and terminate very nearly in a line with the hinder margin of the second true molar. Besides these, there are four (sometimes only two) small round openings in the hinder part of the palatine bone, and two rather larger foramina, one on either side at the back of the palate. The auditory bullæ are rather large, (larger than in *P. myosuros* and *P. fasciata*), having the largest diameter from 6 to $5\frac{1}{3}$ lines in the skulls of which the dimensions are given. The hindermost upper incisor is smaller than the fourth, pointed, and situated nearer to the preceding incisor than in most other species, there being at most a space of three quarters of a line in six skulls examined, but usually about half a line.

	Male, mature, but not old*.		Male, aged.		Female, aged.		Sex? aged: from Western Australia.		Sex? very aged: Western Australia†.		From skeletons in the College of Surgeons, from Van Diemen's Land.				Male, adult, but not old. From Perth, W. Australia‡.		*Per. macroura?*§ Brit. Mus.	
	Ins.	Lines.	Ins.	Lines.	Ins.	Lines.	Ins.	Lines.	Ins.	Lines.	Ins.	Lines.	Ins.	Lines.	Ins.	Lines.	Ins.	Lines.
Length of skull	2	7	2	7¼	2	6	2	7	2	10⅓	2	8½	3	0	2	9	2	8⅓
Width of ditto	1	2	1	2¼	1	1¼	1	1⅓	1	3	1	2¼	1	4⅓	1	2½	1	2½
Length of nasal bones	1	0¾	1	0½	1	0¾	1	0¾	1	2	1	0	1	3½	1	1½	1	1½
Width of ditto behind		2⅓		2½		2¼		2¼		2⅓		2½		3		2⅔		2⅔
From front of foremost incisor to back of last molar	1	5	1	4⅓	1	4	1	4	1	5½	1	5	1	7½	1	5	1	6¼
Length of four true molars, taken together		6		5⅔		5½		5½		4¾		5⅔		6		5¾		7⅓
" of palate	1	6½	1	6	1	5⅛			1	7⅔	1	6⅓	1	9¾	1	6⅔	1	7½
Width between orbits		5⅔		6		6		6		5⅔		5⅔		6		5¾		5¾
Length of lower jaw to angle ...	2	0½	2	1	1	11⅔	1	11½	2	3					2	1	2	1⅓
Height, from apex of coronoid process		8½		8½		8¼		8		9½						9		9¼

* This skull was removed from the specimen the dimensions of which are given in the first of the columns at p. 370.

† It will be observed that, notwithstanding this skull is larger than most others, its molar teeth give a smaller dimension. The skull increases in size, to a certain extent, with the age of the animal, but of course not the teeth; these are not only worn at the crown, but are worn in front and behind by the friction of adjoining teeth—where there are any.

‡ The dimensions of the specimen from which this skull was removed are given in column 8 at p. 370.

§ The skull from which the above measurements are taken I am convinced belongs to an animal distinct from, though nearly allied to, the *Per. obesula*, and as in the structure of its incisor teeth it agrees *only* with the *P. macroura* (a species, it will be remembered, also close to *P. obesula*), I suspect it will prove to belong to that animal; if so, we shall have to add to its characters, that its palate is destitute of any perforations between the principal palatine openings and the incisive foramina; that the principal palatine openings, situated opposite the anterior two true molars, are small, being only 2½ lines in length; and, lastly, that the molar teeth are nearly double the size of those of *P. obesula*, the skulls being of equal size. The posterior incisor of the upper jaw is very nearly in contact with the fourth, and is broader than in *P. obesula*.

Perameles affinis. GRAY.

Is founded upon a small animal from Van Diemen's Land, which appears to me to be a young individual of the *P. obesula:* excepting in size I can perceive no difference: its length from the tip of the nose to the root of the tail is 8 inches. When of this size, the young *P. obesula* has so much the general appearance of an adult animal, in the character of the fur, &c., that I supposed, like Mr. Gray, there really existed a second species resembling the *P. obesula,* but after seeing the skulls removed from two such specimens I was convinced that their small size merely indicated immaturity: they wanted the two or three back molar teeth.

Perameles fusciventer. GRAY.

Two specimens in the British Museum collection are labelled *Perameles fusciventer;* one agrees in every respect with the *P. obesula,* excepting that its head is rather shorter: its dimensions are given in the sixth column of admeasurements in p. 370. The other is considerably smaller than the adult *P. obesula,* and differs in being more strongly pencilled with black on the upper parts of the body, and in having the under parts of the body of a pale brownish yellow, and the hairs on this part are slightly tinted with grey at the root. The head bears the same proportion to the body in length, as in *P. obesula.* Its dimensions are—

	Inches.	Lines.
Length from tip of nose to root to tail ...	9	0
" of tail	3	0
" from nose to ear	2	2
" of ear		7½
" of fore foot and nails		11½
" of hind foot and nails	1	11½

I question much whether the shortness of the head in the larger specimen does not arise from the mode in which the specimen has been stuffed, and with regard to the yellowish tint of the abdomen, I may observe, that in specimens which are undoubtedly the *P. obesula*, the under parts of the body are sometimes tinted with yellow, though less strongly than in the little animal above described. I cannot see any good grounds for regarding the specimens called *fusciventer*, as specifically distinct from the *P. obesula*.

PERAMELES NASUTA.

Long-nosed Perameles.

Perameles nasuta. GEOFFROY, Annales du Muséum, tom. iv. p. 62, Pl. 44.
" *nasuta* and *aurita*, of the Paris Museum.
" " WATERHOUSE, Nat. Libr. (Marsupialia) vol. xi. p. 155. Pl. 13.
? " *Lawsoni.* QUOY et GAIMARD, Voyage de l'Uranie—Zoologie, p. 57, and p. 711.

Muzzle much elongated; ears moderate: general colour pale brown; on the sides of head and body of a very pale vinous red; under parts of body and feet yellowish white; tail brown above, dirty white beneath.

Inhabits New South Wales.

The Long-nosed Perameles, as its name implies, has the muzzle more elongated than in other species of the present section: its fur is almost entirely composed of harsh and flattened hairs, but there is a scanty under fur of fine hairs, which are of a pale grey colour on the upper parts of the body. The longer and coarser hairs on the back have their visible portions pencilled with pale brown, and blackish; on the sides of the body the black is almost entirely absent, and here, as well as on the sides of the head, the general tint is purplish red, but very pale. The under parts of the body are yellowish white, and the hairs on these parts are uniform to the root. The feet are dirty white; the fore leg is greyish at the base externally, and there is a dusky patch immediately above the heel of the hind foot. The ears are of moderate size, broad at the base, and considerably attenuated at the opposite extremity; they are clothed with very small hairs, whitish on the inner side, and dusky on the outer, but pale brown near the anterior angle. The tail is furnished with small stiff hairs, dusky or brownish on the upper surface, and dirty white on the under.

	Mus. Zool. Soc.		Brit. Mus.	
	Inches.	Lines.	Inches.	Lines.
Length from tip of nose to root of tail	16	0	16	0
" of tail	4	6	5	0
" from nose to ear	4	0	4	1
" of ear	1	2	1	4
" of fore foot and nails			1	11*
" of tarsus, including the nails	3	2	3	5

The cranium of the *Perameles nasuta*, according to Geoffroy's figure in the Annales du Muséum, is $3\frac{3}{4}$ inches in length; from the orbit to the apex of the intermaxillaries, 2 inches $1\frac{1}{2}$ lines; from the front of the upper foremost incisor to the canine, half an inch; from the front of the first incisor to the back of the last molar, 1 inch $10\frac{2}{3}$ lines; from ditto to first true molar, 1 inch $3\frac{1}{2}$ lines.

In a specimen in the British Museum, I find the fourth and fifth incisors of the upper jaw separated by a space of two lines, and from the front of the foremost incisor to the same

* The longest nails of the fore foot in this specimen are $10\frac{1}{2}$ lines in length, and of the hind foot $8\frac{1}{2}$ lines.

part of the first true molar measures 1 inch $4\frac{1}{4}$ lines. A naked space on the upper side of the muzzle extends backwards nearly three quarters of an inch.

Perameles Lawsoni. Quoy et Gaimard.

A large species (of Perameles), state the authors above mentioned, was given to us at Bathurst, beyond the Blue Mountains. It was perhaps about two feet in length, from the head to the extremity of the tail: its fur was reddish brown above, and yellowish beneath. It was lost in the shipwreck of the Urania. At p. 711 of the "Zoologie de l'Uranie," the name of *Perameles Lawson* is proposed for this species.

Such are the slender grounds upon which the *Perameles Lawsoni* is inserted into catalogues of the species of Mammalia—a description evidently given from memory; and which, after all, does not include any peculiarities by which the animal may be distinguished from the *Perameles nasuta,* a specimen of which in the British Museum differs only in being three inches short of the *guessed* size of *P. Lawsoni.*

PERAMELES GUNNII.

Gunn's Perameles.

Perameles Gunnii. Gray, Proceedings of the Zoological Soc. for January, 1838, Pt. 6, p. 1; Annals of Nat. Hist. for April, 1838, Vol. i. p. 108.

" " Waterhouse, Naturalists' Library (Marsupialia), p. 156, Pl. 15.

Upper parts of body grey, pencilled with yellow and black; under parts white, hinder part of back blackish, with white bands: ears with yellowish hairs internally, externally dusky, but pale yellow behind, and yellow at the anterior angle; feet and tail white; the latter with a small dusky patch at the base, above.

Inhabits Van Diemen's Land.

Mr. Gunn, who was the discoverer of the present species, and after whom it has been named, informs us, that it is known in Van Diemen's Land as the *Bandicoot*, and states that Bandicoots are numerous in all parts of that island: they burrow in the ground, and live principally upon roots. Certain species of bulbs, cultivated by Mr. Gunn, suffered much from the attacks of the Bandicoots, and a case is mentioned by that gentleman in which a whole collection of Cape bulbs was destroyed by them.

In size and general proportions the ***Perameles Gunnii*** closely resembles the ***P. nasuta***, but it is readily distinguished from that animal by the white bands which adorn the hinder parts of its body, and the dark ground colour of the same parts: the colouring of other parts scarcely differs from that of ***P. nasuta***. Its fur is moderate as to length, and harsh to the touch, though less so than in ***Per. obesula*** or ***P. nasuta***: on the upper parts of the body the hairs are grey at the root, but the visible portion of each hair is pencilled with black and ochreous yellow; on the sides of the body the general hue is somewhat paler than on the upper parts, the hairs having very little of the black pencilling; and here a delicate vinous tint is perceptible: on the hinder third of the back the ground colour may be described as black, there being but little admixture of yellow, but on this part are four broadish white bands; the first of these bands (which are of about equal width to the dark interspaces) crosses the back rather in front of the line of the thigh; the other three radiate, so that the second is most nearly transverse, and the hindermost is longitudinal; a mesial dark line divides, or separates, all the bands. The under parts of the body are pure white, and the hairs are here uniform to the root; the feet and tail are also white, but the latter has a small dusky patch on the upper surface, at the base; and, on the sides of the heel of the hind foot, the hairs are dusky, as are likewise those which form the fringe partly covering the under surface of the foot; a dusky patch is also observable immediately above the heel. The muzzle is slender, and much elongated; the ears are moderate, broad at the base, and attenuated towards the opposite extremity; internally they are clothed with very small pale yellow hairs, and externally they are also chiefly clothed with yellow hairs, but on the hinder part they are nearly white, and a broad dusky mark obliquely crosses the outer surface of the ear, commencing about the middle of the anterior margin, and running obliquely backwards as it descends to the base.

A species greatly resembling the present animal inhabits the main land; it is the *P. fasciata* of Mr. Gray, and at the end of the description of that animal will be found pointed out the distinguishing characters.

	Inches.	Lines.	
Length from tip of nose to root of tail ...	16	6	
" from nose to ear	4	4 or 4	5
" of ear	1	2 or 1	3
Width of ditto		11	
Length of fore-foot and nails	1	6	
(of which the longest nail is		6)	
" of hind-foot and nails	3	0	
(of which the nail of largest toe is ...		6)	
" of tail, about	4	0	
Space between the fourth and fifth upper incisors	0	$1\frac{1}{3}$	
From front of foremost upper incisor to front of first true molar	1	$2\frac{1}{2}$	

PERAMELES FASCIATA.

White-banded Perameles.

Perameles fasciata. GRAY, in the Appendix to Capt. Grey's Journals of Two Expeditions of Discovery in North-west and Western Australia, Vol. ii. p. 407.

Fur moderately long, and harsh to the touch; on upper parts of the body pencilled with black and yellow in about equal proportions; on the sides of the body the yellow prevails, and on the hinder part of the back the black prevails as a ground colour; but here are three broad yellow-white bands, the foremost of which crosses the back, the other two run obliquely downwards and backwards from the mesial line; the hindermost of these two is almost longitudinal, and the one in front of this joins the foremost band—these bands are interrupted on the middle of the back; the feet and under parts of the body are white; the tail is also white, but along the whole upper surface the hairs are partly black and partly yellow, but chiefly the former.

Inhabits Liverpool Plains, and South Australia.

The ears of this species are rather long, very broad at the base, and much attenuated towards the opposite extremity; the posterior margin is indistinctly emarginated; they are clothed with very small adpressed hairs; those on the inner surface are yellowish white, and those on the outer of a pale rusty yellow, but blackish on the fore part, if we except an orange-coloured spot at the base, joining the anterior angle. As in *Per. Gunnii*, the fringes of hairs bordering the sides of the tarsus are dusky. The hairs on the under parts of the body are uniformly white; those on the upper parts are pale grey at the root, rusty yellow near the point, and black at the point. The under, softer fur, is dense on the upper parts of the body, especially over the haunches; but on the under parts the softer fur is scanty.

	Inches.	Lines.
Length from nose to root of tail	11	6
" of tail	4	0
" of tarsus	2	3⅓
" of ear	1	2¾
" from nose to ear	3	1

The *P. fasciata* differs from *P. Gunnii* in being smaller, in having the ears proportionately rather longer, and broader at the base; the tail longer, and dusky along the whole upper surface—(the dark colour not confined to a very small space at the base, as in *P. Gunnii*); the feet more slender, and the muzzle also more slender. The following dimensions, being from two specimens (one of each species) of very nearly equal size, will convey an idea of the amount of difference in the proportions. As the adult of one species is larger than that of the other, of course one of these specimens was younger than the other.

	PERAMELES FASCIATA.		PERAMELES GUNNII.	
	Inches.	Lines.	Inches.	Lines.
Length from nose to tail	11	3	11	9
" of tail	4	0	2	8
" of tarsus and toes, including the nails	2	3½	2	5½
" from nose to ear	3	6	3	1½
" of ear	1	2½		11½
Breadth of ditto at the base		11½		9
Length of fore-foot to end of nails ...	1	2	1	2
" of longest toe of hind-foot, including the nail		10⅔	1	0½
" of outer toe and nail		5		5⅔

A skull of *Perameles fasciata* in the British Museum collection, agrees very closely in size and general proportions with skulls of the *P. myosuros* in the same collection; it is, however, narrower, and the teeth are smaller—the canine is very small: the second pair of palatine openings are rather larger, and the principal opening in the palate is smaller. The lower jaw is more slender. The dimensions are :—

	Inches.	Lines.
Length	2	6
Width	1	0
Length of nasal bones	1	1
Width of ditto, behind		2½
" " in front		1⅓
" of interorbital space		6½
Length of palate	1	6
Space between fourth and fifth upper incisors		1⅓
" " fifth incisor and canine		1¼
" " canine and first premolar ...		2
" from front of foremost incisor to back of last true molar	1	4⅔
" of four true molars, taken together ...		5⅓
" of incisive openings		3¼
" of second pair of palatine openings ...		4
" of principal palatine openings		3⅔
Length of lower jaw...	2	0
Height of ditto from apex of coronoid process ...		6½

PERAMELES MYOSUROS.

Saddle-backed Perameles.

Perameles myosuros. WAGNER, Schreb. Saug., Pl. 155, A. d., Pt. 111, 112, Nov. 1842. GOULD, Mammalia of Australia, Pt. 1, Pl. 2.

" *arenaria.* GOULD, Proceedings of the Zoological Society for June, 1844, Pt. 12, p. 104.

Muzzle very slender, and much elongated ; ears rather large ; tail moderate ; fur rather long : general colour of upper parts brownish ; under parts of the body, feet, and tail, white, indistinctly suffused with yellowish ; the latter with the upper surface dusky ; ears with a broad dusky band externally, a large orange patch in front at the base, and the hinder part whitish yellow: a broadish dusky band crosses the hinder part of the back.

Inhabits the Swan River district, Western Australia, and the neighbourhood of King George's Sound.

The present animal has received the name of the Saddle-backed Perameles from Mr. Gould, on account of its having a large and nearly circular area on the back, which is of a darker hue than other parts, if we except the transverse band which crosses the lumbar region, a little in front of the hind leg: with respect to this dusky patch, it must be remarked, however, that it is extremely indistinct, and it appears to me that it would be more accurate to describe the animal as having a crescent-shaped band, of a rather paler hue than the ground-colour of the back, immediately in front of the dark band on the lumbar region; this pale band, which corresponds to the foremost of the bands in *P. fasciata* or *P. Gunnii*, combined with the sides of the body being paler than the other parts, causes the appearance of the dark area referred to.

The under fur in *P. myosuros* is unusually plentiful, especially on the hinder parts of the back, and is of a whitish grey colour; the longer flattened, harsh hairs, are grey-white at the root, and rusty yellow near or at the point; on the back are many which are black at the point, and these produce the darker general hue of that part: on the under parts of the body the hairs are uniform to the root, white, but slightly suffused with yellow; the feet and tail are of the same colour, but the small hairs on the upper surface of the latter are partly yellowish, but chiefly black. The ears are rather long, very broad at the base, and rather suddenly contracted at the apical portion; their hinder edge is somewhat emarginated; on the inner side they have pale yellow hairs, and externally they are clothed with minute hairs of the same yellow hue on the hinder part; a largish bright rusty yellow, or orange-coloured spot, is observable at the anterior angle, and a broad dusky band crosses the middle, running from the anterior margin, obliquely backwards and downwards. A delicate brownish red tint is observable on the sides of the body. The tail is clothed throughout with very small stiff hairs, which are sufficiently numerous, however, to hide the scaly skin beneath, in the specimens which I have examined, though it appears they are less numerous in the specimen described by Prof. Wagner[1]. The naked portion of the muzzle terminates in a point about $\frac{1}{6}$ of an inch behind the line of the nostril openings.

	Inches.	Lines.
Length from tip of nose to root of tail ...	11	0
" of tail	3	6
" of ear	1	3
" of tarsus, including the nail ...	2	4
" from nose to ear	3	1

The Saddle-backed Perameles, according to Mr. Gould, "inhabits the whole line of coast of the Swan River colony, but is apparently not found to the westward of the Darling range of hills. It resides in the densest scrub; thickets of the seedling *Casuarinæ* being its favourite resort. It makes a compact nest in a hollow on the ground, of grasses and other materials, which assimilate so closely in colour and appearance to the surrounding herbage, that it is very difficult of detection, the difficulty being much increased by the nest having no visible opening for the ingress and egress of the animals. The nests are generally inhabited by pairs: the young are either three or four in number.

"Its food consists of insects, seeds, &c. It excavates holes

[1] D. Wagner says the tail is similar to a rat's tail: the tail of the animals before me will not bear this comparison, being much more densely clothed than in the common rat, which of course must be the one referred to.

in the earth with rapidity and ease, and to these, and the hollow trunks of fallen trees, it flies for shelter when pursued by its enemies.

"Mr. Gilbert remarks, that this species is, without exception, the most difficult to skin of all the Marsupials with which he is acquainted; the skin is, in fact, so tender, that the weight of one of the limbs, if left hanging by the skin, is sufficient to separate it from the body."

I am indebted to Mr. Neill, of King George's Sound, for a drawing, and some notes, of a species of Perameles, which that gentleman informs me is found to the westward of King George's Sound only. The animal (which is there known to the natives by the name of "Gnemmel"), judging from these materials, and also from some parts of the skeleton, also forwarded to me, is clearly the ***Perameles myosuros***. The specimen drawn by Mr. Neill lived for several weeks in confinement: its food, which consisted of raw mutton and insects, Mr. Neill states, was encircled by its long tongue, and conveyed by that organ to the mouth, piece by piece, with the greatest rapidity. The natives state that it makes a nest composed of sticks, straws, &c.

The skull of *Perameles myosuros* is of a more delicate make than that of ***P. obesula***. In this latter animal, the cranium, when old, presents a slight trace of sagittal crest, but in ***P. myosuros*** the temporal ridges can scarcely be traced, and never meet: its form is more elongated; the muzzle is more attenuated, and the zygomatic arch is more slender. As would be anticipated, these differences are accompanied by a reduced size in the teeth; and, the muzzle being longer in ***P. myosuros***, some of the teeth are more widely separated from those which precede, or follow them in the series; thus the fifth incisor is more widely separated from the fourth, or from the canine, and between this latter tooth and the premolars, as well as between the two foremost of the premolars, the spaces are greater. The upper true molars differ from those of *P. obesula* in having the posterior internal lobe distinctly smaller than the anterior one. The lower jaw in the animal under consideration is much more slender than in *P. obesula*, and has the condyloid and coronoid processes less elevated.

	Inches.	Lines.
Length of skull	2	8½
Width of ditto	1	1¼
Length of nasal bones	1	1½
Width of ditto behind		2½
" " in front		1⅓
" " between orbits		7
Length of palate		
Space between fourth and fifth upper incisors		1¼
" " fifth incisor and canine ...		1½
" " canine and fifth premolar ...		1⅓
" from front of foremost incisor to back of last molar	1	5½
Length of four true molars taken together ...		5½
" of incisive openings		4¼
" of second pair of palatine openings		3½
" of principal palatine openings ...		4¾
" of auditory bullæ		4
" of lower jaw	2	0⅓
Height of ditto from the apex of the coronoid process		7¼

PERAMELES BOUGAINVILLEI.

Bougainville's Perameles.

Perameles Bougainvillei. QUOY et GAIMARD, Zoologie du Voyage de l'Uranie, p. 56, Tab. 5. Bulletin des Sci. Nat. 1824, tom. i. p. 270.

Muzzle attenuated; eyes tolerably large; ears ovate; fur moderately harsh; that on the upper parts of the body of a rusty tint, and that on the under parts, and inner side of limbs, rusty grey: tail rusty brown above, and greyish beneath; claws yellowish.

Inhabits Peron's Peninsula—in Shark Bay, Western Australia.

The dimensions of this species as given by Messrs. Quoy and Gaimard, reduced to English measure, are as follow:—

	Inches.	Lines.
Length from tip of nose to root of tail ...	6	6
" of head	1	10½
" of tail	2	8
" of fore legs	1	5
" of hind legs	2	8

The small size of this animal, and its having the teeth but little developed, caused MM. Quoy and Gaimard to suspect it might be young; these authors say, however, that they saw many specimens all of which were of the same size. The skull, represented on the same plate with the animal, and of the natural size, shows the four true molars as perfectly developed, a circumstance of itself sufficient to prove that the animal to which the skull belonged was adult. The cranium has lost the occiput, but, making allowance for this part, it would be considerably smaller than that of any other known species of *Perameles*: its length, when entire, indeed, could not have been more than two inches.

PERAMELES DOREYANUS.

New Guinea Perameles.

Perameles doreyanus. Quoy et Gaimard, Voyage de Découvertes de l'Astrolabe—Zoologie, tom. i. p. 100, Pl. 16, figs. 1—5.

Head conical; the muzzle long, and tolerably thick; eye small, the pupil linear, and in the direction of the muzzle; ears large, and somewhat rounded; limbs short and stout; the outer and inner toes of the fore feet clawless: all the claws strong, short, slightly arched, and rounded above and beneath: fur harsh; of a rusty brown colour on the upper parts of the body—a tint produced by the admixture of blackish brown harsh hairs with others of a finer quality, and golden rusty colour; sides of the body of a pale hue; under parts, and inner sides of the limbs, pale yellowish; ears yellowish, and destitute of hair, excepting on the anterior margin: toes of the hind feet covered with pale yellowish hairs; tail short, and but sparingly furnished with short harsh hairs.

Inhabits New Guinea.

	Inches.	Lines.
Total length	19	5
Length of head	4	4
Distance between ears	1	3
Length of ears	1	1
Width of ditto		$9\frac{3}{4}$
Length of anterior members about	3	3
" of posterior ditto	5	9
" of tail	3	5

The above is the description given by Messrs. Quoy and Gaimard, but slightly abbreviated. With regard to the dentition, it is stated the *Perameles doreyanus*[1] has in all 46 teeth; the two wanting to complete the number found in other species of *Perameles*, are, two incisors of the upper jaw, eight only being enumerated in the dental formula given. Possibly these teeth may at one time have existed in the skull, but had been lost whilst the animal was alive, in which case the sockets would soon become obliterated. A skull of an aged specimen of *Perameles obesula*, now before me, has but eight incisors in the upper jaw, the foremost pair having been lost,

[1] M. Lesson arranges this animal under a new generic title, in his Nouveau Tableau du Règne Animal. It will be found under the name *Echymipera Kalubu*, in that work (see p. 192).

and the sockets filled with bone: from the condition of the second pair it is probable they would soon likewise have been cast.

The specimen described by Messrs. Quoy and Gaimard was procured at Dorey Harbour; hence its name.

Perameles Harveyi and *Perameles Tuckeri*.

The so-called species of Perameles, named *Harveyi*[1] and *Tuckeri*[2], must be expunged from our lists. The former was founded by myself upon an imperfect skin, from South Australia, which I have since discovered appertains to the *Hypsiprymnus Graii*. The latter I had an opportunity of inspecting soon after it was described by Mr. Gray, and found, upon examining its teeth, &c., to be also a species of *Hypsiprymnus;* indeed, I could perceive no difference between this animal and the *Hypsiprymnus murinus*.

Genus, *Chœropus*[3].

Chœropus. OGILBY, Proceedings of the Zoological Society for March, 1838, Pt. 6, p. 26.

Peramelidæ having very slender limbs; the fore feet provided with two toes only, and these small, equal, and furnished with short, compressed nails; the hind feet with but one well developed toe, the joined toes being very small, and far removed from the extremity of the foot; and the outer toe being represented by a mere tubercle, placed about midway between the extremities of the foot. All the toes of the hind foot are provided with nails.

[1] Proceedings of the Zoological Society, Part 10, p. 47.

[2] Annals of Natural History, vol. i. p. 150.

[3] From χοῖρος, a hog; and πους, foot.

The present genus was founded by Mr. Ogilby upon a singular little animal discovered by Sir Thomas Mitchell on the banks of the River Murray. The animal was presented to the Natural History Museum at Sydney, but a careful drawing, and some notes, were brought to England, and upon these Mr. Ogilby's account is drawn up. More recently, specimens of the Chœropus have been procured by Mr. Gould, and by Capt. G. Grey, and are now deposited in the British Museum. From these we learn that Mr. Ogilby's original conjecture as to the affinities of the animal, is correct—that it is allied to Perameles. The structure of the skull and teeth is very nearly the same as in Perameles. In the number of the teeth I have good reason to believe *Chœropus* does not differ from Perameles[1]. The five incisor teeth on each side of the upper jaw are contiguous, and rather less compressed than in Perameles, being slightly convex on the outer surface, and their apices are not truncated as in the species of the genus just mentioned, but terminate in a somewhat obtuse point. The canine is rather small, but comparatively broad from front to back, and is formed nearly like the premolars: it is distant about three quarters of a line from the incisors. The first false molar is separated from the canine by a space of $1\frac{1}{2}$ lines, and from the second false molar by about three quarters of a line; the second and third false molars, and the four true molars, form a continuous (or very nearly continuous) series. These teeth, as represented in the drawing kindly lent me by Prof. Owen, do not appear to differ in structure from those of *Perameles*. With regard to the teeth of the lower jaw, I have only to remark, that the incisors are broader (the hindermost one distinctly so) than in *Perameles*.

The cranium is shorter, and the cerebral portion is broader,

[1] The only teeth which I have not seen are the back molars of the lower jaw.

than in *Perameles;* but the muzzle is very narrow. The ears are larger than usual in the genus last mentioned. The limbs are longer, and remarkable for their slenderness, the fore legs being scarcely as thick as an ordinary sized goose-quill. The two toes with which they are terminated are very small, and provided with small, compressed, and but little curved, nails; they have each an oblong fleshy pad on the under surface, behind which is a second flesby tubercle, but this latter is very small. The two toes most probably correspond to the second and third, since the fourth is the smallest of the three developed toes in Perameles. The hind legs are distinctly longer than the anterior, and almost equally slender; the foot is long, and at first glance appears to have but one large toe, the other toes being very small, and far removed from the extremity of the foot; the little outer toe is placed about midway between the extremities of the foot, and has a small conical nail: the joined inner toes are almost as small; they have hollow nails, and are placed rather in advance of the outer toe. The tarsus is entirely covered with hair beneath, if we except a small tubercle situated at the commencement of the anterior third of the foot. The great developed toe has a large fleshy pad at its extremity on the under surface, and terminates in a nail in the form of a compressed cone. From these circumstances it would appear that the heel is not applied to the ground. The tail is short and slender. The pouch opens backwards, as in *Perameles*.

CHŒROPUS CASTANOTIS.

The Chœropus, or Pig-footed Perameles,

(*Plate* 13, fig. 2.)

Chœropus ecaudatus.	OGILBY, Proceedings of the Zool. Society for March, 1738, Pt. 9, pp. 25-6. Perameles ecaudatus, Ogilby, in Mitcheh's Journal, ii. p. 131, Pl. 27.
" *castanotis.*	GRAY, Annals of Natural History for March, 1842, vol. ix. p. 42.
" "	GOULD, Mammals of Australia, Pt. 1. Pl.

Fur long, loose, and rather soft to the touch; on the upper parts of the body of a brown-grey tint, and on the under parts white, or yellowish white; ears clothed with very small hairs, which are chiefly of a rusty yellow colour, but are dusky towards the point of the ear: fore feet whitish; tarsi pale rufous; the great central toe dirty white. Tail clothed with short hairs, but those on the upper surface distinctly longer than elsewhere; the sides, under surface, and tip, brown-white; the upper surface, black.

Inhabits South Australia.

Such is the colouring of one of two specimens in the collection of the British Museum; the other specimen differs in having the general tint of the body brown, and suffused with rust colour, especially on the flanks: the limbs are also tinted with pale rusty red. The fore feet, and the large toe of the hind foot, are whitish; the tail is of a pale rust colour, if we except the longer hairs of the upper surface, which form a black crest[1]. The under parts of the body are rusty white. The dusky or blackish hue of the apical portion of the ear is more extended on the outer surface of that organ than on the inner, in both specimens, and the fur, both on

[1] These longer hairs are less than a quarter of an inch in length.

the upper and under parts of the body, is grey at the root. The under fur is more abundant than in *Perameles obesula,* and some others, and in this respects resembles that of the *P. myosurus;* the longer interspersed hairs are less harsh than usual in the *Peramelidæ.* The ears are very broad at the base, and much attenuated at the opposite extremity, and their hinder edge is emarginated.

	Inches.	Lines.	Inches.	Lines.
Length from tip of nose to root of tail	9	6	11	0
" of tail	4	0	?	
" from nose to ear	2	7	2	8
" of ear	1	11	1	10
Width of ditto at the base	1	2½	1	3½
Length of fore leg, about	2	6	2	8
" of fore foot	1	0½	1	1½
Of which the two toes and nails of ditto are		5½		5¾
The nails being		2		2½
Length of tarsus	2	7¾	2	8
Of which the great central toe is ...	1	0	1	0¼

Sir Thomas Mitchell, in his notes on the Chœropus, especially mentions the total absence of a tail as a remarkable peculiarity in the animal, and hence the specific name *ecaudatus* was applied to it by Mr. Ogilby; since, however, other specimens, agreeing essentially with the *Chœropus ecaudatus,* have subsequently been discovered—and one of these, I may observe, was found in the same district as Sir Thomas Mitchell's animal—we can scarcely refuse to believe that the latter specimens are specifically identical with the former, and that the peculiarity in question is due to some accident. Two examples of the Chœropus are contained in the British Museum collection, one of which was presented by his Excellency George Grey, Esq., and is from South Australia; the second is from the Swan River district, where, according to Mr. Gould, the species is confined to the interior of the country. Its food is said to consist of insects and vegetable substances; and the author just mentioned states that it forms a nest, composed of leaves, &c.; in these respects resembling the species of *Perameles*.

DASYURIDÆ, OR DASYURUS FAMILY.

Dasyuridæ. WATERHOUSE, Transactions of the Zoological Society, vol. ii. Pt. 2, p. 149.—Naturalists' Library, vol. xi.
Dasyurina. GRAY, Annals of Philosphy, xxvi. 1825.

Marsupialia having the second and third toes of the hind foot disunited, and well developed; the thumb, or first toe, small, or absent; the tail non-prehensile, and hairy; eight incisors in the upper jaw, and six in the lower; the canine teeth well developed; the molar teeth either with trenchant crowns, or with the masticating surface presenting numerous prickly points. Stomach simple: no cœcum marking the division of the large and small intestines.

With respect to the above, which are the chief distinguishing characters of a group of Marsupial animals whose diet is either carnivorous or insectivorous, it is necessary to observe, that the Myrmecobius has yet to be dissected; hence we are not sure that the cœcum is absent in that animal.

Several fossil species of this family have been found in Australia, to which continent the recent species are also confined—with one exception only, the *Phascogale melas*, which is a native of New Guinea.

Genus, *Myrmecobius* [1].

Myrmecobius. WATERHOUSE, Proceedings of the Zoological Society for July, 1836.—Transactions of the Zoological Society, vol. ii. Pt. 2, p. 149.

Teeth small and detached: incisors, $\frac{4-4}{3-3}$; canines, $\frac{1-1}{1-1}$; molars, $\frac{8-8}{9-9} = 52$. The molar teeth provided with prickly points.

Head somewhat depressed above; the muzzle moderately elongated; muffle naked; nostrils lateral: ears of moderate size, and pointed: tongue very long and slender: bony palate very long, and destitute of the ordinary large palatine openings.

Legs rather short and strong; fore feet naked beneath; provided with five toes, having compressed and curved claws: hind feet naked beneath, in front, and along the mesial line behind; provided with four toes having compressed and curved nails.

Tail long and bushy.

The female destitute of pouch, and having, apparently, eight mammæ, arranged in a circle.

The only known species of the genus Myrmecobius is found on the west coast of Australia, chiefly in the Swan River district, but has likewise been met with as far southwards as the Murray Scrubs. One of its most remarkable peculiarities consists in the great number of its teeth, these being at the same time, many of them, of complicated structure [2]. The incisor teeth are arranged

[1] From μύρμηξ, ant; and βιος, life.

[2] In the Porpoises and Armadillos the teeth are often very numerous, but in these animals they are of a very simple form, and cannot by any difference of structure be divided into canines, false and true molars, &c. It has appeared to me probable that the simple teeth in question represent parts only of the more complicated teeth of other mammals.

laterally, rather small, compressed, and pointed, and slightly recurved at the apex: the foremost two incisors of the lower jaw are distinctly the largest, and incline forwards, but are somewhat recurved at the point. Narrow spaces separate all the incisors of either jaw. The canines, although but moderately developed, are the largest of the dental series: they are compressed, rather wide in the antero-posterior direction, and slightly recurved at the apex. The foremost three molar teeth on either side of each jaw present the most common form of premolars in insectivorous or carnivorous mammals; compressed, pointed, and having a small anterior and posterior lobe, which lobes are very distinct in the hindermost of the teeth in question, and almost obliterated in the foremost, which is the largest of the three. The fourth molar of the upper jaw is very small and compressed, and its crown is divided by notches into four small tubercles; the following four molars present each five or six small prickly points, the number of which differs in the corresponding teeth of opposite sides of the jaw; but it is important to notice that these four teeth differ from the molars which precede them, in having a double instead of a single row of tubercles: this difference of structure induces me to believe, that the more complicated four posterior teeth are true molars, that the small fifth molar corresponds to the principal premolar, and that the three foremost molars are premolars; that, in fact, the unusual number of the molar teeth is due to the presence of extra premolars. But if we are guided by the same kind of differences in the structure of the molars of the lower jaw, we find four premolars and five true molars. The premolars are of the same form as those of the upper jaw, and the true molars differ only in having the inner range of tubercles more developed. The ramus of the lower jaw is so twisted, that the true molars are directed inwards.

MYRMECOBIUS FASCIATUS.

The Banded Myrmecobius.

(*Plate* 14, fig. 1.)

Myrmecobius fasciatus. WATERHOUSE, Proceedings of the Zoological Society for July, 1836, Part iv. p. 69; Transactions of the Zoological Society, Vol. ii. Part ii. p. 149, Plates 27 and 28; Naturalists' Library (Marsupialia), Vol. xi. p. 145, Plate 11.

" " GOULD, Mammals of Australia, Part i. Pl. 10.

" *fasciatus* and *Diemenensis.* GRAY, List of the Mammalia in the British Museum (1843), p. 100.

Fur harsh and somewhat adpressed; of a bright rusty red colour, and pencilled with white, but almost black on the hinder parts of the body, and yellowish white on the under parts: numerous transverse, white, or cream-coloured bands cross the back; are distinct on the hinder parts, but become indistinct, and are interrupted, as they approach the shoulders. A black mark runs along the side of the muzzle, and, passing through the eye, terminates near the ear, which latter is rather small, pointed, and well clothed with small hairs, which are almost entirely black on the outer surface, and of a rusty yellow hue on the inner: feet of a pale rust colour. The long hairs of the tail are chiefly black at the basal half, and yellow-white at the terminal half; but the mesial line of the tail is of a bright rust colour on the under surface.

Inhabits Western and Southern Australia.

The Banded Myrmecobius was first discovered by Lieut. Dale, who procured a specimen whilst on an exploring expedition into the interior of the Swan River settlement, about ninety miles to the south-east of the mouth of the river.

Two specimens of this very elegant little animal were seen by Lieut. Dale, both of which fled to hollow trees for shelter upon being pursued. The district in which they were found abounded in decayed trees and ant-hills; and from some peculiarities in the dentition of the animal, combined with its extremely long and slender tongue, the author, when the animal was placed in his hands by the discoverer, to be described, felt certain that its food was not only insects, but consisted of the softer and smaller species, for procuring which, by scratching up the earth, the strong fore feet and claws appeared to be adapted. Indeed, the peculiarities of structure, combined with the fact that the animal was found in the vicinity of ant-hills, suggested that its food in all probability consisted chiefly of ants—and hence the generic name. As yet, however, we have no direct evidence that ants form the chief food of the *Myrmecobius*, though it is stated in Mr. Gould's "Mammals of Australia," that wherever this animal takes up its abode, there ants are found to be very abundant. In the same work the following particulars of the habits of our animal are given, from the pen of Mr. Gilbert:—

"I have seen a good deal of this beautiful little animal. It appears very much like a squirrel when running on the ground, which it does in successive leaps, with its tail a little elevated; every now and then raising its body, and resting on its hind feet. When alarmed it generally takes to a dead tree lying on the ground, and before entering the hollow invariably raises itself on its hind feet, to ascertain the reality of approaching danger. In this kind of retreat it is easily captured, and when caught, is so harmless and tame, as scarcely to make any resistance, and never attempts to bite. When it has no chance of escaping from its place of refuge it utters a sort of half smothered grunt, apparently produced by a succession of hard breathings.

"The female is said to bring forth her young in a hole in the ground, or in a fallen tree, and to produce from five to nine in a litter. I have not myself observed more than seven young attached to the nipples. Like the members of the genus *Antechinus*, this animal has no pouch for the protection of the young; the only protection afforded their delicate offspring is the long hairs which clothe the under surface of the abdomen of the mother."

With regard to the range of the Myrmecobius, Mr. Gould states that it is very generally dispersed over the interior of the Swan River settlement, from King George's Sound on the south, to the neighbourhood of Moore's River on the north, and as far westward as civilized man has yet been able to penetrate; and the author has recently learnt from the same gentleman, that he has received a specimen of this animal from the Murray Scrubs, whence Mr. Gould has also obtained the *Lagorchestes fasciata*—an animal which had hitherto been found only on the west coast.

The Myrmecobius is about equal in size to the common squirrel (*Sciurus vulgaris* of authors); its body is moderately slender, the limbs rather short; the fore legs and feet strong, the latter armed with tolerably large claws, which are compressed and curved—less so, however, than we find the claws of climbing animals, and are more especially adapted to scratching in the soil. The fur is tolerably long, rather glossy, and composed almost entirely of harsh hairs, the softer under fur being very scanty. The head, fore part of the back, and outer side of the limbs, are of a bright rust colour, but more or less pencilled with white; the hinder third of the back is black, slightly pencilled with white, and on this part are usually six transverse white bands. Other bands are observable in front of these, but they are comparatively indistinct, and confined almost entirely to the sides of the body, being interrupted on the back. The more distinct bands, I may observe, are not unfrequently slightly interrupted in the middle of the back. The under parts of the head and body, which are much less densely clothed with fur than the upper, are white, or yellowish white, as are also the inner parts of the limbs. A black mark, commencing about midway between the eye and the tip of the muzzle, runs backwards through the eye, and terminates near the anterior angle of the ear: the dark hairs forming this mark are shorter than on the adjacent parts, and those on either side of the mesial line are pointed in opposite directions. A white mark borders the black one just mentioned both above and below, but the upper white mark, which runs immediately above the eye, is indistinct. The hairs of the moustaches are black, and neither long nor numerous. The ears are rather small, narrow, and pointed; they are well clothed with small hairs, internally of a rusty yellow colour, externally almost black, but sometimes reddish. The feet are of a pale rust colour. The tail is bushy, and about equal to the body in length, and owing to the hairs which spring from its sides being longer than elsewhere, has a flattish appearance. The hairs are rusty red at the root, black beyond, and white at the point, the black and white being in about equal proportions. The hairs springing from the upper surface of the tail display very little of the rust colour, but those on the under surface are so coloured that the whole mesial portion of the tail is of a bright rusty hue.

In young specimens of the Myrmecobius in the British Museum collection, the white bands on the back are less defined than in the adult; the hairs of the tail are comparatively short, and, when in their natural position, the visible portion of each hair is yellow.

The adult specimens are subject to some variety in their colouring: in some, the pale bands on the back are of a cream colour, and the rusty red of the fore parts of the body is much less bright. I have before me a drawing, kindly sent me by Mr. Neill, of a specimen found by that gentlemen at King's George's Sound, in which the whole of the upper parts of the body are much darker than usual. The specimen (a female) was regarded by Mr. Neill as "evidently a variety of the Myrmecobius." This gentleman particularly notices the great length of the tongue of the animal, and the same peculiarity struck Lieut. Dale.

	FEMALE.	
	Inches.	Lines.
Length from tip of nose to root of tail ...	12	0
" from nose to ear	2	6
" of ear		10½
" of tail		7*
" of tarsus and nails	2	1½
" of hind foot and nails	1	3

The skull of the Myrmecobius is of a depressed conical form. Viewing it from above, the most striking points are the almost total absence of crests or muscular ridges; the greatly developed supra-orbital ridge, which is deeply notched in the middle, and sends outwards a large post-orbital process, which is separated but by a narrow space from a corresponding, though smaller, process of the malar bone: the great extent of the lachrymal bones is also striking. The nasal bones are by no means long; narrow in front, and greatly expanded behind; the frontal bones are broad. A semicircular and divided interparietal bone is observable on the hinder part of the skull; the squamous bone is small: the malar bone is broad, being much expanded under the eye, where it inclines outwards, so as to approach to the horizontal; it enters slightly into the composition of the glenoid cavity. The palate is of great extent, especially in the longitudinal direction, and reminds us of the palate of the true ant-eaters. There are no perforations in the palatine bone; the palatine portion of the superior

* The length of the tail, *including* the hair, is about 1¾ inches more than the above.

maxillaries presents two longitudinal ridges, and on the fore part are some oblique ridges, near which some minute perforations for small vessels are observable, as in the Armadillos; and, besides these, there are four very minute openings near the palatine bone. The incisive openings are small, being scarcely one line in length. The auditory bullæ are of moderate size, very convex (almost spherical), and formed by an expansion of the sphenoid alæ. The occipital opening is large, and notched above. All the sutures of the cranium are distinct.

The horizontal rami of the lower jaw are long, and rather slender; the coronoid process is moderately elevated; the condyloid is much elevated; its articular surface is nearly flat, transverse, and approaches to a semicircular form: the angle is slender, by no means long, and is less suddenly bent inwards than is usual in the Marsupialia. The two rami are but loosely joined at the *symphysis menti*.

Of *Vertebræ*, there are—cervical, 7; dorsal, 13; lumbar, 6; sacral, 4; and caudal, 23, or perhaps 24. The ring of the atlas is complete beneath—that is to say, the body is anchylosed to the neural arch, and not separate, as in many Marsupialia. The remaining cervical vertebræ present no characters that need arrest our attention: they all have small spinous processes; that of the vertebra dentata is of great antero-posterior extent, though but little elevated. The dorsal vertebræ have moderately elevated spinous processes: the ribs are slender; the clavicle tolerably strong. The sternum is composed of six bones, of which the foremost, or manubrium sterni, is suddenly dilated rather in front of the middle. The lumbar, as well as the four or five first caudal vertebræ, have a small perforation traversing the body from beneath, which has two outlets on the upper surface: the lumbar vertebræ have the transverse processes greatly developed; the corresponding processes of the four or five first caudal vertebræ are also large, being much expanded from before backwards. The scapula has the spine much elevated; in front it is suddenly bent over the infra-spinal fossa, and this reflected portion of the spine is much dilated in the middle[1]. The humerus is rather short, and powerful—shorter, and with the muscular ridges more developed, than in *Didelphys:* the deltoid ridge is very prominent, projecting boldly about midway between the extremities of the humerus: the internal condyle is perforated.

With respect to the remaining portions of the skeleton I shall only observe, that the ulna and radius are nearly in contact for their whole length, as in the Dasyures, and the same remark applies to the tibia and fibula. The metatarsal bone of the thumb is exceedingly short, and supports but one phalanx, which is very small, and is so enclosed in the integument, in the living animal, as not to project in the form of a free point.

In the mounted skeleton of a Myrmecobius in the College of Surgeons, the inner toe or thumb of the fore foot has three phalanges[2].

Genus, *Phascogale*[3].

Phascogale. TEMMINCK, Monographies de Mammalogie, tom. i. p. 56. 1827.

Dasyuridæ with the two foremost incisors of the upper and lower jaw larger than the others; premolars $\frac{3-3}{3-3}$, true molars $\frac{4-4}{4-4}$, studded with prickly tubercles; those of the upper jaw with triangular crowns; the last tooth very narrow, and transverse. Five toes to the fore, and the same number to the hind feet; the inner toe of the latter in the form of a small,

[1] It nearly resembles the scapula of a Didelphys, but is proportionately shorter, and has the spine much more developed.

[2] In a foot of a bear in the same museum, I find three phalanges to the inner toe, and M. De Blainville represents three to the inner toes of the fore and hind foot of *Ursus ferox*. I have never seen any other instance amongst the Mammalia in which the thumb has more than two phalanges.

[3] From θάσκωλος, a pouch; and γαλῆ, a weasel. *Phascologale* would have been a more correct compound.

naillless, prehensile thumb. Tail either clothed with short hair throughout, or with short hairs only on the basal portion, the apical having long and bushy hair. The females sometimes destitute of pouch; mammæ eight, arranged in a circle.

The differences which present themselves upon comparing the skulls of these small *Dasyuridæ* with those of the larger species of the family, are of the same nature as those to which I called attention at p. 308, when comparing the smaller Phalangers with the larger. In addition to the comparatively large size of the cranial cavity, and of the occipital opening, and the very slight indications of muscular ridges in the skulls of the Phascogales, I may notice that the spinous processes of the cervical vertebræ in their development follow very closely that of the muscular ridges of the skull.

The two foremost incisor teeth of the upper jaw are slightly separated from the rest; the remaining three on either side are closely packed: the canine teeth are tolerably well developed: the premolars are compressed and pointed, and, viewed from the outer side, present nearly a triangular figure; the two foremost have a small anterior and posterior tubercle; in the third, a posterior small basal tubercle only remains. The molars, which, together with the premolars, form a continuous series, have the masticating surface of a triangular form; they each have three external cusps, two in the mesial line of the tooth, and one internal lobe, if we except the last, which has but two external cusps, one mesial, and one internal tubercle. The true molar teeth of the lower jaw have each three principal cusps, two placed on the inner side of the tooth, and one on the outer side; and besides these there is a posterior lobe, which is more or less distinctly divided by a notch into two tubercles; these are, however, but little elevated.

The muzzle is pointed, and but little elongated; the muffle is naked, and the nostrils are lateral. The ears are of moderate size, broadest at the base, and have the hinder edge emarginated, or nearly straight. The limbs are short; the fore feet are provided with five well developed toes, having compressed, curved, and pointed claws: four of the toes of the hind feet have similar claws, but the toe corresponding to the thumb is nailless.

The Phascogales are insectivorous mammals, and climb trees and shrubs in quest of their prey. The largest known species is about equal in bulk to the Common Rat, and most of the species are considerably smaller than that animal.

Within the last four or five years the described species of the present section have increased from two, which have long been known, to about a dozen, and some trivial differences observed among these have given rise to the establishment of new genera; as sections of minor value only can we adopt them.

The dentition of the Phascogales presents some modifications in the different species which are worthy of notice: thus, in ***P. penicillata***, ***P. calura***, and ***P. apicalis***, the third premolar of the lower jaw is very much smaller than the preceding two premolars, but the last mentioned of these three species differs from ***P. penicillata*** and ***P. calura*** in having the third premolar of the upper jaw also very small, the corresponding tooth being distinctly the largest of the premolars in those animals. *P. penicillata* and *P calura* differ from all other species, in having the foremost of the three lateral incisors of the upper jaw the largest: in other Phascogales they are of equal size, or perhaps in some it may be said to be smaller than the others: in ***P. albipes*** and ***P. crassicaudata*** the tooth in question is rather smaller than the other incisors. The lateral upper incisors in all the Phascogales are vertical, whilst the anterior pair, which are larger, are directed forwards.

In *P. Swainsonii*, *P. flavipes*[1], and *P. leucogaster*, the third is the largest of the premolars in the upper jaw, and is rather smaller than the preceding two, in the lower jaw. *P. leucopus*, *P. albipes*, and *P. crassicaudata*, differ from the above only in having the second and third premolars of the lower jaw equal.

With regard to the true molars, those of the upper jaw differ from the true molars of the Marsupial animals already described, in having the crown of a triangular form, a form of the masticating surface in these teeth which runs through the remaining species of the order. In the herbivorous species we have found the crowns of the true molars approaching more or less to a square form, and the same form was found in some of the insectivorous species forming the genus *Perameles;* in other members of that genus, however, owing to the reduced size of the posterior internal lobe, the masticating surface of the molars approximated to the triangular form. Now if we compare the true molar teeth of Phascogale with these species of Perameles, we shall find that the chief difference consists in the absence of the posterior internal lobe in the Phascogale. Another point of distinction in the molar teeth of the two animals is observable in the number of prickly tubercles arranged along the outer side of the tooth, there being four in Perameles, and but three in Phascogale: it is the foremost of the four, in Perameles (which is the smallest), which is lost in Phascogale, or of which there remains only an indistinct trace[2].

[1] In the immature animal, when the skull presents but three true molars on either side of the jaws, the third, or hindermost, of the upper and lower premolars is a minute tooth in *P. flavipes*, and probably in other species; it is then a milk tooth, and therefore to be replaced by another.

[2] In the true molar teeth of the species of Perameles I find very evenly developed all the cusps, or pointed tubercles, which are found in Insectivorous

Section 1. *Phascogale* proper.

Phascogales having the terminal half of the tail clothed with very long and bushy hair.

This section necessarily contains the *Phascogale penicillata,* that being the animal upon which M. Temminck founded the genus. A second species of Bushy-tailed Phascogales has recently been added to our lists by Mr. Gould.

Mammalia: their structure can scarcely be said to differ from that observable in the true molar teeth of the genus *Tupaia,* in the order Insectivora.

The dentition of Phascogale is very interesting, as showing an intermediate condition between what may be called the insectivorous and carnivorous types of dentition, and since it enables us to ascertain what parts are wanting in the more simple carnivorous molar tooth, where we find the remaining parts have a proportionate increase in their development—that such is the case may be clearly seen upon comparing the true molars of the *Perameles obesula* with those of a Phascogale. A transverse indentation divides the molar of the Perameles into two equal parts, and the structure of one of these is a repetition of the other, each part presenting an elevated triangular area externally, the angles of which are marked by three prickly tubercles; and an internal, less elevated lobe. The same transverse indentation exists in the molar of the Phascogale, and the same raised triangular area is found to each half, but the hinder half wants the internal lobe; here the triangular area is more developed, whilst on the anterior half of the tooth, where the inner lobe is larger than in Perameles, the corresponding area is proportionately small, and the foremost of the four outer tubercles of the tooth of the Perameles, belonging to the area in question, is obliterated. The true molar teeth of the lower jaw, when compared with those of *Perameles,* present corresponding differences—they have the anterior half (which presents three much elevated prickly cusps) more developed than in Perameles, and this increase of development is, as it were, at the expense of the hinder half of the tooth, which is proportionately small.

PHASCOGALE PENICILLATA.

Brush-tailed Phascogale.

The *Tapoa Tafa*, or *Tapha*.	WHITE's Journal of a Voyage to New South Wales, p. 281, and Pl.
Didelphis penicillata.	SHAW, General Zool., i. Pt. 2, p. 502, Tab. 113, fig. 1.
Dasyurus penicillatus.	GEOFF. Ann. du Mus. iii. p. 361.
" *Tafa.*	GEOFF., loc. cit. p. 360.
Phascogale penicillata.	TEMMINCK, Monographies de Mammalogie, tom. i. p. 58, Skull Pl. 7, figs. 9—12.
" "	GOULD, Mammals of Australia, Part 1, Pl. 6.

Fur long, and moderately soft; grey, pencilled with white; under parts white; feet grey, toes whitish; ears large, clothed with minute dusky hairs; mesial portion of head dusky; tail bushy, having long black hairs, excepting on the basal third, where the hairs are short, adpressed, and of a grey hue, but black, or dusky on the under side.

Inhabits New South Walcs, South Australia, and Western Australia.

This pretty little animal was first noticed in White's Journal, above quoted, but being there mentioned with the same name, Tapoa Tafa, as *Dasyurus viverrinus*, the two species were confounded together by Shaw, in his account of the last mentioned animal, though he nevertheless characterises the present species in another part of his work, under the specific name *penicillata*, a name still used. Geoffroy Saint-Hilaire perceived that the two animals, confounded by Shaw under the name *Didelphys viverrinus*, were distinct, and proposed for the present animal the name *Dasyurus Tafa*, but he did not suspect that his *D. Tafa* was identical with the *D. penicillatus*, a fact which we have ascertained by an examination of White's specimen still preserved in the Royal College of Surgeons. The Brush-tailed Phascogale is nearly equal in size to the Common Squirrel (*Sciurus vulgaris*); has a long, and moderately soft fur, which, on the upper parts of the head and body, is of a grey colour, and on the under parts, white, or yellow-white. The eyes are encircled with black, and there is a pale spot both above and below the eye: along the middle portion of the head, the hairs are more suffused with black than elsewhere: on the back, the hairs are finely annulated with whitish near the point, and are black at the point, whilst they are of a deep grey colour at the root. The ears are sparingly clothed with small pale hairs internally and externally, but on the fore part of the outer surface they are dusky. The fore legs and feet are of a pale greyish brown colour; the toes almost white; the hind feet are somewhat dusky behind, but pale in front, and the toes are grey-white. The tail is about equal to the body in length; at the root, covered with fur like that of the body; there is then a portion of about two inches in length which is clothed with short stiff hairs, which are of a very pale grey hue (sometimes chiefly yellowish white), but dusky along the middle line beneath; and the remaining portion, being about two-thirds of the whole, is clothed with very long and glossy black hairs, varying from one to two inches in length.

	Inches.	Lines.	Inches.	Lines.
Length from nose to root of tail ...	9	0	9	6
" of tail	7	0*	7	6
" from nose to ear	1	9	1	11
" of ear		10		11
" of tarsus	1	6¾	1	7

The present species has an unusually wide range, being found in New South Wales, and both in Southern and

* About 8½ inches, *including* the long hairs.

Western Australia. It climbs trees in quest of its insect food, and makes its nest in the hollows of the trunks or branches. Mr. Gould states that it enters the stores of the settlers, and that it is accused of attacking their poultry.

PHASCOGALE CALURA.

Handsome-tailed Phascogale.

(*Plate* 14, fig. 2.)

Phascogale calura — Handsome-tailed Phascogale. GOULD, Mammals of Australia, Pt. 1, Pl. 7. Proceedings of the Zoological Society for June, 1844, p. 104.

General colour of the upper parts of the body ashy grey ; of under parts yellowish white : tail at the base with short bright rust coloured hairs ; the apical half with long black hairs.

Inhabits Western Australia.

A beautiful species, discovered by Mr. Gilbert, at Williams' River, Western Australia. Like the *P. penicillata* it has the apical portion of the tail clothed with long black hairs, but it may be distinguished from that animal by its smaller size, and the brilliant rust colour of the basal part of the tail; the black hairs on the remaining portion, moreover, although long, are by no means so long as in *P. penicillata.* The skull, which has been removed from the specimen described by Mr. Gould, and which is now in the British Museum, has all the teeth developed, and its much smaller size, combined with some differences of proportion when compared with that of *P. penicillata*, leave no doubt on my mind as to the distinctness of the two species. The prevailing colour of the

P. calura is ashy grey; the body beneath is white, slightly suffused with yellow, and a delicate yellow tint is observable on the sides of the body. In front of the eye is a black patch, and a whitish spot is situated immediately beneath the eye. The ears are rather sparingly clothed with minute hairs, which are of a yellowish white colour on the inner sides, excepting on the anterior margin, where they are black, and brownish on the outer surface. The feet are of a greyish white hue. The apical half of the tail is densely clothed with glossy black hairs, averaging about half an inch in length, or rather more, and the basal half is covered with short adpressed hairs, which are of a brilliant rust colour, excepting along the mesial line beneath, where, almost to the root of the tail, they are black.

	Male.	
	Inches.	Lines.
Length from tip of nose to root of tail ...	5	3
" of tail	5	2
" of ear		8½
" from nose to ear	1	2
" of hind foot and nails		11

Section 2. *Antechinus.*

Antechinus. MacLeay, Annals, and Magazine of Natural History, for December, 1841, vol. viii. p. 242, and for January, 1842, vol. viii. p. 338.

" Gray, List of the Mammalia in the collection of the British Museum (1843), p. 99.

Phascogales in which the tail is clothed throughout with very short hairs.

With respect to the animal to which the generic title *Antechinus* was given by Mr. MacLeay, that gentleman observes that it differs from Phascogale in having the three lateral incisors of the upper jaw of equal size, and also in having all the premolars equal[1]. We have already pointed out the differences observable in the dentition in the different species of the genus Phascogale (in which I include *Antechinus*), and it is only necessary here to observe, that all the species which we, following Mr. Gray, place in the present section, do not precisely agree amongst each other in their dentition.

PHASCOGALE (*Antechinus*) SWAINSONII.

Swainson's Phascogale.

Phascogale Swainsonii. **WATERHOUSE, Mag. Nat. Hist. for 1840, vol. iv. p. 300.**

Above deep brown, very finely pencilled with rusty brown; beneath grey, obscurely tinted with yellow: tail and feet dusky brown.

This species inhabits Van Diemen's Land, and ranks in size next to the Brush-tailed Phascogales, being larger than most of the species of the present section. Its colouring is much darker than other hitherto discovered species, and is almost destitute of any grey hue. The fur is long, and moderately soft, and is of a deep grey colour next the skin: on the back the hairs are most of them narrowly annulated with rusty yellow, or brownish rust colour, the deeper tint being observable on the hinder parts. The hairs of the under parts of the body are grey, but tipped with yellowish. The tail is clothed throughout with small adpressed hairs of a dusky brown colour, and a trifle paler on the under than on

[1] Annals of Nat. Hist., vol. viii. p. 338.

the upper surface. The feet are uniform dusky brown: the fleshy pads on their under surface are transversely striated, and the remaining naked portion of each foot is apparently smooth. The muzzle is narrower, and more elongated than usual. The specimen from which the original description was drawn up, measured, from the tip of the nose to the root of the tail, five inches and two lines in length, and its tail was three inches and five lines long; the animal, however, attains a larger size, as will be perceived from the following dimensions, taken from a specimen, from Tasman's Peninsula, in the collection of the British Museum:—

	Inches.	Lines.
Length from tip of nose to root of tail	7	0
" of tail	4	0
" of tarsus		11¾
" of ear		4
" from tip of nose to ear	1	4

Mr. Gould imagined this species was identical with the *Dasyurus minimus* of Geoffroy; I have recently compared the two animals together, and find this is not the case.

The skull of *P. Swainsonii* is proportionately narrower, and more elongated, than in other species of its genus, and its upper surface is remarkably flat; the interorbital space is broad; the anterior upper pair of incisors are smaller, and are in contact with the others, not being directed outwards and forwards as in other Phascogales, or at least very slightly so: a space on either side of the foremost upper premolar separates this tooth from the canine, or from the second premolar. The third upper premolar is about equal in size to the second: the corresponding tooth in the lower jaw is smaller than the second lower premolar. The incisive openings of the palate are much larger than usual, extending backwards so as to terminate opposite the second premolar.

PHASCOGALE (*Antechinus*) APICALIS.

Freckled Phascogale.

Phascogale apicalis.	GRAY, Annals and Magazine of Natural History, vol. ix. p, 518. 1842.
Antechinus apicalis.	GRAY, List of the Mammalia in the British Museum, 1843, p. 99.
" "	Freckled Antechinus. GOULD, Mammals of Australia, Pt. 1, Pl. 11.

General colour grey-brown—of a very rich brown hue on the hinder parts of the body; on the head and fore parts of the body distinctly freckled with black and white; under parts dirty yellow-white; fore legs of a bright rust colour; hind legs less distinctly tinted externally with the same colour: tail clothed at the root with hairs like those on the body, but with the hairs becoming gradually shorter towards the apex, where they are black.

Inhabits Western Australia.

This species is found in the Swan River district, and at King George's Sound, and, indeed, appears to be pretty generally distributed over Western Australia. Mr. Neill informs me that it is called the "Dibbler" at King George's Sound, and I perceive, upon consulting Mr. Gould's beautiful work, that that gentleman has received it from the same district with the name just mentioned. Mr. Gilbert, in his notes, quoted by Mr. Gould, states that he found insect remains in the stomachs of specimens which he dissected, and that, while at King George's Sound, he obtained a female specimen having seven young attached; they were little more than half an inch in length, quite naked, and blind. "Above the mammæ of the mother, observes Mr. Gilbert, is a very small fold of skin, from which the long hairs of the under surface spread downwards, and effectually cover and protect the young. This fold in the skin of the abdomen is the only approximation to a pouch which I have found in any of the members of this genus. The young are very tenacious of life: those above mentioned lived nearly two days attached to the mammæ of the dead mother.

The black hairs of the end of the tail form a small pencil or tuft; hence the name *apicalis* was applied to this species, but the tail can scarcely be said to differ in this respect from other species of the Antechinus section; the basal part of the tail, however, is clothed with longer hairs than usual. The English name given by Mr. Gould I have adopted as calling attention to a striking characteristic of the *P. apicalis*, when compared with other Phascogales, viz. its freckled appearance. On the back the hairs are of a very dark grey at the root; each hair is of a yellowish brown colour in the middle, annulated with white or rusty white below the point, and black at the point: on the hinder part of the back the white is replaced by rusty yellow, and hence the general tint is here darker and richer. On the under parts of the body the hairs are yellowish white, but faintly tinted with grey at the root. The ears are rather small, and tolerably well clothed, and the hairs are for the most part black and yellowish. The fore legs are always suffused with rust colour externally—usually very bright: the hind legs are also tinted with rust colour, but less distinctly than the anterior limbs. The feet are of a dirty yellowish white colour: the whole sole of the foot is covered with small but distinct tubercles, with the exception of the larger fleshy pads, which are in part transversely striated, and the under sides of the toes, which present numerous strong transverse incisions.

	Inches.	Lines.	Inches.	Lines.
Length from tip of nose to root of tail	6	0	6	8
" of tail	3	2	4	0
" from nose to ear	1	4	1	6
" of ear		4⅓		5
Width of ditto		6		
Length of hind foot and nails ...		11½	1	0

The dimensions in the first column are from a female specimen in the British Museum; those in the second, are from a very large male in Mr. Gould's collection.

PHASCOGALE (*Antechinus*) FLAVIPES.

Yellow-footed Phascogale.

Phascogale flavipes. WATERHOUSE, Proc. Zool. Soc. for 1837, p. 75.
" *rufogaster.* GRAY, App. Grey's Journal, p. 407.
Antechinus flavipes. GRAY, List of the Mammalia in the British Museum, 1843, p. 99.
? *Antechinus Stuartii.* MACLEAY, Annals and Mag. Nat. Hist., vol. viii. p. 242, and p. 338.

General colour of upper parts grey, but on hinder parts of back tinted with fulvous; sides of body washed, as it were, with bright rusty yellow, or with rust colour; feet, and under parts of body, bright rust colour, or rusty yellow; chin, throat, and chest, whitish: tail black; freckled with yellow towards the base above; under parts, at base, rusty yellow.

Inhabits New South Wales and South Australia.

The fur of *P. flavipes* is moderately long, and soft, of a deep grey colour next the skin, but blackish, and, for the most part, freckled with yellow externally; that on the belly is usually of a brilliant yellow-rust tint, but is sometimes palish ochreous, and the colour of the feet is the same in either case. The tail is clothed throughout with short hairs, but at the point they are a trifle longer, and form a small tuft.

	Inches.	Lines.
Length from tip of nose to root of tail ...	5	4
" of tail	3	2
" from nose to ear	1	1½
" of ear		5½
" of tarsus		9½

The fleshy pads on the under surface of the feet are transversely striated; the outer tarsal pad is very long, extending from the base of the inner toe, or thumb, to the proximal extremity of the metatarsal bones: the remaining naked portions of the foot are apparently smooth, or very indistinctly tuberculated. The skull is rather short, and broad, and very flat above; the palate presents two short incisive foramina, and two posterior palatal openings situated opposite the penultimate and antepenultimate molar teeth. Its dimensions are given hereafter.

Antechinus Stuartii. MacLeay.

Under the above generic and specific names Mr. W. S. MacLeay characterises an animal, obtained at Spring Cove, near Sydney, which, I strongly suspect, will prove to be the *Phascogale flavipes;* the very short specific description given by Mr. MacLeay, however, does not enable me to satisfy myself upon this point. The animal is said to be of a fulvous colour, but having the limbs, internally, and the abdomen whitish: the tail nearly equal to the body in length; and the whole length of the animal, to the tip of the tail, 9½ inches.

Mr. MacLeay's original description appeared in the Annals and Magazine of Natural History for December, 1841, (vol. viii. p. 242), and is drawn up from a figure selected from a number of drawings made by Mr. Stuart (a surgeon in the army), combined with some notes by the same gentleman. From these it appeared the animal agreed with *Phascogale* generally, but was remarkable for the possession of but six incisors in the upper jaw, and thus afforded an exception to the dental formula of other carnivorous or insectivorous Marsupialia. Subsequently, however, Mr. MacLeay had an opportunity of examining the skeleton of the animal in question, and discovered the true dental formula to be as in *Phascogale*[1]. The animal, however, differs, Mr. MacLeay observes, "in the three lateral incisors of the upper jaw being of equal size, and also in the pseudomolars being all of equal size."

PHASCOGALE (*Antechinus*) LEUCOGASTER.

White-bellied Phascogale.

Phascogale leucogaster. GRAY, Append. Grey's Journal, p. 407.
Antechinus leucogaster. GRAY, in List of the Mammalia in the collection of the British Museum, 1843, p.

Grey; hinder part of back tinted with rusty brown: under parts white: feet dusky white; tail dusky above and beneath, but blackish at the apex; ears rather large, and sparingly clothed, for the most part, with minute pale hairs.

This animal so greatly resembles the *Phascogale flavipes* in its proportions, as well as in the structure of its skull and teeth, that it is with considerable hesitation I describe it as

[1] See Annals and Mag. Nat. Hist. for January, 1842, vol. viii. p. 338.

a distinct species. I have seen, however, several specimens from Western Australia which agree perfectly with the individual from which Mr. Gray drew up his original description, and differ from the *P. flavipes* in having the under parts of the body white, and in having little or no rusty yellow on the sides of the body, and on the feet. The general tint of the upper parts of the body likewise differs somewhat, being less grey: on the fore parts of the body it is brownish grey, and on the hinder parts rich brown. The feet are brownish white; not unfrequently suffused with brown behind. The tail is brown above, pale-brown beneath, and dusky towards the point.

A skull, removed from a specimen sent me from King George's Sound by Mr. Neill, differs from a skull of *P. flavipes* in the British Museum collection, in having the muzzle (and consequently the nasal bones) a trifle shorter, but the difference is not more than is often found in individuals of the same species, and I think it by no means improbable that the *P. leucogaster* is a local variety of *P. flavipes*.

	Ins.	Lines*	Ins.	Lines†	Ins.	Lines‡
Length from tip of nose to root of tail ...	4	5	4	9	4	6
" of tail	2	10	3	2	3	2
" from nose to ear		11½		11½		11
" of ear		5		5		5
" of hind foot and nails		8		9		9¼

* From specimen in the British Museum, from the neighbourhood of the Canning River, Western Australia.

† From specimen in the same collection from Victoria Plains.

‡ From a male specimen found at King George's Sound, and presented to the author by Mr. Neill.

PHASCOGALE (*Antechinus*) MINIMA.

Geoffroy's Phascogale.

Dasyurus minimus.	GEOFFROY, Annales du Muséum, tom. iii. p. 362.
" "	SCHREBER, Saugth. Suppl. Tab. 152, B. *e*.
Phascogale minima.	TEMMINCK, Monographies de Mammalogie, tom. i. p. 59.
" *affinis.*	GRAY, Appendix to Grey's Journal of Two Expeditions of Discovery in Australia, vol. ii. p. 406.

Brown, suffused with rusty yellow, the latter tint very distinct on the hinder parts of the body; under parts palish buff-yellow, as well as the feet, inner surface of the ears, and tail beneath; upper surface of the tail dusky brown.

Inhabits Tasmania.

The fur of this animal is long, and moderately soft; the hairs of the back are grey, broadly ringed with rich rusty yellow near the point, and dusky at the point; the longer interspersed hairs have the visible portion black; the general tint produced by this admixture of colours is rich yellowish brown; a brown hue prevails on the upper surface of the head, and the hinder part of the body is of a very deep reddish yellow tint: on the under parts of the body the hairs are grey at the base, and of a palish buff-yellow externally. The ears are small, and have the hinder margin nearly straight; they are densely clothed with minute yellowish hairs, both internally and externally. The tail is about as long as the body, well clothed with short hairs, of a dusky brown colour on the upper surface, but of a dirty yellow hue on the sides and under surface. The feet are very nearly of the same tint; the claws of the fore feet are rather long.

		Inches.	Lines.
Length from tip of nose to root of tail	...	3	10
" of tail	...	2	6
" from nose to ear	...		$10\frac{1}{4}$
" of ear	...		$2\frac{3}{4}$
" of fore foot and claws ...	...		$5\frac{3}{4}$
" of claw of middle toe	...		$1\frac{3}{4}$
" of hind foot and claws ...	...		$9\frac{1}{3}$

Formerly I was led, upon Mr. Gould's authority, to link together, as names of the same species, *Phascogale Swainsonii* and *P. minima;* having since, however, had reason to doubt the accuracy of this identification, I took with me to Paris the animal to which I had given the name *Swainsonii*, and compared it with Geoffroy's *Dasyurus minimus.* I found the two animals to be very distinct, and that the latter was a species exceedingly close to, if not identical with, either the *Phas. flavipes* or the *P. affinis.* From *P. flavipes* it appeared to me to differ in having the ears proportionately smaller, and the claws of the feet rather larger; and this I find, upon comparing dimensions, to be actually the case, and I also find that these differences distinguish the *P. affinis* from *P. flavipes.* In fact, I can discover no points of distinction between the *Dasyurus minimus* of Geoffroy and the *Phascogale affinis*, excepting that the former is of smaller size, and has the fur differently coloured; and when we bear in mind that the little animal described by Geoffroy has been exposed to the action of light in a museum for upwards of forty years, we cannot but suppose its colouring has changed. The difference of size, I need scarcely say, may arise from difference of age in the specimens when captured, and hence it does not appear that we have good grounds for separating, as distinct species, these two animals, which may have been found within about five miles of the same spot, the one being from Tasman's Peninsula, and the other from Maria Island. The following description is from the animal found in Tasman's Peninsula.

Phascogale affinis. GRAY.

Fur long, and moderately soft; on the head and fore part of the back, brownish grey, slightly tinted with yellow; on the hinder parts of the body, of a deep fulvous brown, and on the under parts of a very pale yellowish hue. The feet are of a dusky brown colour, and so is the upper surface of the tail, but near the base it is somewhat pencilled with yellow. The ears are clothed with small hairs, which are for the most part yellowish. The für, both on the upper and under parts of the body, is of a deep slate-grey colour next the skin; the hairs of the back are annulated with yellowish towards the point, and black at the point; or, on the hinder parts of the back, the yellow is replaced by a rich brown colour.

	Inches.	Lines.
Length from tip of nose to root of tail ...	5	6
" of tail	2	10
" from nose to ear	1	3
" of ear		4
" of hind foot and nails		10

PHASCOGALE (*Antechinus*) ALBIPES.

The White-footed Phascogale.

Phascogale albipes. WATERHOUSE, Proceedings of the Zoological Society for March, 1842, Pt. 10, p. 48.

Fur moderate as to length, and very soft; on the upper parts of the body brownish grey, and on the under parts white. Tail dirty white beneath, and of a dusky ash colour above. Feet white, slender, and with the naked portions entirely covered with minute tubercles. Ears large.

Inhabits South Australia (Western Australia and Van Diemen's Land ?)

The White-footed Phascogale was discovered by the late J. B. Harvey, Esq., a very zealous corresponding Member of the Zoological Society, who for many years exerted himself in collecting specimens to enrich that Society's Museum[1]. In size and colouring this little animal greatly resembles the Field Mouse (***Mus sylvaticus*** of authors) : its form is less robust than that of any of the Phascogales already described; its feet are more slender, and a greater portion of the palm of the hind foot is clothed with fur. In neither of the preceding species does the hair on the under side of the foot extend beyond the heel; terminating in front at the proximal extremity of the metatarsal bones. In ***P. albipes*** the hair encroaches on either side of the sole of the foot (and more especially on the outer side), so as to contract the naked portion—most so behind, and gradually less towards the fore part. In all the preceding species, the fleshy pads[2] on the under surface of the foot, at the base of the toes, are transversely or obliquely striated, whilst in the present, and other species about to be described, they are covered with small tubercles, as well as the other naked parts of the foot: on the toes the tubercles are arranged in two longitudinal rows.

[1] Mr. Harvey is also the author of some communications printed in the Zoological Society's publications.

[2] With regard to the pads in question, it will be found that usually in quadrupeds which have five toes to their feet, there are three of these pads, or fleshy tubercles, situated near the base of the four outer toes, and a fourth placed near the root of the inner toe : they are evidently destined to protect the under surface of the toes from pressure, which, by hardening the skin, would impede the motions of those members. Other pads there are on the under side of the foot which serve to protect the blood-vessels. These latter might be called *carpal* or *tarsal pads*, and the former, digital pads.

The fur, both on the upper and under parts of the body, is of a deep slate grey colour next the skin: on the upper parts, the hairs are of a very pale yellow colour near the point, and black at the point; those on the under parts of the body are white at the point. The eyes are encircled with black. The ears, which are large, are clothed throughout with minute hairs, chiefly of a pale hue, but dusky on the outer surface near the anterior margin. The tail is clothed with very small hairs (smaller than in either of the preceding species), and these are of a dirty white colour on the under surface, and partly black and party yellow-white on the upper suface.

	Inches.	Lines.
Length from tip of nose to root of tail ...	3	9
" of tail	3	2
" from nose to ear		10½
" of ear		6
" of tarsus and nails		8½

The skull in *P. albipes* is proportionately narrower, and less depressed, than in *P. flavipes* and its allies, and the palate has an extra pair of openings—they are situated entirely in the palatine bone. The principal palatine openings are situated opposite the first, second, and third true molars.

PHASCOGALE (*Antechinus*) LEUCOPUS.

Tasmanian White-footed Phascogale.

Phascogale leucopus. GRAY, Annals and Magazine of Natural History for December, 1842, vol. x. p. 261.

Fur very soft, and rather long: general colour grey, much suffused with black on the back of the animal, and very finely pencilled with pale yellow; the yellow most distinct on the head and sides of the body: feet and under parts of the body white: ears tolerably large, and clothed with minute hairs, for the most part dusky, but pale at the basal portion of the ear externally: upper surface of the tail nearly black; under surface dirty white.

Inhabits Van Diemen's Land.

The general tint of this animal is somewhat darker than that of *Phascogale albipes;* the upper surface of the tail is almost black, whilst in the species just mentioned it is greyish, and the ears are smaller. Beyond these, I can perceive no other points of distinction between the Van Diemen's Land animal and the continental one, *P. albipes.* Of the former I have seen but one specimen, and I can scarcely satisfy myself, from such imperfect materials as are before me, that these White-footed *Phascogales* are specifically distinct. *P. leucopus* presents the following dimensions: —

	Male.	
	Inches.	Lines.
Length from tip of nose to root of tail ...	4	4
" of tail	3	7
" of ear		5
" of hind foot and nails		8¾

A small Phascogale is found at King George's Sound which agrees very closely with the *P. leucopus,* being of the same dark colour, and having the tail black above, or nearly so. Two specimens in Mr. Gould's collection, thus resembling the Van Diemen's Land animal, differ, however, in having the chest of a dusky grey hue: their dimensions are given in the first and second columns of admeasurements which follow. In the third column I have added the dimensions of a specimen from the same quarter, and which is preserved in spirits[1]; it appears to have had the same colouring. A

[1] In this specimen I have found no trace of a cœcum, but in the *Phascogale leucogaster* a small hemispherical projection seemed to mark the line of

fourth specimen from King George's Sound, and contained in the British Museum collection, differs in having the colouring less dark, and, indeed, very closely resembling that of *Phascogale albipes ;* its size and proportions may be gleaned from an inspection of the fourth column.

	Male.		Female.		Male.			
	Ins.	Lines.	Ins.	Lines.	Ins.	Lines.	Ins.	Lines.
Length from nose to root of tail...	3	9	3	8	3	6	3	10
" of tail	3	2	3	0	3	3	3	1
" from nose to ear	1	0		11½		10½		
" of ear		5		5		5		4¾
" of hind foot and nails ...		8		7⅔		8		8

PHASCOGALE (*Antechinus*) MURINA.

Murine Phascogale.

Phascogale murina. Waterhouse, Proceedings of the Zoological Society for July, 1837, Pt. 5, p. 76. Naturalists' Library, vol. xi. (Marsupialia), p. 143, Pl. 10.

Fur very soft: upper parts of the body ashy grey, under parts white; feet and tail also white; tail silvery white; ears moderate; clothed with minute pale hairs, but brownish in front on the outer surface. Eyes encircled with black.

Inhabits New South Wales.

The *Phascogale murina* is considerably smaller than *P. albipes ;* its general colouring is paler, and its tail is uni-

division between the small and large intestines, and may be regarded as representing a cœcum in a very rudimentary condition. Around the base of the swelling in question, the wall of the intestine was slightly incrassated.

formly white. The tarsi are almost entirely covered with hair on the under side, there being only a very narrow naked space running backwards from the digital pads, which are covered with small tubercles. Two longitudinal rows of small tubercles are observable on the under side of each toe. The fur, both on the upper and under parts of the body, is grey next the skin.

	Inches.	Lines.
Length from tip of nose to root of tail ...	3	3
" of tail	2	10
" of ear		$4\frac{1}{2}$
" of hind foot and nails		$7\frac{2}{3}$

PHASCOGALE (*Antechinus*) MACROURA.

Ash-coloured Phascogale.

Podabrus macrourus. GOULD, Proceedings of the Zoological Society for June, 1845.

Fur moderately long, and soft; on the upper parts of the body ashy grey, somewhat suffused with black on the back and mesial line of the head, on the under parts of the body white, and on the sides of the body tinted with cream-colour: feet white: tail with the basal half much thickened—in the male sex at least; clothed throughout with small hairs, above partly black and partly yellow on the basal portion, but entirely black on the apical portion; beneath dirty white: ears moderately large, clothed with whitish hairs, but dusky on the outer side, in front. Eye encircled with black. The fur on the upper and under parts of the body of a deep grey colour next the skin,

Inhabits New South Wales.

In Mr. Gould's collection are two specimens of this species which appear to me to be decidedly distinct from the *Phasco-*

gale crassicaudata. Like that animal it has the tail incrassated, a character which perhaps may be sexual, or even may not be constant in individuals of the same sex at different periods. Of the two specimens, one is decidedly a male, and this has the tail much incrassated; the other is apparently a female, and here the tail can scarcely be described as incrassated, though it is thicker than in the little slender-footed species, *P. albipes*, &c. From *P. crassicaudata* the present animal may be distinguished by its much larger size, the proportionately greater length of its tail, and its ashy-grey colouring, which presents no admixture of yellow, excepting on the sides of the body, where a delicate cream-coloured band separates the white of the under parts from the grey of the upper. I have assured myself that the difference of size between this species and the *P. crassicaudata* is not dependent upon either a difference of age or sex, having seen adult specimens of both sexes of *P. crassicaudata*, and I have scarcely a doubt that the two specimens of *P. macroura* are of different sexes.

	Male.		Female?	
	Inches.	Lines.	Inches.	Lines.
Length from tip of nose to root of tail	4	2	3	11
" of tail	3	3	2	10
" from nose to ear	1	0		11½
" of ear		5		4¾
" of hind foot and nails		8½		7½

In the male specimen the diameter of the tail is between three and four lines, near the base: from the middle it tapers rather suddenly to the apex. In the female the tail is rather more than two lines in diameter at the base.

PHASCOGALE (*Antechinus*) CRASSICAUDATA.

The Thick-tailed Phascogale.

(*Plate* 15, fig. 2.)

Phascogale crassicaudata. GOULD, Proceedings of the Zoological Society for June, 1844, Pt. 12, p. 105.
Podabrus crassicaudatus. GOULD, Mammals of Australia, Pt. 1, Pl. 5.

Fur long, dense, and extremely soft; on the upper parts of the body grey washed with yellow; on the sides of the body of a delicate yellow, and on the under parts white: feet, and under surface of the tail, white: ears large; sparingly clothed with very minute hairs, which are for the most part whitish, but on the outer surface is a large black patch extending inwards from the anterior margin.

Inhabits Western and Southern Australia.

Mr. Gould, in his Mammals of Australia, places this animal under the new generic name *Podabrus*, on account of its having slender and delicate feet, as compared with the ***Phascogale flavipes***, and its allies. Whatever may be the value of this new section, it certainly should include, as well as the ***P. crassicaudata,*** the ***P. albipes, P. leucopus, P. murina,*** and ***P. macroura*** and the last mentioned animal approaches most nearly, in having the tail incrassated, to the present species. All these ***Phascogales*** have the slender feet, the heel more clothed with hair than in others of the group, and the pads on the under side of the foot covered with minute tubercles, instead of being striated, as represented in the foot of ***P. penicillata*** (see Pl. 12, fig. 2.)

The fur of the Thick-tailed Phascogale is of a deep lead-grey colour next the skin; on the under parts of the body each hair is white at the point; on the sides, the hairs are of a delicate yellow at the point, and on the back they are yellow near the point, and black at the extremity. The ears are large, slightly emarginated behind, and somewhat pointed at the apex: the flesh of the ears appears to have been dark in the living animal, but in the middle of each ear is a spot which was apparently of a pale flesh colour. The eye is encircled with black; and in front of that organ the black is considerably extended. The tail is short, and, in the specimen described by Mr. Gould, is about 2½ lines in diameter in the middle: this specimen is a male, and is from William's River, Western Australia. In a second specimen, apparently a female, and which is from South Australia, the tail is scarcely incrassated. Both specimens are in the British Museum collection: the dimensions of the first mentioned of these, were taken by Mr. Gilbert before the animal was skinned, and are—

	Inches.	Lines.
Length from tip of nose to root of tail ...	3	9¾
" of tail	1	9⅔
" from nose to ear		10½
" from ditto to eye		6
" of ear		9
" of hind foot and nails		7*

PHASCOGALE (*Antechinus*) MELAS.

New Guinea Phascogale.

Phascogale melas. SAL. MÜLLER, Verhandelingen over der Naturalijke Geschiedenis der Nederlandsche overzeeschte Beritingen—Zoologie, Tab. 25, figs. 1-3.

Fur black: ears short, and clothed with small hairs.

Inhabits New Guinea.

* Added from the stuffed specimen.

This animal is said to be of the size of the Black Rat (*Mus Rattus*), and of a nearly uniform black hue: the short and soft fur of the back is rather glossy; on the under parts of the body the fur is of a less pure black colour, assuming a slight rusty hue. The ears, feet, and tail, are clothed with short hairs. The eyes are brown.

According to Dr. Müller's figures, the *P. melas* is about 8 inches in length; its tail $6\frac{1}{4}$ inches; hind foot and nails 1 inch 5 lines; and the ear half an inch. The portion of the skull represented shows that the nasal bones are rather suddenly expanded behind. The third molar tooth of the upper jaw is represented as rather smaller than the preceding premolars, and the corresponding tooth of the lower jaw is very small, as in *Phascogale calura* and *P. penicillata*.

In the following table I have given the dimensions of the skulls, or parts of skulls, of the different species of Phascogale which I have had an opportunity of examining.

	Phascogale penicillata.		*P. calura.*		*P. Swainsonii.*		*P. apicalis.*		*P. flavipes.*		*P. leucogaster,* MALE. King George's Sound.		*P. albipes.*		*P. leucopus.*		*P. murina.*		*P. crassicaudata.*	
	Ins.	Lines.	Ins.	Lines.	Ins.	Lines.	Ins.	Lines.	Ins.	Lines.	Ins.	Lines.	Ins.	Lines.	Ins.	Lines.	Ins.	Lines.	Ins.	Lines.
Length of skull	1	11	1	$2\frac{1}{2}$		*			1	2	1	$1\frac{1}{2}$	1	$0\frac{3}{4}$						
Width of ditto	1	1		$8\frac{1}{2}$						$8\frac{1}{2}$		$8\frac{1}{4}$		$7\frac{1}{4}$						
" between orbits ...		4		3		4		$3\frac{1}{2}$		3		3		$2\frac{2}{3}$		3		$2\frac{1}{2}$		
Length of nasal bones		$8\frac{1}{3}$		$4\frac{3}{4}$		$6\frac{1}{4}$				$5\frac{1}{2}$		$4\frac{3}{4}$		5				$3\frac{2}{3}$		
Width of ditto behind		$3\frac{1}{2}$		2		2				$2\frac{1}{4}$		2		$1\frac{1}{4}$				1		
From front of foremost incisor to back of last molar tooth		$11\frac{3}{4}$		$7\frac{1}{4}$		8		8		$7\frac{1}{2}$		$6\frac{3}{4}$		$6\frac{1}{4}$		6		$5\frac{2}{3}$		$5\frac{3}{4}$
Length of four upper true molar teeth taken together		5		$3\frac{1}{2}$		$3\frac{1}{3}$		$3\frac{3}{4}$		$3\frac{1}{3}$		$3\frac{1}{4}$		$2\frac{3}{4}$		$2\frac{3}{5}$		$2\frac{1}{2}$		$2\frac{1}{2}$
" of palate		$11\frac{1}{2}$		$6\frac{3}{4}$				$8\frac{1}{3}$		$7\frac{1}{3}$		$6\frac{1}{2}$		$6\frac{1}{4}$		6		$5\frac{1}{2}$		
" of lower jaw, to back of condyle	1	$5\frac{2}{3}$		10				$11\frac{3}{4}$		$10\frac{1}{3}$		$9\frac{3}{4}$		$9\frac{1}{2}$		9		$8\frac{1}{4}$		8
" of four true molar teeth of ditto taken together		$5\frac{3}{4}$		$3\frac{3}{4}$		$3\frac{2}{3}$		$4\frac{1}{4}$		$3\frac{2}{3}$		$3\frac{1}{2}$		$2\frac{5}{6}$		3		$2\frac{2}{3}$		$2\frac{4}{5}$

* The only skull of *P. Swainsonii* I have seen has the hinder part imperfect; hence I cannot give its exact length: but, taking the proportions of the other parts as my guide, it must be *very* nearly 1 inch 4 lines; from the tip of the intermaxillaries to the back of the frontal bones, it measures $11\frac{1}{4}$ lines: the width of the palate between the last molars is $2\frac{3}{4}$: the width of the skull must be about 7 lines.

Genus *Dasyurus*.

Dasyurus. GEOFFROY, Annals du Müséum, tom. iii. p. 333. 1804.
" TEMMINCK, Monographies de Mammalogie, tom. i. p. 66. 1827.

Dasyuridæ with equal incisor teeth, $\frac{8}{6}$; canines, $\frac{1-1}{1-1}$; premolars, $\frac{2-2}{2-2}$; true molars, $\frac{4-4}{4-4} = 42$. Tail usually long, and well clothed with long, or moderately long hairs[1].

This section contains flesh-eating animals, all of which are of moderate size, and nearly all have the fur spotted. They have been compared to the Weasels and Martins of the order *Carnivora*, and, indeed, such is the general resemblance of these carnivorous Marsupialia to the ordinary carnivora, that some naturalists have associated them with that group; the resemblance, however, is a superficial one, and when the species of the two groups are compared in detail, many important differences present themselves. Many of these differences have already been alluded to, since they include characters which separate the order Marsupialia from other orders. I will now notice one or two points connected with the structure of the feet, as serving to illustrate the foregoing remarks.

In the order Carnivora I am acquainted with no species in which the scaphoid and lunare (two of the upper row of wrist bones) are not joined so as to form a single bone: the carnivorous Marsupials differ from the ordinary carnivora in having the bones in question distinct. In the true carnivora the astragalus articulates with the heel bone, or calcaneum, by two surfaces, between which there is a strong groove; this groove is not found in the Dasyurus, where the articular surface is continuous[2].

[1] Hence the generic name; from δασὺς, and οὔρος.

[2] The Thylacinus, which is merely a modified Dasyure, and which has been compared to the Dog, differs more from the Dog in the structure of the astragalus, than does the Dog from a Monkey.

The teeth of the Dasyuri differ but little from those of the Phascogales, which have been described in detail. The most striking difference consists in the reduced number of premolars, of which there is one less on each side of each jaw. The incisors form a continuous series in either jaw, the foremost pair of the upper jaw not being thrown outwards and forwards as in the Phascogales; the incisors, moreover, in the Dasyuri are very nearly equal in size. With regard to the true molars, the cusps are rather less sharply pointed than in the Phascogales; I can perceive no other difference in the molars of the species which I shall first describe; in the *Dasyurus ursinus* the teeth present some modifications which will be noticed when that animal comes under consideration. Some of the Dasyuri have no inner toe to the hind feet; we will commence with those which, like the Phascogales, have a small, nailless, and thumb-like inner toe to the hind foot.

DASYURUS HALLUCATUS.

North Australian Dasyurus.

Dasyurus hallucatus. GOULD, Proceedings of the Zoological Society for February, 1842, Pt. 10, p. 41.

Fur of moderate length, and rather harsh ; general colour of the upper parts of the body dusky brown, much pencilled with yellowish, and having numerous irregular white spots ; under parts white, suffused with yellow : ear pale ; the fur which covers the root of the ear externally, whitish ; tail but little bushy, cylindrical, the apical half, or more, black ; fleshy pads on the under surface of the feet with minute oblique striæ.

Inhabits North Australia—Port Essington.

This is the smallest species of the true Dasyures, being a trifle less than the *D. Maugei* or *D. hallucatus* With the latter animal it might be confounded, having, like it, a thumb to the hind foot; upon a close examination, however, I discovered several characters by which it may be easily distinguished. It is of smaller size[1] than *D. Geoffroyi*, of a darker colour, the ears have the flesh of a paler colour, and they are clothed with pale hairs; the longer hairs which cover the root of the ear outside are whitish; the toes of the hind foot are longer, since I found them to be seven lines in length in *hallucatus*, and only 6¼ in a specimen of *Geoffroyi* which was of the same sex, and of considerably larger size; and, lastly, I find the whole sole, both of fore and hind feet in *D. Geoffroyi*, covered with minute, but distinct fleshy tubercles, (as is also the case in *D. viverrinus*), whilst in *D. hallucatus* I could scarcely perceive a trace of tubercles, and the fleshy pads at the base of the toes, and elsewhere, on which the tubercles were most distinct in *Geoffroyi*, are covered with numerous oblique or transverse grooves. The pads, moreover, at the base of the toes, were much narrower, and proportionately longer. I was led to examine these parts upon finding similar little striæ on the pads of the feet of many of the *Phascogales*.

The following description was drawn up from a female specimen in Mr. Gould's collection:—

Upper parts of body dusky brown, inclining to black, but pretty freely pencilled with yellowish, and having numerous irregular, and moderate sized, white spots, which extend likewise on the sides of the body; on the crown of the head are a few very small white spots; the under parts of the body are white, but suffused with yellowish; most distinctly so

[1] I have examined three specimens, and both sexes, of *D. hallucatus*, and many specimens of the *D. Geoffroyi*, also of both sexes.

about the throat; the cheeks, a large patch above the eye, and the sides of the body, are greyish; ears very pale (probably whitish flesh-colour in the living animal), thinly clothed with small pale hairs; immediately at the base externally the hairs are longer and dense, and of a yellowish white colour, and the part of the head immediately adjoining the root of the ear has similar pale hairs. The tail is immaculate, cylindrical; clothed throughout with longish harsh hairs, but by no means bushy; about one-third (basal portion) is brownish, but considerably pencilled with black, and the remaining portion is almost entirely black. The feet are brownish, and the region of the pouch is clothed with very dark red hairs, appearing as if stained with blood. The fur is less dense, and harsher than in *D. Geoffroyi*.

A female specimen in the British Museum collection differed from the above in having the general hue of the upper parts of the body yellowish brown—the ground colour being of a tolerably bright brown, but much pencilled with yellow; on the head and sides of the body the yellow prevails. A male specimen in the same collection agrees with my first description, in having the ground colour of the back almost black, and a similar specimen in the national collection, there are reasons to believe, is from Western Australia.

	Inches.	Lines.
Length from tip of nose to root of tail ...	12	0
" of tail	10	0
" of tarsus and nails	1	0
" of the longest toes and nails of the hind foot		7
" of ear	1	10
" from tip of nose to ear	2	3

DASYURUS GEOFFROYI.

Geoffroy's Dasyurus.

Dasyurus Geoffroyi. GOULD, Proceedings of the Zoological Society for November, 1840, Pt. 8, p. 151.

Fur moderate; general colour of the upper parts yellowish, pencilled with black, and having numerous irregular white spots; body beneath white; tail immaculate, black at the apex: hind foot with a distinct, but small inner toe, or thumb.

Inhabits Western Australia, Southern Australia, and New South Wales.

The present species somewhat resembles the pale variety of *Dasyurus viverrinus*, to which Geoffroy applied the specific name *Maugei*, but is distinguished most readily by its tail being much less bushy, and the hind foot being furnished with an inner toe, or thumb. The fur is moderately long, rather soft, and on the upper parts of the body of a greyish hue, but much suffused with yellow, and pencilled with black; and these parts, moreover, as well as the sides of the body, are adorned with numerous irregular white spots: the head has a few small white spots only, and is often of a greyer hue than other parts, but the muzzle is somewhat tinted with brownish, and in front of the eye is a dusky patch; the ears are dusky brown, and clothed with minute blackish brown hairs externally; internally, with longish, pencilled black and grey hairs, at and near the anterior margin, but towards the apex, and on the hinder parts, the hairs are minute and brownish: the under parts of the body are white, or very nearly so; the fore feet are brownish—sometimes brown-white; the hind feet are nearly white, or greyish suffused with yellow: the tail is yellowish at the base, but much pencilled with black, the ends of the hairs being of that colour; the black gradually increases towards the tip of the tail, and usually about one-third is entirely black.

	Male.		Female.		Male.	
	Ins.	Lines.	Ins.	Lines.	Ins.	Lines.
Length from tip of nose to root of tail ...	14	0	13	0	16	0
" of tail	10	0	8	6	12	0
" of tarsi, including toes and nails ...	2	4	2	3	2	2
" from nose to ear	2	5	2	4	2	7
" of ear	1	2	1	2	1	0

The British Museum collection contains specimens of this species from Liverpool Plains, from the brushes on the Murray, and from the neighbourhood of Perth. The dimensions in the third column are taken from an unstuffed specimen, from Liverpool plains, in Mr. Gould's collection: a female specimen, also belonging to Mr. Gould, and which appeared to be adult, the teats being very large, was rather smaller. The mammæ were six in number.

The anterior portion of a skull of the *Dasyurus Geoffroyi* contained in the British Museum collection, agrees very nearly in size with the corresponding part of the skull of *D. viverrinus*, but differs in having the zygomatic arches thrown more boldly outwards, and in having the muzzle and nasal bones shorter. Its dimensions are as follows:—

	Inches.	Lines.
Length from the hinder root of the zygoma to the apex of the intermaxillary bones ...	2	5¼
Greatest width—which is rather behind the middle of the zygomatic arch	1	9
Depth of the zygoma behind		3¾

Width between orbits		8
" behind orbits		5
Length from anterior angle of orbit to apex of intermaxillaries	1	1¼
" of nasal bones		11⅔
Width of ditto behind		5½
" in front		2⅔
Length of palate	1	6⅓
Width of ditto between the hindermost molars		7⅔
Length of the four true molars taken together	0	8¼
Length of lower jaw, measuring to the back part of the condyle	2	3⅓
Height of ditto from the apex of the coronoid process		10¾

DASYURUS MACULATUS.

Spotted-tailed Dasyurus.

The Spotted Martin. PHILLIP's Voyage to Botany Bay, p. 276. Martin, Cat. Pl. 46. 1789.

Viverra maculata. SHAW, General Zoology, vol. i. Pt. 2, p. 433. 1800.

Dasyurus macrourus. GEOFFROY, Ann. du Mus. tom. iii. p. 358. 1804.

" " PERON et LESUEUR, Voyage aux terres Australes, Pl. 33.

" " TEMMINCK, Monog. de Mammalogie, tom. i. p. 69. 1827.

" " WATERHOUSE, Naturalists' Library (Marsupialia), vol. xi. p. 130, Pl. 6.

" *maculatus*, Spotted-tailed Dasyure. GRAY, List of the Mammalia in the British Museum, 1843, p. 98.

Fur rather coarse, and by no means long: general colour reddish brown, or deep brown, pencilled with yellowish: body beneath yellowish: body and tail spotted with white.

Inhabits Van Diemen's Land.

In no known species of Dasyurus besides the present is the tail spotted as well as the body; hence the English name given by Mr. Gray to this animal at once draws attention to its most striking characteristic. Its muzzle is proportionately shorter, and more obtuse, than in most of the preceding species; the ears are also shorter; and the tail is longer. The superior size and more robust form will also help to distinguish the *D. maculatus* from other species in which the hind foot is furnished with a rudimentary inner toe.

The fur in the present animal is harsh to the touch, and rather short; its colour varies from a very deep brown to a rich red-brown: the head is always paler than the back, and sometimes assumes a yellowish hue, being much pencilled with this pale tint; other parts of the body are more or less pencilled with yellowish, and the whole under parts of the body, as well as the fore legs and feet, are of a dirty yellow; the upper lip, chin, and throat, are of a more pure yellow tint: the toes of the fore feet are yellowish. The hind legs externally, and the hind feet, scarcely differ in tint from the upper parts of the body. The tail is nearly equal in length to the head and body taken together; cylindrical, and clothed with tolerably long and harsh hairs; its general colour is the same as that of the body, or very nearly so. The ears are short, clothed internally for the most part with small yellowish hairs, but at the margin the hairs are longer, and near the anterior angle they are tolerably long; on the outer side the ears are of the same colour as the crown of the head. With regard to the white spots with which this animal is adorned, they vary considerably in different individuals, and are very irregular in size and form; they are observed on the whole of the upper parts and sides of the body; some few are also visible on the under parts and on the legs; the head is usually immaculate, or presents but two or three very small spots. The spots on the tail are often large, and never numerous.

	Inches.	Lines.	Inches.	Lines.
Length from tip of nose to root of tail	17	0	24	0
" of tail	15	0	20	0
" from nose to ear	3	0	4	3
" of ear	1	0	1	2
" of hind foot and nails ...	2	0	3	7

The admeasurements given in the first column are from a specimen in the museum of the Zoological Society; and a very large individual, apparently a male, furnished the dimensions in the second column. I will now add the dimensions of two skulls of the *D. maculatus* contained in the museum of the Royal College of Surgeons.

	Adult.		Aged.	
	Inches.	Lines.	Inches.	Lines.
Total length of skull			3	6½
Length from hinder root of zygomatic arch to apex of intermaxillaries ...	3	1	3	0
Width	2	2	2	2½
" between orbits		10		9¾
" immediately behind orbits ...		5½		5
Length of nasal bones	1	3		
Width of ditto behind		5¾		
" " in front		3¾		
Length of palate	1	11	1	10½
Width of ditto between hindermost molars		9⅓		
Length of four true molars taken together		9¾		9½
" of lower jaw			2	11
Height of ditto from apex of coronoid process			1	2½

The Spotted-tailed Dasyurus appears to be confined to Van Diemen's Land. The earliest notice of the animal is that contained in Phillip's Voyage, under the name of the Spotted Martin, a name which, it appears, misled Shaw in his location of the animal, it being described by him under the generic title *Viverra*. Geoffroy's specific name *macrourus* is generally adopted for this animal, but as I do not perceive that there were sufficient grounds for rejecting the earlier name of *maculatus*, I have followed Mr. Gray in restoring that appellation.

An account of the internal anatomy of the *D. maculatus*, or *macrourus*, will be found in the Proceedings of the Zoological Society for January, 1835, Part 3, p. 7. The specimen, which was dissected by Prof. Owen, was a female, and the Professor found the mammæ to be six in number, arranged three on either side, describing three quarters of a circle, and seated within a slight fold of integument of a corresponding shape.

DASYURUS VIVERRINUS.

The Viverrine Dasyurus.

The Spotted Opossum. PHILLIP, Voyage to Botany Bay, p. 147, Pl. 15. 1789.

Topoa Tafa. WHITE, Journal of a Voyage to New South Wales, p. 285, and Plate. 1790.

Didelphis viverrina. SHAW, General Zoology, vol. i. Pt. 2, p. 491, Pl. 111. 1800.

Dasyurus viverrinus, et *D. Maugei.* GEOFFROY, Annales du Muséum, tom. iii. pp. 359 and 360. 1804.

" " TEMMINCK, Monogr. de Mamm. pp. 71 and 72, Pl. 7, Figs. 1—8 ; skull and lower jaw.

Ears moderately large ; fur long, and rather soft ; tail bushy ; head and body spotted with white : general colour black, or brown-black—(*Dasyurus viverrinus* of authors) ; or grey, washed with yellow ; under parts of body and feet white ; tail whitish at the extremity—(*Das. Maugei* of authors).

In all systematic works up to a very recent period, the animal under consideration appears under two names—

Dasyurus Maugei and *D. viverrinus:* these names having been applied to animals differing much in the colouring of their fur, it seems never to have been suspected that they belonged to the same species, as has been proved by the fact that Mr. Gould found young individuals presenting both conditions in the same litter—that is, black and grey specimens[1]. As no individuals presenting an intermediate condition of colouring are found, I at first suspected the difference might be sexual, but such is not the case, since I have seen male specimens both of the black and grey varieties. The former vary only from brown-black to black; the under parts of the body, and the feet, are generally brownish. The fur on the back is grey next the skin, and that on the abdomen is also grey, but of a paler hue. The white spots on the body vary in size, some being very small, and others more than half an inch in diameter; on the head there are a few small white spots. The tail is bushy, being provided with long hairs averaging on the basal portion about an inch in length, but of double that length at the point; on the under surface they are, however, comparatively short: in length, the tail is about equal to the body. The ears are tolerably large, and somewhat attenuated at the apex; they are clothed with short black hairs; these are most abundant on the outer surface, but are also plentiful on the inner surface at the point and near the anterior angle, in which latter part the hairs are considerably longer than elsewhere. The flesh of the ear is of a pale pink colour in the living animal, as is also the naked tip of the nose and soles of the feet, the latter being also destitute of hair, but covered with small fleshy tubercles. There is no trace of an inner toe to the hind foot, unless it be a slight swelling of the flesh, marking the

[1] See Proceedings of the Zoological Society for November, 1840, Pt. 8, p. 151.

situation of the rudimentary metacarpal bone beneath[1]. The fore foot is provided with a remarkably large and prominent tubercle on the under side of the wrist, immediately behind which some long and bristly hairs have their origin.

I have before me very copious notes made upon a fine male specimen of the present species, which had lived for a considerable time in the menagerie of the Zoological Society; from these I have already made some extracts; I will now add the dimensions taken immediately after its death in December, 1839.

	Inches.	Lines.
Length from tip of nose to root of tail ..	15	0
" from nose to ear	3	0
" of ear	1	6
" of tail	8	6
" of fore foot and toes (the nails not included)	1	7
(or, with the nails)	1	9½
" of hind foot and toes	2	8
" of ditto, including the nails	2	10¾
Circumference of the body at the chest ...	8	6
Height of the animal at the shoulders, about	6	0

Such are the characters of the black variety of our animal, which is the *D. viverrinus* of authors; I will now briefly notice the grey variety.

Dasyurus Maugei. GEOFFROY.

The general colour of the fur is here greyish, but much suffused with yellow. Each hair of the ordinary fur on the upper parts of the body is of a pale grey colour at the root,

[1] In *D. maculatus* (and no doubt in *D. Geoffroyi* and *D. hallucatus*), where a small inner toe is visible, the skeleton of the foot presents not only the inner metatarsal bone, but this is provided with the ordinary two phalanges.

pale yellow near the point, and black at the point, and the coarser interspersed hairs have their visible portions almost entirely black; on the feet and under parts of the body the hairs are of an uniform yellowish white tint. The sides of the face are almost of an uniform pale yellow. The ears are for the most part rather sparingly clothed with pale hairs: at their base externally is a white spot. The tail is bushy, of the same general hue as the body at the base, but becomes gradually paler towards the opposite extremity, and is terminated with white or dirty yellow-white hairs.

	Inches.	Lines.
Length from tip of nose to root of tail ...	12	0
" of tail (not including the long hair) ...	9	0
" from tip of nose to ear	2	8
" of ear	1	3
" of hind foot (not including the nails) ...	2	1½

I have seen specimens whose dimensions slightly exceeded the above, which are taken from a stuffed specimen in the museum of the Zoological Society. Upon comparing the skins of the black and the grey specimens together, I could perceive no constant difference of size, as M. Temminck found in the specimens which came under his notice[1].

Both varieties of the Viverrine Dasyurus are common in New South Wales and Van Diemen's Land.

Subjoined are the dimensions of some skulls of the grey variety.

[1] This author observes, that the *D. viverrinus* is *constantly* smaller than the *D. Maugei*, but the dimensions taken by myself from a recent specimen of the former animal, denote a larger size than those given by M. Temminck from the *D. Maugei*.

	Adult, probably a Female.		Aged, probably a Male.		Adult.		Adult.	
	Ins.	Lines	Ins.	Lines	Ins.	Lines	Ins.	Lines
Length of skull	2	11½	3	2	2	11	2	10½
Width of ditto	1	8	1	10½	1	8	1	7⅓
" between orbits		8¾		9¾		8¾		7¾
" immediately behind the orbits		6		6		6		5⅔
Length of nasal bones	1	1¾	1	3½	1	0½	1	0¼
Width of ditto behind		5½		5¾		5		5¼
" " in front ...		2		3		2¼		2½
From anterior angle of orbit to apex of intermaxillaries .	1	3⅓	1	4½	1	3½	1	3
Length of palate	1	7	1	8⅔	1	7	1	7
Width of ditto between posterior molars		7½		8⅓		7⅔		
Length of four true molars taken together		8⅓		9		8½		8½
" of lower jaw			2	6			2	4
Height of ditto from apex of coronoid process			0	11½				10

Sub-genus, *Sarcophilus*.

Sarcophilus[1]. F. CUVIER, Mammifères, 70me livr. 1838.

Diabolus. GRAY, in Appendix to Grey's Journals of Two Expeditions in North-West, and Western Australia. 1841.

Dasyuri with the body stout; the head short, and very broad; the tail shorter than the body; premolar teeth with the antero-posterior and transverse diameters equal, or nearly so.

Unfortunately I am not aware what are the points which induced M. F. Cuvier[2], or Mr. Gray, to separate the *Dasyurus*

[1] From σὰρξ, flesh; and φίλος, loving.

[2] The great work of MM. F. Cuvier and Geoffroy, in which the section *Sarcophilus* is proposed, although in our National Library, is not complete, and is deficient of the part containing the account of the *Dasyurus ursinus*. The section *Diabolus* of Mr. Gray I do not find anywhere characterised.

ursinus of authors, from the other Dasyuri, but I presume they are derived from the proportions of the animal, which are more bulky than in other Dasyuri, and perhaps from the comparatively short tail. The form of the skull, and some slight modifications observable in the structure of the teeth, no doubt were also taken into consideration. With regard to the latter, the following points struck me upon comparing the teeth of *D. ursinus* with those of *D. maculatus.* The premolars, owing to the shortness of the muzzle in *D. ursinus,* are crowded together, whilst in Dasyurus proper they are somewhat isolated, and they differ, moreover, in having the transverse diameter as great as the longitudinal; the inner lobe of the upper true molar teeth is less developed, and the anterior lobe is smaller in proportion to the hinder one: in the foremost true molar this lobe presents but one point; in the second it is partially divided, and in the third true molar the anterior lobe is distictly divided The hindermost of the two external lobes differs from the corresponding lobe in the true molar of *D. maculatus,* in being considerably compressed, and the small tubercle, which, in the typical Dasyuri, is situated on the outer side of the tooth, and about midway between the anterior and posterior angles, is here brought near the mesial tubercle: in the hindermost molar, the tubercle in question is obliterated, as is the case in the molars of the Thylacinus. These observations refer only to the three foremost of the upper true molars; the hindermost molar is a narrow, transverse, and very simple tooth. The true molars of the lower jaw differ in having the posterior lobe, or heel, less developed, and these teeth want the inner tubercle, which is found midway between the extremities of the tooth of *Dasyurus viverrinus, D. Maugei,* &c.

DASYURUS (*Sarcophilus*) URSINUS.

Ursine Dasyurus[1].

Didelphis ursina.	HARRIS, Linnean Transactions, vol. ix. p. 176, Pl. 19, fig. 2.
Dasyurus ursinus.	GEOFFROY, Annales du Muséum, tom. xv. p. 305.
" "	TEMMINCK, Monographies de Mammalogie, tom. i. p. 69.
Diabolus ursinus.	GRAY, in Appendix to Grey's Journal. List of the Mammalia in the British Museum, 1843, p. 97.

Fur coarse, moderate as to length, and black, excepting on the head, tail, and under parts of the body, where it is brown-black: a broad white band usually crosses the chest, and

[1] The Ursine Dasyurus is called by the colonists "The Devil," or "Native Devil;" I see no necessity for adopting such barbarous names, nor do I think it desirable to adopt other names given by our colonists, when they convey an erroneous impression of the nature of the animals which have received them.

extends backwards on either side, more or less over the base of the fore leg; and a second band crosses the back, near the root of the tail.

Inhabits Van Diemen's Land.

The *Dasyurus ursinus* may be compared to a bear, in the general proportions of its body and limbs, as well as in the texture of its fur: in many of its actions, and in its gait, moreover, the original describer of this species, Mr. Harris, was struck with a resemblance between the two animals. The tail of the Dasyurus, however, is long, when compared with that of a bear. In size it is about equal to a badger.

To the short diagnosis already given, little need be added with regard to the specific characters of the animal. The white markings noticed, vary in extent in different individuals, and not unfrequently on opposite sides of the body of the same individual. Of three specimens contained in the British Museum collection, one is entirely black, with the exception of a white spot on the chest; the second is also black, with the exception of a white mark on the chest, and a white patch immediately behind the base of the fore leg; and the third presents the more common condition of these markings, having the chest band entire, and running back over the base of the legs on to the sides of the body; the transverse band on the hinder part of the back is also distinct. The ears are rather short, but very broad: the tail is about half as long as the head and body taken together, and clothed with tolerably long, coarse hairs.

A very fine specimen of the Ursine Dasyurus, which died in the menagerie of the Zoological Society, furnished the subjoined dimensions.

	Inches.	Lines.
Length from tip of nose to root of tail ...	23	9
" of tail	11	0
Circumference of body at the chest, about ...	20	0

	Inches.	Lines.
Length from nose to ear	5	3
" of ear	1	6
" of hind foot	4	7

To these admeasurements I will add those of a skull in the collection of the Royal College of Surgeons:—

	Inches.	Lines.
Total length of skull	4	6
Width	3	6
" between orbits	1	$6\frac{1}{2}$
" in the temporal region		9
Length of nasal bones	1	8
Width of ditto behind		$7\frac{1}{2}$
" in front		$4\frac{1}{3}$
Length of palate	2	6
Width of ditto between the hindermost molars	1	$9\frac{2}{3}$
Length of posterior palatine openings ...		5
" of four true molar teeth taken together	1	$5\frac{2}{3}$

The skull of the *D. ursinus* is short, very broad, and somewhat depressed; the great breadth is produced by the bold arch formed by the zygoma, near the posterior root of which the skull is broadest, and from that point it gradually tapers to the front. The temporal region is much contracted, and the sagittal crest considerably elevated, circumstances which, combined with a strong zygoma, much arched in the horizontal direction, and also in the vertical, indicate great power in the muscles which work the lower jaw. The interorbital space is broader than in other Dasyuri, and the frontals throw out a distinct post-orbital process; a smaller post-orbital process arises from the malar bone[1]. The posterior palatine openings are of moderate size, and remote from the hinder boundary of the palate, being situated opposite the interspace of the second and third true molars.

[1] In having a small post-orbital process to the frontal and malar bones, *Dasyurus maculatus* approaches most near to the Ursine species.

In a skeleton of the present animal, in the Royal College of Surgeons, I find one the of humeri imperforated at the lower extremity but the opposite humerus has the inner condyle perforated[1]

The Ursine Dasyurus is found only in Van Diemen's Land, and it is to Mr. Harris that we are indebted for the earliest account of this singular animal: but little has been added to that gentleman's observations upon its habits, which appeared in the Linnean Transactions.

"These animals," Mr. Harris observes, "were very common on our first settling at Hobart Town, and were particularly destructive to poultry, &c. They, however, furnished the convicts with a fresh meal, and the taste was said to be not unlike veal. As the settlement increased, and the ground became cleared, they were driven from their haunts near the town to the deeper recesses of the forests yet unexplored. They are, however, easily procured by setting a trap in the most unfrequented parts of the woods, baited with raw flesh, all kinds of which they eat indiscriminately and voraciously; they also, it is probable, prey on dead fish, blubber, &c., as their tracks are frequently found on the sands of the sea-shore.

"In a state of confinement they appear to be untameably savage, biting severely, and uttering at the same time a low yelling growl. A male and female, which I kept for a couple of months chained together in an empty cask, were continually fighting; their quarrels began as soon as it was dark (as they slept all day), and continued throughout the night almost without intermission, accompanied by a kind of hollow barking, not unlike that of a dog, and sometimes a sudden kind of snorting, as if the breath was retained a

[1] In having the humerus imperforated on the inner condyle, the true Dasyures differ from the Phascogales, and, indeed, from all other *Marsupiata*.

considerable time, and then suddenly expelled. They frequently sat on their hind parts, and used their fore paws to convey food to their mouths. The muscles of the jaws were very strong, as they cracked the largest bones with ease asunder."

According to Mr. Gunn[1], these animals commit great havoc among the sheep, and, notwithstanding their comparatively small size, they are so fierce, and bite so severely, that they are a match for any ordinary dog.

Dasyurus laniarius (Fossil).

Although the section *Sarcophilus* is at present confined to Van Diemen's Land, such was not always the case, since remains of a species nearly allied to the *Sarcophilus ursinus* have been found on the main land. These remains, consisting of portions of both jaws, and exhibiting nearly the whole of the molar teeth, appertained, however, to an animal of a larger size than the recent species. Two premolar, and two true molar teeth, in a fragment of an upper jaw, in the Museum of the College of Surgeons, measure together one inch and a half, whilst the corresponding teeth in the *Dasyurus*, or *Sarcophilus ursinus*, occupy an extent of an inch and a quarter only; and a last molar tooth of a lower jaw of the fossil Dasyure, measures from front to back seven and a half lines, being two and a half lines more than the hindermost molar of *D. ursinus*.

All the fragments referred to, were found in the caves of Wellington Valley, and are described by Prof. Owen, some of them in Mitchell's Eastern Australia, vol. ii. p. 363, and the remainder in the Catalogue of the Fossil Organic Remains contained in the Museum of the Royal College of Surgeons.

[1] Annals of Natural History, vol. i. p. 101.

Genus, *Thylacinus*.

Thylacinus.[1] TEMMINCK, Monographies de Mammalogie, tom. i. p. 60, 1827.

Peracyon[2]. GRAY, Annals of Philosophy, for November, 1825, vol. x. of New Series, p. 344.

Dasyuridæ with the outermost incisors slightly exceeding the others in size; the premolars separated from each other, and $\frac{3-3}{3-3}$ in number; the three foremost of the upper true molars with a much elevated central cusp, an anterior and posterior cusp but little elevated, and an internal lobe; the hindermost upper true molar transverse; the true molars of the lower jaw nearly resembling those of the upper jaw, but destitute of internal lobe, and with the central cusp more elevated; the humerus with the inner condyle perforated; the hind foot destitute of inner toe; marsupial bones wanting; the females with a distinct pouch, and provided with four mammæ.

The premolars are more numerous in *Thylacinus* than in *Dasyurus*, there being three, instead of two, of these teeth on either side of each jaw: the teeth, indeed, agree in number with those of *Phascogale*, though in other respects they differ much: the incisors differ, inasmuch as the outermost, instead of the innermost pair, are the largest; the premolars differ in being isolated, and the true molars are of a more simple form. The canine teeth are of large size, of a simple, elongated conical form, and are slightly recurved at the apex; those of the upper jaw are separated from the incisor teeth by a

[1] From θύλᾶκος, a pouch.

[2] From pera, a pouch; and κῦων, a dog (?) Although Mr. Gray proposed the above name for the present genus in 1825, so far as I can learn the section was first characterized by M. Temminck.

large and deep concavity into which the apex of the canine of the lower jaw is lodged when the jaws are closed[1]. The premolar teeth present a triangular outline when viewed from the outer or inner side, and have a small posterior basal cusp: the two foremost of the upper premolars are separated from each other, or from the canine; the third upper premolar is in contact with the true molars; but in the lower jaw all the premolars are isolated. The crowns of the three foremost upper true molars are in the form of a right-angled triangle, of which one of the sides is about one-fourth shorter than the others; this short side is in front. Viewed from the outer side, these molars present a large central cusp, and two smaller and much less elevated cusps, placed one behind, and the other in front of the principal cusp: the foremost cusp is separated by a distinct transverse notch from the body of the tooth. The inner lobe of the tooth is but little elevated. The hindermost true molar is nearly of the same structure as the others, but of smaller size, and so placed that its greatest diameter is in the opposite direction to the long diameter of the other molars. The true molars of the lower jaw have the crown divided by transverse incisions into three cusps, of which the central one is the largest, pointed, and much elevated.

But one recent species of this genus is known, and that is confined to Van Diemen's Land; but amongst the fossil remains found in the caves in Wellington Valley already alluded to, Prof. Owen has detected some portions of lower jaws, which he regards as belonging to a species of *Thylacinus*. An anterior extremity of one of these fossils is figured in Sir T. Mitchell's Australia (vol. ii. pl. 31, fig. 7), and other portions subsequently sent to England are de-

[1] The Dasyuri all present this character, which is not found in the ordinary *Carnivora*; in those animals the lower canine passes on the outer side of the upper jaw, when the mouth is closed.

scribed in the Catalogue of the College of Surgeons, under the name

Thylacinus spelæus. OWEN.

The principal known differences between this fossil species and the recent one are, that the lower jaw is deeper, being nine lines below the first premolar tooth, whilst in *Thylacinus cynocephalus* the depth of the jaw at the same point is but seven lines. A penultimate molar tooth of the lower jaw in the College of Surgeons' Museum, when compared with the corresponding tooth of the recent species, " differs, moreover, in having a small accessary cusp on the inner side of the large middle compressed cusp, which cusp is also less deeply and angularly divided from the anterior lobe of the tooth. [1]" In this last mentioned character the fossil agrees more nearly with *Dasyurus* proper, than with *Thylacinus*.

[1] Owen, in Catalogue of the Fossil Remains contained in the Museum of the College of Surgeons, p. 336.

THYLACINUS CYNOCEPHALUS.

Dog-headed Thylacinus.

(*Plate* 16, fig. 2).

Didelphis cynocephala. HARRIS, Transactions of the Linnean Society, vol. ix. p. 174, Pl. 19. 1807.

Dasyurus cynocephalus. GEOFFROY, Annales du Muséum, tom. xv. p. 304.

Thylacinus " FISCHER, Synopsis Mammalium, p. 270.

A. WAGNER, in Schreb. Saug. Suppl. 109—110 Heft, p. 19.

WATERHOUSE, Nat. Library (Marsupialia), vol. xi. p. 123, Pl. 5.

Thylacinus Harrisii. TEMMINCK, Monographies de Mammalogie, vol. i. p. 63, Pl. 7, figs. 1—4,—the skull and lower jaw.

Peracyon cynocephalus, the Tasmanian Wolf. GRAY, List of the Mammalia in the British Museum (1843), p. 97.

Tiger, Hyæna, Zebra-Opossum, Zebra-Wolf, and Dog-headed Opossum of the colonists.

About equal in size to the Common Wolf; head formed like that of a Dog; tail about half the length of the body; fur short, and closely applied to the skin; general colour grey-brown; the back with from about twelve to fourteen transverse black bands, narrow and short on the fore parts of the back, longer and broader on the hinder parts; region of the eye pale; tail with short fur, nearly like that of the body, excepting on the under side of the apical portion, and at the tip, where the hairs are comparatively long.

Inhabits Van Diemen's Land.

The general resemblance which the *Thylacinus* bears to a Wolf or large Dog, has struck many, and, indeed, has caused it to be, by some, arranged amongst the ordinary Carnivora [1].

[1] The Thylacinus is arranged by Mr. Swainson amongst the *Felidæ*, or Cat Family, and in support of his views, the author quotes some observations from Temminck with regard to the dentition of the animal; Mr. Swainson,

Its legs, however, are proportionately shorter than in the Wolf, and, judging from the structure of its foot, its body must be brought much nearer to the ground, in walking, than that of the Wolf, it being what may be termed a semi-plantigrade animal. The muzzle is more elongated and narrower than in other *Dasyuridæ*. The ears are rather short, very broad at the base, and somewhat pointed at the opposite extremity; they are well clothed with hairs, both internally and externally; on the outer side the hairs are coloured like those on the upper part of the head, excepting towards the tip of the ear, where they are paler; on the inner side the hairs are of a brownish white hue, slightly inclining to yellow; near the anterior angle they are very long. The eyes, according to Mr. Harris, are large and full, of a black colour, and provided with a nyctitant membrane. Long black bristles spring from the upper lip; a few are also observed on the cheeks, and above the eye. The fur of the animal is short, somewhat closely applied to the skin, though of a slightly woolly texture, owing to each of the hairs of which it is

however, restricts his quotation to those parts only, of M. Temminck's account, in which certain resemblances existing between the teeth of the Thylacinus and those of some of the true Carnivora are pointed out. Certain teeth in the animal under consideration, it is true, will bear a close comparison with certain teeth found in Cats and Dogs, but striking differences are observable when the whole series of the teeth of the carnivorous marsupial quadruped is compared with that of any mammal belonging to the true *Carnivora*. The increase in the number of the incisors, and true molars, in the Thylacinus, becomes important when it is found that these teeth are implanted in a skull and lower jaw which in every respect are conformable to the marsupial type of structure, and that in that type only is this increase found. The passage quoted from M. Temminck, however, will not bear the construction which Mr. Swainson has put upon it; M. Temminck compares the true molars of the Thylacinus with the principal *false molar* ("carnassièr") of the Cats and Dogs. Now, the *true* molars of the last mentioned animals differ much in their structure from the *corresponding* teeth in the pouched Thylacinus; and hence the arguments of Mr. Swainson, founded upon a presumed resemblance in the dentition of the animals mentioned, are not valid.

composed being waved. The general tint of the animal is greyish brown, but faintly suffused with yellowish; on the under parts of the body of a paler hue than the upper. The fur on the back is of a deep brown colour next the skin, and each hair (excepting those which form the transverse black bands) is yellowish brown towards the point, and dusky at the point; on the abdomen the hairs are of a paler brown at the root, and brown-white externally. The black bands alluded to are usually about fourteen in number; they commence immediately behind the shoulders, and are at first narrow and confined to the back, but, proceeding towards the tail, they become gradually broader, and are more extended in the lateral direction; those on the haunches are the longest, and are often forked at their extremities. The general tint of the head is rather paler than that of the body, and the region of the eye is of a whitish hue, but a dark spot is observable at the anterior angle of the eye, and a narrow dark line runs over the eye: the muzzle is dusky; the edge of the upper lip white. The limbs, externally, and the feet, scarcely differ in colour from the body. The large pads at the base of the toes of the fore foot are naked, and exceedingly rough, and a narrow naked mark runs backwards from these pads to the wrist; a similar narrow naked mark runs along the under side of the hind foot, from the heel to the great rough pads at the base of the toes. The claws of the fore and hind feet are nearly equal in size; short, thick, but slightly compressed, and solid; and they are of a brown colour. The tail is about half as long as the body; thick at the base, where it is covered with somewhat woolly fur, like that on the body, but at about the commencement of the second fourth of the tail the hairs become short and harsh, and are closely applied to the skin; they are brown on the upper surface, and pale brown on the under: on the under surface of the apical portion of the tail, however, the hairs are comparatively long.

as well as at the point, where they are blackish: about three or four black bands are observable on the basal part of the tail above.

The region of the pouch in the female is clothed with rusty red hairs.

The dimensions of the female specimen from which the above description is taken, will be found in the first of the columns of admeasurements; those added in the second are from a male specimen in the British Museum collection.

	Female.		Male.	
	Inches.	Lines.	Inches.	Lines.
Length from tip of nose to root of tail	33	0	45	0
" of tail	14	0	20	0
" from tip of nose to ear ...	6	0	7	9
" from ditto to eye, about ...	3	0	4	3
" of ear	2	0	2	2
Width of ditto at the base	2	6	3	3
Length of hind foot	5	3	6	7
Height at shoulders, about ...			18	6

A skeleton of an adult male Thylacinus in the Museum of the College of Surgeons measures in total length about five feet, and its height at the shoulders is about 21¼ inches. The subjoined dimensions are taken from skulls in the British Museum, and that of the Royal College of Surgeons.

	MALE? British Museum.		FEMALE? College of Surgeons.		MALE. College of Surgeons.		FEMALE. College of Surgeons.	
	Ins.	Lines.	Ins.	Lines.	Ins.	Lines.	Ins.	Lines.
Total length of skull	9	6	7	6	9	0	7	9
Width	5	7	4	1½	5	6	4	3
" between orbits	2	1	1	7				
" in the temporal region ...	1	5	1	3½				
Length of nasal bones	3	8	3	0				
With of ditto behind	1	3	1	0				
" " in front		8¾		6				
Length of palate	4	9½	3	11				
Width of ditto between last molars	1	10	1	6½				
Length of posterior palatine openings	1	1½	1	0½				
" of four upper true molar teeth, taken together ...	1	10	1	8				
" of lower jaw	7	9	6	3				
Height of ditto from apex of coronoid process	2	11	2	1½				

The auditory bullæ are small, little convex, and formed, as in nearly all the other Marsupialia, of the alæ of the sphenoid: the palatine openings are tolerably large, and situated for the most part in the palatine bone. The nasal process of the intermaxillary bones notches into the nasal bones in a very unusual manner. The facial portion of the skull is very narrow, and considerably elongated; the skull is again much contracted in the temporal region, but broad between the orbits, which latter are more than three parts enclosed, there being a distinct post-orbital process to the frontal, and a corresponding process to the malar bone. The zygoma is thrown boldy outwards, leaving a large temporal fossa, and it is also arched upwards to furnish greater resistance to the muscles of the lower jaw.

Mr. Harris, who was the first to make this animal known, states that it inhabits amongst caverns and rocks in the deep and almost impenetrable glens in the neighbourhood of the highest mountains of Van Diemen's Land The specimen from which his description was taken was caught in a trap baited with Kangaroo's flesh : it remained alive but a few hours, having received some internal hurt whilst being secured. From time to time it uttered a short guttural cry, and it appeared exceedingly inactive and stupid, and, like the owl, had an almost continual motion of the nyctitant membrane of the eye. Remains of an Echidna were found in the stomach of the animal. Mr. Gunn informs us that these animals are common only in the remoter parts of the colony, and are frequently caught at Woolnooth and the Hampshire Hills. They attack the sheep at night, but are occasionally seen during the day-time; upon which occasions, perhaps from imperfect vision, their pace is very slow. Mr. Gunn also observes that the Thylacinus sometimes attains so large and formidable a size that a number of dogs will not face it. That gentleman denies that the tail of the animal is compressed, as has been stated by some authors, and his observations do not confirm the aquatic habits which have been attributed to it.

Prof. Owen, who has prepared a memoir upon the internal anatomy of the Thylacinus, found no marsupial bones in three of the specimens which he dissected, two of which were full-grown females, and the third a male; but in a large and old male he detected a few particles of the bone-salts in the centre of the fibro-cartilage. The pouch, Prof. Owen observes, is well developed in the female Thylacine, and in one of the specimens dissected, four well developed teats, each four inches long, indicated that it had contained four young ones when, or shortly before, it was killed[1]

[1] See Proceedings of the Zoological Society for December, 1843, p. 148.

DIDELPHIDÆ ; or Opossum Family.

Didelphidæ. Waterhouse, Nat. Libr. vol. xi. (Marsupialia), p. 75. 1841.
Didelphina. Gray, Annals of Philosophy, vol. x. (New Series), p. 340. 1825.

Marsupialia with incisor teeth, $\frac{5-5}{4-4}$; canines, $\frac{1-1}{1-1}$; premolars, $\frac{3-3}{3-3}$; molars, $\frac{4-4}{4-4}$: feet five-toed, plantigrade: cæcum moderate.

The family *Didelphidæ* is composed of numerous Marsupial animals, which at the present time are confined to the American continents. They are of small size, the largest known species being scarcely equal to the Common Cat in bulk, whilst by far the greater portion of them would bear a closer comparison with the Common Rat in this respect. Their food consists chiefly of insects; but small reptiles, as well as birds and their eggs, are attacked by the larger species.

Some of the Didelphidæ, or Opossums, have no pouch, or at least this receptacle for the young is found only in a very rudimentary condition in certain species, and the young, which at first remain firmly attached to the nipples, are subsequently carried upon the back of the parent, where they retain their position by entwining their tails round that of the mother. The mammæ are very numerous, varying from nine to thirteen, the odd one being in the centre of a circle formed by the other nipples. One of the species of the present family (the Yapok) lives in the water, and differs from others of its group in having the feet webbed; the remaining species live for the most part in trees, in the hollows or amongst the foliage of which, they remain concealed during the day-time, sallying forth in the night only to procure their food.

The Opossums may at once be distinguished from other Mammalia by the great number of their incisor teeth, there being ten in the upper, and eight in the lower jaw. These teeth are arranged in each jaw nearly in the form of a semicircle: in the upper jaw the two foremost incisors are rather longer than the rest, and are generally separated from them by a narrow space; their form is nearly cylindrical, but at the apex they are slightly dilated. The canine teeth are well developed, and those of the upper jaw are the larger. The premolars are two-rooted, compressed, and pointed, and have a small posterior talon. The true molars of the upper jaw are three-rooted, and have the crown of a triangular form, and tubercular: those of the lower jaw are longer than broad, and, when viewed from either the outer, or inner side, they present three prickly cusps, of which the middle cusp is the highest; the foremost of these cusps is single, whilst the others are double, and there are therefore five prickly cusps to each molar. With regard to the general form, these animals, for the most part, have very nearly the proportions of the Common Rat, excepting that they have the muzzle more elongated, and terminated in a distinct naked muffle. In the larger species the body is proportionately stout. The tail is almost always very long, nearly destitute of hair, excepting at the root, and covered by a scaly skin[1]: this organ is here (as in many other of the Marsupialia) prehensile. The feet are naked beneath, five-toed, and all the toes are furnished with moderate-sized claws, excepting the inner toe of the hind foot, which is clawless.—(See Pl. 10, fig. 3). In the possession of a cœcum, and in having ten incisor teeth in the upper jaw, the *Didelphidæ* agree with the *Peramelidæ*; the incisor teeth of the Opossums, however,

[1] Very minute scattered hairs only, are found on the tail of the Opossums, excepting at the root, where the tail is usually clothed with fur like that on the body.

agree more closely in their form and arrangement with those of the Dasyuri. The molar teeth of the Opossums differ from those of the Perameles group in wanting the posterior inner lobe, and in this and every other respect they agree with the molar teeth of the Dasyuri. The structure of the skull, and of the extremities, is nearly the same in the last-mentioned group and the Opossums; the only difference worthy of notice consists in the thumb of the hind foot being more developed in the Opossums; but, compared with the *Peramelidæ*, the animals under consideration differ much in external characters, and very considerably in the structure of the skull. The situation of the *Didelphidæ* in a natural system, then, should be between the *Dasyuridæ* and the *Peramelidæ*, but nearer to the first of these two families.

Genus, *Didelphys*.

Didelphis[1]. LINN. Syst. Nat. i. p. 71.
Philander. BRISS. Règn. Anim. 1756.

Didelphidæ without cheek pouches, and in which the toes of the feet are free—that is, not joined by a web.

Section 1. *Opossums in which the pouch is well developed.*

We shall commence our descriptions with those species which are of large size, and have very long hairs interspersed with those which form the ordinary fur, as is the case in the Common Opossum of North America.

[1] From δὶς and δελφὺς, double uterus.

DIDELPHYS VIRGINIANA.

Virginian Opossum.

Virginian Opossum. PENNANT, Synopsis of Quadrupeds, p. 204, Pl. 31, f. 1. 7171.
Didelphis Virginiana. SHAW, General Zool. vol. i. Pt. 2, p. 473, Pl. 107. 1800.
" " TEMMINCK, Monogr. de Mammalogie, tom. i. p. 27.
Sarigue des Illinois et Sarigue à longs poils. BUFFON, Quad. Suppl. vol, 7, Pls. 33 and 34.
Sarigue à oreilles bicolores. CUVIER, Règne Animal, tom. i. p. 175.

Fur long, loose, and somewhat woolly ; white, the tips of the hairs more or less suffused with brown-black : numerous long interspersed white hairs, mixed with those of the ordinary fur : a dusky patch at the anterior angle of the eye : ears naked, black, with the tip white : tail having the naked portion at first black, but terminated with white : legs and feet brown, or brown-black.

Inhabits North America.

The Virginian Opossum is one of the largest species of its genus, being about equal in bulk to the Common Cat. It has a large, elongated, and pointed head, terminated by a naked, and flesh-coloured muffle. The ears are tolerably large, oval-shaped, and naked, and from their black colour form a strong contrast with the white head of the animal. The tail is nearly equal to the body in length ; at the base it is clothed with fur like that on the body, this occupying one-fourth of the entire length ; the remaining portion is protected by a scaly skin, and between the small scales, which nearly resemble those of the tail of the Rat, are some very small hairs : of the scaly portion, the first, or basal half, is black, and the terminal half, white ; or sometimes about one third is black, and the rest white. The fur of the animal is somewhat woolly, very long, loose, and moderately soft; it is of a dirty yellowish white hue, and the hairs of the ordinary fur are more or less suffused with black or brown at the extremity; most distinctly so on the back; numerous very long silky hairs are interspersed with those which form the chief clothing of the animal, the visible portions of which are white. The head is white, but usually the hairs on the occipital portion are dusky at the point, and an indistinct mark running forwards from this part, and terminating between the eyes, is formed by the hairs being similarly suffused with brownish at the point on the middle part of the head. The eye is narrowly encircled with dusky brown, excepting in front, where the dark hue is more extended, and forms a tolerably conspicuous spot. The ears are sometimes entirely black, but usually they are tipped with white, the white in some individuals forming a narrow edging, and in others, is rather more extended. The legs and feet are of a brownish black colour.

A specimen of the Virginian Opossum in the British Museum is entirely white, with the exception of the ears, which are black, but margined with white at the apex, and a small patch in front of the eye, which is brown.

			Female.		Male.	
	Ins.	Lines.	Ins.	Lines.	Ins.	Lines.
Length from tip of nose to root of tail ...	22	0	21	0	22	0
" of tail	15	0	12	0	14	0
" from nose to ear	4	5	4	9	4	9
" of ear	1	5	1	6	1	6
" of hind foot	2	5	2	7	2	7

The structure of the skull of the *D. Virginiana* indicates great power in the action of the jaws, the temporal ridges meeting so as to form a much-elevated sagittal crest; the interorbital space is moderately broad, but in the temporal region the skull is much contracted: an obtuse post-orbital process is thrown out from the frontal bone, and the malar bone is also produced into an angular process, so as partially to separate the temporal from the orbital fossa. The zygomatic arch is curved both outwards and upwards. The nasal bones are broad towards the hinder part, and are produced posteriorly so as to encroach much upon the frontal bones. This is one of the points of distinction which strikes one upon comparing the skulls of the Opossum with that of a Dasyurus, the nasal bones in the animals of that genus being much less produced in the mesial line of the skull; their greatest diameter is at the hinder part, whilst in the Opossum the broadest part of the nasal bones is always considerably in advance of that point. Another distinction observable in these bones, when the *Dasyuridæ* and *Didelphidæ* are compared, consists in their being pointed in front and produced considerably over the entrance of the nasal cavity in the latter group, whilst in the Dasyures the nasal bones are truncated or emarginated in front, so that the upper boundary of the nasal cavity terminates in a line with the lateral boundaries formed by the intermaxillaries, or behind that line. The facial portion of the skull is larger, the cerebral portion smaller in proportion than in the Dasyuri [1], and the palate is longer.

[1] The brain, as in a great measure indicated by the structure of the skull, is remarkable for the proportionately large size of the olfactory lobes, and the small size of the cerebral hemispheres; these latter are much contracted in front, and destitute of convolutions. As compared with the brain of the Mammalia of the higher classes, that of the Opossum furnishes the most remarkable contrast in its small size in proportion to the bulk of the animal, and the small development of the cerebrum, which is here distinctly separated from the cerebellum, and leaves exposed the optic lobes above, as well as the *crura cerebri* below. The Dasyuri come next in grade, and their brain con-

	Inches.	Lines.
Total length of skull	4	6
Width	2	6
" between orbits	1	0
" in the temporal region		5½
Length of nasal bones	2	2
Greatest width		9½
Width in front		3½
Length of palate	2	5
" of four true molar teeth taken together		9½
" of principal palatine openings ...		6

With regard to other parts of the skeleton, there is one point which I should not omit to notice, and that relates to the structure of the neck, the second, third, fourth, fifth, and sixth vertebræ of which are remarkable for the great development of the spinous processes, and more particularly for the great thickness, or transverse diameter, of these processes, and their mode of junction with each other, which is such as to preclude any upward flexure of the neck in the region of these vertebræ. I have met with no very satisfactory explanation of the use of this structure, which is found not only in the Virginian Opossum, but likewise in the *D. cancrivora*, and no doubt in the other large species of *Didelphys*.

The Virginian Opossum, according to M. Temminck, is found from Mexico to the southern provinces of the United States; according to Pennant, it occurs also in Brazil and Peru, but in making this statement he has undoubtedly confounded some other nearly allied species with the *D. Virginiana*. This animal is said to be very destructive to poultry, sucking their blood, but not eating the flesh; it feeds

sequently differs widely from that of the placental Carnivora, in the points just alluded to, as well as in that remarkable character of the Marsupial brain—the almost total absence of *corpus callosum*—which was first pointed out by Prof. Owen, and has since been confirmed by the able editors of Cuvier's Anatomie Comparée—See p. 102 of vol. iii.

upon roots and fruits, is very expert in climbing trees, from the branches of which it suspends itself by the tail, and, by swinging its body, reaches the boughs of neighbouring trees; it hunts eagerly after birds and their nests; when pursued and overtaken, it will feign death, and I am informed will bear much torture without evincing the slightest sign of life upon these occasions. It is extremely tenacious of life. The female Opossum brings forth from twelve to sixteen young at a time; her nest, which is formed of dry grass, is usually deposited at the root of a tree, or in some close bush. At the time of their birth, the young are described as being scarcely more than a grain in weight, blind, naked, and shapeless; nevertheless they find the teats in the pouch of the parent, to which they attach themselves so firmly that they cannot be separated without difficulty. When the young have attained the size of a mouse, and all their parts are developed, which takes place in about five days, they then leave the pouch, but return to suckle, and when danger threatens. During this time the female shows an excessive attachment to her young, and will suffer any torture rather than permit the pouch to be opened. The flesh of the Opossum is said to be well flavoured, resembling that of a sucking pig. The skin is very fœtid. The hair is dyed by the Indians, and woven into girdles, &c.

DIDELPHYS AZARÆ.

Azara's Opossum.

(*Plate* 18, fig. 2).

Didelphis Azaræ.	TEMMINCK, Monogr. de Mammalogie, tom. i. p. 30.
" "	WATERHOUSE, Nat. Library (Marsupialia), vol. xi. p. 83, Pl. 1.
" *aurita.*	PR. MAXIM, Beitr. ii. 392.
Micouré premier.	AZARA, Quad. de Paraguay, i. p. 244.

Fur long and loose, suffused with black; head white, with a central black streak running backwards from the forehead, and a streak on each side running through the eye; ears white, more or less clouded with black at the base; legs and feet black; basal half of the naked portion of tail also black, the terminal half white: very long white hairs are interspersed with those of the ordinary fur on the body.

Inhabits Brazil, Paraguay, Bolivia, &c.

Didelphys Azaræ greatly resembles the Common Opossum of North America, but may be distinguished by the three distinct black marks on the head, and by its longer tail. The head and neck are white; the mesial black stripe commences between the eyes, and is at first narrow, but as it approaches the back part of the head it becomes broader, and on the occiput it is much dilated; the lateral stripe commences considerably in front of the eye, and encloses the eye as it passes backwards almost to the ear; this latter is tolerably large, ovate, and sometimes entirely white, but usually clouded with black at the base. The bristly hairs of the moustaches are very long; those nearest the mouth are white, and the remainder are black; a few similar long bristly hairs spring from above the eye, and there is a small tuft of long hairs on the cheeks The throat and chest are yellowish, or sometimes rusty brown. The fur on other parts of the body is of a dirty yellowish white hue next the skin, and more or less suffused with sooty black externally; on the back very little of the pale colour is perceptible, but on the sides and under parts of the body the black is less extended on each hair, and the pale hue is seen when the fur is in its ordinary position—most so, on the under parts of the body. The visible portions of the very long interspersed hairs, which are abundant on the back and sides of the body, are pure white. The legs and feet are sooty black. About one-third of the tail is clothed with fur like that on the body; on the remaining portion there are but minute scattered hairs springing from the interstices of the small scales, with which this part of the tail is covered; both the scales and the small hairs are black on the second third of the tail, and white on the terminal third.

	Inches.	Lines.
Length from tip of nose to root of tail ...	19	0
" of tail	19	0
" from nose to ear	4	3
" of ear	1	9
" of fore foot	1	11
" of hind foot	2	6

A specimen of the *D. Azaræ*, brought by Mr. Darwin from Maldonado, La Plata, measured seventeen inches in length from the tip of the nose to the root of the tail, and its tail was about thirteen and a half inches in length. These admeasurements slightly exceed those given by M. Temminck, which are as follows:—

	Adult, Female.		Male.		Male.	
	Ins.	Lines.	Ins.	Lines.	Ins.	Lines.
Head and body	16	3	14	7	14	0
Tail	14	7	13	6	13	0

Azara's Opossum has a wide geographical range. It was first noticed by Azara in Paraguay; in the British Museum collection are specimens from Brazil, and St. Fé de Bogota. Mr. Darwin found it at Maldonado, and Mr. Bridges has forwarded us specimens from Bolivia. In Guiana it is replaced by a nearly allied species, which we shall next describe—the *D. cancrivora*. The *Didelphys Azaræ* is a nocturnal animal, lies concealed during the day in burrows in the ground, or in thickets, and climbs trees in the night to feed upon fruits, and birds' eggs. Like the Weasel it sucks the blood of birds, and is very destructive to poultry. Azara states that it is found in the open parts of the country as well as in the thickets. The body of this animal, Mr. Darwin informs us, has a very offensive odour after death. Mr. Bridges found eight young ones in a litter of one of these animals.

Didelphys albiventris.

Didelphys albiventris. Lund, Det. K. Danske Vidensk. Selsk. Afh. viii. p. 236.

Fur of a pale Isabella-yellow colour, but with the hairs black at the point on the back and sides of the body: head with a longitudinal black mark running through each eye, and a third dark mark on the forehead; ears grey at the base, and whitish at the point; legs and feet black; tail with the basal half black, and the terminal half pale Isabella-yellow.

Inhabits Brazil.

The above description is taken from Dr. Lund's account of a species of *Didelphys*, which that author regards as distinct from others with which he is acquainted. I strongly suspect, however, it will prove to be specifically identical with the *D. Azaræ*. It is said to be 22 inches in total length, of which the body constitutes one-half; the ears are $2\frac{1}{4}$ inches in length. Compared with the *Did. Virginiana*, Dr. Lund observes, it differs in being of smaller size, in having the tail longer, the ears larger, and the abdomen white.

DIDELPHYS CANCRIVORA.

The Crab-eating Opossum.

Didelphis cancrivora.	GMEL. Linn. Syst. i. p. 108, sp. 7.
" *marsupialis.*	Id. p. 105, sp. 1.
" *cancrivora.*	TEMMINCK, Monogr. de Mammalogie, tom. i. p. 32, Pl. 5—skeleton.

Fur long, loose, and rather glossy, of a dirty yellowish white colour next the skin, but with the visible portions of the hairs of a black or brown-black colour, and this dark hue is nearly uniform on all parts of the body: ears black; tail also black, with the exception of the apical half, which is white, or nearly so.

Inhabits chiefly the northern parts of South America.

The nearly uniform dark hue of this animal at once distinguishes it from the other large species of *Didelphys*, to which it is nearly allied, such as the *D. Virginiana* and *D. Azaræ*. In these two species the long hairs of the fur are white, whilst in *D. cancrivora* they are black—at least the visible portions of these hairs, for they are whitish at the root. In size the Crab-eating Opossum is rather inferior to the Common Opossum of North America, and its tail is proportionately longer than in that animal.

The following description is drawn up from specimens of *D. cancrivora* contained in the collection of the British Museum.

	Female.		Male ?		Male.	
	Ins.	Lines.	Ins.	Lines.	Ins.	Lines.
Length from tip of nose to root of tail ...	17	0	14	0	15	2
" of tail, about	15	0	13	0	16	3
" from nose to ear	3	8	2	7	3	6
" of ear	1	3	1	3		
" of fore foot and nails	1	7	1	6	1	6
" of hind foot and nails	2	3	1	11	2	2
" of inner toe of ditto, about ...				7		
" of skull	3	10*			3	9½
" of ditto from hinder root of zygoma to apex of intermaxillary bones ...	3	4½			3	5
Width of ditto	1	10½			1	10¼
Length of nasal bones	1	10			1	9½
Width of ditto towards the hinder part ...		6½				
" " in front		3				
Length of palate	2	4			2	2½
Width of ditto between the hindermost molars		9½				9
Length of posterior palatine openings† ...		6				5
" of four upper true molar teeth taken together		9¼				9¾
" of lower jaw	3	0½			3	1
Height of ditto from apex of coronoid process	1	4½			1	1

The fur is glossy, rather harsh to the touch, and by no means dense; on all parts of the body it is long, but on

* Allowance being made for a small portion which is lost, the skull being fractured behind.

† These openings are long and narrow, and situated partly in the palatine portion of the superior maxillary bones, and partly in the palatines; in these latter bones are four other openings of about $\frac{1}{12}$ of an inch in diameter; they are all nearly in the same transverse line, and near the termination of the palate.

the back it is very long, the longest hairs being as much as three inches in length. The general hue of these animals is sooty black, but the hairs are of a dirty yellow-white hue next the skin, and this pale colour is not altogether hidden by the long black points of the hairs: on the under parts of the body, where there are but few of the coarser and longer hairs, the colour may be described as yellowish white, suffused with black; and on the throat a rusty hue is visible: the tip of the muzzle is brown; the ears are black. About one-fourth of the tail is clothed with fur like that on the body; the remaining portion is curved with small scales, between which spring very minute hairs; the first half of the scaly portion of the tail is black, and the apical half is white.

I have given the dimensions of two specimens alluded to as I find them, but the admeasurements in the third column are more to be depended upon, since they were taken by M. Temminck from a specimen preserved in spirits, whilst the others are from skins. I have added, also, in the third column, the dimensions of a skull taken from Temminck's figure. The author just mentioned states that in upwards of thirty specimens of *D. cancrivora* which had come under his notice, the only variation in colouring which he perceived was, that some specimens had the free points of the hairs of the fur perfectly black, whilst in others they were of a blackish chestnut hue.

The Crab-eating Opossum extends into Brazil, but is chiefly found in Guiana; like its congeners, it climbs trees, and it is said to be a bad runner. It prefers swampy situations where small crabs abound, these forming its food, which consists likewise of small birds, reptiles, and insects. The Indians eat its flesh, which is said to resemble in flavour that of the Hare.

DIDELPHYS CALIFORNICA.

Californian Opossum.

Didelphis	*Californica.*	BENNETT, Proceedings of the Zoological Society for March, 1833, p. 40.
? "	*pruinosa.*	WAGNER, in Wiegmann's Archiv. 1842, p. 358.

Fur long, somewhat woolly, and with very long interspersed harsher hairs; next the skin the hairs are dirty white, but externally those of the ordinary fur are black, or brown-black, and the longer hairs are white; head brown, with a darker mark running through the eye, and a white band immediately beneath this: ears black; tail with nearly the whole of the scaly portion white: legs and feet brown-black.

This species agrees with the *Did. virginiana* and *D. Azaræ* in having the long bristly hairs on the upper parts of the body of a white colour; but it differs from both those animals in having the upper surface of the head of a nearly uniform dusky brown hue; the mesial portion of the head is rather darker than other parts, if we except the black mark, which, commencing considerably in front of the eye, extends backwards to the ear. From both the animals mentioned the present species differs, moreover, in having the ears entirely black: as in the other large species of Opossums just described, the fur is of a dirty yellowish white colour next the skin, and this pale hue is not perfectly hidden by the overlaying dark points of the hairs; indeed, on the abdomen the pale colour is but little concealed. The somewhat bristly, long, interspersed white hairs, are exceedingly abundant on the back of the animal. The root of the tail is clothed with fur, like that on the body; the remaining portions are scaly, at first black, but the greater part appears to have been white in the living animal.

	Inches.	Lines.
Length of head and body, together	17	0
" of tail	14	0
" from nose to ear	3	11
" of fore foot	1	8
" of hind foot	2	1
" of ear	1	5

Two specimens of the present species were obtained by the Zoological Society, from that part of California which adjoins to Mexico; and it is upon these that Mr. Bennett's description was drawn up.

Didelphys pruinosa. WAGNER.

The colouring of the fur of the animal described by Dr. Wagner, under the above name, can scarcely leave a doubt but that it is specifically identical either with the *D. californica* or the *D. breviceps;* but, unfortunately, the skin upon which Dr. Wagner founds his species had had the skull removed before it came into his possession; hence the proportions of the head could not be ascertained with any degree of accuracy. I have been induced, however, to associate the *D. pruinosa* with the *D. californica,* rather than the *D. breviceps,* from the circumstance of its agreeing with the former in having the ears totally black. The length of the tail in *D. pruinosa* is said to be about equal to that of the head and body, taken together, which is about one foot. The ears are nearly an inch and a half in length.—It is from Mexico.

DIDELPHYS BREVICEPS.

Short-headed Opossum.

Didelphis breviceps. Bennett, Proceedings of the Zoological Society for March, 1833, p. 40.

Head short; fur long and somewhat woolly, whitish next the skin, and black externally; upper parts and sides of the body with very long interspersed white hairs; head with the upper surface brown, a black mark running through the eye, and extending to the ear, and a white mark immediately below this: ears black, but mottled with white at the point: tail with the scaly portion at first black, but with the apical part white: legs and feet black.

Inhabits California.

The specific name given by Mr. Bennett to this animal, calls attention to its chief distinguishing character, as compared with either of the foregoing species. In its colouring it can scarcely be said to differ from the *D. californica*. The long bristly hairs on the upper parts of the body are remarkably abundant, and many of them measure as much as three and a half inches in length It was obtained from the same part of California as the preceding species.

	Inches.	Lines.
Length of head and body	14	0
" of tail	12	0
" from nose to ear	3	2
" of ear	1	1
" of fore foot	1	6
" hind foot	2	0

Besides the foregoing species, belonging to the first division of the Opossums in which the pouch is well developed, the British Museum Collection contains two specimens of an animal which resembles the *Didelphys cancrivora* in having the body and limbs almost of an uniform black colour, but which, like the *D. Azaræ*, has a white head adorned with three black marks, one of which commences in a point between the eyes, and runs backwards on to the occiput, where it is much expanded; of the others, one is placed on each side of the head, commencing on the muzzle, and running back to the ear; they inclose the eye, as in *D. Azaræ*. From this species they differ, not only in having the body black, but in having the ears uniformly white. The tail is black at the base, and has the apical half white. The fur is white next the skin, or nearly so, as usual, and on the sides and under parts of the body the pale parts of the hairs are not altogether hidden by the overlapping points. In size these animals nearly agree with the *D. cancrivora*.

	MALE. Inches.	Lines.
Length of head and body, together	15	6
" of tail	15	0
" from nose to ear	3	2
" of ear	1	2
" of fore foot	1	5
" of hind foot	2	0

It is not known from what part of America these specimens were procured.

The following species of Opossums are of smaller size, and of a more slender form, than those already described; they differ, moreover, in having the fur short, and destitute of the long bristly interspersed hairs.

DIDELPHYS QUICA.

Quica Opossum.

Didelphys Quica. (NATTERER) TEMMINCK, Monographies de Mammalogie, tom. i. p. 36. 1827.
" " DESMAREST, Dict. des Sci. Nat. tom. xlvii. p. 387.
" " WAGNER, in Schreber's Saugethière Suppl. 109—110 Heft, p. 42.

Fur short, dense, and rather harsh to the touch; general hue of the upper parts and sides of the body ashy grey, indistinctly pencilled with silvery white, and, on the back, somewhat suffused with black; under parts white: head, on the upper surface of a sooty black hue, but with two largish white spots, situated rather behind the line of the eyes: ears large, oval, and of a brown hue: tail about equal in length to the head and body taken together; blackish, excepting at the extremity, which is white; about two inches of the basal portion clothed with fur.

Female with the upper parts of the body of a dusky brown colour.

Inhabits Brazil, Guiana, and Surinam.

Three male specimens of the *D. Quica* in the British Museum collection can be scarcely said to vary in their colouring or size. Their fur is dense, rather harsh to the touch, and by no means closely applied to the skin; it averages at about a quarter of an inch in length. On the upper parts and sides of the body the hairs are grey at the root, of a glistening silvery white near the point, and dusky at the point, and the general hue produced may be described as ashy grey, slightly suffused with blackish, and having at the same time a very faint purple tint. On the sides of the body the black is less extended on each hair, and hence these parts are paler. The whole upper surface of the head is sooty black, with the exception of two largish white spots on the forehead, which are separated from each other by a space of a quarter of an inch in width; a small rusty spot is observable near the anterior angle of the ear, and immediately in front of the ear-opening the fur is pale grey or whitish. The upper lip, lower part of the cheeks, and the whole of the under parts of the body, are white, or cream-coloured; and so are likewise the inner side of the hind and fore legs, and the hinder part of the latter; the outer surface of the legs, and the upper surface of the feet, are grey. The ears are large, of an oval form, and apparently were of a darkish brown colour in the living animal, but paler at the root. The tail is about equal to the head and body in length, thick at the root, where it is clothed with fur like that on the body; at about two inches from the root the fur terminates, and the remaining portion is scaly; at first, of a dusky brown colour, but the apical portion is white, and usually the white extends back about four or five inches from the point.

The female of *D. Quica* I have not seen, but it is said to differ somewhat in its colouring from the male M. Temminck describes it as of a blackish fawn colour, with a slight silvery hue; the flanks and limbs of a bright ash-colour, and on the abdomen and region of the pouch of a rusty hue: the head and muzzle black or blackish on the upper surface, and with three large, white or whitish spots, on each side— one above, one beneath, and one behind each eye.

The fur of the young is said to be more tinted with fawn colour, or with brown, than the adult.

The subjoined dimensions are from two male specimens.

	Inches.	Lines.	Inches.	Lines.
Length from tip of nose to root of tail	12	0	14	0
" of tail	11	6	13	6
" from nose to ear	2	6	2	8
" of ear	1	1½	1	0
" of fore foot	1	2	1	2
" of hind foot	1	7	1	6½
" of skull	2	8½		
Width of ditto	1	5½		
" between orbits		6		
" in the temporal region ...		4¼		
Length of nasal bones	1	4		
With of ditto towards the hinder part		4		
" in front		2		
Length of palate	1	7		
" of four true molar teeth taken together		7		
" of lower jaw	2	2		
Height of ditto from apex of coronoid process		10		

The Quica, as this animal is called by the natives of Brazil, is found in the Surinam district, and, no doubt, in other parts of Guiana, but appears to be most abundant in Brazil. It lives upon trees, and feeds upon small birds, insects, and fruits. During the day-time, like others of its group, it sleeps, having its body then rolled into a ball.

DIDELPHYS NUDICAUDATA.

Naked-tailed Opossum.

Didelphis nudicaudata. (Geoffroy), Desmarest, Nouv. Dict. Hist. Nat. tom. ix. p. 424.—Mammologie, p. 257.
" *myosuros* Temminck, Monogr. de Mammal. p. 48.

General hue brownish yellow, or sienna-yellow, brightest on the sides of the neck, and somewhat suffused with blackish on the back; under parts yellowish white; head dusky above, and with two small white spots on the forehead : tail with the fur less extended on the basal portion than usual, about half an inch only being clothed.

Inhabits Brazil and Guiana.

This species is nearly allied to the *D. Quica* and the *D. Opossum*, but in both those animals fur like that of the body is extended for at least two inches on to the root of the tail, whereas in the present species not more than half an inch of the tail is clothed with fur. Its feet are rather longer than in *D. Quica.* The fur is very short, and dense; its prevailing hue on the back of the animal is brownish, but it is slightly pencilled with white; on the sides of the body the fur assumes a brighter hue, being suffused with sienna yellow—most distinctly so on the sides of the neck; the under parts of the body are white or cream-colour, but here and there tinted with yellow. The upper surface of the head, and the sides of the muzzle, are brownish black; the upper part of the latter is brown; on the forehead are two small white spots, which are separated from each other by a space of about half an inch in width [1]. A small rusty yellow spot is observed near the anterior angle of the ear; the hairs near the ear-opening are dusky: the cheeks are yellowish white, and so are the chin and throat. The ears are large, ovate, and of a brownish colour. The hairs of the moustaches are black. The limbs are coloured externally like the sides of the body; internally they are white, or nearly so. The tail is about equal in length to that of the head and body taken together, and of a brown colour, excepting the apical portion, which is white; the white occupies as much as five inches of the tail, but often is less extended.

[1] The large pale spots on the forehead of *D. Quica* have a dark space between them of not more than a quarter of an inch in width, in the full-grown animal.

	Inches.	Lines.	Inches.	Lines.
Length from tip of nose to root of tail	12	6	13	0
" of tail	13	6	12	6
" from nose to ear	2	5	2	4
" of ear	1	0	1	0
" of fore foot	1	3		
" of hind foot	1	10	1	10½
Length of skull, about	2	6		
" from posterior root of zygomatic arch to apex of intermaxillaries	2	1⅓		
Width	1	3		
" between orbits		6½		
" in temporal region		4⅓		
Length of nasal bones	1	2⅓		
Width of ditto towards the hinder part		4¾		
" in front		2¼		
Length of palate	1	4½		
Width of ditto between the hindermost molars		6½		
Length of four upper true molars taken together		6		
" of lower jaw	1	10½		
Height of ditto, from apex of coronoid process, about		9		

The skull from which the above dimensions are taken has been removed from the skin of an adult male animal, and, compared with a skull of the *D. Quica,* also belonging to an adult male, differs in being considerably smaller, and in having the nasal bones more expanded behind. The palate in *D. Quica* has two longish palato-maxillary openings, and four other openings behind these, placed nearly in the same transverse line: the middle two are rather longer than broad, the breadth being about one line. In *D. nudicaudata* there are but two of the four openings just alluded to, the middle pair being absent; and the palato-maxillary openings are considerably smaller. The foremost upper premolar, moreover, in this animal, is placed nearer to the second premolar than in *D. Quica,* the muzzle being rather less elongated.

The Naked-tailed Opossum is said to be very abundant in Brazil; and, according to M. Temminck, is found also in Guiana, but is rarely received from Surinam.

DIDELPHYS OPOSSUM.

Four-spotted Opossum.

Didelphis Opossum.		LINN., Syst. Ed. xii. vol. i. p. 72.
"	"	SCHREB., Säug. iii. p. 537, Tab. 146 A and B.
"	"	TEMM., Monogr. de Mamm. tom. i. p. 41.
"	"	WAGNER, SCHREB., Säug. Suppl. iii. p. 44.

Fur short; general colour rusty yellow; the lower part of the cheeks, and the whole under parts of the animal, yellowish white; head with two large white spots on the forehead, and with two similar white spots, placed one behind each ear: ears naked, oval; tail as long as the body, or rather longer; the fur of the basal portion considerably extended; the naked portion at first brown, but terminated with white.

Inhabits Guiana and Brazil.

In size this animal rather exceeds the Common Squirrel; the fur on the upper parts of the head and body, as well as that which covers the base of the tail, is of a rusty red, or reddish cinnamon colour; less pure on the lower parts of the limbs; above each eye is a large whitish spot, and a similar spot is placed behind each ear; the lower part of the cheeks is also whitish; the eye is encircled with the same rusty red hue as that on the back; the whole of the under parts of the body, as well as the inner sides of the limbs, are yellowish white: the naked portion of the tail is for the most part brown, but the apex is whitish. The female is larger than the male.

	Male.		Female.	
	Inches.	Lines.	Inches.	Lines.
Length from tip of nose to root of tail	10	3	10	3
" of tail	9	2	10	3
" of hairy portion of ditto ...	2	2	2	3
Width of the tail at the root ...	1	7		
From tip of nose to eye ..	1	1	1	3

The above account is taken from Temminck, who moreover states that the skull of the *D. Opossum* greatly resembles that of *D. Quica,* but that in the former animal the upper surface of the cranium forms a straight and descending line, whilst in *D. Quica* the central portion of the skull is elevated.

The Four-spotted Opossum[1] is very common in Guiana, and is found also in Brazil, but apparently in less abundance in the last mentioned district.

[1] The French call this animal *Le Didelphe Quatre-œil,* the conspicuons white spots on the forehead being compared to a second pair of eyes. I have applied the English name, Four-spotted, on account of there being two pairs of spots on the head ; the second pair are wanting, or scarcely traceable, in the other Opossums, which, like the present species, have spots above the eyes.

DIDELPHYS PHILANDER.

The Philander Opossum.

Didelphis Philander.	Linn. Syst. i. p. 72, sp. 2. 1766.
" "	Shreb. Säug. vol. iii. p. 541, Pl. 147. 1778.
" "	Temm. Monog. de Mammal. vol. i. p. 43, 1827, Pl. 6 (skeleton).
" "	Wagner, Shreb., Säug. Suppl. vol. iii. p. 45. 1842.
" *Cayopollin.*	Desm. Nouv. Dict. d'Hist. Nat. tom. ix. p. 147, 1817. Mammalogie, p. 257. 1820.
Philander Marsupialis[1].	Gray, List of Mammal. in British Museum, 1843, p. 101.

Upper parts of body of a yellowish rust colour (ashy grey, suffused with rusty yellow, in the female) ; under parts yellow ; upper surface of head pale silvery grey, but with a slender, dusky, longitudinal mark ; region of the eye dusky, and this tint extended considerably in front of the eye : tail with about one-fifth of its length clothed with fur like that on the body ; the remaining, scaly portion, at first brown, then clouded, or spotted with brown and yellowish white, and entirely of the latter hue at the distal extremity. Ribs very broad.

Inhabits northern parts of South America.

[1] With respect to the *Didelphis marsupialis* of the older authors, there can be no doubt that the descriptions and figures generally refer to one or other of the large species allied to the *Didelphis Virginiana*, or to that animal. Linnæus refers to Seba's plate and description as his *D. marsupialis*, and that plate, &c. is clearly taken from the animal called *D. cancrivora* by modern authors, though Linnæus notices in his description that the ears are black tipped with white—a character found in the *D. Virginiana*, but, to my experience, never found in *D. cancrivora*. I find no notice in Seba's account of the ears being tipped with white. The *D. marsupialis*, according to Schreber's description, must be the *D. cancrivora* of Temminck, but his figure, although copied from Saba's, it would appear is coloured from one of the Opossums belonging to the second division. On the whole, as there appears to be more than one species con-

This species agrees very closely in size with the *D. Quica, D. nudicaudata*, and *D. Opossum*, but may readily be distinguished by its short head, and the absence of the white spots on the forehead: on this part is a narrow, dusky longitudinal mark, which runs forwards nearly to the end of the muzzle, and is rendered conspicuous by the parts of the head immediately adjoining it, on either side, being of a silvery grey hue. The region of the eye is dusky brown, and this dark colour is much extended in front of the eye. The upper lip is edged with whitish, and the cheek is yellow. The hinder portion of the head and upper parts of the body are of a yellowish rust colour, and the sides of the body and outer side of the limbs are rusty yellow; the limbs are, however, somewhat tinted with grey, the pale grey of the fur next the skin not being altogether hidden by the brighter coloured points of the hairs on these parts, and the sides of the body near the limbs is of a brighter hue than elsewhere, assuming a golden yellow tint: the same rich yellow is observed on the sides of the neck: the fur covering the basal portion of the tail differs from that of the body only in being of a less bright tint. The whole of the under parts of the animal, as well as the inner surface of the legs, is of a bright yellow hue. The feet are sparingly clothed with pale hairs. The naked portion of the tail is at first brown; it is then spotted, or clouded with brown and yellowish white, and at the extremity, and, indeed, for a considerable distance from the point, is entirely yellow-white. The long bristly hairs, which as usual spring from the sides of the muzzle, from above the eye, and from the cheek, are brown.

The above description is taken from a male specimen

founded under the name *D. marsupialis* by Linnæus, I have followed those authors who reject the name altogether.

The term Philander is used by Mr. Gray in a generic sense for nearly all the Opossums in which the fur is short, and destitute of long bristly hairs.

preserved in the National Collection ; this specimen is stuffed, and, unfortunately, has lost its ears. It measures—

	Inches.	Lines.
From tip of nose to root of tail	14	0
Tail, about	14	0
Of which the hairy portion is	3	0
From nose to ear	2	3
Fore foot	1	1
Hind foot	1	8

In the same collection are some specimens of the *D. Opossum,* together with a second male Philander, which, being preserved in spirits, affords us a favourable opportunity of comparing the proportions of the two animals. The most striking difference which presented itself upon laying the two species side by side, was that observable, as already noticed, in the porportions of the head. With the body but one-third of an inch longer in *D. Opossum* than in *D. Philander,* the head was nearly three-quarters of an inch longer. Other differences were more or less marked : thus the tail of the *D. Opossum* was considerably shorter and thicker than in the *Philander,* and differed, moreover, in having the scales larger, and consequently more distinct : when the limbs are compared, the proportions of length and thickness are reversed in the two animals, the legs being shorter and stouter in the Philander than in the *D. Opossum*[1] In the latter animal the vertical groove on the naked muffle is moderately indented, and, besides the notch formed by the termination of this groove on the lower edge of the muffle, there are two other notches, one on each side of the mesial indentation. In *D. Philander* the nasal groove is more deeply impressed, and there are two distinct notches on either

[1] Judging from stuffed specimens, the shorter and thicker legs will also distinguish the *D. Philander* from the *D. nudicaudata* and the *D. Quica.*

side. The extremity of the tongue, as in others of the genus, is fringed with sharply-pointed fleshy tubercles; in *D. Opossum* I could perceive but one distinct row[1] of these tubercles, whilst in *D. Philander* there were several rows. The subjoined dimensions were taken from the two animals above alluded to. I regret it has not been in my power to compare individuals of the same sex.

	D. Opossum. Female.		*D. Philander.* Male.	
	Inches.	Lines.	Inches.	Lines.
Length from tip of nose to root of tail	11	0	10	0
" of tail	10	0	12	6
" of hairy portion of ditto ...	1	2	2	0
Diameter of ditto at the root ...		7½		6
Length from tip of nose to ear ...	2	7	1	11
" from ditto to eye	1	3½		10½
" of ear	1	1½	1	2
Width of ditto	1	0		11
Length of fore leg, from tip of elbow to the wrist	2	2	1	9
" of fore foot		11½		11
" of hind leg, from the knee to the ankle	2	7½	2	1
" of hind foot	1	7	1	4

Schreber's description of *D. Philander* agrees so closely in all essential points with the two specimens from which my account is taken, that I cannot doubt their being the same animal; and moreover I cannot but think my *D. Philander* is specifically identical with the animal figured by Seba, and which is referred to by Linnæus as his ***D. Philander***[2]. M. Temminck's description of the male Philander

[1] This row of tubercles edges the tongue; behind it were some others, but these could only be perceived with a strong lens: the difference consisted in the tubercles forming the hinder rows in *D. Philander* being more developed.

[2] See Seba's *Locupletissimi Rerum Naturalium Thesauri accurata descriptio*, tom. i. Plate 36, fig. 4. Seba states in his text (p. 57) that the female animal described by him has no pouch. Schreber describes the Philander

agrees perfectly, as regards the colouring, with the specimen described by myself, excepting that in that individual the under parts of the body are yellow, whilst the author just mentioned states that they are white in the specimens which he examined. The specimen preserved in spirit, of which I have given the dimensions, agrees more nearly with M. Temminck's account in this respect, the abdomen being nearly white. These male specimens, on the other hand, differ much from those noticed by M. Temminck, with regard to size; that author gives thirteen and a half inches (French measure) as the total length of three male specimens in his possession, of which, he states, the tail measured eight and a half inches, and the head one inch and ten lines[1].

The female Philander differs from the male in being less bright in its colouring; it has the same narrow dusky mark on the forehead, bordered on either side by pale grey, or whitish, and the tail spotted in parts with brown on a whitish ground colour, but the upper parts of the body are ashy grey, slightly suffused with rusty yellow; the under parts are yellowish white, and so are the cheeks and inner side of the limbs.

According to Schreber and Temminck, the Philander Opossum inhabits Surinam; the last mentioned of these authors states that he had never seen specimens in collections

as possessing a pouch, but his observations tend to show that it is more open than in the species hitherto described. Temminck states that the pouch is perfect in the female Philander. I cannot put much faith in Seba's statement, since I do not find his descriptions by any means remarkable for accuracy or precision: he appears, as Schreber remarks, to have drawn up his descriptions chiefly from his plates.

[1] The head and body, taken together, would then be only five Freneh inches, and yet one inch and ten lines is given as the length of the head, thus making it more than one-third of the length of the trunk. If these dimensions be correct, the *D. Philander* of Temminck would have a head larger in proportion to the body than the *D. Opossum*, whilst his description states that it is remarkable for the shortness of its head.

formed in Brazil. In the stomachs of specimens dissected, the remains of birds have been found.

	Specimen in Mus. Zool. Soc.		From Temminck		Brit. Mus.	
	Ins.	Lines.	Ins.	Lines.	Ins.	Lines.
Length from tip of nose to root of tail ...	9	0	9	9	10	6
" of tail	13	0	14	0	13	6
" of hairy portion of ditto	2	9	2	11	2	2
" from nose to ear	1	9			2	0
" of ear		10				9

Didelphys dichura.

Didelphys dichura. (NATTERER), WAGNER, in Wiegmann's Archiv. fur Naturgeschichte, vol. viii. Pt. 6 for 1842, p. 358.

General tint reddish grey; under parts yellowish white; tail equal to the body in length, and of a whitish flesh-colour, spotted with brown above, and immaculate beneath. Length of head and body 8½ inches; of tail, 9 inches.

Inhabits Brazil.

Dr. Wagner observes, that both *D. dichura* and *D. affinis* are liable to be confounded with the *Did. Philander*, and that the diagnosis of the last mentioned species might be as follows:—above, reddish, or hoary grey; beneath, yellow white; tail much longer than the body, the naked portion at first brown, and then white, spotted both above and below with brown, and with a considerable portion at the terminal extremity of an immaculate fleshy white tint; female with a distinct pouch. Length of head and body, 11 inches and 2 lines; tail, 15 inches, of which the hairy portion measures 2¾ inches.

Section II. *Opossums in which the pouch is rudimentary, or entirely wanting.*

Didelphys affinis.

Didelphys affinis. (NATTERER) WAGNER, in Wiegmann's Archiv. fur Naturgesch. vol. viii. Pt. 6 for 1842, p. 358.

General tint reddish; under parts yellowish white; tail longer than the body, spotted with white, both above and beneath; pouch wanting. Length of head and body 9 inches; of tail, 10 inches.

Inhabits Brazil (Matto grosso).

DIDELPHYS DERBIANUS.

Derby's Opossum.

Didelphys Derbiana. WATERHOUSE, Naturalist's Library (Marsupialia), vol. xi. p. 97, Pl. 2. Aug. 1841.

Body stout; tail considerably longer than the head and body taken together, and covered with fur for more than one-third of its length; fur of moderate length; general colour bright brownish rust; under parts dirty white; head greyish, and with a longitudinal dusky line above; lower half of fore legs white; hind feet dusky; naked portion of tail of a pale tint, but spotted with dark brown; a short silvery-grey stripe along the back.

Inhabits —— ?[1]

In having a dark mark on the head, and the naked portion

[1] Dr. Schinz states that Dr. Tschudy, in a recently published account of Peru, has described a species of Opossum, under the name *bicolor*; the work in question has not yet found its way into our National Library, and I have not been able to procure it.

of the tail spotted, this species agrees with the *D. Philander*, but it is readily distinguished from that, and other species, by the great extent of the tail, which is clothed with fur, combined with the silvery grey stripe on the back, the white fore legs, and the dusky hue of the hind feet. Its body seems to have been, moreover, considerably stouter than in other species of the short-furred section. The upper parts, and sides of the body, as well as the outer surface of the hind legs, are of a bright, brownish rust colour, and the under parts of the head and body are dirty white. The general tint of the head is brownish grey; a faint brown streak runs along the upper surface, commencing near the extremity of the muzzle, and terminating in a line with the anterior portion of the ears, and near this dark line the fur is of a purer and paler grey than on other parts of the head; the region of the eye is brown; the edge of the upper lip is white: the ears are tolerably large, naked, and of a pale colour, as are also the naked soles of the feet. The grey stripe on the back commences in a line with the shoulders, and terminates about midway between that point and the root of the tail; a second pale grey mark runs upwards on the side of the body, immediately behind the fore leg, and a third grey mark is observable on the outer surface of the hind leg, running obliquely upwards and backwards from the knee. The lower half of the fore leg is white, and the scanty hairs which cover the upper surface of the foot are also white; the hind feet are dusky. The fur covering the basal portion of the tail is of a dull brown colour, excepting on the under surface, where it is dirty white, and it is paler on the upper surface at the root than on other upper parts: the naked portion of the tail is of a pale pinkish hue, but is spotted with dark brown; the spots are numerous near the hairy portion, and gradually decrease in number as they become more remote from that part.

	Inches.	Lines.
Length from tip of nose to root of tail ...	13	6
" of tail, about	17	0
" of hairy portion of ditto	7	6 [1]
" from nose to ear	2	6
" from ditto to eye	1	1
" of ear	1	2
Breadth of ditto		11
Length of hind foot	1	9

I have seen but one specimen of this species, and that forms part of the collection of the Earl of Derby, after whom I have taken the liberty of naming it. In my notes (made some years ago) I omitted to notice the sex, and cannot therefore state whether it possessed a pouch. Its habitat is unknown.

DIDELPHYS LANIGERA.

Woolly Opossum.

Didelphis lanigera. Desm., Mammalogie, Pt. 1, p. 258, sp. 395.
" " Rengger, Naturgesch. der Säugth. von Paraguay, p. 225. 8vo. Basel, 1830.
Micouré second, ou Micouré laineux. Azara, Ess. sur l'Hist. Nat. des Quadr. du Paraguay, tom. i. p. 275.

Fur rather long and woolly; prevailing colour bright brown, on the sides of body, and outer sides of limbs, reddish brown; under parts reddish white; head with a narrow longitudinal dusky mark above; tail with the terminal third only, entirely destitute of fur: ears moderate, and of a livid violet tint.

Inhabits Paraguay.

[1] On the under surface of the tail the hair does not extend so far by about an inch and a half.

According to Azara this animal has a soft and woolly fur, which, on the back, is about one inch in length; its general colour is compared to that of Spanish tobacco, deeper on the back, and bright on the sides of the body; the under parts, and inner surface of the legs, whitish. On the head is a narrow brown line, commencing on the muzzle, and running backwards to the occiput. The ears are somewhat pendulous, naked within, and with the terminal half also naked externally; in the living animal they are of a livid violet hue: the edge of the lower lip is white. Fur, like that on the body, covers an unusually large portion of the tail, extending on the upper surface to the commencement of the terminal third part, but on the under surface it is less extended, terminating at the commencement of the second third; the naked portion of the tail is whitish. Both Azara and Rengger agree in stating that the tail has a triangular form, arising, according to the latter author, from the great development of the spinous and transverse processes of its vertebræ.

Three specimens examined by Dr. Rengger are described as of a light brown colour, passing into reddish brown around the eyes, on the sides of the neck, and outer surface of the extremities; as having a black stripe on the upper part of the head, which runs forwards to the point of the nose; the edge of the lower lip white, the bristles on the face black, the point of the muzzle flesh-coloured, the abdomen and inner side of the limbs reddish white, the ears and soles of the feet violet blue mingled with grey, and the naked portion of the tail reddish white.

Both sexes are said to agree in their colouring, and the female has no true pouch, this being represented only by two folds of the integument.

	Azara. Inches.	Azara. Lines.	Rengger. Inches.	Rengger. Lines.
Length from tip of nose to root of tail	8	8	8	10
" of tail	13	6	13	6
" of ear	1	0	1	0
Width of ditto		6	(nearly)	6
Length of head	2	3	2	4

Our English collections do not yet possess this species, nor did I find it in the French National Museum; M. Temminck, moreover, states, in his Monograph on the present genus, that he had never seen the *Micouré laineux* of Azara in any of the collections which he had examined. It approaches most nearly in its characters to the *Did. Derbiana.*

DIDELPHYS CRASSICAUDATA.

Thick-tailed Opossum.

Didelphis crassicaudata. Desm., Nouv. Dict. d'Hist. Nat. tom. ix. p. 425.—Mammalogie, p. 527, sp. 393.
" " Rengger, Nat. der Säugeth. von Paraguay, p. 226.
" " Waterh., Zool. of H.M.S. Beagle—Mammalia, p. 94, Pl. 30. Skull, &c. 25 *a* to *d*.
" *mustelina.* Geoff., Paris Museum.
Micouré troisième, ou *M. à queue grosse.* Azara, Essais sur l'Histoire Nat. des Quad. de la Province de Paraguay, vol. i. p. 284.

Muzzle short and obtuse; ears short, and clothed with small hairs; tail very thick at the base, rather shorter than the head and body taken together, with the basal third clothed with fur like that on the body, the remaining portion with short hairs; fur somewhat harsh, its general colour yellow-brown; under parts of body dirty yellow; the region of the eye, and the muzzle, brownish: tail with the terminal two-thirds black, with the exception of a smallish space at the end, which is white. Feet short.

Inhabits Brazil and Paraguay, and extends southwards as far as the River Plata.

The *D. crassicaudata* is not only remarkable for the great thickness of its tail, but differs from all the preceding species in having small ears, and these tolerably well clothed with small hairs; its tail, moreover, has the part which is usually naked. or nearly so, almost covered by small semi-adpressed hairs; its feet are very short, and so is its head. The fur of the animal, moreover, is remarkable for being composed chiefly of somewhat harsh hairs, and these pretty closely applied to the body, the under fur being rather scanty. In colouring it varies considerably. A specimen brought home by Mr. Darwin from Maldonado, La Plata, has the upper parts of the head and body of a brown-yellow hue, and the under parts dirty yellow; the muzzle, as well as the space round the eye, is dusky brown; the tip of the chin, and also the tip of the muzzle, on either side, whitish; the cheeks yellow; the small hairs covering the ears on the outer side are brownish, and those on the inner side are yellow, but brown towards the outer margin: about one-third of the tail is clothed with fur like that on the body: beyond this the tail is black, excepting a small portion of about one inch in length at the apex, which is white: these black and white parts of the tail are clothed with hairs which are nearly a quarter of an inch in length (shorter near the point), and are almost sufficiently numerous to hide the scaly skin beneath. The hairs on the back (which are moderately long) are grey next the skin, ochreous yellow towards the point, and yellowish brown at the point: on the under parts of the body they are faintly tinted with grey at the root, and of a dirty yellow colour at the point.

A specimen labelled *Didelphis mustelina* in the French national museum, but evidently belonging to the present species, is almost of an uniform pale yellow tint, but slightly suffused with brown on the back, and on the upper surface of the head; the muzzle is brown, and the same colour surrounds the eye; the lips and chin are whitish; the whole of the under parts of the body are of a straw yellow, the under surface of the neck is somewhat more distinctly tinted with yellow; the fore feet and legs are yellowish white; the hind feet brownish: about three inches of the apical portion of the tail is white; the middle third of the tail is black.

Rengger states that occasionally the prevailing hue of the upper parts of the body in this animal is brown-red[1], and that of the under parts, reddish grey.

	From Azara.							
	Male.		Female.		Male.			
	Ins.	Lines.	Ins.	Lines.	Ins.	Lines.	Ins.	Lines.
Length from tip of nose to root of tail	13	2	12	4	12	1	11	3
" of tail	12	1	9	7	11	0	10	3
Circumference of ditto at the root	3	10	2	2	2	9		
" of the body immediately behind the fore legs	7	5			5	8		
Length from nose to ear							2	1½
" of ear							1	5½
" of hind foot								6

The dimensions in the fourth column are taken from Mr.

[1] I may observe, that upon separating the fur on the back of the specimen brought home by Mr. Darwin, I perceived that at the time of its death the animal was about to shed its fur, and that the new coat would have been of a reddish hue, for numerous young hairs of this colour were visible.

Darwin's specimen already alluded to: the admeasurements of the cranium of the same specimen are—

	Inches.	Lines.
Total length of skull	2	4
Width	1	3
Length of nasal bones		$9\frac{1}{2}$
" of palate	1	$2\frac{3}{4}$
Width of ditto between the posterior molar teeth		5
Distance between fore part of front incisors and fore part of canine tooth ...	2	$0\frac{3}{4}$
" between fore part of canine and hinder part of last molar tooth	1	0
Length of lower jaw	1	$10\frac{1}{2}$

This specimen weighed, according to Mr. Darwin, $14\frac{1}{2}$ oz.

Dr. Rengger states that the animal is found throughout Paraguay, but not very abundantly. The specimen in the Paris Museum is labelled as coming from Brazil. A specimen which Azara saw in captivity allowed itself to be handled, although it was adult, and had only been caught a few days previously. He describes it as being very stupid, and in this respect resembling the *D. Azaræ*. It was fed upon raw meat, and a parrot happening to approach, it was captured and killed in a moment by the Opossum. The author just alluded to found six mammæ in a female which he examined; there were two folds of skin on the lower part of the abdomen, but no true pouch.

Didelphys ochropus.

Didelphys ochropus. (Natterer) Wagner, in Wiegmann's Archiv für Naturgesch. vol. viii. Pt. 6 for 1842, p. 539.

All that we learn from Dr. Wagner relating to this species is, that it resembles the *Didelphys lanigera*, but is of smaller size, of a more rufous tint, has the sides of the head, neck, and body, suffused with hoary grey, and the tail with nearly one-half naked. It inhabits Brazil (Barra).

DIDELPHYS CINEREA.

Ash-coloured Opossum.

Didelphis cinerea. TEMMINCK, Mongraphies de Mammalogie, tom. i. p. 46.
" " PR. MAXIM. Beiträge zur Naturgeschicte Braziliens, ii. p. 406.
" " WAGNER, in Schreber's Säugth. Suppl. iii. p. 47.
" " CUVIER, Règne Animal (1829), p. 177.

Fur short, dense, and of a cotton-like texture; general colour ashy grey, the back slightly suffused with black; under parts impure white; eyes encircled with black; tail with rather more than one-fifth of its length clothed with fur like that on the body, the naked portion at first brown, but with the greater portion white. The female somewhat suffused with rusty yellow.

Inhabits Brazil.

M. Temminck describes this animal as follows:—Size about equal to that of the Black Rat (*Mus Rattus*); head small, muzzle very short; ears somewhat contracted at the base, naked; tail much longer than the head and body taken together, very slender, and well clothed with fur at the base, the remaining portions perfectly naked; the extremity white, the white occupying more than half of the length of the tail. Females destitute of pouch.

The fur is dense, but short, and of a cotton-like texture; in the males it is of an ashy grey colour, but the extreme points of the hairs are suffused with blackish; the under parts of the body, as well as the inner surface of the limbs, are whitish, and the throat and chest are rusty white; the colouring of the fur on the head and muzzle does not differ from that of the back, and exhibits no trace of a mark on the forehead, or of pale spots above the eyes; the eyes are encircled with black, and the black is most extended in front of the eyes; the tail is covered at the base by a dense fur of the same ashy grey hue as that on the body; the remaining portion is covered by a scaly skin, but the scales are by no means distinct: this hairless part of the tail is at first brown, but the greater half, which is the terminal one, is white.

The females have the fur somewhat suffused with rusty yellow, and a yellowish tint is perceptible in the region of the ears, and on the cheeks; the white of the under parts of the body is less pure than in the males; the region of the mammæ, and of the longitudinal abdominal fold of skin, is of a rusty hue; the black band which encircles the eye is less broad, and less defined, than in the males.

	Temminck.				Male, Brit. Mus.	
	Ins.	Lines	Ins.	Lines	Ins.	Lines
Length from nose to root of tail	6	7	7	2	7	6
" of tail	9	10	10	5	9	6
" of hairy portion of ditto	2	2	2	2	1	4
Diameter of ditto at the base						3¾
From tip of nose to eye		7		7		8½
" " to ear					1	6½
Length of fore leg, from elbow to wrist ...					1	3
" of fore foot						7½
" of hind leg, from knee to ankle ...					1	6
" of hind foot						11½
" of ear						10
Width of ditto						8½
From front of foremost incisor tooth to back of last molar						10¼

I have added to the dimensions given by M. Temminck,

the admeasurements of an animal preserved in spirits in the collection of the British Museum: in all essential points it agrees with M. Temminck's description of *D. cinerea*, having the upper parts of the body grey, and the under parts white, or nearly so, and the eyes encircled with black, but the portion of the tail which is clothed with fur is less than in the specimens examined by the author just mentioned; it is, however, greatly extended, as compared with many of the species of the present section: such as *D. murina*, &c. The muffle in the specimen under consideration has a double cleft on its lower edge, on either side of the notch formed by the termination of the mesial groove, as in *D. Philander*, and I find the same character in *D. murina* and *D. pusilla*. Three very small Opossums accompanied the specimen in the British Museum, which appear to me to be clearly the young of the same species: they agree with the adult in colouring, and upon comparing them with the young of *D. murina* of the same size, they differed in having the tail longer in proportion to the body; the former being 2 inches, and the tail $2\frac{1}{4}$ inches, whilst in *D. murina* the tail was equal to the head and body in length. The scales on the tail of the adult animal are very indistinct.

DIDELPHYS INCANA.

Hoary-grey Opossum.

Didelphis incana. LUND, Det. K. Danske Vidensk. Selsk. Afh. viii. p. 236.
" " SCHINZ, Synopsis Mammalium, i. p. 503.

Upper parts grey, under parts white; a grey-black band passes through the eye; tail light brown, whitish at the extremity, and naked to the root. Length of body 4 inches; of tail, 5 inches.

Inhabits Brazil, in the province of Minas Geraes.

This animal, briefly described by Dr. Lund in the Transactions of the Society of Sciences at Copenhagen, it would appear, differs from the *D. cinerea* of Temminck in being of smaller size, and in having the tail destitute of fur at the base: this organ being of two colours, would serve to distinguish the *D. incana* from *D. murina* and *D. dorsigera*, and in the grey colouring of the fur on the upper parts of the body it likewise differs from the two species just mentioned. Whilst, however, there appear to be good grounds for regarding the *D. incana* as distinct from the species alluded to, that animal must, on the other hand, be exceedingly close to, if not identical with, Azara's *Micouré à queue longue*—a species only known to us through description.

DIDELPHYS GRISEA.

Grey Opossum.

Micouré quatrième, ou *M. à queue longue*. AZARA, Essais sur l'Hist des Quadr. de la Province du Paraguay, tom. i. p. 290. 1801.

Didelphis grisea. DESMAREST, in Dict. des Sciences Naturelles, tom. xlvii. p. 393.

Fur short and soft; general colour mouse-grey; sides of body pale; under parts dirty white; eye encirled with black, and external to this dark ring is a second, which is whitish.

Inhabits Paraguay.

Azara describes this species as having the fur as short and

as soft as that of the Common Mouse, and on the back of the head and upper parts of the body of the same colour as in that animal, but the sides of the body are said to be paler; the under side of the lower jaw, and fore part of the anterior limbs, nearly white, and the under parts of the body dirty white: a narrow dark line on the upper surface of the head separates the pale hue which adjoins the black eye-ring: the tail is of the same colour as the upper parts of the body.

	Inches.	Lines.
Length from tip of nose to root of tail ...	3	11
" of tail	5	4

The *D. grisea* is placed by authors in the pouchless division of the Opossums; but I find that Azara merely surmised that the animal was destitute of that appendage, and had been positively informed by a friend that the female possessed a pouch: perhaps it has a rudimentary pouch, like many of the species of the second section. Dr. Lund, in his description of *D. incana*, makes no mention of any pale ring encircling the black one which surrounds the eye, but in other respects his description agrees very closely with Azara's *Micouré à queue longue.*

I will here notice a small species of Opossum contained in the British Museum collection, which will perhaps prove distinct from either of the species briefly characterised by Dr. Wagner, and which appears to me to be undoubtedly distinct from either of the Opossums more fully described by M. Temminck, in his Monograph upon the group.

Didelphys ——— ?

Fur very soft, long and loose; general colour of the upper parts ashy grey, of the under parts white, faintly tinted with yellow; cheeks, and sides of the neck and body, suffused with yellow; eyes encircled with black, the black much extended in front of the eye: ears naked, of a brown colour, and tolerably large size: tail rather longer than the head and body taken together, with the basal half brown, and the terminal half whitish; about half an inch at the base clothed with fur like that on the body.

In size and general colouring this species approaches the *Didelphys cinerea,* but in that animal the fur is described as short, and of a cotton-like texture, whilst in the present species the hairs of the fur are longer than usual, being more than half an inch in length, and they are less dense. Both on the upper and under parts of the body they are grey next the skin, whilst in most of the small Opossums the fur on the abdomen is uniform in tint throughout. As compared with *D. cinerea,* another point of distinction is presented in the British Museum animal, its tail having a very small space at the root which is clothed with fur—a space of at most half an inch in length, whilst in *D. cinerea* a space of upwards of two inches on the tail (according to Temminck) is clothed with fur.

	Male.	
	Inches.	Lines.
Length from tip of nose to root of tail ...	6	9
" of tail	7	3
" from tip of nose to ear	1	6
" of fore foot		8
" of hind ditto	1	0
" of ear		8

DIDELPHYS DORSIGERA.

Merian's Opossum.

Didelphis dorsigera. LINN., Syst. (ed. 12), i. p. 72.
" " SCHREBER, Säugeth. iii. p. 546, tab. 150.
" " TEMMINCK, Monographies de Mammalogie, tom. i. p. 48.

Fur short, and of a grey-brown, or brownish yellow hue, on the upper parts of the animal, and impure white on the hinder parts; eye encircled with dark brown, and this dark tint extended along the sides of the muzzle; the intervening space, as well as the forehead, yellow-white: tail distinctly longer than the head and body together, and of an uniform brown colour; the fur at the base moderately extended.

Inhabits Surinam.

Although the Opossums of the section now under consideration differ from most other Marsupialia in having no pouch, they agree with the animals of their order in that remarkable peculiarity of being comparatively little advanced in development at the time of their birth.

The young, when sufficiently advanced to leave the teats of the parent (to which they are at first firmly attached), are carried by the parent on the back, where they retain their position by means of their prehensile tails, which are entwined round the tail of the mother. It is this habit of carrying the young on the back which gave rise to the name *dorsigera*, applied by Linnæus to the present species; the habit in question being, it was supposed, peculiar to it.

The present species was described as early as the year 1719, by Madame Merian, and is introduced in one of the plates of her large work upon the insects of Surinam, where it is represented with the young upon the back.

In size the *D. dorsigera* is rather inferior to the common black rat; its fur is short and scanty, and the hairs of which it is composed are of a deep grey colour at the root, and of a grey brown, or brownish yellow tint, at the point, and the general tint resembles that of the common brown rat. The eyes are surrounded by a deep brown ring, which is narrow both above and below the eye, but the dark hue is much extended in front: the forehead is yellowish white, as are also the cheeks, the outer surface of the fore legs, and the feet. The tail is of a uniform brown colour, and perfectly destitute of hairs, excepting at the root.

	Inches.	Lines.
Length from tip of nose to root of tail ...	6	0
" of tail	7	8½
" of hairy portion of ditto	1	0
" from tip of nose to eye		8

DIDELPHYS MURINA.

Murine Opossum.

Didelphis murina.	LINN., Syst. (ed. 12), i. p. 72.
" "	SCHREBER's Säugeth. iii. p. 545, tab. 149.
" "	TEMMINCK, Monographies de Mammalogie, tom. i. p. 50.
Murine Opossum.	PENNANT, Quadr. p. 207.
" "	SHAW, General Zoology, Vol. i. Pt. 2, p. 427.

Fur short and soft, of an ochreous and sometimes rusty yellow hue on the upper parts of the body, and yellow-white on the under parts; eyes surrounded by black, which is much extended in front of the eye; tail rather longer than the head and body together, and of an uniform yellow tint.

Inhabits Guiana, Brazil, Peru, and Mexico.

The short and soft fur on the back of this animal is of a palish grey hue on the upper parts and sides of the body, but the hairs are of a more or less rich yellow hue (sometimes inclining to rusty) at or near the point: on the middle of the back the extreme points are brownish, and the yellow tint is therefore less pure than on the sides of the body, where the hairs are of an ochreous yellow at the point: the grey colour of the basal part of each hair is not altogether hidden by the overlapping yellow points. The outer surface of the limbs is of the same colour as the upper parts of the body. The fur on the under parts of the head and body, as well as the inner surface of the limbs, is uniformly yellowish white to the root. The black ring which surrounds the eye is narrow, both above and beneath that organ, rather broad behind it, and in front is much extended; the mesial portion of the muzzle above, and the forehead, are of a very pale yellow tint. The edge of the upper lip and the cheeks are yellow-white. The ears are brown. The feet are whitish.

	From Temminck.				Paris Museum.		Male. Brit. Mus.		Male. Brit. Mus.		Female. Brit. Mus.		Female, preserved in spirit, Brit. Mus.	
	Ins.	Lines.	Ins.	Lines.	Ins.	Lines.	Ins.	Lines.	Ins.	Lines.	Ins.	Lines.	Ins.	Lines.
Length of head and body	5	3	5	4	5	0	5	9	5	9	5	4½	5	4
" of tail	5	9	5	10	6	0	6	6	6	6	5	8	6	0
" of hairy portion of ditto		7		7				5		5		4½		5
" from nose to eye		7		7		7⅓		7						6
" from ditto to ear					1	2	1	2⅓	1	3¼	1	3	1	0½
" of fore leg, from elbow to wrist ...														11
" of fore foot								7		7		6		6
" of hind leg, from knee to ankle ...													1	2
" of hind foot						9¼		9½		10		9		8⅓
" of ear						6½		7		7		6		7½

With the exception of some slight differences which will be seen in the dimensions, and which differences are no doubt partly attributable to the measurements being taken, most of them, from stuffed specimens, the above description accords perfectly with the "Marmose" (*Didelphys murina*) of the French authors[1], and with four specimens contained in the British Museum collection, one of which is preserved in spirits. These specimens, moreover, accord perfectly with M. Temminck's account of the Murine Opossum; but, on the other hand, they cannot be regarded as the *D. murina* of Dr. Wagner, inasmuch as they have large ears, and these organs are described by that author as small in the animal which he regards as *D. murina.*

The Murine Opossum is found in Brazil, but is apparently most abundant in the northern portions of South America. Among my notes I have a description of a little Opossum, brought from Mexico, which differs only from the *D. murina* in being a trifle smaller: its dimensions were—

	Inches.	Lines.
From tip of nose to root of tail	4	10
Length of tail	5	6
" from nose to ear	1	0
" of ear		$6\frac{1}{2}$
" of fore foot		5
" of hind foot		$7\frac{1}{2}$

The specimen was a female, and had nine very distinct mammæ. I found the same number in the female *D. murina* which is preserved in spirits in the British Museum

Small as the present animal is, it attacks birds, their feathers being found, mixed with numerous insect remains, in the stomachs of specimens dissected by M. Temminck. The author just mentioned states that it burrows in the ground, and climbs trees, and, like others of the genus, will eat fruits.

Dr. Wagner describes two small species of Brazilian Opossums in Wiegmann's Archiv[2], which he suspects may have been confounded with the *D. murina*. The first—

[1] I have descriptions before me which I drew up from the specimens in the French National Museum.

[2] See vol. viii. (1842), p. 359.

Didelphys macrotarsus,

Dr. Wagner states, resembles the *D. murina*, excepting in having the ears much larger, its tail entirely naked, and of a deep reddish ash colour, and its feet stout and elongated.

The second species—

Didelphys microtarsus,

Also resembles *D. murina*, but has the ears much larger, the tail of a deep reddish grey, and with minute hairs on the under side, and the tarsi short and slender.

I will here introduce some other species of Opossums described by the same author in the work mentioned. The species in question were collected by Dr. Natterer during his long sojourn in Brazil; some of them are stated to be destitute of a pouch, and therefore belong to the present section, and it is highly probable that the females of the remainder are also pouchless. The shortness of the tail in the following four species would seem to indicate that they are very nearly allied to the *Didelphys tricolor*, and the *D. brachyura*.

Didelphys domestica. (Natt.) Wagner.

Wiegmann's Archiv für Naturgeschichte, viii. (1842), p. 359.

General tint yellowish grey, beneath pale yellowish; ears moderately large; head without any stripe; tail short and thickish, with minute white scattered hairs. The female has no abdominal pouch. Length of the body 7 inches; of tail, 2 inches 4 lines.

From Cuyaba.

Didelphys glirina. (NATT.) WAGNER.

WIEGM., Archiv, *l. c.*

Fur ashy grey; under parts of the body hoary grey, washed with yellowish; the sides ochreous; head short; ears moderate; tail not quite equal to half the body in length, hairy at the base; beyond, almost naked. Length of the body 6½ inches; of tail, 2 inches 7 lines.

From Mamoré.

Didelphys velutina. (NATT.) WAGNER.

WIEGM., Archiv, *l. c.* p. 360.

General tint mouse-grey, under parts whitish, the white separated from the grey by a well marked line; tail, excepting at the base, beset throughout with minute, adpressed, glossy hairs, and these of a brown colour. Length of body 3¾ inches; of tail, 2¾ inches.

From Ypanema.

Didelphys unistriata. (NATT.) WAGNER.

WIEG., Archiv, *l. c.*

This species has the upper parts reddish mixed with grey, and on the back is a darker stripe; the under parts are reddish; the tail short, and slightly hairy. Length of the body 5¼ inches; of tail, 2½ inches.

From Ytarare.

All the above new species of Didelphys, described by Dr. Wagner, are contained in the Museum at Vienna, and were collected in Brazil by the late Dr. Natterer.

DIDELPHYS PUSILLA.

Diminutive Opossum.

Didelphis pusilla. DESMAREST, Nouv. Dict. d'Hist. Nat. tom. ix. p. 430. 1817. Mammalogie, Pt. 1, p. 261, sp. 399. 1820.
Micouré sixième, ou *Micouré nain.* AZARA, Essai sur l'Hist. Nat. des Quadrupèdes du Paraguay, tom. i. p. 304.

Fur short, and soft; on the upper parts of the animal, mouse-grey, and on the under, whitish; eye surrounded with black, the black most extended in front; the region above the eye, whitish, and a yellowish white spot is situated beneath the eye: tail naked.

Inhabits Paraguay.

A small Opossum in the British Museum Collection bearing the name *D. pusilla*, agrees closely with Azara's description and dimensions. The specimen is preserved in spirits, and in its size and proportions, and (so far as one may judge from a specimen in the condition mentioned), in the character of the fur and its colouring, greatly resembles the Common Mouse. The under side of the tail at the extremity is perfectly naked, and the callous skin of this part shows that it is used for prehension; the naked part extends backwards from the point of the tail about half an inch; the remaining portion is covered with small scales, from between which small hairs spring, as in the common mouse. A narrow black ring surrounds the eye, and the sides of the muzzle, near the anterior angle of the eye, are dusky; the upper lip and the cheeks are yellowish white. There are two distinct grooves on either side of the mesial notch formed by the termination of the nasal groove, in the upper lip. The ears are clothed throughout with small hairs. I add the dimensions of this animal (which I think is without doubt the true *D. pusilla* of Desmarest) to those given by Azara.

	From Azara.		From B. M. specimen.	
	MALE.		MALE.	
	Inches.	Lines.	Inches.	Lines.
Length from tip of nose to root of tail	3	6	3	3
" of tail	3	10	3	9½
" from nose to eye				4⅓
" from ditto to ear				10
" of ear		6		5½
Width of ditto		4⅓		5½
Length of fore foot				4½
" of hind foot				6

DIDELPHYS ELEGANS.

Elegant Opossum.

(*Plate* 16, fig. 1).

Didelphys elegans. WATERHOUSE, Zoology of the Voyage of H.M.S. Beagle, Mammalia, p. 95, Pl. 31. — Naturalists' Library, (Marsupialia), Vol. xi. p. 106.

" *hortensis.* REID, Proceedings of the Zoological Society for January, 1837, Pt. 5, p. 4.

Thylamys elegans. GRAY, List of the Mammalia in the British Museum, p. 101. 1843.

Fur long, and very soft, of a delicate yellowish ash colour on the upper parts, slightly suffused with black, or brown, on the back, and white on the under parts; ears rather large; eyes surrounded with black; tail about equal to the head and body in length, and usually very thick; feet small.

Inhabits Chili.

The fur of this little animal is long, and remarkably soft, and composed of hairs of one kind only; presenting none of that woolly or cotton-like texture so generally found in the small Opossums; the hairs on the back are of a darkish slate-grey next the skin, and annulated near the point with ash colour, and sometimes with yellowish, and tipped with rusty black; on the sides of the body they are also of a deep slate-grey at the root, but at the point they are of a delicate yellow tint, or sometimes of a deep cream colour; on the under parts of the body the hairs are of a pure white colour throughout their whole length. The general tint of the upper parts varies somewhat, being sometimes ashy grey, slightly suffused with black along the middle of the back, and sometimes yellowish ash colour, suffused with brown. The white of the under parts is separated by a tolerably well defined line from the darker hue of the upper, and extends about half way up the sides of the body; the flanks, immediately above the white parts, are always more or less tinted with yellow. The head is rather short; the ears tolerably large, and of a greyish brown hue, but pale at the root; they are clothed throughout with very minute hairs: a broadish black ring encircles the eye, and the sides of the muzzle are suffused with blackish. There is but one indentation on the lower margin of the muffle, on either side of the mesial groove. The legs and feet are white; the latter are smaller than usual. The tail is remarkably thick in most of the specimens which have come under my notice, but I have seen some in which it was scarcely thicker in proportions than in most other Opossums; when incrassated the thickest part is near the termination of the first quarter of the entire length; from that point it rather suddenly diminishes in diameter, and at the extremity it terminates in a fine point; at the base it is distinctly constricted; about a quarter of an inch of the tail, at the apex, is bare beneath, and there is a longitudinal groove on this part; the remaining portions are covered with minute adpressed hairs, which, however, are not sufficiently numerous to hide the skin: the scales on the skin are very indistinct.

	Female?		Male.				Male, preserved in spirits, Brit. Mus.		Female?[1]	
	Ins.	Lines.	Ins.	Lines.	Ins.	Lines.	Ins.	Lines.	Ins.	Lines.
Length from tip of nose to root of tail	4	1	5	5	4	6	4	1	5	3
" of tail	3	6	4	2	4	4	4	3	5	2
Diameter of ditto near the root		5						4¾		
Length from nose to eye		6½		7				5½		
" " nose to ear		9¼	1	0½	1	1½		11½	1	0½
" of ear		7¾		7½		7¼		7¾		7¾
Width of ditto ...		7½				7½		7		
Length of fore foot ...		5½						6		5½
" of hind foot ...		7½		7		7½		7¼		7⅓
" of skull ...	1	1	1	2¼	1	2½				
Width of ditto		7⅙		8⅓		8				
" " between orbits		2¼				2½				
Length of nasal bones[2]		5⅓								
" of palate ...		7				7¼				
" from front of foremost incisor to back of last molar tooth		7¼								
" of lower jaw ...		10				10½				

[1] In this specimen, which is apparently a female, the tail is not incrassated, but I am satisfied that the thick tail is not peculiar to either sex.

[2] The nasal bones are not expanded behind, as is usual in the Opossums, but are of very nearly the same width throughout.

The *Didelphys elegans* is found nearly throughout Chili; Mr. Bridges informs me that he has found it as far north as Cobija, and that in the opposite direction it extends to Curico, in the province of Colchagua. Mr. Darwin, who first brought specimens to Europe, states that these little animals are frequent in the thickets growing on the rocky hills, near Valparaiso. They are exceedingly numerous, and are easily caught in traps baited either with cheese or meat. The tail appeared to be scarcely at all used as a prehensile organ: they are able to run up trees with some degree of facility. Larvæ of beetles were found in the stomachs of specimens dissected.

DIDELPHYS TRISTRIATA.

Three-striped Opossum.

Didelphys tristriata.	KUHL, Beiträge, p. 63. 1820.
" "	FISCHER, Synopsis Mammalium, p. 269. 1829.
" "	WATERHOUSE, Nat. Libr. (Marsupialia), Vol. xi. p. 107, Pl. 3.
Sorex Brazilensis.	ERXLEBEN, Systema Regni Animalis, p. 127. 1777.
" "	SCHREB., Säug. p. 577. 1778.
" "	GMEL. LINN., Syst. i. p. 115. 1788.
La Musaraigne du Brézil.	BUFF., Hist. Nat. tom. xv. p. 160. 1767.
Brazilian Shrew.	PENNANT, Synopsis of Quadrupeds, p. 309. 1771.

Fur very short, of a rich brown hue on the upper parts of the body, and suffused with rust colour on the hinder parts; abdomen rusty yellow: three longitudinal black stripes on the back. Ears small; tail short.

Inhabits Brazil.

The small size, and comparatively short ears and tail,

combined with a sharply-pointed muzzle, give to the present species the aspect of a Shrew-mouse, and hence by the older authors we find it arranged amongst the Shrews; no animal belonging to the order *Insectivora*, however, has yet been found in South America: the true Insectivora are there replaced by the Opossums. The species already described all climb trees more or less, and possibly the Three-striped Opossum may also climb, but I suspect that this animal lives chiefly upon the ground, and that it seeks shelter in burrows, as is the case with the short-tailed species, which remain to be noticed. Our little animal is found in but few museums, having probably escaped collectors from its small size, it being not larger than the common mouse. The following description is drawn up from a specimen contained in the museum of the Zoological Society. It was purchased from a dealer in Liverpool, who did not know from what part of America it was procured.

The fur of the Three-striped Opossum is short, moderately soft, somewhat closely applied to the body, and of a grey colour next the skin, both on the upper and under parts of the animal. The general tint of the upper parts is rich brown, the fur being pencilled with deep yellow and black on the back; on the sides of the body the black gradually disappears towards the lower parts, where a rich rusty yellow tint prevails, and the same yellow hue is observed on the whole of the under parts of the animal. The three black marks on the back are rather broad; the central one commences near the tip of the muzzle, and extends backwards along the head and back, and on to the basal portion of the tail: on the middle of the back this stripe is fully a quarter of an inch in width; the other two dark stripes are less broad, they commence immediately behind the ears, and extend to the root of the tail. The ears are small, rounded, and nearly naked, but some very minute brown hairs are scattered over both the outer and inner surfaces. The tail, in like manner, has minute scattered hairs, and these are of a blackish brown hue on the upper surface of the organ in question, and brown on the under: fur, like that on the body, extends on the basal portion of the tail, covering a space of about one-third of an inch in length.

		Inches.	Lines.
Length	from tip of nose to root of tail ...	4	3
"	of tail	2	1
"	from tip of nose to ear	1	$0\frac{1}{2}$
"	of ear		3
"	of hind foot and claws		$8\frac{3}{4}$

DIDELPHYS TRICOLOR.

Tricoloured Opossum.

Didelphis tricolor.	(GEOFF.) DESM., Nouv. Dict. d'Hist. Nat., ix. p. 429.
" "	TEMM. Monogr. i. p. 52.
" *brachyura.*	PALLAS, Act. Petrop. Ann. 1780, ii. p. 235, Tab. 5.
Le Touan.	BUFF., Hist. Nat. Suppl. vii. p. 252, Tab. 61.

Head large; tail about equal in length to half that of the trunk of the animal: fur short; on the upper parts of the head and body, black, obscurely pencilled with white; on the sides of the body deep rusty red, and on the under parts, white.

Inhabits Guiana.

The Tricoloured Opossum is readily distinguished by its short tail, combined with the three distinct colours of black (or nearly black), rusty red, and white, with which its body is adorned. Its fur is dense, and very short, and of a grey hue next the skin, both on the upper and under parts of the body; on the back the visible portions of the hairs are black, but they have most of them a white ring near the point, sometimes very narrow and indistinct, and sometimes broader, in which case the general tint produced by the mixture of black and white is grey-black, or dark grey. The chin, throat, the sides of the face (extending above the eye), and the sides of the body, as well as the outer surface of the legs, are of a deep rusty red hue; the chest, and the whole of the abdomen, are pure white, or sometimes white faintly tinted with yellow. The ears are rather smaller than usual, and of a dark hue. The tail is short, and thick at the root; the basal half, or rather more, is clothed with fur like that on the body, excepting on the under surface, which is naked, or very nearly so; here, as well as on the remaining portion, the terminal half above, is clothed with short black hairs. The feet are black, slightly freckled with rust colour.

	Inches.	Lines.
Length from tip of nose to root of tail ...	5	6
" of tail	2	8
" from nose to ear	1	2½
" of ear		5
" of hind foot		10

According to M. Temminck, adult specimens of the present species vary from nine, to nine and a quarter inches, or rather more, in entire length, of which the tail is about three and a quarter inches.

My description is taken from specimens in the Paris Museum, one of which is from Cayenne, where it is said to be most abundant. M. Desmarest states that, whilst in Paris, Azara recognized, in a specimen of the *D. tricolor* contained in the National Museum, his *Micouré à queue courte*; hence we find in systematic works, published since that time, the Three-coloured Opossum is given as a native of Paraguay as well as Guiana. Azara's description, however, agrees perfectly with the ***Didelphys brachyura*** of authors, a nearly allied species, and which we know extends as far south as the River Plata ; and hence I am induced to believe that Azara, who was possibly not acquainted with this second short-tailed species, and who, it must be borne in mind, was trusting to his memory, was mistaken in the identification.

DIDELPHYS BRACHYURA.

Short-tailed Opossum.

(*Plate* 16, fig. 2).

Didelphis brachyuros.	SCHREB., Säug. iii. p. 548, Pl. 151. 1778.
" *brachyura.*	GMEL. LINN., Syst. i. p. 108.
" "	TEMM., Monogr. i. p. 53.
" "	WATERH., Zoology of H.M.S. Beagle (Mammalia), p. 97, Plate 22.
Short-tailed Opossum.	PENNANT, Quadr. p. 208.

Tail short ; fur short, grey on the upper parts of the head and body, and rusty yellow on the sides and under parts, rather paler on the abdomen than elsewhere : feet yellowish.

Inhabits Guiana and Brazil, and extends southwards as far as the River Plata.

The head of this animal is larger than usual, in proportion to the body ; the canine teeth are large ; the tail is not quite equal to half the body in length. The fur is short and dense, and moderate as to texture, of an ashy grey hue on the upper surface of the head, and on the back, and of a rusty yellow on the sides of the head and body, as well as on the under parts ; but here the hue is rather paler than elsewhere. The tail is tolerably well clothed with short stiff hairs, which, excepting on the basal portion, are not sufficiently numerous to hide the scaly skin beneath: at the tip of the tail beneath is a small naked space of about a quarter of an inch in length. The ears, which are very short as compared with those of most other Opossums, are clothed with small yellowish hairs.

It is necessary to observe that this animal is subject to some slight variation in its colouring. Sometimes the upper parts of the body are suffused with yellowish. A specimen in the British Museum collection has the sides and under parts of the body unusually pale, the former being tinted with ochreous yellow, and the latter yellow-white: its back is ashy grey, pencilled with black; about half an inch of the tail at the base is covered with fur like that on the body, and the remaining portions are well clothed with short black hairs; the sides of the muzzle are brown; and the feet are of a pale dirty yellow tint. It is from Brazil.

	Zoological Society.		Male, Brit. Mus.		Paris Museum.	
	Ins.	Lines.	Ins.	Lines.	Ins.	Lines.
Length from tip of nose to root of tail ...	6	0	6	6	6	9
" of tail	2	8	2	7	2	6
" from nose to ear	1	6	1	4½	1	3
" of ear		3¼		4		3¼
" of fore foot and nails				8		
" of hind foot		8¾		9½		8½

The first of the above columns contains the dimensions of a specimen brought by Mr. Darwin from Maldonado, La Plata.

Didelphys Hunteri.

Didelphys Hunteri. **Waterh., Naturalist's Library (Marsupialia), Vol. xi. p. 110.**

The above name was proposed for a small species of Opossum, agreeing in size and proportions very nearly with the *D. tricolor* and the *D. brachyura*, but which differs from those animals in its colouring, the upper parts of its body being of an uniform brown-black tint, and the under parts pale brown: it presents no trace of the rusty hue on the flanks, which characterizes the *D. tricolor*. The specimen is preserved in spirits in the Museum of the College of Surgeons, and formerly formed part of the collection of the celebrated John Hunter, after whom the animal is named. Of course this cannot be regarded as a well-established species.

Didelphys pæcilotus.

This species, described by Dr. Wagner in the volume of Wiegmann's Archive, already quoted (at p. 358 of Vol. viii.), should have been introduced next to the *Didelphys cancrivora*, agreeing with that animal in its general characters, and no doubt possessing a pouch. It is said to be nearly equal in size to the *D. cancrivora*. Its woolly fur is whitish, and the long interspersed silky hairs are white at the base, and black at the point, or some few of them are white throughout: the head is white, and has three narrow black stripes; the ears are of a whitish flesh colour, spotted with black. Inhabits Brazil—Agaba.

Since the preceding pages were printed, copies of the published parts of Dr. Tschudi's *Fauna Peruana* have arrived in England, and the author, having been favoured by Mr. Gould with the loan of this work, is enabled to notice certain Peruvian species of Opossum described by Dr. Tschudi, which that naturalist regards as new to science. These are respectively named *Didelphys ornata*, *D. noctivaga*, and *D. impavida*.

Didelphys ornata.

Didelphys ornata. TSCHUDI, Fauna Peruana, Pt. 2, Pl. 7, and Pt. 4, p. 146. 1845.

Fur dense, and somewhat woolly; on the upper parts of the body of a rusty brown colour, suffused with black; on the under parts grey-white; chest pure white: head with a blackish mesial line, on each side of which are two longitudinal grey lines; muzzle dusky; ear encircled with brown, the brown extended forwards on to the sides of the muzzle, and backwards to the ear: between the shoulders is an oblong, but nearly heart-shaped, white patch. About half the tail is clothed above with fur like that on the body; the naked portions are yellowish: ears large and naked, and of a brown-red colour.

	Inches.	Lines.
Length from nose to root of tail	11	0
" of tail	12	0
" of hairy portion of ditto, above ...	6	6
" of ditto, below	3	6
" of head	2	6
" of ears	1	0
Width of ditto		9
Length of fore foot	1	0
" of hind ditto	1	3

The description, it will be seen, accords in all essential points with my account of *D. Derbiana*. The pale mark on the back, it would appear, is more expanded in Dr. Tschudi's animal than in that described by myself, but I see no reason to doubt the specific identity of *D. ornata* and *D. Derbiana*.

Didelphys impavida.

Didelphys impavida. TSCHUDI, Fauna Peruana, Pt. 4, p. 149; Pt. 2, Pl. 9.

Fur moderate as to length; on the upper parts of the body reddish brown, suffused with black; on the sides of the body of a fulvous yellow, and on the under parts white; eyes encircled with black, the black prolonged in front almost to the extremity of the muzzle; a longitudinal pale stripe on the forehead: tail grey-brown, and with small silvery white hairs springing from between the scales. Female with abdominal folds of skin.

	Inches.	Lines.
Length from tip of nose to root of tail ...	6	3
" of tail	5	4
" from nose to ear	1	6
" of ear		9

Like others of its group, this species is nocturnal, and during the day remains hidden in holes in the earth, or under the roots of trees. The strong light of the sun, Dr. Tschudi states, so dazzles this animal, that it is easily captured. Of the seven species of Didelphys noticed in Peru by the author just mentioned, six were met with in the thickest forests, and one in the open fields: several live exclusively on vegetable substances, and prefer the fruits of the bananas to all others. The *D. noctivaga* was frequently caught in the huts on the bananas hung up to ripen. By using a dark lantern a strong light was suddenly thrown upon them when committing their nightly attacks upon the fruits in question, and this so stupified them that they were readily captured by the hand. The *D. impavida* also frequently visited the huts of the travellers in the night, and, indeed, was exceedingly troublesome, running, as the Doctor observes, over their faces and hands, and always returning after being driven away.

Didelphys noctivaga.

Didelphys noctivaga. Tschudi, Fauna Peruana, Pt. 4, p. 148, and Pt. 2, Pl. 8.

Fur soft and long; upper parts of the body greyish brown, suffused with dusky; sides of the body reddish yellow; under parts yellowish white; along the mesial line of the belly the hairs uniform in colour to the root; on other parts of the body the fur grey next the skin: a pale, yellowish, longitudinal stripe on the upper surface of the head; the eyes encircled with black-brown, and this dark colour well defined; lower region of the ear reddish yellow, and the throat and breast suffused with the same colour: a very small portion of the tail clothed with fur; the naked portion red-brown: ears naked, and reddish: feet light brown; the toes whitish yellow; the soles flesh coloured. Female provided with abdominal folds of the integument.

	Inches.	Lines.
Length from tip of nose to root of tail ..	6	9
" of tail	7	4
" from nose to ear	1	3
" of ear		9½

With respect to the geographical ranges of the preceding three species, and some other Opossums found in Peru, Dr. Tschudi furnishes us with the following particulars:—

Didelphys Azaræ, Dr. Tschudi observes, has a very extensive range, being met with throughout the whole of South America, to 45° south latitude: in Peru it occurs in all the districts, and particularly in the western district. A female, with young, was killed at a spot having an elevation of more than 12,500 feet above the sea: in the hot, woody parts, it disappears, and it is very rare on the east slope of the coast Cordilleras at an elevation of 6,000 feet above the sea. *Didelphys myosuros* (or *D. nudicaudata*) and *D. Opossum*, live in the districts of the coast, and the forest; the former also occurs in the Sierra region at an elevation of from 9,000 to 10,000 feet above the sea; the latter has not been found above the Ceja region. The three species *D. noctivaga*, *D. impavida*, and *D. ornata*, have hitherto been exclusively found in the middle and deep forest regions, and the climate at an elevation above 3,500 seems to be unfitted for them. Their horizontal range is also small, since they have only been met with between the 10th and 12th degrees of south latitude. *D. murina*, on the other hand, has a wide horizontal range, being found throughout the whole of the northern parts of South America, from the Atlantic to the foot of the inner Cordilleras.

Didelphys Cuvieri (Fossil).

Didelphys Cuvieri. **FISCHER, Synopsis Mammalium, p. 268.**

Dr. Fischer gives the above name to the animal whose skeleton was discovered by Cuvier in the Eocene Tertiary formation of the Paris basin. The skeleton in question agrees most nearly in size with that of the *Didelphys murina*, but the separate bones differ in their proportions. It exhibits distinct marsupial bones; the angle of the lower jaw was found to be bent inwards, as in the Marsupialia, and the true molar-teeth remaining in the fractured skull, agree with those of the Opossums, both in number and form. In fact, very satisfactory evidence is deduced by Cuvier from this skeleton, to the effect that a small insectivorous Marsupial Mammal coexisted with the Anoplotheria, Palæotheria, and other now extinct Mammals of the Paris Gypsum Quarries, but that that animal was a member of the genus *Didelphys*, as now restricted, could not be satisfactorily ascertained, inasmuch as the specimen did not exhibit the incisor and premolar teeth. A copy of Cuvier's figure of this skeleton will be found at p. 13; figs. A and B.

Fossil Didelphidæ of the Brazilian Caverns.

I have before me fragments, from the above caverns, which are clearly referrible to seven or eight distinct species of *Didelphys*. The specimens are chiefly rami of lower jaws, which differ in size and proportions, and correspond very closely, as regards size, with the different species now inhabiting Brazil. Indeed, there are no fragments appertaining to the marsupiate division of Mammalia, in the extensive collection of remains, procured from the Brazilian caverns, now in the British Museum, which are not referrible to the true Opossums, nor do I find any, the size and proportions of which would lead me to suspect they belong to species which are not known to exist at the present day in Brazil.

Sub-genus, *Chironectes.*

Chironectes. Illiger, Prodromus Systematis Mammalium, &c. p. 76. 1811.

Opossums having large hind feet, the toes of which are united by a web; the fore feet moderate, and with an unusual elongation of the pisiform bone. Females provided with a perfect pouch. Habits aquatic.

The present section contains but one known species, the *Chironectes variegatus*, or *Chir. Yapock*, of modern authors, an animal rarely found in museums, and of which we have only had an opportunity of examining an imperfect specimen. A skin, having the skull *in sitû*, was placed in Mr. Ogilby's hands for examination by Dr. Natterer, and in a communication published in the Proceedings of the Zoological Society will be found a detailed account of such characters as could be gleaned from this specimen. Amongst other peculiarities of the Chironectes, pointed out by Mr. Ogilby[1], that gentleman notices the existence of large cheek-pouches, extending far back into the mouth, and of which the opening was very apparent[2]. Mr. Ogilby's remarks upon the dentition of the Chironectes, as the only detailed account of these organs hitherto published, I will give in that gentleman's own words. "The teeth of this animal," Mr. Ogilby states, "are altogether different from those of the Opossums (Didelphys), and I am at a loss to reconcile my own observations with those of M. F. Cuvier upon this subject, as given in the Dents des Mammifères, p. 73, unless by supposing there must have been some mistake about the skull referred by M. Cuvier to the *Yapock*. For my own part, I could not be deceived in this matter, as the skull which I examined had never been extracted from the specimen. The incisors and canines are of the same form and number as in the true Opossums; the two middle incisors above, being rather longer than the lateral, those below broader, and a little separate. The molars are five on each side, two false, and three real, both in the upper and under jaws. The first false molar is rather small, and in contact with the canine both above and below: the second is half

[1] See Proceedings of the Zool. Soc. for May, 1836, p. 57, Part 4.

[2] Dr. Wagner remarks, that unless Mr. Ogilby's observations upon this point were made upon a specimen preserved in spirits, they require confirmation.

as large again, and both are of a triangular form, with apparently two roots. The three real molars are of the normal form of these teeth among the Opossums. The first of the upper jaw is longer than broad, and has four sharp, elevated tubercles, with a low heel projecting backwards; the second resembles it in general form, but is larger and broader; the third is small, and resembles the tuberculous molars of the true *Carnivora*. In the lower jaw the three real molars do not materially differ in point of size: they are narrower than those of the upper, have their tubercles arranged in a single longitudinal series, a single large one in the centre, and a smaller one on each side." This last phrase points out a difference in the structure of the molars of the Chironectes and true Opossums, since the latter animals present two longitudinal series of tubercles; beyond this it does not appear that there is any difference in the structure of the teeth of the animal under consideration, and the species of *Didelphys*, unless it be in the last upper true molar, which is compared to the tuberculous molars of the true *Carnivora*—teeth which differ very much in different species of that group. With regard to the number of the teeth, which we find in the animal examined by Mr. Ogilby was less than in the Opossums, it is necessary to observe, that such a difference might arise from a difference of age; and on this point I have to remark, that the *Yapock*, upon which the foregoing notes were drawn up, was inferior in size to others brought to Europe by Dr. Natterer, and that that gentleman informed me it was a young animal. Now, we know that in Opossums which have nearly attained their full size, but which are still immature, the hindermost molar tooth, on either side of the jaws, is but partially developed, and, were the skin not removed from the skull (as I suspect was the case in the Yapock described by Mr. Ogilby), would not be visible; and beyond this, we find that at the time that the last true molar of the Opossum is in this undeveloped condition, the hindermost of the milk teeth is shed to make way for a permanent false molar which takes its place; there would then, however, be a vacant space between the second false molar and the first true molar, and such a vacancy is not noticed by Mr. Ogilby[1].

CHIRONECTES VARIEGATUS.

Water-Opossum.

(*Plate* 17, fig. 1).

Lutra minima.	ZIMMERMAN, Geogr. Geschicht. ii. p. 317. 1780.
" "	BODDAERT, Elenchus Animalium, i. p. 165. 1785.
" *Sarcovienna.*	SHAW, General Zool. vol. i. Pt. 2, p. 447. 1800.
Chironectes variegatus.	ILLIGER, in Abh. der Berl. Acad. 1811, p. 107.
" *Yapock.*	DESMAREST, Mammalogie, p. 261, 1820. Dict. des Sci. Nat. tom. xlvii. p. 400. 1827.
Didelphys palmata.	(GEOFF.) FISCHER, Synopsis Mammalium, p. 266. 1829.
Petite Loutre de la Guyane.	BUFFON, Hist. Nat. Suppl., tom. iii. p. 159, Tab. 22. 1776.
Saricovienne.	PENNANT, Quad., vol. ii. p. 355. 1781.
Yapock.	CUV., Règne Animal (ed. 1817), tom. i. p. 174.

Ears large and naked; tail longer than the head and body taken together; fur short, dense, and somewhat woolly; upper

[1] I have thought it desirable to make these remarks, because I find Dr. Schinz, in his recently published work upon Mammalia, has adopted as the true dental formula of the Chironectes, that which has been found in one specimen only—the specimen examined by Mr. Ogilby, and for the reasons stated I am by no means satisfied that this animal, when adult, does not present (as M. F. Cuvier has stated it actually does) the same dentition as the true Opossums.

surface of the head, and the back, sooty black ; sides of body grey, the grey running upwards in three places, so as nearly to divide the black of the back into separate patches ; a grey mark behind the ear, and a transverse, but somewhat curved band, of the same colour, on the crown of the head ; under parts of head and body white : tail naked and scaly (excepting at the root, which is clothed with fur like that on the body), black, but terminated with white.

Inhabits Guiana and Brazil.

The following description is taken from a specimen of the Yapock, or Water-Opossum, contained in the collection of the Zoological Society. The specimen is rather larger than the common Brown Rat (*Mus decumanus*), and somewhat resembles that animal in its proportions. The fur is soft, dense, and somewhat woolly; white on the under parts of the head and body, and on the upper parts grey, but with large, sooty-black patches : black is the prevailing hue on the back, and grey on the sides of the body. The muzzle, and, indeed, the whole upper surface of the head, is black, if we except a curved grey band (having its convex side forwards) which crosses the forehead ; the black runs backwards from behind the eye on to the sides of the neck; the upper lip is white. The hairs of the moustaches are for the most part black, but some of them are white. On the back are four large black patches, joined by a broad black dorsal line. The foremost of the patches crosses the shoulders, and extends a short distance on the outer surface of the fore leg, leaving, however, the anterior and lower parts grey. The second patch is placed near the middle of the back, and is almost circular; the third is situated near the hinder part of the back, and the fourth crosses the rump, runs down the back of hind legs, and extends also on to the hairy portion of the tail above. About two and a half inches of the tail is clothed with fur like that on the body; the remaining portion is covered with scales, between which spring short bristly hairs; the scaly part of the tail is chiefly black, but the apical portion is white, the white extending backwards about four inches from the point. The upper suface of the feet is brownish: the toes of the fore feet are long, united at the base to the end of the first phalanx: on the outer side of the foot is an elongated tubercle, having the appearance of a rudimentary sixth toe; the fleshy pads on the under surface of the foot are very large and rough; the nails are small, and partially embedded in the large fleshy pads with which the toes are terminated. The hind feet are very large; the toes are long, and tied together by an ample web, which extends to the base of the claws. The opposable, thumb-like, inner toe, as usual, has no nail; the claws of the other toes are of moderate size, curved, and compressed; those of the two inner toes are more compressed than the others.

	Inches.	Lines.
Length from tip of nose to root of tail ...	10	6
" of tail	13	6

In the specimen examined by Mr. Ogilby, which has already been referred to, the head and body measured ten inches, and the tail was of the same length. Dr. Natterer, however, informed Mr. Ogilby that he had specimens which measured fourteen or fifteen inches. The specimen described by Buffon was a young animal, and was supposed by him to be a species of Otter, and indeed for a long period the Yapock was classed by mammalogists amongst the Otters.

In its habits the Water Opossum greatly resembles the animals just mentioned. Buffon's specimen was found in Cayenne, and, according to Dr. Natterer, it is likewise a native of Brazil, occurring in all the smaller streams of that district. Two of Dr. Natterer's specimens, that gentleman informed me, were caught near water not far from Rio Janeiro, and a third was captured in the water, alive, near Para, in a basket similar to those used for catching eels in this country: it had made its way through the funnel-shaped opening, and could not return; thus proving that the animals are good divers. They feed upon crustaceous, and no doubt upon other aquatic animals.

Distribution of the Species of Didelphys.

United States.	*California.*	*Mexico.*	*Peru.*	*Guiana.*	*Brazil.*	*Paraguay.*	*Banda oriental.*	*Chili.*
Virginiana								
			Azaræ		Azaræ	Azaræ	Azaræ	
	Californica	pruinosa			pœcilotus			
	breviceps	Californica			albiventris			
				cancrivora	cancrivora			
				Quica	Quica			
			nudicaudata	nudicaudata	nudicaudata			
			Opossum	Philander	Opossum			
			Derbiana		dichura			
					ochropus	lanigera		
					cinerea	crassicaudata	crassicaudata	
					macrotarsus	grisea		elegans
			noctivaga		microtarsus			
			impavida		domestica			
					glirina			
		murina	murina	murina	murina			
					pusilla	pusilla		
					velutina			
					unistriata			
					tricolor			
				brachyura	brachyura	brachyura	brachyura	
					tristriata			
				variegata	variegata			

The preceding table shows that the great metropolis of the Opossums is Brazil, and that in proportion as the various districts mentioned are more widely separated from that province, so do the species decrease in number. The natural barrier of the Cordilleras, it seems, has prevented the passage of a single species from the eastern, to the western side of those mountains. The Great River Plata forms the southern boundary of the province of the Opossums [1].

With respect to the distribution of the Marsupiata inhabiting Australia and the neighbouring islands, some remarks will be found in the introductory observations upon the group; and at page 3 is a table showing the distribution of the species of *Cuscus* which inhabit the islands north of Australia; in that table the name *chrysorrhos* should be inserted in the column for New Guinea, a specimen of that species of Cuscus having been shot by Mr. Jukes in that island [2]. The species of Marsupiata now known to be natives of New Guinea are as follows:—*Cuscus maculatus, Cuscus chrysorrhos, Dendrolagus ursinus, Dendrolagus inustus, Macropus Brunii, Petaurus* (*sciureus ?*), *Phascogale melas,* and *Perameles Doreyanus.*

Australia may be conveniently divided into five principal divisions or districts, of which the East, West, North, and South portions of the main land will each form one province, and Van Diemen's Land the fifth. Of these provinces, the northern one has the greatest number of species peculiar to it, since out of ten species discovered in that part of Australia,

[1] With regard to the numerous so-called species forming the first section of the genus Didelphys, I have to observe, that they have been founded for the most part upon very trivial characters, and I think it highly probable that when sufficient materials are collected for a rigorous examination of the group, it will be found to be composed of but one or two variable species, instead of seven or eight, as now supposed.

[2] The specimen has recently been presented by Mr. Jukes to the British Museum.

eight are not found elsewhere. The Marsupiata of the Eastern district are for the most part distinct from those of the opposite side of the continent, there being but eight species, out of upwards of sixty inhabiting the two provinces, which are found in both. But if the three districts mentioned are characterized by the few species which they have in common, South Australia must be characterized by an opposite quality, that of having a comparatively large proportion of species identical with those of other districts; indeed, I know of but four species which are peculiar to this district: it possesses sixteen species in common with Western Australia, and fifteen in common with Eastern Australia. Western Australia possesses one genus (*Tarsipes*) which is peculiar to it, and one sub-genus (*Macrotis*); none of the other districts of continental Australia possess any genera which are not found elsewhere. About half of the species found in Van Diemen's Land are peculiar to that island—in fact, nine out of twenty: of the remainder, the greater portion are found on the eastern part of the main land. This island, moreover, possesses one genus (*Thylacinus*), and one sub-genus (*Sarcophilus*), which are now peculiar to it. Examples of both these sections have, however, been found in a fossil state on the main land.

Subjoined is a list of the species inhabiting each of the districts mentioned.

DISTRIBUTION OF THE MARSUPIATA IN AUSTRALIA.

South Australia.

Macropus giganteus.
,, lunatus.
,, leporoïdes.
,, fasciatus.
,, rufus.
,, Greyi.
,, Eugenii (Nuyt's Archipelago).
,, Derbianus (from islands off the coast).
Hypsiprymnus Grayi.
,, penicillatus.
,, Gaimardi.
,, campestris.
Phascolomys Wombat.
,, latifrons.
Phascolarctos cinereus.
Phalangista vulpina.
,, Cookii.
,, concinna.
Petaurus Australis.
Perameles obesula.
,, fasciata.
Chœropus castanotis.
Myrmecobius fasciatus.
Phascogale penicillata.
,, flavipes.
,, albipes.
,, crassicaudata.
Dasyurus Geoffroyi.

North Australia.

Macropus unguifer.
Macropus antilopinus.
,, agilis.
,, brachiotis.
,, inornatus.
,, concinnus.
Phalangista vulpina.
Petaurus breviceps var. Ariel
Perameles macroura.
Dasyurus hallucatus.

Western Australia.

Echidna aculeata.
Macropus giganteus.
,, (giganteus) ocydromus.
,, lunatus
,, isabellinus.
,, fasciatus.
,, hirsutus.
,, conspicillatus, Barrow Island.
,, Irma.
,, Derbianus.
,, brachyurus, K. George's Sound.
,, lateralis.
Hypsiprymnus Grayi.
,, penicillatus.
,, Gilbertii (King George's Sound).
,, platyops.
Phalangista vulpina.
,, ,, var. xanthopus.
,, Cookii.
,, concinna.
,, Neillii (King George's Sound).
Tarsipes rostratus (Swan River and King George's Sound).

Perameles (Macrotis) lagotis (ditto, ditto).
,, Bougainvillii (Peron's Peninsula—Shark's Bay).
,, obesula.
,, myorsuros (Swan River and King George's Sound).
Chœropus castanotis (Swan River).
Myrmecobius fasciatus (Swan River and King George's Sound).
Phascogale penicillata.
,, calura.
,, apicalis.
,, leucogaster, var. flavipes ?
,, albipes (?)
Phascogale crassicaudata.
Dasyurus Geoffroyi.

Van Diemen's Land.

Ornithorhynchus paradoxus.
Echidna (aculeata) setosa.
Macropus giganteus.
,, ruficollis, var. Bennettii.
,, Billardierii.
Hypsiprymnus cuniculus.
,, murinus.
Phascolomys Wombat.
Phalangista vulpina.
,, Cookii.
,, nana.
Perameles obesula.
,, Gunnii.
Phascogale Swainsonii.
,, minima.
,, leucopus.
Dasyurus maculatus.
,, viverrinus.
,, ursinus.
Thylacinus cynocephalus.

Bass' Strait.

Echidna.
Macropus ruficollis (King Island).
Phascolomys Wombat (ditto)
Dasyurus maculatus (ditto).

New South Wales.

Ornithorhynchus paradoxus.
Echidna aculeata.
Macropus giganteus.
,, frænatus (interior).
,, leporoïdes.
,, robustus (interior).
,, rufus.
,, Parryi.
,, Ualabatus.
,, Thetidis.
,, Parma.
,, dorsalis (interior).
,, penicillatus.
Hypsiprymnus rufescens.
,, penicillatus.
,, Gaimardi.
,, murinus.
Phascolomys Wombat.
Phascolarctus cinereus.
Phalangista vulpina.
,, canina (interior).
,, Cookii.
Petaurus taguanoïdes.
,, Australis.
,, sciureus.
,, breviceps.
,, (acrobata) pygmæus.
Perameles obesula.
,, nasuta.
,, fasciata.
Phascogale penicillata.
,, flavipes.
,, murina.
,, macroura.
Dasyurus Geoffroyi.
,, viverrinus.

INDEX TO THE TECHNICAL NAMES.

E.

H.

I.

K.

L.

M.

N.

O.

P.

INDEX TO THE ENGLISH NAMES.

EXPLANATION OF THE PLATES.

In most cases the names of the objects figured, are added to the plates, but in some instances they were too numerous; these latter are all that it is necessary here to explain.

Plate 2, Figs. 1 to 6. Parts of the Ornithorhynchus; explained in detail at p. 39.

Figs. 7—9. Parts of the Echidna, explained at p. 49.

Plate 3, Fig. 1. Skull and teeth of the Wombat—see p. 244.

Figs. 2—5. Skull and teeth of *Macropus Thetidis*, and teeth of *Macropus giganteus*[1]—see p. 61.

Plate 8. Skulls of species of *Hypsiprymnus;* the letters are explained at p. 195.

Plate 12. Fore and hind feet of various Marsupial animals: the fore feet are represented in the upper row, and the hind, in the lower. The inner toe of the hind foot is marked * in the figures.

Fig. 1. Fore and hind feet of *Dasyurus viverrinus.*
2. ,, ,, of *Phascogale penicillata.*
3. ,, ,, of *Didelphys cancrivora.*
4. Hind foot of *Phalangista Cookii.*
5. Fore and hind feet of *Hypsiprymnus penicillatus.*
6. ,, ,, of *Perameles obesula.*
7. ,, ,, of *Phascolomys Wombat.*

Plate 19, Fig. 1. Skull of *Phalangista vulpina,* viewed from beneath.

Fig. 1*a*. A molar tooth of the same, magnified.

[1] By mistake "cigantus" on the Plate.

Fig. 2. Skull of *Phalangista Cookii,* viewed from above.

Fig. 3. Skull of *Petaurus taguanoïdes.*

Fig. 4. Skull of *Petaurus pygmæus,* magnified: 4 *a,* side view of the upper and lower teeth much enlarged.

Fig. 5. Skull of *Phalangista Neillii,* magnified: 5 *a,* side view of the teeth of the upper and ower jaws much magnified.

Fig. 6. Upper view of the skull of *Tarsipes rostratus,* magnified: 6 *a,* side view of the same: 6 *b,* side view of the lower jaw.

Plate 20, Fig. 1. Skull of *Perameles lagotis.*

2. Ditto of *Perameles obesula,* a large male: 2 *a,* side view of the teeth of the upper jaw: 2 *b,* side view of the lower jaw, from the skull of a female?

3. Ditto of *Perameles myosuros;* 3 *a,* side view of the teeth of the upper jaw.

4. Ditto of *Chœropus castanotis;* 4 *a,* side view of the same. From a drawing belonging to Prof. Owen.

5. Ditto of *Phascolarctus cinereus,* half the natural size; 5 *a,* side view of the same.

Plate 21. Skulls of the various genera of the family *Dasyuridæ.*

Fig. 1. *Myrmecobius fasciatus;* 1 *a,* side view of the teeth of the upper jaw, magnified; 1 *b,* the lower jaw viewed from the inner side.

Fig. 2. *Dasyurus maculatus,* half the natural size.

Fig. 3. *Thylacinus cynocephalus,* half the natural size.

Fig. 4. *Dasyurus ursinus,* half the natural size.

Fig. 5. *Phascogale penicillata,* and side view of the same.

Plate 22. Fig. 1. Skull of *Didelphys cancrivora.*

Fig. 2. Fore part of the skull of a Dasyurus, to show the differences of form and proportions of the nasal bones in this genus, as compared with those of the *Didelphys,* where they are produced in front, and much expanded behind.

Fig. 3. Skull of *Didelphys elegans*; 3 *a* and 3 *b*, other views of the same; 3 *c*, lower jaw; 3 *d*, the same magnified, and viewed from the inner side.

Fig. 4. Molar tooth of the upper jaw of *Thylacinus cynocephalus*; 4 *a*, a true molar tooth of the lower jaw of the same.

Fig. 5. Molar tooth of a Dasyurus; 5 *a*, ditto of lower jaw.

Fig. 6. „ of *Perameles myosuros*.

Fig. 7. „ of *Perameles obesula*.

Fig. 8. „ of a species of *Hypsiprymnus*.

Fig. 9. „ of *Phalangista Cookii*.

Fig. 10. „ of *Phalangista vulpina*.

Fig. 11. „ of a species of Macropus.

The above figures, from No. 4 to No. 11 inclusive, are more or less magnified representations of the penultimate true molar tooth of the left side of the upper jaw, as found in the principal genera of the Marsupiata.

The four principal cusps, or tubercles, observable on the crown of the more complicated of these teeth, are marked A, B, C, and D; the anterior pair, A and C, are partially joined, so as to form a slightly interrupted transverse ridge in the teeth, fig. 10, and a second similar ridge is formed by the junction of the posterior pair of principal cusps, B and D. The principal cusps are also similarly united in the tooth, fig. 8, but in *Macropus* (fig. 11), they are most perfectly united, and form the high transverse ridges observed in the molar teeth of the species of that genus. In the tooth, fig. 7, all the cusps are most evenly developed, and there are besides these, four smaller cusps, 1, 2, 3, and 4, arranged in a line along the outer side of the tooth, and, if I am not mistaken in my identification of the corresponding parts in these molars, they become very important portions of the tooth in the carnivorous type of dentition, fig. 4; here, it appears as if the cusps 2 and 3 were much developed, and bent inwards so as to be united to the principal cusps, A and B[1]; the cusp, D, is wanting,

[1] It appeared to me, upon first comparing the molar teeth of the Thylacinus with those of the Dasyurus, that the chief difference consisted in the absence of the small cusps, 2 and 3, in the molars of the former animal: but observing

as will be rendered pretty evident upon inspecting the tooth 6, where, owing to the comparatively small development of the cusp D, the crown approaches to the triangular form which is so characteristic of figs. 4 and 5. The part of the tooth which I have called the band, is most developed in the fore part of the molar, fig. 11, and is apparent in the teeth figured, where marked *.

that these small cusps are approximated to the principal cusps in the second true molar of Dasyurus (or at least are much less widely separated than in the third molar), and that in the first molar the cusps 3 and B, are still more approximated, and the cusps 2 and A are partially joined, I have since thought it more probable that in *Thylacinus* the small and principal cusps are united.

END OF VOL. I

Wilson and Ogilvy, 57, Skinner Street, Snowhill, London.

MACROPUS PENICILLATUS. DENDROLAGUS URSINUS.

London, Published by H. Baillière, 219, Regent St. 1845

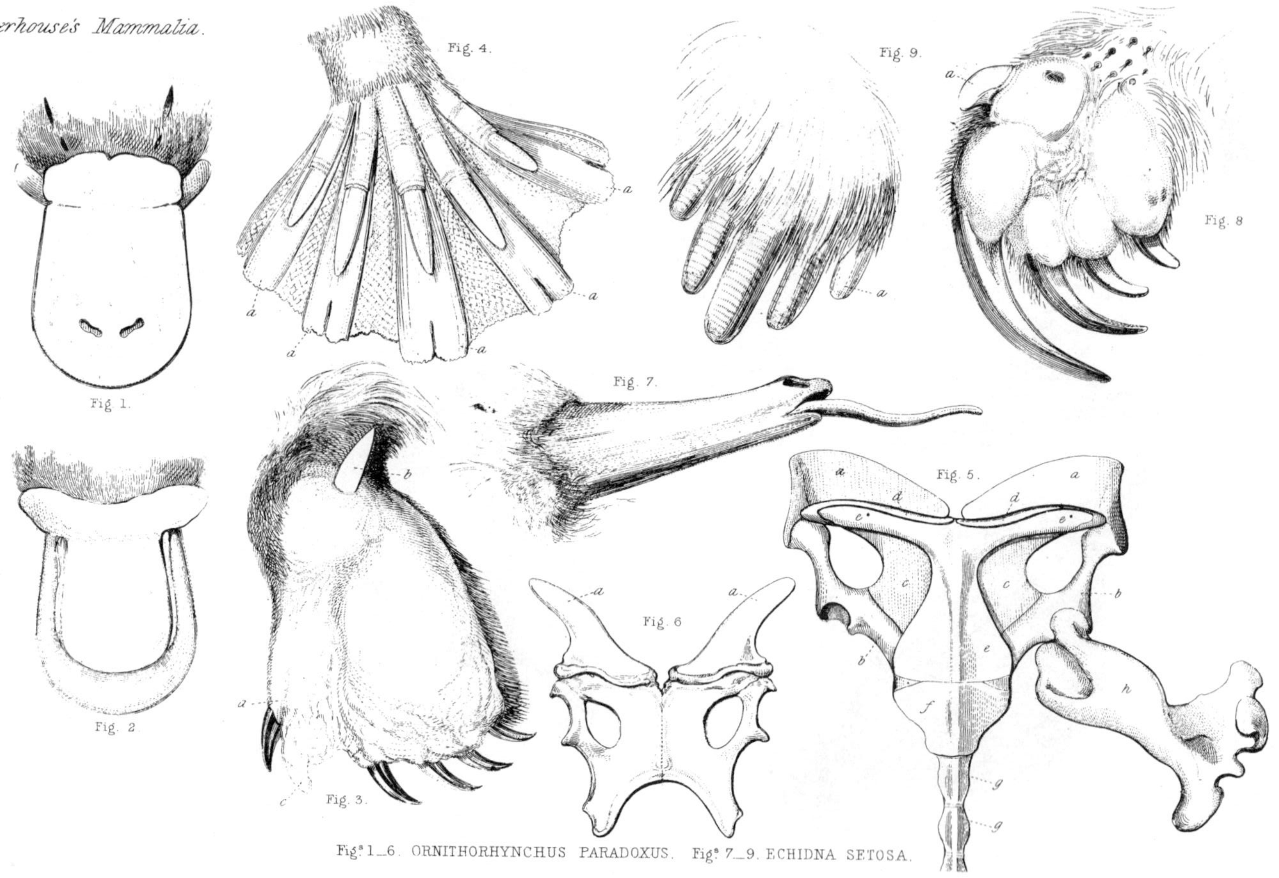

Figs. 1_6. ORNITHORHYNCHUS PARADOXUS. Figs. 7_9. ECHIDNA SETOSA.

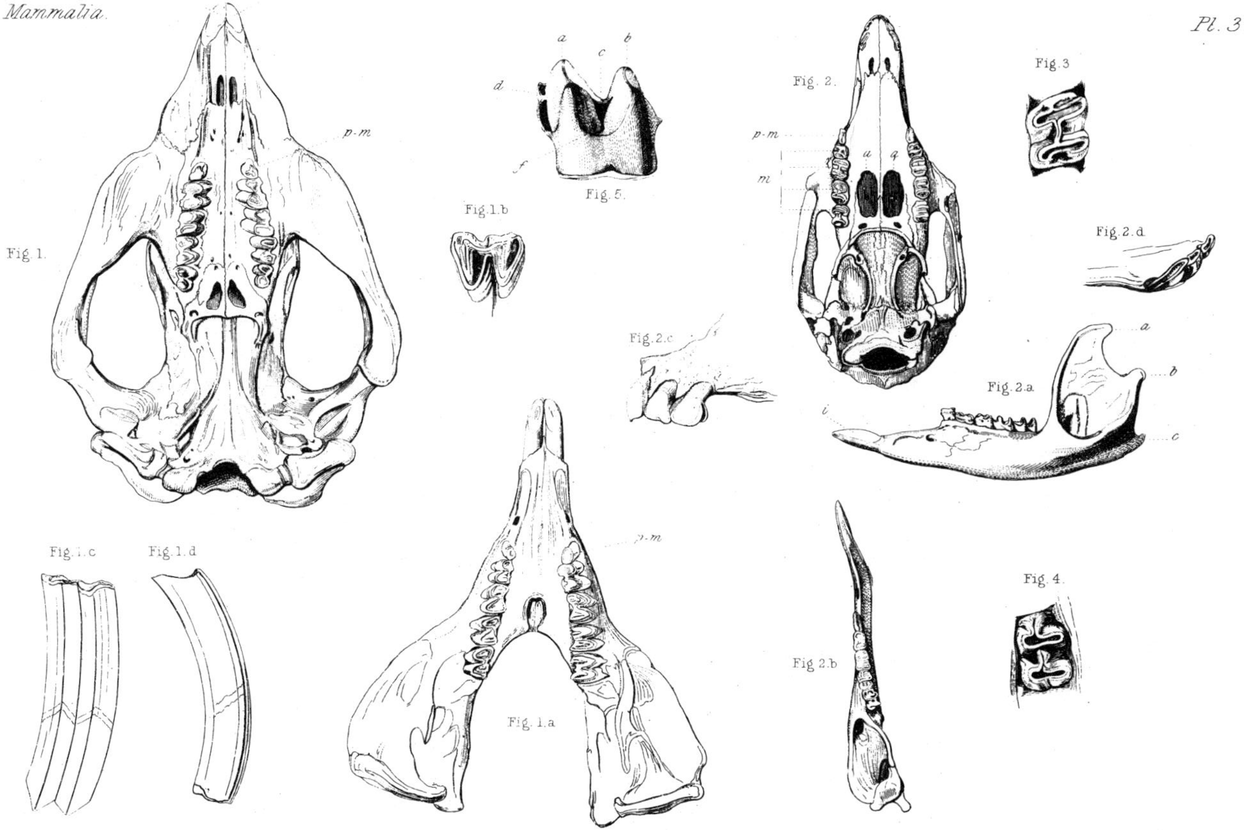

Figs. 1. PHASCOLOMYS WOMBATUS. Figs. 2. MACROPUS · THETIDIS. Figs. 3. 4. & 5. MACROPUS CICANTUS

London, Published by H. Baillière, 219 Regent St. 1845.

MACROPUS FRÆNATUS. (from Gould.) MACROPUS FASCIATUS. (from Gould.)

Engraved by H. S. Beckwith. *London, Published by H. Baillière, 219, Regent St. 1846* *Printed by Brooker & Harrison.*

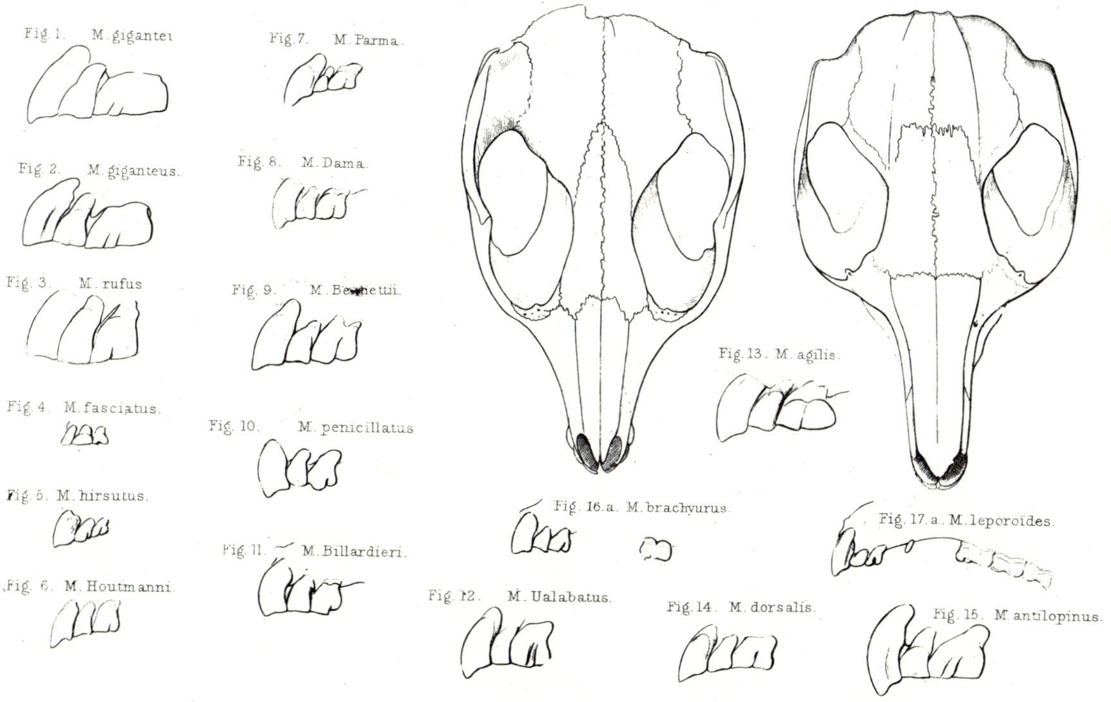

Figs. 1_15. UPPER INCISOR TEETH of VARIOUS SPECIES of MACROPUS. Fig. 16. SKULL of M. BRACHYURUS Fig. 17. SKULL of M. LEPOROÏDES.

Engraved by H.S. Beckwith

London, Published by H. Baillière, 219 Regent St. 1845.

Printed by Brooker & Harrison

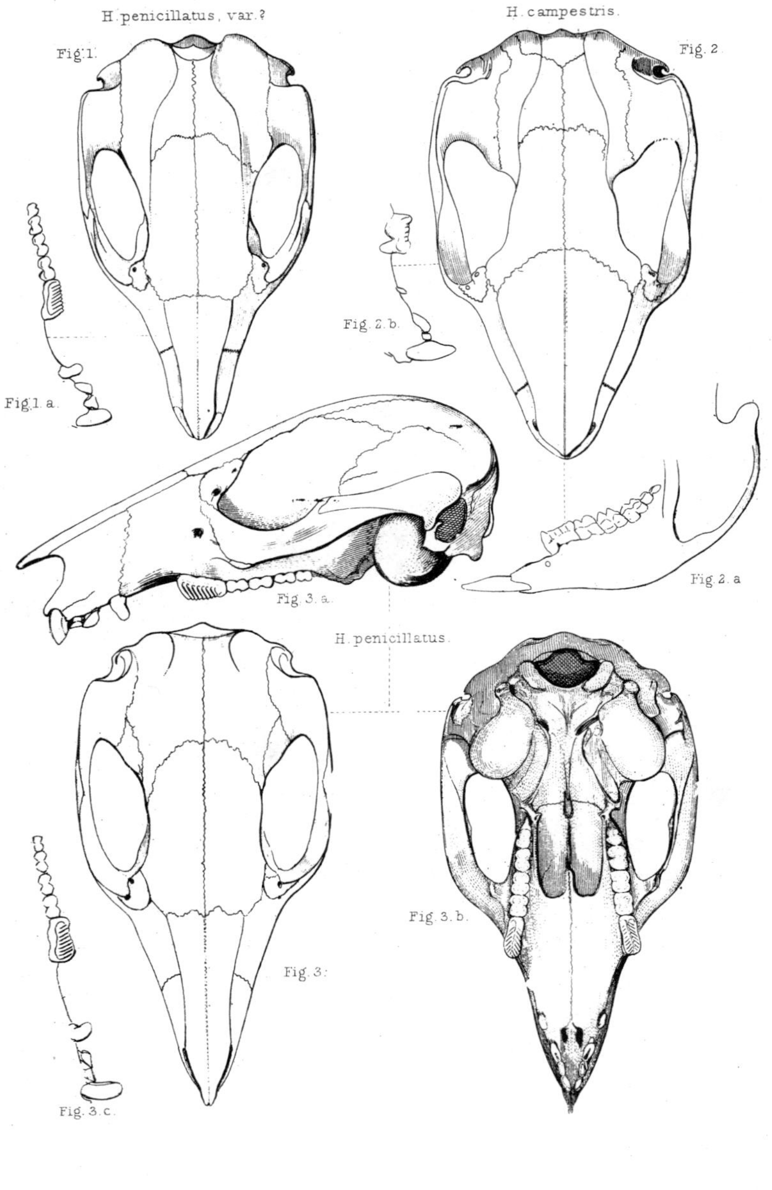

SCULLS OF VARIOUS SPECIES OF HYPSIPRYMNUS.

Engraved by H. S. Beckwith. London. Published by H. Baillière. 219. Regent St. 1845. Printed by Brooker & Harrison.

HYPSIPRYMNUS PENICILLATUS. (Gould)

MACROPUS ROBUSTUS. (Gould)

London, Published by H. Baillière, 219, Regent St. 1845.

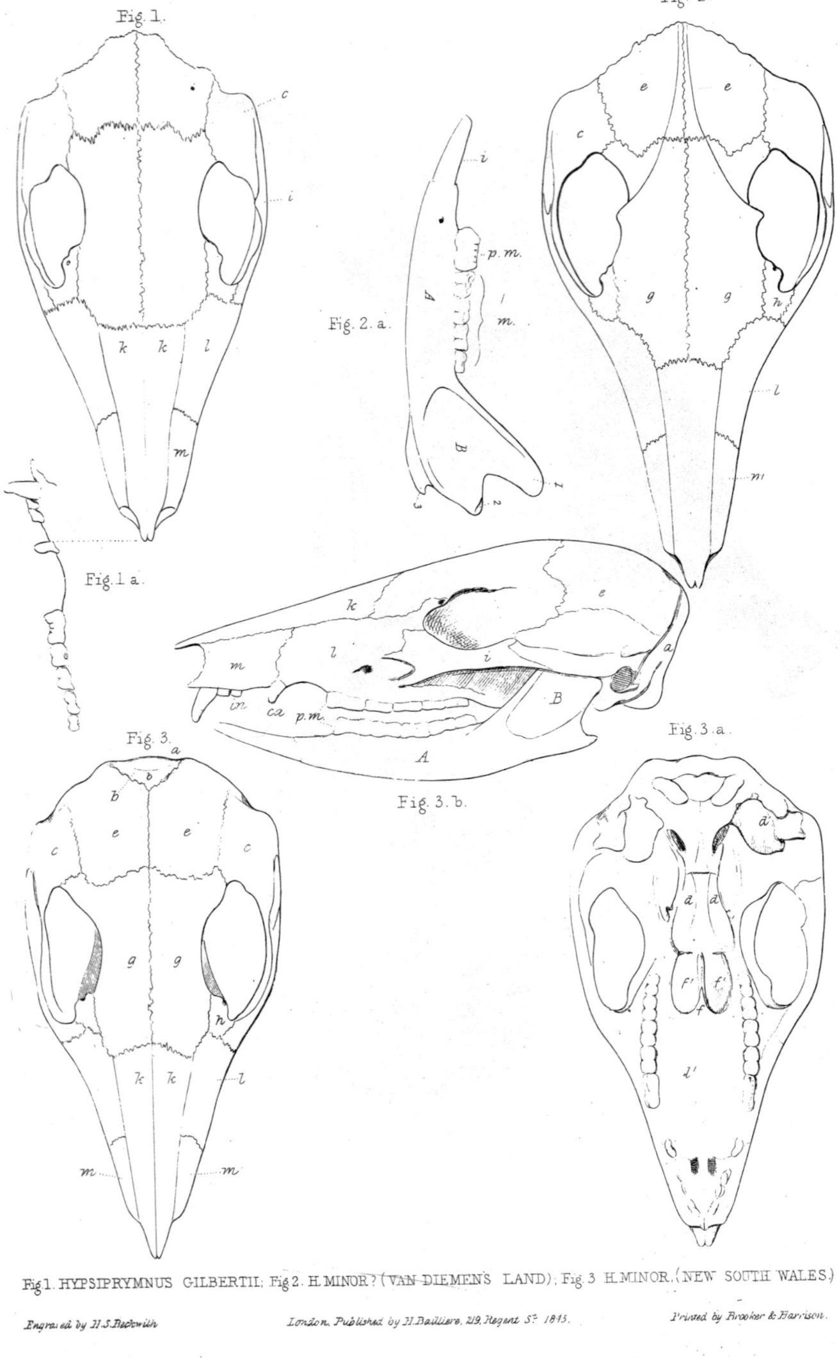

Fig 1. HYPSIPRYMNUS GILBERTII; Fig 2. H. MINOR? (VAN DIEMEN'S LAND); Fig 3 H. MINOR, (NEW SOUTH WALES.)

Engraved by H. S. Beckwith London, Published by H. Bailliere, 219, Regent St. 1845. Printed by Brooker & Harrison.

Fig. 1.

Fig. 2.

Fig. 1 PHALANGISTA VULPINA; Fig. 2 PHASCOLARCTUS FUSCUS.

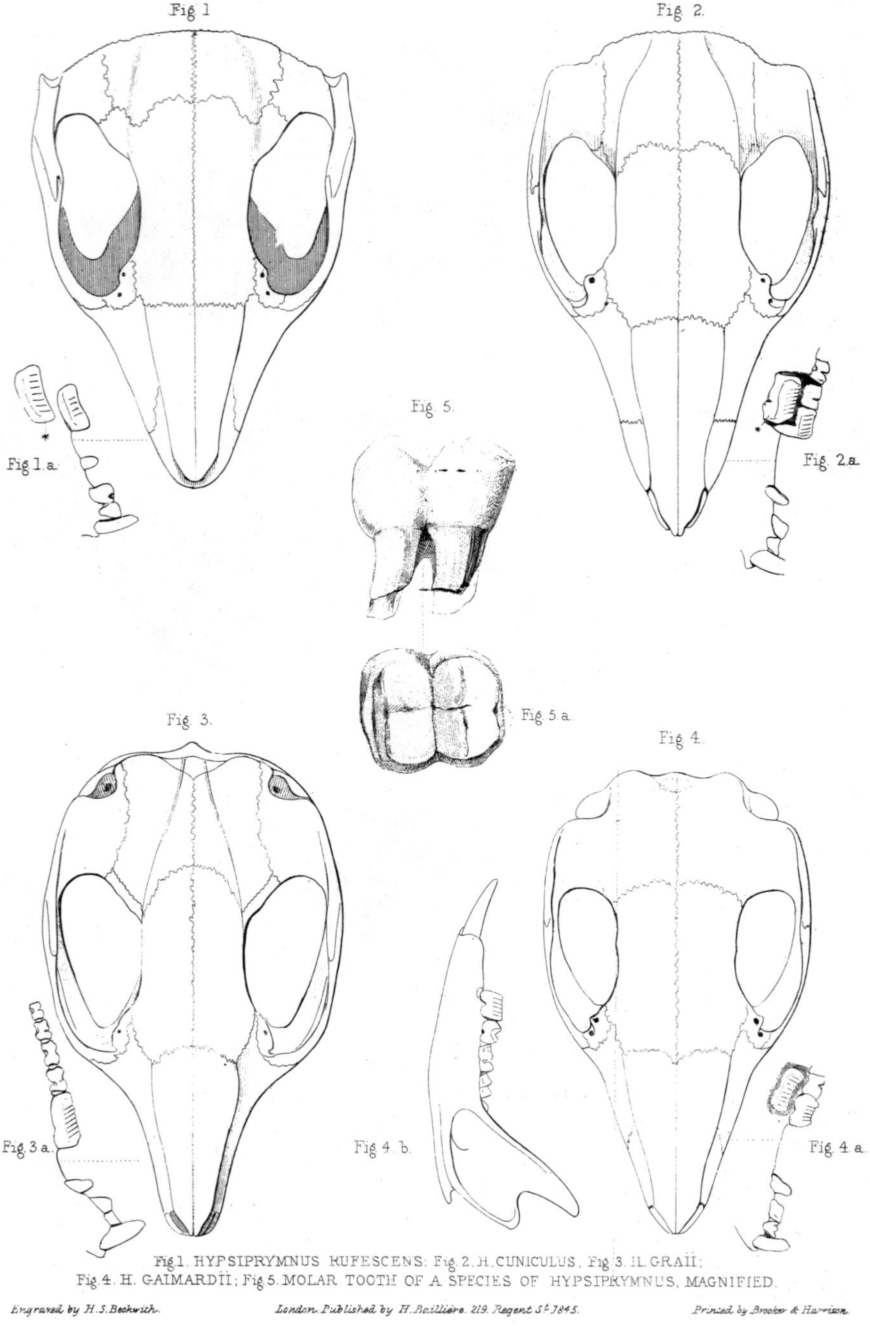

Fig. 1. HYPSIPRYMNUS RUFESCENS; Fig. 2. H. CUNICULUS, Fig. 3. H. GRAII;
Fig. 4. H. GAIMARDII; Fig. 5. MOLAR TOOTH OF A SPECIES OF HYPSIPRYMNUS, MAGNIFIED.

Engraved by H. S. Beckwith. London. Published by H. Baillière. 219. Regent St. 1845. Printed by Brooker & Harrison.

Fig. 1.

Fig. 1. TARSIPES ROSTRATUS. Fig. 2. PHALANGISTA CONCINNA. (Gould)

Engraved by [illegible] London Published by H. Baillière 219. Regent St. 1846. Printed by Bro[illegible]

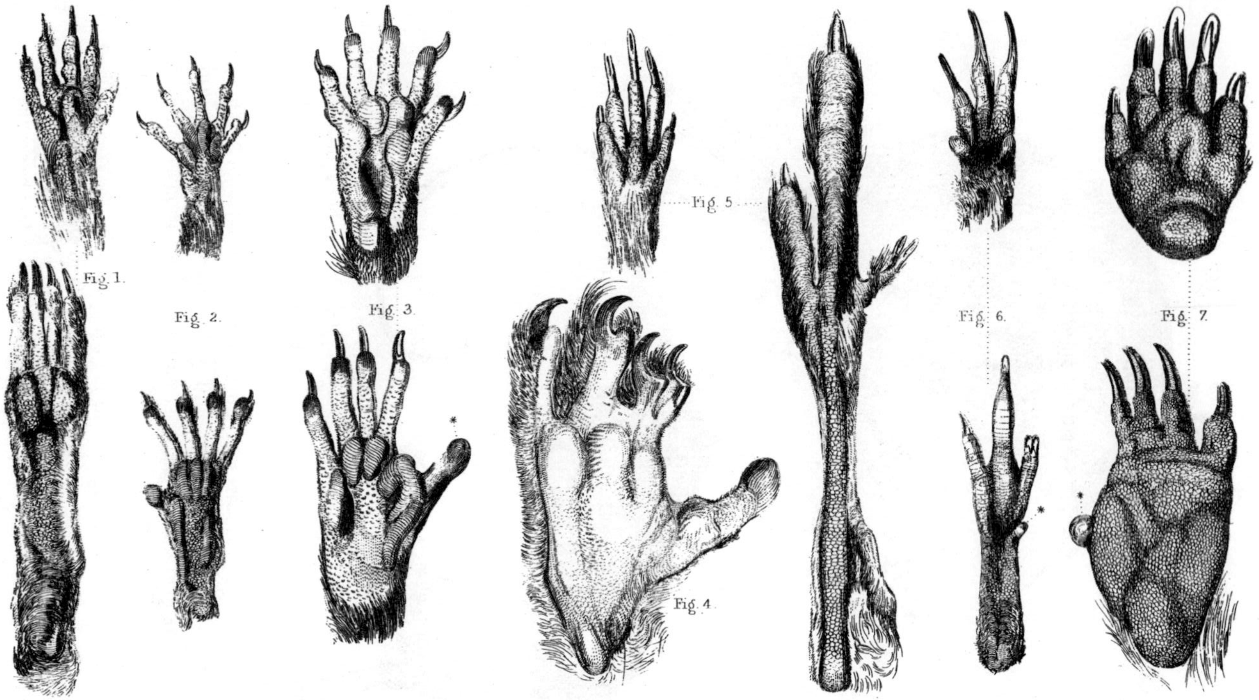

FORE & HIND FEET OF VARIOUS MARSUPIAL MAMMALS; THE FORE FEET ARE REPRESENTED IN THE UPPER ROW. * THUMB OF HIND FOOT

Fig. 1.

Fig. 2.

Fig. 1. PERAMELES LAGOTIS, Fig. 2. CHOEROPUS CASTANOTIS (from Gould.

London, Published by H. Baillière, 219 Regent St.

Fig. 1

Fig. 2.

Fig. 1. MYRMECOBIUS FASCIATUS. Fig. 2. PHASCOGALE CALURA. (Gould)

Engraved by H. Beckwith. London. Published by H. Baillière. 219. Regent St. 1846. Printed by Brooker & Harrison

Fig. 1.

Fig. 2.

Fig. 1. PERAMELES OBESULA. Fig. 2. PHASCOGALE CRASSICAUDATA. (Gould.)

Engraved by H. S. Beckwith. *London: Published by H. Bailliere, 219, Regent S^t. 1846.* *Printed by Brooker & Harrison.*

Fig. 1.

Fig. 2.

Fig. 1. DIDELPHYS ELEGANS. Fig. 2. DIDELPHYS BRACHYURA.

Engraved by H. S. Beckwith. London. Published by H. Baillière, 219 Regent St. 1846. Printed by Brooker & Harrison.

Fig

Fig. 2.

Fig. 1. CHIRONECTES VARIEGATUS   Fig. 2. THYLACINUS CYNOCEPHALUS.

Engraved by H. S. Beckwith.   *London. Published by H. Baillière, 219, Regent St. 1846.*   *Printed by Brooker & Harrison.*

Fig. 1.

Fig. 2.

Fig. 1. DIDELPHYS DORSIGERA. Fig. 2. DIDELPHYS AZARÆ.

Engraved by H. S. Beckwith. *London: Published by H. Bailliere, 219, Regent St. 1846.*

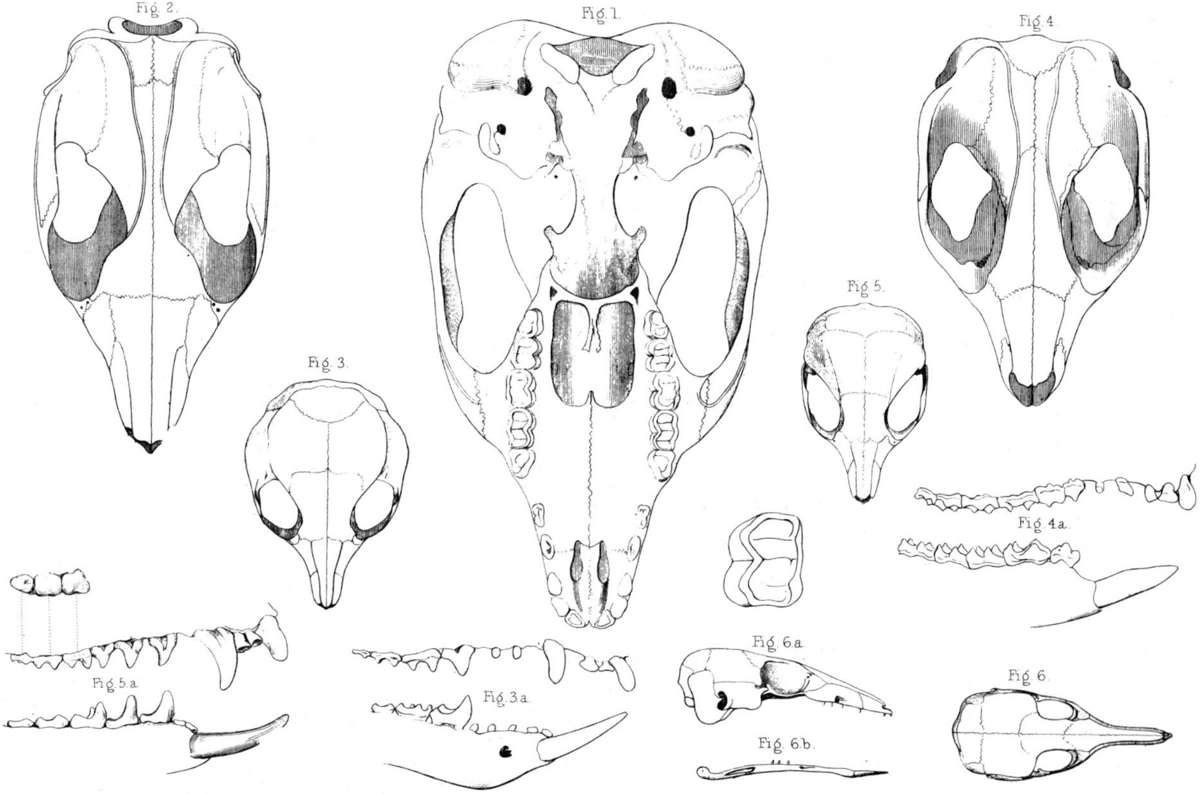

Fig. 1. PHALANGISTA VULPINA. Fig. 2. PHAL. COOKII. Fig. 3. PHAL. NEILLII. Fig. 4. PETAURUS TAGUANOIDES. Fig. 5. PET. PYGMÆUS. Fig. 6. TARSIPES ROSTRATUS.

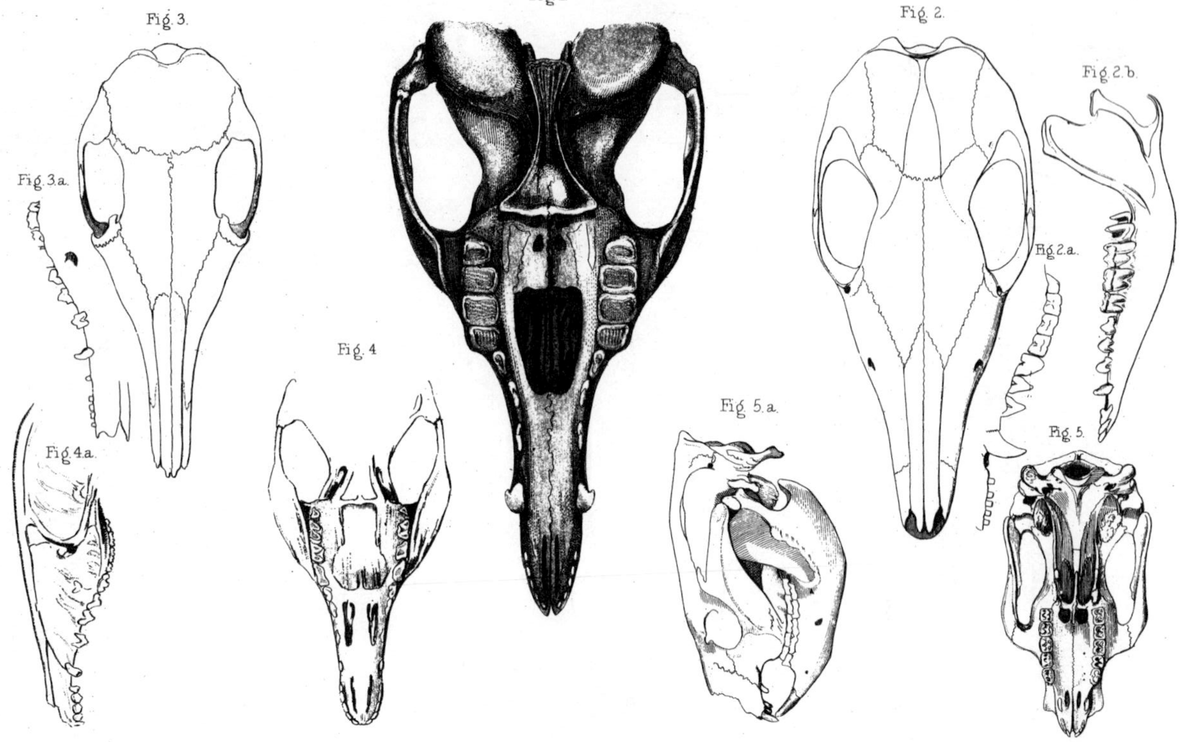

Fig. 1. PERAMELES LAGOTIS. Fig. 2. PER. OBESULA. Fig. 3. PER. MYOSUROS. Fig. 4. CHŒROPUS CASTANOTIS. Fig. 5. PHASCOLARCTUS CINEREUS.

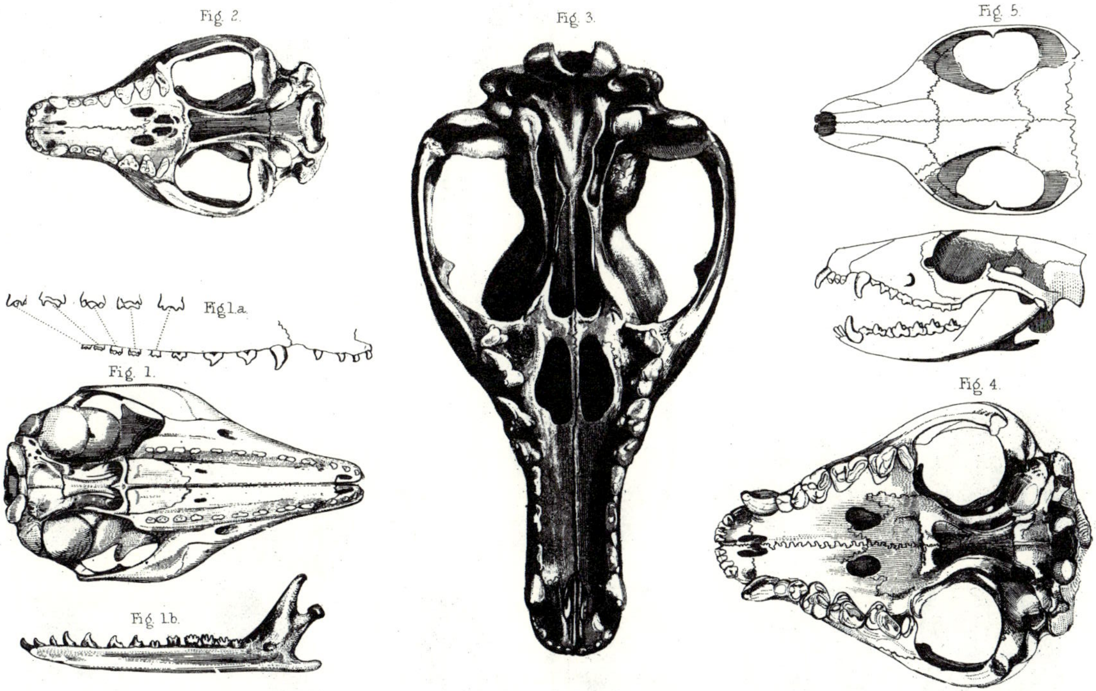

Fig. 1. SKULL OF MYRMECOBIUS. Fig. 2. DASYURUS. Fig. 3. THYLACINUS. Fig. 4. SARCOPHILUS. Fig. 5. PHASCOGALE.

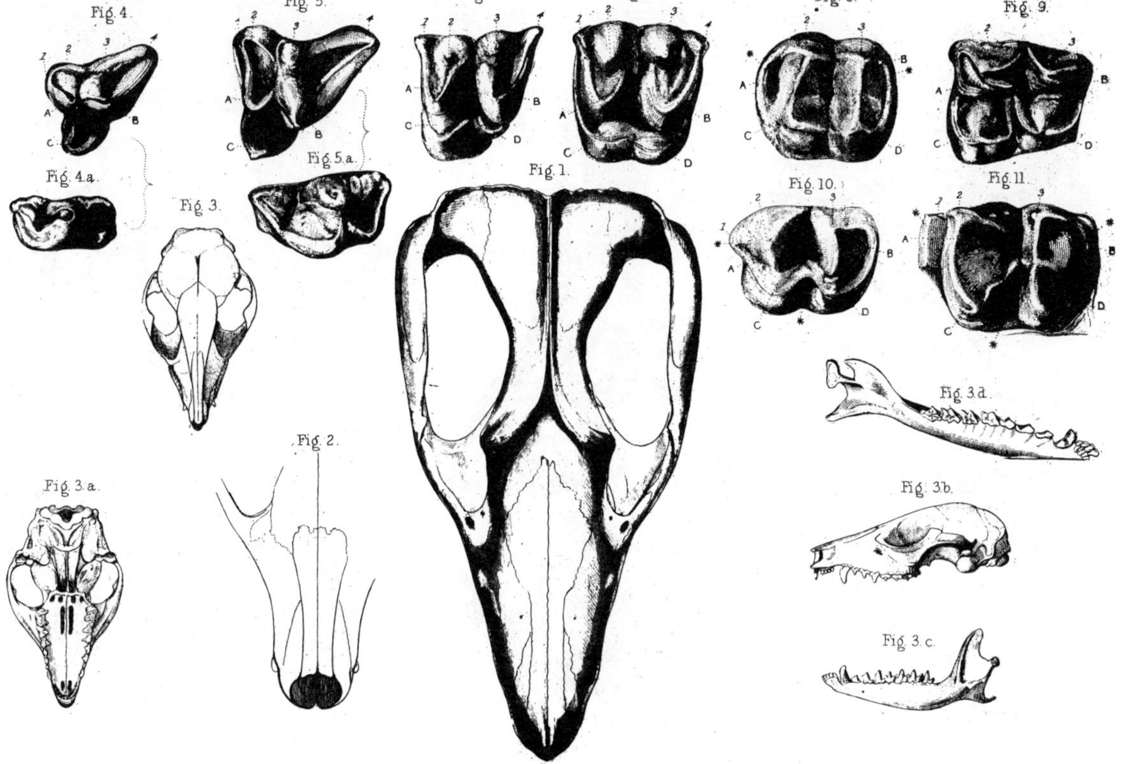

Fig. 1. DIDELPHYS CANCRIVORA. Fig. 2. DASYURUS. Fig. 3. DIDELPHYS ELEGANS. Fig. 4-11. MOLAR TEETH OF VARIOUS MARSUPIATA.

BIOLOGISTS AND THEIR WORLD

An Arno Press Collection

Adler, Kraig, editor. **Early Herpetological Studies and Surveys in the Eastern United States.** 1978

Adler, Kraig, editor. **Herpetological Explorations of the Great American West.** Two vols. 1978

Agassiz, [Jean] Louis [Rodolphe]. **Contributions to the Natural History of the United States of America.** Four vols. in two. 1857/1860/1862

Allard, Dean Conrad, Jr. **Spencer Fullerton Baird and the U.S. Fish Commission.** 1978

Altum, Bernard. **Der Vogel Und Sein Leben.** 1868

Azara, Don Felix de. **Apuntamientos Para La Historia Natural De Los Quadrúpedos Del Paragüay Y Rio De La Plata.** Two vols. in one. 1802

Baer, Karl Ernst v[on]. **Reden Gehalten in Wissenschaftlichen Versammlungen und Kleinere Aufsaetze Vermischten Inhalts.** Three vols. in two. 1864/1876/1873

Barrett-Hamilton, Gerald E. H. and Martin A. C. Hinton. **A History of British Mammals.** Two vols. 1910-1921

Boulenger, G[eorge] A[lbert]. **The Tailless Batrachians of Europe.** Two parts in one. 1897/1898

Brocchi, [Paul]. **Mission Scientifique Au Mexique Et Dans L'Amérique Centrale,...** Étude Des Batraciens De L'Amérique Centrale. 1882

Buffon, [Georges L. L.]. **The History of Singing Birds.** Translated from the French. 1791

Buffon, [Georges L. L.]. **The Natural History of Oviparous Quadrupeds and Serpents.** Translated by Robert Kerr. Four vols. in two. 1802

The Cabinet of Natural History and American Rural Sports. Three vols. in one. 1830/1832/1833

Candolle, A[ugustin] P[yramus] de and K[urt] Sprengel. **Elements of the Philosophy of Plants.** Translated from the German. 1821

Cassin, John. **United States Exploring Expedition During the Years 1838, 1839, 1840, 1841, 1842 Under the Command of Charles Wilkes, U. S. N.:** Mammalogy and Ornithology. Two vols. 1858

Chapman, Frank M. **Essays in South American Ornithogeography.** Edited by Keir B. Sterling. 1978

Cope, Edward D[rinker]. **The Vertebrata of the Tertiary Formations of the West.** One vol. in two. 1883

Cuvier, [Georges]. **The Class Mammalia.** Five vols. 1827

Donovan, E[dward]. **The Natural History of British Fishes.** Five vols. in two. 1802/1803/1804/1806/1808

Duméril, Auguste [Henri André], [Marie-Firmin] Bocourt and [François] Mocquard. **Mission Scientifique Au Mexique Et Dans L'Amérique Centrale,...** Étude Sur Les Reptiles. Two vols. 1870-1909

Flower, William Henry and Richard Lydekker. **An Introduction to the Study of Mammals, Living and Extinct.** 1891

Forbush, Edward Howe. **Birds of Massachusetts and Other New England States.** Three vols. 1925/1927/1929

Girard, Charles. **United States Exploring Expedition During the Years 1838, 1839, 1840, 1841, 1842 Under the Command of Charles Wilkes, U. S. N.:** Herpetology. Two vols. 1858

Grinnell, Joseph. An Account of the Mammals and Birds of the Lower Colorado Valley. 1914

Howard, H[enry] Eliot. **Territory in Bird Life.** 1920

Hume, Edgar Erskine. **Ornithologists of the United States Army Medical Corps.** 1942

Huxley, Leonard. **Life and Letters of Sir Joseph Dalton Hooker.** Two vols. 1918

LeConte, John L. and George H. Horn. **Classification of the Coleoptera of North America.** 1883

Linsley, E. Gorton, editor. **Beetles From the Early Russian Explorations of the West Coast of North America, 1815-1857.** 1978

Linsley, E. Gorton, editor. **The Principal Contributions of Henry Walter Bates to a Knowledge of the Butterflies and Longicorn Beetles of the Amazon Valley.** 1978

Murray, Andrew. **The Geographical Distribution of Mammals.** 1866

Osborn, Henry Fairfield. **Cope:** Master Naturalist. 1931

Peale, Titian R. **United States Exploring Expedition During the Years 1838, 1839, 1840, 1841, 1842 Under the Command of Charles Wilkes, U. S. N.:** Mammalogy and Ornithology. 1848

Phillips, John C. **American Game Mammals and Birds.** 1930

Rafinesque: Autobiography and Lives. 1978

Ray, John. **Synopsis Methodica Animalium Quadrupedum et Serpentini Generis.** 1693

Ray, John. **Synopsis Methodica Avium & Piscium.** [Edited by William Derham]. 1713

Richardson, John. **Fauna Boreali-Americana;** Or the Zoology of the Northern Parts of British America, Part Third: The Fish. 1836

Richardson, John, William Swainson and William Kirby. **Fauna Boreali-Americana;** Or the Zoology of the Northern Parts of British America, Part Four: Insecta, 1837

Riley, Charles V[alentine]. **[Nine] Annual Report[s] on the Noxious, Beneficial and Other Insects of the State of Missouri, 1869-1877** *and* **General Index and Supplement.** Ten vols. in three. 1869-1877/1881

Say, Thomas. **The Complete Writings of Thomas Say on the Entomology of North America.** Edited by John L. LeConte. Two vols. 1859

Schuchert, Charles and Clara Mae LeVene. **O. C. Marsh:** Pioneer in Paleontology. 1940

Sclater, William Lutley and Philip Lutley Sclater. **The Geography of Mammals.** 1899

Seton, Ernest Thompson. **Trail of an Artist-Naturalist:** The Autobiography of Ernest Thompson Seton. 1940

Smith, James Edward. **A Selection of the Correspondence of Linnaeus and Other Naturalists, From the Original Manuscripts.** Two vols. 1821

Stuckey, Ronald L., editor. **Development of Botany in Selected Regions of North America Before 1900.** 1978

Stuckey, Ronald L., editor. **Essays on North American Plant Geography From the Nineteenth Century.** 1978

Stuckey, Ronald L., editor. **Scientific Publications of Charles Wilkins Short.** 1978

Swammerdam, John. **The Book of Nature.** 1758

Wadland, John Henry. **Ernest Thompson Seton:** Man in Nature and the Progressive Era, 1880-1915. 1978

Waterhouse, G[eorge] R[obert]. **A Natural History of the Mammalia.** Two vols. 1846/1848

Weiss, Harry B. and Grace M. Ziegler. **Thomas Say:** Early American Naturalist. 1931

Wheeler, Geo[rge] M., [editor]. **[Reports Upon Insects Collected During Geographical and Geological Explorations and Surveys West of the One Hundredth Meridian During the Years 1872, 1873, and 1874].** 1875

Willughby, Francis. **De Historia Piscium** *and* **Icthyographia Ad Amplisimum Virum Dnum., Samuelem Pepys, Presidem Soc. Reg.** Two vols. in one. 1686/1685

Youmans, William Jay, editor. **Pioneers of Science in America.** Revised edition. 1896